Classical Mathematical Physics

Springer
New York
Berlin
Heidelberg
Hong Kong
London
Milan
Paris
Tokyo

Walter Thirring

Classical Mathematical Physics

Dynamical Systems and Field Theories

Third Edition

Translated by Evans M. Harrell II

With 146 Illustrations

Springer

Walter Thirring
Institute for Theoretical Physics
University of Vienna
Vienna, A-1090
Austria

Library of Congress Cataloging-in-Publication Data
Thirring, Walter E., 1927–
Classical mathematical physics: dynamical systems and field theories/Walter Thirring. — 3rd ed.
p. cm.
Rev. ed. of: A course in mathematical physics 1 and 2. 2nd ed. 1992
Includes bibliographical references and index.
1. Mathematical physics. 2. Dynamics. 3. Field Theory (Physics).
I. Thirring, Walter E., 1927– . Lehrbuch der mathematischen Physik. English. II. Title.
QC20.T4513 1997
530.1′5—dc20 96-32751

ISBN 0-387-40615-8 Printed on acid-free paper.

First softcover printing, 2003.

Printed in the United States of America.

Volume 1 (now Part I) 1992, 1978 by Springer-Verlag/Wien.
Volume 2 (now Part II) 1986, 1979 by Springer-Verlag/Wien.

9 8 7 6 5 4 3 2 1 SPIN 10947258

www.springer-ny.com

Springer-Verlag New York Berlin Heidelberg
A member of BertelsmannSpringer Science+Business Media GmbH

Preface to the Third Edition

This edition combines the earlier two volumes on Classical Dynamical Systems and on Classical Field Theory, thus including in a single volume the material for a two-semester course on classical physics.

In preparing this new edition, I have once again benefited from valuable suggestions and corrections made by M. Breitenecker.

Vienna, Austria, February 1997 Walter Thirring

Preface to the Second Edition: Classical Dynamical Systems

The last decade has seen a considerable renaissance in the realm of classical dynamical systems, and many things that may have appeared mathematically overly sophisticated at the time of the first appearance of this textbook have since become the everyday tools of working physicists. This new edition is intended to take this development into account. I have also tried to make the book more readable and to eradicate errors.

Since the first edition already contained plenty of material for a one-semester course, new material was added only when some of the original could be dropped or simplified. Even so, it was necessary to expand the chapter with the proof of the K–A–M theorem to make allowances for the current trend in physics. This involved not only the use of more refined mathematical tools, but also a reevaluation of the word *fundamental*. What was earlier dismissed as a grubby calculation is now seen as the consequence of a deep principle. Even Kepler's laws, which determine the radii of the planetary orbits, and which used to be passed over in silence as mystical nonsense, seem to point the way to a truth unattainable by superficial observation: The ratios of the radii of Platonic solids to the radii of inscribed Platonic solids are irrational, but satisfy algebraic equations of lower order. These irrational numbers are precisely the ones that are the least well approximated by rationals, and orbits with radii having these ratios are the most robust against each other's perturbations, since they are the least affected by resonance effects. Some surprising results about chaotic dynamics have been discovered recently, but unfortunately their proofs did not fit within the scope of this book and had to be left out.

In this new edition, I have benefited from many valuable suggestions of colleagues who have used the book in their courses. In particular, I am deeply grateful to H. Grosse, H.-R. Grümm, H. Narnhofer, H. Urbantke, and above all

M. Breitenecker. Once again the quality of the production has benefited from drawings by R. Bertlmann and J. Ecker and the outstanding word processing of F. Wagner. Unfortunately, the references to the literature have remained sporadic, since any reasonably complete list of citations would have overwhelmed the space allotted.

Vienna, Austria, July 1988 Walter Thirring

Preface to the Second Edition: Classical Field Theory

In the past decade, the language and methods of modern differential geometry have been increasingly used in theoretical physics. What seemed extravagant when this book first appeared 12 years ago, as lecture notes, is now a commonplace. This fact has strengthened my belief that today students of theoretical physics have to learn that language—and the sooner the better. After all, they will be the professors of the twenty-first century, and it would be absurd if they were to teach then the mathematics of the nineteenth century. Thus, for this new edition I did not change the mathematical language. Apart from correcting some mistakes, I have only added a section on gauge theories. In the last decade, it has become evident that these theories describe fundamental interactions, and on the classical level, their structure is sufficiently clear to qualify them for the minimum amount of knowledge required by a theoretician. It is with much regret that I had to refrain from incorporating the interesting developments in Kaluza–Klein theories and in cosmology, but I felt bound to my promise not to burden the students with theoretical speculations for which there is no experimental evidence.

I am indebted to many people for suggestions concerning this volume. In particular, P. Aichelburg, H. Rumpf, and H. Urbantke have contributed generously to corrections and improvements. Finally, I would like to thank Dr. I. Dahl-Jensen for redoing some of the figures on the computer.

Vienna, Austria, December 1985 Walter Thirring

Preface to the First Edition

This textbook presents mathematical physics in its chronological order. It originated in a four-semester course I offered to both mathematicians and physicists, who were only required to have taken the conventional introductory courses. In order to be able to cover a suitable amount of advanced material for graduate students, it was necessary to make a careful selection of topics. I decided to cover only those subjects in which one can work from the basic laws to derive physically relevant results with full mathematical rigor. Models that are not based on realistic physical laws can at most serve as illustrations of mathematical theorems, and theories whose predictions are only related to the basic principles through some uncontrollable approximation have been omitted. The complete course comprises the following one-semester lecture series:

I. Classical Dynamical Systems

II. Classical Field Theory

III. Quantum Mechanics of Atoms and Molecules

IV. Quantum Mechanics of Large Systems

Unfortunately, some important branches of physics, such as the relativistic quantum theory, have not yet matured from the stage of rules for calculations to mathematically well-understood disciplines, and are therefore not taken up. The above selection does not imply any value judgment, but only attempts to be logically and didactically consistent.

General mathematical knowledge is assumed, at the level of a beginning graduate student or advanced undergraduate student majoring in physics or mathematics.

Some terminology of the relevant mathematical background is collected in the Glossary near the beginning of the book. More specialized tools are introduced as they are needed; I have used examples and counterexamples to try to give the motivation for each concept and to show just how far each assertion may be applied. The best and latest mathematical methods to appear on the market have been used whenever possible. In doing this, many an old and trusted favorite of the older generation has been forsaken, as I deemed it best not to hand dull and worn-out tools down to the next generation. It might perhaps seem extravagant to use manifolds in a treatment of Newtonian mechanics, but since the language of manifolds becomes unavoidable in general relativity, I felt that a course that used them right from the beginning was more unified.

References are cited in the text in square brackets [ ] and collected near the end of the book. A selection of the more recent literature is also to be found there, although it was not possible to compile a complete bibliography.

I am very grateful to M. Breitenecker, J. Dieudonné, H. Grosse, P. Hertel, J. Moser, H. Narnhofer, and H. Urbantke for valuable suggestions. F. Wagner and R. Bertlmann have made the production of this book very much easier by their greatly appreciated aid with the typing, production, and artwork.

Vienna, Austria, February 1977 Walter Thirring

Note About the Translation

In the English translation, we have made several additions and corrections to try to eliminate obscurities and misleading statements in the German text. The growing popularity of the mathematical language used here has caused us to update the Bibliography. We are indebted to A. Pflug and G. Siegl for a list of misprints in the original edition. The translator is grateful to the Navajo Nation and to the Institute for Theoretical Physics of the University of Vienna for hospitality while he worked on this book.

Atlanta, Georgia, USA — Evans M. Harrell II

Vienna, Austria — Walter Thirring

Contents

Preface to the Third Edition v

Preface to the Second Edition: Classical Dynamical Systems vii

Preface to the Second Edition: Classical Field Theory ix

Preface to the First Edition xi

Note About the Translation xiii

Glossary xix

Symbols Defined in the Text xxv

Part I Classical Dynamical Systems

1 Introduction 3
- 1.1 Equations of Motion 3
- 1.2 The Mathematical Language 6
- 1.3 The Physical Interpretation 7

2 Analysis on Manifolds 11
- 2.1 Manifolds 11
- 2.2 Tangent Spaces 23

2.3 Flows . . . 36
2.4 Tensors . . . 45
2.5 Differentiation . . . 65
2.6 Integrals . . . 77

3 Hamiltonian Systems **89**
3.1 Canonical Transformations . . . 89
3.2 Hamilton's Equations . . . 96
3.3 Constants of Motion . . . 105
3.4 The Limit $t \to \pm\infty$ . . . 122
3.5 Perturbation Theory: Preliminaries . . . 145
3.6 Perturbation Theory: The Iteration . . . 157

4 Nonrelativistic Motion **169**
4.1 Free Particles . . . 169
4.2 The Two-Body Problem . . . 173
4.3 The Problem of Two Centers of Force . . . 182
4.4 The Restricted Three-Body Problem . . . 190
4.5 The N-Body Problem . . . 204

5 Relativistic Motion **213**
5.1 The Hamiltonian Formulation of the Electrodynamic Equations of Motion . . . 213
5.2 The Constant Field . . . 219
5.3 The Coulomb Field . . . 226
5.4 The Betatron . . . 232
5.5 The Traveling Plane Disturbance . . . 237
5.6 Relativistic Motion in a Gravitational Field . . . 242
5.7 Motion in the Schwarzschild Field . . . 248
5.8 Motion in a Gravitational Plane Wave . . . 257

6 The Structure of Space and Time **265**
6.1 The Homogeneous Universe . . . 265
6.2 The Isotropic Universe . . . 267
6.3 M_e According to Galileo . . . 269
6.4 M_e as Minkowski Space . . . 271
6.5 M_e as a Pseudo-Riemannian Space . . . 277

Part II Classical Field Theory

7 Introduction to Classical Field Theory **285**
7.1 Physical Aspects of Field Dynamics . . . 285
7.2 The Mathematical Formalism . . . 294
7.3 Maxwell's and Einstein's Equations . . . 312

8 The Electromagnetic Field of a Known Charge Distribution **329**

8.1 The Stationary–Action Principle and Conservation Theorems 329
8.2 The General Solution 340
8.3 The Field of a Point Charge 352
8.4 Radiative Reaction 370

9 The Field in the Presence of Conductors **383**

9.1 The Superconductor 383
9.2 The Half-Space, the Wave-Guide, and the Resonant Cavity 393
9.3 Diffraction at a Wedge 405
9.4 Diffraction at a Cylinder 417

10 Gravitation **433**

10.1 Covariant Differentiation and the Curvature of Space 433
10.2 Gauge Theories and Gravitation 452
10.3 Maximally Symmetric Spaces 468
10.4 Spaces with Maximally Symmetric Submanifolds 481
10.5 The Life and Death of Stars 499
10.6 The Existence of Singularities 512

Bibliography **529**

Index **539**

Glossary

Logical Symbols

$\forall$	for every
$\exists$	there exist(s)
$\nexists$	there does not exist
$\exists!$	there exists a unique
$a \Rightarrow b$	if a then b
iff	if and only if

Sets

$a \in A$	a is an element of A
$a \notin A$	a is not an element of A
$A \cup B$	union of A and B
$A \cap B$	intersection A and B
CA	complement of A (In a larger set B: $\{a : a \in B, a \notin A\}$)
$A \setminus B$	$\{a : a \in A, a \notin B\}$
$A \Delta B$	symmetric difference of A and B: $(A \setminus B) \cup (B \setminus A)$
$\emptyset$	empty set
$C\emptyset$	universal set
$A \times B$	Cartesian product of A and B: the set of all pairs (a, b), $a \in A$, $b \in B$

Important Families of Sets

open sets	contains Ø and the universal set and some other specified sets, such that the open sets are closed under union and finite intersection
closed sets	the complements of open sets
measurable sets	contains Ø and some other specified sets, and closed under complementation and countable intersection
Borel-measurable sets	the smallest family of measurable sets that contains the open sets
null sets, or sets of measure zero	the sets whose measure is zero. "Almost everywhere" means "except on a set of measure zero."

An equivalence relation is a covering of a set with a nonintersecting family of subsets. $a \sim b$ means that a and b are in the same subset. An equivalence relation has the following properties: (i) $a \sim a$ for all a; (ii) $a \sim b \Rightarrow b \sim a$; (iii) $a \sim b$, $b \sim c \Rightarrow a \sim c$.

Numbers

$\mathbb{N}$	natural numbers
$\mathbb{Z}$	integers
$\mathbb{R}$	real numbers
$\mathbb{R}^+$ ($\mathbb{R}^-$)	positive (negative) numbers
$\mathbb{C}$	complex numbers
sup	supremum, or lowest upper bound
inf	infimum, or greatest lower bound
I	any open interval
(a, b)	the open interval from a to b
$[a, b]$	the closed interval from a to b
$(a, b]$ and $[a, b)$	half-open intervals from a to b
$\mathbb{R}^n$	$\underbrace{\mathbb{R} \times \cdots \times \mathbb{R}}_{N \text{ times}}$ This is a vector space with the scalar product $(y_1, \ldots, y_N \mid x_1, \ldots, x_N) = \sum_{i=1} y_i x_i$

Maps (= Mappings, Functions)

$f : A \to B$	for every $a \in A$ an element $f(a) \in B$ is specified
$f(A)$	image of A, i.e., if $f : A \to B$, $\{f(a) \in B : a \in A\}$
$f^{-1}(b)$	inverse image of b, i.e., $\{a \in A : f(a) = b\}$
f^{-1}	inverse mapping to f. *Warning*: (1) it is not necessarily a function, and (2) distinguish from $1/f$ when $B = \mathbb{R}$.

$f^{-1}(B)$	inverse image of $B : \bigcup_{b\in B} f^{-1}(b)$
f is injective (one-to-one)	$a_1 \neq a_2 \Rightarrow f(a_1) \neq f(a_2)$
f is surjective (onto)	$f(A) = B$
f is bijective	f is injective and surjective. Only in this case is f^{-1} a true function
$f_1 \times f_2$	the function defined from $A_1 \times A_2$ to $B_1 \times B_2$, so that $(a_1, a_2) \to (f_1(a_1), f_2(a_2))$
$f_2 \circ f_1$	f_1 composed with f_2: if $f_1 : A \to B$ and $f_2 : B \to C$, then $f_2 \circ f_1 : A \to C$ so that $a \to f_2(f_1(a))$
1	identity map, when $A = B$; i.e., $a \to a$. *Warning*: do not confuse with $a \to 1$ when $A = B = \mathbb{R}$.
$f\big\|_U$	f restricted to a subset $U \subset A$
$f\big\|_a$	evaluation of the map f at the point a; i.e., $f(a)$
f is continuous	the inverse image of any open set is open
f is measurable	the inverse image of any measurable set is measurable
supp f	support of f: the smallest closed set on whose complement $f = 0$
C^r	the set of r times continuously differentiable functions
C_0^r	the set of C^r functions of compact (see below) support
χ_A	characteristic function of $A : \chi_A(a) = 1 \ldots$

Topological Concepts

topology	any family of open sets, as defined above
compact set	a set for which any covering with open sets has a finite subcovering
connected set	a set for which there are no proper subsets that are both open and closed
discrete topology	the topology for which every set is an open set
trivial topology	the topology for which the only open sets are $\emptyset$ and $C\emptyset$
simply connected set	a set in which every closed path (loop) can be continuously deformed to a point
(open) neighborhood of $a \in A$	any open subset of A containing a. Usually denoted by U or V
(open) neighborhood of $B \subset A$	any open subset of A containing B
p is a point of accumulation (= cluster point) of B	for any neighborhood U containing p, $U \cap B \setminus \{p\} \neq \emptyset$
$\bar{B}$	closure of B: the smallest closed set containing B
B is dense in A	$\bar{B} = A$

B is nowhere dense in A	$A \setminus \bar{B}$ is dense in A
metric (distance function) for A	a map $d : A \times A \to \mathbb{R}$ such that $d(a, a) = 0$; $d(a, b) = d(b, a) > 0$ for $b \neq a$; and $d(a, c) \leq d(a, b) + d(b, c)$ for all a, b, c in A. A metric induces a topology on A, in which all sets of the form $\{b : d(b, a) < \eta\}$ are open
separable space	a space with a countable dense subset
homeomorphism	a continuous bijection with a continuous inverse
product topology on $A_1 \times A_2$	the family of open sets of the form $U_1 \times U_2$, where U_1 is open in A_1 and U_2 is open in A_2, and unions of such sets

Mathematical Conventions

$f_{,i}$	$\partial f/\partial q_i$
$\dot{q}(t)$	$dq(t)/dt$
$\det \lvert M_{ij} \rvert$	determinant of the matrix M_{ij}
$\operatorname{Tr} M$	$\sum_i M_{ii}$
δ^i_j, δ_{ij}	1 if $i = j$, otherwise 0
$\varepsilon_{i_1,\ldots,i_m}$	the totally antisymmetric tensor of degree m, with values ± 1
M^t	transposed matrix: $(M^t)_{ij} = M_{ji}$
M^*	Hermitian conjugate matrix: $(M^*)_{ij} = (M_{ji})^*$
$\mathbf{v} \cdot \mathbf{w}$, $(\mathbf{v} \mid \mathbf{w})$, or $(\mathbf{v} \cdot \mathbf{w})$	scalar (inner, dot) product
$\mathbf{v} \times \mathbf{w}$ or $[\mathbf{v} \wedge w]$	cross product
∇f	gradient of f
$\nabla \times \mathbf{f}$	curl of $\mathbf{f}$
$\nabla \cdot \mathbf{f}$	divergence of $\mathbf{f}$
$\lVert\mathbf{v}\rVert$ (in three dimensions, $\lvert\mathbf{v}\rvert$)	length of the vector $\mathbf{v}$: $\lVert\mathbf{v}\rVert = (\sum_{i=1} v_i^2)^{1/2} = d(\mathbf{0}, \mathbf{v})$
ds	differential line element
$d\mathbf{S}$	differential surface element
$d^m q$	m-dimensional volume element
$\perp$	is perpendicular (orthogonal) to
$\parallel$	is parallel to
$\angle$	angle
$d\Omega$	element of solid angle
$\mathrm{Mat}_n(\mathbb{R})$	the set of real $n \times n$ matrices
$O(x)$	order of x

The summation convention for repeated indices is understood except where it does not make sense. For example, $L_{ik}x_k$ stands for $\sum_k L_{ik}x_k$.

Groups

GL_n	group of $n \times n$ matrices with nonzero determinant
O_n	group of $n \times n$ matrices M with $MM^t = \mathbf{1}$ (unit matrix)
SO_n	subgroup of O_n with determinant 1
E_n	Euclidean group
S_n	group of permutations of n elements
U_n	group of complex $n \times n$ matrices M with $MM^* = \mathbf{1}$ (unit matrix)
Sp_n	group of symplectic $n \times n$ matrices

Physical Symbols

m_i	mass of the ith particle
$\mathbf{x}_i$	Cartesian coordinates of the ith particle
$t = x^\circ/c$	time
s	proper time
q_i	generalized coordinates
p_i	generalized momenta
e_i	charge of the ith particle
κ	gravitational constant
c	speed of light
$\hbar = h/2\pi$	Planck's constant divided by 2π
F^α_β	electromagnetic field tensor
$g_{\alpha\beta}$	gravitational metric tensor (relativistic gravitational potential)
$\mathbf{E}$	electric field strength
$\mathbf{B}$	magnetic field strength in a vacuum
$\sim$	is on the order of
$\gg$	is much greater than

Symbols Defined in the Text

Df	derivative of $f : \mathbb{R}^n \to \mathbb{R}^m$	(2.1.1)
(V, Φ)	chart	(2.1.3)
T^n	n-dimensional torus	(2.1.7; 2)
S^n	n-dimensional sphere	(2.1.7; 2)
∂M	boundary of M	(2.1.20)
$\Theta_C(q)$	mapping of the tangent space into $\mathbb{R}^m$	(2.2.1)
$T_q(M)$	tangent space at the point q	(2.2.4)
$T_q(f)$	derivative of f at the point q	(2.2.7)
$T(M)$	tangent bundle	(2.2.12)
Π	projection onto a basis	(2.2.15)
$T(f)$	derivative of $f : M_1 \to M_2$	(2.2.17)
$T_0^1(M)$	set of vector fields	(2.2.19)
Φ_*	induced mapping on T_s^r	(2.2.21)
L_X	Lie derivative	(2.2.25; 1), (2.5.7)
∂_i	natural basis on the tangent space	(2.2.26)
Φ_t^x	flow	(2.3.7)
τ_t^x	automorphism of a flow	(2.3.8)
W	action	(2.3.16)
L	Lagrangian	(2.3.17)
H	Hamiltonian	(2.3.26)
$T_q^*(M)$	cotangent space	(2.4.1)
e_i^*	dual basis	(2.4.2; 1)
df	differential of a function	(2.4.3; 1)
$T_{qs}^r(M)$	space of tensors	(2.4.4)

$\otimes$	tensor product	(2.4.5)
$\wedge$	wedge (outer, exterior) product	(2.4.7)
i_X	interior product	(2.4.9), (2.4.16)
$*$	*-mapping	(2.4.18)
$T_s^r(M)$	tensor bundle	(2.4.25)
g	pseudo-Riemannian metric	(2.4.27)
$\mathcal{T}_s^r(M)$	set of tensor fields	(2.4.28)
$E_p(M)$	set of p-forms	(2.4.28)
$\overset{\pi}{\times}$	fiber product	(2.4.34)
$T^*(\Phi)$	transposed derivative	(2.4.34)
Φ^*	pull-back, or inverse image of the covariant tensors	(2.4.41)
d	exterior derivative	(2.5.1)
$[\,]$	Lie bracket	(2.5.9; 6)
Θ, ω	canonical forms	(3.1.1)
Ω	Liouville measure	(3.1.2; 3)
X_H	Hamiltonian vector field	(3.1.9)
b	bijection associated with ω	(3.1.9)
$\{\,\}$	Poisson brackets	(3.1.11)
M_e	generalized configuration space	(3.2.12)
$\mathcal{H}$	Hamiltonian on M_e	(3.2.12)
(I, φ)	action-angle variables	(3.3.14)
$\Omega_\pm$	Møller transformations	(3.4.4)
S	scattering matrix	(3.4.9)
$d\sigma$	differential scattering cross-section	(3.4.15)
$\mathbf{L}$	angular momentum	(4.1.3)
$\mathbf{K}$	boost	(4.1.9)
$\eta_{\alpha\beta}$	Minkowski space metric	(5.1.2)
γ	$1/\sqrt{1-v^2/c^2}$ (relativistic dilatation)	(5.1.4; 2)
F	electromagnetic 2-form	(5.1.10; 1)
A	1-form of the potential	(5.1.10; 1)
Λ	Lorentz transformation	(5.1.12)
r_0	Schwarzschild radius	(5.7.1)
$e^{i_1 i_2 \ldots i_p}$	basis of the p-forms	(7.2.3)
$E_p(M)$	linear space of the p-forms	(7.2.5; 2)
d	exterior differential	(7.2.6)
$\omega_{\mid N}$	restriction of a form	(7.2.7; 3)
$E_m^0(U)$	space of m-forms with compact support	(7.2.9)
$\langle e^i(x) \mid e^k(x) \rangle$	scalar product	(7.2.14)
i_v	interior product	(7.2.16)
$*$	isomorphism between E_p and E_{m-p}	(7.2.17)
δ	codifferential	(7.2.19)
Δ	Laplace–Beltrami operator	(7.2.20)
L_v	Lie derivative	(7.2.23)

ω_k^i	affine connection	(7.2.25)
ω_{ik}	affine connection	(7.2.25)
$\Theta(x)$	Heaviside step function	(7.2.31)
$\delta(x)$	Dirac delta function	(7.2.31)
$\delta_{\bar{x}}$	Dirac delta form	(7.2.33)
$G_{\bar{x}}$	Green function	(7.2.35)
$\mathbf{E}, \mathbf{B}, \mathbf{F}$	electric and magnetic fields	(7.3.1)
A	vector potential	(7.3.7)
Λ	gauge function	(7.3.10; 1)
J	current	(7.3.12)
Q	total charge	(7.3.18; 2)
$T^{\alpha\beta}$	energy-momentum tensor	(7.3.20)
P^α	total energy-momentum	(7.3.21)
T^α	energy-momentum form of the field	(7.3.22)
$z(s)$	world-line	(7.3.25; 2)
t^α	energy-momentum form of matter	(7.3.25; 2)
$\mathcal{L}$	Lagrangian	(8.1.1)
W	action	(8.1.1)
$\mathcal{S}$	Poynting's vector	(8.1.13)
$D^\pm(N)$	domains of influence	(8.1.15)
$D_{\bar{x}}$	Green function	(8.2.5)
$D^{\text{ret}}(x)$	retarded Green function	(8.2.7)
$G_{\bar{x}}^{\text{ret}}$	retarded Green function (form)	(8.2.7)
F^{ret}	retarded field strength	(8.2.9)
F^{in}	incoming field strength	(8.2.15)
F^{out}	outgoing field strength	(8.2.15)
F^{rad}	radiation field	(8.2.21)
$D(x)$	D-function	(8.2.22)
δE	energy loss per period	(8.4.4; 2)
j	specified current	(9.1.7)
$\varepsilon(k)$	dielectric constant	(9.1.19; 3)
S	superpotential	(9.1.21; 1)
$F(z)$	Fresnel's integral	(9.3.10)
$\langle\mid\rangle$	scalar product	(10.1.3)
S_p	Sections	(10.1.9)
D	exterior covariant derivative	(10.1.10)
D_X	covariant derivative	(10.1.15)
Ω	curvature form	(10.1.19)
R	curvature in space-time	(10.1.20; 1)
Γ_{ijk}	Christoffel symbol	(10.1.36)
R_{ijkm}	Riemann–Christoffel tensor	(10.1.44; 2)
C_{jk}	Weyl forms	(10.1.44; 3)
K	curvature parameter	(10.4.42)
c	rate of convergence	(10.6.8)

$J^+(x)$	future of x	(10.6.18(a))
$J^-(x)$	past of x	(10.6.18(a))
$C(x, S)$	set of causal curves	(10.6.18(b))
$C^1(x, S)$	set of differentiable causal curves	(10.6.18(b))
$d(\lambda)$	length of λ	(10.6.18(c))

Part I

Classical Dynamical Systems

1

Introduction

1.1 Equations of Motion

The foundations of the part of mechanics that deals with the motion of point-particles were laid by Newton in 1687 in his *Philosophiae Naturalis Principia Mathematica*. This classic work does not consist of a carefully thought-out system of axioms in the modern sense, but rather of a number of statements of various significance and generality, depending on the state of knowledge at the time. We shall take his second law as our starting point: "Force equals mass times acceleration." Letting $\mathbf{x}_i(t)$ be the Cartesian coordinates of the ith particle as a function of time, this means

$$(1.1.1) \qquad m_i \frac{d^2\mathbf{x}_i(t)}{dt^2} = \mathbf{F}_i(\mathbf{x}_i), \qquad i = 1, 2, \ldots, N,$$

where $\mathbf{F}_i$ denotes the force on the ith particle. In nature, so far as we know, there are just four fundamental forces: the strong, weak, electromagnetic, and gravitational forces. In physics books there are in addition numerous other forces, such as friction, exchange forces, forces of constraint, fictitious forces (centifugal, etc.), and harmonic forces, with which we shall only be peripherally concerned. The first two fundamental forces operate at the subatomic level, outside the realm of classical mechanics, so in fact we shall only discuss gravitation and electromagnetism.

The exact expressions for these forces are rather complicated in their full generality, but, surprisingly, they both simplify greatly in the limit where the velocities of the particles are much less than the speed of light. They are the gradients of the

Newtonian and Coulombic potentials, i.e.,

$$\mathbf{F}_i(\mathbf{x}_i) = \sum_{j \neq i} \frac{\mathbf{x}_j - \mathbf{x}_i}{|\mathbf{x}_j - \mathbf{x}_i|^3}(\kappa m_i m_j - e_i e_j), \tag{1.1.2}$$

where κ is the gravitational constant and e_i is the charge of the ith particle.

For the elementary constituents of matter, e^2 and κm^2 are of very different orders of magnitude: for protons, $e^2 \sim 10^{36}\ \kappa m_p^2$. The reason that gravitation is nonetheless significant is that all masses are positive and add constructively in a large object, whereas the overall charge can be neutral. In astronomical bodies ($N \sim 10^{57}$ for the sun), only gravitation contributes significantly to (1.1.2). One might hestitate to apply (1.1.1) to such bodies, because a star is hardly a point-particle, and it is unclear what meaning should be attached to $\mathbf{x}_i$. But it is noteworthy that (1.1.1) also applies to the center of mass of the whole body, which moves according to Newton's law in response to the net force. In practice there is no difficulty with the meaning of $\mathbf{x}_i$, since heavenly bodies are usually rather small, compared with typical distances between them.

To get a feeling for the meaning of the constants of nature just introduced, let us look at their orders of magnitude in the framework of (1.1.1) and (1.1.2). Suppose a particle orbits a star with $N \sim 10^{57}$ protons, with period τ at radius R. Then from (1.1.1) and (1.1.2), essentially

$$\frac{R^3}{\tau^2} = N\kappa m_p, \tag{1.1.3}$$

in which the mass of the orbiting particle has dropped out, with a purely gravitational force. In cgs units, $\kappa m_p \sim 10^{-32}$, so for a given R we expect period $\tau \sim 10^{16}\ R^{3/2}\ N^{-1/2}$ and velocity $v \sim 10^{-16}\ R^{-1/2}\ N^{1/2}$. For typical cosmic distances and $N \sim 10^{57}$:

	R (cm)	τ (sec)	v (cm/sec)
Earth's orbit	10^{13}	10^{7}	10^{6}
double star	10^{11}	10^{4}	10^{7}
black hole	10^{5}	10^{-5}	10^{10}

We see that in a planetary system typical speeds are 10–100 km/sec, which may seem rather fast, but is modest compared with the speed of light. It is only when the dimensions are roughly those of a black hole, in which the mass of a star is compressed to within a few kilometers, that gravitation can lead to speeds approaching the speed of light. At that point the equations of motion (1.1.1) lose their validity and must be replaced with their relativistic version, discussed below.

As already noted, the electrical force between protons is 10^{36} times stronger than their gravitational force. For a proton–electron system this number is raised by three orders of magnitude, the ratio of the proton's mass to the electron's mass, giving 10^{39}. Correspondingly, the relationships between R, τ, and v become $\tau \sim 10^{-7/2}\ R^{3/2}\ N^{-1/2}$ and $v \sim 10^{7/2}\ R^{-1/2}\ N^{1/2}$. On the atomic scale ($R \sim 10^{-8}$ cm), and for $N \sim 1$, we now find impressive speeds, $v \sim 10^{7.5}$ cm/sec and $\tau \sim 10^{-15.5}$ sec. It is thus relatively easy to accelerate charged elementary particles to nearly the speed of light, which necessitates a generalization of Newton's equation of motion.

The law that replaces (1.1.1) and (1.1.2) in these cases is best formulated if one regards ct and $\mathbf{x}$ as dependent variables x^α, $\alpha = 0, 1, 2, 3$, and introduces a parameter s, the proper time, as the independent variable, defined so that $ds^2 = c^2\,dt^2 - |d\mathbf{x}|^2$. The electromagnetic field is no longer a vector, but a tensor field of the second degree. The equation of motion generalizing (1.1.1) for a charged particle in an electromagnetic field then reads

$$m\frac{d^2x^\alpha}{ds^2} = eF^\alpha_\beta(x)\frac{dx^\beta}{ds}, \tag{1.1.4}$$

where by convention the repeated index β is summed over.

The force in (1.1.2) can be written as the gradient of a potential. In the relativistic case the electromagnetic field may be expressed with derivatives of a vector potential as

$$F_{\alpha\beta} = \frac{\partial}{\partial x^\alpha}A_\beta - \frac{\partial}{\partial x^\beta}A_\alpha. \tag{1.1.5}$$

Since A_α depends on the positions, or more precisely on the trajectories, of the charged particles, the relativistic formula (1.1.5) is rather more complicated than (1.1.2) and requires the use of field theory. At present we must content ourselves with the restricted problem of a particle in a specified external field $F_{\alpha\beta}$.

The utility of (1.1.4) is further reduced, because macroscopic objects rarely approach c, while the motion of elementary particles actually belongs to the quantum theory. Nonetheless, the classical equation (1.1.4) gives the essential behavior in many cases.

The equations of motion which generalize (1.1.1) for fast-moving bodies in a gravitational field are even more complicated than (1.1.4). As in the nonrelativistic theory the force is proportional to the mass, but one now needs an equation with three indices:

$$\frac{d^2x^\alpha}{ds^2} = -\Gamma^\alpha_{\beta\gamma}(x)\frac{dx^\beta}{ds}\frac{dx^\gamma}{ds}. \tag{1.1.6}$$

Gravitation is generalized through $\Gamma^\alpha_{\beta\gamma}$, which again can be written with derivatives of a potential, though now a symmetric tensor of the second degree:

$$\Gamma^\alpha_{\beta\gamma} = \tfrac{1}{2}(g^{-1})^{\alpha\sigma}\left(\frac{\partial g_{\sigma\beta}}{\partial x^\gamma} + \frac{\partial g_{\sigma\gamma}}{\partial x^\beta} - \frac{\partial g_{\beta\gamma}}{\partial x^\sigma}\right). \tag{1.1.7}$$

Once more we must resort to field theory at this point if we wish to determine $g_{\alpha\beta}(x)$ for a given distribution of mass. We shall only study these equations of motion for certain g's; it turns out that despite a mathematical structure similar to (1.1.1), the physics enters a completely different world.

1.2 The Mathematical Language

Formula (1.1.1) is an ordinary differential equation of second order for a vector in $\mathbb{R}^{3N}$. However, since the forces (1.1.2) have a singularity when $\mathbf{x}_i = \mathbf{x}_j, i \neq j$, it is advisable to remove those points and work in an open subset of $\mathbb{R}^{3N}$. In doing this one gives up all information about what happens after a collision, but that is just as well, for otherwise the equations would undoubtedly be pushed beyond their physical validity. The equations could in fact be regularized by the introduction of another variable in place of t, so that the solutions would extend beyond the collision (see [6] and [7]). There is indeed some physical interest in these regularizations, but only in the possibility of more accurate numerical analyses of near misses; they cannot describe true catastrophes.

We shall, however, broaden the mathematical domain of definition of the equations of motion on open sets of $\mathbb{R}^{3N}$ somewhat further. The process of differentiation depends only on local properties of a Euclidean space, and thus carries over to anything that looks just like a Euclidean space to a near-sighted observer. In this way we are led to introduce differentiable manifolds, for the following reasons:

1. When one deals with a three-dimensional space with the origin removed, polar coordinates are preferable to Cartesian coordinates for many purposes. The space does not then appear as an open subset of $\mathbb{R}^3$, but as (positive numbers) $\times$ (surface of a sphere). Hence it is desirable to formulate a differential calculus for spherical surfaces, which are not open subsets of $\mathbb{R}^n$.

2. If we know a constant of the motion K, we may restrict the equations of motion to the surface $K =$ const., which is a manifold. This might typically be motion on a torus, which has quite different properties from free motion in $\mathbb{R}^n$.

3. Equation (1.1.6) and problems with constraints are generally set up on manifolds in the first place.

4. It is essential to distinguish local and global quantities in order to understand the mathematical structure of classical mechanics. A Hamiltonian system with n degrees of freedom will always locally have $2n - 1$ time-independent constants. The crucial question is how many of these may be defined globally. The concept of manifold serves to clarify this distinction.

In the second chapter we shall develop the necessary mathematical methods. The almost infinitesimal ratio of the number of propositions to the number of definitions is plain evidence that it is less a question of obtaining deep results than of generalizing and sharpening our knowledge of elementary mathematics, or simply common sense. Elementary mechanics gets extended to a more flexible scheme. The various infinitely small quantities like "infinitesimal variations" and "virtual displacements" disappear and are replaced more precisely with mappings of the tangent spaces. The tangent spaces and their associated bundles are the real stage for dynamics, where, roughly speaking, the tangent bundle is the space of

q and $\dot{q}$, and the cotangent bundle is the space of q and p, that is, phase space. After a little necessary preparation we thus arrive at Cartan's symbolism, in which all the rules of elementary differential and integral calculus are written down with a very few symbols. At first it may seem only an exercise in the abstract style of writing. But the reward is that this abstract notation succeeds in reducing the general assertions of classical mechanics to trivialities.

1.3 The Physical Interpretation

In order to interpret the formalism it must first be agreed what the observable quantities are. The observables generally correspond to the coordinates and momenta of the particles. There is of course no reason that the coordinate system should necessarily be Cartesian; for example, in astronomy it is usually angles that are directly measured. We should therefore allow arbitrary functions of coordinates and momenta as observables, subject only to boundedness and, for mathematical convenience, differentiability. Such functions form an (Abelian) algebra, and the time-evolution defined by the equations of motion gives an automorphism of the algebra, since sums are transformed to sums and products to products. It is well to distinguish this algebra of observables conceptually from the state in which a particular specimen of the system is to be found; the state has nothing to do with the laws of nature, but only reflects our knowledge of the initial conditions that happened to be realized.

Whereas the observables are functions on phase space, the states are construed as probability measures on it. For each state there is a probability distribution $\rho(q, p)$ such that the average of many measurements of an observable $f(q, p)$ is predicted to be

$$\bar{f} = \int d^{3N}q\, d^{3N}p\rho(q, p) f(q, p). \tag{1.3.1}$$

Note that $\overline{f+g} = \bar{f} + \bar{g}$, but $\overline{f \cdot g} \neq \bar{f} \cdot \bar{g}$. This means that fluctuations arise so that $\overline{(f - \bar{f})^2} \neq 0$, unless the measure $d^{3N}q\, d^{3N}p\rho$ is concentrated at a point. Such "extremal" states amount to complete knowledge of all coordinates and momenta. With the solution $q(0), p(0) \to q(t), p(t)$ of the equations of motion, the automorphism mentioned above is $f(q(0), p(0)) \to f(q(t), p(t))$.

Although this conceptual distinction between observables and states is avoidable until one encounters quantum mechanics, it draws attention to the essential nature of the problem even in classical mechanics. It is not sufficient to solve the equations of motion for a few initial conditions which happen to arise; instead, they must be solved for arbitrary initial conditions. In particular, the stability of the solutions under small perturbations of the initial conditions, which are never exactly known in reality, becomes an essential question. Above all, this point of view is well suited to statistical physics, where only a small amount of information is given for a system of many degrees of freedom, and the critical facts are the absence of stability and mixing properties of the time-evolution.

To be sure, the execution of this program for realistic forces (1.1.2) creates some difficulties. As mentioned, when there is a collision the trajectory leaves the domain of definition of the problem, at which point we can look no further into the time-evolution. Since initial conditions can always be found so that a collision takes place within an arbitrarily short time, we do not really have an automorphism of the algebra. In the two-body problem it happens that the situation may be remedied by removing the region with angular momentum zero from phase space, since in the rest of phase space no collisions can occur. However, in a three-body system this only avoids triple collisions, and it is necessary to regularize the equations of motion with a new time variable if one wants to get an automorphism. In the relativistic case (1.1.4, 6) the situation is even more hazardous, and even in the two-body problem particles that have nonzero angular momentum can be pulled into the singularity. Popularly speaking, there is a black hole and not just a black point. Hence we must moderate our demands and be contented to examine smaller pieces of phase space. The central questions become: Which configurations are stable? Will collisions ever occur? Will particles ever escape to infinity? Will the trajectory always remain in a bounded region of phase space? The words "always" and "ever" make it hard to give exact answers. Computer calculations and, often, mathematical existence theorems provide answers only for the not-too-distant future, and predictions for longer times are notoriously inaccurate. In any case, an assertion that something will happen loses its interest for physics when the time in question is longer than the age of the universe.

For (1.1.1; 2) with two particles it is known that all finite orbits are periodic. But this is a degenerate case, which does not hold relativistically (1.1.4; 6) or when there are three particles. Instead, almost-periodic orbits are more typical, where the system returns arbitrarily close to the starting point, but the orbits are not closed. Rather, they intertwine densely in some higher-dimensional shape (a Lissajou figure). Between these almost-periodic orbits are no doubt imbedded an infinite number of others that are strictly periodic. For (1.1.2), $e_j = 0$, and more than two particles, there is a strong suspicion that the trajectories for which particles are sent off to infinity fill up most of phase space for all energies. This is certainly energetically possible, since the remaining particles can use potential energy to compensate for the loss. In fact, computer studies [8] show that fairly soon two particles will come so near that they can release enough energy to accelerate one of them off to infinity. It is apparent that this process is of great significance for planetary and stellar systems.

The book closes with an investigation of how the physical space–time manifold is determined by the laws of mechanics. At first the structure of space and time appears to be given a priori. Yet it is determined by real rulers and clocks, which are themselves subject to the equations of motion.[1] Thus it will be necessary to

---

[1] Of course, real matter is governed by the quantum theory, so we must anticipate some later material.

study whether the relationship between rulers and clocks that comes out of the equations of motion is consonant with our original assumptions about space and time. We shall see, for example, that space–time loses its pseudo-Euclidean nature through equations (1.1.6) and gets in its place a Riemannian structure. In other words, gravitation affects rulers so that the space they measure appears curved.

The attraction of the mechanics of point particles is that despite the simplicity of the basic laws, the trajectories that are possible produce such a large and complex picture that it is difficult to survey it all. It is already evident that the consequences of these laws of nature, which can be expressed so briefly, are hugely complicated.

2

Analysis on Manifolds

2.1 Manifolds

The intuitive picture of a smooth surface becomes analytic with the concept of a manifold. On the small scale a manifold looks like a Euclidean space, so that infinitesimal operations like differentiation may be defined on it.

A function f from an open subset U of $\mathbb{R}^n$ into $\mathbb{R}^m$ is differentiable at a point $x \in U$ if it may be approximated there with a linear mapping $Df: \mathbb{R}^n \to \mathbb{R}^m$. We can make this notion more precise by requiring that for all $\varepsilon > 0$ there exists a neighborhood U of x such that

$$\|f(x') - f(x) - Df(x)(x' - x)\| < \varepsilon \|x - x'\| \quad \forall x' \in U.$$

Here x and f are, respectively, vectors in $\mathbb{R}^n$ and $\mathbb{R}^m$, and $\|v\|$ is the length of the vector v. (We shall always make use of vector and matrix notation with the indices dropped, unless there is some reason to write them out.) Written out in components, Df is the matrix of the partial derivatives,

$$(2.1.1) \qquad (Df)_{ij} = \frac{\partial f_i}{\partial x_j}, \qquad i = 1, \ldots, m, \; j = 1, \ldots, n.$$

(2.1.2) **Remarks**

1. The function f must be given in a neighborhood of x. If we speak simply of differentiability (at all points), we have to deal with a mapping of some open set.

2. At every point the derivative Df is a linear mapping $\mathbb{R}^n \to \mathbb{R}^m$, which has the following significance: if the curve $u: I \to \mathbb{R}^n$ passes through x, then Df transforms the direction of the curve into the direction of the image of the curve under f. $(df_i(x(t))/dt = f_{i,j}\, dx_j/dt)$.
3. Df can also be regarded as a function, specifically as a mapping into the linear transformations. As such it can itself be differentiable, which simply means that the $f_{i,j}$ are further differentiable. We denote the set of p-times continuously differentiable functions by C^p, the set of infinitely often differentiable functions by C^∞, and the set of C^∞-functions of compact support by C_0^∞.

In this section we extend the idea of differentiability to sets M which resemble open sets in $\mathbb{R}^n$ only locally. In §2.2 we can then look for the spaces which are mapped linearly by the derivative. First we introduce some concepts which should be perfectly clear due to their geographical flavor.

(2.1.3) **Definition**

Let M be a topological space. A **Chart** (V, Φ) is a homeomorphism Φ of an open set V (the **domain** of the chart) of M to an open set in $\mathbb{R}^m$. Two charts are **compatible** in case $V_1 \cap V_2 = \emptyset$, or if the mappings $\Phi_1 \circ \Phi_2^{-1}$ and $\Phi_2 \circ \Phi_1^{-1}$, restricted in the obvious way, are C^∞-mappings of open sets in $\mathbb{R}^m$ (Figure 2.1).

(2.1.4) **Definition**

An **atlas** is a set of compatible charts that cover M. Two atlases are called **compatible** if all their charts are compatible.

(2.1.5) **Remarks**

1. Compatibility of atlases is obviously an equivalence relationship: every atlas is compatible with itself, and the definition is symmetric. Suppose $\bigcup_i (V_{1i}, \Phi_{1i})$ is compatible with $\bigcup_i (V_{2i}, \Phi_{2i})$, which is compatible with $\bigcup_i (V_{3i}, \Phi_{3i})$. Cover $V_{1i} \cap V_{3j}$ with the V_{2k}, and recall that $f \circ g$ is differentiable when f and g are.
2. Assuming that all the charts map M into an $\mathbb{R}^m$ with the same m, m is called the **dimension** of M. Occasionally this definition is also used when $m = 0$, although $\mathbb{R}^0$ is a point, for which there can be nothing to differentiate.
3. If the V's are chosen small enough, we can suppose that they are all connected sets.

(2.1.6) **Definition**

A differentiable **manifold** is a separable, metrizable space M with an equivalence class of atlases.

(2.1.7) **Examples**

1. $M = \mathbb{R}^n = V$. $\Phi = \mathbf{1}$. Only one chart is necessary in this case. This is also true for the somewhat more general case of an open subset of $\mathbb{R}^n$.

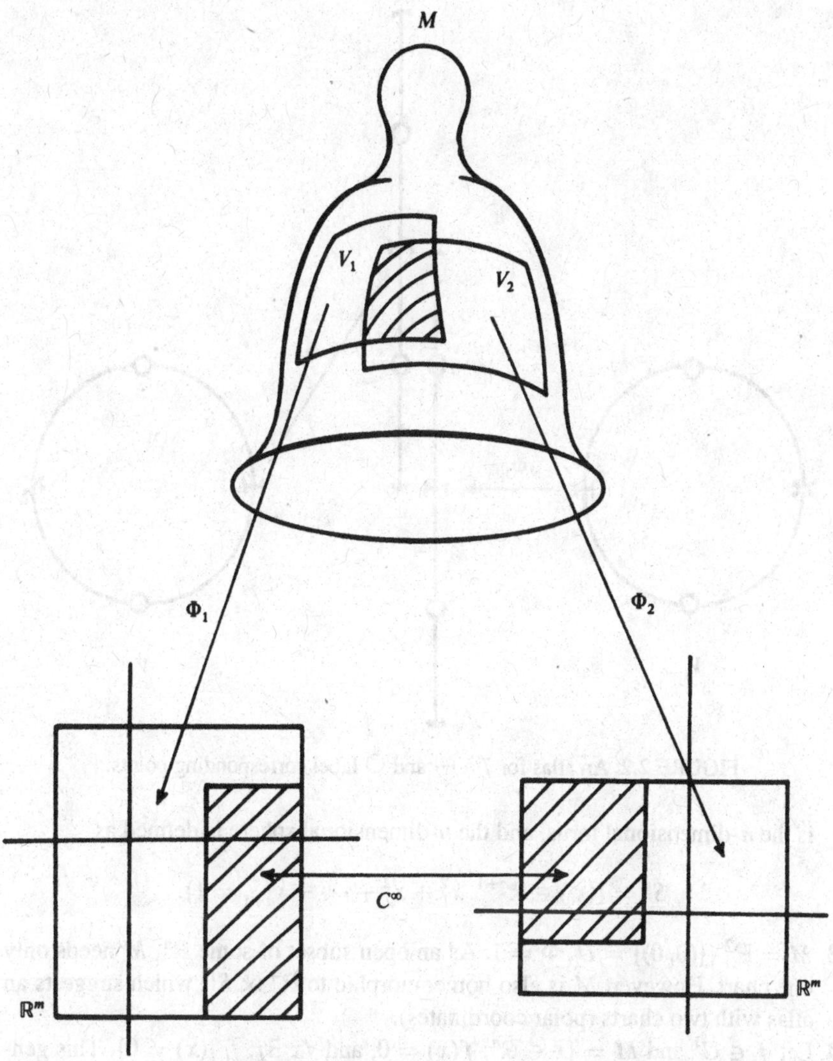

FIGURE 2.1. Compatibility of two charts.

2. $M = \{(x_1, x_2) \in \mathbb{R}^2 : x_1^2 + x_2^2 = 1\}$ is called the one-dimensional **sphere** S^1 or the one-dimensional **torus** T^1. M is compact and therefore not homeomorphic to an open subset of $\mathbb{R}$. At least two charts are needed:

$$V_1 = T^1 \backslash \{(-1, 0)\}, \qquad \Phi_1^{-1}: \varphi \to (\cos\varphi, \sin\varphi), \qquad -\pi < \varphi < \pi,$$

$$V_2 = T^1 \backslash \{(1, 0)\}, \qquad \Phi_2^{-1}: \varphi \to (\cos\varphi, \sin\varphi), \qquad 0 < \varphi < 2\pi.$$

The compatibility of Φ_1 and Φ_2 is trivial (Problem 1); but they can not be replaced with a single mapping (see Figure 2.2).

$$\underbrace{T^1 \times T^1 \times \cdots \times T^1}_{n \text{ times}} = T^n$$

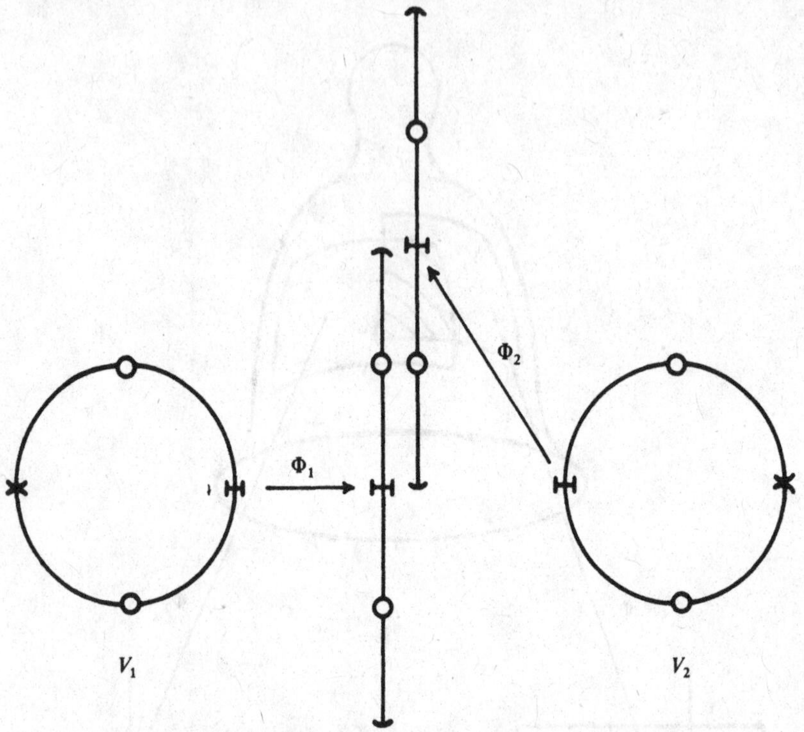

FIGURE 2.2. An atlas for T^1. $\mapsto$ and O label corresponding points.

is the n-dimensional torus, and the n-dimensional sphere is defined as

$$S^n = \{(x_i) \in \mathbb{R}^{n+1}: x_1^2 + x_2^2 + \cdots + x_{n+1}^2 = 1\}.$$

3. $M = \mathbb{R}^2 \setminus \{(0,0)\} = U$, $\Phi = 1$. As an open subset of some $\mathbb{R}^n$, M needs only one chart. However, M is also homeomorphic to $\mathbb{R}^+ \times S^1$, which suggests an atlas with two charts (polar coordinates).
4. Let $f \in C^1$ and $M = \{x \in \mathbb{R}^n: f(x) = 0, \text{ and } \forall x\ \exists j: f_{,j}(x) \neq 0\}$. This generalizes Example 2, and the implicit function theorem guarantees the existence of suitable charts so that M becomes an $(n-1)$-dimensional manifold. The condition on the derivative is obviously necessary, for suppose f is a constant function; then the inverse image of $\mathbf{0}$ is either the empty set or all of $\mathbb{R}^n$.
5. The n^2 elements of an $(n \times n)$-matrix define a point in $\mathbb{R}^{n^2}$. Hence the $(n \times n)$-matrices may be identified with $\mathbb{R}^{n^2}$ and inherit its structure as a manifold (and also as a vector space). The invertible matrices M, $\det M \neq 0$, are an open subset, and form the group $GL(n)$. The unimodular matrices M, $\det M = 1$, are characterized by a condition as in Example 4, and are thus an $(n^2 - 1)$-dimensional submanifold.
6. $M = \{\mathbf{x} \in \mathbb{R}^2: |x_1| = |x_2|\}$. This cannot be a manifold, since every neighborhood of $(0, 0)$ decomposes M without that point into four rather than two components, and consequently cannot be mapped homeomorphically onto an

open interval.

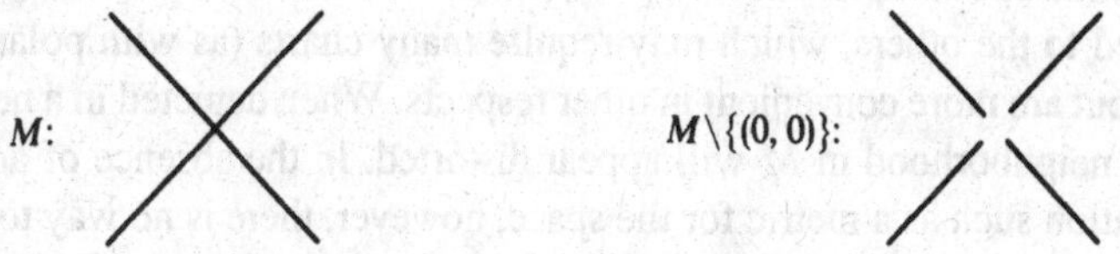

7. $M = \bigcup_{n=1,2,\dots} \{(1/n, \mathbb{R})\} \cup \{(0, \mathbb{R})\} \subset \mathbb{R}^2$ is certainly no manifold, since it is not locally connected at (0, 0).
8. Given two manifolds one can define the **product manifold** $M_1 \times M_2$ (cf. Examples 2 and 3). This set comes equipped with the product topology and the product chart $(V_1, \Phi_1) \times (V_2, \Phi_2) = (V_1 \times V_2, \Phi_1 \times \Phi_2)$ uses the mapping $(q_1, q_2) \to (\Phi_1(q_1), \Phi_2(q_2))$ into $\mathbb{R}^{m_1+m_2}$. It is clear that the product of two atlases is another atlas, since the conditions of covering and compatibility are fulfilled.

(2.1.8) **Remarks**

1. In Examples 1 to 4, M is given directly as a subset of $\mathbb{R}^n$ with the induced topology. It is not always done this way. More obviously, manifolds can be constructed by piecing overlapping regions together. This determines the global structure, while locally everything is determined by the dimension. However, it can be shown [1, Chapters 16 and 25] that every m-dimensional manifold is homeomorphic to a subset of $\mathbb{R}^{2m+1}$.
2. It must be assumed that M is separable in order to exclude a number of pathologies; it is not implied by M's being locally Euclidean. This is why we require a topology on M, rather than simply defining one with the charts.
   Example: $M = \mathbb{R} \times (\mathbb{R}$ with the discrete topology), $V_y = \mathbb{R} \times \{y\}$, $\Phi_y: (x, y) \to x$. By this devious construction a plane becomes a one-dimensional manifold.
3. We shall usually suppose that the manifolds are C^∞, which is not an excessively burdensome restriction. Of course, many results could be obtained with fewer assumptions, but it is not our goal to figure out what the optimal assumptions are. Moreover, in the future we will not always check whether all the assumptions of differentiability are satisfied. This is left to the conscientious reader, who will find that there are no real difficulties, since for these local questions everything works as in $\mathbb{R}^n$. For this same reason we shall simply say "manifold" rather than "C^∞-manifold."
4. Since in the cases that will concern us, most of the functions that crop up are analytically continuable, it is sometimes convenient to work with complex manifolds. In this $\mathbb{C}^n$ is substituted for $\mathbb{R}^n$, and analyticity for all degrees of differentiability. For an example of a complex manifold, think of the Riemann surface for $\sqrt{z}$ or for $\ln z$.
5. Physicists are used to the terms "**local coordinate system**" or "parametrization" instead of charts. That M is not defined with any particular atlas, but with an equivalence class of atlases, is a mathematical formulation of "general covari-

ance." Every suitable coordinate system is equally good. A Euclidean chart may well suffice for an open subset of $\mathbb{R}^n$, but this coordinate system is not to be preferred to the others, which may require many charts (as with polar coordinates), but are more convenient in other respects. When depicted in a new chart, a given neighborhood in M will appear distorted. In the absence of additional information such as a metric for the space, however, there is no way to say that one or other chart gives the "true picture."

As we have seen (Examples 6 and 7), not all subsets of $\mathbb{R}^n$ may be used as manifolds. They need not necessarily be open subsets of $\mathbb{R}^n$, but one should at least be able to define differentiation on them. The question now arises of when a subset can inherit the structure of a manifold.

(2.1.9) **Definition**

$N \subset M$ is an n-dimensional **submanifold** iff for all $q \in N$ there exists a chart (V, Φ), where $q \in V \subset M$ and $\Phi(V) \subset \mathbb{R}^m$, such that for all $q' \in N \cap V$, $\Phi(q') = (x_1, \ldots, x_n, 0, \ldots, 0)$.

(2.1.10) **Examples**

1. N is an open subset of M. This is the trivial case with $m = n$.
2. $N = S^1$, $M = \mathbb{R}^2$. The charts in (2.1.7; 2) are not of the form (2.1.9), but charts of that form are easy to find (Problem 2).
3. Let f_i, $i = 1, \ldots, k \leq m$, be differentiable functions $\mathbb{R}^m \rightarrow \mathbb{R}$ such that the vectors Df_i at each point where $f_i = 0, i = 1, \ldots, k$, are linearly independent, or, equivalently, the rank of the matrix $f_{i,j}$, $i = 1, \ldots, k$ and $j = 1, \ldots, m$, is maximal. Then according to the rank theorem [1, Section X.3], $N = \{x \in \mathbb{R}^m: f_i(x) = 0\ \forall i\}$ is a closed submanifold of dimension $m - k$ of $\mathbb{R}^m$. In particular the orthogonal matrices M, $MM^t = \mathbf{1}$, are a submanifold of the invertible matrices (cf. (2.1.7; 5)).
4. $M = \mathbb{R}^2$, $N = \{x \in M: x_2 = |x_1|\}$ can be equipped with a manifold structure but is not a submanifold of M.[1] There is a kink in N, which cannot be put into the differentiable form required in the definition, even with a new set of charts. Yet the atlas $(U = N, \Phi: (x_1, x_2) \rightarrow x_1)$ makes N a manifold.

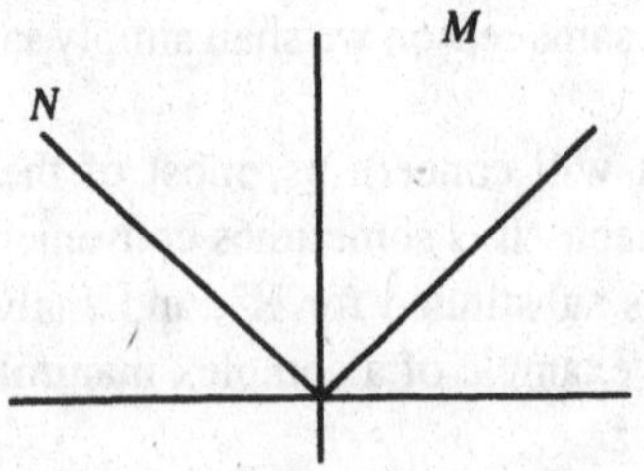

[1] The set N is, however, the union of three submanifolds.

(2.1.11) **Remarks**

1. It is easy to see that the atlas in (2.1.9) gives N the structure of a differentiable manifold: the differentiability required for compatibility is unaffected, since only the first n coordinates vary.
2. The last example shows that even when a manifold M is a set-theoretical and topological subspace of $\mathbb{R}^n$, it does not necessarily have to be a submanifold of $\mathbb{R}^n$. However, with the imbedding of M in $\mathbb{R}^{2m+1}$ mentioned in (2.1.8; 1), M is in fact a submanifold of $\mathbb{R}^{2m+1}$.
3. We produced submanifolds of $\mathbb{R}^n$ by requiring that $f_i(x) = 0$, $f_i \in C^\infty$, and that the Df_i were linearly independent. The existence of such functions f is implicit in the definition, at least locally.
4. The following proposition can be proved: Let Y be a submanifold of X, and Z a subset of Y. Then Z is a submanifold of Y iff it is a submanifold of X [1, 16.8.7].

Now we are ready to generalize the concept of a differentiable mapping of open sets in $\mathbb{R}^n$ to manifolds by following the usual custom in physics: something is called differentiable when it is differentiable in local coordinates.

(2.1.12) **Definition**

A mapping $f: M_1 \to M_2$ is **p-times differentiable** iff for all charts of an atlas for M_1 and of an atlas for M_2, the obvious restriction of $\Phi_2 \circ f \circ \Phi_1^{-1}$ is a p-times differentiable mapping from $\Phi_1(U_1 \cap f^{-1}(U_2)) \subset \mathbb{R}^{m_1}$ to $\mathbb{R}^{m_2}$ (see Figure 2.3).

(2.1.13) **Examples**

1. If M_1 is a submanifold of M_2, then the natural injection is infinitely-often differentiable, because a projection in $\mathbb{R}^n$ is (cf. (2.1.9)).
2. Let $\mathrm{Mat}_n(\mathbb{R})$ denote the manifold of real $n \times n$ matrices. Addition and multiplication of two $n \times n$ matrices are C^∞ mappings $\mathrm{Mat}_n(\mathbb{R}) \times \mathrm{Mat}_n(\mathbb{R}) \to \mathrm{Mat}_n(\mathbb{R})$.
3. If f_1 and $f_2 \in C^p$, then their composition $f_1 \circ f_2: M_3 \stackrel{f_2}{\to} M_2 \stackrel{f_1}{\to} M_1$, is also a C^p-mapping (Problem 6).
4. If M is the product manifold $M_1 \times M_2$ and $f = f_1 \times f_2$, then when the f_i are C^p-mappings, so is f.
5. $M_1 = I \subset \mathbb{R}$. Thinking of I as an interval of time and M_2 as space, we will refer to the function f as a **curve** and to $f(I)$ as its **trajectory**.
6. $M_2 = \mathbb{R}$. The p-times differentiable functions in this case are denoted by $C^p(M_1)$. They form an algebra with the usual product in $\mathbb{R}$ because of the elementary rules for differentiation of sums and products. For any f which never vanishes, $1/f$ also belongs to the algebra.

(2.1.14) **Remarks**

1. Definition (2.1.12) dealt with only one atlas. The condition of compatibility guarantees that differentiability is defined equivalently for all atlases of an equiv-

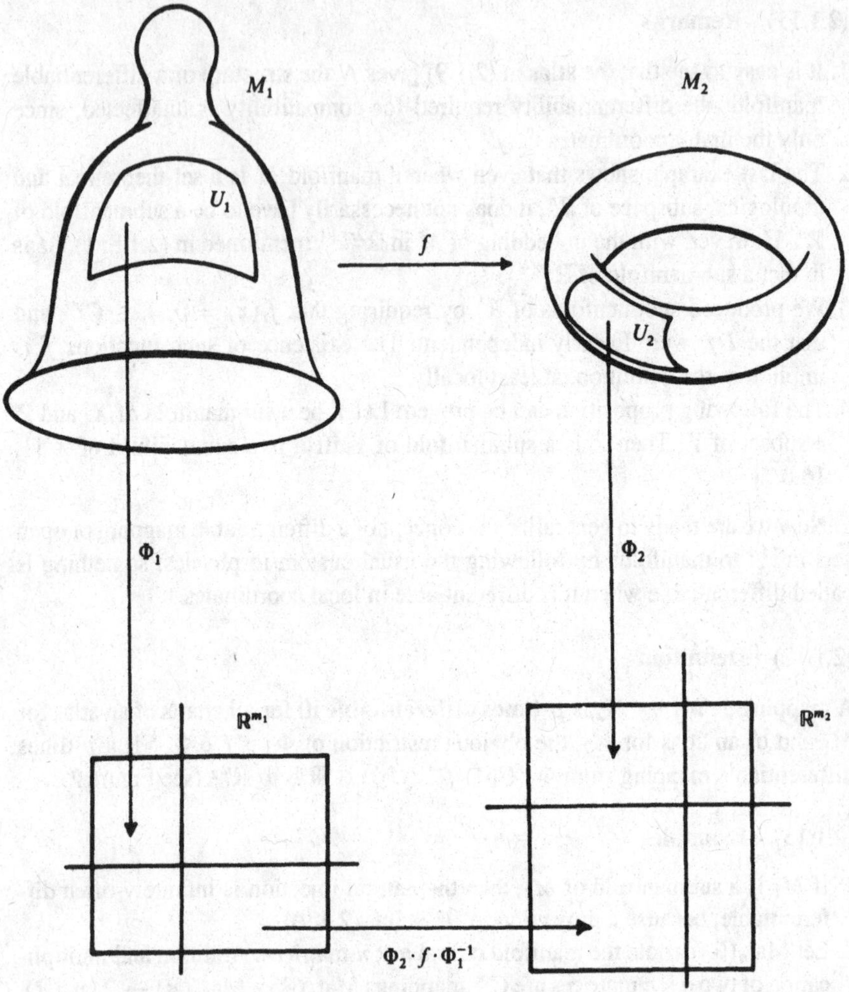

FIGURE 2.3. Differentiability of a mapping of manifolds.

alence class. This means that a differentiable mapping remains differentiable under a change of charts, and a nondifferentiable mapping can never become differentiable (cf. (2.1.10; 4)).

2. If N_1 is a submanifold of M_1, then $f_{|N_1}$ is as differentiable as f. On the chart (2.1.9) the restriction only amounts to holding the last $m - n$ coordinates fixed, which does not adversely affect differentiability.

The topological concept of homeomorphism can now be somewhat sharpened for manifolds.

(2.1.15) Definition

A **diffeomorphism** f of two manifolds is a bijection for which both f and f^{-1} are C^∞. Two manifolds are **diffeomorphic** iff there exists a diffeomorphism between them.

(2.1.16) Examples

1. A chart (U, Φ) provides a diffeomorphism Φ between the submanifold U and $\Phi(U) \subset \mathbb{R}^m$, since $\mathbf{1} \in C^\infty$.
2. $M_1 = \mathbb{R}$ topologically, but has the atlas $U = \mathbb{R}$, $\Phi: x \to x^3$ (only one chart). According to Example 1, Φ is a diffeomorphism $M_1 \overset{\Phi}{\to} \mathbb{R}$.
3. $\Phi: x \to x^3$ is not a diffeomorphism $\mathbb{R} \to \mathbb{R}$, since $\Phi^{-1} \notin C^\infty$.
4. Two manifold structures on the same set M are identical (i.e., are defined with equivalent atlases) iff $\mathbf{1}$ is a diffeomorphism.

(2.1.17) Remarks

1. These examples show that over $\mathbb{R}$ there exist diffeomorphic manifold structures that are not identical, since $M_1 \overset{\mathbf{1}}{\to} \mathbb{R}$ is not a diffeomorphism. It should be borne in mind, if one wants to identify manifolds related by a diffeomorphism, that they are not necessarily the same manifold in the sense of the definition. When we refer to $\mathbb{R}^n$ as a manifold without qualification, we mean $\mathbb{R}^n$ with the standard chart $(\mathbb{R}^n, \mathbf{1})$.
2. On complicated topological spaces, and even on $\mathbb{R}^4$, there are distinct manifold structures that are not even diffeomorphic. But of the connected one-dimensional manifolds the compact ones are all diffeomorphic to S^1 and the noncompact ones to $\mathbb{R}$.

Although an open interval (a, b) and $\mathbb{R}$ can be made into diffeomorphic manifolds, care must be taken in integrating by parts, since (a, b) has boundary points $\{a, b\}$, but $\mathbb{R}$ has none. To emphasize this distinction and highlight the boundary points, we introduce the somewhat more general concept of a manifold with a boundary. It is modeled on a **half-space** with its boundary.

(2.1.18) Definition

$\mathbb{R}^n_+ := \{x \in \mathbb{R}^n : x_1 \geq 0\}$, $\partial\mathbb{R}^n_+ = \{x \in \mathbb{R}^n : x_1 = 0\}$. A mapping f of an open subset $U \subset \mathbb{R}^n_+$ into $\mathbb{R}^m$ is **differentiable** iff there exist an open subset $\tilde{U}$ of $\mathbb{R}^n$ which contains U, and a differentiable mapping $\tilde{f}: \tilde{U} \to \mathbb{R}^m$, such that $\tilde{f}_{|U} = f$ (see Figure 2.4).

(2.1.19) Remarks

1. U need not be open in $\mathbb{R}^n$ and can contain parts of $\partial\mathbb{R}^n_+$.
2. $\mathbb{R}^n_+$ is not a submanifold of $\mathbb{R}^n$, though $\partial\mathbb{R}^n_+$ is.

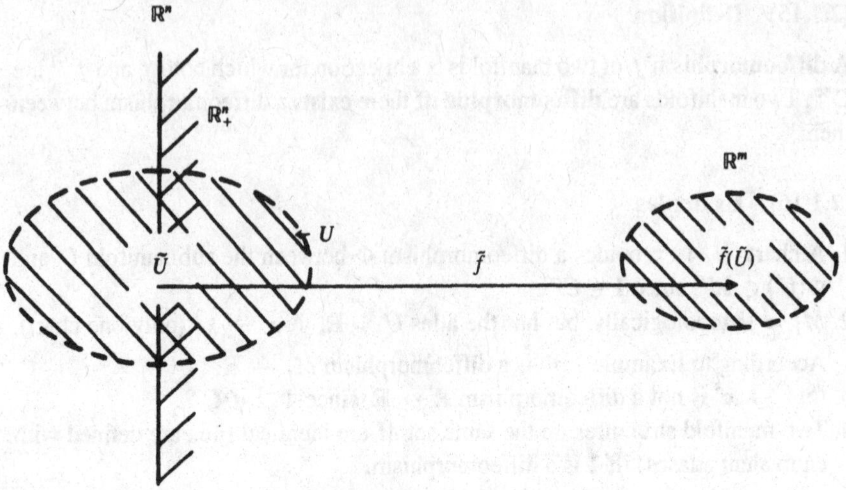

FIGURE 2.4. A differentiable mapping of $\mathbb{R}^n_+$.

Just as a manifold is composed of the inverse images of open sets of $\mathbb{R}^n$, a manifold with a boundary is likewise composed of inverse images from $\mathbb{R}^n_+$. The concept of an atlas generalizes to this situation.

(2.1.20) **Definition**

Let M be a separable metrizable space. M has the structure of a **manifold with a boundary** when there exist an open covering $\{U_i\}$ and homeomorphisms $\Phi_i: U_i \to$ open subsets of $\mathbb{R}^n_+$, for which $\forall i, j$, $\Phi_i \circ \Phi^{-1}_{j|\Phi_j(U_i \cap U_j)} \in C^\infty$. The **boundary** of M is $\partial M = \bigcup_i \Phi_i^{-1}(\Phi_i(U_i) \cap \partial \mathbb{R}^n_+)$ (see Figure 2.5).

(2.1.21) **Examples**

1. $M = [a, b]$: $U_1 = [a, b)$, $\Phi_1: x \to x - a$, $U_2 = (a, b]$, $\Phi_2: x \to b - x$; then $\partial M = \{a\} \cup \{b\}$.
2. $M = \{x \in \mathbb{R}^2: x_1^2 + x_2^2 \le 1\}$: $U_1 = \{x_1^2 + x_2^2 < 1\}$, $\Phi_1: (x_1, x_2) \to (x_1 + 1, x_2)$, and let charts be introduced on $U_2 = \{\frac{1}{2} < x_1^2 + x_2^2 \le 1\} = T^1 \times (\frac{1}{2}, 1]$ as in Example 1 and (2.1.7; 2). $\partial M = T^1$.
3. $M = \{x \in \mathbb{R}^2: |x_1| \le 1, |x_2| \le 1\}$. This is not a manifold with a boundary, because it has corners. However, its interior is a manifold, while its boundary alone is not.[2]

(2.1.22) **Remarks**

1. The boundary ∂M is to be distinguished from a topological boundary, which depends on an imbedding. Topologically $\partial \mathbb{R}^n_+$ as a subset of $\mathbb{R}^n$ is its own

[2]For manifolds with boundaries, see [19].

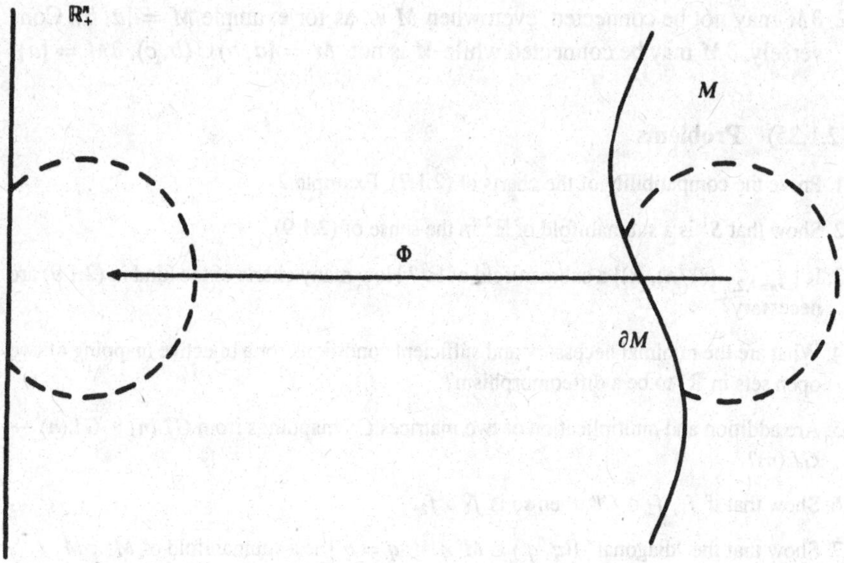

FIGURE 2.5. A manifold with a boundary.

boundary, but as $\mathbb{R}^{n-1}$ it has no boundary. In $\mathbb{R}^n$ the topological boundary of $\mathbb{R}^n_+$ is $\partial\mathbb{R}^n_+$.

2. Once again, a system of charts such that $M = \bigcup_i U_i$ suffices to determine a manifold, but this system is not to be preferred to other compatible ones. Structures which possess compatible atlases will henceforth be identified.
3. If $\partial M = \emptyset$, this definition reduces to (2.1.6), and in this case we speak simply of a manifold, and only otherwise of a manifold with a boundary.
4. Since the $\Phi_i \circ \Phi^{-1}_{j|\Phi_j(U_i \cap U_j)}$ are homeomorphisms, boundary points of $\mathbb{R}^n_+$ are mapped to boundary points, and thus compatible atlases define the same boundary.
5. A manifold with a boundary need not be compact (e.g., $(0, 1]$) and a compact manifold may not have a boundary (e.g., T^1).

At the same time the U_i in (2.1.20) provide charts both for the interior, $M\backslash\partial M$, and for the boundary ∂M. These are mapped respectively into open subsets of $\mathbb{R}^n$ and of $\mathbb{R}^{n-1}$, and it is easy to prove that the condition of compatibility continues to hold under these restrictions. Thus we conclude

(2.1.23) **Proposition**

Both $M\backslash\partial M$ and ∂M have the structure of manifolds (without boundary).

(2.1.24) **Remarks**

1. That a boundary has no boundary will also follow from a generalization of Stokes's integral theorem (cf. (2.6.6; 3)).

2. ∂M may not be connected, even when M is, as for example $M = [a, b]$. Conversely, ∂M may be connected while M is not: $M = [a, b) \cup (b, c)$, $\partial M = \{a\}$.

(2.1.25) Problems

1. Prove the compatibility of the charts of (2.1.7), Example 2.
2. Show that S^1 is a submanifold of $\mathbb{R}^2$ in the sense of (2.1.9).
3. Is $\bigcup_{n=1,2,\ldots}\{((1/n), \mathbb{R})\}$ a submanifold of $\mathbb{R}^2$? How many charts of the kind in (2.1.9) are necessary?
4. What are the minimal necessary and sufficient conditions for a bijective mapping of two open sets in $\mathbb{R}^n$ to be a diffeomorphism?
5. Are addition and multiplication of two matrices C^1-mappings from $GL(n) \times GL(n) \to GL(n)$?
6. Show that if $f_1, f_2 \in C^p$, then so is $f_1 \circ f_2$.
7. Show that the "diagonal" $\{(q, q') \in M \times M : q = q'\}$ is a submanifold of $M \times M$.
8. Let $g: \mathbb{R}^m \to \mathbb{R}$ be C^∞ with $Dg \neq 0$ for all x at which $g(x) = 0$. Show that $\{x \in \mathbb{R}^m : g(x) \leq 0\}$ is a manifold with a boundary.
9. Show: $q \in \partial M \Leftrightarrow \exists$ a chart (U, Φ) such that $\Phi(q) = (0, 0, \ldots, 0)$.

(2.1.26) Solutions

1. $V_1' \cap V_2' = T^1 \setminus \{(-1, 0)\} \setminus \{(0, 1)\} \xrightarrow{\Phi_2} (0, \pi) \cup (\pi, 2\pi) \xrightarrow{\Phi_1 \circ \Phi_2^{-1}} (0, \pi) \cup (-\pi, 0)$, by $\varphi \to \varphi$, or as the case may be $\varphi \to \varphi - 2\pi$ (see Figure 2.2). Likewise for $\Phi_2 \circ \Phi_1^{-1}$.
2. Use four charts:
$$U_{\frac{1}{2}} = \{(x_1, x_2) \in S^1 : x_2 \gtrless 0\}, \qquad \Phi_{\frac{1}{2}} : (x_1, x_2) \to x_1,$$
$$U_{\frac{3}{4}} = \{(x_1, x_2) \in S^1 : x_1 \gtrless 0\}, \qquad \Phi_{\frac{3}{4}} : (x_1, x_2) \to x_2.$$
Compatible, since for $|x| \neq 1$, $x \to \sqrt{1 - x^2}$ is C^∞.
3. Yes. Infinitely many: $U_n = I_n \times \mathbb{R}$, where I_n is an open interval containing $1/n$, and $I_m \cap I_n = \emptyset$. $\Phi_n : (x, y) \to (y, x - 1/n)$.
4. Let the mapping be $x \to y(x)$. Minimal conditions are that $\forall x, y_{i,j} \in C^\infty$ and $\det y_{i,j} \neq 0$ (see [1, Chapter 10, 2.5]).
5. Yes, assuming $\det(M_1 + M_2) \neq 0$, since they are restrictions (2.1.14; 2) of differentiable mappings (2.1.13; 2).
6. $\Phi_1 \circ f_1 \circ \Phi_2^{-1} \circ \Phi_2 \circ f_2 \circ \Phi_3^{-1} = \Phi_1 \circ f_1 \circ f_2 \circ \Phi_3^{-1}$.
7. Introduce the variables $q_k^{\pm} = q_k \pm q_k'$ on the product chart $m_1 \times m_2 \to (q_1, \ldots, q_m, q_1', \ldots, q_m')$; this gives a chart of the type of (2.1.9).
8. On the charts for which the submanifold $g = 0$ of $\mathbb{R}^m$ has the form (2.1.9), $g \leq 0$ can be put into the form (2.1.20).

9. If $\Phi(q) = (0, 0, \dots, 0)$, then q is a boundary point by (2.1.20). Conversely, if $q \in \partial M$, then on every chart, $\Phi(q) = (0, x_2, \dots, x_m)$. By changing the chart so that $\Phi \to \Phi - (0, x_2, \dots, x_m)$, q is mapped to the origin.

2.2 Tangent Spaces

A smooth surface may be approximated at a point by the tangent plane. The generalization of this concept is the tangent space. The derivative of a mapping of a manifold acts as a linear transformation on its tangent space.

Mechanics is concerned with the trajectories of particles, which were introduced in (2.1.13, 5) in the guise of mappings u from an interval $I \subset \mathbb{R}$ to a manifold $M: t \to u(t) \in M$. One's next inclination after this would be to introduce a velocity vector $\dot{u}$, but this does not work because M does not have a linear structure. If M were imbedded in $\mathbb{R}^n$, then $\dot{u}$ would lie in the tangent hyperplane and would stick out from M. What we shall do is to associate velocity vectors with trajectories without reference to imbeddings. In a chart $C = (V \ni q = u(0), \Phi)$, the image $\Phi \circ u$ of the trajectory at the point q defines the velocity vector[3] $D(\Phi \circ u)\big|_{t=0} \in \mathbb{R}^m$. In a coordinate system its components are as usual $(\partial/\partial t)x^i(u(t))\big|_{t=0}$. Even though velocities at different points can only be compared with reference to a chart, the statement that two velocities at a single point are equal is independent of the chart: Suppose that u and v are two trajectories $I \to M$, and $D(\Phi_1 \circ u)\big|_{t=0} = D(\Phi_2 \circ u)\big|_{t=0}$. Then with a different chart,

$$D(\Phi_2 \circ u)\big|_{t=0} = D(\Phi_2 \circ \Phi_1^{-1}) \cdot D(\Phi_1 \circ u)\big|_{t=0} = D(\Phi_2 \circ \Phi_1^{-1}) \cdot D(\Phi_1 \circ v)\big|_{t=0}$$
$$= D(\Phi_2 \circ v)\big|_{t=0}.$$

There is thus a chart-independent way to apportion the trajectories passing through q into equivalence classes of **tangential** trajectories. To each such class is associated the tangent vector of the image of the trajectories through $\Phi(q)$, i.e., $D(\Phi \circ u)\big|_{t=0} \in \mathbb{R}^m$. In any chart $C = (V \ni q, \Phi)$ this creates a bijection $\Theta_C(q)$ between the equivalence classes of trajectories and vectors in $\mathbb{R}^m$:

(2.2.1) **Definition**

The **mapping** $\Theta_C(q)$ sends a curve u that passes through q to the following vector:

$$u \xrightarrow{\Theta_C(q)} D(\Phi \circ u)(0) \in \mathbb{R}^m.$$

Conversely, for any $v \in \mathbb{R}^m$ the inverse mapping defines a representative curve u of the appropriate class ($0 \in I \subset \mathbb{R}$):

(2.2.2)
$$v \xrightarrow{\Theta_C^{-1}(q)} u := \{t \in \mathbb{R} \to \Phi^{-1}(\Phi(q) + tv) \in M\}.$$

---

[3] An $(m \times 1)$-matrix is regarded as equivalent to a vector in $\mathbb{R}^m$.

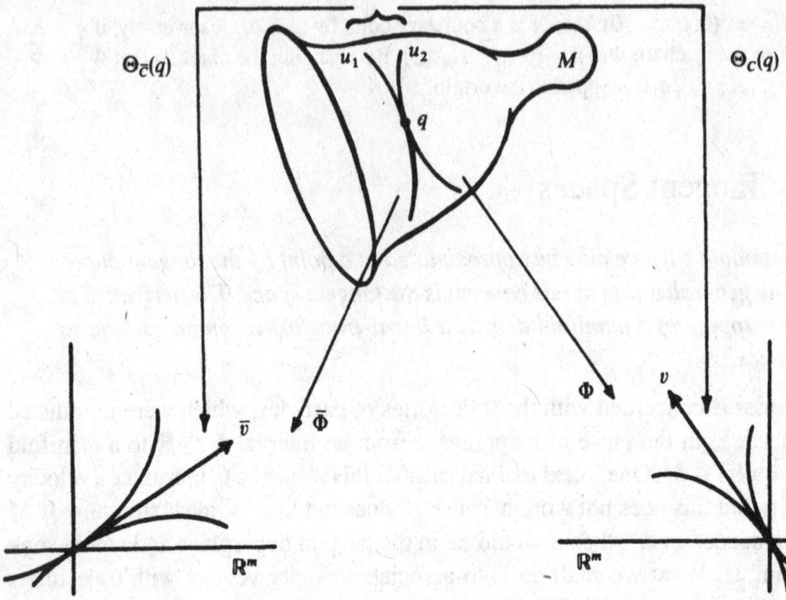

FIGURE 2.6. Action of the bijection $\Theta_C(q)$.

(2.2.3) **Remarks**

1. Different curves of one class correspond to the same vector by $\Theta_C(q)$, but on different charts the same class is associated to different vectors (Figure 2.6).
2. It might be supposed that a tangent vector directed along the curve u could be defined simply as $\lim_{n\to\infty}(n(u(1/n) - u(0))$, thereby avoiding abstract mental acrobatics. The unfortunate drawback is that this difference is undefined for finite n.
3. The mapping $\Theta_C(q)$ provides the equivalence classes with the structure of a vector space. This is independent of the choice of charts, since under a change of charts $\Theta_C(q)$ is multiplied by $D(\bar{\Phi}\circ\Phi^{-1})(q)$. The chain rule for D then implies that, as the derivative of a bijection, this is an invertible linear transformation (Problem 1), and thus preserves the vector-space structure. Hence the bijection Θ_C allows the desired characteristics of a tangent plane to be preserved, although in fact in the absence of a canonical imbedding of M into $\mathbb{R}^m$ no tangent plane is defined.

(2.2.4) **Definition**

The space of equivalence classes of curves tangent at q is called the **tangent space** of M at q and denoted $T_q(M)$. It has the structure of a vector space when by definition for $v, w \in T_q(M)$ and $\alpha, \beta \in \mathbb{R}$ we set

$$\alpha v + \beta w = \Theta_C^{-1}(q)(\alpha\Theta_C(q)(v) + \beta\Theta_C(q)(w)),$$

in which Θ_C is given by (2.2.1); this structure is chart-independent.

(2.2.5) Examples

1. $M = \mathbb{R}^n$, $\Phi = \mathbf{1}$. In this case to any vector v, $\Theta_C^{-1}(q)$ assigns the line which passes through q and is parallel to v. $T_q(M)$ may be identified naturally with M, crudely writing $\Theta = \mathbf{1}$.[4]
2. Suppose a surface F in $\mathbb{R}^3$ is given by a parametrization $g\colon \mathbb{R}^2 \to \mathbb{R}^3$, $(u, v) \overset{g}{\to} (x(u, v), y(u, v), z(u, v))$. If $g^{-1}\big|_F$ is used as a chart, then the coordinate lines $u =$ const. and $v =$ const. are just sent to the two axes in $\mathbb{R}^2$ by $\Theta_C(q)$.

(2.2.6) Remarks

1. Definition (2.2.4) may seem abstract, but it really only formalizes the intuitive notion of vectors in a tangent plane as arrows pointing in the directions of the curves passing through the point. This makes them elements of $\mathbb{R}^n$ on the charts used to make M a manifold.
2. It is clear that with a change of charts Θ is multiplied by $D(\bar{\Phi} \circ \Phi^{-1})$, which is the matrix that specifies how the images of the curves u are twisted around when put in different charts. This is closely related to the usual transformation relations for vectors under a coordinate transformation; if $\bar{x} \overset{\bar{\Phi}}{\leftarrow} q \overset{\Phi}{\rightarrow} x$, then $D(\bar{\Phi} \circ \Phi^{-1})$ is simply expressed as the matrix $\partial \bar{x}_i / \partial x_j$, and a vector of the tangent space is transformed as $v_i \to \bar{v}_i = v_j \, \partial \bar{x}_i / \partial x_j$. This property is frequently made the definition of a vector, in the sense of "a vector is a vector that transforms like a vector."
3. If there is no distinguished coordinate system for M, then there is also no distinguished basis for $T_q(M)$ (and hence no scalar product, either). It is only due to their structure as vector spaces that we can identify $\mathbb{R}^n$ and $T_q(\mathbb{R}^n)$, as we do from now on.
4. If N is a submanifold of M, then for any $q \in N$, $T_q(N)$ may be identified with a subspace of $T_q(M)$. The members of the set $T_q(N)$ correspond to trajectories in N.

According to (2.1.1), the derivative of a mapping $\mathbb{R}^{m_1} \to \mathbb{R}^{m_2}$ can be interpreted as a linear transformation, which turns trajectories in the direction of their images under this transformation. This notion can be carried over to manifolds by using the linear structure of $T_q(M)$: In local coordinates, D is a linear transformation and Θ transcribes it to the more abstract setting of a tangent space.

(2.2.7) Definition

Let $f\colon M_1 \to M_2$ be a C^1 transformation. The **derivative** of f, written $T_q(f)$, is defined as $T_q(f) = \Theta_{C_2}^{-1}(f(q)) \circ D(\Phi_2 \circ f \circ \Phi_1^{-1}) \circ \Theta_{C_1}(q)$. It is a linear transformation $T_q(M_1) \to T_q(M_2)$, and can be written in a chart-independent manner as $T_q(f) \cdot [u] = f \circ [u] = [f \circ u]$ where equivalence classes $[u]$ of trajectories are identified with vectors (cf. Problem 4).

---

[4] This notation is deprecated by pedants but all right among friends.

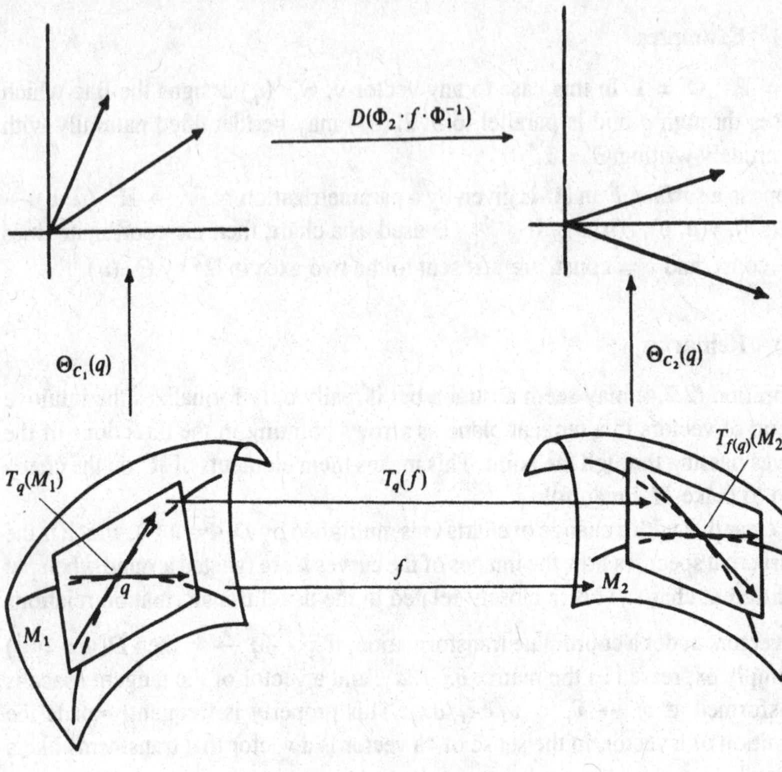

FIGURE 2.7. The action of $T_q(f)$.

(2.2.8) Remarks

1. Figure 2.7 is an attempt to clarify this difficult notation for the action of $T_q(f)$.
2. Although this definition makes reference to a chart, it is in fact chart-independent, since the tangent spaces and $D(\Phi_2 \circ f \circ \Phi_1^{-1})$ transform so as to compensate. If, for example, the chart on M_1 is changed to $\bar{\Phi}_1$, $D(\Phi_2 \circ f \circ \Phi_1^{-1})$ is multiplied on the right by $D(\Phi_1 \circ \bar{\Phi}_1^{-1})$ by the chain rule, while, as we already know, $\Theta_{C_1}(q)$ gains a factor $D(\bar{\Phi}_1 \circ \Phi_1^{-1})$ on the left.

(2.2.9) Examples

1. $M_i = \mathbb{R}^{n_i}$, $i = 1, 2$. The canonical identification of the tangent spaces implies $\Theta = \mathbf{1}$, and so $T_q(f) = D(f)(q)$. This test case brings us back to the old definition.
2. $M_1 = \mathbb{R} = T_0(M_1)$, $\Phi_1 = \mathbf{1}$. This provides a canonical way to identify an equivalence class with one of its vectors. With this identification we denote the basis vector of $T_0(\mathbb{R})$ by $\mathbf{1}$. Then $T_0(f) \cdot \mathbf{1} = \Theta_C^{-1} \circ D(\Phi_2 \circ f)(0) = \Theta_C^{-1} \circ \Theta_C[f] = [f] \in T_{f(0)}(M_2)$ is just the representative vector of the equivalence

class of f, which in this case consists of a single curve. Schematically,

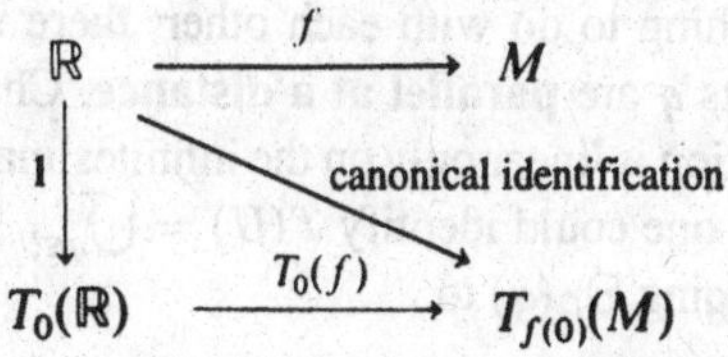

3. If $v \in T_q(M_1)$ is determined by the curve u, then $T_q(f) \cdot v$ is determined by $f \circ u$, because

$$T_q(f) \cdot v = \Theta_{C_2}^{-1}(f(q))D(\Phi_2 \circ f \circ \Phi_1^{-1})D(\Phi_1 \circ u) = \Theta_{C_2}^{-1}(f(q))D(\Phi_2 \circ f \circ u).$$

In words: f transforms the curve u into $f \circ u$, and $T_q(f)$ maps the tangent vectors to the curve u at the point q to the tangent vectors to $f \circ u$ at the point $f(q)$. This statement reads the same in all charts, showing that $T_q(f)$ is chart-independent. Schematically,

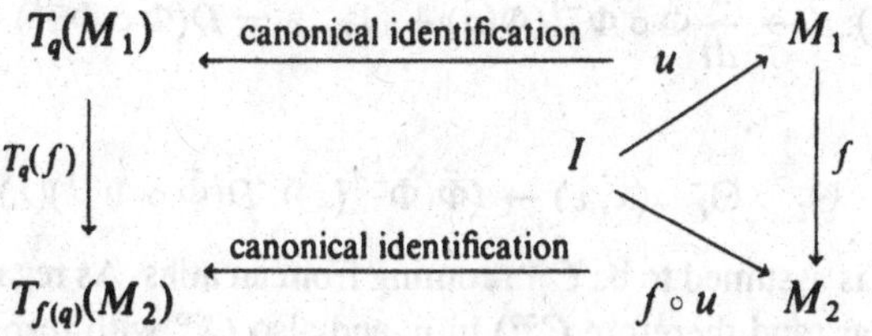

4. $M_2 = \mathbb{R}$. The f's form an algebra, and the derivative behaves as usual for the algebraic operations: from $(f_1 + f_2) \circ \Phi^{-1} = f_1 \circ \Phi^{-1} + f_2 \circ \Phi^{-1}$ and $(f_1 \cdot f_2) \circ \Phi^{-1} = (f_1 \circ \Phi^{-1}) \cdot (f_2 \circ \Phi^{-1})$ it follows that:

   (a) $T_q(\text{const.}) = 0$;

   (b) $T_q(f_1 + f_2) = T_q(f_1) + T_q(f_2)$; and

   (c) $T_q(f_1 \cdot f_2) = f_1(q) \cdot T_q(f_2) + f_2(q) \cdot T_q(f_1)$.

5. M_1 is a submanifold of M_2 and f is the natural injection. By using a single chart it can be shown (Problem 6) that $T_q(f)$ is injective, and we can identify $T_q(M_1)$ as a subspace of $T_q(M_2)$.
6. We concluded in (2.1.13; 3) that the composition of two differentiable maps is differentiable. This can also be verified for T (Problem 7):

$$T_q(f_1 \circ f_2) = T_{f_2(q)}(f_1) \circ T_q(f_2),$$

using the chain rule for D.

These examples illustrate how the essential rules of differentiation for manifolds can be expressed independently of the charts.

In order to progress from evaluating the derivative at a point to treating it as a function of q, we have to connect the tangent spaces at different points. At this stage they still have nothing to do with each other; there is no way to say that vectors at different points q are **parallel at a distance.** Changes of chart distort domains; the transformation is linear only on the infinitesimal scale. Yet within the domain of a single chart one could identify $T(U) = \bigcup_{q\in U} T_q(M)$ with $U \times \mathbb{R}^m$, and then extend the mapping $\Theta_C(q)$ to

$$\Theta_C: T(U) \to \Phi(U) \times \mathbb{R}^m, \qquad (q, v) \to (\Phi(q), \Theta_C(q) \cdot v). \tag{2.2.10}$$

It is possible to compare tangent vectors at different points within this "tangent bundle" over U. The mapping Θ_C is plainly a bijection, and $T(U)$ can be topologized so that it becomes a homeomorphism. One can even make it into a diffeomorphism and thereby confer a manifold structure on $T(U)$. The atlas then has only one chart, so there are no conditions of compatibility to verify. To extend the tangent bundle over all of M, one constructs it for all U_i of an atlas $\bigcup_i (U_i, \Phi_i)$. It suffices to show the compatibility of these charts, which also verifies the compatibility of the product topologies on the individual $T(U_i) = U_i \times \mathbb{R}^m$. Now,

$$\Theta_{\bar{C}}(q) \circ \Theta_C^{-1}(q): v \to \frac{d}{dt} \bar{\Phi} \circ \Phi^{-1}(\Phi(q) + vt)_{|t=0} = D(\bar{\Phi} \circ \Phi^{-1}) \cdot v, \quad v \in \mathbb{R}^m,$$

and thus

$$\Theta_{\bar{C}} \circ \Theta_C^{-1}: (x, v) \to (\bar{\Phi}(\Phi^{-1}(x)), D(\bar{\Phi} \circ \Phi^{-1})(x) \cdot v), \tag{2.2.11}$$

where $\bar{\Phi} \circ \Phi^{-1}$ is assumed to be C^∞, coming from an atlas. As regards the second factor, it is linear (and therefore C^∞) in v, and also C^∞ with respect to x by the assumption on $\bar{\Phi} \circ \Phi^{-1}$. This proves compatibility of the Θ's, allowing us to make

(2.2.12) Definition

$T(M) = \bigcup_{q\in M} T_q(M)$ is called the **tangent bundle** of M. It is a manifold with the atlas $\bigcup_i (U_i \times \mathbb{R}^m, \Theta_{C_i})$.

(2.2.13) Examples

1. $M = U \subset \mathbb{R}^m$, $C = (U, 1)$, $\Theta_C: (x_i, x_i + tv_i) \to (x_i, v_i)$, where the second argument stands for the curve $t \to x + tv$. $T(M) = U \times \mathbb{R}^m$. As we see, for open subsets of a Euclidean space the tangent bundle is a Cartesian product.
2. $M = S^1$, charts as in (2.1.7; 2). $\Theta_{C_{1,2}}: (\cos\varphi, \sin\varphi; \cos(\varphi+\omega t), \sin(\varphi+\omega t)) \to (\varphi, \omega)$. The two charts C_i are simply the products of charts C_i of S^1 and the identity chart of $\mathbb{R}$. Combined, they give a diffeomorphism between $T(S^1)$ and $S^1 \times \mathbb{R}$. The tangent bundle is again a product.
3. $M = S^2 = \{x \in \mathbb{R}^3: x_1^2 + x_2^2 + x_3^2 = 1\}$, $C_\pm = (S^2 \setminus (0, 0, \pm 1); (x_1, x_2, x_3) \to (x_1, x_2)/(1 \mp x_3))$ (**stereographic projection**),

$$\Theta_{C_\pm}: (x_1, x_2, x_3; x_1 + v_1 t, x_2 + v_2 t, x_3(t))$$
$$\to \left(\frac{(x_1, x_2)}{1 \mp x_3}; \frac{(v_1, v_2)}{1 \mp x_3} \mp \frac{(x_1, x_2)}{(1 \mp x_3)^3} \frac{x_1 v_1 + x_2 v_2}{x_3} \right).$$

Interchange of the coordinates x_1 and x_3 removes the singularity at $x_3 = 0$, but at $x_3 = \pm 1$ the mappings do not agree on their region of overlap, and they cannot be extended continuously. For example, as $x_3 \to 1$, in the second position Θ_{C_-} becomes the mapping: (left side) $\to$ (; $(v_1, v_2)/2$); thus it acts essentially like the identity. On the other hand, Θ_{C_+} becomes the mapping

$$\text{(left side)} \to (\ ; (2/(x_1^2 + x_2^2))((v_1, v_2) \\ - 2(x_1, x_2)(x_1 v_1 + x_2 v_2)/(x_1^2 + x_2^2))),$$

acting like a dilatation by $1/|\mathbf{x}|^2$ followed by a reflection about $\mathbf{x}$. This is singular at $\mathbf{x} = \mathbf{0}$, and cannot be continuously joined to Θ_{C_-}. No product structure can be identified, as it would contradict (2.6.15; 6), and as a matter of fact $T(S^2) \neq S^2 \times \mathbb{R}^2$.

4. Obviously, $T(M_1 \times M_2) = T(M_1) \times T(M_2)$, with the product charts.

(2.2.14) **Remarks**

1. At present $T(M)$ is defined abstractly and not given concretely as a submanifold of some $\mathbb{R}^n$. The meaning of the tangent bundle becomes more intuitive, however, if we think of it as the space of the positions and velocities of particles.
2. If $T_q(M)$ is thought of as the pair $\{q\} \times \mathbb{R}^m$, then in a purely set-theoretical sense $T(M) = \bigcup_{q \in M} T_q(M) = \bigcup_{q \in M}(\{q\} \times \mathbb{R}^m) = (\bigcup_{q \in M}\{q\}) \times \mathbb{R}^m = M \times \mathbb{R}^m$ is always a product. However, with the Θ_{C_i} it could be topologized as, say, a **Möbius strip** (cf. (2.2.16; 3)), so that $T(M) \neq M \times \mathbb{R}^m$ topologically. If, however, $T(M)$ is diffeomorphic as a manifold to $M \times \mathbb{R}^m$, then we say that M is **parallelizable**, since the product chart makes it possible to define what is meant by saying that tangent vectors at different points are parallel, or for that matter equal. The only n-spheres that are parallelizable are S^1, S^3, and S^7. Locally, $T(M)$ is always a product manifold.
3. M may be identified with the submanifold of $T(M)$ corresponding to the point $\{\mathbf{0}\}$ in $\mathbb{R}^m$, because of which there exists, even globally, a projection $\Pi: T(M) \to M, (q, v) \to q$, onto a distinguished submanifold. Note that for a Cartesian product $(q, v) \to v$ would also be given canonically, but here it is chart-dependent.
4. As a manifold $T(M)$ admits various other charts, though it is for the so-called bundle chart used up to now that Π has the simple form $(q, v) \to q$. A change of charts for M induces a transition from one bundle chart to another on $T(M)$.

In differential geometry it is always necessary to distinguish among the infinitesimal (q), local (U), and global (M) levels. The following definition abstracts from the tangent bundle the property that a product structure may be defined at least at the local level, although not necessarily at the global level.

(2.2.15) **Definition**

A **vector bundle** consists of a manifold X, a submanifold M (known as the **basis**), and a surjection $\Pi: X \to M$. Furthermore, for each $q \in M$, the **fibers** $\Pi^{-1}(q)$ are assumed to have the structure of vector spaces, which are all isomorphic to a fixed vector space F. **Bundle atlases** are assumed given on X with domains $\Pi^{-1}(Ui)$, where U_i are neighborhoods in M. The corresponding chart mappings Φ_i are not only diffeomorphisms on $U_i \times F$, but also map the fibers linearly to F. If X is diffeomorphic to $M \times F$, then X is said to be **trivializable**, and **trivial** iff it is given as a Cartesian product.

(2.2.16) **Examples**

1. $X = \mathbb{R} \times \mathbb{R}$, $M = F = \mathbb{R}$, $\Pi: (x, y) \to x$. $X = M \times F$ is trivial. There are many coordinate systems for $\mathbb{R} \times \mathbb{R}$ as a manifold, though the product structure distinguishes the Cartesian one.
2. $X = T(M)$, $F = \mathbb{R}^m$, $\Pi: (q, v) \to q$. The fibers are the tangent spaces $T_q(M)$. Trivializable iff parallelizable.
3. $X = [0, 2\pi) \times \mathbb{R}$ (as sets), with two charts $C_i = (U_i, \Phi_i)$, which also define the topology on X:

$$C_1: ((0, 2\pi) \times \mathbb{R}, 1),$$

$$C_2: \left([0, \pi) \cup (\pi, 2\pi) \times \mathbb{R}, (\varphi, x) \to \begin{matrix} (\varphi, x) & \text{if } 0 \le \varphi < \pi \\ (\varphi - 2\pi, -x) & \text{if } \pi < \varphi < 2\pi \end{matrix} \right),$$

$$M = S^1, \qquad F = \mathbb{R}, \qquad \Pi: (\varphi, x) \to \varphi.$$

With C_2, $\phi \to 2\pi$ is joined to $\varphi = 0$, but in such a way that the sign of x is reversed. X is then an infinitely wide Möbius strip (see Figure 2.8). It is not trivializable, since if there were $\Phi: X \to S^1 \times \mathbb{R}$, then the image of the unit vector in $\mathbb{R}$ would change sign under Φ^{-1} during a circuit of S^1, which is not possible without its vanishing somewhere by continuity.

The derivative defined at a point (2.2.7) can now be generalized as a mapping of the tangent bundle.

(2.2.17) **Definition**

The mapping $T(M_1) \to T(M_2): (q, v) \to (f(q), T_q(f) \cdot v)$ is called the **derivative** $T(f)$ of the function $f: M_1 \to M_2$.

(2.2.18) **Remarks**

1. If $f \in C^r$, then $T(f) \in C^{r-1}$.
2. If f is a diffeomorphism, then so is $T(f)$ (cf. Problem 1).
3. The diagram

$$\begin{array}{ccc} M_1 & \xrightarrow{f} & M_2 \\ \uparrow \Pi_1 & & \uparrow \Pi_2 \\ T(M_1) & \xrightarrow{T(f)} & T(M_2) \end{array}, \text{ which means } \begin{array}{ccc} q & \xrightarrow{f} & f(q) \\ \uparrow \Pi_1 & & \uparrow \Pi_2 \\ (q, v) & \xrightarrow{T(f)} & (f(q), T_q(f) \cdot v) \end{array}$$

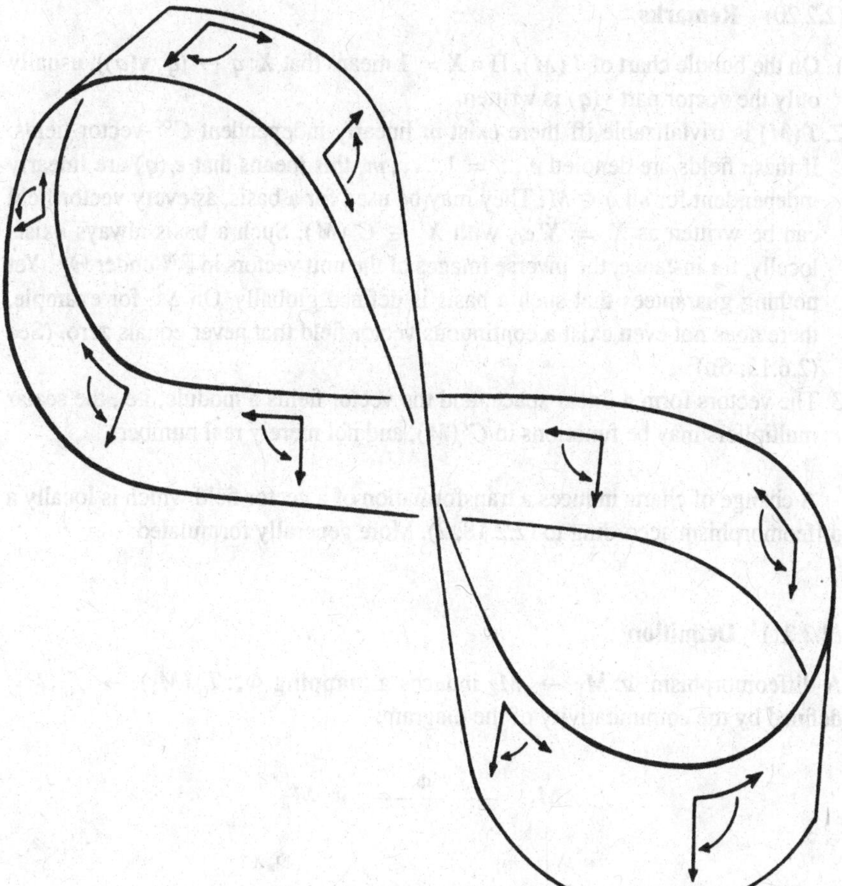

FIGURE 2.8. A nontrivializable vector bundle.

commutes.

4. The chain rule (2.2.9; 6) is now written more conveniently $T(f \circ g) = T(f) \circ T(g)$ (Problem 7).
5. With Cartesian products everything factorizes, including the derivative: $T(f \times g) = T(f) \times T(g)$.

The manifold structure of $T(M)$ defines when the transition between vectors at neighboring points is continuous This allows

(2.2.19) Definition

A C^r-**vector field** X is a C^r-mapping $M \to T(M)$ such that $\Pi \circ X = \mathbf{1}$. The set of all vector fields for M is denoted $T_0^1(M)$.

(2.2.20) Remarks

1. On the bundle chart of $T(M)$, $\Pi \circ X = \mathbf{1}$ means that $X: q \to (q, \mathbf{v}(q))$; usually only the vector part $\mathbf{v}(q)$ is written.
2. $T(M)$ is trivializable iff there exist m linearly independent C^∞-vector fields. If these fields are denoted e_i, $i = 1, \ldots, m$, this means that $e_i(q)$ are linearly independent for all $q \in M$. They may be used for a basis, as every vector field can be written as $X = X^i e_i$, with $X^i \in C^r(M)$. Such a basis always exists locally, for instance, the inverse images of the unit vectors in $\mathbb{R}^m$ under Θ_C. Yet nothing guarantees that such a basis is defined globally. On S^2, for example, there does not even exist a continuous vector field that never equals zero. (See (2.6.15; 6).)
3. The vectors form a linear space, and the vector fields a module, i.e., the scalar multipliers may be functions in $C^r(M)$, and not merely real numbers.

A change of charts induces a transformation of a vector field which is locally a diffeomorphism according to (2.2.18; 2). More generally formulated:

(2.2.21) Definition

A diffeomorphism $\Phi: M_1 \to M_2$ induces a **mapping** $\Phi_*: T_0^1(M_1) \to T_0^1(M_2)$ defined by the commutativity of the diagram:

$$\begin{array}{ccc} M_1 & \xrightarrow{\Phi} & M_2 \\ {\scriptstyle X}\downarrow & & \downarrow{\scriptstyle \Phi_* X} \\ T(M_1) & \xrightarrow{T(\Phi)} & T(M_2) \end{array}$$

That is, $\Phi_* X = T(\Phi) \circ X \circ \Phi^{-1}$. This clearly means that Φ_* turns the vector fields in just the same way as Φ turns the curves that define the direction.

(2.2.22) Examples

1. $M = \mathbb{R}^n$, $\Phi: \mathbf{x} \to \mathbf{x} + \mathbf{a}$, $T(\Phi): (\mathbf{x}, \mathbf{u}) \to (\mathbf{x} + \mathbf{a}, \mathbf{u})$, $X: \mathbf{x} \to (\mathbf{x}, \mathbf{v}(\mathbf{x}))$, $\Phi_* X: \mathbf{x} \to (\mathbf{x}, (\mathbf{v}(\mathbf{x} - \mathbf{a}))$. A vector remains unchanged under a displacement, but one must take care to talk only about a vector at one particular point, which has different coordinates in the new system.
2. $M = \mathbb{R}^n$, $\Phi: x_i \to L_{ik} x_k$, $T(\Phi): (x_i, u_j) \to (L_{ik} x_k, L_{jm} u_m)$,[5] $\Phi_* X: x_i \to (x_i, L_{ik} v_k(L^{-1} x))$. Under linear transformations v transforms like x.

[5] Here x_i stands for $x_i e_i \in \mathbb{R}^n$, where $\{e_i\}$ is a basis for $\mathbb{R}^n$, and similarly for v.

3. The following diagram of a transformation holds in general:

$$\begin{array}{ccc}
q_i & \xrightarrow{\quad\Phi\quad} & \bar{q}_i \\
\Big\downarrow X & & \Big\downarrow \Phi_* X \\
(q_i, v_i(q)) & \xrightarrow{\quad T(\Phi)\quad} & \left(\bar{q}_i, \dfrac{\partial \bar{q}_i}{\partial q_j} v_j(q)\right)
\end{array}$$

(2.2.23) **Remark**

A vector field X induces a transformation sending a function $f \in C^1(M)$ to the **Lie derivative**

$$L_X(f) \equiv I \cdot T(f) \cdot X \colon M \xrightarrow{X} T(M) \xrightarrow{T(f)} T(\mathbb{R}) \xrightarrow{I} \mathbb{R},$$

in which I denotes the projection onto the second factor for $T(\mathbb{R}) = \mathbb{R} \times \mathbb{R}$. Given a bundle chart in which $q \to (q, \mathrm{v})$, this function becomes $L_X(f) = v^i(q)\, \partial f / \partial q^i$, i.e., the rate of change of f along the direction defined by X. The Lie derivative has the properties:

(a) $L_X(f + g) = L_X(f) + L_X(g)$ for all $f, g \in C(M)$;

(b) $L_X(f \cdot g) = f \cdot L_X(g) + g \cdot L_X(f)$; and

(c) $L_{\alpha X_1 + \beta X_2}(f) = \alpha L_{X_1}(f) + \beta L_{X_2}(f)$ for all $\alpha, \beta \in \mathbb{R}$.

Indeed, Properties (a) and (b) characterize the vector fields. It is thus possible to define a direction on a manifold which determines the rate of change of the C^1-functions. As our main concern is the geometric intuition, instead of pursuing this line of thought further (cf. Problem 8), we merely quote

(2.2.24) **Theorem**

A mapping $L \colon C^\infty(M) \to C^\infty(M)$ with the properties:

(a) $L(f + g) = L(f) + L(g)$; *and*

(b) $L(f \cdot g) = f \cdot L(g) + g \cdot L(f)$,

known as a **derivation**, *determines a unique C^∞ vector field such that $L = L_X$.*

(2.2.25) **Remarks**

1. In mechanics L_X is known as the **Liouville operator**, so the notation L is doubly justified.

2. Since L_X is defined with a local operation, it suffices to know the action of L_X on the C^∞-functions of compact support in order to determine X.
3. Since the diagram

$$\begin{array}{ccccc} M_1 & \xrightarrow{\Phi} & M_2 & \xrightarrow{f} & \mathbb{R} \\ \downarrow X & & \downarrow \Phi_* X & & \\ T(M_1) & \xrightarrow{T(\Phi)} & T(M_2) & \xrightarrow{T(f)} & T(\mathbb{R}) \end{array}$$

commutes, that is, $T(f) \circ T(\Phi) \circ X = T(f) \circ \Phi_* X \circ \Phi$, we conclude that

$$L_X(f \circ \Phi) = (L_{\Phi_* X}(f)) \circ \Phi.$$

Thus the image of X acts on a function as X acts on its inverse image. This fact becomes obvious when one thinks of L_X (and $L_{\Phi_* X}$) as the rate of change along a curve in the direction of X (resp. along the image of the curve).
4. The module structure of $\mathcal{T}_0^1(M)$ extends to L_X when the latter acts on $C^r(M)$. Property (c) holds even for $\alpha, \beta \in C(M)$.
5. If we apply Property (b) with $f = 1$, we see that $L_X(1) = 0$, and hence with (a) that $L_X(k) = 0$ for any constant k.

(2.2.26) **The Natural Basis**

According to (2.2.5; 2) the coordinate grid of $\mathbb{R}^n$ equips the domain V of a chart with a basis for $\mathcal{T}_0^1(V)$, known as the **natural basis.** It is often denoted $\{\partial/\partial x_i\}$ or simply $\{\partial_i\}$ for the following reason: Let $\{e_i\}$ be the basis for $\mathbb{R}^m$ and $\Phi: q \to \sum_i e_i x^i \in \mathbb{R}^m$. For any function $g \in C^\infty(M)$ we have a mapping on the image $\Phi(V)$ of the chart given by $g \cdot \Phi^{-1}$. According to (2.2.23), the Lie derivative along e_i is the derivative $\partial/\partial x^i$, so

$$L_{\Theta_C^{-1} e_i} g = \frac{\partial}{\partial x_i} g(q(x)).$$

(2.2.27) **Problems**

1. Show that for a diffeomorphism Ψ, $T(\Psi^{-1}) = (T(\Psi))^{-1}$.
2. Show that in the natural basis of a chart $\Phi_* X$ (2.2.21) produces the usual transformation law for vectors (2.2.6; 2).
3. Write $L_X g$ out explicitly on a chart.
4. Show the chart-independence of (2.2.4).
5. Show that if $M_1 \xrightarrow{f} M_2 \supset N_2$, $T(f)$ is surjective, and N_2 is a submanifold of M_2, then $f^{-1}(N_2)$ is a submanifold of M_1. (If N_2 is a one-point space, this reduces to (2.1.10; 3).)
6. Show that for the natural injection to a submanifold, $T_q(f)$ is injective $\forall q$.
7. Verify the chain rule.

8. Show that a mapping $L: C^\infty \to C^\infty$ with the properties: (i) $L(f_1+f_2) = L(f_1)+L(f_2)$; and (ii) $L(f_1 \cdot f_2) = L(f_1)\cdot f_2 + f_1 \cdot L(f_2)$ must be of the form $L(f)(q) = (df|X)_{|q}$. For the definition of df, see (2.4.3).

(2.2.28) Solutions

1. This follows from the chain rule applied to $\Psi \circ \Psi^{-1} = 1$ and from $T(1) = 1$.
2. Let $X: q \to (q, v^i(q)\partial_i)$ and $\Phi: q \to \bar{q}(q)$. Then $\Phi_* X: \bar{q} \to (\bar{q}, v^j(q)(\partial\bar{q}_i/\partial q_j)\bar{\partial}_{i|\bar{q}})$. Observe that the components v^i transform the same way as the differentials dq_i and the other way around from the basis $\frac{\partial}{\partial q_i} = \frac{\partial \bar{q}_j}{\partial q_i}\frac{\partial}{\partial \bar{q}_j}$.
3. Let $g: q \to g(q)$. Then $L_X g: q \to v^i(q)(\partial g(q)/\partial q_i)$ when $X: q \to (q, v(q))$.
4.

$$\Theta_C^{-1}(q)(\alpha\Theta_C(q)(v) + \beta\Theta_C(q)(w)) = \Theta_{\bar{C}}^{-1}(D(\Phi \circ \bar{\Phi}^{-1}))^{-1}(\alpha D(\Phi \circ \bar{\Phi}^{-1})\Theta_{\bar{C}}(v) + \beta D(\Phi \circ \bar{\Phi}^{-1})\Theta_{\bar{C}}(w)) = \Theta_{\bar{C}}^{-1}(\alpha\Theta_{\bar{C}}(v) + \beta\Theta_{\bar{C}}(w)).$$

5. Use a chart homeomorphism Φ on M_2 of the type (2.1.9). If $\Phi \circ f = \sum e_i f_i$, then $f^{-1}(N_2) = \{x \in M_1 : f_i(x) = 0, i = n_2+1, \ldots, m_2\}$. That $T(f)$ is surjective means that $f_{i,k}$, $i = 1, \ldots, m_2$, $k = 1, \ldots, m_1 \geq m_2$, has maximal rank, that is, the vectors $f_{i,k}$, $i = n_2+1, \ldots, m_2$ must be linearly independent. Then we have the situation of (2.1.10; 3).
6. On the chart of (2.1.9), $f: (x_1, \ldots, x_m) \to (x_1, \ldots, x_n, 0, \ldots, 0)$

$$T(f): m\left\{\overbrace{\begin{bmatrix} \left.\begin{matrix} 1 & & & & & & \\ & 1 & & & & & \\ & & 1 & & & & \\ & & & 1 & & & \\ & & & & 1 & & \\ & & & & & 1 & \\ & & & & & & 1 \end{matrix}\right\}n \\ \\ \\ \\ \end{bmatrix}}^{n}\right. \qquad T(f)v = 0 \Leftrightarrow v = 0.$$

7.

$$\begin{aligned} T_q(f_2 \circ f_1) &= \Theta_{C_3}^{-1}(f_2 \circ f_1(q)) \circ D(\Phi_3 \circ f_2 \circ f_1 \circ \Phi_1^{-1})\Theta_{C_1}(q) \\ &= \Theta_{C_3}^{-1}(f_2 \circ f_1(q))D(\Phi_3 \circ f_2 \circ \Phi_2^{-1})\Theta_{C_2}(f_1(q)) \\ &\quad \times \Theta_{C_2}^{-1}(f_1(q))D(\Phi_2 \circ f_1 \circ \Phi_1^{-1}) \circ \Theta_{C_1}(q) = T_{f_1(q)}(f_2) \circ T_q(f_1). \end{aligned}$$

8. On a chart which maps q to the origin, f is of the form

$$f(x) = f(0) + x_i f_{,i}(0) + \tfrac{1}{2}x_i x_j f_{,ij}(0) + \cdots.$$

From (i) and (ii) it follows that $L(f(0)) = L(\text{constant}) = 0$, and then from (ii), $L(x_i f_{,i}(0)) = L(x_i) f_{,i}(0)$ and $L(x_i x_j f_{,ij})/2 = x_i L(x_j) f_{,ij}(0)$. Therefore $L(f)(0) = f_{,i} L(x_i)$. The $L(x_i)$ are the components of the vector field X. It is obviously necessary that $L(f) \in C^\infty$ for the components to be C^∞.

2.3 Flows

A vector field X defines a motion in the direction of X at all points of a manifold. Under the right circumstances this defines a flow, *that is, a one-parameter group of diffeomorphisms of M.*

A vector field X is to be regarded as a field of direction indicators; to every point of M it assigns a vector in the tangent space at that point. A curve $u: I \to M$, $t \to u(t)$ will be called an **integral curve**[6] of X if it has the same direction as X at every point; or more precisely if the tangent vector determined by it equals X at every point. This is shown pictorially in Figure 2.9.

To express this in a formula, let $e: t \to (t, 1)$ be the unit vector field on I, and let $\dot{u} = T(u) \cdot e$:

(2.3.1) $$\dot{u} = X \circ u.$$

Expressed schematically it is the following commutative diagram:

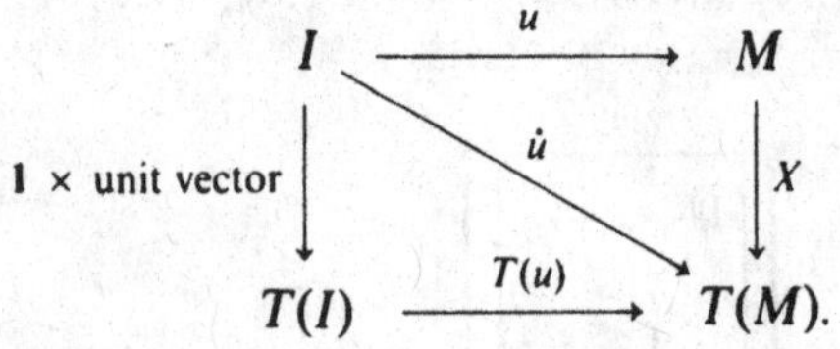

On a chart for which $\Phi \circ u: t \to (u_i(t))$ and $T(\Phi) \circ X \circ \Phi^{-1}: q_i \to (q_i, X_i(q))$, this is commonly written in the form

(2.3.2) $$\dot{u}_i(t) = X_i(u(t)), \qquad i = 1, \ldots, n,$$

an ordinary differential equation of first order in m dimensions.

(2.3.3) **Remarks**

1. Because of our physical orientation we refer to the parameter t as time, so that (2.3.2) describes motion on M.
2. It is no real restriction that (2.3.2) is of first order, since a higher-order equation can always be reduced to a first-order equation (on a space of higher dimension) by introducing new variables.

---

[6]When an integral curve is the path a physical system follows, i.e., the solution of the equations of motion, we shall generally refer to it as a **trajectory**.

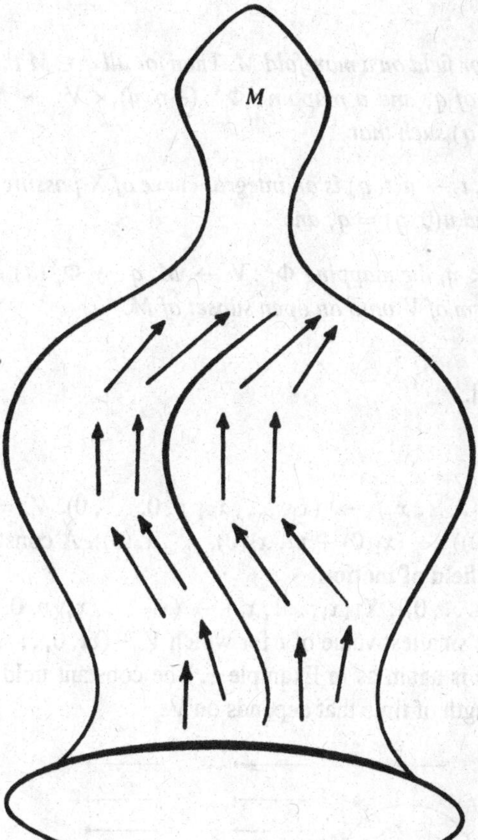

FIGURE 2.9. An integral curve of a vector field.

3. It is also no loss of generality that the independent variable t only occurs implicitly on the right side of (2.3.2) through the dependent variables. One could easily take t as a dependent variable and introduce a new independent variable s; we shall look into this possibility later. In the meantime let us discuss the general properties of (2.3.1).

Recall that the existence of a solution is guaranteed by the continuity of $X_i(q)$, and local uniqueness by a Lipschitz condition. With these assumptions, there is precisely one integral curve through each point, and integral curves can never cross. Since we shall be concerned only with C^∞-vector fields, for which these remarks apply, the basic results from the theory of differential equations as to the solution with arbitrary initial conditions are expressed in

(2.3.4) Theorem

Let X be a C^∞-vector field on a manifold M. Then for all $q \in M$ there exist $\eta > 0$, a neighborhood V of q, and a mapping $\Phi^X: (-\eta, \eta) \times V \to M$, $(t, q(0)) \to \Phi^X(t, q(0)) =: u(t, q)$ such that

1. *for all $q \in V$, $t \to u(t, q)$ is an integral curve of X passing through q, i.e., $\dot{u} = X \cdot u$, and $u(0, q) = q$; and*

2. *for all t, $|t| < \eta$, the mapping $\Phi_t^X: V \to M$, $q \to \Phi_t^X(q) \equiv \Phi^X(t, q)$ is a diffeomorphism of V onto an open subset of M.*

Proof: See [1, 10.8].

(2.3.5) Examples

1. $M = \mathbb{R}^n$, $X: (x_1, \ldots, x_n) \to (x_1, \ldots, x_n; v, 0, \ldots, 0)$. $V = \mathbb{R}^n$, $\eta = \infty$, $u(t, x(0)): (t, x_i(0)) \to (x_1(0) + vt, x_2(0), \ldots, x_n(0))$. A constant vector field induces a **linear field of motion.**
2. $M = \mathbb{R}^n \setminus \{(0, 0, \ldots, 0)\}$, $X: (x_1, \ldots, x_n) \to (x_1, \ldots, x_n; v, 0, \ldots, 0)$, V arbitrary, but η is the smallest value of t for which $V + (vt, 0, \ldots, 0)$ contains the origin. $u(t, x(0))$ is again as in Example 1. The constant field of motion may leave M, in a length of time that depends on V.

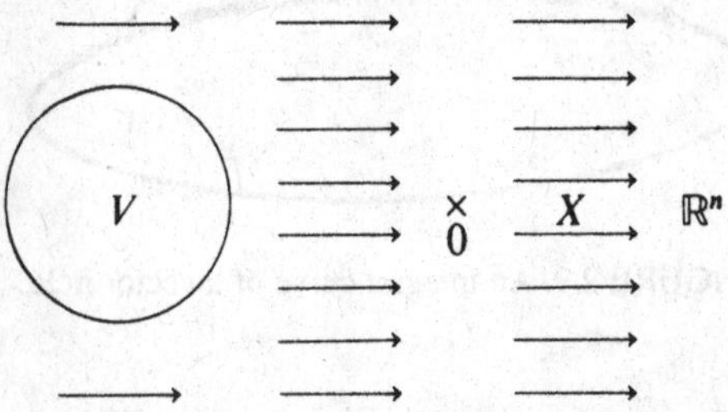

3. $M = \mathbb{R}^n \setminus \{(x, 0, \ldots, 0) \mid x \in \mathbb{R}\}$, that is, the x_1-axis is removed; and X is as above. Once again $V = M$, $\eta = \infty$.
4. $M = \mathbb{R}$, $X: x \to (x, x^3/2)$, V arbitrary, $\eta = \inf_{x \in V} 1/x^2$, $u: (t, x) \to x(1 - tx^2)^{-1/2}$. For large x the vector field becomes so strong that every point except the origin is sent to infinity in a finite time.

(2.3.6) Remarks

1. Theorem (2.3.4) states that trajectories that are near neighbors cannot suddenly be separated. We shall see that neighboring points may not diverge faster than exponentially in time, provided that the derivative of X remains bounded in a suitable norm.
2. In Example 1, X provides a one-parameter group of diffeomorphisms Φ_t^X on M. Because $u(t_1 + t_2, q(0)) = u(t_2, u(t_1, q(0)))$, its existence is equivalent to the

possibility of letting $V = M$ and $\eta = \infty$. It can be shown that this is possible, for instance, when X is of compact support. This is intuitively clear, since the worst eventuality is for some trajectories to leave M in a finite time. But if X equals zero outside some compact subset of M, the trajectories can not leave M [1, 18.2.11].

3. In Example 2 there is no diffeomorphism of all of M, and in Example 3 we saved the group of diffeomorphisms by getting rid of the trajectories that go through the origin. This is not always possible; in Example 4 only one point of the manifold would be left after a similar operation.

These possibilities are delineated by

(2.3.7) **Definition**

If the diffeomorphisms Φ_t^X of Theorem (2.3.4) form a one-parameter group of bijections $M \to M$, X is said to be **complete** and the group is called a **flow**. If the relationship

$$\Phi_{t_1}^X \circ \Phi_{t_2}^X = \Phi_{t_1+t_2}^X$$

holds only for sufficiently small neighborhoods of any point and sufficiently short times, Φ_t^X is called a **local flow** (a local group of diffeomorphisms).

As mentioned in §1.3 we would like to construe time-evolution as a group of automorphisms of the algebra of observables. Choosing the algebra as C_0^∞, the C^∞-functions of compact support, the local flow of a vector field provides an automorphism for short times by

(2.3.8) $$\tau_t^X(f) := f \circ \Phi_t^X, \qquad f \in C_0^\infty.$$

If X is complete, the τ_t^X are a one-parameter group:

(2.3.9) $$\tau_{t_1}^X \circ \tau_{t_2}^X = \tau_{t_1+t_2}^X, \qquad \forall t_1, t_2 \in \mathbb{R}.$$

In any case, the mapping $t \to \tau_t^X(f)(q)$ is differentiable for t in some neighborhood of 0, the size of which depends on f. As can be seen by using a chart (Problem 4), the time-derivative is the same as the Lie derivative associated with X (cf. (2.5.7)).

(2.3.10) $$\frac{d}{dt}\tau_t^X(f)_{|t=0} = L_X f, \qquad \forall f \in C_0^\infty.$$

(2.3.11) **Remarks**

1. Thus a vector field determines a local flow, which then determines the automorphisms of C_0^∞ given by (2.3.8). By (2.3.10) and (2.2.24) the automorphisms determine in turn a vector field, so we can combine the three concepts into one.
2. If M, X, and f are all analytic, then $t \to \tau_t^X(f)\big|_q$ is analytic in a complex neighborhood of 0. The power series in t may be written as

$$\tau_t^X(f) = e^{tL_X} f := \sum_{n=0}^{\infty} \frac{t^n}{n!}(L_X)^n f.$$

3. It may happen that the flows of two vector fields $\tilde{X}$ and X approach each other asymptotically, i.e., that for every $q \in M$ there exists a $p \in M$ such that the flows $\Phi_t^{\tilde{X}}(q)$ and $\Phi_t^X(p)$ converge together. Since, however, they will not converge individually, what we require is that the limit

$$\lim_{t\to\infty} \Phi_{-t}^X \circ \Phi_t^{\tilde{X}} = \Omega$$

exists. A (pointwise) limit of diffeomorphisms might not be a diffeomorphism; e.g., the limit of the mappings $x \to x/t$ on $\mathbb{R}$ is $\mathbb{R} \to \{0\}$. However, if Ω is a diffeomorphism, then it follows from the group property that $\Omega \circ \Phi_\tau^{\tilde{X}} = \lim_{t\to\infty} \Phi_{-t}^X \circ \Phi_{t+\tau}^{\tilde{X}} = \lim_{t\to\infty} \Phi_\tau^X \circ \Phi_{-t}^X \circ \Phi_t^{\tilde{X}} = \Phi_\tau^X \circ \Omega$, so

$$\Phi_t^X = \Omega \circ \Phi_t^{\tilde{X}} \circ \Omega^{-1}$$

for all t. Therefore the flows induced by X and $\tilde{X}$ must also be diffeomorphic. According to (2.2.25; 3) and (2.2.24), taking the time-derivative (at $t = 0$) of $f \circ \Phi_t^X \circ \Omega = f \circ \Omega \circ \Phi_t^{\tilde{X}}$, $f \in C^\infty$, yields $L_{\Omega_* \tilde{X}} = L_X$, or $\Omega_* \tilde{X} = X$: The diffeomorphism Ω transforms the vector fields into each other. Thus asymptotically equal flows may be regarded as the same flow expressed in different coordinate systems.

The trivial case (2.3.5; 1) is typical in that in the neighborhood of any point q where $X(q) \neq 0$ (i.e., other than at a point of equilibrium) the general case may be reduced to it by a suitable change of coordinates, by letting the lines of flow become the new coordinate lines.

(2.3.12) **Theorem**

At every point $q \in M$ where $X(q) \neq 0$ there exists a chart (U, Φ) such that $\Phi(U) = I \times V$, $V \subset \mathbb{R}^{m-1}$; for all $x \in V$, $t \to \Phi^{-1}(t \times \{x\})$ $\forall t \in I$ is an integral curve for X; and $\Phi_* X: (x_1, \ldots, x_m) \to (x_1, \ldots, x_m; 1, 0, \ldots, 0)$.

Proof: Since $X(q) \neq 0$, a chart (U_1, ψ) with $\psi(q) = 0 \in \mathbb{R}^m$ can be found such that $\psi_* X(0) = (1, 0, \ldots, 0)$. Since $\psi_* X \in \mathcal{T}_0^1(\Psi(U_1))$ is continuous, there is an open, relatively compact neighborhood U_2 of $\mathbf{0}$ on which the first component of the image of X is greater than $\frac{1}{2}$: $(\psi_* X)^1(x) > \frac{1}{2}$ $\forall x \in U_2$. If $X_0 \in \mathcal{T}_0^1(\mathbb{R}^m)$: $x \to (x; 1, 0, \ldots, 0)$, and $U \subset U_2$ is an open set containing 0, then, given a function $f \in C^\infty(\mathbb{R}^m)$ with $0 \le f \le 1$,

$$f = \begin{cases} 0 & \text{on } CU_2\,, \\ 1 & \text{on } U, \end{cases}$$

we may define the interpolating vector field

$$\tilde{X} = f \cdot \psi_* X + (1 - f) X_0 \in \mathcal{T}_0^1(\mathbb{R}^m).$$

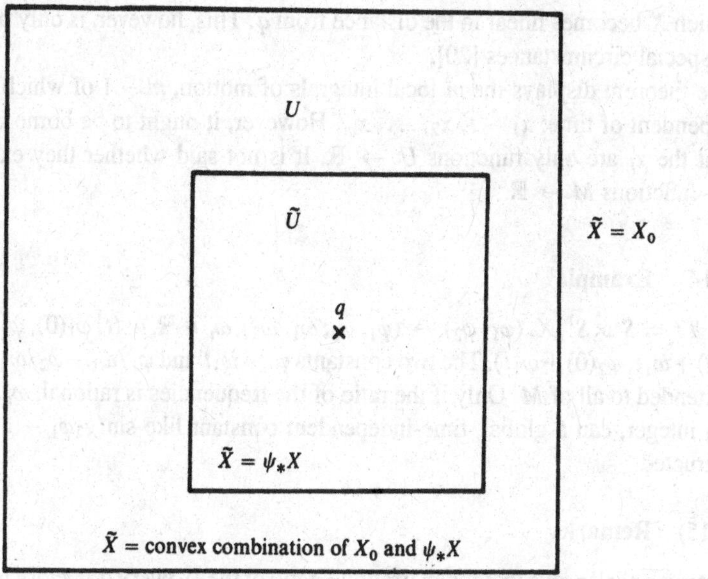

FIGURE 2.10. The interpolating vector field $\tilde{X}$.

A well-known theorem guarantees the existence of functions f of this sort. Clearly $(\tilde{X})^1(x) > \frac{1}{2}\ \forall x \in \mathbb{R}^m$, and $\tilde{X}$ induces a flow, because it agrees with X_0 outside some compact set (see Figure 2.10). Hence

$$\Omega = \lim_{t\to\infty} \Phi^{X_0}_{-t} \circ \Phi^{\tilde{X}}_t$$

also exists, for $(\Phi^{\tilde{X}}_t(x))^1 \geq x^t + t/2$, and $\Phi^{X_0}_t$ and $\Phi^{\tilde{X}}_t$ are identical on

$$\left\{x \in \mathbb{R}^m : x^1 > \sup_{\bar{x}\in U_2} \bar{x}^1\right\}.$$

If for $t > \tau$, $\Phi^{\tilde{X}}_t$ has mapped some point out of U_2, then at that point,

$$\Phi^{X_0}_{-\tau-t_1} \circ \Phi^{\tilde{X}}_{\tau+t_1} = \Phi^{X_0}_{-\tau} \circ \Phi^{X_0}_{-t_1} \circ \Phi^{\tilde{X}}_{t_1} \circ \Phi^{\tilde{X}}_{\tau} = \Phi^{X_0}_{-\tau} \circ \Phi^{\tilde{X}}_{\tau}$$

for all $t_1 > 0$. Therefore the limit is attained on compact sets after a finite time, and Ω is a diffeomorphism. According to (2.3.11; 3) Ω transforms $\tilde{X}$ into X_0, and $\tilde{X}$ and $\psi_* X$ are equal on U. The mapping Φ of the theorem is $\Omega \circ \psi$. □

(2.3.13) **Remarks**

1. The idea of a comparison diffeomorphism used in the proof plays an important role in physics. For a direct proof see Problem 5 and (2.5.12; 10).
2. Points q at which $X(q) = 0$ are fittingly called critical points; they are fixed points of the flow. In §3.4 we investigate what happens in their vicinity. It might be conjectured that it was always possible to find a chart containing q, for

which X becomes linear in the distance from q. This, however, is only possible in special circumstances [20].

3. The theorem displays the m local integrals of motion, $m-1$ of which are independent of time: $x_1 - t, x_2, \ldots, x_m$. However, it ought to be borne in mind that the x_i are only functions $U \to \mathbb{R}$. It is not said whether they extend to C^r-functions $M \to \mathbb{R}$.

(2.3.14) **Example**

$M = T^2 = S^1 \times S^1, X: (\varphi_1, \varphi_2) \to (\varphi_1, \varphi_2; \omega_1, \omega_2), \omega_i \in \mathbb{R}. u: (t, \varphi_1(0), \varphi_2(0)) \to (\varphi_1(0) + \omega_1 t, \varphi_2(0) + \omega_2 t)$. The two constants $\varphi_1 - \omega_1 t$ and $\varphi_1/\omega_1 - \varphi_2/\omega_2$ cannot be extended to all of M. Only if the ratio of the frequencies is rational, $\omega_i = g_i \omega$, g_i an integer, can a global, time-independent constant like $\sin(g_2\varphi_1 - g_1\varphi_2)$ be constructed.

(2.3.15) **Remarks**

1. Later we shall prove that for an irrational ratio of the ω_i every trajectory is dense. Because of this there can be no C^∞-function K which is constant in time and for which $T_q(K) \neq 0$ for all q. The set $K =$ const. would be the one-dimensional submanifold containing the trajectory, which cannot be dense in M. This is clear from Definition (2.1.9).
2. The phrase "integrals of motion" will be reserved for C^r-functions ($r \geq 1$) $K: M \to \mathbb{R}$ with $L_X K = 0$. At this stage the existence of one or more integrals of the motion is an open question.
3. Example (2.3.14) is typical in that it can be shown [3, 25.17] that every vector field can be approximated arbitrarily well by one that does not have any constants (= integrals) of motion. This fact is of physical interest only when the uncertainty of our knowledge allows appreciable alterations of the solutions in physically relevant times.

The differential equations of mechanics, (1.1.1) through (1.1.6), are somewhat special, as they are the Euler–Lagrange equations of a variational problem, to wit, the requirement that the (Fréchet) derivative DW of a functional

$$W = \int dt\, L(x(t), \dot{x}(t)) \tag{2.3.16}$$

of $x(t)$ vanishes. This has the advantage of a coordinate-free formulation, since the requirement that $DW = 0$ does not single out any particular coordinate system. We will not delve further into this matter, because later we shall prove the more general invariance of the equations of motion under canonical transformations. For that end we need only the elementary fact that with the Lagrangian

$$L = \sum_{i=1}^{N} m_i \frac{|\mathbf{x}_i|^2}{2} - \sum_{i>j} (e_i e_j - \kappa m_i m_j) |\mathbf{x}_i - \mathbf{x}_j|^{-1}, \tag{2.3.17}$$

the Euler–Lagrange equations

$$\frac{d}{dt}\frac{\partial L}{\partial \dot{\mathbf{x}}_i} = \frac{\partial L}{\partial \mathbf{x}_i}, \qquad i = 1, \ldots, N, \tag{2.3.18}$$

produce the equations of motion (1.1.1) and (1.1.2). With generalized coordinates $q_i(x)$, $i = 1, \ldots, 3N$, L may be written

$$L = \sum_{i,k=1}^{3N} m_{ik}(q)\frac{\dot{q}_i \dot{q}_k}{2} - V(q). \tag{2.3.19}$$

Because m_{ik} is a nonsingular matrix for all q, the q_i can be expressed with the conjugate momenta $p_i = \partial L/\partial \dot{q}_i = m_{ik}(q)\dot{q}_k$, and the Euler–Lagrange equations

$$\frac{d}{dt}\frac{\partial L}{\partial \dot{q}_i} = \frac{\partial L}{\partial q_i}, \qquad i = 1, \ldots, 3N, \tag{2.3.20}$$

can equally well be written in the Hamiltonian form

$$\frac{dq_i}{dt} = \frac{\partial H}{\partial p_i}, \qquad \frac{dp_i}{dt} = -\frac{\partial H}{\partial q_i}, \tag{2.3.21}$$

where

$$H(q, p) = \sum_i p_i \dot{q}_i - L = \sum_{i,k} \frac{p_i p_k}{2}(m^{-1}(q))_{ik} + V(q). \tag{2.3.22}$$

The Legendre transformation leading from L to the Hamiltonian H is invertible:

$$L = \sum_i p_i \frac{\partial H}{\partial p_i} - H. \tag{2.3.23}$$

To see these equations in the framework of the structure we have constructed up to now we need the concept of a cotangent bundle, which we develop in the next chapter. Briefly, L furnishes a vector field on the tangent bundle (coordinates $(q, \dot{q})$) and H furnishes one on the cotangent bundle (coordinates (q, p)), which will be called phase space, whereas the underlying manifold will be called configuration space.

(2.3.24) Problems

1. In what sense is equation (2.3.1) formulated "invariantly" (or "covariantly")?
2. Discuss the integrals of motion for the one-dimensional and two-dimensional harmonic oscillators:

$$H_1 = p^2 + \omega^2 q^2, \qquad H_2 = p_1^2 + p_2^2 + \omega_1^2 q_1^2 + \omega_2^2 q_2^2.$$

3. Same problem for $M = T^2$, $X: (\varphi_1, \varphi_2) \to (\varphi_1, \varphi_2; \omega, \alpha \sin \varphi_1)$.

4. Derive (2.3.10).

5. Prove (2.3.12) by using the streamlines of X as coordinate lines.

6. Given an example of a vector field for $M = \mathbb{R}$, which is continuous but not C^1, such that (2.3.1) has more than one solution for a given initial value.

(2.3.25) Solutions

1. With a diffeomorphism $\Phi: M_1 \to M_2$, the commutativity of the diagram (cf. (2.2.21))

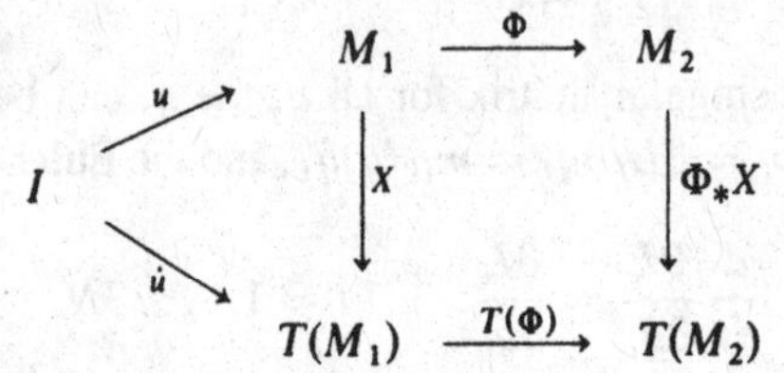

implies for $\tilde{u} := \Phi \circ u$ that $\dot{\tilde{u}} = \Phi_* X \circ \tilde{u}$.

2. The general solution of the equation of motion is

$$(q_i(t), p_i(t)) = (A_i \sin(\omega_i t + \varphi_i), A_i \omega_i \cos(\omega_i t + \varphi_i)).$$

For H_1 the constant $A^2 = p^2/\omega^2 + q^2$ is defined globally, but $\varphi = \arctan(q/p) - \omega t$ only locally. Similarly for H_2 there are two integrals, A_1^2 and A_2^2, and again the φ_i exist only locally, and likewise for the third time-independent constant, $\varphi_1/\omega_1 - \varphi_2/\omega_2 = (1/\omega_1)\arctan(q_1/p_1) - (1/\omega_2)\arctan(q_2/p_2)$. If $\omega_i = g_i\omega$, with g_i integral, then once more there is a global constant $\sin(\varphi_1 g_2 - \varphi_2 g_1)$.

3. Locally, $\varphi_1 - \omega t$ and $\varphi_2 + (\alpha/\omega)\cos\varphi_1$ are constant. In this case there is a global time-independent constant, $\sin(\varphi_2 + (\alpha/\omega)\cos\varphi_1)$.

4. Let $q(t) = u(t, q)$ be the solution of (2.3.1). Then $\tau_t^X f_{|q} = f \circ \Phi^X_{t|q} = f(q(t))$. Consequently,

$$\frac{d}{dt}\tau_t^X f|_{t=0} = \frac{\partial f}{\partial q_i}\frac{\partial q_i}{\partial t}\bigg|_{t=0} = \frac{\partial f}{\partial q_i} X_i(q) = L_X f.$$

5. In the notation of the proof of Theorem (2.3.12), let U_1 be the domain of ψ and let $\psi(U_1) = I_1 \times V_1$, $I_1 \subset \mathbb{R}$, $V_1 \subset \mathbb{R}^{m-1}$. Theorem (2.3.4) guarantees the existence of a local solution $u(t; x_1, \ldots, x_m)$ of the equation $\psi_* X \circ u = \dot{u}$, using this chart. At the origin the function $f(t, x_2, \ldots, x_m) := u(t, 0, x_2, \ldots, x_m)$: $I_2 \times V_2 \to \mathbb{R}^m$, $I_2 \subset I_1$, $V_2 \subset V_1$, has the derivative $Df(0) = \mathbf{1}: \mathbb{R}^m \to \mathbb{R}^m$, because the components f_i are found to satisfy

$$\frac{\partial f_i}{\partial t}\bigg|_0 = X_i(0) = (1, 0, 0, \ldots, 0), \quad \frac{\partial f_i}{\partial x_2}\bigg|_0 = \delta_{i2}, \text{ etc.}$$

Therefore Df is invertible on a neighborhood $I_3 \times V_3$, where $I_3 \subset I_2$ and $V_3 \subset V_2$, and consequently f is a diffeomorphism there [1, 10.2.5]. Because $f(0, x_2, \ldots, x_m) =$

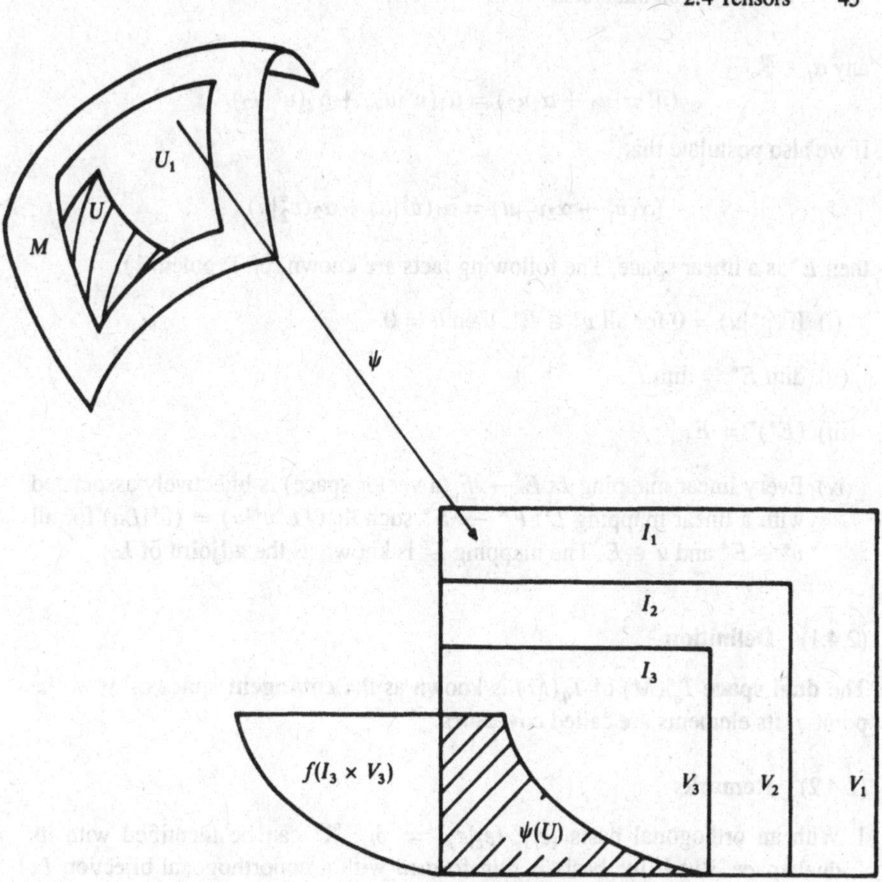

FIGURE 2.11. The relationship of the domains.

$(0, x_2, \ldots, x_m)$, $\psi(U) = I_3 \times V_3 \cap f(I_3 \times V_3) \neq \emptyset$, it is possible to introduce $(U, f^{-1} \circ \psi_{|U})$ as a new chart. For a picture of this see Figure 2.11. On this chart the vector field has the form $\Phi_* X = T(\Phi) \circ X \circ \Phi^{-1} = T(f^{-1}) \circ \psi_* X \circ f = T(f^{-1}) \circ \dot{f} = T(f^{-1}) \circ T(f) \circ (1, 0, \ldots, 0) = (1, 0, 0, \ldots, 0)$. Therefore the $I \times \{x\}$ are integral curves.

6. $X: x \to (x, \sqrt{x}$ for $x > 0$, and otherwise 0). For $u(0) = 0$ there are two solutions, $u(t) = 0$ and $u(t) = t^2/4$.

2.4 Tensors

Multilinear algebra defines algebraic structures on a vector space. In differential geometry these are extended to the local and global levels.

If E is a (finite-dimensional) vector space, then the space of linear mappings $E \to \mathbb{R}$ (or $\mathbb{C}$) is called its dual space E^*. We may write these mappings as scalar products: To any $v^* \in E^*$ there corresponds a mapping $u \to (v^*|u)$, such that for any $\alpha_i \in \mathbb{R}$,

$$(v^*|\alpha_1 u_1 + \alpha_2 u_2) = \alpha_1(v^*|u_1) + \alpha_2(v^*|u_2).$$

If we also postulate that

$$(\alpha_1 v_1^* + \alpha_2 v_2^*|u) = \alpha_1(v_1^*|u) + \alpha_2(v_2^*|u),$$

then E^* is a linear space. The following facts are known (cf. Problem 1):

(i) If $(v^*|u) = 0$ for all $v^* \in E^*$, then $u = 0$.

(ii) $\dim E^* = \dim E$.

(iii) $(E^*)^* = E$.

(iv) Every linear mapping $L: E \to F$ (a vector space) is bijectively associated with a linear mapping $L^t: F^* \to E^*$ such that $(L^t v^*|u) = (v^*|Lu)$ for all $v^* \in F^*$ and $u \in E$. The mapping L^t is known as the **adjoint** of L.

(2.4.1) Definition

The **dual space** $T_q^*(M)$ of $T_q(M)$ is known as the **cotangent space** of M at the point q. Its elements are called **covectors**.

(2.4.2) Remarks

1. With an orthogonal basis $\{e_i\}$, $(e_i|e_j) = \delta_{ij}$, $\mathbb{R}^n$ can be identified with its dual space. But if the basis is transformed with a nonorthogonal bijection L, $e_i \to Le_i$, the **dual basis** $\{e^{*i}\}$ must be transformed with $(L^{-1})^t$ so as to satisfy $(e^{*i}|e_j) = \delta_{ij}$. Since the transformation $T_q(\Phi)$ induced on $T_q(M)$ by a diffeomorphism Φ is not generally orthogonal, and since no coordinate system is distinguished, the statement that a vector in $T_q(M)$ is the same as one in $T_q^*(M)$ has no chart-independent meaning. With a change of charts they transform differently and become unequal. Hence it is necessary to distinguish $T_q(M)$ from $T_q^*(M)$ unless M possesses additional structure, such as a Riemannian metric.
2. The identification of the dual space $T_q^{**}(M)$ of $T_q^*(M)$ with $T_q(M)$ is unaffected by a change of charts, since $(((L^{-1})^t)^{-1})^t = L$.
3. Any $v^* \in T_q^*(M)$ defines an $(m-1)$-dimensional hyperplane, consisting of all v orthogonal to v^*. Since $(v^*|v) = 0$, we see that the scalar product measures the component of a vector pointing out of the hyperplane. If v is thought of as an arrow and v^* as a row of parallel hyperplanes, then $(v^*|v)$ tells how many of them are penetrated by the arrow. Note, however, that v^* does not define a unique decomposition $v = v_\parallel + v_\perp$ with $(v^*|v_\perp) = 0$.

(2.4.3) Examples

1. In elementary vector calculus one encounters the gradient of a function f as an example of a vector. Here we shall recognize it as an element of $T_q^*(M)$.

A function $f \in C^\infty(M)$ defines a mapping $T_q(f): T_q(M) \to T_{f(q)}(\mathbb{R}) = \mathbb{R}$, which is therefore an element of $T_q^*(M)$. We may denote this mapping by $df_{|q}$ and call it the **exterior derivative** of f at the point q. On a chart we write the usual formula

$$df_{|q}(v) = v^i \left.\frac{\partial f}{\partial q^i}\right|_q, \qquad \forall v \in T_q(M).$$

If a vector of $T_q(M)$ is specified by the vector field X, this mapping becomes $df_{|q}(X(q)) = (L_X f)(q)$. Thus the effect of df is to assign to any vector X the rate of change of f in the direction of X.

2. Given a chart $C: (U, \Phi)$, $\Phi(q) = \sum e_i q^i \in \mathbb{R}^m$, the inverse mapping $\Theta_C^{-1}(q)$ transplants the basis $\{e_i\}$ from $\mathbb{R}^m$ to $T_q(U)$, where we write it symbolically as $\{\partial/\partial q^i\}$. Similarly, $\Theta_C^t(q)$ transforms the e^{*i} into the basis of $T_q^*(U)$ dual to the $\partial/\partial q^i$. In the notation of Example 1, this is written as dq^i, if we consider q^i as a C^∞-function on M:

$$(dq^i|\Theta_C^{-1}(q)e_j) = dq^i|_q(\Theta_C^{-1}(q)e_j) = L_{\Theta_C^{-1}(q)e_j}(q^i) = \frac{\partial q^i}{\partial q^j} = \delta_{ij}.$$

The dq^i, the differentials of the coordinates,[7] are referred to as the **natural basis** of $T_q^*(U)$ (cf. (2.2.26)). According to (1)

$$df = \frac{\partial f}{\partial q^i} dq^i.$$

Taking these algebraic considerations a step further produces the idea of the space of tensors at a point. If $T_q(M)$ is identified with $T_q^{**}(M)$, then $T_q(M)$ can be considered as a linear mapping $T_q^*(M) \to \mathbb{R}$. A mapping

$$\underbrace{T_q^*(M) \times T_q^*(M) \times \cdots \times T_q^*(M)}_{r \text{ times}} \to \mathbb{R}$$

which is linear in every factor is called a contravariant tensor of degree r. Similarly, a covariant tensor is a multilinear mapping $T_q(M) \times \cdots \times T_q(M) \to \mathbb{R}$, and more generally we make

(2.4.4) Definition

A mapping

$$\underbrace{T_q^*(M) \times T_q^*(M) \times \cdots \times T_q^*(M)}_{r \text{ times}} \times \underbrace{T_q(M) \times T_q(M) \times \cdots \times T_q(M)}_{s \text{ times}} \to \mathbb{R}$$

[7] Following the usual convention we use subscripts for the bases and superscripts for the components in a tangent space, and do it the other way around in a cotangent space. This does not fix what to do about coordinates, which are not vectors.

at the point q, which is linear in every factor is a **tensor** which is **contravariant of degree** r **and covariant of degree** s. This multilinear mapping will be written as a scalar product

$$(v_1^*, \ldots, v_r^*; v_1, \ldots, v_s) \to (t|v_1^*, \ldots, v_r^*; v_1, \ldots, v_s) \in \mathbb{R},$$

where

$$\begin{aligned}(t|v_1^* + u_1^*, v_2^*, \ldots, v_r^*; v_1, \ldots, v_s)\\ = (t|v_1^*, v_2^*, \ldots, v_r^*; v_1, \ldots, v_s) + (t|u_1^*, v_2^*, \ldots, v_r^*; v_1, \ldots, v_s),\end{aligned}$$

etc. If we also postulate a distributive law for the first factor,

$$(\alpha_1 t_1 + \alpha_2 t_2|\ldots) = \alpha_1(t_1|\ldots) + \alpha_2(t_2|\ldots), \qquad \alpha_k \in \mathbb{R},$$

then the space of multilinear mappings also inherits a linear structure. This space will be designated $T_{qs}^r(M)$. (Hence $T_{q0}^1(M) = T_q(M)$ and $T_{q1}^0(M) = T_q^*(M)$.)

In order to construct a basis for this vector space we shall have to recall some of the concepts of multilinear algebra. Since they apply to arbitrary vector spaces, we shall drop the q and M.

(2.4.5) Definition

The **tensor product** $u_1 \otimes u_2 \otimes \cdots \otimes u_r \otimes u_1^* \otimes u_2^* \otimes \ldots u_s^* \in T_s^r$ of r vectors and s covectors is defined by

$$\begin{aligned}(u_1 \otimes u_2 \otimes \cdots \otimes u_r \otimes u_1^* \otimes u_2^* \otimes \ldots u_s^*|v_1^* \ldots v_r^*; v_1 \ldots v_s)\\ = \prod_{i=1}^{r}(v_i^*|u_i)\prod_{j=1}^{s}(u_j^*|v_j).\end{aligned}$$

(2.4.6) Remarks

1. The distributive law $u_1 \otimes (u_2 + u_3) = u_1 \otimes u_2 + u_1 \otimes u_3$ follows from the definition. In this way, the tensor product $\otimes$ can be interpreted as a distributive and associative mapping from $T_s^r \times T_{s'}^{r'}$ to $T_{s+s'}^{r+r'}$. It is not commutative. With this mapping $\oplus_{r,s} T_s^r$ becomes an algebra, known as the **tensor algebra** T. Each space T_s^r is a linear subspace.
2. Although not every tensor can be written as a tensor product of vectors, T_s^r is clearly spanned by linear combinations of such expressions. If $\{e^{*i}\}$ and $\{e_j\}$ are bases, respectively, for T_1^0 and T_0^1, then any $t \in T_s^r$ may be written as

$$t = \sum_{(i),(j)} t_{i_1 \ldots i_s}^{j_1 \ldots j_r} e^{*i_1} \otimes \cdots \otimes e^{*i_s} \otimes e_{j_1} \otimes \cdots \otimes e_{j_r}.$$

The quantities $t_{i_1 \ldots i_s}^{j_1 \ldots j_r}$ are the components of t, and are what physicists usually refer to as tensors.

3. Considered as a vector space, T^r_s has dimension m^{r+s}. Note that with the Cartesian product the dimensions add, but with the tensor product they multiply.
4. While the total tensor algebra T is infinite-dimensional, with the aid of ideals finite-dimensional quotient algebras can be generated. The set

$$I = \left\{ \sum_i t \otimes t_i \otimes \bar{t}, \text{ with } t_i \in T_1, t, \bar{t} \in T \right\}$$

is known as the two-sided ideal generated by the subset T_1 of T, and the quotient algebras are the equivalence classes modulo I. Sums and products are defined on them in the natural way.

The quotient algebra of the tensor algebra by the ideal generated by $u^* \otimes u^*$, with $u^* \in T^0_1$ will earn a special status.

We will construct it explicitly in different but equivalent notation using

(2.4.7) **Definition**

The **exterior product**, or **wedge product**, $v_1^* \wedge v_2^* \wedge \cdots \wedge v_p^* \in T^0_p$ of p covectors is defined by $(v_1^* \wedge v_2^* \wedge \cdots \wedge v_p^* | u_1, u_2, \ldots, u_p) = \det(v_i^* | u_j)$.

(2.4.8) **Remarks**

1. The connection with the notation $\otimes$ is that for any two covectors, $v_1^* \wedge v_2^* = v_1^* \otimes v_2^* - v_2^* \otimes v_1^*$. More generally, if $t_{i_1 \ldots i_p}$ is totally antisymmetric,

$$t_{i_1 \ldots i_p} e^{*i_1} \otimes \cdots \otimes e^{*i_p} = \frac{1}{p!} t_{i_1 \ldots i_p} e^{*i_1} \wedge \cdots \wedge e^{*i_p}.$$

In this expression, $t_{P_{i_1} \ldots P_{i_p}} = (-1)^P t_{i_1 \ldots i_p}$ for any permutation $(i_1 \ldots i_p) \to (P_{i_1 \ldots i_p})$, where $(-1)^P$ is the signature of the permutation.
2. The linear space consisting of the covariant, totally antisymmetric tensors of pth degree is denoted Λ_p. Its elements are of the form

$$\sum_{(i)} \frac{1}{p!} e^{*i_1} \wedge \cdots \wedge e^{*i_p} \omega_{i_1 \ldots i_p},$$

and its dimension is $\binom{m}{p}$. If the associative and distributive laws for the wedge product $\wedge$ are extended to all of Λ_p, then the set $\oplus_{p=0}^m \Lambda_p$ becomes a graduated algebra known as the **exterior algebra** (let Λ_0 be the scalars). It is a **Grassmann algebra**, since it is generated by the e^{*i}, for which the product is anticommutative.
3. The reason that Λ_p is of interest is that its elements measure the volumes spanned by p vectors. Of course, it is not a question here of a positive measure, as the volume is measured according to a particular orientation. We have already seen that the covectors $\in \Lambda_1$ define the lengths of components of a vector. In $\mathbb{R}^n$, where

the natural scalar product identifies e^{*i} with e_i, $(e^{*1} \wedge e^{*2}|u, v) = u^1v^2 - u^2v^1$ is just the area of the parallelogram spanned by the projections of u and v on the $(1-2)$-plane. This clearly generalizes to the situation of p vectors, $1 \leq p \leq m$.

Until now we have worked on the level of general geometry, where length and orthogonality of vectors have not been defined. The only general axioms at hand for measuring volumes are as follows: Letting $\mu(u, v, w, \dots)$ be the volume spanned by the vectors $u, v, w, \dots$:

(i) $\mu(\alpha u, v, w, \dots) = \mu(u, \alpha v, w, \dots) = \dots = \alpha\mu(u, v, w, \dots)$;

(ii) $\mu(u_1 + u_2, v, w, \dots) = \mu(u_1, v, w, \dots) + \mu(u_2, v, w, \dots)$, and analogously for $v = v_1 + v_2$, etc.; and

(iii) $\mu(u, u, w, \dots) = \mu(u, v, v, \dots) = \dots = 0$.

These requirements can be depicted schematically as follows:

(i) 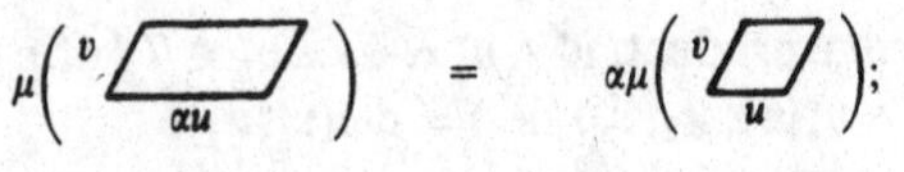

(ii) 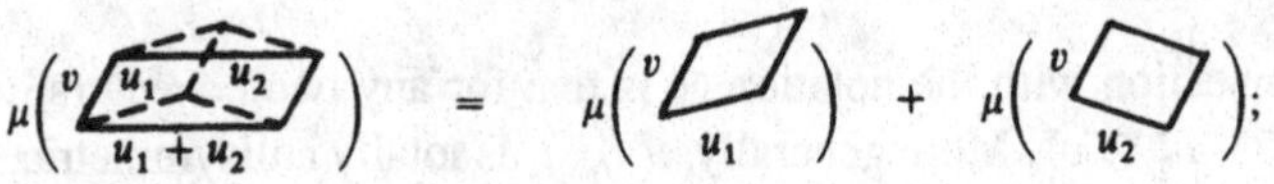

(iii) 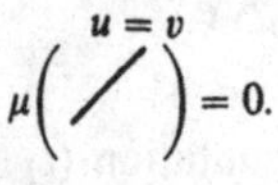

These requirements are equivalent to the statement that for some $t \in \Lambda_p$, $\mu(u, v, w, \dots) = (t|u, v, w, \dots)$: Axioms (i) and (ii) produce the multilinear structure, and Axiom (iii) implies the total antisymmetry. Total antisymmetry is equivalent to a change of sign upon interchange of two arguments, and (iii) leads to

$$0 = \mu(u + v, u + v, w, \dots) = \mu(u, v, w, \dots) + \mu(v, u, w, \dots).$$

The inner product $(|)$ as a mapping $\Lambda_1 \times T_0^1 \to \mathbb{R}$ can be generalized:

(2.4.9) Definition

The **interior product** $(\omega, X) \to i_X\omega$ is a mapping from $\Lambda_p \times T_0^1$ to Λ_{p-1}. It is linear in both factors and determined by:

(i) $i_X\omega = (\omega|X)$ for $\omega \in \Lambda_p$; and

(ii) $i_X(\omega \wedge \nu) = (i_X\omega) \wedge \nu + (-1)^p\omega \wedge i_X\nu$ for $\omega \in \Lambda_p$.

(By convention we set $\Lambda_p = 0$ for all $p < 0$, so $i_X\omega = 0$ when $\omega \in T_0^0$. The notation $X\rfloor\omega$ is frequently encountered for $i_X\omega$.)

A tensor $g \in T_2^0$ maps $(u, v) \in T_0^1 \times T_0^1$ to $g(u, v) \in \mathbb{R}$. In components this may be written $g(u, v) = u^i v^k g_{ik}$, where $u = u^i e_i$, $v = v^i e_i$, and $g = e^{*i} \otimes e^{*k} g_{ik}$. If the matrix g_{ik} is strictly positive, i.e., $g_{ik} = g_{ki}$ and all its eigenvalues are positive, then this bilinear mapping will be written as $g(u, v) =: \langle u, v\rangle$. It has the properties of a scalar product:

(2.4.10) $\langle u|v\rangle = \langle v|u\rangle, \quad \langle v|v\rangle \geq 0 \quad$ and $\quad \langle v|v\rangle = 0 \quad$ iff $\quad v = 0.$

If the eigenvalues of g_{ij} are not necessarily positive but none of them is zero, then we obtain the weaker statement that

(2.4.11) $$g(u, v) = 0 \quad \text{for all} \quad v \quad \text{iff} \quad u = 0.$$

Such a g is referred to as **nondegenerate**. We shall continue to use the notation $g(u, v) = \langle u, v\rangle$ even in the case when only (2.4.11) applies. This relationship guarantees that each $v \in T_0^1$ can be bijectively assigned a $v^* = e^{*i} g_{ik} v^k$ by the formula

$$\langle v|w\rangle = (u^*|w) \qquad \text{for all} \quad w.$$

(2.4.12) **Remarks**

1. The tensor g thus equips the space with an additional structure allowing an identification of T_0^1 and T_1^0. We may then speak simply of a vector and refer to v^k as its contravariant components and $v_i = g_{ik} v^k$ as its covariant components. With this identification the star can be dropped from the cobasis, which can be recognized from the position of the indices:

$$u = u^i e_i = u_i e^i, \quad \text{with} \quad e^i g_{ik} = e_k,$$

so

$$(e^i)^* = e^{*i},$$

since

$$\delta^i_j = \langle e^i|e_j\rangle = ((e^i)^*|e_j) = (e^{*i}|e_j).$$

2. The inverse matrix to g is written with raised indices: $g_{ik}g^{kj} = \delta_i^j$. Other indices can likewise be "raised and lowered" with g^{ik} and g_{ik}:

$$e^i = g^{ij} e_j \quad \text{and} \quad u^i = g^{ij} u_j.$$

3. If (2.4.10) holds, then $\langle v|v\rangle^{1/2}$ may be interpreted as the length of a vector. Neither u^i nor u_i is, however, the length of the ith component of u, as, for instance, $\langle u^1 e_1|u^1 e_1\rangle^{1/2} = |u^1|\sqrt{g_{11}}$. Only components in an **orthogonal basis**, for which $g_{ik} = \delta_{ik}$, can be so interpreted.

If vectors and covectors are identified, then all T_s^r with the same $\sigma := r + s$ may be treated as identical by, say, mapping them all bijectively to T_σ^0 by

$$t = t^{i_1 \ldots i_r}_{j_1 \ldots j_s} e^{j_1} \otimes \ldots e^{j_s} \otimes e_{i_1} \otimes \ldots e_{i_r} \tag{2.4.13}$$
$$= t_{i_1 \ldots i_r j_1 \ldots j_s} e^{i_1} \otimes \ldots e^{i_r} \otimes e^{j_1} \otimes \ldots e^{j_s},$$

with

$$t_{i_1 \ldots i_r j_1 \ldots j_s} = t^{k_1 \ldots k_r}_{j_1 \ldots j_s} g_{i_1 k_1} \ldots g_{i_r k_r}.$$

In this manner, for a given g we may extend the scalar product $\langle | \rangle$ to all of T_s^r.

(2.4.14) Definition

The scalar product of two tensors is a bilinear mapping $T_s^r \times T_s^r \to \mathbb{R}$, $(\bar{t}, t) \to \langle \bar{t} \mid t \rangle \in \mathbb{R}$, defined by

$$\langle \bar{t} | t \rangle = \bar{t}^{i_1 \ldots i_r}_{j_1 \ldots j_s} t^{m_1 \ldots m_r}_{n_1 \ldots n_s} g_{i_1 m_1} \ldots g_{i_r m_r} g^{j_1 n_1} \ldots g^{j_s n_s}.$$

(2.4.15) Remarks

1. If g is positive and $t \in \Lambda_p$, then $\langle t | t \rangle^{1/2}$ provides a measure for the p-dimensional volume defined by t.
2. Treating $T_s^r \times T_s^r$ as a subset of T_{r+s}^{r+s}, the scalar product can be viewed as the mapping of this set to $\mathbb{R}$ by "contracting the upper and lower indices." Analogously, if $r \geq s$, T_s^r can be mapped to T_0^{r-s}, though the result depends on which indices get contracted and which are left alone. If, however, T_s^r comes from $\Lambda_r \times \Lambda_s$, then the result is uniquely determined up to a sign because of the antisymmetry of the tensors. This is the motivation for the following definition.

(2.4.16) Definition

The **interior product** on Λ_p is defined as the bilinear mapping $\Lambda_p \times \Lambda_q \to \Lambda_{p-q}$, $p \geq p$, $(\omega, \nu) \to i_\nu \omega$ determined by the rules:

(i) $i_\nu \omega = \langle \omega | \nu \rangle$ for $p = q = 1$;

(ii) $i_\nu(\omega_1 \wedge \omega_2) = (i_\nu \omega_1) \wedge \omega_2 + (-1)^{p_1} \omega_1 \wedge i_\nu \omega_2$ for $\nu \in \Lambda_1$ and $\omega_i \in \Lambda_{p_i}$; and

(iii) $i_{\nu_1 \wedge \nu_2} = i_{\nu_2} \circ i_{\nu_1}$.

(2.4.17) Remarks

1. In order to express i in components, we introduce the abbreviation

$$e^{j_1 \ldots j_p} = e^{j_1} \wedge e^{j_2} \wedge \cdots \wedge e^{j_p}$$

for the basis of Λ_p. With

$$\omega = \frac{1}{p!}\omega_{j_1\ldots j_p}e^{j_1\ldots j_p}, \qquad \nu = \frac{1}{q!}\nu_{k_1\ldots k_q}e^{k_1\ldots k_q},$$

we find

$$i_\nu \omega = \frac{1}{q!(p-q)!}\nu^{j_1\ldots j_q}\omega_{j_1\ldots j_p}e^{j_{q+1}\ldots j_p}.$$

2. Definition (2.4.16) generalizes (2.4.9), when T_0^1 is identified with Λ_1.
3. The tensor g allows an identification of the spaces Λ_p with their dual spaces. If exterior multiplication by some $\nu \in \Lambda_q$ is interpreted as a mapping $\Lambda_p \to \Lambda_{p+q}$, then interior multiplication by ν is the adjoint mapping (with respect to $\langle | \rangle$):

$$\langle \nu \wedge \omega | \mu \rangle = i_{\nu\wedge\omega}\mu = i_\omega(i_\nu \mu) = \langle \omega | i_\nu \mu \rangle = \langle (i_\nu)^t \omega | \mu \rangle.$$

4. If $p = q$, then

$$i_\nu \omega = \frac{1}{p!}\langle \omega | \nu \rangle = \frac{1}{p!}\omega^{j_1\ldots j_p}\nu_{j_1\ldots j_p} = i_\omega \nu.$$

In particular, the canonical m-form $\varepsilon := |g|^{1/2}e^{1\ldots m}$ is normalized by $i_\varepsilon \varepsilon = (-1)^s$. Here $g := \det(g_{ik})$ and $(-1)^s = g/|g|$. Both spaces Λ_p and Λ_{m-p} have dimension $\binom{m}{p}$, and hence they may also be identified by the use of a g with property (2.4.11).

(2.4.18) Definition

The **duality mapping**, or **star operation**, is the linear bijection $\Lambda_p \xrightarrow{*} \Lambda_{m-p}$ defined by $\omega \to {}^*\omega := i_\omega \varepsilon$.

(2.4.19) Remarks

1. It is clear that * is injective. For linear mappings of the same finite dimension injectivity implies surjectivity.
2. In components,

$$^*\omega = \frac{\omega^{j_1\ldots j_p}}{p!(m-p)!}e^{j_{p+1}\ldots j_m}\varepsilon_{j_1\ldots j_m}.$$

(2.4.20) Properties of the Star Operation

(i) $\varepsilon = {}^*1$, ${}^*\varepsilon = (-1)^s$;

(ii) ${}^{*\circ *} = (-1)^{p(m-p)+s}$;

(iii) $i_\nu {}^*\omega = {}^*(\omega \wedge \nu)$; and

(iv) for all ν and ω in Λ_p, $\nu \wedge {}^*\omega = \varepsilon i_\nu \omega = \omega \wedge {}^*\nu = \varepsilon(-1)^s i_{*\nu}{}^*\omega$.

(2.4.21) Remarks

1. Property (i): Λ_0 and Λ_m are both one-dimensional and hence isomorphic to $\mathbb{R}$. The canonical m-form is dual to the number 1.
2. Property (ii): By this property, * is its own inverse, up to a sign. It is unfortunately not possible to modify the definition of the sign of * in order to dispense with the possible negative sign.
3. Property (iii): This means that the interior product turns out to be dual to the exterior product.
4. Property (iv): Herein lies the origin of the term "duality." The space Λ_{m-p} can be treated as the dual space of Λ_p, by defining a scalar product $\{\ ,\ \}$ via $\nu \wedge \omega = \varepsilon\{\nu, \omega\}$ for $\nu \in \Lambda_p$ and $\omega \in \Lambda_{m-p}$. It is connected with i through the formula $\{\nu, \omega\} = (-1)^{p(m-p)+s} i_\nu{}^*\omega$.

Proof of (2.4.20)

(i) $i_\varepsilon \varepsilon$ was normalized as $(-1)^s$.

(ii) This follows from (i) together with

$$\varepsilon_{i_1 \ldots i_m} = (-1)^{p(m-p)} \varepsilon_{i_{p+1} \ldots i_m i_1 \ldots i_p}.$$

(iii) $i_\nu{}^*\omega = i_\nu i_\omega \varepsilon = i_{\omega \wedge \nu} \varepsilon$.

(iv) This follows from $i_\nu \omega = i_\omega \nu$ when $p = q$, together with Properties (ii) and (iii).

(2.4.22) Example

$m = 3$ and $g_{ik} = \delta_{ik}$. In this case $^*\circ^* = \mathbf{1}$. The definition of the basis vectors specializes to

$$p = 0, 3: \quad \mathbf{1} \overset{*}{\leftrightarrow} dx^1 \wedge dx^2 \wedge dx^3;$$

$$p = 1, 2: \quad (dx^1, dx^2, dx^3) \overset{*}{\leftrightarrow} (dx^2 \wedge dx^3, dx^3 \wedge dx^1, dx^1 \wedge dx^2).$$

For the components this means that

$$\omega \in \Lambda_0: \quad ({}^*\omega)_{ijk} = \omega \varepsilon_{ijk},$$

$$\omega \in \Lambda_1: \quad ({}^*\omega)_{ij} = \omega_k \varepsilon_{kij},$$

$$\omega \in \Lambda_2: \quad ({}^*\omega)_i = \tfrac{1}{2} \omega_{kj} \varepsilon_{kji},$$

$$\omega \in \Lambda_3: \quad {}^*\omega = \tfrac{1}{3!} \omega_{ijk} \varepsilon_{ijk}.$$

In elementary vector calculus the exterior product in three dimensions is ordinarily expressed as the cross product $^*(v \wedge w) = [v \wedge w]$, frequently written $[v \times w]$.

Brackets are used because the cross product is nonassociative. The scalar product is connected with * and $\wedge$ by $(v \cdot w) = {}^*({}^*v \wedge w)$.

The foregoing construction proceeded on the infinitesimal level. We next investigate how it carries over to the local and global levels. The extension to the domain of a chart is straightforward, whereas the global existence of e_i, g, and ε is open to question.

The next step is to collect all the tensors at different points into a bundle over M. We see again that on the domain U of a chart C of M, $(\Theta_C^{-1})^t$ provides a chart for the **cotangent bundle**,

(2.4.23)
$$T^*(U) = \bigcup_{q \in U} T_q^*(M)$$

through

(2.4.24)
$$T^*(U) \rightarrow \Phi(U) \times \mathbb{R}^m : (q, v^*) \rightarrow (\Phi(q), (\Theta_C^{-1}(q))^t v^*).$$

This follows from: $\Theta_C(q): T_q(M) \rightarrow \mathbb{R}^m$ implies $(\Theta_C^{-1}(q))^t: T_q^*(M) \rightarrow \mathbb{R}^{m*} \equiv \mathbb{R}^m$, hence $(v^*|U) = ((\Theta_C^{-1}(q)^t v^* | \Theta_C(q)U)$. As noted above (2.2.11), for different U's these charts are compatible; $D(\bar{\Phi} \circ \Phi^{-1})$ is merely replaced with $D(\Phi \circ \bar{\Phi}^{-1})^t$, which does not destroy the required differentiability. The bundle structure carries over directly to the tensors. The mapping

$$\underbrace{\Theta_C(q) \otimes \Theta_C(q) \otimes \cdots \otimes \Theta_C(q)}_{r \text{ times}} \otimes \underbrace{(\Theta_C^{-1}(q))^t \otimes (\Theta_C^{-1}(q))^t \otimes \cdots \otimes (\Theta_C^{-1}(q))^t}_{s \text{ times}}$$

sends T_{qs}^r at every point $q \in U$ into $\mathbb{R}^{m(s+r)}$. As a bijection this mapping can be used for the charts of the tensor bundle, leading us to make a general

(2.4.25) Definition

Let M be a manifold with the atlas $\bigcup_i C_i = \bigcup_i (U_i, \Phi_i)$. The vector bundle over M defined by the atlas $\bigcup_i (\bigcup_{q \in U_i} T_q(M)_s^r, (q; u_1 \otimes u_2 \otimes \cdots \otimes u_r \otimes v_1 \otimes v_2 \otimes \cdots \otimes v_s)) \rightarrow (\Phi_i(q); \Theta_{C_i}(q) u_1 \otimes \cdots \otimes \Theta_{C_i}(q) u_r \otimes (\Theta_{C_i}^{-1}(q))^t v_1 \otimes \cdots \otimes (\Theta_{C_i}^{-1}(q))^t v_s))$ on $T_s^r(M) = \bigcup_q T_{qs}^r(M)$ is called the **bundle of r-fold contravariant and s-fold covariant tensors.**

(2.4.26) Remarks

1. With this definition $T(M) \equiv T_0^1(M)$ and $T^*(M) \equiv T_1^0(M)$.
2. The linear structure required by Definition (2.2.15) is that of the tensors, and the projection is $\Pi: (q; u_1, \ldots, u_r, v_1, \ldots, v_s) \rightarrow (q; 0, \ldots, 0)$. As with $T(M)$, the topology used on $T_s^r(M)$ is the product topology of $U \times \mathbb{R}^{m(s+r)}$.
3. Since the chart is a linear mapping on each fiber, it suffices to specify the images of the bases, which is what was done in Definition (2.4.25) by writing out the way r contravariant and s covariant vectors are transformed.

(2.4.27) Examples

1. Suppose M is an m-dimensional, linear topological space: $T^r_s(M) = M \times \mathbb{R}^{m(r+s)}$. Then $T^*(M)$ and $T(M)$ are both of the form $M \times \mathbb{R}^m$, but cannot be identified, because no basis has been provided that is distinguished as orthogonal. If $M = \mathbb{R} \times \cdots \times \mathbb{R}$, an orthogonal basis would exist, because of the additional Riemannian structure we discuss later.
2. $M = S^1$, $T^r_s(M) = M \times \mathbb{R}$. Although the angle φ of (2.1.7; 2) cannot be defined globally as continuous, the bases $d\varphi$ and $\partial/\partial\varphi$ can be. Again no canonical identification of $T^*(M)$ is given; a canonical identification of $T^*(M)$ and $T(M)$ would amount to a specification of arc length, but as a manifold S^1 could be a circle of any radius.
3. $M = S^2$: $T^r_s(M)$ is not a Cartesian product, any more than $T(M)$ is.

There is an immediate generalization of the concept of a vector field.

(2.4.28) Definition

A C^∞-mapping $t: M \to T^r_s(M)$ such that $\Pi \circ t = 1$ is an **r-fold contravariant and s-fold covariant tensor field.** The set of all such tensor fields is denoted by $\mathcal{T}^r_s(M)$. The p-fold covariant, totally antisymmetric tensor fields are called **p-forms.** The set of p-forms is denoted $E_p(M)$, $p = 0, 1, \ldots, m$.

(2.4.29) Remarks

1. The terms "vector field" and "1-fold contravariant tensor field" are synonymous, as are "covariant vector field" and "1-fold covariant tensor field" and "1-form."
2. A tensor field can be written locally in the natural basis of a chart as

$$\sum_{(i)(j)} c^{i_1,\ldots,i_r}_{j_1,\ldots,j_s}\, \partial_{i_1} \otimes \cdots \otimes \partial_{i_r} \otimes dq^{j_1} \otimes \cdots \otimes dq^{j_s},$$

with $c^{(i)}_{(j)} \in C^\infty(M)$ (cf. (2.2.26) and (2.4.3; 2)). In the physical literature the components $c^{(i)}_{(j)}$ are referred to as the tensor fields. In this basis a p-form is written

$$\frac{1}{p!}\sum_{(j)} c_{j_1\ldots j_1}\, dx^{j_1} \wedge dx^{j_2} \wedge \cdots \wedge dx^{j_p}.$$

3. If there exist global bases $\{e_i\}$ and $\{e^i\}$ for T^1_0 and T^0_1, then there is also obviously a global basis for all T^r_s. The manifold is then parallelizable (cf. (2.2.14; 2)), the desired diffeomorphism $M \times \mathbb{R}^m \to T(M)$ being $(x, v) \to (x, e_i(x)v^i)$. It then also follows that all bundles T^r_s are trivializable. If there is an atlas consisting of a single chart, then the natural basis is defined globally. This condition is, however, not necessary; S^1 is also parallelizable. In contrast, on S^2 there does not even exist a nowhere-vanishing smooth vector field.
4. If there exists a globally nonvanishing m-form, then M is said to be **orientable.** Any parallelizable M is also orientable, since $e^1 \wedge e^2 \wedge \cdots \wedge e^m$ never vanishes.

This is again not a necessary condition, as, for example, S^2 is orientable. On the other hand, the Möbius strip (2.2.16; 3) is not orientable, as the 2-form $d\varphi \wedge dx$, which is defined on both charts, cannot be extended continuously to the whole manifold. Since every manifold can be equipped with a Riemannian structure (see (2.4.31) and (2.4.32; 2)), such a structure clearly does not guarantee orientability. In contrast, a symplectic space (2.4.31) is always orientable, since the m-form

$$\underbrace{g \wedge g \wedge \cdots \wedge g}_{m/2 \text{ times}} = dq_1 \wedge \cdots \wedge dq_m \sqrt{\det(g)}$$

is nonvanishing by assumption. We shall encounter a criterion for orientability in (2.6.15; 4).

(2.4.30) **Examples**

1. In (2.4.3) there was an example of a covariant vector field (= 1-form), the differential $df \in \mathcal{T}_1^0(M)$. Thus df symbolizes the rate of change of f in some direction (to be specified later), and is not some infinitely small quantity.
2. If $g \in \mathcal{T}_2^0(M)$ satisfies (2.4.11) at all points of M, then the construction made there may be extended to all of M and creates an additional structure on M. The important cases for us will be when either g_{ik} is symmetric or else is antisymmetric and satisfies a differential condition, which cannot be formulated at a single point. If (2.4.11) holds everywhere, the tensor field is said to be **nondegenerate**.

(2.4.31) **Definition**

If a manifold M is given a nondegenerate, symmetric tensor field $g \in \mathcal{T}_2^0(M)$, it is called a **pseudo-Riemannian** space. If g is in fact positive, M is a **Riemannian** space, and g is called its **metric**. If $g \in E_2$ is nondegenerate (which requires m to be even), and there is a natural basis dq_j, so that

$$g = \sum_{j=1}^{m/2} dq_j \wedge dq_{j+m/2},$$

then M is called a **symplectic space**.

(2.4.32) **Examples**

1. $\mathbb{R}^n$ becomes a pseudo-Riemannian space with $g = \sum_{i,k} dx^i \otimes dx^k g_{ik}$, where g_{ik} is a constant symmetric matrix with all nonzero eigenvalues. The matrix g can be diagonalized with some orthogonal transformation $x^i \to m^{ij} x^j$, and then the eigenvalues can all be normalized to $g_{ii} = \pm 1$ with a dilatation $x^i \to x^i/(|g_{ii}|)^{1/2}$. These charts have a special status, because they are determined

up to pseudo-Euclidean transformations. (For $n = 4$ and $g_{ii} = (-1, 1, 1, 1)$, the transformations would form the **Poincaré group**.) When all $g_{ii} = 1$, $\mathbb{R}^n$ becomes a Riemannian space. On other charts the g_{ij} of this space do not have to be either diagonal or constant. For example, in the Riemannian case on $\mathbb{R}^2$, and using polar coordinates, $g = dr \otimes dr + r^2\, d\varphi \otimes d\varphi$.

2. If N is a submanifold of M, and therefore $T(N)$ is a submanifold of $T(M)$, a nondegenerate $g \in \mathcal{T}_2^0(M)$, $g > 0$, induces a Riemannian structure on N, because g also provides a nondegenerate mapping $T_q(N) \times T_q(N) \to \mathbb{R}$. The metric $g_{ik} = \delta_{ik}$ on $\mathbb{R}^m$ induces the usual metric on S^n or $T^n \subset \mathbb{R}^{n+1}$. Since every m-dimensional manifold can be imbedded as a submanifold of $\mathbb{R}^{2m+1}$, it is always possible to find a Riemannian structure for any manifold.
3. The Riemannian structure of $\mathbb{R}^n$ shows up in mechanics because of the kinetic energy, which we wrote as $m_{ik}(q)\dot{q}_i\dot{q}_k/2$ in (2.3.19). Up to a factor, this mapping $T(M) \times T(M) \to \mathbb{R}$ is exactly the metric. In the last chapter of the book we shall discuss why a concept of length that was introduced purely mathematically should be the physically measured interval. The bijection $T(M) \to T^*(M)$ mentioned above (2.4.12; 1), induced by the metric, sends $\dot{q}_i$ to $m_{ik}(q)\dot{q}_k = \partial L/\partial \dot{q}_i$, that is, to the canonically conjugate momentum p_i: (q, p) represents a point of $T^*(M)$.
4. Only manifolds of even dimensionality may have symplectic structures. We shall see in §3.1 that $T^*(M)$ always has a symplectic structure.

(2.4.33) **Remarks**

1. We take over the notation of (2.4.12) through (2.4.20) at individual points for the corresponding tensor fields (* is globally definable only if M is orientable):

$$\begin{aligned} \langle | \rangle &: \mathcal{T}_s^r \times \mathcal{T}_r^s \to C^\infty(M), \\ i &: E_p \times \mathcal{T}_1^0 \to E_{p-1}, \\ \wedge &: E_p \times E_q \to E_{p+q}, \\ i &: E_p \times E_q \to E_{p-q}, \\ * &: E_p \qquad\;\; \to E_{m-p}. \end{aligned}$$

2. Since g_{ik} is diagonable on a pseudo-Riemannian space, it is possible at least locally to introduce an orthogonal basis $\{e^i\}$, in which g has the normal form

$$g = e^i \otimes e^k \eta_{ik}, \qquad \eta_{ik} = \begin{cases} 0 & \text{if } i \neq k, \\ \pm 1 & \text{if } i = k. \end{cases}$$

Unlike with the normal form (2.4.31) of a symplectic matrix, $\{e^i\}$ is not necessarily a natural basis.

A diffeomorphism $\Phi: M_1 \to M_2$ induces a diffeomorphism $T(\Phi): T(M_1) \to T(M_2)$. Now we introduce another diffeomorphism $T^*(M_1) \to T^*(M_2)$, such that the scalar product, $(|): T^*(M) \overset{\pi}{\times} T(M) \to M \times \mathbb{R}$, remains invariant,[8] and consequently dual bases are mapped to dual bases.

(2.4.34) Transformations of Covectors

For any diffeomorphism $\Phi: M_1 \to M_2$, we define another diffeomorphism $T^*(\Phi): T^*(M_1) \to T^*(M_2)$, so as to make the diagrams

$$\begin{array}{ccc} M_1 & \xrightarrow{\quad\Phi\quad} & M_2 \\ \uparrow{\scriptstyle \Pi_1} & & \uparrow{\scriptstyle \Pi_2} \\ T^*(M_1) & \xrightarrow{\quad T^*(\Phi)\quad} & T^*(M_2) \end{array}$$

$$\begin{array}{ccc} M_1 \times \mathbb{R} & \xrightarrow{\quad\Phi \times 1\quad} & M_2 \times \mathbb{R} \\ \uparrow{\scriptstyle (|)} & & \uparrow{\scriptstyle (|)} \\ T^*(M_1) \overset{\pi}{\times} T(M_1) & \xrightarrow{\quad T^*(\Phi) \times T(\Phi)\quad} & T^*(M_2) \overset{\pi}{\times} T(M_2) \end{array}$$

commute.

(2.4.35) Remark

On the domain of a chart, $T^*(\Phi)$ is evidently $(q, u) \to (\Phi(q), (T(\Phi^{-1}))^{t}(q) \cdot u)$, the differentiability and uniqueness of which are obvious. We shall often denote this simply by $T^*(\Phi) = T(\Phi^{-1})^{t}$, with the understanding that t refers to the transposition of a linear mapping at constant q.

Constructing the tensor product

$$\underbrace{T(\Phi) \otimes T(\Phi) \otimes \cdots \otimes T(\Phi)}_{r \text{ times}} \otimes \underbrace{T^*(\Phi) \otimes T^*(\Phi) \otimes \cdots \otimes T^*(\Phi)}_{s \text{ times}}$$

shows how $T_s^r(M)$ is altered under a diffeomorphism. The vectors of the basis $\partial_{i_1} \otimes \cdots \otimes \partial_{i_r} \otimes dq^{j_1} \otimes \cdots \otimes dq^{j_s}$ must be transformed by $T(\Phi)$ or, respectively, $T^*(\Phi)$, and linearity extends this to all of $T_s^r(M)$. In this way we obtain the transformation law for tensor fields under diffeomorphisms of a manifold, and in particular under a change of charts.

---

[8] The "fiber product" notation $\overset{\pi}{\times}$ means that the factors are to be taken at the same base point.

(2.4.36) Definition

A diffeomorphism $\Phi\colon M_1 \to M_2$ induces a mapping $\Phi_*\colon T_s^r(M_1) \to T_s^r(M_2)$ defined by the permutability of the diagram:

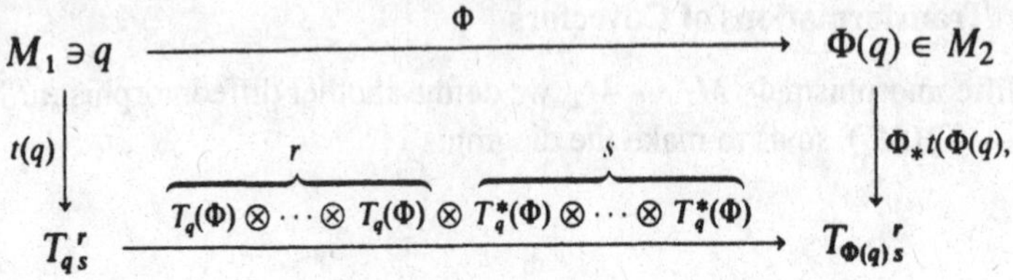

$$
\begin{array}{ccc}
M_1 \ni q & \xrightarrow{\quad\quad\quad \Phi \quad\quad\quad} & \Phi(q) \in M_2 \\[2pt]
{\scriptstyle t(q)}\Big\downarrow & \overbrace{\underbrace{T_q(\Phi) \otimes \cdots \otimes T_q(\Phi)}_{r} \otimes \underbrace{T_q^*(\Phi) \otimes \cdots \otimes T_q^*(\Phi)}_{s}} & \Big\downarrow{\scriptstyle \Phi_* t(\Phi(q)),} \\[2pt]
T_{q\,s}^{\,r} & \xrightarrow{\hspace{4cm}} & T_{\Phi(q)\,s}^{\,r}
\end{array}
$$

where $t \in T_s^r(M)$:

$$
\Phi_* t = \underbrace{T(\Phi) \otimes \cdots \otimes T(\Phi)}_{r \text{ times}} \otimes \underbrace{T^*(\Phi) \otimes \cdots \otimes T^*(\Phi)}_{s \text{ times}} \circ t \circ \Phi^{-1}.
$$

(2.4.37) Examples

1. $T_0^0\colon \Phi_* f = f \circ \Phi^{-1}$. The value of the transformed function at a given point is the same as that of the original function at the inverse image of the point.
2. $M = \mathbb{R}^n$, $\Phi\colon \mathbf{x} \to \mathbf{x} + \mathbf{a}, \mathbf{a} \in \mathbb{R}^n$. $T(\Phi)\colon (\mathbf{x}, \mathbf{v}) \to (\mathbf{x} + \mathbf{a}, \mathbf{v})$, and $T(\Phi)$ has already been given in (2.2.22; 1). Under displacements the components of a tensor t transform like scalar functions (see Part I).

$$
t\colon \mathbf{x} \to (\mathbf{x}, t_{j_1,\ldots,j_s}^{i_1,\ldots,i_r}(\mathbf{x}) \partial_{i_1} \otimes \cdots \otimes \partial_{i_r} \otimes dx^{j_1} \otimes \cdots \otimes dx^{j_s}),
$$

$$
\Phi_* t\colon \mathbf{x} \to (\mathbf{x}, t_{j_1,\ldots,j_s}^{i_1,\ldots,i_r}(\mathbf{x} - \mathbf{a}) \partial_{i_1} \otimes \cdots \otimes \partial_{i_r} \otimes dx^{j_1} \otimes \cdots \otimes dx^{j_s}).
$$

3. $M = \mathbb{R}^n$, $\Phi\colon x_i \to L_{ik} x_k$, $T^*(\Phi)\colon (x_i; dx^j v_j) \to (L_{ik} x_k, dx^j L_{kj}^{-1} v_k)$. For a transformation of $\partial_i v^i$, $(L^{-1})^t$ must be replaced with L. With t as above,

$$
\Phi_* t\colon \mathbf{x} \to (\mathbf{x}, \partial_{i_1} \otimes \cdots \otimes \partial_{i_r} \otimes dx^{j_1} \otimes \cdots
$$
$$
\otimes dx^{j_s} L_{i_1 m_1} \ldots L_{i_r m_r} L_{n_1 j_1}^{-1} \ldots L_{n_s j_s}^{-1} t_{n_1,\ldots,n_s}^{m_1,\ldots,m_r}(L^{-1} x)).
$$

4. $g \in C^\infty(M_1)$: $\Phi_* dg = d(g \circ \Phi^{-1}) = d(\Phi_* g)$. It is intuitively clear that the image of the exterior derivative of a function must be the exterior derivative of the image of the function. When dg is applied to a vector v determined by a curve $u\colon I \to M_1$, it yields the rate of change of g along u. But that is the same as the rate of change of $g \circ \Phi^{-1}$ along $\Phi \circ u$,

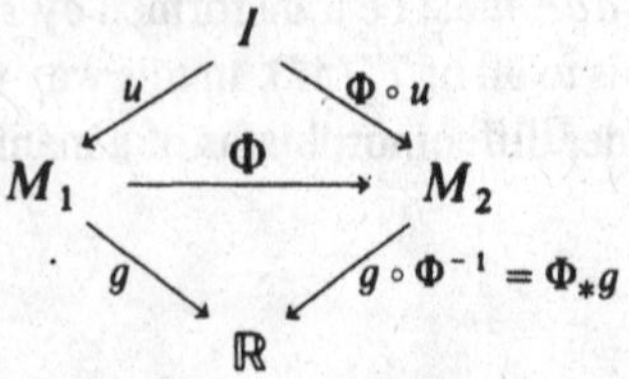

and the latter curve determines the image of the vector v under $T(\Phi)$. Let us summarize

(2.4.38) The Properties of the Mapping Φ_*

The image of a vector is determined by the images of the curves that define it. The image of a covector is such that its product with the image of any vector equals the original product of the vector and the covector. These conditions fix the relationships among the bases and, because of the permutability of Φ_* with algebraic operations:

$$\Phi_*(t_1 + t_2) = \Phi_*(t_1) + \Phi_*(t_2), \qquad \Phi_*(t_1 \otimes t_2) = \Phi_*(t_1) \otimes \Phi_*(t_2),$$

among all tensors. As for the tensor fields, at every point they transform in the same way as the tensors at that point.

For compositions, $(\Phi_1 \circ \Phi_2)_* = \Phi_{1*} \circ \Phi_{2*}$.

(2.4.39) Remarks

1. Although the scalar product $(|)$ is invariant under diffeomorphisms Φ, the same is true of $\langle|\rangle$ or more generally the interior product i provided that Φ leaves the metric g invariant. Diffeomorphisms that leave a pseudo-Riemannian structure invariant are called **isometries**, and those that leave a symplectic structure invariant are called **canonical transformations.** We shall return later in detail to these transformations and their generating vector fields, known, respectively, as the **Killing** and **Hamiltonian vector fields.**
2. Until now we have only investigated Φ_* for diffeomorphisms. In case Φ is not bijective, it is only possible to define the inverse images of covariant tensor fields. Even if Φ is injective, for example if it is the injection j of a submanifold, $j: N \to M \supset N$, $T(j): T(N) \to T(M) \supset T(N)$ (cf. (2.2.27; 6)), neither the image nor the inverse image of a vector field is defined; the image fails to be defined everywhere, and the inverse image lacks a distinguished subspace of $T_q(M)$ complementary to $T_q(N)$ unless M is given a metric.

(2.4.40) Example

$M = \mathbb{R}^2$ without a scalar product, $N = \mathbb{R}^1$, $j: x \to (x, 0)$, $T(j): (x, v) \to (x, 0; v, 0)$, and $X \in \mathcal{T}_0^1(M)$: $(x, y) \to (x, y; 1, 1)$. What could $j_*^{-1}X$, the inverse image of X under j, possibly be? While $Y \in \mathcal{T}_0^1(N): x \to (x, 1)$ reproduces the components of $X_{|N}$ in $T(N)$, if we look at the basis: $e_1 = (1, 0)$, $e_2 = (1, 1)$, and we let (v_1, v_2) mean the vector $v_1 e_1 + v_2 e_2$, then we see that j is still $x \to (x, 0)$; but X: $(x, y) \to (x, y; 0, 1)$ has the component 0 in $T(N)$. A glance at Figure 2.12 shows that $j_*^{-1}X = X_{|N}$ is not uniquely determined. The inverse image of X under j would be a restriction $X_{|N}$, which is undefined, unless on N the vector field X takes its values in $T(N)$. These difficulties do not occur for covariant tensors, because they are defined as multilinear mappings of $T(M)$. Their restrictions are merely the mappings of the vectors of $T(N) \subset T(M)$. By (2.4.36) the inverse image of a covariant vector field can be depicted as follows:

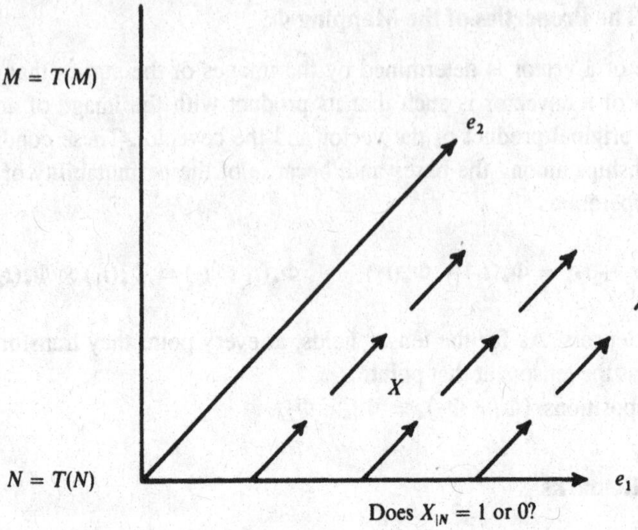

FIGURE 2.12. Restriction of a vector field to a subspace.

(2.4.41) **Definition**

Let $\Phi: N \to M$ be a differentiable mapping of two manifolds. The **inverse image,** or **pull-back,** $\Phi^*: T_s^0(M) \to T_s^0(N)$ of covariant tensor fields is defined by the formula

$$\Phi^* X = (T(\Phi))^t \circ X \circ \Phi.$$

This is equivalent to the commutativity of the diagram:

$$\Phi^* X = (T(\Phi))^t \circ X \circ \Phi, \qquad \begin{array}{ccc} N & \xrightarrow{\;\Phi\;} & M \\ {\scriptstyle \Phi^* X}\downarrow & & \downarrow{\scriptstyle X} \\ T^*(N) & \xleftarrow{\;T(\Phi)^t\;} & T^*(M) \end{array}$$

For diffeomorphisms, $\Phi^* = (\Phi^{-1})_*$.

(2.4.42) **Remark**

The pull-back $\Phi^* t$ of a covariant tensor field t acts on vectors exactly as t acts on their images under $T(\Phi)$:

$$\Phi^* t(v_1, \ldots, v_s) = t(T(\Phi)v_1, \ldots, T(\Phi)v_s).$$

(2.4.43) **Examples**

1. Let us return to Example (2.4.40) and reinterpret the vector fields introduced at the beginning as $X \in T_1^0(M)$ and $Y \in T_1^0(N)$. This time $(Y(q)|v) =$

$(X(q)|T_q(j)v)\ \forall v \in T_q(N)$. Suppose we transform the coordinates linearly with the matrix

$$L = \begin{bmatrix} 1 & -1 \\ 0 & 1 \end{bmatrix}.$$

Then X transforms with

$$(L^{-1})^t = \begin{bmatrix} 1 & 0 \\ 1 & 1 \end{bmatrix},$$

becoming $(x, y) \to (x, y; 1, 2)$. The components in the direction of $T(N)$ remain unchanged.

2. If $t = dg \in T_1^0(M)$, then according to (2.4.37; 4), $\Phi^* dg = d(g \circ \Phi)$, where Φ does not have to be either injective or surjective.

(2.4.44) **Problems**

1. Show that the adjoint $L \to L^t$ introduced at the beginning of the section is bijective.
2. Show that *: $E_p \to E_{n-p}$ is bijective.
3. Calculate the explicit form of $\Phi^*\omega$, $\omega \in T_1^0$, in local coordinates. How is this related to the elementary transformation of differentials of coordinates (or of the gradient)?
4. Calculate the components of the gradient df in spherical and cylindrical coordinates in $\mathbb{R}^3$.
5. Show that $T^*(M_1 \times M_2) = T^*(M_1) \times T^*(M_2)$.
6. Show

   (a) $(i_X\omega|X_1, \ldots, X_{p-1}) = p(\omega|X, X_1, \ldots, X_{p-1})$, $\omega \in E_p(M)$;

   (b) $i_{fX}\omega = f i_X\omega$, $f \in E_0(M)$, $\omega \in E_p(M)$; and

   (c) $i_X \circ i_Y = -i_Y \circ i_X$.

7. What is the transformation law for an m-form in m dimensions?

(2.4.45) **Solutions**

1. $v \to (w^*|Lv)$ is a linear functional on E, and can therefore be written $(L^t w^*|v)$, where L^t is a linear mapping $F^* \to E^*$. The association $L \to L^t$ is injective, for

$$\begin{aligned} L_1^t = L_2^t &\Leftrightarrow L_1^t w^* = L_2^t w^*\ \forall w^* \in F^* \Leftrightarrow \\ \langle L_1^t w^*|v\rangle &= \langle L_2^t w^*|v\rangle\ \forall v \in E \Leftrightarrow \\ \langle w^*|L_1 v\rangle &= \langle w^*|L_2 v\rangle\ \forall v \in E,\ \forall w^* \in F^* \Leftrightarrow L_2 = L_2. \end{aligned}$$

Since $E^{**} = E$ and $F^{**} = F$ (remember, that E and F are finite-dimensional), for any $L: E \to F$

$$\langle L^{tt} v|w^*\rangle = \langle v|L^t w^*\rangle \equiv \langle L^t w^*|v\rangle = \langle w^*|Lv\rangle,$$

whence $L^{\iota\iota} = L$. Thus for any $L^*: F^* \to E^*$,

$$L := L^{*\iota}: E \to F, \quad \langle Lv|w^*\rangle := \langle v|L^*w^*\rangle \ \forall v \in E,\ \forall w^* \in F^*$$

and therefore $L^{\iota} = (L^{*\iota})^{\iota} \equiv L^{*\iota\iota} = L^*$, i.e., the association is also surjective.

2. A linear mapping $\omega \to {}^*\omega$,

$$({}^*\omega)_{i_{p+1},\ldots,i_n} = \frac{1}{p!(n-p)!}\omega^{i_1,\ldots,i_p}\varepsilon_{i_1,\ldots,i_n},$$

was defined in (2.4.18). It is injective because $\omega \neq 0 \Rightarrow {}^*\omega \neq 0$. But for linear mappings of finite-dimensional spaces of equal dimension, injectiveness is equivalent to surjectiveness.

3. For $\Phi: q^i \to \bar{q}^i(q)$, we have $(T(\Phi))_{ij} = \partial\bar{q}^i/\partial q^j$. Therefore $(T(\Phi^{-1}))^{\iota}_{ij} = (T^*(\Phi))_{ij} = \partial q^j/\partial\bar{q}^i$. Let $\omega: q \to (q, \omega_i(q)\, dq^i)$. Then $\Phi_*\omega: \bar{q} \to (\partial q^j/\partial\bar{q}^i)\omega_j(\Phi^{-1}(\bar{q}))\, d\bar{q}^i)$. The covariant components transform in the same way as the basis ∂_i of $T(M)$, and hence as the gradient $(\partial/\partial\bar{q}^i)f(q(\bar{q})) = (\partial q^j/\partial\bar{q}^i)(\partial f/\partial q^j)$. On the other hand, one can write the differential using $dq^i = (\partial q^i/\partial\bar{q}^j)\, d\bar{q}^j$ and leave the components unchanged.

4. With a change of charts, $(x, y, z) = (r\sin\vartheta\cos\varphi, r\sin\vartheta\sin\varphi, r\cos\vartheta)$ (or, respectively, $(\rho\cos\varphi, \rho\sin\varphi, z)$), the metric $dx^2 + dy^2 + dz^2$ becomes

$$dr^2 + r^2\, d\vartheta^2 + r^2\sin^2\vartheta\, d\varphi^2$$

(or $d\rho^2 + \rho^2\, d\varphi^2 + dz^2$). (Pedantically, dx^2 should be written $dx \otimes dx$, etc.) Following Remark (2.4.12; 3), from the covariant components $(f_{,r}, f_{,\vartheta}, f_{,\varphi})$ (or $(f_{,\rho}, f_{,\varphi}, f_{,z})$) we obtain the components $v_i/(g_{ii})^{1/2}$:

$$\left(f_{,r}, \frac{1}{r}f_{,\vartheta}, \left(\frac{1}{r\sin\vartheta}\right)f_{,\varphi}\right) \left(\text{or } \left(f_{,\rho}, \frac{1}{\rho}f_{,\varphi}, f_{,z}\right)\right).$$

5. Use a product chart; everything factors out.

6. (a) Following Remark (2.4.17; 1), on a chart containing q,

$$\begin{aligned}(i_X\omega|X_1, \ldots, X_{p-1}) &= \frac{1}{(p-1)!}X^i\omega_{ii_1\ldots i_{p-1}}X_1^{i_1}\ldots X_{p-1}^{i_{p-1}}\\ &= p(\omega|X, X_1, \ldots, X_{p-1}).\end{aligned}$$

(b) This follows from (a), since $\omega(fX, X_1, \ldots, X_{p-1}) = f\omega(X, X_1, \ldots, X_{p-1})$.
(c) This follows from (a) and $(\omega|X, Y, \ldots) = -(\omega|Y, X, \ldots)$.

7. Let

$$\begin{aligned}\omega &= \omega_{1,\ldots,m}dx^1 \wedge \cdots \wedge dx^m\\ &= \omega_{1,\ldots,m}\frac{\partial x^1}{\partial\bar{x}^{j_1}}\cdots\frac{\partial x^m}{\partial\bar{x}^{j_m}}\, d\bar{x}^{j_1} \wedge \cdots \wedge d\bar{x}^{j_m}\\ &= \omega_{1,\ldots,m}\det\left|\frac{\partial x^i}{\partial\bar{x}^j}\right| d\bar{x}^1 \wedge \cdots \wedge d\bar{x}^m.\end{aligned}$$

2.5 Differentiation

The only generalization of the elementary operation of differentiation for a manifold with no additional structure is the exterior differential of a form. If a local flow is given by some vector field, then it defines the Lie derivative of an arbitrary tensor field.

The exterior derivative d (2.4.3) generalizes to a mapping $d\colon E_p(M) \to E_{p+1}(M)$, which contains the differentiation operations of elementary vector calculus as special cases.

(2.5.1) **Definition**

Let ω be a p-form, which is written as

$$\omega = \frac{1}{p!}\sum_{(i)} c_{(i)}\, dq^{i_1} \wedge \cdots \wedge dq^{i_p}, \qquad c_{(i)} \in C^\infty(U),$$

on some chart. Then the $(p+1)$-form,

$$d\omega = \frac{1}{p!}\sum_{(i)} dc_{(i)} \wedge dq^{i_1} \wedge \cdots \wedge dq^{i_p},$$

is known as its **exterior differential.**

From the definition follow the

(2.5.2) **Rules of Exterior Differentiation**

(a) $d(\omega_1 + \omega_2) = d(\omega_1) + d(\omega_2)$, $\omega_i \in E_p(M)$.

(b) $d(\omega_1 \wedge \omega_2) = (d\omega_1) \wedge \omega_2 + (-1)^p \omega_1 \wedge d\omega_2$, $\omega_1 \in E_p$, $\omega_2 \in E_q$.

(c) $d(d\omega) = 0$, $\omega \in E_p$, $p = 0, 1, \ldots, m$.

Rules (a) and (b) are obvious. Rule (c) follows from the symmetry of the partial derivative:

$$d(d\omega) = \sum_{(i)}\sum_{k,j} \frac{1}{p!}\frac{\partial^2 c_{(i)}}{\partial q^k\, \partial q^j} dq^k \wedge dq^j \wedge dq^{i_1} \wedge \cdots \wedge dq^{i_p} = 0.$$

(2.5.3) **Remarks**

1. Since we wish Definition (2.5.1) to be independent of the coordinate system, it is essential for d to be natural with respect to diffeomorphisms, a phrase which means that for a diffeomorphism $\Phi\colon M_1 \to M_2$,

$$\Phi_* d\omega = d\Phi_*\omega;$$

or, equivalently, that the diagram

$$\begin{array}{ccc} E_p(M_1) & \xrightarrow{\ \Phi_*\ } & E_p(M_2) \\ \Big\downarrow d & & \Big\downarrow d \\ E_{p+1}(M_1) & \xrightarrow{\ \Phi_*\ } & E_{p+1}(M_2) \end{array}$$

is permutable. This follows from the special case proved above (2.4.37; 4), by which

$$\begin{aligned} \Phi_*\omega &= \sum_{(i)} \Phi_*(c_{(i)})\Phi_*(dq^{i_1}) \wedge \cdots \wedge \Phi_*(dq^{i_p}) \\ &= \sum_{(i)} c_{(i)} \circ \Phi^{-1} d(q^{i_1} \circ \Phi^{-1}) \wedge \cdots \wedge d(q^{i_p} \circ \Phi^{-1}) \end{aligned}$$

and

$$\Phi_*(d\omega) = \sum_{(i)} d(c_{(i)} \circ \Phi^{-1}) \wedge d(q^{i_1} \circ \Phi^{-1}) \wedge \cdots \wedge d(q^{i_p} \circ \Phi^{-1}).$$

If in particular Φ is the diffeomorphism of a change of charts, then $d\omega$ is constructed in the new coordinate system exactly as in the old one, except that everything is expressed in the new coordinates.

2. The relationship $\Phi_* d\omega = d\Phi_*\omega$ does not hold only for diffeomorphisms, but for all inverse images of forms (2.4.41). Given any mapping $\Psi\colon M_2 \to M_1$, we can carry forms over from M_1 to M_2 with Ψ^*. If M_2 happens to be a submanifold of M_1 and Ψ is the natural injection, and thus Ψ^* is the restriction to M_2, then the relationship merely means that the exterior derivative of the restriction is the restriction of the exterior derivative.
3. Linearity (a) and Leibniz's rule (b) are properties of every derivative. The sign in (b) for a derivative $E_p \to E_{p+1}$ arises from the requirement of consistency with the formula $\omega_1 \wedge \omega_2 = (-1)^{pq}\omega_2 \wedge \omega_1$. By rule (c), d is determined by its action on E_0. The covariant exterior derivative that arises in field theory does not satisfy (c), and so the action on E_1 needs to be an extra postulate.

(2.5.4) Examples

Let $M = \mathbb{R}^3$. As in (2.4.33; 1) we identify E_0 with E_3 and E_1 with E_2. The connection between our notation and that of vector calculus is: $(df)_i = (\nabla f)_i = (\text{grad } f)_i$; ${}^*(d\mathbf{v})_i = (\nabla \times \mathbf{v})_i = (\text{curl }\mathbf{v})_i$; and ${}^*({}^*d\mathbf{v}) = \nabla \cdot \mathbf{v} = \text{div }\mathbf{v}$. The rules (2.5.2) contain the following special cases:

(b) $p = q = 0$: $\nabla(f \cdot g) = f\nabla g + g\nabla f$;

(b) $p = 0, q = 1$: $\nabla \times (f \cdot \mathbf{v}) = [\nabla f \times \mathbf{v}] + f\nabla \times \mathbf{v}$;

(b) $p = q = 1$: $\nabla \cdot [\mathbf{v} \times \mathbf{w}] = {}^*(d(v \wedge w)) = {}^*(dv \wedge w) - {}^*(v \wedge dw) = (w \cdot \nabla \times v) - (v \cdot \nabla \times w)$;

(c) $p = 0$: $\nabla \times \nabla f = 0$;

(c) $p = 1$: $\nabla \cdot (\nabla \times \mathbf{v}) = 0$; and

(b) and (c) $\nabla \cdot (f \cdot \nabla \times \mathbf{v}) = (\nabla f \cdot \nabla \times \mathbf{v})$.

In vector calculus one learns that curl-free vectors ($\nabla \times \mathbf{v} = 0$ everywhere) can be written as gradients, and divergence-free vectors can be written as curls. In order to state the analogous fact for manifolds, we make use of

(2.5.5) Definition

A p-form ω is said to be **closed** iff $d\omega = 0$ and **exact** iff $\omega = d\nu$ for some $\nu \in E_{p-1}(M)$.

(2.5.6) Remarks

1. By (2.5.2(c)), exact $\Rightarrow$ closed, and the exact forms are a linear subspace of the closed forms.
2. The exact forms are in general a proper subspace. Consider on $M = \mathbb{R}^2\backslash\{\mathbf{0}\}$ the 1-forms
$$\omega_i = \frac{-y\,dx + x\,dy}{x^2 + y^2} = \operatorname{Im}\frac{dz}{z}, \qquad z = x + iy,$$
and
$$\omega_r = \frac{x\,dx + y\,dy}{x^2 + y^2} = \operatorname{Re}\frac{dz}{z}.$$
Certainly $d\omega_i = d\omega_r = 0$, and locally $\omega_r + i\omega_i = d \ln z$. But since $\ln z$ is not defined continuously on M, the forms are not exact. Here it is crucial that we have removed the origin of $\mathbb{R}^2$, at which point the forms are singular and their differentials by no means zero.
3. If M is a **starlike**[9] open set in $\mathbb{R}^n$ (see Figure 2.13), then there exists a mapping $A: E_p \to E_{p-1}$ such that $A \circ d + d \circ A = \mathbf{1}$ (Problem 7). It follows that for manifolds M diffeomorphic to starlike sets, $d\omega = 0$ implies $\omega = d(A\omega)$ (**Poincaré's lemma**). Since in $\mathbb{R}^n$ every neighborhood contains a convex set, closed $\Rightarrow$ exact on small enough subsets. That is, locally (2.5.6; 1) holds the other way around.
4. Since p-forms with $p < 0$ are identically zero by definition, it would seem that $df = 0$ implies $f = 0$. But this is the degenerate case, and in fact it only implies that f is locally constant.
5. At every point x, any set of $m - n$ linearly independent 1-forms ω_j defines an n-dimensional subspace of $T_x(M)$, $\mathcal{N}_x = \{v \in T_x(M)\colon (\omega_j|v) = 0 \text{ for all } j\}$.

---

[9]A set $S \subset \mathbb{R}^n$ is starlike with respect to a point P iff the line connecting any point of S with P lies wholly within S. A convex set is starlike with respect to all its points.

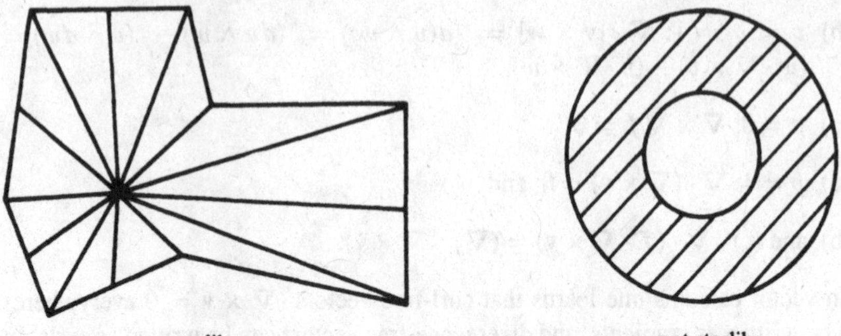

FIGURE 2.13. Starlikeness in the plane.

The question arises, whether it is possible to find n-dimensional submanifolds N such that $T_x(N) = \mathcal{N}_x$ for all x. This is referred to as "integrability" of the ω_j. By Definition (2.1.9), N is locally given by a set of equations of the form $f_j = 0$, $j = 1, \ldots, m - n$, for $f_j \in C^\infty(M)$. The 1-forms df_j satisfy $df_j|_N = 0$, and thus have the desired property that $(df_j|v) = 0$ for all $v \in T_x(N)$. They form a basis of forms of this kind, and locally we may write $\omega_j = c_{jk}\, df_k$. Hence $d\omega_j = \nu_{jk} \wedge \omega_k$, for $\nu_{ij} = dc_{ji} \cdot (c^{-1})_{ik} \in E_1$, so $d\omega_j \wedge \omega_1 \wedge \omega_2 \wedge \cdots \wedge \omega_{m-n} = 0$ for all j. This means that N can exist only if the exterior derivatives of the ω's contain at least one factor lying in the space that they span. If $d\omega_j = 0$, then, locally, $\omega_j = df_j$ because of Remark 3. The ω_j are then integrable, and N is given by $f_j = \text{const}$. By a theorem of Frobenius [1, §10.9], the apparently more general case where $d\omega_j = \nu_{jk} \wedge \omega_k$ may be reduced to this by the use of appropriate linear combinations, and this condition is both necessary and sufficient for local integrability. It guarantees that area elements may be extended from the infinitesimal to the local level, although it remains open whether this may be done globally. If $n = 1$, the condition for integrability is always satisfied, as the only thing missing from the basis for $T_x^*(M)$ is a single ω_0, and since $\omega_0 \wedge \omega_0 = 0$, every $d\omega_j$ must contain at least one factor ω_j. In this case, N is characterized by a single $X \in \mathcal{T}_0^1$, and as we saw in §2.3, there must then exist local integral curves. In any case, the n-dimensional submanifolds are invariant under the local flows Φ^X generated by the X such that $(\omega_j|X) = 0$, and are moreover locally generated by allowing the flows Φ^X to act at a point. We shall soon encounter an integrability condition for flows which is dual to the foregoing. When one tries to extend the flows arbitrarily far, however, it may happen as in (2.3.14) that trajectories return arbitrarily near their initial points, and that the set of points actually attained is dense in a higher-dimensional region, and is not a submanifold at all.

It is not always possible to attribute a coordinate-independent sense to the derivative of a tensor field $\mathcal{T}$. One would have to compare $\mathcal{T}(q)$ and $\mathcal{T}(q + \delta q)$, but the relative orientation of the tangent spaces depends on the coordinate system (2.2.10). Taking as an example $v^* \in \mathcal{T}_1^0$, the derivative $v^*_{i,k}$ does not transform as a tensor of degree two, although the unwanted terms cancel out in the transformation of the

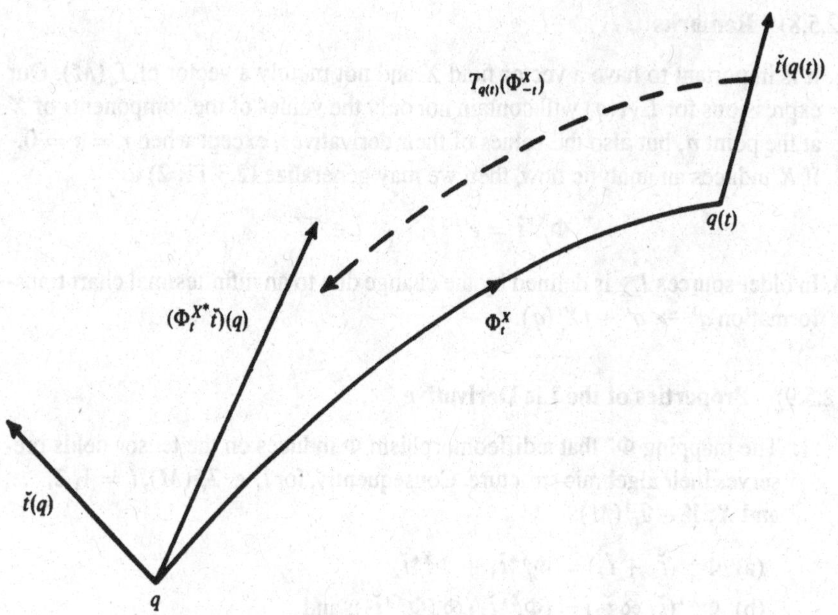

FIGURE 2.14. The Lie derivative.

combination $v^*_{i,k} - v^*_{k,i}$ that comes from the exterior differential. Yet if a vector field X is given on M, it induces a local flow Φ_t^X, and in order to define the derivative of another vector field $\check{t}$ at the point q, one could map the tangent vectors along the path through q, $q(t) = \Phi_t^X(q)$, back into $T_q(M)$ by using $T_{q(t)}(\Phi^X_{-t})$. Then both vectors, $\check{t}(q)$, which is the value of the vector field at q, and the vector generated from $\check{t}(q(t))$ by running time backwards, can be compared at the same point q. The second of these vectors can also be written as $(\Phi_t^{X*}\check{t})(q)$ (cf. (2.4.41) and Figure 2.14).

The corresponding derivative

$$\frac{d}{dt}(\Phi_t^{X*}\check{t})(q)$$

is independent of the coordinate system used, since it involves vectors in a single tangent space $T_q(M)$, and because differentation of vectors commutes with linear transformations. This line of reasoning applies as well to arbitrary tensor fields, leading us to make (cf. (2.2.23)).

(2.5.7) **Definition**

The Lie derivative $L_X \colon T^r_s \to T^r_s$ is defined by

$$L_X\check{t} = \frac{d}{dt}\Phi_t^{X*}\check{t}_{|t=0}, \qquad \check{t} \in T^r_s.$$

(2.5.8) Remarks

1. It is important to have a vector field X and not merely a vector of $T_q(M)$. Our expressions for $L_X \check{t}(q)$ will contain not only the values of the components of X at the point q, but also the values of their derivatives, except when $r = s = 0$.
2. If X induces an analytic flow, then we may generalize (2.3.11; 2) to

$$\Phi_t^{X*}\check{t} = e^{tL_X}\check{t}, \qquad \check{t} \in \mathcal{T}_s^r.$$

3. In older sources L_X is defined by the change due to an infinitesimal chart transformation $q^i \to q^i + tX^i(q)$.

(2.5.9) Properties of the Lie Derivative

1. The mapping Φ^* that a diffeomorphism Φ induces on the tensor fields preserves their algebraic structure. Consequently, for $\check{t}_i \in \mathcal{T}_s^r(M), i = 1, 2, \ldots$, and $X, Y \in \mathcal{T}_0^1(M)$:

   (a) $\Phi_t^{X*}(\check{t}_1 + \check{t}_2) = \Phi_t^{X*}\check{t}_1 + \Phi_t^{X*}\check{t}_2$;
   (b) $\Phi_t^{X*}(\check{t}_1 \otimes \check{t}_2) = (\Phi_t^{X*}\check{t}_1) \otimes (\Phi_t^{X*}\check{t}_2)$; and
   (c) $\Phi_t^{X*}V\check{t} = V\Phi_t^{X*}\check{t}$, for V a contraction. (See (2.5.15).)

   For infinitesimal t it follows that:

   (a) $L_X(\check{t}_1 + \check{t}_2) = L_X\check{t}_1 + L_X\check{t}_2$;
   (b) $L_X(\check{t}_1 \otimes \check{t}_2) = (L_X\check{t}_1) \otimes \check{t}_2 + \check{t}_1 \otimes L_X\check{t}_2$; and
   (c) $L_X V\check{t} = V L_X\check{t}$.

   For isometric (resp. canonical) transformations Φ, $\Phi_* g = g$, i.e., $\Phi_*\langle\check{t}_i|\check{t}_k\rangle = \langle\Phi_*\check{t}_i|\Phi_*\check{t}_k\rangle$. It follows for the associated Killing (resp. Hamiltonian) vector fields X that $L_X\langle\check{t}_i|\check{t}_k\rangle = \langle L_X\check{t}_i|\check{t}_k\rangle + \langle\check{t}_i|L_X\check{t}_k\rangle$.

2. The permutability of the diagram

$$\begin{array}{ccc}
\mathcal{T}_s^r(M_1) & \xrightarrow{\Psi_*} & \mathcal{T}_s^r(M_2) \\
\Big\downarrow \Phi_{t*}^X & & \Big\downarrow (\Phi_t^{\Psi^* X})_* \\
\mathcal{T}_s^r(M_1) & \xrightarrow{\Psi_*} & \mathcal{T}_s^r(M_2)
\end{array}$$

implies the permutability of its infinitesimal version

$$\begin{array}{ccc}
\mathcal{T}_s^r(M_1) & \xrightarrow{\Psi_*} & \mathcal{T}_s^r(M_2) \\
\Big\downarrow L_X & & \Big\downarrow L_{\Psi_* X}, \\
\mathcal{T}_s^r(M_1) & \xrightarrow{\Psi_*} & \mathcal{T}_s^r(M_2)
\end{array}$$

that is, $L_{\Psi_* X}\Psi_* \check{t} = \Psi_* L_X \check{t}$. This means that the flow on the transformed system is determined by the transformed vector field, so that the image with Ψ_* of the Lie derivative of a tensor field is the same as the Lie derivative with respect to the image of the vector field of the image of the tensor field. This naturalness of L_X with respect to diffeomorphisms is a consequence of the chart-independence of its definition. The same is true for the pull-back Ψ^*.

3. Since d is natural with respect to diffeomorphisms (2.5.3), it commutes with L_X. Formally, it is like this: $(\omega \in E_p)$

$$L_X\, d\omega = \frac{d}{dt}\Phi_t^{X*}\,\omega_{|t=0} = \frac{d}{dt} d\Phi_t^{X*}\,\omega_{|t=0}$$
$$= d\frac{d}{dt}\Phi_t^{X*}\omega_{|t=0} = dL_X\omega.$$

Expressed diagrammatically:

$$\begin{array}{ccc} E_p(M) & \xrightarrow{\;L_X\;} & E_p(M) \\ \Big\downarrow d & & \Big\downarrow d \\ E_{p+1}(M) & \xrightarrow{\;L_X\;} & E_{p+1}(M) \end{array}$$

commutes.

4. On E_p, L_X may be expressed in terms of d and the interior product i_X (2.4.33),

$$L_X = i_X \circ d + d \circ i_X, \qquad i_X \circ L_X = L_X \circ i_X.$$

For the proof, see Problem 6. This also shows that $dL_X = d \circ i_X \circ d = L_X d$, $i_X \circ L_X = L_X \circ i_X$.

5. The Lie derivative L_X is consistent with linearity of X:

(a) $L_{X_1+X_2} = L_{X_1} + L_{X_2}$; and

(b) $L_{cX} = cL_X$, c any constant.

As for the module structure of the vector fields, it follows from Property 4 that

$$L_{fX} = fL_X + df \wedge i_X \qquad \text{on} \quad E_p,$$

since

$$i_{fX}\, d\omega + d i_{fX}\omega = f i_X\, d\omega + d(f i_X \omega)$$
$$= f(i_X d + d i_X)\omega + df \wedge i_X\omega, \quad \forall \omega \in E_p.$$

The extra term containing df reflects the presence of the derivatives of X in L_X. Roughly speaking, X and fX displace the beginning and end points of a vector differently.

6. According to 1(iii) and Property 3, $(df|L_X Y) = L_X(df|Y) - (dL_X f|Y) = (L_X L_Y - L_Y L_X)f$. Let us denote the vector field $L_X Y$ by the **Lie bracket** $[X, Y]$. The calculation just done implies

$$L_{L_X Y} = L_X L_Y - L_Y L_X \quad \text{on } \mathcal{T}_0^0,$$

from which it follows that $L_X Y = -L_Y X$, because a vector field is completely characterized by its action on $\mathcal{T}_0^0$. The relationship

$$L_{[X,Y]} = L_X L_Y - L_Y L_X$$

can be extended to all $\mathcal{T}_s^r$. If $L_X L_Y - L_Y L_X$ is applied to $\check{t} \in \mathcal{T}_s^0$:

$$\check{t} = \sum_{(i)} c_{(i)}\, dq^{i_1} \otimes \cdots \otimes dq^{i_s},$$

using the rules for sums and tensor products, only the terms in which one factor is differentiated twice remain. The others cancel out because of antisymmetry. For the remaining terms, the relationship in question holds by Property 3. Then by Property 1 it must hold for all $\mathcal{T}_s^r$.

(2.5.10) **Examples**

1. $r = s = 0$. $\Phi_t^{X*} f = f \circ \Phi_t^X = e^{tL_X} f = \tau_t^X f$. In this case (2.5.7) coincides with the earlier definition (2.3.11; 2). If on some chart X is $X^i \partial_i$, (2.5.7) yields

$$L_X f = X^i f_{,i}.$$

Observe that Φ_t^{X*} induces the automorphism τ_t.

2. $r = 0$, $s = 1$: $\omega = \omega_i\, dq^i$. The rules imply that

$$\begin{aligned} L_X \omega &= (L_X \omega_i)\, dq^i + \omega_i\, d(L_X q^i) \\ &= (X^k \omega_{i,k} + \omega_k X^k_{,i})\, dq^i. \end{aligned}$$

3. $r = 1$, $s = 0$. For $\omega \in \mathcal{T}_1^0$, $Y = Y^i \partial_i \in \mathcal{T}_0^1$, we calculate

$$\begin{aligned} L_X(\omega|Y) &= \omega_i Y^i_{,k} X^k + \omega_{i,k} Y^i X^k \\ &= \omega_i (Y^i_{,k} X^k - X^i_{,k} Y^k) + (\omega_{i,k} X^k + \omega_k X^k_{,i}) Y^i \\ &= (\omega|L_X Y) + (L_X \omega|Y). \end{aligned}$$

Thus the ith component of the Lie derivative of Y is

$$Y^i_{,k} X^k - X^i_{,k} Y^k.$$

4. $X = \partial_i$ and $Y = \partial_j$. Then $[X, Y] = 0$. The vanishing of the Lie bracket of the natural basis vectors means that the partial derivatives commute.

(2.5.11) Integrability of Vector Fields

A question that might well be asked at the end of (2.5.6; 5) is: Given vector fields X_j defining an n-dimensional subspace $\mathcal{N}_x \subset T_x(M)$ at each point x, j running from 1 to at least n, when do there exist n-dimensional submanifolds N such that $T_x(N) = \mathcal{N}_x$ for all $x \in N$? (If so, the X_j are referred to as **surface-forming**, or **integrable.**) On the chart (2.1.9), $N = \{x_1, \ldots, x_m;\ x_{n+1} = \cdots = x_m = 0\}$, and $T_x(N)$ is spanned by $\{\partial/\partial x_k,\ k = 1, \ldots, n\}$. Hence the X_j must be of the form $c_j^k\, \partial/\partial x_k$, where c is of rank n. According to (2.5.10; 3),

$$[X_i, X_\ell] = \left(c_i^j \frac{\partial}{\partial x_j} c_l^k - c_l^j \frac{\partial}{\partial x_j} c_i^k \right) \frac{\partial}{\partial x_k}, \qquad i, l, j, k = 1, \ldots, n,$$

and $\partial/\partial x_k$ may be expressed as a linear combination of the X_j. If the X_j are to form a surface, their Lie brackets must also belong to $\mathcal{N}_x$. According to (2.5.6; 5), this condition is also sufficient to guarantee local integrability: Let $\mathcal{N}_x^\perp$ denote the linear space of all ω such that $(\omega(x)|X_j(x)) = 0$ for all j and all x in a neighborhood. Local integrability means that $d\omega$ has at least one factor belonging to $\mathcal{N}_x^\perp$. In Problem 9 it is found that $(d\omega|X, Y) = -(\omega|[X, Y])$ for all $\omega \in \mathcal{N}_x^\perp$ and $X, Y \in \mathcal{N}_x$. Hence $d\omega$ has a factor in $\mathcal{N}_x^\perp$ iff $(d\omega|X, Y) = 0$ iff $(\omega|[X, Y]) = 0$ iff $[X, Y] \in \mathcal{N}_x$. If one would rather prove integrability beginning with the vector fields, that can be accomplished by following the path laid out in (2.5.6; 5).

The following argument shows that $[X_j, X_k]$ must belong to $\mathcal{N}_x$: The flows $\Phi_{\tau_j}^{X_j}$ must leave N invariant. Therefore

$$\Phi_\tau^{X_j} \circ \Phi_\tau^{X_k} \circ \Phi_{-\tau}^{X_j} \circ \Phi_{-\tau}^{X_k}$$

maps N into itself, and $[X_j, X_k]$ can be obtained from it:

$$[L_{X_j}, L_{X_k}] = \lim_{\tau \to 0} \tau^{-2} (\exp[\tau L_{X_j}] \exp[\tau L_{X_k}] \exp[-\tau L_{X_j}] \exp[-\tau L_{X_k}] - 1).$$

For the X_j of every natural basis:

$$\Phi_{\tau_j}^{X_j} \circ \Phi_{\tau_k}^{X_k} = \Phi_{\tau_k}^{X_k} \circ \Phi_{\tau_j}^{X_j}$$

and the chart (2.1.9) for N is

$$\Phi_{\tau_1}^{\partial_1} \circ \Phi_{\tau_2}^{\partial_2} \circ \cdots \circ \Phi_{\tau_n}^{\partial_n} x \to (\tau_1, \ldots, \tau_n) \in \mathbb{R}^n.$$

By reversing the argument, if we are given n independent X_i such that $\Phi_{\tau_j}^{X_j} \circ \Phi_{\tau_k}^{X_k} = \Phi_{\tau_k}^{X_k} \circ \Phi_{\tau_j}^{X_j}$, then we obtain an N and a chart in which $\{X^j\}$ is the natural basis $\{\partial/\partial \tau_j\}$ (Problem 10). Therefore, $[X_j, X_k] = 0$ is a sufficient condition for local integrability. The apparently more general case where $[X_j, X_k] = c_{jkm} X_m$ reduces to this case by the introduction of linear combinations X_j of the X_k satisfying $[\bar{X}_j, \bar{X}_k] = 0$.

(2.5.12) Examples

1. $M = \mathbb{R}^3$. Let $\omega = dx + z\,dy \in E_1(\mathbb{R}^3)$: $\omega \wedge d\omega = dx \wedge dz \wedge dy \neq 0$, so ω is not integrable. The space orthogonal to ω is spanned by

$$X = -z\frac{\partial}{\partial x} + \frac{\partial}{\partial y} \quad \text{and} \quad Y = \frac{\partial}{\partial z}.$$

It is easy to see that $[X, Y] = \partial/\partial x$ cannot be written as $aX + bY$, and, indeed, when the flows generated by X and Y:

$$\Phi^X_{\tau_1}: (x, y, z) \to (x - \tau_1 z, y + \tau_1, z)$$

and

$$\Phi^X_{\tau_2}: (x, y, z) \to (x, y, z + \tau_2),$$

are applied to an initial point, the result is not a surface but rather an open set in $\mathbb{R}^3$:

$$\Phi^Y_{\tau_3} \circ \Phi^X_{\tau_2} \circ \Phi^Y_{\tau_1}(0, 0, 0) = (-\tau_2\tau_1, \tau_2, \tau_1 + \tau_3);$$

and as (τ_1, τ_2, τ_3) runs over, say $\mathbb{R} \times \mathbb{R}^+ \times \mathbb{R}$, the image of the origin covers this same region. This example also reveals that there need not exist hypersurfaces which are orthogonal to a vector field in the sense of a particular metric. For example, consider $\mathbb{R}^3$ with the Euclidean metric. The vectors orthogonal to $X = \partial/\partial x + z\,\partial/\partial y$ are annihilated by $\omega = dx + z\,dy$, and hence do not compose surfaces. Yet by Theorem (2.3.12) there are hypersurfaces N transversal to X in the sense that $T_x(N)$ and $X(x)$ span all of $T_x(\mathbb{R}^m)$.

2. $M = \mathbb{R}^3$. Let X_j denote the generator of the rotations about the jth coordinate axis:

$$X_1 = y\frac{\partial}{\partial z} - z\frac{\partial}{\partial y}, \qquad X_2 = z\frac{\partial}{\partial x} - x\frac{\partial}{\partial z}, \qquad X_3 = x\frac{\partial}{\partial y} - y\frac{\partial}{\partial x}.$$

The three vector fields generate only a two-dimensional $\mathcal{N}_x$, and are orthogonal to dr, where $r = \sqrt{x^2 + y^2 + z^2}$: $(dr|X_j) = L_{X_j} r = 0$. Since dr is integrable, $d\,dr = 0$, the X_j must likewise be integrable. It turns out in fact that $[X_i, X_j] = \varepsilon_{ijk} X_k$. The flows generated by the X_j map the two-dimensional submanifolds $r =$ const. into themselves. In this example, the X_j form surfaces that are diffeomorphic to S^2. Mappings like $\Phi^{X_1}_{\tau_1} \circ \Phi^{X_2}_{\tau_2}$ can of course only generate a local chart on such a surface.

The Lie bracket provides the vector fields with an additional algebraic structure. It distributes over addition, and instead of the associative law it satisfies

(2.5.13) Jacobi's Identity

$$[X, [Y, Z]] + [Y, [Z, X]] + [Z, [X, Y]] = 0.$$

Proof: Follows from the identity

$$L_{[X,[Y,Z]]} + L_{[Y,[Z,X]]} + L_{[Z,[X,Y]]} = L_X(L_Y L_Z - L_Z L_Y)$$
$$- (L_Y L_Z - L_Z L_Y)L_X + L_Y(L_Z L_X - L_X L_Z) - (L_Z L_X - L_X L_Z)L_Y$$
$$+ L_Z(L_X L_Y - L_Y L_X) - (L_X L_Y - L_Y L_X)L_Z = 0,$$

because $L_X = 0$ (even if only on $\mathcal{T}_0^0(M)$) implies $X = 0$. □

(2.5.14) **Problems**

1. Why is it not possible to define d on all of $\mathcal{T}_s^0$?
2. How are the covariant components of $d\omega$ written in the notation of (2.5.1)?
3. Show explicitly that $\Phi_* d\omega = d\Phi_*\omega$ for $M = \mathbb{R}^n$, $p = 1$.
4. Show directly for $f \in C(M)$ that $L_X\, df = d(L_X f)$.
5. Calculate the components of $L_X Y$, X and $Y \in \mathcal{T}_0^1$, and of $L_X\alpha$, $\alpha \in \mathcal{T}_1^0$. Prove that $L_X(\alpha \mid Y) = (L_X\alpha \mid Y) + (\alpha \mid L_X Y)$.
6. Show that $L_X = i_X \circ d + d \circ i_X$ on E_p (2.5.9; 4).
7. Define the mapping A: $E_p \to E_{p-1}$ of Remark (2.5.6; 3) as follows: Let U be starlike with respect to the origin and h: $(0, 1) \times U \to U$ be the mapping $(t, x) \to tx$. For $\omega \in E_p(U)$ we may decompose the inverse image of ω under h into one part with dt and another without dt:

   $$h^*\omega = \omega_0 + dt \wedge \omega_M, \qquad \omega_0 \in E_p((0, 1) \times U), \qquad \omega_M \in E_{p-1}((0, 1) \times U).$$

   Then

   $$A\omega \equiv \int_0^1 dt \wedge \omega_M \in E_{p-1}(U).$$

   Show that $A \circ d + d \circ A = \mathbf{1}$, and calculate $A\omega$ in $\mathbb{R}^3$ for $\omega \in E_1$ and E_2.
8. Find an example of a vector field $\mathbf{E}$ that is divergence-free on $\mathbb{R}^3\backslash\{\mathbf{0}\}$, but which cannot be written $\mathbf{E} = \nabla \times A$.
9. Show that for $\omega \in E_1$,

   $$(d\omega|X, Y) = L_X(\omega|Y) - L_Y(\omega|X) - (\omega|[X, Y]).$$

10. Let X_j, $j = 1, \ldots, m$, be vector fields satisfying $[X_i, X_j] = 0$, with $\{X_j(q)\}$ linearly independent. Show that in the vicinity of q there exists a chart (U, Φ) such that $\Phi_* X_j = \partial/\partial x_j$. (*Hint*: Consider the m-parameter group of diffeomorphisms generated by the X_j.) Investigate what happens in a simple case when $[X_i, X_j] \neq 0$.

(2.5.15) **Solutions**

1. Antisymmetry is the key to the proof that $d(d\omega) = 0$.
2. $(d\omega)_{i_1,\ldots,i_{p+1}} = \sum_{\ell=1}^{p+1} \frac{\partial}{\partial x_{i_\ell}} (\omega)_{i_1,\ldots,i_{\ell-1},i_{\ell+1},\ldots,i_{p+1}} \cdot (-1)^\ell.$

3.

$$\omega = c_k(x)\,dx^k, \qquad d\omega = c_{k,i}\,dx^i \wedge dx^k,$$
$$\Phi_*\omega = c_k(x(\bar{x}))\frac{\partial x^k}{\partial \bar{x}^j}\,d\bar{x}^j,$$
$$d\Phi_*\omega = \left(c_{k,i}\frac{\partial x^i}{\partial \bar{x}^r}\frac{\partial x^k}{\partial \bar{x}^j} + c_k\frac{\partial^2 x^k}{\partial \bar{x}^r \partial \bar{x}^j}\right) d\bar{x}^r \wedge d\bar{x}^j$$
$$= c_{k,i}\frac{\partial x^i}{\partial \bar{x}^r}\frac{\partial x^k}{\partial \bar{x}^j}\,d\bar{x}^r \wedge d\bar{x}^j \equiv \Phi_*\,d\omega.$$

4. $L_X L_Y f = L_X(df|Y) = (df|L_X Y) + (L_X\,df \mid Y) = L_{[X,Y]}f + (L_X\,df|Y)$, so $(L_X\,df|Y) = L_Y L_X f = (d(L_X f)|Y)\ \forall Y \in \mathcal{T}_0^1(M) \Rightarrow L_X\,df = d(L_X f)$, i.e., $dL_X = d_{i_X}d = L_X d$ on f.

5. For $X\colon q \to (q_i, X^i)$, etc.,

$$(L_X\alpha|Y) + (\alpha|L_X Y) = (L_X\alpha)_i Y^i + \alpha_i(L_X Y)^i = (\alpha_{i,k}X^k + \alpha_k X^k_{,i})Y^i$$
$$+\ \alpha_i(Y^i_{,k}X^k - X^i_{,k}Y^k) = \alpha_i Y^i_{,k}X^k + \alpha_{i,k}Y^i X^k = L_X(\alpha|Y).$$

(And similarly.)

6. Proof of induction: For $p = 0$, $i_X f = 0$ by definition, and $i_X\,df = (df|X) = L_X f$. Every $(p+1)$-form may be written as

$$\sum_i df_i \wedge \omega_i, \qquad \omega_i \in E_p,\ f \in C^\infty.$$

Now

$$\begin{aligned}(i_X \circ d + d \circ i_X)(df \wedge \omega) &= i_X \circ (-df \wedge d\omega) + d \circ ((i_X\,df)\omega - df \wedge i_X\omega)\\ &= -(i_X\,df) \wedge d\omega + df \wedge (i_X\,d\omega) + (d(i_X\,df)) \wedge \omega\\ &\quad + (i_X\,df) \wedge d\omega + df \wedge d(i_X\omega)\\ &= df \wedge L_X\omega + (L_X df) \wedge \omega = L_X(df \wedge \omega).\end{aligned}$$

Since both sides of the equation in (2.5.9; 4) are linear operators, this relationship also holds for $\sum_i df_i \wedge \omega_i$, and consequently on $E_{p-1}(M)$.

7. For

$$\omega = \frac{1}{p!}\omega_{(i)}(x)\,dx^{i_1} \wedge \cdots \wedge dx^{i_p},$$

we find

$$h^*\omega = \omega_{(i)}(xt)(t\,dx^{i_1} + x^{i_1}\,dt) \wedge \cdots \wedge (t\,dx^{i_p} + x^{i_p}\,dt)\frac{1}{p!}$$
$$= \omega_0 + dt \wedge \omega_M.$$

Let us designate the exterior derivative with t held constant by d'; then

$$dA\omega = \int_0^1 dt\,d'\omega_M, \qquad A\,d\omega = \int_0^1 dt\left(\frac{\partial \omega_0}{\partial t} - d'\omega_M\right).$$

As defined above, $\omega_{0|t=1} = \omega$ and $\omega_{0|t=0} = 0$ and so $dA\omega + A\,d\omega = \omega$.

$$p = 1: \quad A\omega = \int_0^1 dt\, x_i v_i(xt) = \int_0^{\mathbf{x}} d\mathbf{s}\cdot\mathbf{v};$$

$$p = 2: \quad \omega_{ij} = \varepsilon_{ijk} B_k, \qquad (A\omega)_i = \int_0^1 dt\, t B_k x_j \varepsilon_{kji}.$$

8. $\mathbf{E} = \mathbf{x}/|\mathbf{x}|^3$. It is impossible that $\mathbf{E} = \nabla\times\mathbf{A}$, for then we would have

$$4\pi = \int_{S^2} d\mathbf{S}\cdot\mathbf{E} = \int_{S^2} d\mathbf{S}\cdot\nabla\times\mathbf{A} = \int_{\partial S^2} d\mathbf{s}\cdot\mathbf{A} = 0,$$

because $\partial S^1 = \emptyset$.

9.

$$\begin{aligned} L_X(\omega|Y) &= ((i_X\circ d + d\circ i_X)\omega|Y) + (\omega|L_X Y) \\ &= (d\omega|X, Y) + L_Y(\omega|X) + (\omega|[X, Y]). \end{aligned}$$

10. The mapping $(\tau_j) \to \Phi^{X_m}_{\tau_m}\circ\cdots\circ\Phi^{X_1}_{\tau_1}(q) =: q(\tau_j)$ is a diffeomorphism of a neighborhood of $0 \in \mathbb{R}^m$ to a neighborhood of q: $\partial q(\tau)/\partial\tau_j = X_j(q(\tau))$, and by assumption $\det(X^k_j(q))$ is different from 0 at $\tau = 0$ and hence also throughout a neighborhood of 0. On $\mathbb{R}^2$, let $X_1: (x, y) \to (x, y; 1, 0)$ and $X_2: (x, y) \to (x, y; 0, 1 + x)$, and take $q = 0$. Then

$$q(\tau) = \Phi^{X_2}_{\tau_2}\circ\Phi^{X_1}_{\tau_1}(0) = (\tau_1, \tau_2(1 + \tau_1))$$

is a diffeomorphism, but

$$\frac{\partial}{\partial\tau_1} q(\tau) = (1, \tau_2) \neq X_1(q(\tau)).$$

2.6 Integrals

An m-form defines a measure on a manifold. Its integral is an inverse of the exterior derivative in the sense that integration by parts can be generalized as Stokes's theorem.

In (2.4.8; 3) we interpreted p-forms as measures for p-dimensional volume elements. It is possible to define a coordinate-independent integral over p-forms by applying them to a coordinate basis and then integrating over those coordinates in the usual manner. Once again we begin by extending our earlier analysis from a single point to a neighborhood. Suppose that Ω is an m-form the support of which lies in a chart (U, Φ). Its image under the chart is of the form

$$\Phi_*(\Omega) = w(x)\, dx^1\wedge dx^2\wedge\cdots\wedge dx^m, \qquad w = (\Omega|\partial_1, \partial_2, \ldots, \partial_m),$$

and if U is relatively compact we define

$$\int \Omega := \int_{-\infty}^{\infty} dx^1 \cdots \int_{-\infty}^{\infty} dx^m \, w(x). \tag{2.6.1}$$

The value of this integral does not change under a diffeomorphism, since by (2.4.44; 7) w gets multiplied by $\det(\partial x^i/\partial \bar{x}^j)$, and

$$\int d^m \bar{x} \det\left(\frac{\partial x^i}{\partial \bar{x}^j}\right) \omega(x(\bar{x})) = \int d^m x \omega(x). \tag{2.6.2}$$

(See [1, 16.22.1].)

Here we encounter a choice of sign, a choice that must be confronted even in the simplest case: Should $\int_{(ab)}$ mean $\int_a^b$ or $\int_b^a$? The difficulty is that a consistent sign convention for the whole manifold M is possible only if M is orientable, for only in that case does there exist a nowhere-vanishing m-form Ω that can determine the sign.

In order to avoid questions of convergence we shall integrate only forms that vanish outside a finite region. As usual, on manifolds the notion of a finite region is replaced by that of a compact set, as compactness is invariant under diffeomorphism. Hence we will consider forms that can be written as $f\Omega$, where $f \in C(M)$ such that the support supp(f) is compact. The set of these forms is denoted E_p^0. Given any atlas, it is then always possible to choose finitely many charts (U_i, Φ_i) such that

$$\operatorname{supp}(f) \subset \bigcup_i^{\text{finite}} U_i.$$

By the use of a partition of unity (see, e.g., [1, 12.6.4]), f may be written as $f = \sum_i f_i$, where supp $f_i \subset U_i$. This enables us to make

(2.6.3) Definition

The **integral** of an m-form $f\Omega$ with compact support on an orientable manifold M is

$$\int_M \Omega f = \sum_i \int \Phi_{i*}(\Omega f_i),$$

where $f = \sum_i f_i$, f_i is of compact support on the domain of the chart (U_i, Φ_i), and the integrals summed over are given by (2.6.1), assuming the use of charts for which $(\Omega|\partial_1 \ldots \partial_m) > 0$.

(2.6.4) Remarks

1. The integral is independent of the choice of charts provided that all the charts have the same orientation $(\Omega|\partial_1 \ldots \partial_m) > 0$, so that under a change of charts it remains true that $\det(\partial x^i/\partial \bar{x}^j) > 0$. If we use the charts $(\partial \bar{U}_i, \Phi_i)$, then by introducing a partition of unity $\sum_{i,j} f_{ij} = f$ such that $\operatorname{supp}(f_{ij}) \subset U_i \cap \bar{U}_j$,

we will find that the value of $\int f\Omega$ is unchanged. The change from Φ_i to $\bar{\Phi}_j$ on $U_i \cap \bar{U}_j$ is a change of variables of the type mentioned earlier, and since $\sum_{i,j} f_{ij} = f$ is always a finite sum without convergence problems, interchange of the order of summation is allowed.

2. For all C_0^∞-functions f, $\int \Omega f$ is a linear functional bounded by $\sup|f|$, some constant depending only on supp f, and so defines a measure on M. The linear functional may then be extended to a larger class of functions, which need not have compact support, but must only fall off sufficiently rapidly.
3. If ω is a p-form and N an orientable p-dimensional submanifold of M, then $\int_N \omega$ is defined by (2.6.3) with $\omega_{|N}$.
4. There is no meaning independent of the charts for an integral over other tensor fields.

If $M = (a, b)$ and ω is the 1-form df with supp $f \subset M$, then

$$\int df = \int_a^b dx \frac{\partial f}{\partial x} = 0,$$

because f vanishes at the boundary. Without the condition on the support of f, $\int df = f(b) - f(a)$. If we make the immediate extension of Definition (2.6.3) to manifolds with boundaries, this rule generalizes to

(2.6.5) Stokes's Theorem

Let M be an orientable m-dimensional manifold with a boundary and let ω be an $(m-1)$-form with compact support. Then

$$\int_M d\omega = \int_{\partial M} \omega.$$

(2.6.6) Remarks

1. It does not need to be assumed that ∂M is orientable, since the orientation of M induces one on ∂M. Indeed, it is a consequence of the proof of the theorem that if on some chart of the form (2.1.20) the orientation of M is given by $w(x)\, dx^1 \wedge dx^2 \wedge \cdots \wedge dx^m$, $w > 0$, then we ascribe the orientation $-dx^2 \wedge \cdots \wedge dx^m$ to ∂M. The sign is important, for, if it were reversed, (2.6.5) would be false: for $M = [0, \infty)$,

$$\int_0^\infty \frac{df}{dx} dx = -f(0).$$

2. The requirement of a compact support is necessary even if M is a finite part of $\mathbb{R}^n$. For example, $M = (a, b)$, $\partial M = \emptyset$, $f = x$, and

$$\int_a^b df = b - a \neq \int_{\partial M} f = 0.$$

3. Note that the rule $d \circ d = 0$ follows from the fact that a boundary has no boundary: Let V be a compact submanifold of M with a boundary. Then

$$\int_V d \circ d\omega = \int_{\partial V} d\omega = \int_{\partial\partial V} \omega = 0.$$

It is easy to convince oneself that an m-form vanishes if its integral over every compact submanifold with a boundary vanishes, and hence that $d \circ d = 0$.

Proof: Let us again write $\int d\omega = \sum_i \int d\omega_i$, where each ω_i has compact support in the domain U_i of a chart of the form of (2.1.20); then it suffices to show that $\int_M d\omega_i = \int_{\partial M} \omega_i$. On a chart of the type (2.1.18),[10]

$$\Phi_i^* \omega_i = \sum_{j=1}^m g_j \, dx^1 \wedge \cdots \wedge \widehat{dx^j} \wedge \cdots \wedge dx^m,$$

and we choose $dx^1 \wedge dx^2 \wedge \cdots \wedge dx^m$ as the orientation. Then

$$\int_M d\omega_i = \sum_{j=1}^m (-1)^{j+1} \int_0^\infty dx^1 \int_{-\infty}^\infty dx^2 \cdots \int_{-\infty}^\infty dx^m \frac{\partial g_j}{\partial x^j}$$

$$= -\int_{-\infty}^\infty dx^2 \cdots \int_{-\infty}^\infty dx^m g_i(0, x^2, \ldots, x^m).$$

On the other hand, we know (cf. (2.6.6; 1)) that

$$\int_{\partial M} \omega_i = -\int_{-\infty}^\infty dx^2 \cdots \int_{-\infty}^\infty dx^m g_1(0, x^2, \ldots, x^m),$$

because the restriction of dx^1 to ∂M vanishes, so that

$$\omega_{i|\partial M} = g_1 \, dx^2 \wedge \cdots \wedge dx^m. \qquad \square$$

(2.6.7) Examples

1. $M = \{(x, y) \in \mathbb{R}^2, \frac{1}{2} \le x^2 + y^2 \le 1\}$, $\omega = -y\,dx + x\,dy/x^2 + y^2$, $d\omega = 0$ (cf. (2.5.6; 2))

$$0 = \int_{x^2+y^2=1} \omega - \int_{x^2+y^2=1/2} \omega = 2\pi - 2\pi.$$

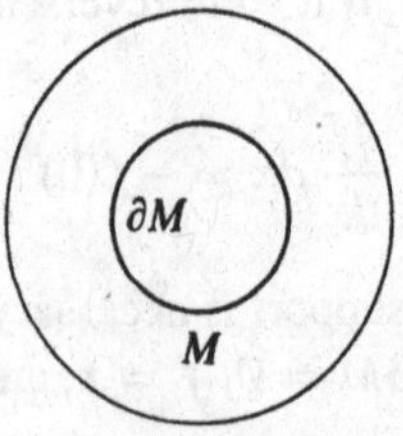

[10] The symbol $\widehat{dx^j}$ indicates that the jth differential is missing.

It is again apparent that the compact support of ω is essential, as otherwise one could take ω on $M = \{(x, y) \in \mathbb{R}^2, 0 < x^2 + y^2 \le 1\}$, $\partial M = S^1$, and get the contradiction $0 = 2\pi$. We also see that ω cannot be exact, as $\omega = dv$ would imply

$$2\pi = \int_{S^1} \omega = \int_{S^1} dv = \int_{\partial S^1} v = 0, \quad \text{since } \partial S^1 = \emptyset.$$

2. C = any one-dimensional submanifold with a boundary in $\mathbb{R}^3$, $\partial C = \{a, b\}$,

$$\int_C df = \int_{\partial C} f, \quad \text{or} \quad \int_C d\mathbf{s} \cdot \nabla f = f(b) - f(a).$$

3. M = a two-dimensional submanifold with a boundary of $\mathbb{R}^3$ (not necessarily a part of a plane), and ω is the 1-form **w**. In vector notation (2.6.5) reads

$$\int_M d\mathbf{S} \cdot \nabla \times \mathbf{w} = \int_{\partial M} d\mathbf{s} \cdot \mathbf{w}.$$

4. M = a three-dimensional submanifold with a boundary of $\mathbb{R}^3$, and ω is the 2-form $^*\mathbf{w}$. We obtain Gauss's theorem:

$$\int_M dV \nabla \cdot \mathbf{w} = \int_{\partial M} d\mathbf{f} \cdot \mathbf{w}.$$

In order to discover the relationship of the Lie derivative to integration, recall that the integral is invariant under diffeomorphisms:

(2.6.8) $$M_1 \overset{\Phi}{\to} M_2 : \int_{M_1} \omega = \int_{M_2} \Phi_* \omega.$$

If Φ is specifically a flow on $M = M_1 = M_2$, then the infinitesimal version of (2.6.8) is

(2.6.9) $$\int_M L_X \omega = 0, \qquad X \in \mathcal{T}_0^1, \quad \omega \in E_m(M).$$

These facts have physically interesting formulations when we consider an m-form Ω which is invariant under the flow. This is the situation for one-parameter groups of orientation-preserving isometries and canonical transformations, which leave ^{*1} or $g \wedge g \wedge \cdots \wedge g \wedge g \wedge$ invariant. On phase space the latter is known as the **Liouville measure**, and is commonly written $dq_1 \ldots dq_m \, dp_1 \ldots dp_m$.

(2.6.10) Incompressibility of the Flow

Let Φ_t be a flow on M and Ω an m-form such that $\Phi_{t*}\Omega = \Omega$. Then $\forall f \in C_0^\infty(M)$,

$$\int \Omega \cdot f = \int \Omega \cdot (f \circ \Phi_t).$$

Proof: Use (2.6.8) and the fact that

$$\Phi_{-t*}(\Omega \cdot f) = \Phi_{-t*}\Omega \cdot \Phi_{-t*}f = \Omega \cdot (f \circ \Phi_t).$$

This holds for all measurable functions. If f is the characteristic function χ_A of a set A, then the equation states that the volume of the set, as measured by Ω, stays unchanged during the time-evolution. □

The motion thus resembles that of an incompressible fluid. An incompressible flow cannot have sinks. Indeed, if the phase space is compact, an incompressible flow always maps trajectories back onto themselves. The precise statement of this fact is known as

(2.6.11) **Poincaré's Recurrence Theorem**

Let $A \subset M$, $\Phi_t(A) \subset A\ \forall t \in \mathbb{R}$, and $\Omega(A) := \int \Omega\chi_A < \infty$. If $\Phi_{t*}\Omega = \Omega$, then for almost every point p of any measurable subset $B \subset A$, the trajectory through p returns infinitely often to B.

Proof: Let $B \subset A$ be an arbitrary measurable set, $\Omega(B) > 0$, and let $\tau \in \mathbb{R}^+$ be a unit of time. $K_n = \bigcup_{j=n}^{\infty} \Phi_{-j\tau}(B)$, j and $n \in \mathbb{Z}^+$, is the set of points that enter B after n or more time units (and possibly earlier as well). We clearly have the inclusions $B \subset K_0 \supset K_1 \supset \cdots \supset K_{n-1} \supset K_n$. The set of points of B that return after arbitrarily long times is $B \cap (\bigcap_{n\geq 0} K_n)$. This is disjoint from the set of points which do not return infinitely often, but, instead, are in B for a last time, and never come back. We want to show that the measure of the first set equals the measure of B. By assumption,

$$\Omega(K_n) = \Omega(\Phi_\tau K_n) = \Omega(K_{n-1}) \leq \Omega(A) < \infty,$$

because of the ordering by inclusion of the K_n's; and

$$\lim_{m\to\infty} \Omega\left(B \cap \left(\bigcap_{n\geq 0}^{m} K_n\right)\right) = \Omega(B \cap K_0) - \sum_{j=1}^{\infty} \Omega(B \cap (K_{i-1} \backslash K_i)) = \Omega(B),$$

since $B \cap K_0 = B$, and since $K_{n-1} \supset K_n$ and $\Omega(K_n) = \Omega(K_{n-1}) \Rightarrow \Omega(K_{n-1} \backslash K_n) = 0$. Hence the measure of the arbitrary measurable set B equals that of the set of its points that return to B infinitely often. □

Under the right circumstances conservation of energy provides a time-invariant submanifold of finite volume in phase space, for which the theorem applies. However, invariant regions of finite measure for unbounded forces (1.1.2) and more than two particles are not known, as the trajectories for which particles escape to infinity fill up a large portion of phase space. Theorem (2.6.11) does not apply to these **escape trajectories**, but since the flow cannot be compressed, most trajectories that come in from infinity must go back there:

(2.6.12) Schwarzchild's Capture Theorem

Again assume that $\Phi_t \cdot \Omega = \Omega$, and let A be a measurable set of finite measure, $\Omega(A) < \infty$. Then the set of points of A which were in A infinitely long in the past but will leave A forever in the future has measure zero. The same is true of the set of points that first enter A at some finite time and remain in A forever in the future.

Proof: Let $A_\pm = \bigcap_{\tau \gtrless 0} \Phi_\tau(A)$, the set of points which will remain in A forever, or, respectively, which have always been in A. Then for any time t,

$$\Omega(A_+) = \Omega(\Phi_{-t} A_+) = \Omega\left(\bigcap_{\tau > -t} \Phi_\tau(A)\right)$$

$$= \Omega\left(\bigcap_{-\infty < \tau < \infty} \Phi_\tau(A)\right) = \Omega(A_+ \cap A_-) = \Omega(A_-);$$

hence

$$\Omega(A_+ \backslash A_+ \cap A_-) = \Omega(A_- \backslash A_+ \cap A_-) = 0.$$

Trajectories that come from infinity and get bound in A, or, conversely, those that leave A forever, having formerly always been in A, can thus compose at most a subset of A of measure zero. Of course, the system could be unstable and $\Omega(A_+) = 0$. □

(2.6.13) Remarks

1. The invariance of Ω is the basis of what is called ergodic theory, which is concerned with establishing the general existence of the time-average

$$f_\infty = \lim_{T \to \infty} \left(\frac{1}{T}\right) \int_0^T dt\, \tau_1 f \text{ a.e.}$$

for functions f, required merely to be measurable rather than C^∞ (the **ergodic theorems of Birkhoff and von Neumann**). Because the spectral theory of operators in Hilbert space is used in the proof, we defer it until Volume IV, *Quantum Mechanics of Large Systems*, where it appears as a special case.

2. The utility of these theorems for physics is limited, in that they do not state how long it will be before a trajectory returns or before a time-average approaches its asymptotic value. The surprising fact is that these theorems have global consequences, which are not easy to read off the differential equations directly.
3. Ω serves only to measure the probabilities of various configurations. This interpretation of Ω seems well justified for the Liouville measure. Every open set is Liouville measurable, and $B \neq \emptyset$ iff $\Omega(B) > 0$. As a consequence, Theorems (2.6.11) and (2.6.12) apply to arbitrary open sets B.
4. The Recurrence Theorem gave rise to all sorts of confusion, because people got the impression that whenever a flow was measure-preserving on a compact set, an initial state must always recur. States, however, are probability measures.

Moreover, if the illusion of infinite accuracy of measures is abandoned, they can never be point measures, but rather must be of the form $f\Omega$, with f measurable and > 0, $\int f\Omega = 1$. In Volume IV we shall learn about mixing systems, for which the states tend toward the uniformly distributed state $\Omega/\int\Omega$. What the apparent contradiction overlooks is that while the individual points of the measure of f must return to any neighborhood of their initial positions, they do not do so all at the same time. Instead, the trajectories spread out over all of M so that the effect of $f\Omega$ on an integral tends to that of $\Omega/\int\Omega$, and time averages of functions converge to their averages with respect to the measure $\Omega/\int\Omega$.

We sum up by collecting the important formulas for differentiation and integration:

$$\int_M d\omega = \int_{\partial M} \omega,$$

$$d(\alpha_i\omega_i) = \alpha_i\, d\omega_i, \qquad \alpha_i \in \mathbb{R},\ \omega_i \in E_p, \qquad \int \alpha_i\omega_i = \alpha_i \int \omega_i,$$

$$d(\omega_1 \wedge \omega_2) = d\omega_1 \wedge \omega_2 + (-1)^p \omega_1 \wedge d\omega_2, \qquad \partial(M_1 \times M_2) = \partial M_1 \times M_2 \cup M_1 \times (-)^{\dim M_1} \partial M_2$$

($-M$ denotes M with opposite orientation)

$$d \circ d\omega = 0, \qquad \partial\partial M = \emptyset$$

$$d\Phi_*\omega = \Phi_*\, d\omega, \qquad \int_{\Phi(M)} \Phi_*\omega = \int_M \omega$$

(2.6.14) $$\int_M L_X\omega = 0 \quad \text{if} \quad \Phi_t^X M = M,$$

as the formulas on the left are in correspondence to the formulas on the right.

(2.6.15) **Problems**

1. Show that the Liouville measure $d^m q\, d^m p \equiv dq_1 \wedge \cdots \wedge dq_m \wedge dp_1 \wedge \cdots \wedge dp_m$ is invariant under a point transformation $q \to \bar{q}$.
2. Work through Example (2.6.7; 2), following the steps of the definition of the integral.
3. Using (2.6.5), prove the following theorem: $\oint_C dz\, f(z) = 0$ if the path of integration C encircles a region in a neighborhood of which f is a meromorphic function without poles of the first order.
4. Show that orientability is equivalent to the existence of an atlas $\bigcup_i (U_i, \Phi_i)$ with $d_{ij} \equiv \det D(\Phi_i \circ \Phi_j^{-1}) > 0\ \forall i, j$ such that $U_i \cap U_j \neq \emptyset$.
5. Two C^∞-mappings f and $g: M \to N$ of two manifolds are said to be diffeotopic to each other iff there exists a C^∞-mapping $F: [0, 1] \times M \to N$, such that $f = F \circ i_0$ and

$g = F \circ i_1$, where i_0 and i_1 are the imbeddings $i_0: M \to \{0\} \times M$ and $i_1: M \to \{1\} \times M$. Show that if M and N are orientable, compact, and n-dimensional, then $\forall \omega \in E_n(N)$,

$$\int_M f^*\omega = \int_M g^*\omega,$$

if f and g are diffeotopic. (First show that $\varphi \in E_n$ and closed $\Rightarrow g^*\varphi - f^*\varphi$ is exact.)

6. Use Problem 5 to prove the theorem that you can't comb a hedgehog: If n is even, then every C^∞-vector field X on S^n has at least one point where it vanishes.

7. The hydrodynamic equations, $\dot{v}_i + v_k v_{i,k} = -p_{,i}$ are written as $\dot{v} + L_v v = d(v^2/2 - p)$ (cf. (2.5.14; 5)), if we construe v as a 1-form on $\mathbb{R}^3$ and denote the Lie derivative with respect to v by L_v, by making the covariant vector field contravariant with the metric $g_{ik} = \delta_{ik}$. Hence the vorticity $w = dv$ satisfies the equation $\dot{w} + L_v w = 0$. If at any time w vanishes everywhere, it remains zero afterwards. Derive the following integrated version of conservation of vorticity, known as **Thompson's theorem**: Let C_t be a closed curve that follows the flow of v. Then

$$\frac{d}{dt}\int_{C_t} v = 0.$$

8. For a divergence-free vector field ($E \in \mathcal{T}_0^1(\mathbb{R}^3)$, $g_{ik} = \delta_{ik}$, $d * E = 0$), show that the field strength is proportional to the density of the lines of force in the following sense: Lines of force are the trajectories of Φ_t^E, so that the same number of lines of force pass through both N and $\Phi_t^E N$. For $E \| d\mathbf{f}$, where $d\mathbf{f}$ is the surface area element,

$$\int_N {}^*E = \int_N \mathbf{E} \cdot d\mathbf{S} \propto \frac{\text{field strength}}{\text{density of lines of force}},$$

because (surface area) $\times$ (density of lines of force) is constant. Prove the invariance of this relationship under Φ_t^E.

(2.6.16) Solutions

1. $$d^m q = d^m \bar{q} \det\left(\frac{\partial q_i}{\partial \bar{q}_j}\right), \qquad d^m p = d^m \bar{p} \det\left(\frac{\partial \bar{q}_i}{\partial q_j}\right).$$

2. The 1-form df should be integrated along the trajectory of the curve $u: I = (a, b) \to \mathbb{R}^3$. To do this we use the chart $(u(I), u^{-1})$ on this one-dimensional submanifold. Then

$$\int_{u(I)} df = \int_I (u)^* df = \int_I d((u)^* f) = \int_I d(f \circ u) = \int_{\partial I} (f \circ u)$$
$$= f(u(b)) - f(u(a)).$$

3. It follows from the assumptions that $f(z) = (\partial/\partial z)F(z)$, so

$$\int_C dz\, f(z) = \int_C dF = \int_{\partial C} F = 0,$$

since $\partial C = \emptyset$ for a closed path C.

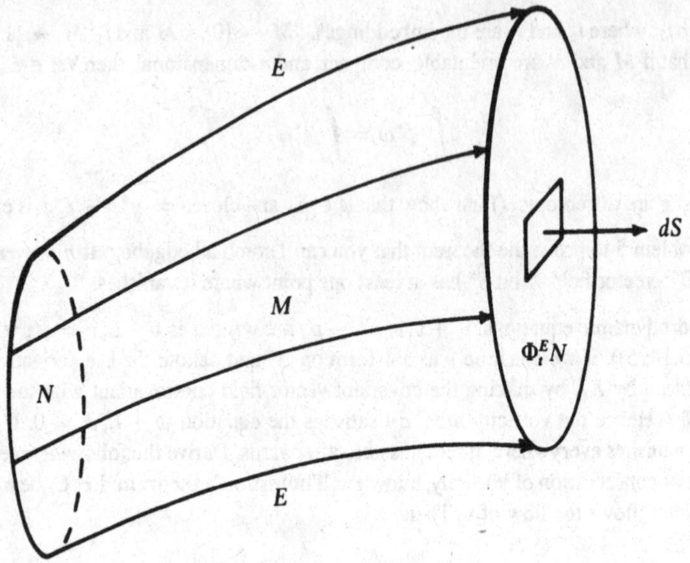

FIGURE 2.15. The cylinder spanned by the lines of force.

4. $\Rightarrow$: Choose Φ_i such that in the coordinate system $(x_1, \ldots, x_m) = \Phi_i(x)$, $\Omega = g_i\, dx_1 \wedge \cdots \wedge dx_m$, $g_i > 0$. Then $d_{ij} = g_j/g_i > 0$.
   $\Leftarrow$: Let $\omega_i = dx_1 \wedge \cdots \wedge dx_m$ on U_i, and let x_k be as above. Construct $\sum_i \omega_i f_i$, where f_i is a partition of unity as in (2.6.3). Given $x \in M$, let I be the set of i such that $f_i(x) \neq 0$, and fix $i_0 \in I$. Then since f_i and $d_{ii_0}(x) > 0$ by assumption,
$$\omega(x) = \sum_{i \in I} [f_i(x)\, d_{ii_0}(x)]\omega_{i_0}(x) \neq 0.$$

5. In analogy with (2.5.14; 7); define a mapping $K: E_{p+1}(I \times M) \to E_p(M)$ such that $\omega_0 + dt \wedge \omega_M \to \int_0^1 dt\, \omega_M$. This mapping satisfies $d \circ K + K \circ d = i_1^* - i_0^*$. Hence $d \circ K \circ F^* + K \circ F^* \circ d = (d \circ K + K \circ d) \circ F^* = (i_1^* - i_0^*) \circ F^* = (F \circ i_1)^* - (F \circ i_0)^* = g^* - f^*$. If $d\varphi = 0$, then $g^*\varphi - f^*\varphi = d \circ K \circ F^* \varphi$ is exact. But then $\int_M (g^*\omega - f^*\omega) = 0$, because ω, as an n-form, is closed, and M has no boundary.

6. Imagine S^n imbedded in $\mathbb{R}^{n+1}$, where $S^n = \{x \in \mathbb{R}^{n+1}: \|x\|^2 := \langle x|x\rangle = 1$ and $T(S^n)$ equipped with the corresponding Riemannian structure, and let $X(x) \neq 0\ \forall x \in S^n$. If we replace X with $X/\|X\|$, then we can treat it as a mapping from S^n to S^n. Let $F: [0, 1] \times S^n \to S^n$ be given by $(t, x) \to x \cos \pi t + X(x) \sin \pi t$ (note that $X(x) \perp x$); this furnishes a homotopy between $x \to x$ and $x \to -x$ (the antipodal mapping a). On S^n let us consider the n-form given by
$$\omega_j = \frac{(-1)^{j+1}}{x_j} dx_1 \wedge \cdots \wedge \widehat{dx_j} \wedge \cdots \wedge dx_{n+1}$$
in the chart given by the imbedding $\{x_j \neq 0\}$. Since when restricted to S^n, $\sum_{i=1}^{n+1} x_i\, dx_i|_{S^n} = 0$, all the ω_j agree on S^n and define the volume element ω on S^n. Now note that $a_*\omega = (-1)^{n+1} V(S^n)$, so
$$V(S^n) \equiv \int_{S^n} \omega = \int_{S^n} a_*\omega = (-1)^{n+1} V(S^n),$$

which is impossible when n is even.

7. Since C_t follows the flow of v, $\int_{C_{t+\tau}} v = \int_{C_t} e^{\tau L_v} v$, which contributes the term $\int_{C_t} L_v v$ to the derivative:

$$\frac{d}{dt}\int_{C_t} v = \int_{C_t} (\dot{v} + L_v v) = \int_{C_t} d\left(\frac{v^2}{2} - p\right) = 0.$$

8. Let M be the cylinder spanned by N and $\Phi_t^E N$ (see Figure 2.15). $\partial M = N \cup \Phi_t^E N \cup$ the outer surface. Since $\mathbf{E} \perp d\mathbf{f}$ on the outer surface,

$$0 = \int_M d^*E = \int_{\Phi_t^E N} {}^*E - \int_N {}^*E.$$

3

Hamiltonian Systems

3.1 Canonical Transformations

A 2-form is canonically defined on the cotangent bundle of a manifold. Diffeomorphisms leaving this 2-form invariant are called canonical transformations.

The Lagrangian (2.3.23) defines a bijection $T(M) \to T^*(M)$: $\dot{q} \to p$, and the corresponding local flow on $T^*(M)$ satisfies Hamilton's equations. The flow has the special property of preserving the symplectic structure of $T^*(M)$, which is determined by the canonical 2-form:

(3.1.1) Definition

Let Π denote the projection of the cotangent bundle $T^*(M)$ such that $T(\pi)$: $T(T^*M)) \to T(M)$, and define $\Theta \in E_1(T^*(M))$ by $(\Theta|v)(z) = (z|T(\Pi)v(z))$ for all $z \in T^*(M)$ and $v \in \mathcal{T}_0^1(T^*(M))$. The 1-form Θ and its differential $\omega := -d\Theta$ are called the **canonical forms** on $T^*(M)$.

(3.1.2) Remarks

1. On a bundle chart $T(\Pi)$ acts by

$$\left(\frac{\partial}{\partial q}, \frac{\partial}{\partial p}\right) \to \frac{\partial}{\partial q}.$$

If in the same chart v is such that

$$(q, p) \to \left(q, p; v^i \frac{\partial}{\partial q_i} + w^i \frac{\partial}{\partial p_i}\right),$$

then $(\Theta|v)(q, p) = \sum_i p_i v_i(q, p)$. Hence, in the sense of Example (2.4.3; 2) one often writes $\Theta = p_i\, dq^i$ and $\omega = dq^i \wedge dp_i$. The form ω is not only closed, but also nondegenerate, and thus defines a symplectic structure on $T^*(M)$.

2. The forms are chart-independent. More generally, under diffeomorphisms Φ: $M_1 \to M_2$, for which $\Psi \equiv T^*(\Phi)$: $T^*(M_1) \to T^*(M_2)$, the canonical forms $\Theta_{1,2}$ and $\omega_{1,2}$ are transformed into each other: $\Psi_*\Theta_1 = \Theta_2$, and $\Psi_*\omega_1 = \omega_2$ (Problem 8).
3. The 1-form Θ vanishes at the origin of $T^*(M)$, but ω is always nonzero. The p-fold exterior product,

$$\Omega_p := \frac{(-1)^{(p-1)p/2}}{p!} \underbrace{\omega \wedge \omega \wedge \cdots \wedge \omega}_{p \text{ times}}$$

is a $2p$-form which likewise never vanishes. In particular, on a bundle chart $\Omega \equiv \Omega_m$ it is $dq_1 \wedge dq_2 \wedge \cdots \wedge dq_m \wedge dp_1 \wedge \cdots \wedge dp_m$ (Liouville measure). Since ω is exact, so are its powers. Specifically,

$$\Omega = -\frac{(-1)^{(m-1)m/2}}{m!} d(\Theta \wedge \omega \wedge \cdots \wedge \omega).$$

As we have seen, the transformations induced by diffeomorphisms of M, known as point transformations, leave the canonical forms invariant. However, we can ask about more general transformations of $T^*(M)$ in which the new q depends on the old q and p, but which leave ω invariant.

(3.1.3) **Definition**

A diffeomorphism $T^*(M_1) \supset U_1 \xrightarrow{\Psi} U_2 \subset T^*(M_2)$ that takes the canonical 2-form $\omega_{|U_1}$ to $\omega_{|U_2}$ is called a **local canonical transformation.** If $U_i = T^*(M_i)$, then Ψ is called a **canonical transformation.**

(3.1.4) **Remarks**

1. Point transformations are canonical.
2. An example of a canonical transformation that is not a point transformation is the interchange $(q, p) \to (p, -q)$ on $M = \mathbb{R}$, where $T^*(M) = \mathbb{R} \times \mathbb{R}$ and $\omega = dq \wedge dp$.
3. The interchange $(q, p) \to (p, q)$ changes the sign of ω; thus not every linear transformation is canonical.
4. Because $(\Psi_1 \circ \Psi_2)_* = \Psi_{1*} \circ \Psi_{2*}$, the canonical transformations form a group.
5. Since $\Psi_*(\omega_1 \wedge \omega_2) = (\Psi_*\omega_1) \wedge (\Psi_*\omega_2)$, canonical transformations also leave the $2p$-form Ω_p invariant.

In general the 1-form Θ will change under a canonical transformation. However, it is still true that $\Psi_* d\Theta - d\Theta = d(\Psi_*\Theta - \Theta) = 0$, so that, at least in a neighborhood U, $\Psi_*\Theta = \Theta + d\bar{f}$, for some $\bar{f} \in C^\infty(U)$. Letting $\Psi: (\bar{q}, \bar{p}) \to (q, p)$, on some chart, this formula may be written

$$p_i\, dq^i = \bar{p}_i\, d\bar{q}^i + d\bar{f} = -\bar{q}^i\, d\bar{p}_i + df, \qquad (3.1.5)$$
$$f(\bar{q}, \bar{p}) = \bar{f}(\bar{q}, \bar{p}) + \bar{q}^i \bar{p}_i.$$

If we want an explicit expression for Ψ, we must evaluate (3.1.5) on a basis. If

$$\det\left(\frac{\partial q^i}{\partial \bar{q}^j}\right)_{|\bar{p}\ \text{constant}} \neq 0,$$

which holds, for instance, for a point transformation or for Ψ sufficiently close to **1**, it will suffice to express everything in terms of dq^i and $d\bar{p}_j$ and to equate coefficients. Then we may write $\bar{q}$ locally as a function of q and $\bar{p}$ by inverting $q(\bar{p}, \bar{q})$. If we also call $f(\bar{q}(q, \bar{p}), \bar{p})$ simply $f(q, \bar{p})$ and plug it into (3.1.5), we obtain

(3.1.6) **Lemma**

A local canonical transformation $\Psi: (\bar{q}, \bar{p}) \to (q, p)$ *with*

$$\det\left(\frac{\partial q^i}{\partial \bar{q}^j}\right)_{|\bar{p}\ \text{constant}} \neq 0$$

may be written locally as

$$p_i = \frac{\partial f}{\partial q^i}, \qquad \bar{q}^i = \frac{\partial f}{\partial \bar{p}_i}, \qquad f(q, \bar{p}) \in C^\infty.$$

The function f is known as the local generator. Conversely, if $f(q, \bar{p}) \in C^\infty(U)$ such that $\det(\partial^2 f/\partial q^i\, \partial \bar{p}_j) \neq 0$ *is given, then the above equations define a local canonical transformation.*

(3.1.7) **Remarks**

1. The canonical transformation $q^i = \bar{p}_i$, $p_i = -\bar{q}^i$ on $T^*(M) = \mathbb{R}^{2m}$ is not induced in this way. The construction fails because $\partial q^i/\partial \bar{q}^j_{|\bar{p}} = 0$.
2. If $\det(\partial q^i/\partial \bar{p}_j) \neq 0$, then $\bar{p}(q, \bar{q})$ may be calculated locally. Substituting into $\bar{f}$, we obtain the alternative form

$$p_i = \frac{\partial \bar{f}}{\partial q^i}, \qquad \bar{p}_i = -\frac{\partial \bar{f}}{\partial \bar{q}^i}.$$

Point transformations cannot be written like this. It is clear that $\bar{f}$ is additive with respect to composition, i.e., $\bar{f}_1 \circ \Psi_2^{-1} + \bar{f}_2$ generates $\Psi_2 \circ \Psi_1$.

3. We learn from integration theory (§2.6) that integrals over ω and Ω are left invariant by canonical transformations, where of course the new integral is taken over the image of the original integration region:

$$\int_N \omega = \int_{\Psi N} \omega, \qquad \int_U \Omega = \int_{\Psi U} \Omega,$$

in which N and U are, respectively, two-dimensional and $2m$-dimensional submanifolds of M. Nothing like this is necessarily true for Θ. However, if C is a one-dimensional submanifold without a boundary, contained in some neighborhood in which equation (3.1.5) holds, that is, C is a sufficiently small closed curve, then

$$\int_{\Psi^{-1}C} \Theta = \int_C \Theta + \int_{\partial C} \bar{f} = \int_C \Theta.^1$$

It is easy to go astray for arbitrary closed curves (Problem 6).

4. Care should be taken to distinguish between the generator of a finite transformation and the generating vector field of a transformation group. The connection between them will be examined in (3.2.9).

(3.1.8) **Examples**

1. $p = \sqrt{2\omega\bar{p}}\cos\bar{q}$, $q = \sqrt{2\bar{p}/\omega}\sin\bar{q}$, $\omega \in \mathbb{R}^+$, is a local canonical transformation from $\mathbb{R}^+ \times S^1 \subset T^*(S^1)$ to $\mathbb{R}^2\backslash\{0\} \subset T^*(\mathbb{R})$. Calculating

$$dp = \tfrac{1}{2} d\bar{p}\cos\bar{q}\sqrt{2\omega/\bar{p}} - d\bar{q}\sqrt{2\omega\bar{p}}\sin\bar{q},$$
$$dq = \tfrac{1}{2} d\bar{p}\sqrt{2/\omega\bar{p}}\sin\bar{q} + d\bar{q}\sqrt{2\bar{p}/\omega}\cos\bar{q},$$

we see

$$dq \wedge dp = d\bar{q} \wedge d\bar{p}.$$

This can obviously not be continued to a canonical transformation $T^*(S^1) \to T^*(\mathbb{R})$.

2. We would like to determine when a linear transformation on $T^*(\mathbb{R}^m) = \mathbb{R}^{2m}$ is canonical. Let us treat $(q^1, \ldots, q^m, p_1, \ldots, p_m)$ as a single vector $(x_1, \ldots, x_{2m})$

---

[1]The forms ω and Ω are called **integral invariants,** and Θ is called a **relative integral invariant.**

and write

$$\omega = dx_i\, W_{ik}\, dx_k, \qquad W = \frac{1}{2}\;\begin{array}{|c|c|l}
\multicolumn{1}{c}{} & \multicolumn{1}{c}{m} & \\ \hline
0 & \begin{matrix} 1 & & & \\ & 1 & & \\ & & \ddots & \\ & & & 1 \end{matrix} & m \\ \hline
\begin{matrix} -1 & & & \\ & -1 & & \\ & & \ddots & \\ & & & -1 \end{matrix} & 0 & \\ \hline
\end{array}$$

A linear transformation $x_k = L_{kj}\bar{x}_j$ is canonical in case $L^t W L = W$, that is, L is a symplectic matrix. Symplectic matrices have the following properties (see Problem 4):

(a) $\det L$ must be either $+1$ or -1.

(b) If λ is an eigenvalue, then so are $1/\lambda$, λ^*, and $1/\lambda^*$.

It follows that in fact $\det L = 1$, and L preserves volumes. If L expands in one direction, it must contract in another.

The canonical 2-form is given everywhere as the invertible matrix W. Hence ω, just like g from (2.4.12; 2), defines a nondegenerate bilinear form, and consequently a bijection from $\mathcal{T}_1^0(T^*(M))$ to $\mathcal{T}_0^1(T^*(M))$.

(3.1.9) Definition

The canonical form ω defines a bijection b from $\mathcal{T}_1^0(T^*(M))$ to $\mathcal{T}_0^1(T^*(M))$ by $i_{\ell_v}\omega = v$, $v \in \mathcal{T}_1^0(T^*(M))$. The vector fields $X_H \equiv b(dH)$, $H \in C^\infty(T^*(M))$ are said to be **Hamiltonian.** If H can only be defined locally, X_H is said to be **locally Hamiltonian.**

(3.1.10) Remarks

1. The action of b can be expressed schematically, in that the diagram

$$\begin{array}{ccc}
\mathcal{T}_0^1(T^*(M)) \times \mathcal{T}_0^1(T^*(M)) & \overset{\omega}{\searrow} & \\
\qquad {\scriptstyle b\times 1}\Big\uparrow & & C^\infty(T^*(M)) \\
\mathcal{T}_1^0(T^*(M)) \times \mathcal{T}_0^1(T^*(M)) & \underset{(|)}{\nearrow} &
\end{array}$$

commutes.

2. On a chart, $b(v_{q^i}\, dq^i + v_{p_i}\, dp_i) = v_{p_i}\, \partial q^i - v_{q^i}\, \partial p_i$, and

$$X_H\colon (q, p) \to \left(q, p; \left(\frac{\partial H}{\partial p_i}\right) \partial q^i, -\left(\frac{\partial H}{\partial q^i}\right) \partial p_i\right).$$

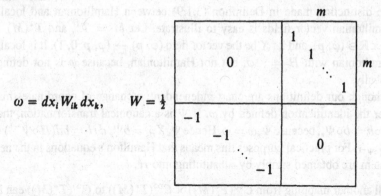

Note that X_H is exactly of the form of the vector field that came up in (2.3.21).

3. The distinction made in Definition (3.1.9) between Hamiltonian and locally Hamiltonian vector fields is easy to illustrate. Let $M = T^1$, and $T^*(M) = T^1 \times \mathbb{R} \ni (\varphi, p)$, and let X be the vector field $(\varphi, p) \to (\varphi, p; 0, 1)$. It is locally Hamiltonian with $H = -\varphi$, but not Hamiltonian, because φ is not defined globally.
4. Although our definitions are chart-independent, a change of coordinates may alter the identification defined by ω. If Ψ is a canonical transformation, then $\Psi_* \circ b = b \circ \Psi_*$, because $\Psi_* \omega = \omega$. Hence $\Psi_* X_H = b\Psi_* dH = bd(H \circ \Psi^{-1}) = X_{H \circ \Psi^{-1}}$. For practical purposes this means that Hamilton's equations in the new system are obtained simply by substituting into H.

Still another mapping from $C^\infty(T^*(M)) \times C^\infty(T^*(M))$ to $C^\infty(T^*(M))$ can be defined using ω, by applying it to the vector fields associated with two functions.

(3.1.11) **Definition**

The **Poisson bracket** of two functions F and G in $C^\infty(T^*(M))$ is defined by

$$\{G, F\} \equiv \omega(X_G, X_F) = i_{X_F} i_{X_G} \omega = i_{X_F} dG = L_{X_F} G = -L_{X_G} F.$$

(3.1.12) **Remarks**

1. The antisymmetry of ω leads to $\{G, F\} = -\{F, G\}$, accounting for the final equality.
2. On a bundle chart,

$$\{G, F\} = \sum_{i=1}^{m} \left(\frac{\partial G}{\partial q^i} \frac{\partial F}{\partial p_i} - \frac{\partial F}{\partial q^i} \frac{\partial G}{\partial p_i} \right).$$

and in particular, $\{q^i, q^j\} = \{p_i, p_j\} = 0$, and $\{q^i, p_i\} = \delta_{ij}$.
3. Poisson brackets are invariant under canonical transformations Ψ in the sense that in the new system they just become the Poisson brackets of the new functions: By (3.1.10; 4), $\{F, G\} \circ \Psi = \{F \circ \Psi, G \circ \Psi\}$. Conversely, if Ψ is a diffeomorphism of $T^*(M)$ that satisfies this equation for all functions F and G, then Ψ is canonical. To see this, choose functions that equal q^i and p_i on the domain of a chart; then the transformed coordinates are $\bar{q}^i \equiv q^i \circ \Psi$ and $\bar{p}_i \equiv p_i \circ \Psi$, and they also satisfy $\{\bar{q}^i, \bar{p}_j\} = \delta_{ij}$ and $\{\bar{q}^i, \bar{q}^j\} = \{\bar{p}_i, \bar{p}_j\} = 0$. Hence the transformed $\omega = d\bar{q}^i \wedge d\bar{p}_i$ and Ψ is canonical.
4. The rules (2.2.24) for the Lie derivative have the consequences that $\{F + G, H\} = \{F, H\} + \{G, H\}$ and $\{F \cdot G, H\} = G\{F, H\} + F\{G, H\}$.

(3.1.13) **Problems**

1. Find the generator $\bar{f}$ of the local canonical transformation $q = \sqrt{2\bar{p}/\omega} \sin \bar{q}$, $p = \sqrt{2\omega\bar{p}} \cos \bar{q}$.

2. For real α and β, calculate $\{e^{\alpha q}, e^{\beta p}\}$, $\alpha, \beta \in \mathbb{R}$.

3. Let us write the matrix L of the linear canonical transformation (3.1.8; 2) in block form

$$L = \left|\begin{array}{c|c} A & B \\ \hline C & D \end{array}\right|.$$

What are the conditions for L to be symplectic, and what is the generator $\bar{f}$ of the transformation?

4. Let λ be an eigenvalue of the symplectic matrix L. Show that $1/\lambda$ (and hence also λ^* and $1/\lambda^*$) is an eigenvalue. If L is an element of a one-parameter group generated by a function F, $L = e^{tF}$, then what does this imply about the eigenvalues of F?

5. Consider the flow on $T^*(M)$ generated by the canonical vector field $b\Theta$. Is it canonical?

6. Construct a canonical transformation Ψ for which $\Psi_*\Theta \neq \Theta + d\bar{f}$, and hence $\int_C \Theta \neq \int_{\Psi C} \Theta$ for a closed curve C.

7. What is the form of the generator f (cf. (3.1.5)) of a point transformation $\psi = T^*(\Phi)$, $\Phi: q \to \bar{q}(q)$?

8. Show that $\Psi_*\Theta = \Theta$ (and hence $\Psi_*\omega = \omega$ and $\Psi_*\Omega = \Omega$) for the point transformation $\Psi = T^*(\Phi)$, where Φ is a diffeomorphism of neighborhoods in M. Is this the most general local canonical transformation that leaves Θ invariant and which has a domain of definition including the submanifold $\bar{p}_i = 0$?

(3.1.14) Solutions

1. $\bar{f} = \bar{p} \cos \bar{q} \sin \bar{q}$, because

$$\begin{aligned} p\,dq &= \sqrt{2\omega \bar{p}} \cos \bar{q} \left(d\bar{p} \sqrt{\frac{1}{2\bar{p}\omega}} \sin \bar{q} + d\bar{q} \sqrt{\frac{2\bar{p}}{\omega}} \cos \bar{q} \right) \\ &= \bar{p}\, d\bar{q} + d(\bar{p} \cos \bar{q} \sin \bar{q}). \end{aligned}$$

Note that $\bar{f}$ is defined globally on $T(S^1)$, but even so it generates only a local canonical transformation.

2.

$$\begin{aligned} \{e^{\alpha q}, p\} &= \sum_{n=0}^{\infty} \frac{\alpha^n}{n!} \{q^n, p\} = \alpha \sum_{n=0}^{\infty} \frac{\alpha^n}{n!} q^n = \alpha e^{\alpha q}, \\ \{e^{\alpha q}, p^m\} &= p^{m-1} \{e^{\alpha q}, p\} + \{e^{\alpha q}, p^{m-1}\} p = m\alpha p^{m-1} e^{\alpha q}, \\ \{e^{\alpha q}, e^{\beta p}\} &= \sum_{m=0}^{\infty} \frac{\beta^m}{m!} \{e^{\alpha q}, p^m\} = \sum_{m=0}^{\infty} \frac{\beta^m}{m!} \alpha\beta e^{\alpha q} p^m = \alpha\beta e^{\alpha q + \beta p}. \end{aligned}$$

3. The matrices $A^t C$ and $B^t D$ must be symmetric, and $A^t D - C^t B = 1$.

$$\bar{f} = \bar{p} B^t C \bar{q} + \tfrac{1}{2} (\bar{q} C^t A \bar{q} + \bar{p} D^t B \bar{p}).$$

4. (a) Take the determinant of $L^t W L = W$, to conclude that $(\det L)^2 = 1$.

(b) $\det(L^{-1}-\lambda) = \det(W^{-1}L^t W - \lambda) = \det(L^t - \lambda) = \det(L-\lambda) = 0$, which implies that λ is an eigenvalue of L^{-1}. Therefore if φ is an eigenvalue of F, then so are $-\varphi$, φ^*, and $-\varphi^*$, since L is real.

5. $b\Theta = -p_i\,\partial/\partial p_i$ generates the flow $q^i(t) = q^i(0)$, $p_i(t) = e^{-t}p_i(0)$. It is not canonical, yet it does characterize the bundle structure of $T^*(M)$: Its flow lines are the fibers and its fixed points are the basis.

6. $M = S^1$, $T^*(M) = S^1 \times \mathbb{R}$, and $\Psi: (\varphi, p) \to (\varphi, p+a)$, $a \in \mathbb{R}$, for both charts (2.1.7; 2) on S^1. Locally, $\bar{f} = a\varphi$, but $\varphi \notin C^\infty(T^*(M))$. So for $C: t \in [0, 2\pi) \to (t, p) \in T^*(M)$,

$$\int_C \Theta = 2\pi p \neq \int_{\Psi C} \Theta = 2\pi(p+a).$$

Note that C is not the boundary of a compact surface. If S^1 were imbedded in $\mathbb{R}^2$, then C would be the boundary of a circular disc D, and one might be led to the false conclusion that

$$\int_C \Theta = -\int_D \omega = -\int_{\Psi D} \omega = \int_{\Psi C = \partial\Psi D} \Theta.$$

The trick is that Θ would be singular at the origin, but the region D would not be compact with the origin deleted. This is not mere mathematical nit-picking; it shows up in the quantum theory of the Aharonov–Bohm effect, and thus has observable consequences.

7. $f(\bar{p}, q) = \bar{p}_i \bar{q}^i(q)$.

8. By Problem 7, $f = \bar{p}\bar{q}(q)$, and so $\bar{f} = 0$ and $\Psi_*\Theta = \Theta$. If on the other hand Θ is invariant, then so is $f = \bar{p}\bar{q}(q)$. According to (3.1.6), $\bar{q}^i = \partial f/\partial \bar{p}_i$, so $f = \bar{p}_i\,\partial f/\partial \bar{p}_i$, and thus f is homogeneous of degree 1 in the variables p_i. If $\{\bar{p}_i = 0\}$ belongs to the domain of definition of f, then f must be linear, and consequently $\bar{q}^i$ is a function of q^i alone.

3.2 Hamilton's Equations

Hamiltonian vector fields generate local flows that leave ω invariant. These flows are determined by Hamilton's equations.

A local flow Ψ_t on $T^*(M)$ that leaves ω invariant, $\Psi_t^*\omega = \omega$, has a vector field X which generates it and satisfies $L_X\omega = 0$. It turns out that Hamiltonian vector fields (3.1.9), X_H, $H \in C^\infty(T^*(M))$, have this very property. To see that this is so, recall that the inverse of the bijection from (3.1.9), $b^{-1}: \mathcal{T}_0^1(T^*(M)) \to \mathcal{T}_1^0(T^*(M))$ is the mapping $X \to i_X\omega$ (2.4.33), and consequently one can write $dF = i_{X_F}\omega$ (see (3.1.1; 1)). Then using Property (2.5.9; 4) and the relationship $d\omega = -d\,d\Theta = 0$,

$$L_{X_F}\omega = d i_{X_F}\omega = d\,dF = 0. \tag{3.2.1}$$

Moreover, locally the argument goes the other way, too, since $0 = L_X\omega = d i_X\omega$, which implies $i_X\omega = dF$, for $F \in C^\infty(T^*(M))$, which then implies $X = X_F$. This proves

(3.2.2) Theorem

Every locally Hamiltonian vector field generates a local flow of canonical transformations. Conversely, any local flow of canonical transformations has a locally Hamiltonian generator.

(3.2.3) Remark

The above statement is false if the qualification "local" is dropped. A Hamiltonian vector field need not be complete even if H is defined globally. And conversely, the flow $\Psi_t: (\varphi, p) \to (\varphi, p+t)$ on $T^*(S^1) = S^1 \times \mathbb{R}$ has the generator φ, which is only defined locally (cf. (3.1.10; 3)).

Let us consolidate the foregoing results.

(3.2.4) Definition

For all $H \in C^\infty(T^*(M))$, **Hamilton's equations,**

$$\dot{u} = X_H \circ u,$$

define the **local canonical flow** Φ_t, and (M, Φ_t) is called a **Hamiltonian system.**

(3.2.5) Remarks

1. The variation in time of an observable is determined by its Poisson bracket with H:

$$\frac{d}{dt} F \circ \Phi_t = \{F \circ \Phi_t, H\}.$$

2. According to Remark (3.1.10; 4), Hamilton's equations are invariant under canonical transformations in the sense that in the new system it is necessary merely to use the transformed H. More explicitly, let $\Psi: (\bar{q}, \bar{p}) \to (q(\bar{q}, \bar{p}), p(\bar{q}, \bar{p}))$, and let us call $H_T(q, p) = H(\bar{q}(q, p), \bar{p}(q, p))$. Then the pairs of equations

$$\dot{\bar{q}}^i = \partial H / \partial \bar{p}_i, \quad \dot{\bar{p}}_i = -\partial H / \partial \bar{q}^i \quad \text{and} \quad \dot{q}^i = \partial H_T / \partial p_i, \quad \dot{p}_i = -\partial H_T / \partial q^i$$

are equivalent.
3. Because of the antisymmetry of the Possion bracket, H is constant in time. For the same reason, any quantity that generates a local canonical flow that does not change H is constant in time.
4. Time-evolution leaves invariant not only ω, but also, as in Remark (3.1.4; 5), the phase-space volume Ω. (This is Liouville's theorem. In the framework of classical mechanics the proof of this theorem requires some effort. But modern concepts are so formulated that there is really nothing to prove.)
5. If H and F are analytic, and X_H is complete, then the canonical flow may be written more explicitly following Remark (2.3.11; 2) as

$$F \circ \Phi_t = \sum_{n=0}^{\infty} \frac{t^n}{n!} \underbrace{\{\{\cdots\{\{F, H\}, H\}\cdots\}, H\}}_{n \text{ times}}.$$

(3.2.6) Examples

1. The canonical transformation of Example (3.1.8; 1) transforms the Hamiltonian of the harmonic oscillator,

$$H = \tfrac{1}{2}(p^2 + \omega^2 q^2),$$

into

$$H_T = \bar{p}\omega.$$

With these coordinates the time-evolution is $\bar{p} = \text{const.}$, $\bar{q}(t) = \bar{q}(0) + \omega t$. With the old coordinates, $p(t) = \sqrt{2\omega\bar{p}}\cos(\bar{q}(0)+\omega t)$ and $q(t) = \sqrt{2\bar{p}/\omega}\sin(\bar{q}(0)+\omega t)$, which is precisely the solution of the equations for p and q using H. Note that: (i) Time-evolution is a rotation in phase space, (q, p), and therefore it leaves phase-space volumes invariant. (ii) This canonical transformation can be used even though it is only local, because it maps time-invariant regions into one another.

2. $M = \mathbb{R}^3$, $T^*(M) = \mathbb{R}^3 \times \mathbb{R}^3$, $H = |\mathbf{p}|^2$, and $G_1 = p_1$. The vector field X_{G_1} generates the flow $(q_1, q_2, q_3; p_1, p_2, p_3) \to (q_1 + \lambda, q_2, q_3; p_1, p_2, p_3)$. This leaves H invariant, which is equivalent to $G_1 = \text{const.}$ (**conservation of momentum** for a free particle).

3. M and H as in Example 2, but $G_2 = p_1 q_2 - q_1 p_2$. The vector field X_{G_2} generates the flow $(q_1, q_2, q_3; p_1, p_2, p_3) \to (q_1 \cos\lambda + q_2 \sin\lambda, -q_1 \sin\lambda + q_2 \cos\lambda, q_3; p_1 \cos\lambda + p_2 \sin\lambda, -p_1 \sin\lambda + p_2 \cos\lambda, p_3)$. The flow leaves H invariant, which is equivalent to $G_2 = \text{const.}$ (**conservation of angular momentum** for a free particle).

The fact that Hamiltonian vector fields generate (locally) canonical transformations establishes a connection between the Lie and Poisson brackets.

(3.2.7) Theorem

The Lie bracket (2.5.9; 6) *of two Hamiltonian vector fields is the Hamiltonian vector field of their Poisson bracket*: $[X_H, X_G] = X_{\{G,H\}}$.

Proof: Suppose that X_H generates the local flow Ψ_t. Differentiating $\Psi_t^* X_G = X_{G\circ\Psi_t}$ by time at $t = 0$ and using (2.5.7), we obtain $L_{X_H} X_G = X_{L_{X_H} G}$. But $L_{X_H} X_G = [X_H, X_G]$, while $L_{X_H} G = \{G, H\}$. □

(3.2.8) Remarks

1. This can also be expressed as the commutativity of a diagram:

$$\begin{array}{ccc}
\mathcal{T}^1_0(T^*(M)) \times \mathcal{T}^1_0(T^*(M)) & \xrightarrow{[\ ]} & \mathcal{T}^1_0(T^*(M)) \\
\uparrow{\scriptstyle b\circ d \times b\circ d} & & \uparrow{\scriptstyle b\circ d} \\
C^\infty(T^*(M)) \times C^\infty(T^*(M)) & \xrightarrow{-\{\ \}} & C^\infty(T^*(M))
\end{array}$$

2. As with the Lie bracket, the Poisson bracket is not associative, but instead Jacobi's identity,

$$\begin{aligned}\{F, \{G, H\}\} + \{G, \{H, F\}\} + \{H, \{F, G\}\} &= L_{X_F} L_{X_G} H \\ &\quad - L_{X_G} L_{X_F} H + L_{X_{\{F,G\}}} H \\ &= (L_{[X_F, X_G]} + L_{X_{\{F,G\}}}) H = 0\end{aligned}$$

holds. Cf. (2.5.9; 6).

3. The flows that leave H invariant form a group, the center of which includes the time-evolution. If the Poisson brackets $\{K_i, K_j\}$ are not expressible in terms of the K_i, then the group must contain additional elements. More specifically, it follows from Remark 2 that the Poisson bracket of two constants of motion is itself a constant, as $\{G, H\} = \{F, H\} = 0 \Rightarrow \{H, \{F, G\}\} = 0$. For instance, in Examples (3.2.6; 2 and 3), the momentum $p_2 = \{G_1, G_2\}$ is a constant.

Let us next try to derive the generator $\bar{f}_t$ of a group of canonical transformations Φ_t from the associated generating vector field. To this end, we regard the finite transformation Φ_t as the composition of many infinitesimal transformations:

$$\Phi_t = \lim_{n\to\infty} \underbrace{\Phi_{t/n} \circ \Phi_{t/h} \circ \cdots \circ \Phi_{t/n}}_{n \text{ times}},$$

recalling the additivity of generators under composition of mappings (3.1.7; 2). If we write $\Phi_{t/n}$ as $\exp((t/n)L_{X_H})$, then we see that in the limit $n \to \infty$ it transforms Θ into $\Theta + (t/n)d(-H + i_{X_H}\Theta)$, since

$$L_{X_H}\Theta = i_{X_H}\, d\Theta + d i_{X_H}\Theta$$

and

$$dH = i_{X_H}\omega = -i_{X_H}\, d\Theta.$$

On a bundle chart,

$$i_{X_H}\Theta = \sum_i p_i (dq^i | X_H) = \sum_i p_i \frac{\partial H}{\partial p_i},$$

so for infinitesimal time t the generator $i_{X_H}\Theta - H$ turns out to be the Lagrangian $\mathcal{L}$ defined in (2.3.23). In the limit as $n \to \infty$, the sum

$$\sum_{k=0}^{n-1} \frac{t}{n} \mathcal{L} \circ \Phi_{tk/n}$$

turns into an integral, namely the action introduced in (2.3.16).

(3.2.9) The Generator of the Canonical Flow

Locally,

$$\Phi^*_{-t}\Theta = \Theta + d\bar{f}_t,$$

where in their explicit forms, Φ_t: $(q(0), p(0)) \to (q(t), p(t))$ *and*

$$\bar{f}_t = W(q(0), p(0), t) = \int_0^t dt' \, \mathcal{L}(q(t'), p(t')).$$

Explanatory Comment. Here we consider the Lagrangian as a function on $T^*(M)$, which is made possible by the diffeomorphism $T^*(M) \leftrightarrow T(M)$ (2.4.32; 3):

$$L = \sum_i p_i \frac{\partial H}{\partial p_i} - H.$$

Of course L can just as well be expressed in terms of q and $\dot{q}$; in either case, the integration is along the trajectory that passes through $(q(0), p(0))$.

(3.2.10) **Remarks**

1. This $\bar{f}_t$ is a time-dependent system of generators of the transformation $(q(0), p(0)) \to (q(t), p(t))$, for brevity written $(\bar{q}, \bar{p}) \to (q, p)$. In writing this we have treated $\bar{f}_t$ as a function of $\bar{q}$ and $\bar{p}$, but, as in (3.1.7; 2), it is more convenient to use the variables $\bar{q}$ and q. If $\det(\partial q^i / \partial \bar{p}_j^1)_{|\bar{q}} \neq 0$, then we can consider the initial momentum as a function $\bar{p}(\bar{q}, q, t)$. If we define the **action**

$$W(\bar{q}, q, t) = \bar{f}(\bar{q}, \bar{p}(\bar{q}, q, t), t),$$

then by (3.1.7; 2),

$$p_i = \frac{\partial W}{\partial q^i} \qquad \text{and} \qquad \bar{p}_i = -\frac{\partial W}{\partial \bar{q}^i}.$$

The time-dependence is affected in the following way: Taking a partial derivative, with $\bar{q}$ and q fixed, yields

$$\frac{\partial W}{\partial t} = \frac{\partial \bar{f}}{\partial t} + \frac{\partial \bar{f}}{\partial \bar{p}^j} \frac{\partial \bar{p}_i}{\partial t},$$

where $\bar{p}_i$ stands for $\bar{p}_i(\bar{q}, q, t)$. But since the initial conditions do not depend on time,

$$\frac{\partial \bar{p}_i}{\partial t} + \frac{\partial \bar{p}_i}{\partial q^j} \frac{dq^j}{dt} = 0.$$

Finally, because

$$\frac{\partial W}{\partial q^j} = \frac{\partial \bar{f}}{\partial \bar{p}_i} \frac{\partial \bar{p}_i}{\partial q^j}$$

we conclude that

$$\frac{\partial W}{\partial t} = L - p_j \frac{dq^j}{dt} = -H.$$

When written out explicitly, the action satisfies the **Hamilton–Jacobi partial differential equation** with $H(q^i, p_j)$,

$$\frac{\partial}{\partial t} W(\bar{q}, q, t) + H\left(q^i, \frac{\partial}{\partial q^j} W(\bar{q}, q, t)\right) = 0. \tag{3.2.11}$$

2. This whole treatment is as yet only local; in particular, it shows that local solutions of the Hamilton–Jacobi equations exist. Whether there exist global solutions is a problem of a much higher level of difficulty.

It is often convenient to introduce time as a dependent variable. The formalism is then changed as follows.

(3.2.12) **Definition**

We shall call $M_e \equiv M \times \mathbb{R}$ **extended configuration space** and $T^*(M_e)$ **extended phase space**. Let t and $-E$ be the coordinates of the final Cartesian factor, so that the canonical 1-form becomes $\Theta_e = p_i\, dq^i - E\, dt$ on $T^*(M_e)$. Now

$$\mathcal{H} \equiv H(p, q; t) - E \in C^\infty(T^*(M_e))$$

generates a local canonical flow (with the parameter s), for which Hamilton's equations are as in (3.2.4), and

$$\frac{dt}{ds} = 1 \quad \text{and} \quad \frac{dE}{ds} = \frac{\partial H}{\partial t}. \tag{3.2.13}$$

(3.2.14) **Remarks**

1. As always, $\mathcal{H}$ is a constant (since we do not consider the case where it depends explicitly on s), and we may restrict ourselves to the submanifold $\mathcal{H} = 0$, where $E = H$, i.e., the actual energy.
2. It is possible for H to depend explicitly on t, in which case (3.2.13) says that energy is conserved iff H is invariant under the transformation $t \to t + c$.
3. The invariance of the equations of motion under displacements in time is irrelevant for conservation of energy. For example, the equation of the damped oscillator, $\ddot{x} = -\mu\dot{x} - \omega^2 x$, is invariant under time-displacements, although its energy is not conserved, because its Hamiltonian,

$$H = e^{-\mu t}\frac{p^2}{2} + \omega^2 \frac{x^2}{2} e^{\mu t},$$

depends explicitly on t (Problem 3).
4. If a potential is turned on, so that $H = H_0 + V(q)e^{\alpha t}$, it causes a change in the energy between $t = -\infty$ and $t = 0$:

$$\delta E = \int_{-\infty}^{0} \alpha\, ds\, e^{\alpha s} V(q(s)),$$

which is the **Cesàro average** of V.

5. From equations (3.2.13), t and s are equal up to a constant; but time-dependent coordinate transformations, such as the passage to an accelerated frame of reference, are point transformations on M_e.
6. For many purposes it is desirable to choose $t \neq s$ (cf. (1.1.6) and (1.1.4)). For example, the Hamiltonian $\mathcal{H} = f(q, p)(H(q, p) - E)$ with f positive yields the equations

$$\frac{dq}{ds} = fH_{,p} + f_{,p}(H - E), \qquad \frac{dp}{ds} = -fH_{,q} - f_{,q}(H - E),$$
$$\frac{dt}{ds} = f.$$

These equations are equivalent to (2.3.21) on the invariant surface $\mathcal{H} = 0$. Thus if the canonical equations can be solved after a factor has been separated off from H, then the above equations solve the problem with another parameter in place of t—their solution gives the trajectories directly, and it only remains to integrate $dt/ds = f(q(s), p(s))$ in order to calculate the time-evolution.

(3.2.15) **Examples**

1. **Free fall.** $M = \mathbb{R}$, $M_e = \mathbb{R}^2$, and $\mathcal{H} = p^2/2 + gx - E$.

(a) One might at first consider changing to a co-moving coordinate system. The transformation

$$\Phi\colon x = \bar{x} - \frac{g}{2}\bar{t}^2, \qquad t = \bar{t},$$

is a point transformation on M_e, and with

$$T_{(\bar{x},\bar{t})}(\Phi) = \begin{vmatrix} 1 & -g\bar{t} \\ 0 & 1 \end{vmatrix}, \qquad T_{(\bar{x},\bar{t})}(\Phi)^{t-1} = \begin{vmatrix} 1 & 0 \\ g\bar{t} & 1 \end{vmatrix},$$

it induces the canonical transformation $T^*(\Phi)$:

$$(x, t; p, -E) = \left(\bar{x} - \frac{g}{2}\bar{t}^2, \bar{t}; \bar{p}, -\bar{E} + g\bar{t}\bar{p}\right):$$
$$\mathcal{H} = \frac{\bar{p}^2}{2} + g\left(\bar{x} + \bar{p}\bar{t} - \frac{g}{2}\bar{t}^2\right) - \bar{E}.$$

This indeed produces an equivalent set of equations of motion, but makes no real advance.

(b) If we wish to separate off the influence of the gravitational field g, we can make use of a canonical transformation that agrees with the above Φ on M, but acts differently on the fibers $(p, -E)$:

$$(x, t; p, -E) = \left(\bar{x} - \frac{g}{2}\bar{t}^2, \bar{t}; \bar{p} - g\bar{t}, -\bar{E} + g(\bar{t}\bar{p} - \bar{x})\right):$$
$$\mathcal{H} = \frac{\bar{p}^2}{2} - \bar{E}.$$

With these coordinates the motion is the same as free motion without gravitation.

(c) It is always possible to transform the system canonically to equilibrium, so that everything but t becomes a constant:

$$(x, t; p, -E) = \left(\bar{x} + \bar{p}\bar{t} - \frac{g}{2}\bar{t}^2, \bar{t}; \bar{p} - g\bar{t}, -\bar{E} - g\bar{x} - \frac{\bar{p}^2}{2}\right):$$

$$\mathcal{H} = -\bar{E}.$$

Then $\dot{\bar{x}} = \dot{\bar{p}} = \dot{\bar{E}} = 0$, $s = \bar{t} = t$ (cf. (2.3.12)).

2. **Rotating system.** $M = \mathbb{R}^2$, $M_e = \mathbb{R}^3$, and $\mathcal{H} = |\mathbf{p}|^2/2 + V(|\mathbf{x}|) - E$. With the canonical transformation

$$x = \bar{x}\cos\omega\bar{t} + \bar{y}\sin\omega\bar{t}, \qquad p_x = \bar{p}_x\cos\omega\bar{t} + \bar{p}_y\sin\omega\bar{t},$$
$$y = -\bar{x}\sin\omega\bar{t} + \bar{y}\cos\omega\bar{t}, \qquad p_y = -\bar{p}_x\sin\omega\bar{t} + \bar{p}_y\cos\omega\bar{t},$$
$$t = \bar{t}, \qquad E = \bar{E} - \omega(\bar{x}\bar{p}_y - \bar{y}\bar{p}_x),$$

(x and p are transformed the same way by orthogonal transformations) there results

$$\mathcal{H} = \frac{|\bar{\mathbf{p}}|^2}{2} + \omega(\bar{x}\bar{p}_y - \bar{y}\bar{p}_x) + V(|\bar{\mathbf{x}}|) - \bar{E}.$$

The extra term contains the **Coriolis** ($\sim \omega$) and **centrifugal** ($\sim \omega^2$) forces,

$$\ddot{\bar{x}} = -V_{,x} - 2\omega\dot{\bar{y}} + \omega^2\bar{x},$$

3. The situation of Example 1(c) can be formulated generally. Let a canonical transformation on $T^*(M_e)$ be given by

$$\sum_i p_i\,dq^i - E\,dt = \sum_i \bar{p}_i\,d\bar{q}^i - \bar{E}\,d\bar{t}$$
$$+ \left(f(q, \bar{p}, t) + \bar{E}(\bar{t} - t) - \sum_i \bar{p}_i\bar{q}^i\right).$$

This implies that

$$\bar{t} = t, \quad E = \bar{E} - \frac{\partial f}{\partial t}, \quad p_i = \frac{\partial f}{\partial q^i} \quad \text{and} \quad \bar{q}^i = \frac{\partial f}{\partial \bar{p}_i}.$$

Now if f satisfies the **Hamilton–Jacobi equation**

$$\frac{\partial f}{\partial t} + H\left(q, \frac{\partial f}{\partial q}, t\right) = 0, \tag{3.2.16}$$

then it follows that $\mathcal{H} = -\bar{E}$, and thus $\bar{t} = s$ and $\bar{q}$, $\bar{p}$, and $\bar{E}$ are constant. Hence equation (3.2.16) determines the generator of the transformation on $T^*(M_e)$ that always keeps the system in equilibrium.

(3.2.17) Problems

1. Using (2.3.23), calculate L in the rotating system (3.2.15; 2).

2. Check whether the transformations given in (3.2.15; 1) are canonical on extended phase space.

3. Verify Hamilton's equations for

$$H = e^{-\mu t}\frac{p^2}{2} + e^{\mu t}\omega^2\frac{x^2}{2}$$

(cf. Remark (3.2.14; 3)).

4. The transformation $t \to t - a, x \to xe^{\mu a/2}, p \to pe^{-\mu a/2}$ leaves $\mathcal{H} = H - E$ invariant, where H is as in Problem 3. Show that its generator is constant.

5. Show that the function f introduced in equation (3.1.5) also satisfies the Hamilton–Jacobi equation (3.2.11) with suitable variables.

6. Derive the equations $p = \partial W/\partial q$ and $\bar{p} = -\partial W/\partial \bar{q}$ with a variational argument, where $q \to q + \delta q$, by using equation (2.3.20).

7. Verify the calculations of (3.2.10; 1) explicitly for the action W of the harmonic oscillator, $H = (p^2 + q^2)/2$. Verify the same equations for the f defined in (3.1.5).

8. Show that $dF = i_{X_F}\omega$ by using the expression for ω on a bundle chart.

(3.2.18) Solutions

1. L can be written as

$$\tfrac{1}{2}(|\dot{\bar{\mathbf{x}}}| - [\omega \cdot \bar{\mathbf{x}}])^2 - V(|\bar{\mathbf{x}}|),$$

(where ω points in the z-direction), which also follows from direct subsitution into $L(\dot{x}, x)$.

2. Calculate the Poisson brackets.

3. $$-e^{\mu t}\omega^2 x = \dot{p} = \frac{d}{ds}(\dot{x}e^{\mu t}) = e^{\mu t}(\ddot{x} + \dot{x}\mu).$$

4. The generator is $E + (\mu/2)xp$, and

$$\dot{E} = \dot{H} = -\mu\left(e^{-\mu t}\frac{p^2}{2} - e^{\mu t}\frac{\omega^2 x^2}{2}\right) = -\frac{d}{ds}\frac{\mu}{2}xp.$$

5. With $f(\bar{p}, q, t) = W(\bar{q}(\bar{p}, q, t), q, t) + \sum_i \bar{q}^i(\bar{p}, q, t)\bar{p}_i$, we find that

$$\frac{\partial f}{\partial q^i} = p_i, \qquad \frac{\partial f}{\partial \bar{p}_i} = \bar{q}^i, \qquad \frac{\partial f}{\partial t} = -H(q, p(\bar{p}, q, t)),$$

so

$$\frac{\partial}{\partial t}f(\bar{p}, q, t) + H\left(q, \frac{\partial}{\partial q}f(\bar{p}, q, t)\right) = 0.$$

6. $W = \int_{t_0}^{t_1} dt\, L(q, \dot{q})$, and for

$$q^i(t) \to q^i(t) + \delta q^i(t), \qquad \dot{q}^i(t) \to \dot{q}^i(t) + \frac{d}{dt}\delta q^i(t),$$

$$\delta W = \delta q^i \left.\frac{\partial L}{\partial \dot{q}^i}\right|_{t_0}^{t} + \int_{t_0}^{t} dt\, \delta q^i \left(\frac{\partial L}{\partial q^i} - \frac{d}{dt}\frac{\partial L}{\partial \dot{q}^i}\right).$$

Along the trajectory, $(\cdots) = 0$, and $q^i(t_0) = \bar{q}^i$ and $q^i(t) = q^i$, and so $dW = p_i\, dq^i - \bar{p}_i\, d\bar{q}^i$.

7. $q(t) = \bar{q}\cos t + \bar{p}\sin t$, $p(t) = -\bar{q}\sin t + \bar{p}\cos t$,

$$\bar{f}(\bar{q}, \bar{p}, t) = -\bar{q}\bar{p}\sin^2 t + \frac{\sin t\cos t}{2}(\bar{p}^2 - \bar{q}^2),$$

$$\frac{\partial \bar{f}}{\partial t} = \tfrac{1}{2}(p(t)^2 - q(t)^2) = L,$$

$$W(\bar{q}, q, t) = -\bar{q}q\sin t + \frac{\sin t\cos t}{2}\bar{q}^2 + \frac{\cos t}{2\sin t}(q - \bar{q}\cos t)^2,$$

$$\frac{\partial W}{\partial q} = -\bar{q}\sin t + \frac{q - \bar{q}\cos t}{\sin t}\cos t = p,$$

$$\frac{\partial W}{\partial \bar{q}} = -q\sin t + \sin t\cos t\,\bar{q} - \frac{\cos^2 t}{\sin t}(q - \bar{q}\cos t) = -\bar{p},$$

$$\frac{\partial W}{\partial t} = -\frac{1}{2}\left(\bar{q}^2 + \left(\frac{\bar{q}\cos t - q}{\sin t}\right)^2\right) = -H,$$

$$f(q, \bar{p}, t) = q\bar{p}\cos t - \frac{\bar{p}^2}{2}\sin t\cos t - \frac{\sin t}{2\cos t}(q - \bar{p}\sin t)^2,$$

$$\frac{\partial f}{\partial q} = \bar{p}\cos t - \frac{q - \bar{p}\sin t}{\cos t}\sin t = p,$$

$$\frac{\partial f}{\partial \bar{p}} = q\cos t - \bar{p}\sin t\cos t + \frac{\sin^2 t}{\cos t}(q - \bar{p}\sin t) = \bar{q},$$

$$\frac{\partial f}{\partial t} = -\frac{1}{2}\left(\left(\frac{q - \bar{p}\sin t}{\cos t}\right)^2 + \bar{p}^2\right) = -H.$$

8. $i_{X_F}\omega = (i_{F_{,p}\partial_q} - i_{F_{,q}\partial_p})\, dq \wedge dp = F_{,p}\, dp + F_{,q}\, dq = dF.$

3.3 Constants of Motion

Constants of motion divide phase space into time-invariant submanifolds. A Hamiltonian system always has at least one constant of motion. A trajectory is completely determined by 2m − 1 constants; yet m constants are often sufficient for a solution of the problem.

As we saw with Theorem (2.3.12), all flows are locally diffeomorphic to linear fields of motion (2.3.5; 1), except at points of equilibrium. This leaves two kinds of questions open, dealing on the one hand with the behavior in the vicinity of an equilibrium position, and on the other with global characteristics of the trajectories. In this section we shall be concerned with the latter questions, and leave the former to §3.4.

We know that a trajectory cannot fill up all of M; on the contrary, it always remains on the **energy shell H** = const., which, as long as $dH \neq 0$, is a $(2m-1)$-dimensional submanifold. More generally, if r independent constants, $K_1 = H$, and $K_2, \ldots, K_r$, are known, then the motion must take place only on N: $K_i =$ const. for all i, which is a $(2m-r)$-dimensional submanifold closed in M. By "independent" we mean that for all q and p in N, the dK_i are independent vectors in $T^*_{(q,p)}(T^*(M))$ (cf. (2.1.10; 3)). If it happens that $f(K) = 0$ for some differentiable f, then they are dependent, because $0 = df = \sum_i dK_i\, \partial f / \partial K_i$.

The K_i reduce the problem by allowing the motion to be determined by the restriction of X_H to N. It was noted above (2.4.40) that it is not always possible to speak of the restriction of a vector field. It is possible in the present case, because the values of X_H lie within $T(N)$. It is intuitively clear that the streamlines of X_H must lie in N, since N is time-invariant. Formally, vectors of $T_q(N)$ can be characterized as being perpendicular to the dK_i, in the sense that the dK_i vanish on them, because the derivatives of the K_i in the direction of $T(N)$ all vanish, as the K_i are constant on N. Now, for $i = 1, \ldots, r$,

$$(dK_i | X_H) = L_{X_H} K_i = \{K_i, H\} = \frac{d}{dt} K_i = 0,$$

so $X_H(q) \in T_q(N)$. If $r = 2m - 1$, then N is precisely the trajectory. When $m = 1$, it suffices to write $H = p^2 + V(q)$ in order to solve for the trajectory without further integration.

(3.3.1) **Examples**

1. $M = \mathbb{R}$, $T^*(M) = \mathbb{R} \times \mathbb{R}$, and $V = q^2$.
2. $M = \mathbb{R}$, $T^*(M) = \mathbb{R} \times \mathbb{R}$, and $V = -q^2 + q^4$.
3. $M = \mathbb{R}^+$, $T^*(M) = \mathbb{R}^+ \times \mathbb{R}$, and $V = 1/q$.
4. $M = \mathbb{R}^+$, $T^*(M) = \mathbb{R}^+ \times \mathbb{R}$, and $V = -1/q + 1/q^2$.

(3.3.2) **Remarks**

1. The equilibrium positions (where $dH = 0$) are zero-dimensional trajectories. In Examples 1, 2, 3, and 4 there are, respectively, 1, 3, 0, and 1 such points.
2. The one-dimensional trajectories of Example 1 are all diffeomorphic to T^1, and those of Example 3 are diffeomorphic to $\mathbb{R}$. In Example 2 there are two trajectories diffeomorphic to $\mathbb{R}$, namely the ones where $H = 0$, and the rest are diffeomorphic to T^1. In Example 4 there are infinitely many of both kinds.

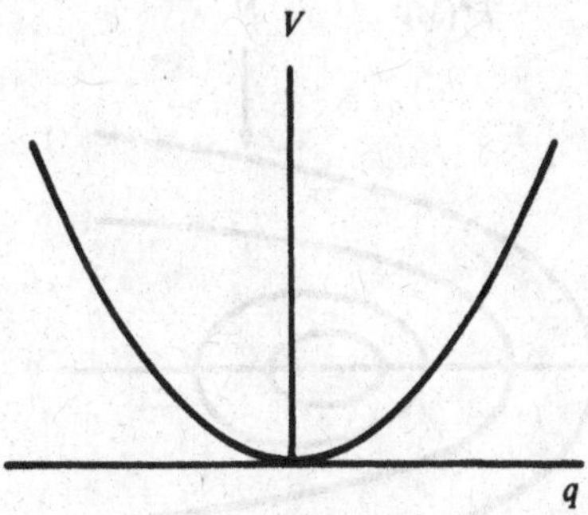

FIGURE 3.1. Harmonic oscillator potential.

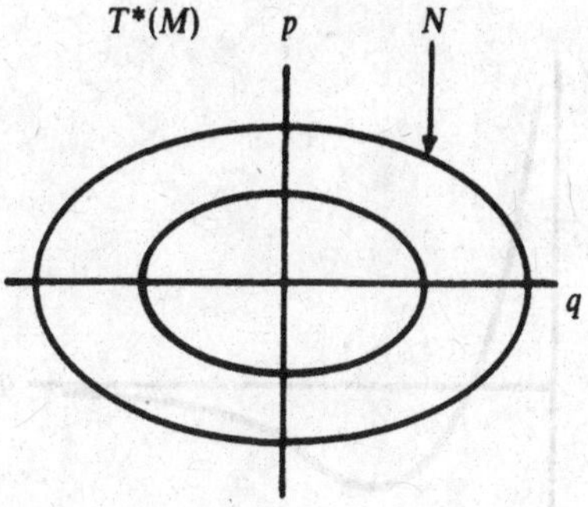

FIGURE 3.2. The trajectories of the harmonic oscillator.

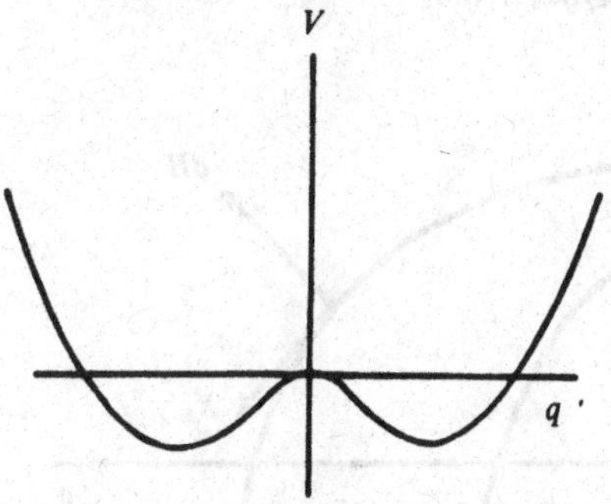

FIGURE 3.3. Potential with two wells.

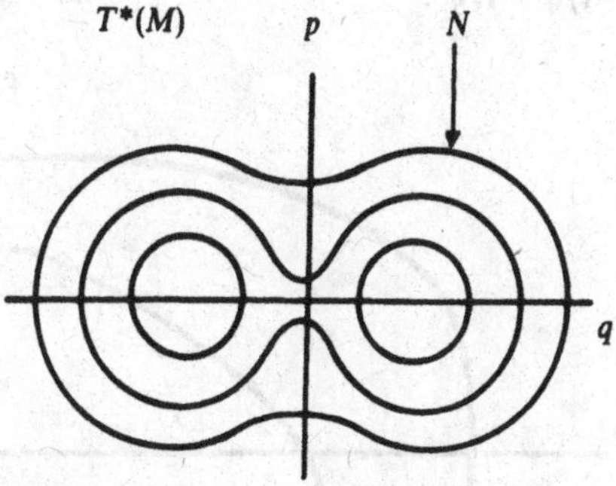

FIGURE 3.4. The trajectories of the two-well potential.

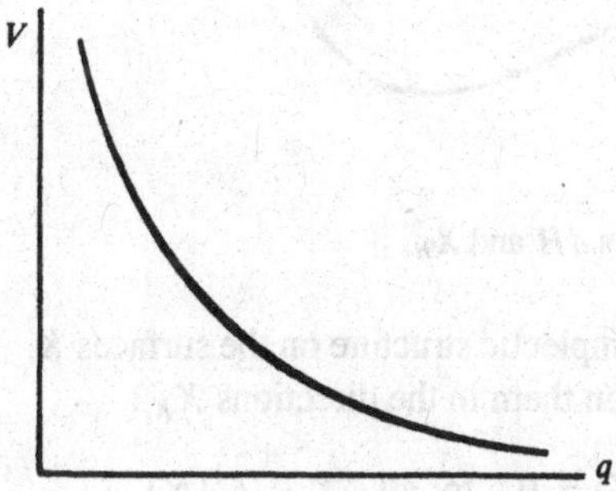

FIGURE 3.5. A $1/q$ potential.

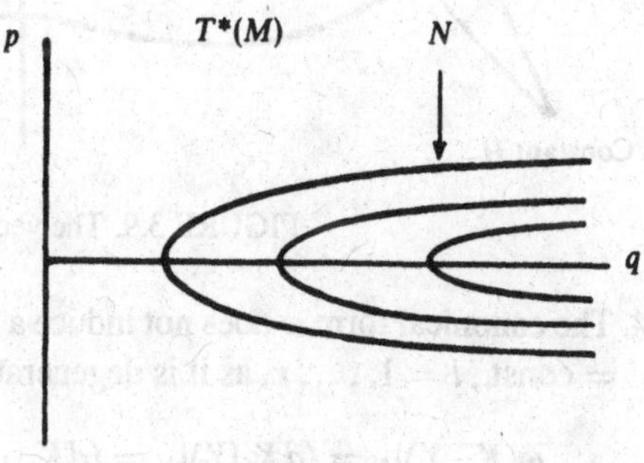

FIGURE 3.6. Trajectories of the $1/q$ potential.

3. The restriction of X_H is not simply $b \cdot$ (the restriction of dH), which would make it zero. If $T^*(\mathbb{R}) = \mathbb{R}^2$, then because $(dH|X_H) = 0$, dH and X_H are as in Figure 3.9.

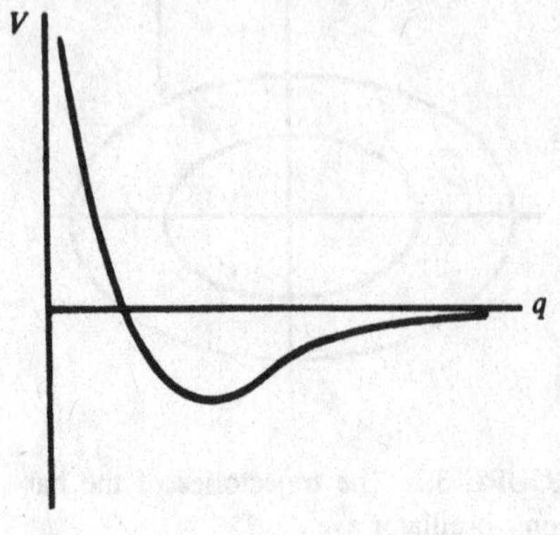

FIGURE 3.7. Potential of the form $-1/q + 1/q^2$.

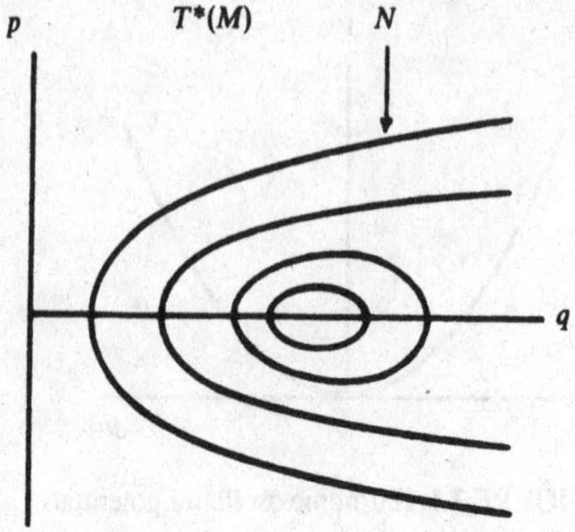

FIGURE 3.8. Trajectories of the potential $-1/q + 1/q^2$.

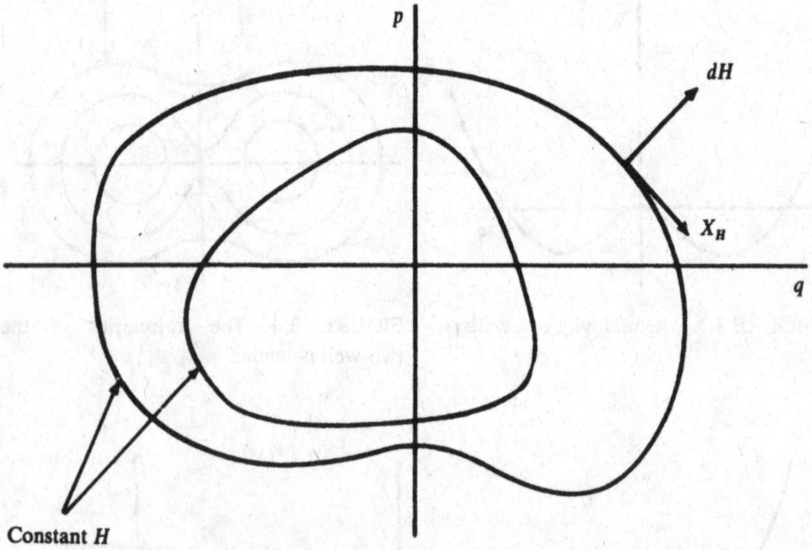

FIGURE 3.9. The vectors dH and X_H.

4. The canonical form ω does not induce a symplectic structure on the surfaces $K_i = \text{const.}$, $i = 1, \ldots, r$, as it is degenerate on them in the directions X_{K_i}:

$$\omega(K_i, Y)|_N = (dK_i|Y)|_N = (dK_i|_N|Y) = 0 \quad \text{for all} \quad Y \in \mathcal{T}_0^1(N).$$

It follows that $X_{K_i|N}$ are not canonical vector fields. An invariant volume form Ω_N can however be defined on N by $\Omega = dK_1 \wedge \cdots \wedge dK_r \wedge \Omega_N$.

Recall that the question being posed here differs from that of (2.5.6; 5). There the issue was whether differential forms $\nu_i \in \mathcal{N}^\perp$ locally form surfaces, which are invariant under the flows associated with $X \in \mathcal{N}$. Here that question is trivially answered in the affirmative, since $\nu_i = dK_i$ are exact. Regardless of that, the local question is answered by Theorem (2.3.12); locally, there exist even one-dimensional manifolds such that $X(q) \in T_q(N)$ for all $q \in N$. The crucial problem here is the global one, and it will be answered if the K_i are defined globally and are independent.

It should be emphasized again that it is essential that the K_i be defined on all of M, as locally it is always possible to find $2m - 1$ time-independent constants of motion (2.3.12). For most problems, the local constants cannot be extended continuously to all of M and do not define a closed $(2m - 1)$-dimensional submanifold. On compact manifolds Hamiltonian systems with additional global constants are exceptional [18]. It then happens that the trajectory is dense in a submanifold of dimension > 1, as the following rather typical case shows.

(3.3.3) **Lemma**

Let Φ_t be the flow generated by $H = \frac{1}{2}(p_1^2 + p_2^2 + \omega_1^2 q_1^2 + \omega_2^2 q_2^2)$ on

$$(T^*(\mathbb{R})\backslash\{0\}) \times (T^*(\mathbb{R})\backslash\{0\}).$$

The functions

$$K_i := \frac{\omega_i^2}{2} q_i^2 + \frac{p_i^2}{2}, \qquad i = 1, 2,$$

are constant, i.e., $\Phi_t^* K_i = K_i$, and are independent on this manifold. If the frequencies ω_i have a rational ratio, then all trajectories are submanifolds diffeomorphic to T^1. If their ratio is irrational, then every trajectory is dense in some two-dimensional submanifold defined by the K_i.

Proof: Map $(\mathbb{R}^2\backslash\{0\}) \times (\mathbb{R}^2\backslash\{0\})$ onto $\mathbb{R}^+ \times \mathbb{R}^+ \times T^1 \times T^1$ with the transformation $(q_i, p_i) = (\sqrt{K_i/\omega_i} \sin\varphi_i, \sqrt{K_i\omega_i} \cos\varphi_i)$, $i = 1, 2$ (cf. (3.1.8; 1)). On this chart the time-evolution is given by

$$\Phi_t: (K_1, K_2, \varphi_1, \varphi_2) \to (K_1, K_2, \varphi_1 + \omega_1 t, \varphi_2 + \omega_2 t)$$

(cf. (3.2.6; 1)). Let $\Psi_n \equiv \Phi_{2\pi n/\omega_1}$, $n \in \mathbb{Z}$, and consider its restriction to the last T^1 factor; on the other factor, $\Psi_n = 1$.

1. Suppose $\omega_1/\omega_2 = g_1/g_2$, where $g_i \in \mathbb{Z}$. Then $\Psi_{g_1} = \mathbf{1}$, and to each value of φ_1 on the trajectory there correspond g_2 values of φ_2. (See figure.)

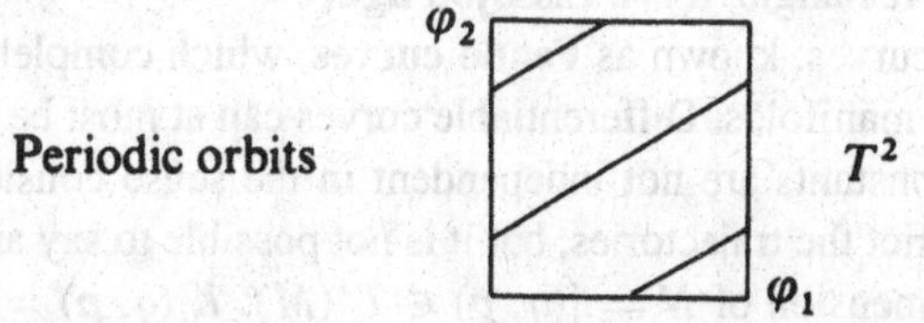

cf. (2.3.14). Trajectories like these, which return to their initial points (in $T^*(M)$), are called **periodic, or closed, orbits**. They are closed submanifolds diffeomorphic to T^1.

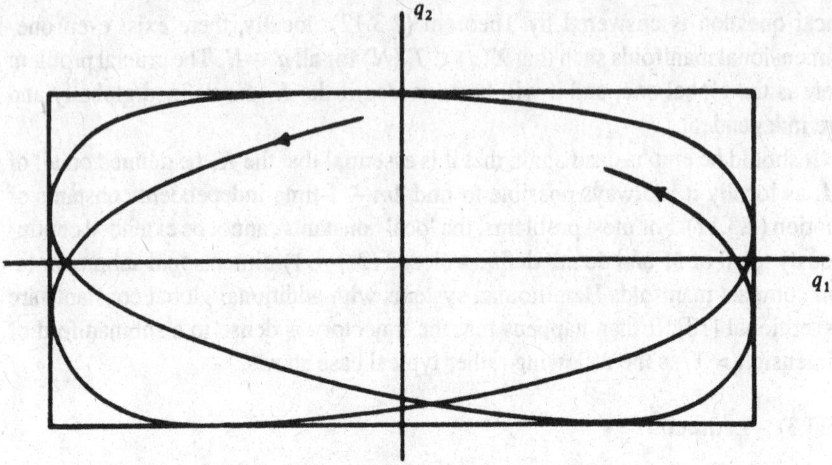

FIGURE 3.10. Quasi-periodic orbits.

2. Suppose ω_1/ω_2 is irrational. Then there is no value g_1 other than 0 for which $\Psi_{g_1} = \mathbf{1}$. Since T^1 is compact, there must be a point of accumulation; i.e., $\forall \varepsilon > 0$, there exist integers g_1 and g_2 such that

$$|\Psi_{g_1}(\varphi_2) - \Psi_{g_2}(\varphi_2)| < \varepsilon,$$

and thus $\Psi_{g_1-g_2}(\varphi_2) = \varphi_2 + \eta$, where $|\eta| < \varepsilon$. Therefore the set

$$\{\Psi_{g(g_1-g_2)}(\varphi_2),\, g \in \mathbb{Z}\}$$

fills T^1 with points that are only some small ε apart. This means that for all φ_1 the points attained by the trajectory are dense in the second factor. Since the trajectory obviously takes on every value of φ_1, it is dense in T^2. Such trajectories are called **quasi-periodic orbits.** □

(3.3.4) **Remarks**

1. This lemma can be generalized to the case of the n-dimensional harmonic oscillator by iterating the proof.
2. Projected onto configuration space (q_1, q_2) for ω_1/ω_2 irrational, the trajectory is dense in a rectangle; it is a **Lissajou figure.**
3. There exist curves, known as **Peano curves**, which completely fill up higher-dimensional manifolds. Differentiable curves can at most be dense in them.
4. When the constants are not independent in the sense considered above, they can still restrict the trajectories, but it is not possible to say anything in general about the dimension of $N = \{(q, p) \in T^*(M): K_i(q, p) = \alpha_i \in \mathbb{R}\}$, or even whether it is a manifold. Recall Example (3.3.1; 1). If $H = 0$, then $dH = 0$ also, and N is a point; thus a single constant reduces the dimension by 2. In (3.3.1; 2) the energy surface $H = E < 0$ divides into two pieces. By choosing

$f(q) \in C^\infty(T^*(M))$ such that f equals different constants on the two pieces, and then multiplying by $g(H) \in C^\infty(T^*(M))$ such that $g = 0$ for $H > E$, we can produce a constant that is not a function of H alone and that forces the trajectory to stay in one part of the energy surface. The constant is not independent of H in our sense, because its differential is proportional to dH. (Of course, this is only on an energy surface $H = E < 0$.) In this case two constants only reduce the dimension by one. And lastly, in Example (3.3.1; 2) the set where $H = 0$ is not a manifold at all; near the origin it has the structure of Example (2.1.7; 6).

5. As mentioned in Example (2.3.14), when $\omega_1/\omega_2 = g_1/g_2$, where the g_i are integers, there exists a constant, $\sin(2\pi(\varphi_1 g_2 - \varphi_2 g_1))$. One might suspect that whenever the trajectories remain restricted to a submanifold there are always additional constants. That this is not generally true is shown by the following non-Hamiltonian example: $M = \mathbb{R}$, $X \in T_0^1(M): x \to (x, x)$; for the time-evolution, $\dot{x} = x \Rightarrow x(t) = x(0)e^t$. There are three trajectories, $\mathbb{R}^-$, 0, and $\mathbb{R}^+$, each of which is a submanifold of M, but none of which is dense. Yet there is no time-independent constant in $C^\infty(M)$; it would have to be constant on the trajectories, and therefore, as a continuous function, on all $\mathbb{R}$. Consequently, it would have a vanishing differential.
6. Suppose that M is an open set of $\mathbb{R}^{2m}$ on which the $2m$ independent coordinate functions z_k, $k = 1, 2, \ldots, 2m$, are defined globally as in (2.1.7; 1). If there exists a function $J \in C^\infty(T^*(M))$ that increases sufficiently fast along the trajectories (specifically, for all $z \in T^*(M)$, $\exists c > 0$: $(d/dt)J \circ \Phi_t(z) > c$ $\forall t$), then the trajectories are one-dimensional submanifolds. In this case there are necessarily $2m - 1$ independent constants of motion $\in C^\infty(T^*(M))$ (see Problem 6). On the extended phase space $T^*(M_e)$ with the independent variable s, the time-coordinate t satisfies the equation $dt/ds = 1$ (cf. (3.2.13)) and plays the rôle of the function J. Thus if $T^*(M_e)$ is an open subset of $\mathbb{R}^{2m+2}$, then every system has $2m + 1$ independent constants of motion on $T^*(M_e)$. In particular, even an ergodic system (i.e., any trajectory is dense on the energy shell) can be considered as a subsystem of a system with one more degree of freedom and the maximal number of constants of motion. Hence the existence of constants does not imply that the motion is simple. The projection of a (one-dimensional) trajectory onto a subsystem may be quite complicated.

The existence of the time-average

$$f_\infty := \lim_{T\to\infty} \frac{1}{T} \int_0^T dt\, \tau_t f$$

was alluded to in (2.6.13). Then

$$\tau_t f_\infty = \lim_{T\to\infty} \frac{1}{T} \int_t^{t+T} dt'\, \tau_{t'} f = f_\infty \quad \forall f \in C^\infty(T^*(M)),$$

and thus f_∞ is constant in time. If it is also spatially constant for all f on $T^*(M)$, that means that with the passage of time the points of any arbitrarily small neighborhood travel throughout the whole manifold. One might hope to always find a nonconstant function f_∞ on the energy shell N whenever the trajectory does not fill N densely. This line of reasoning fails, however, because there is no guarantee that f_∞ is continuous, even for $f \in C^\infty(T^*(M))$.

(3.3.5) Example

$$M = T^2, \qquad T^*(M) = T^2 \times \mathbb{R}^2 \quad \text{and} \quad H = p_1^2/2 + p_2^2/2.$$

The time-evolution is $(\varphi_1, \varphi_2; p_1, p_2) \to (\varphi_1 + p_1 t, \varphi_2 + p_2 t; p_1, p_2)$. If

$$f = g(\varphi_1) \cdot g(\varphi_2), \qquad g(\varphi) = \begin{cases} e^{-1/\varphi - 1/(\varphi - 2\pi)} & \text{for } \varphi \neq 0, \\ 0 & \text{for } \varphi = 0, \end{cases}$$

then

$$f_\infty(\varphi_1, \varphi_2) = \lim_{T\to\infty} \frac{1}{T} \int_0^T dt\, f(\varphi_1 + p_1 t, \varphi_2 + p_2 t)$$
$$= \frac{1}{4\pi^2} \int_0^{2\pi} d\varphi_1\, d\varphi_2\, g(\varphi_1) g(\varphi_2),$$

when p_1/p_2 is irrational; but on the other hand

$$f_\infty = \frac{1}{2\pi} \int_0^{2\pi} d\varphi\, |g(\varphi)|^2$$

when $p_1 = p_2$ and $\varphi_1 = \varphi_2$. Hence f_∞ is quite discontinuous. Constants of this sort can always be found; it is only necessary to assign arbitrary numbers to the trajectories. However, these constants are uninteresting because they do not define submanifolds.

Global generators of groups of canonical transformations that leave H invariant are constants of the motion. To each parameter of the group there corresponds one generator, yet the generators are not generally independent of H. For example, in (3.2.6; 2), p_1, p_2, and p_3 are constant, and $H = |\mathbf{p}|^2$. The generators themselves are certainly linearly independent, but may depend on each other algebraically.

(3.3.6) Example

The m-dimensional oscillator: $M = \mathbb{R}^m$, $m > 1$, and $H = \frac{1}{2}\sum_{i=1}^m (p_i^2 + x_i^2)$. The functions $M_{ik} = p_i p_k + x_k x_i$ and $L_{ik} = p_i x_k - p_k x_i$, where $i, k = 1, \ldots, m$, are constant, as is easily verified. $\{M_{ik}, H\} = \{L_{ik}, H\} = 0$. The Poisson brackets $\{\ \}$ of the M's and L's can be expressed as linear combinations of M's and L's, and provide complete vector fields, which generate a group of canonical transformations that is isomorphic to U_m. There are $(m(m+1)/2)$ M's and $(m(m-1)/2)$ L's, a total of m^2 generators. There is no way that they could all be algebraically independent, as phase space has only $2m$ dimensions. For example, if $m = 2$, then $2H = M_{11} + M_{22}$, and $M_{12}^2 + L_{12}^2 = M_{11} \cdot M_{22}$.

(3.3.7) **Remarks**

1. The group mentioned in this example is far from the largest group that leaves H invariant. The largest such group is generated by the functions $K \in C^\infty(T^*(M))$ for which $\{K, H\} = 0$ and X_K is complete. It does not depend on only a finite number of parameters, and, consequently, it is not even locally compact. Even in the trivial example, $M = \mathbb{R}$, $H = p^2$, the functions $f(p)$, where $f \in C^\infty(\mathbb{R})$, generate the groups $(x, p) \to (x + \lambda f'(p), p)$, which are different unless the f's differ only by a constant. All together, the largest group that leaves H invariant has infinitely many generators, which of course are not all independent.
2. We can just as well pose the opposite question, of what group gives the greatest number of constants with the fewest parameters. Although there can be no more than $2m - 1$ independent constants K_i, it may happen that their Poisson brackets are not expressible as linear combinations of the K_i. In that case, any group that gives all the constants must have more than $2m - 1$ parameters. It can also happen that the minimal group is not unique. In the above example, all the minimal groups have one parameter, and the groups generated by $f(p) = (p^2 + c^2)^{1/2}$, $c \in \mathbb{R}$, are equally good for all $c \neq 0$.

It is not often that one is lucky enough to find $2m - 1$ constants, but it often suffices to find m of them. This situation occurs frequently enough that it is given a name of its own:

(3.3.8) **Definition**

A Hamiltonian system is said to be **integrable on U** iff there exist $m \equiv \dim M$ functions K_i on a time-invariant neighborhood $U \subset T^*(M)$ such that:

(a) $\{K_i, H\} = 0$;

(b) $\{K_i, K_j\} = 0$; and

(c) the dK_i are linearly independent on U.

(3.3.9) **Remarks**

1. The Hamiltonian H and the K_i depend on one another.
2. It is very common to find treatments of integrable systems; in fact most books on mechanics, including this one, are basically catalogs of them. This can lead to the wrong opinion that most systems are integrable on some U that is dense in $T^*(M)$. It is in fact exceptional for such cases to occur, and they are only popular because they are soluble.

The first interesting fact about integrable systems is that the K_i can be used as new coordinates:

(3.3.10) **Theorem (Liouville)**

Consider an integrable system. For all $(q, p) \in U$:

(a) *there exist* $U_1 \subset U$ *with* $(q, p) \in U_1$ *and* $\varphi_i \in C^\infty(U_1)$, $i = 1, \ldots, m$, *such that* $\{K_i, K_j\} = \{\varphi_i, \varphi_j\} = 0$ *and* $\{K_i, \varphi_j\} = \delta_{ij}$ *on* U_1; *and*

(b) *all other sets of variables that satisfy the relationships in* (a) *are of the form*

$$\bar{\varphi}_i = \varphi_i + \frac{\partial \mathcal{X}(K)}{\partial K_i}, \qquad \mathcal{X} \in C^\infty(\mathbb{R}^m).$$

Proof: According to (2.5.11), there exist coordinates $(\tilde{\varphi}_1, \ldots, \tilde{\varphi}_m; K_i, \ldots, K_m)$ for which X_{K_j} generate the flows $\tilde{\varphi}_j \to \tilde{\varphi}_j + \tau_j$, $K_i \to K_i$:

$$\{\tilde{\varphi}_i, K_j\} = L_{X_{K_j}} \tilde{\varphi}_i = \delta_{ij}.$$

In these coordinates, ω is of the form $d\tilde{\varphi}_j \wedge dK_j + F_{ij}\, dK_i \wedge dK_j$, since terms $\sim d\tilde{\varphi}_i \wedge d\tilde{\varphi}_j$ drop out because

$$0 = \{K_j, K_i\} = i_{X_{K_i}} i_{X_{K_j}} \omega = i_{\partial/\partial\tilde{\varphi}_i} i_{\partial/\partial\tilde{\varphi}_j} \omega.$$

Since X_{K_j} generate canonical flows, and hence ω is invariant under the transformation $\tilde{\varphi}_j \to \tilde{\varphi}_j + \tau_j$, the F_{ij} must depend only on the K variables. Moreover, as ω and $d\tilde{\varphi}_j \wedge dK_j$ are exact, $F := F_{ij}\, dK_i \wedge dK_j$ must be exact as well. Therefore $F = df$, where $f = f_j(K)\, dK_j$. Then

$$\omega = d(\tilde{\varphi}_j + f_j(K)) \wedge dK_j = d\varphi_j \wedge dK_j, \quad \text{where } \varphi_j = \tilde{\varphi}_j + f_j(K).$$

This means that if the origin of the coordinates φ on the surfaces $K_j =$ const. is gauged appropriately, then they satisfy (a).

Statement (b) follows because f is only determined up to the addition of an exact form $d\chi$. □

(3.3.11) **Corollary**

By (b), *the canonical transformation corresponding to time-evolution must be of the form*

$$(K_i, \varphi_j) \to \left(K_i, \varphi_j + \frac{\partial}{\partial K_j} \mathcal{X}(t, K)\right).$$

The group property (2.3.7) *implies that*

$$\frac{\partial}{\partial K_j} \mathcal{X}(t_1 + t_2, K) = \frac{\partial}{\partial K_j} \mathcal{X}(t_1, K) + \frac{\partial}{\partial K_j} \mathcal{X}(t_2, K).$$

Since $\mathcal{X}$ *depends continuously on* t *and any contribution to* $\mathcal{X}$ *that is independent of the* K_i *is irrelevant, the time-evolution is*

$$(K_i(t), \varphi_j(t)) = \left(K_i(0), \varphi_j(0) + t \frac{\partial}{\partial K_j} H(K)\right).$$

Because of this, there is again locally a linear field of motion (2.3.5; 1) on N_α. However, among other problems the time-independent constants $\varphi_i\, \partial H/\partial K_j - \varphi_j\, \partial H/\partial K_i$ cannot generally be extended to all of N_α and they do not restrict the motion to a manifold of dimension less than m. Until now all the statements we have made about the motion have only been local, and they contain no information that might answer global questions. Thus we have not progressed much beyond Theorem (2.3.12). With fairly harmless additional assumptions, though, some light is cast on the global structure by

(3.3.12) Theorem (Arnold)

Suppose that for an integrable system on $T^*(M)$

$$N_\alpha = \{(q,p) \in U,\ K_i(q,p) = \alpha_i \in K_i(U) \subset \mathbb{R}\}$$

is compact and connected. Then N_α is diffeomorphic to the torus T^m.

Proof: As we have already seen, on N_α all the $X_{K_i} \in T_0^1(N_\alpha)$. By assumption,

$$L_{X_{K_i}} L_{X_{K_j}} = L_{X_{K_j}} L_{X_{K_i}},$$

and the X_{K_i} are complete since N_α is compact (2.3.6; 2). Consequently,

$$\exp\left(\sum_j \tau_j L_{X_{K_j}}\right), \qquad (\tau_j) \in \mathbb{R}^m,$$

generates an m-parameter group of diffeomorphisms Φ_τ of N_α, and the mapping

$$\Phi: \mathbb{R}^m \to N, \qquad (\tau) \to \Phi_\tau(z), \qquad z \in N \text{ fixed},$$

is locally a diffeomorphism (cf. (2.5.11)). In the vicinity of $\mathbf{0} \in \mathbb{R}^m$, Φ is

$$\tau_j \to z_j + \tau_i (dz_j | X_{K_i}) + 0(\tau^2)$$

(understanding the z_i as coordinates on some chart), and the matrix of derivatives is thus composed of the components of the X_{K_i}, as a result of which it is nonsingular (see [1, 16.5.6]). The group property transfers this to all $(\tau) \in \mathbb{R}^m$; the image of Φ in N_α is both open and closed, and therefore all of N_α. However, Φ is not injective. Because $\Phi_{\tau+\tau'} = \Phi_\tau \circ \Phi_{\tau'}$, the stabilizer $G \equiv \{\tau: \Phi(\tau)z = z\}$ is a subgroup of the additive group $\mathbb{R}^m$, and the mapping of the factor group $\mathbb{R}^m/G$ to N_α is a diffeomorphism [1, 16.10.8]. The discrete subgroups of $\mathbb{R}^m$ are lattices; $\mathbb{R}^m/G$ is diffeomorphic to $T^r \times \mathbb{R}^{m-r}$ for some r, $0 \le r \le m$, and since N_α was assumed to be compact, r must equal m. □

(3.3.13) Remarks

1. Since the trajectory always remains in a connected component of N_α, it is no real restriction to consider only connected N_α's (cf. (3.3.1; 2) for $H < 0$).

2. If N_α is not compact, then it is necessary to add the requirement that all the X_{K_i} are complete. In that case, the above argument shows that N_α is diffeomorphic to some $T^r \times \mathbb{R}^{m-r}$, $0 \le r \le m$.
3. The foregoing considerations remain valid for nonfixed K_i; the K_i may be allowed to vary in a sufficiently small neighborhood $U_1 \subset \mathbb{R}^m$. Then (K_i, φ_j) are locally canonical coordinates for $U_1 \times T^m$, which is a time-invariant submanifold of the U of (3.3.8). If the N_α are compact, connected, and cover all of U, then U becomes a bundle with base U_1 and fibers T^m. It is, however, not necessarily trivial. Under some circumstances, it is not possible to define $(K_i, {}^{\cos}_{\sin}\varphi_j)$ globally on U.

If N_α is compact, then it can be parametrized in accordance with the standard charts of T^1, and it is possible to define a normal form for the coordinates for (3.3.10).

(3.3.14) Definition

Let N_α be diffeomorphic to T^M and C_j a curve in N_α that encircles the jth torus, and whose part on the other tori is continuously contractible to a point. The curve C_j is thus homotopic to $(0 \times 0 \times \cdots \times \underbrace{T}_{j\text{th position}} \times \cdots \times 0)$. Then we call

$$I_j(\alpha) := \frac{1}{2\pi}\int_{C_j} \Theta$$

an **action variable**. If $\det(\partial I_j/\partial \alpha_k) \neq 0$, then the α's can be expressed locally in terms of the I's, and the generator

$$S := \int_{q_0}^{q} \Theta$$

can be treated as a function of the I's and the old coordinates q. This transforms q^i and $p_i = \partial S/\partial q^i$ into I_j and $\varphi_j := \partial S/\partial I_j$. The φ_j are known as the **angle variables** belonging to the I_j.

(3.3.15) Remarks

1. Since the path of integration lies within the N_α, the I_j depend on α_i. It must be established, however, that I_j has the same value for all C_j with the same specifications. In particular, this shows the constancy of the I_j in time. The change in I_j between two times is

$$\frac{1}{2\pi}\int_F d\Theta,$$

where C_j and C_j' compose a surface F, which lies in N_α (see Figure 3.11). But $d\Theta = 0$ on N_α, and hence

$$I_j - I_j' = \frac{1}{2\pi}\int_{\partial F} \Theta = \frac{1}{2\pi}\int_F d\Theta = 0.$$

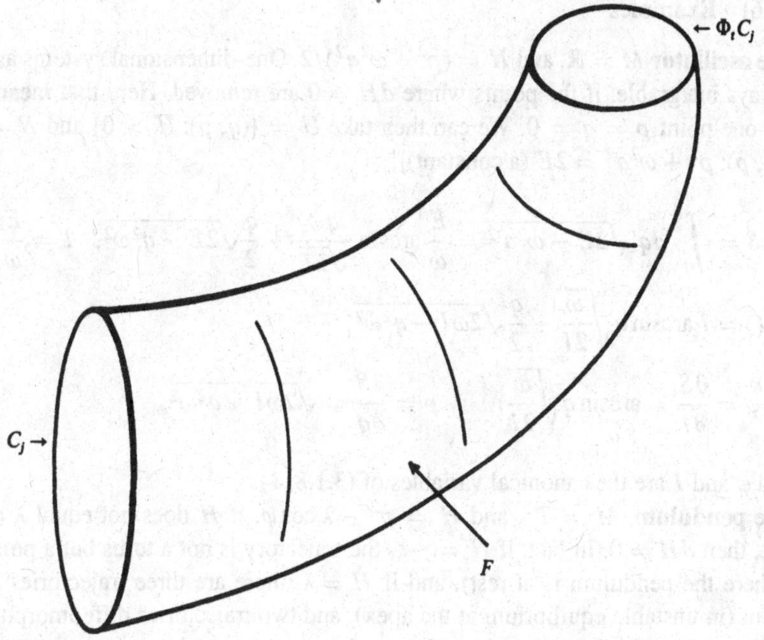

FIGURE 3.11. The paths of integration in the definition of I_j.

2. With the same argument it is possible to show that S is independent of the path, provided that it remains in N_α, so that for fixed q_0 it is a function only of q and I. Nonetheless, S is not defined globally, as it changes by precisely $I_j \cdot 2\pi$ after a circuit of C_j.
3. It follows likewise that the angle variables $\varphi_i = \partial S/\partial I_i$ fail to be defined globally, but instead change by $2\pi\, \partial I_j/\partial I_i = 2\pi\delta_i^j$ after a circuit of C_j. This means that they are the standard angle variables for T^m. The arbitrariness in the choice of q_0 amounts to changing $\varphi_i \to \varphi_i + \partial f(I)/\partial I_j$, i.e., the change of gauge allowed by Liouville's Theorem (3.3.10; 2).
4. Since I and φ are canonical coordinates, the time-evolution is

$$(I_j(0), \varphi_j(0)) \to \left(I_j(0), \varphi_j(0) + t\frac{\partial H(I)}{\partial I_j} \right),$$

according to (3.3.11). The frequencies $\partial H(I)/\partial I_j$ depend continuously on the I's and as a rule do not always have rational ratios, which means that, usually, the trajectory covers N densely.
5. The variables (I, φ) are global if the domain of this chart is a $2m$-dimensional, time-invariant submanifold. The domain of the chart cannot necessarily be extended to all of $T^*(M)$; in the first place, by (3.3.8) U need not be all of $T^*(M)$, and, secondly, U need not be a trivial T^m bundle.

(3.3.16) Examples

1. The **oscillator** $M = \mathbb{R}$, and $H = (p^2 + \omega^2 q^2)/2$. One-dimensional systems are always integrable, if the points where $dH = 0$ are removed. Here that means the one point $p = q = 0$. We can then take $U = \{(q, p): H > 0\}$ and $N = \{(q, p): p^2 + \omega^2 q^2 = 2E \text{ (a constant)}\}$;

$$S = \int_0^q dq' \sqrt{2E - \omega^2 q'^2} = \frac{E}{\omega} \arcsin \frac{q\omega}{\sqrt{2E}} + \frac{q}{2}\sqrt{2E - q^2\omega^2}, \quad I = \frac{E}{\omega},$$

$$S(q, I) = I \arcsin q \sqrt{\frac{\omega}{2I}} + \frac{q}{2}\sqrt{2\omega I - q^2\omega^2},$$

$$\varphi = \frac{\partial S}{\partial I} = \arcsin q \sqrt{\frac{\omega}{2I}}, \qquad p = \frac{\partial S}{\partial q} = \sqrt{2\omega I - q^2 \omega^2},$$

and φ and I are the canonical variables of (3.1.8; 1).

2. The **pendulum.** $M = T^1$, and $H = p^2 - \lambda \cos\varphi$. If H does not equal λ or $-\lambda$, then $dH \neq 0$. In fact, if $H = -\lambda$, the trajectory is not a torus but a point (where the pendulum is at rest), and if $H = \lambda$, there are three trajectories: a point (in unstable equilibrium at the apex), and two trajectories diffeomorphic to $\mathbb{R}$ (asymptotically approaching the apex).

$$I(E) = \frac{1}{2\pi} \oint d\varphi \sqrt{E + \lambda \cos\varphi},$$

$$\frac{\partial I}{\partial E} = \frac{1}{\omega(E)} = \frac{1}{2\pi} \oint \frac{d\varphi}{2\sqrt{E + \lambda\cos\varphi}}.$$

If $E \gg \lambda$, the potential energy makes little difference, and the trajectories are roughly $p = \text{const}$. For $-\lambda < E < \lambda$, the trajectory returns to its starting point when $\cos\varphi_m = -E/\lambda$ (Figure 3.12), so the integral $\oint d\varphi$ runs only between $-\varphi_m$ and φ_m. $\omega(E)$ is an elliptic integral.

3. **Small oscillations.** $M = \mathbb{R}^m$, and

$$H = \sum_{i=1}^{m} \frac{p_i^2}{2m_i} + V(x).$$

Suppose that V has an equilibrium point, which we take as the origin of the coordinate system, so $dV(0) = 0$. Now replace V with the first three terms of its Taylor expansion,

$$V(x) \to V(0) + \tfrac{1}{2} x_i x_k V_{,ik}(0),$$

though as yet we cannot tell how valid this replacement is (see §3.5). In any case, the kinetic and potential energy are turned into quadratic forms, and the system becomes integrable. The matrices of the quadratic forms do not commute, so they cannot be simultaneously diagonalized by an orthogonal matrix. But we

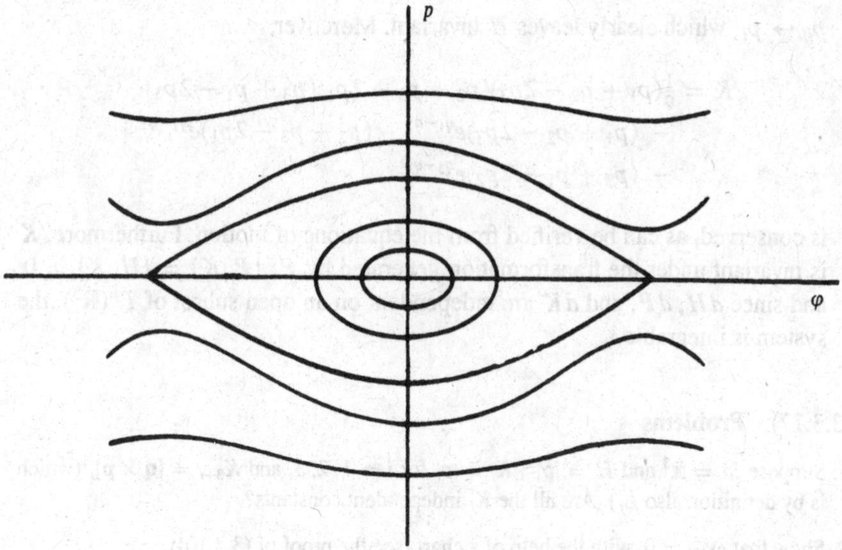

FIGURE 3.12. The trajectories of a pendulum.

can put H in the form

$$H = \sum_{i=1}^{m} \frac{p_i^2}{2} + \frac{1}{2}\sum_{i,k} x_i x_k v_{ik},$$
$$v_{ik} = V_{,ik}(0)/\sqrt{m_i m_k},$$

with the point transformation $p_i \rightarrow p_i\sqrt{m_i}$, $x_i \rightarrow x_i/\sqrt{m_i}$, and then diagonalize v: $(M^t v M)_{ik} = \delta_{ik} v_k$, where $M^t M = 1$. In the canonical coordinates $(\bar{x}, \bar{p})$: $x_i = M_{ik}\bar{x}_k$, and $p_i = M_{ik}\bar{p}_k$,

$$H = \frac{1}{2}\sum_i (\bar{p}_i^2 + v_i \bar{q}^{i2}),$$

and the m constants we are looking for are $\bar{p}_i^2 + v_i(\bar{q}^i)^2$. The N_α are diffeomorphic to $T^r \times \mathbb{R}^{m-r}$, where r is the number of the v_i that are positive. For stable equilibrium ($r = m$ and $V_{,ik}$ is a positive matrix), there are action and angle variables which can be constructed as in Example 1.

4. The "**Toda Molecule**." In problems with several particles, a replacement of harmonic potentials with other functions generally destroys the integrability of the system. But sometimes a miracle occurs and one actually finds additional constants. The following model of a linear molecule with three identical particles:

$$H = \tfrac{1}{2}(p_1^2 + p_2^2 + p_3^2) + e^{q_1 - q_2} + e^{q_2 - q_3} + e^{q_3 - q_1}.$$

has, in addition to H, another constant, the momentum of the center of mass, $P = p_1 + p_2 + p_3$. This constant generates the transformation $q^i \rightarrow q^i + \lambda$,

$p_i \to p_i$, which clearly leaves H invariant. Moreover,

$$\begin{aligned} K = \tfrac{1}{9}&(p_1 + p_2 - 2p_3)(p_2 + p_3 - 2p_1)(p_3 + p_1 - 2p_2) \\ &- (p_1 + p_2 - 2p_3)e^{q_1 - q_2} - (p_2 + p_3 - 2p_1)e^{q_2 - q_3} \\ &- (p_3 + p_1 - 2p_2)e^{q_3 - q_1} \end{aligned}$$

is conserved, as can be verified from the equations of motion. Furthermore, K is invariant under the transformation generated by P; $\{P, K\} = \{H, K\} = 0$, and since dH, dP, and dK are independent on an open subset of $T^*(\mathbb{R}^3)$, the system is integrable.

(3.3.17) Problems

1. Suppose $M = \mathbb{R}^3$ and $H = |\mathbf{p}|^2$; $K_i = p_i$ for $i = 1, 2, 3$, and $K_{3+i} = [\mathbf{q} \times \mathbf{p}]_i$ (which is by definition also L_i). Are all the K_i independent constants?

2. Show that $\omega_{|N_\alpha} = 0$, with the help of a chart (see the proof of (3.3.10)).

3. Show that the harmonic oscillator with a periodic external force,

$$\mathcal{H} = \tfrac{1}{2}(p^2 + q^2) + \lambda q \cos \omega t - E, \qquad \omega \neq 1,$$

is an integrable system in extended phase space $\mathbb{R}^4$.

4. Show that K in (3.3.16; 4) is a constant.

5. Calculate the frequency of vibration of the H_2O molecule, in one dimension and linearized:

$$H = \frac{1}{2m}(p_1^2 + p_3^2) + \frac{1}{2M}p_2^2 + \frac{K}{2}((q_2 - q_1)^2 + (q_3 - q_2)^2).$$

What are the normal modes like?

6. Let the function $J \in C^\infty(T^*(M))$ be such that $\forall z \in T^*M\ \exists c > 0$ with $(d/dt)J \circ \Phi_t(z) > c$. Suppose there exist $2m$ independent functions $\bar{z}_k$ on $T^*(M)$ (e.g., the coordinate functions if $T^*(M)$ is an open set of $\mathbb{R}^{2m}$), and construct $2m - 1$ independent constants of the motion. (*Hint*: To each trajectory assign its initial value.)

(3.3.18) Solutions

1. No; otherwise every trajectory would be a point. From $(\mathbf{L} \cdot \mathbf{p}) = 0$ it follows that $\mathbf{L} \cdot d\mathbf{p} + \mathbf{p} \cdot d\mathbf{L} = 0$.

2. Let A_q and A_p be the $(m \times m)$ matrices of partial derivatives of the K_i by the q^j and, respectively, the p_j. Choose the coordinates so that $\det A_p \neq 0$, and hence it is locally possible to write $p_j(q, K)$. Let P_q be the partial derivative of p_i by q^j, with K fixed. Then $A_p P_q + A_q = 0$. The vanishing of the Poisson bracket implies that $A_p A_q^t = A_q A_p^t$. Thus $A_p P_q A_p^t + A_p A_q^t = A_p(P_q A_p^t + A_q^t) = 0$. Since A_p^{-1} exists, we conclude that $A_p P_q^t + A_q = 0$, and so $P_q^t = P_q$. This is exactly the condition that $\sum p_i(q, \alpha)\, dq^i$ is closed on N_α.

3. In addition to $\mathcal{H}$ there is the constant

$$K = ((\omega^2 - 1)q - \lambda \cos \omega t)^2 + ((\omega^2 - 1)p + \lambda\omega \sin \omega t)^2;$$

$$\frac{dK}{ds} = 2[(\omega^2 - 1)p + \lambda\omega \sin \omega t][(\omega^2 - 1)q - \lambda \cos \omega t - (\omega^2 - 1)(q + \lambda \cos \omega t) + \lambda\omega^2 \cos \omega t] = 0.$$

Here N is not compact, but instead is diffeomorphic to $T^1 \times \mathbb{R}$.

4. Periodic: $q_{i+2} = q_{i-1}$, etc., and

$$\begin{aligned} \frac{d}{dt}\frac{1}{9}\prod_{i=1}^{3}(p_{i+1} + p_{i-1} - 2p_i) &= \frac{1}{3}\sum_{i=1}^{3}(p_{i+1} + p_{i-1} - 2p_i)(p_{i-1} + p_i - 2p_{i+1})(e^{q_{i+1}-q_{i-1}} - e^{q_{i-1}-q_i}) \\ &= -\sum_{i=1}^{3}(p_{i+1} + p_{i-1} - 2p_i)(p_{i-1} - p_{i+1})e^{q_{i+1}-q_{i-1}} \\ &= \frac{d}{dt}\sum_{i=1}^{3}(p_{i+1} + p_{i-1} - 2p_i)e^{q_{i+1}-q_{i-1}}. \end{aligned}$$

5.

$$v_{ik} = K \begin{vmatrix} \frac{1}{m} & -\frac{1}{\sqrt{mM}} & 0 \\ -\frac{1}{\sqrt{mM}} & \frac{2}{M} & -\frac{1}{\sqrt{mM}} \\ 0 & -\frac{1}{\sqrt{mM}} & \frac{1}{m} \end{vmatrix}$$

$$\det(v - 1 \cdot \omega^2) = \omega^2 \left(\frac{K}{m} - \omega^2\right)\left(\omega^2 - \frac{K}{m} - \frac{2K}{M}\right),$$

Frequency	$\omega = 0$	$\omega = \sqrt{\frac{K}{m}}$	$\omega = \sqrt{\frac{K}{m} + \frac{2K}{M}}$
	H O H	H O H	H O H
mode	∘ ∘ ∘	∘ ∘ ∘	∘ ∘ ∘
	→ → →	← ∘ →	→ ← →

6. By assumption, the mapping $\mathbb{R} \times T^*(M) \to \mathbb{R}$ given by $(t, z) \to J \circ \Phi_t(z)$ is strongly monotonic in t for each fixed $z \in T^*(M)$, and therefore invertible. The inverse image of, say, $0 \in \mathbb{R}$ under this mapping assigns a unique value of t to each $z \in T^*(M)$. Letting this function $T^*(M) \to \mathbb{R}$ be called τ, we find that $\tau \circ \Phi_t(z) = \tau(z) - t \; \forall z \in T^*(M)$, since $0 = J \circ \Phi_{\tau(z)}(z) = J \circ \Phi_{\tau(z)-t}(\Phi_t(z))$. Hence the mapping $T^*(M) \to T^*(M)$: $z \to \Phi_{\tau(z)}(z)$ is time-invariant: $\Phi_{\tau(\Phi_t(z))}(\Phi_t(z)) = \Phi_{\tau(z)} \circ \Phi_{-t} \circ \Phi_t(z) = \Phi_{\tau(z)}(z)$. Composing this with the coordinate functions $\bar{z}_k$, $k = 1, \ldots, 2m$, produces the $2m$ constants $\bar{z}_k \circ \Phi_{\tau(z)}(z)$. Define $\tilde{z}(t, z) := \bar{z} \circ \Phi_t(z)$, $\tilde{z}_{i,k} := \partial\tilde{z}_i/\partial z_k$, and $\dot{\tilde{z}}_i := \partial\tilde{z}_i/\partial t$; then the differentials of the constants are

$$d(\bar{z}_k \circ \Phi_{\tau(z)}(z)) := d\tilde{z}_k(\tau(z), z) = \left(\tilde{z}_{k,i} + \dot{\tilde{z}}_k \frac{\partial \tau}{\partial z_i}\right) dz_i.$$

Since Φ_t is a diffeomorphism, the matrix $(\partial \tilde{z}_k/\partial z_i)$ has rank $2m$. The matrix $\dot{\tilde{z}}_k(\partial\tau/\partial z_i)$ has rank 1, so the sum has at least rank $2m-1$. The condition $J \circ \Phi_{\tau(z)}(z) = 0$ implies that the rank is in fact equal to $2m-1$, which is then the number of independent constants.

3.4 The Limit $t \to \pm\infty$

Often the time-evolution of a system approaches that of an integrable system asymptotically. If so, its behavior after long times can be discovered.

Theoretical predictions usually become less precise for longer times, and the future of a system as $t \to \pm\infty$ may be wholly unknown. However, if the potential is of finite range, then particles that escape eventually act like free particles and their time-evolution becomes simple. As we shall see, on the part of phase space filled by the trajectories of escape, Φ_t is diffeomorphic to Φ_t^0, free time-evolution, and there are $2m-1$ constants of motion.

We start by looking for quantities which are not necessarily constant, but approach limiting values.

(3.4.1) **Definition**

Let

$$\begin{aligned} \mathcal{A} &:= \{f \in C^\infty(T^*(M))\text{: the pointwise limit } \lim_{t\to\pm\infty} \tau_t f \text{ exists and} \\ &\in C^\infty(T^*(M))\} := \{\text{the } \textbf{asymptotic constants} \text{ of the motion}\}; \\ \{H\}' &:= \{f \in C^\infty(T^*(M))\text{: } \tau_t f = f\} := \{\text{the } \textbf{constants} \text{ of the motion}\}; \end{aligned}$$

and

$$\tau_\pm\colon \mathcal{A} \to \{H\}', \quad \text{such that} \quad f \to \lim_{t\to\pm\infty} \tau_t f.$$

(3.4.2) **Remarks**

1. Since τ_t commutes with the algebraic operations, $\mathcal{A}$ and $\{H\}'$ are algebras, and $\tau_\pm$ are homomorphisms.
2. Timewise limits are constants in time, and since $\tau_{\pm|\{H\}'} = \mathbf{1}$, $\tau_\pm$ are mappings from $\mathcal{A}$ onto $\{H\}'$.
3. It is not necessary for $\tau_\pm$ to be injective; $\{H\}'$ may be a proper subset of $\mathcal{A}$.
4. $\mathcal{A}$ (or its quantum-mechanical generalization) is of especial interest in atomic physics, where only the asymptotic parts of the trajectories can be measured directly. The **deflection angle** of the particles is given by the difference between $\tau_-\mathbf{p}$ and $\tau_+\mathbf{p}$.

(3.4.3) **Examples**

1. $M = T^1$ and $H = \omega p$ (an oscillator). Since the time-evolution is given by $(\varphi, p) \to (\varphi + \omega t, p)$, a function $f(\varphi, p) \in C^\infty(T^*(T^1))$ is an asymptotic

constant of the motion iff f depends only on p: this case is trivial, as $\mathcal{A} = \{H\}'$ and $\tau_\pm = \mathbf{1}$.

2. $M = \mathbb{R}^+$ and $H = p^2/2 + \gamma/r^2$, $\gamma > 0$. This system is integrable, because $p = \dot{r} = \sqrt{2(E - \gamma/r^2)}$; therefore

$$t = \int \frac{dr\, r}{\sqrt{2Er^2 - 2\gamma}} = \frac{1}{2E}\sqrt{2Er^2 - 2\gamma},$$

so

$$r = \sqrt{\frac{\gamma}{E} + 2Et^2},$$

and

$$p = t\sqrt{2E}\left(\sqrt{\frac{\gamma}{2E^2} + t^2}\right)^{-1}.$$

Thus $r(t)$ is a hyperbola, and the trajectories $H =$ const. are asymptotically horizontal in phase space (see Figure 3.13). If we express E in terms of the initial values (r, p) we obtain the time-evolution

$$\Phi_t : \begin{pmatrix} r \\ p \end{pmatrix} \to$$

$$\begin{pmatrix} \left[p^2 + \frac{2\gamma}{r^2}\right]^{-1/2}\left[2\gamma + \left(t\left(p^2 + \frac{2\gamma}{r^2}\right) + rp\right)^2\right]^{1/2} \\ \left[p^2 + \frac{2\gamma}{r^2}\right]^{1/2}\left[2\gamma + \left(t\left(p^2 + \frac{2\gamma}{r^2}\right) + rp\right)^2\right]^{-1/2}\left(t\left(p^2 + \frac{2\gamma}{r^2}\right) + rp\right) \end{pmatrix}.$$

Observe that

$$p \overset{t\to\pm\infty}{\longrightarrow} \pm\sqrt{p^2 + \frac{2\gamma}{r^2}} \quad \text{and} \quad \frac{1}{r} \to 0,$$

so

$$\mathcal{A} = \left\{ f\left(p, \frac{1}{r}\right), f \in C^\infty(\mathbb{R} \times (0, \infty)) \text{ such that the limit exists as } \frac{1}{r} \to 0 \right\},$$

$$\tau_\pm : f\left(p, \frac{1}{r}\right) \to f\left(\pm\sqrt{p^2 + \frac{2\gamma}{r^2}}, 0\right).$$

This time $\{H\}'$ is just $\{f(H)\}$, and is a proper subset of $\mathcal{A}$.

3. $M = \mathbb{R}^2 \setminus \{\mathbf{0}\}$ and $H = p^2/2 + \alpha/r^2$. The point transformation $(x, y) = (r\cos\varphi, r\sin\varphi)$ generates a canonical transformation $(x, y; p_x, p_y) \to (r, \varphi; p_r, L)$, by which H becomes $p_r^2/2 + \gamma/r^2$, where $\gamma = \alpha + L^2/2$. Consequently, the radial motion is as in Example 2, and for $\gamma > 0$ the equation $\dot{\varphi} = \partial H/\partial L = L/r^2$ can be integrated by substituting $r(t)$ in, yielding

$$\Phi_t : (\varphi, L) \to \left(\varphi + \frac{L}{\sqrt{2\alpha + L^2}}\left[\arctan \frac{rp_r + t\left(p_r^2 + \frac{L^2 + 2\alpha}{r^2}\right)}{\sqrt{2\alpha + L^2}} - \arctan \frac{rp_r}{\sqrt{2\alpha + L^2}}\right], L\right).$$

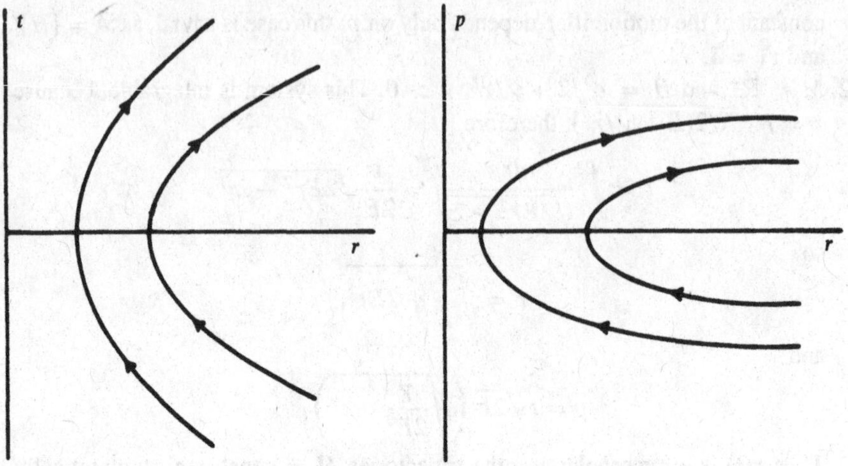

FIGURE 3.13. Trajectories of a $1/r^2$ potential.

The functions $f(\varphi, L)$ now also belong to $\mathcal{A}$, because angular momentum is conserved, and the particle escapes at a definite angle:

$$\tau_\pm f(\varphi, L) = f\left(\varphi + \frac{L}{\sqrt{2\alpha + L^2}}\left[\pm\frac{\pi}{2} - \arctan\frac{rp_r}{\sqrt{2\alpha + L^2}}\right], L\right)$$

(see Figure 3.14). Note that in this case a third constant independent of H and L also appears,

$$\tau_+ p_x = \sqrt{2H}\cos\left(\varphi + \frac{L}{\sqrt{2\alpha + L^2}}\left[\frac{\pi}{2} - \arctan\frac{rp_r}{\sqrt{2\alpha + L^2}}\right]\right).$$

In physics the connection between the observables as $t \to \pm\infty$ is quite important, and one would like to know what the mapping $\tau_+ \circ \tau_-^{-1}$ is. Unfortunately, τ_-^{-1} is not uniquely defined, since τ_- is not injective. One can get around this problem by choosing a subalgebra of $\mathcal{A}$ on which $\tau_\pm$ are injective. The mapping of the asymptotic quantities then depends on what subalgebra has been chosen; if the subalgebra is $\{H\}'$, for instance, then $\tau_+ \circ \tau_-^{-1} = \mathbf{1}$. If the time-evolution Φ_t asymptotically approaches that of a simple reference system Φ_t^0, then as in (2.3.11; 3) it is possible to construct a limiting diffeomorphism, which reproduces the action of the τ's on a subalgebra, and makes them invertible.

(3.4.4) Definition

Let H and H_0 be two Hamiltonians that generate the flows Φ_t and Φ_t^0 on $T^*(M)$. If:

(a) $\lim_{t\to\pm\infty} \Phi_{-t} \circ \Phi_t^0 =: \Omega_\pm$ exist pointwise on some open sets $D_\pm$;

(b) $\Omega_\pm$ are local canonical transformations from $D_\pm$ onto neighborhoods $\mathcal{R}_\pm$; and

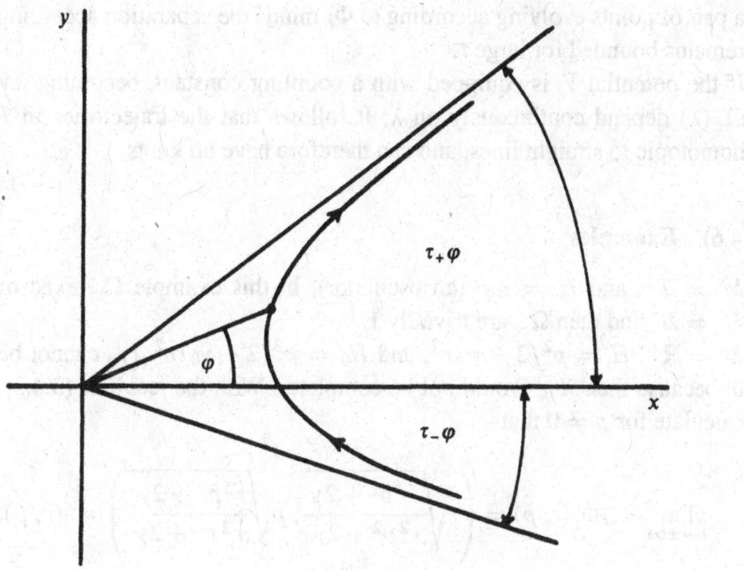

FIGURE 3.14. The trajectories in a plane.

(c) $\lim_{t\to\pm\infty} \Phi^0_{-t} \circ \Phi_t$ exist on $\mathcal{R}_\pm$ and equal $\Omega_\pm^{-1}$,

then we say that the **Møller-transformations** $\Omega_\pm$ exist.

(3.4.5) **Remarks**

1. From (2.3.11; 3), $\Phi_t \circ \Omega_\pm = \Omega_\pm \circ \Phi^0_t$. According to (2.5.9; 2), $L_{\Omega_{\pm *}, X_{H_0}} = L_{X_H}$, and therefore $H_0 \circ \Omega_\pm^{-1} = H_{|\mathcal{R}_\pm}$ (only up to a constant, of course). What is more, $\forall t_0$, $\Phi_{-t} \circ \Phi^0_t$ also converges on $\Phi^0_{t_0} D_\pm$, so we may assume that $D_\pm$ are Φ^0_t-invariant, making $\mathcal{R}_\pm$ invariant under Φ_t. Then the flow that H creates on $\mathcal{R}_\pm$ is diffeomorphic to the flow that H_0 creates on $D_\pm$.
2. For an observable f, $f \circ \Omega_\pm^{-1} \circ \Phi_t \circ \Omega_\pm = f \circ \Phi^0_t$ implies that the image under $\Omega_{\pm *}$ of the time-evolution according to H_0 is the same as the time-evolution according to H of the image: $\tau_t(\Omega_{\pm *} f) = \Omega_{\pm *}(\tau^0_t f)$. Thus $\{H_0\}'$ gets mapped into $\{H\}'$. If H_0 has $2m - 1$ independent constants (as for free motion), then so does H on $\mathcal{R}_\pm$. In particular, such a system is integrable on $\mathcal{R}_\pm$. Theorem (2.3.12) then becomes global in the sense that $\mathcal{R}_\pm$ are time-invariant, although the neighborhood U constructed in the proof of (2.3.12) need not be.
3. The transformations that $\Omega_\pm$ generate on $\{H_0\}'$ are exactly $\tau_\pm$, since $\forall f \in \{H_0\}'$, $f \circ \Phi^0_{-t} \circ \Phi_t = f \circ \Phi_t \to f \circ \Omega_\pm^{-1}$ as $t \to \pm\infty$, i.e., $\tau_\pm(f) = \Omega_{\pm *}(f)$. Hence $\mathcal{A}$ contains $\Omega_{\pm *}\{H_0\}'$, and is usually larger than $\{H\}'$.
4. If H and H_0 are invariant under $T: (x, p) \to (x, -p)$ (**reversal of the motion**, which is not a canonical transformation), then $\Phi_t \circ T = T \circ \Phi_{-t}$ and $\Phi^0_t \circ T = T \circ \Phi^0_{-t}$; so the existence of Ω_+ on D_+ implies that of Ω_- on $T(D_+)$.
5. Points that are initially close together in phase space can spread far apart after long times. The fact that $\Omega_\pm$ are diffeomorphisms means that the separation of

a pair of points evolving according to Φ_t minus the separation according to Φ_t^0 remains bounded for large t.

6. If the potential V is equipped with a coupling constant, becoming λV, then $\Omega_\pm(\lambda)$ depend continuously on λ. It follows that the trajectories in $\mathcal{R}_\pm$ are homotopic to straight lines, and can therefore have no knots.

(3.4.6) Examples

1. $M = T^1$, and $H = \omega p$ (an oscillator). In this example $\Omega_\pm$ exist only for $H_0 = H$, and then $\Omega_\pm$ are trivially $\mathbf{1}$.
2. $M = \mathbb{R}^+$, $H = p^2/2 + \gamma/r^2$, and $H_0 = p^2/2 + \gamma_0/r^2$. ($\gamma_0$ cannot be set to 0, because then X_{H_0} would not be complete.) With the result of (3.4.3; 2), we calculate for $p \neq 0$ that

$$\lim_{t\to\pm\infty} \Phi_{-t}\Phi_t^0(r,p) = \left(r\sqrt{\frac{r^2p^2+2\gamma}{r^2p^2+2\gamma_0}}, p\sqrt{\frac{r^2p^2+2\gamma_0}{r^2p^2+2\gamma}} \right) =: (\bar{r}, \bar{p}).$$

In fact, $(r, p) \to (\bar{r}, \bar{p})$ is a canonical transformation (Problem 3), so that $H_0 = H(\bar{r}, \bar{p})$; that is, $H_0 = H \circ \Omega_\pm$. The domains of $\Omega_\pm$ and their ranges are $D_\pm = \mathcal{R}_\pm = T^*(\mathbb{R}^+)$. All H's of this form produce diffeomorphic flows for all $\gamma > 0$, and $\Phi_{-t}^0 \circ \Phi_t$ always converges to $\Omega_\pm^{-1}$. So in this example, the Møller transformation exists and is different from $\mathbf{1}$.
3. $M = \mathbb{R}^2\backslash\{\mathbf{0}\}$, $H = |\mathbf{p}|^2/2 + \alpha/|\mathbf{x}|^2$, and $H_0 = |\mathbf{p}|^2/2$. Using polar coordinates as in (3.4.3; 3), the radial problem reduces to Example 2, and for the angles we find

$$\Omega_\pm(\varphi, L) = \left(\varphi + \arctan\frac{rp_r}{L} - \frac{L}{\sqrt{2\alpha+L^2}} \arctan\frac{rp_r}{\sqrt{2\alpha+L^2}} \right.$$
$$\left. \mp \frac{\pi}{2}\left(\frac{L}{\sqrt{2\alpha+L^2}} - 1\right), L \right).$$

It is not hard to convince oneself that the $\Omega_\pm$ transform H_0 canonically into H. $D_\pm$ and $\mathcal{R}_\pm$ are $T^*(M^2)\backslash\{\{\mathbf{0}\} \times M^2\}$, and the Møller transformations exist, and can be extended to all of $T^*(M^2)$ and satisfy (a)–(c). $\mathbf{p} \in \{H_0\}'$, and in fact $\mathbf{p} \in \mathcal{A}$.

Some of the properties of the above example hold for a wider class of potentials:

(3.4.7) Theorem

Let $M = \mathbb{R}^m$, $H = |\mathbf{p}|^2/2 + V(\mathbf{x})$, where $V \in C_0^\infty(\mathbb{R}^m)$, and $H_0 = |\mathbf{p}|^2/2$. Then:

(a) $\exists\Omega_\pm$, $D_\pm = T^*(\mathbb{R}^m)\backslash\{\mathbb{R}^m \times \{0\}\}$; *and*

(b) $C\mathcal{R}_\pm = \bigcup_n b_n^\pm$, where $b_n^\pm = \{z \equiv (\mathbf{x}, \mathbf{p}) \in T^*(\mathbb{R}^m): \|\pi_1 \circ \Phi_{\pm t} z\| < n$ for all $t > 0$, or $\lim_{t\to\pm\infty} \|\pi_2 \circ \Phi_{\pm t} z\| = 0\}$.

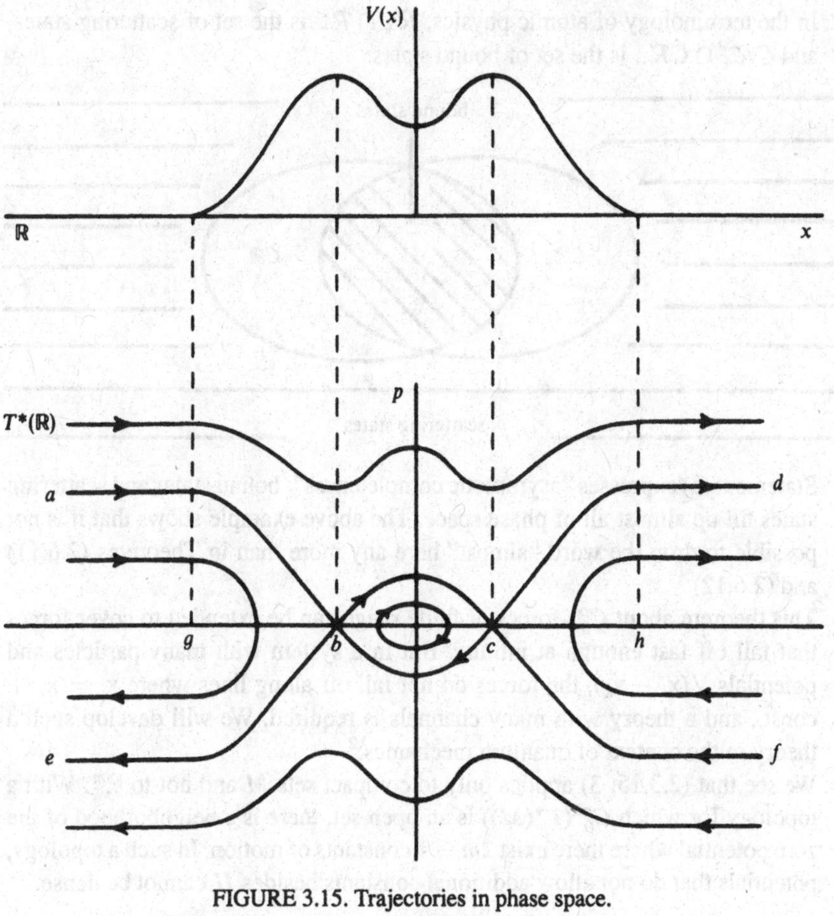

FIGURE 3.15. Trajectories in phase space.

Here π_1 *is the projection* $(\mathbf{x}, \mathbf{p}) \to \mathbf{x}$, π_2 *is the projection* $(\mathbf{x}, \mathbf{p}) \to \mathbf{p}$, *and* $\| \;\|$ *denotes the norm on* $\mathbb{R}^m$.

(c) $\Omega(\mathcal{R}_+ \Delta \mathcal{R}_-) = 0$ Δ *is the symmetric difference and* Ω *is the* **Liouville measure.**

(3.4.8) Remarks

1. The significance of (b) is that $\mathcal{R}_\pm$ are the complements of the trajectories that remain in compact sets for all $t \lessgtr 0$. Yet $\mathcal{R}_\pm$ need not be the same as $\mathcal{R}_-$. To see that, consider the following one-dimensional example: $C\mathcal{R}_+ \cap C\mathcal{R}_- =$ (a closed set bounded by the trajectories bc and cb) $\cup \{(-\infty, g) \times \{\mathbf{0}\}\} \cup \{(h, \infty) \times \{\mathbf{0}\}\}$. In addition, $C\mathcal{R}_+$ (resp. $C\mathcal{R}_-$) contains the trajectories ab and fc (resp. be and cd).

2. In the terminology of atomic physics, $\mathcal{R}_+ \cap \mathcal{R}_-$ is the set of scattering states, and $C\mathcal{R}_+ \cap C\mathcal{R}_-$ is the set of bound states:

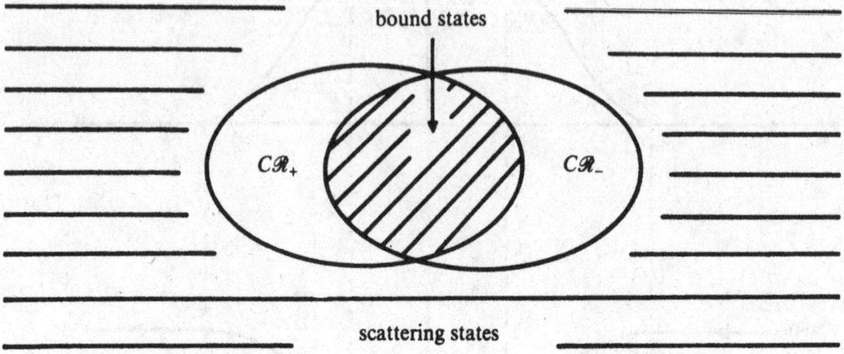

Statement (c) expresses "asymptotic completeness": bound states and scattering states fill up almost all of phase space. The above example shows that it is not possible to drop the word "almost" here any more than in Theorems (2.6.11) and (2.6.12).

3. This theorem about C_0^∞-forces of finite range can be extended to cover forces that fall off fast enough at infinity. But in a system with many particles and potentials $V(\mathbf{x}_i - \mathbf{x}_k)$, the forces do not fall off along lines where $\mathbf{x}_i = \mathbf{x}_k +$ const., and a theory with many channels is required. We will develop such a theory in the context of quantum mechanics.[2]
4. We see that (2.3.15; 3) applies only to compact sets M and not to $\mathbb{R}^m$. With a topology for which $C_0^\infty(T^*(M))$ is an open set, there is a neighborhood of the zero potential where there exist $2m-1$ constants of motion. In such a topology, potentials that do not allow additional constants besides H cannot be dense.

Proof:

(a) $\Phi_t^0 : (\mathbf{x}, \mathbf{p}) \to (\mathbf{x}+\mathbf{p}t, \mathbf{p})$, and if $V(\mathbf{x}) = 0\ \forall \|\mathbf{x}\| > \rho$, then $\Phi_t^0 = \Phi_t\ \forall \|\mathbf{x}\| > \rho$. Hence $\forall (\mathbf{x}, \mathbf{p}) \in D_+$ there exists some T such that $\Phi_{-t} \circ \Phi_{T+t}^0(\mathbf{x}, \mathbf{p}) = \Phi_T^0(\mathbf{x}, \mathbf{p})\ \forall t > 0$. Consequently, $\Phi_{-T-t} \circ \Phi_{T+t}^0(\mathbf{x}, \mathbf{p}) = \Phi_{-T} \circ \Phi_T^0(\mathbf{x}, \mathbf{p})$ $\forall t > 0$, and so for all points of D_+ the limit is reached after a finite time, and is $\Omega_+ = \Phi_{-T} \circ \Phi_T^0$. To understand why the limit is also a diffeomorphism, note that for any compact subset K of D_+ there exists a T such that $\Omega_{+|K} = \Phi_{-T} \circ \Phi_{T|K}^0$, and that $\Phi_{-t} \circ \Phi_t^0$ is a diffeomorphism for all t (X_H and X_{H_0} are certainly complete). Similarly for Ω_- and $\Omega_\pm^{-1}$.

(b) $\forall z \in \mathcal{R}_+, \exists z_0 \in D_+ : z = \Phi_{-t} \circ \Phi_t^0 z_0\ \forall t > T$, which $\Leftrightarrow \Phi_t z = \Phi_t^0 z_0 = (\mathbf{x}_0 + \mathbf{p}_0 t, \mathbf{p}_0)$ for all $t > T$. Hence the spatial part gets arbitrarily large while the momentum remains nonzero, so $\mathcal{R}_+$ lies in the complement of $\bigcup_n b_n^+$. Similarly for $\mathcal{R}_-$. Conversely, if $V = 0$ for all z with $\|\pi_1 z\| > n$

[2] *Quantum Mechanics of Atoms and Molecules.*

and $\|\pi_1 \Phi_t z\| > n$, and if $\pi_2 \Phi_t z \neq 0$ for all $t > T$, then there exists a z_0 such that

$$\Phi_{T+\tau} z = \Phi_\tau^0 \circ \Phi_T z = \Phi_{\tau+T}^0 z_0 \qquad \text{for all } \tau > 0.$$

Hence $z = \Omega_+ z_0$, $z \in \mathcal{R}_+$.

(c) Let $b_n^\pm = \{(\mathbf{x}, \mathbf{p}) \in T^*(\mathbb{R}^m): \|\Phi_{\pm t} \mathbf{x}\| < n \ \forall t > 0\}$. By Theorem (2.6.12), $\Omega(b_n^+ \cap b_n^-) = \Omega(b_n^+) = \Omega(b_n^-) \ \forall n \in \mathbb{Z}^+$, so $\Omega(b_n^\pm \cap Cb_n^\mp) = 0$: because $C\mathcal{R}_\pm = \bigcup_n b_n^\pm$, and noting the monotonicity of $b_n^\pm$ in n, we conclude that

$$\Omega(C\mathcal{R}_+ \cap \mathcal{R}_-) = \lim_{n\to\infty} \Omega(b_n^+ \cap \mathcal{R}_-) \leq \lim_{n\to\infty} \Omega(b_n^+ \cap Cb_n^-) = 0,$$

and hence $\Omega(\mathcal{R}_+ \Delta \mathcal{R}_-) = 0$. □

If only the asymptotic parts of the trajectories are observed, one would like to know the relationships between them. For that purpose we make

(3.4.9) **Definition**

Suppose that $\Omega_\pm$ exist and $\mathcal{R}_+ = \mathcal{R}_-$. Then the local canonical transformation $S = \Omega_+^{-1} \circ \Omega_-$ from D_- to D_+ is called the **scattering transformation.**

(3.4.10) **Remarks**

1. Since S may be written as $\lim_{T\to\infty} \Phi^0_{-T} \circ \Phi_{2T} \circ \Phi^0_{-T}$, it has the following simple interpretation: Proceed along the free trajectory backward for a time T, then follow it through the interaction region for a time $2T$, and finally go backward along another free path for a time T (see Figure 3.16). If the trajectories at great distances are identical to free trajectories, then increasing the value of T will not affect this transformation, since it involves moving back and forth along the same pieces of trajectories. The effect of S is to transform the free trajectory that equals a given trajectory asymptotically as $t \to -\infty$ into the free trajectory that it resembles as $t \to \infty$. Hence S commutes with free time-evolution, $\Phi_t^0 \circ S = S \circ \Phi_t^0$. This can also be seen from the definition

$$\Phi^0_{-T_1+t} \circ \Phi_{T_1+T_2} \circ \Phi^0_{-T_2-t} = \Phi^0_{-\bar{T}_1} \circ \Phi_{\bar{T}_1+\bar{T}_2} \circ \Phi^0_{-\bar{T}_2},$$

where $T_1 = \bar{T}_1 + t$ and $T_2 = \bar{T}_2 - t$.
2. Since S commutes with Φ_t^0, it leaves the algebra $\{H_0\}'$ invariant, but only elements of $\{H_0\}' \cap \{H\}'$ are individually invariant. If $H_0 = |\mathbf{p}|^2/2$, then $\mathbf{p} \in \{H_0\}'$ and $S(\mathbf{p}) = \lim_{T\to\infty} \Phi_t(\mathbf{p})$ gives the change in the momentum over the course of the scattering. Yet S need not always express the long-time change in other observables under Φ_t: For instance, there can be functions, which are invariant under Φ_t, but are changed by S; or the radial momentum p_r may be left unchanged, although it undergoes a change of sign after long times under both Φ_t and Φ_t^0.

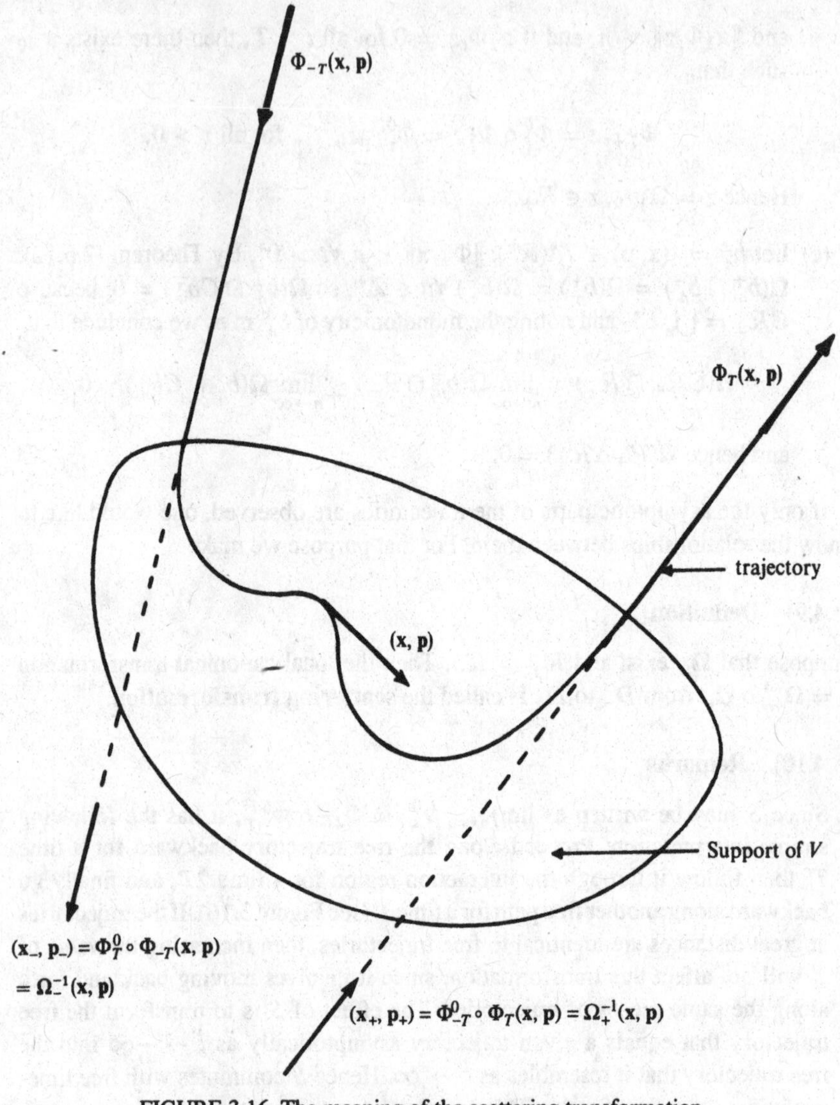

FIGURE 3.16. The meaning of the scattering transformation.

3. Another frequently studied object is the scattering transformation in the Heisenberg representation, $S_H = \Omega_+ \circ \Omega_-^{-1}$. It differs from S by the interchange of H and H_0. Consequently, it commutes with Φ_t and, rather than transforming lines into lines, it transforms trajectories of Φ_t into other such trajectories. If $D_+ = D_-$, then S_H exists and maps $\mathcal{R}_-$ and $\mathcal{R}_+$.

(3.4.11) **Examples**

1. *One-dimensional motion.* Let $H_0 = p^2/2$ and $H = p^2/2 + V(x)$, where V decreases as $|x|^{-1-\varepsilon}$, $\varepsilon > 0$, as $|x| \to \infty$. Then

$$\Phi_t^0(x, p) = (x + pt, p), \quad \Phi_t(x, p) = (x(t), \{p^2 + 2(V(x) - V(x(t)))\}^{1/2}),$$

where $x(t)$ is determined implicitly by

$$t = \int_x^{x(t)} \frac{d\alpha}{\{p^2 + 2(V(x) - V(\alpha))\}^{1/2}},$$

and we assume that $p^2 + 2V(x) > 2\sup_\alpha V(\alpha)$ to ensure that $p(t) > 0$ and the particle is not reflected at the potential. With the notation $(x_-, p_-) = \Phi^0_{-t/2}(x, p) = (x - pt/2, p)$, $(x_+, p_+) = \Phi_t(x_-, p_-)$ (cf. Figure 3.17), and $V_\pm = V(x_\pm)$, the sequence of mappings becomes $\Phi^0_{-t/2} \circ \Phi_t \circ \Phi^0_{-t/2}$,

$$(x, p) \xrightarrow{\Phi^0_{-t/2}} \left(x - \frac{pt}{2}, p\right) \xrightarrow{\Phi_t} \left(x - \frac{pt}{2} + \int_{x_-}^{x_+} d\alpha, \{p^2 + 2V_- - 2V_+\}^{1/2}\right)$$

$$\xrightarrow{\Phi^0_{-t/2}} \left(x - \frac{t}{2}(p + \{p^2 + 2V_- - 2V_+\}^{1/2}) + \int_{x_-}^{x_+} d\alpha, \{p^2 + 2V_- - 2V_+\}^{1/2}\right).$$

As $t \to \infty$, $x_- \to -\infty$, $x_+ \to +\infty$, $V_\pm \to 0$, and, using

$$t = \int_{x_-}^{x_+} \frac{d\alpha}{\{p^2 + 2V_- - 2V(\alpha)\}^{1/2}},$$

we get

$$S(x, p) = \left(x - p\int_{-\infty}^{\infty} d\alpha\, [\{p^2 - 2V(\alpha)\}^{-1/2} - \{p^2\}^{-1/2}], p\right)$$
$$=: (x - p\tau, p) \equiv (\bar{x}, \bar{p}).$$

Since the trajectories of Φ_t^0 are horizontal lines, S can only be of the form $x \to x + f(p)$, $p \to p$. The quantity τ defined above is known as the **delay time.** It is given by an integral that converges by assumption, and can be understood intuitively as the difference in the times required by Φ_t and Φ_t^0 to send (x_-, p_-) to (x_+, p_+) as $t \to \infty$. Since S is a canonical transformation $(x, p) \to (\bar{x}, \bar{p})$, it has a generator (3.1.6) of the form

$$f(\bar{x}, p) = \bar{x}p - 2\delta(p) : (\bar{x}, \bar{p}) = \left(x - 2\frac{\partial \delta(p)}{\partial p}, \bar{p}\right),$$

where

$$\delta(p) = \frac{1}{2}\int_{-\infty}^{\infty} d\alpha\, \{[p^2 - 2V(\alpha)]^{1/2} - [p^2]^{1/2}\}.$$

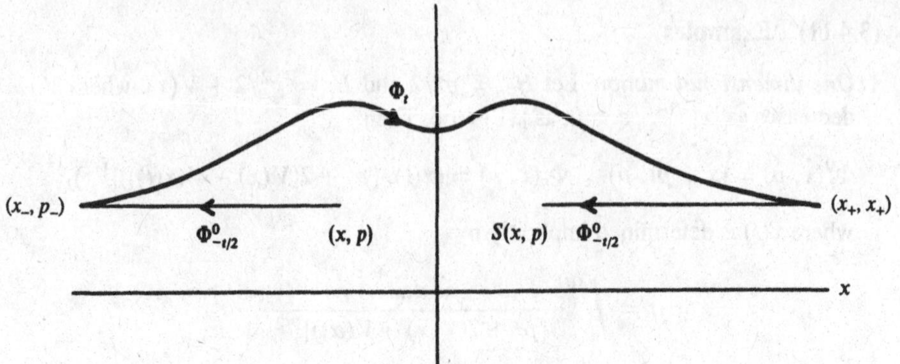

FIGURE 3.17. The scattering transformation on two-dimensional phase space.

2. *Radial force in two dimensions.*

$$H_0 = \frac{|\mathbf{p}|^2}{2}, \qquad H = \frac{|\mathbf{p}|^2}{2} + V(|\mathbf{x}|).$$

In this case a point transformation can be used to switch to polar coordinates,

$$r = |\mathbf{x}|, \quad \varphi = \arccos\frac{x}{r}, \quad p_r = \frac{(\mathbf{p}\cdot\mathbf{x})}{r}, \quad L = |\mathbf{x}\wedge\mathbf{p}|,$$

or, by interchanging **x** and **p**, the variables

$$a = \frac{(\mathbf{p}\cdot\mathbf{x})}{p}, \quad \chi = \arccos\frac{p_x}{p}, \quad p = |\mathbf{p}|, \quad L = |\mathbf{x}\wedge\mathbf{p}|,$$

can be introduced. The latter is of course also a canonical transformation, even though not a point transformation, so we always have two pairs of canonical variables $(r, \varphi; p_r, L)$ and $(a, \chi; p, L)$. By drawing points of $T^*(\mathbb{R}^2)$ as vector arrows in $\mathbb{R}^2$, we discover the geometric significance of these variables (cf. Figure 3.18). Since

$$H_0 = \frac{p_r^2}{2} + \frac{L^2}{2r^2} = \frac{p^2}{2}, \quad H = \frac{p_r^2}{2} + \frac{L^2}{2r^2} + V(r) = \frac{p^2}{2} + V(\sqrt{a^2 + L^2/p^2}),$$

Φ_t^0 is particularly simple when written in the second set of coordinates: $(a, \chi; p, L) \to (a + pt, \chi; p, L)$. Since S commutes with Φ_t^0 and $L \in \{H_0\}' \cap \{H\}'$, S must be of the form $(a, \chi; p, L) \to (a - \tau(p, L)p, \chi; p, L)$. As a canonical transformation, it may be rewritten as

$$(a, \chi; p, L) \to \left(a - 2\frac{\partial\delta(p, L)}{\partial p}, \chi - 2\frac{\partial\delta(p, L)}{\partial L}; p, L\right).$$

If, in order to determine the generator δ, we look again into the transformations $\Phi^0_{-t/2} \circ \Phi_t \circ \Phi^0_{-t/2}$, then we see that

$$\tau p = 2\frac{\partial\delta(p, L)}{\partial p} = \lim_{|a_\pm|\to\infty}\left(tp - \int_{a_-}^{a_+} da\right),$$

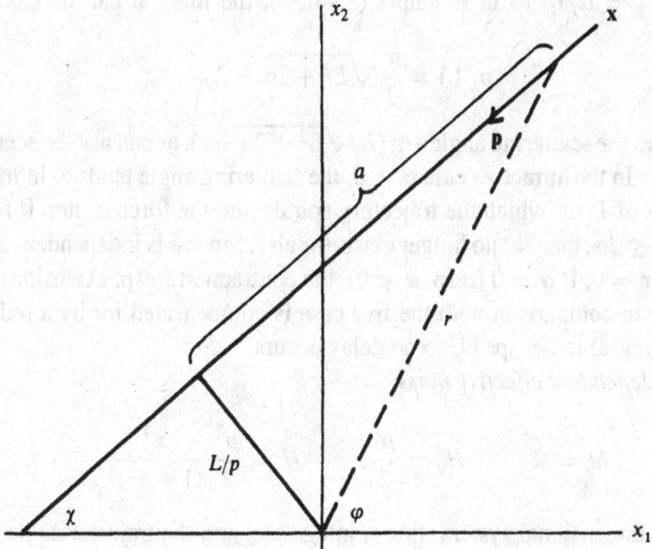

FIGURE 3.18. The geometric significance of a and χ.

as in Example 1. Here t is the time taken on the trajectory of Φ to go from $(\mathbf{x}_-, \mathbf{p}_-)$ to $(\mathbf{x}_+, \mathbf{p}_+)$. It is most conveniently expressed in polar coordinates. It follows from $\dot{r} = p_r$ and $H = \text{const.} \to p^2/2$ as $V(x_\pm) \to 0$ that

$$t = \int_{r_0}^{r_-} \frac{dr}{\sqrt{p^2 - L^2/r^2 - 2V(r)}} + \int_{r_0}^{r_+} \frac{dr}{\sqrt{p^2 - L^2/r^2 - 2V(r)}},$$

where $\sqrt{\cdot}$ is defined as 0 when $r = r_0$. In order to express the integral $\int_{a_-}^{a_+} da$ likewise in these variables, use the relationship $a_\pm = \sqrt{r_\pm^2 - L^2/p_\pm^2}$ or, in case $p_\pm$ has already attained its asymptotic value,

$$\int_{a_-}^{a_+} da = \left(\int_{L/p}^{r_+} + \int_{L/p}^{r_-}\right) \frac{p\,dr}{\sqrt{p^2 - L^2/r^2}}.$$

Thus, all told,

$$\tau = 2 \lim_{R\to\infty} \left\{ \int_{r_0}^{R} \frac{dr}{\sqrt{p^2 - L^2/r^2 - 2V(r)}} - \int_{L/p}^{R} \frac{dr}{\sqrt{p^2 - L^2/r^2}} \right\}.$$

This determines $\partial\delta/\partial p$. Since the scattering angle $2\partial\delta/\partial L$ tends to 0 as $p \to \infty$, it has no contribution that depends on L alone, and

$$\delta(p, L) = \lim_{R\to\infty} \left\{ \int_{r_0}^{R} dr\sqrt{p^2 - L^2/r^2 - 2V(r)} - \int_{L/p}^{R} dr\sqrt{p^2 - L^2/r^2} \right\}.$$

With $V(r) = \alpha/r^2$ as in Example (3.4.6; 3), the integral can be calculated explicitly:

$$\delta(p, L) = \frac{\pi}{2}(\sqrt{L^2 + 2\alpha} - L).$$

This makes the scattering angle $-\pi(L/\sqrt{L^2 + 2\alpha} - 1)$, as can also be seen from $\Omega_+^{-1} \circ \Omega_-$. In the attractive case $\alpha < 0$, the scattering angle tends to infinity for the values of L for which the trajectory spirals into the force center. It follows that if $L^2 \leq 2\alpha$, then Φ_t no longer exists for all t. Since δ is independent of p, in this case $\tau = 0$. If $\alpha > 0$ (resp. $\alpha < 0$), the contraction (resp. extension) of the trajectory in comparison with the free case is compensated for by a reduction (resp. increase) in the speed, so no delay occurs.

3. *Position-dependent effective mass.*

$$M = \mathbb{R}, \qquad H_0 = \frac{p^2}{2}, \qquad H = \frac{p^2}{2}\frac{x^4}{(1+x^2)^2}.$$

As a one-dimensional system, this is integrable, and the trajectories lie on the curve $H = E =$ const. in $T^*(M)$, i.e., $p = \pm\sqrt{2E}(1+1/x^2)$. Since the effective mass becomes infinite at $x = 0$, trajectories have a dead end there. It is also easy to work Φ_t out. If $p > 0$, $\dot{x}(1 + 1/x^2) = \sqrt{2E}$, which integrates to

$$x(t) = \frac{1}{2}\left\{x_0 - \frac{1}{x_0} + t\sqrt{2E} \pm \sqrt{\left(x_0 - \frac{1}{x_0} + t\sqrt{2E}\right)^2 + 4}\right\} \quad \text{for } x \gtrless 0.$$

If $p < 0$, $\sqrt{2E}$ becomes $-\sqrt{2E}$. In quadrants I ($x > 0$, $p > 0$) and III ($x < 0$, $p < 0$) the time-evolution tends to free motion $x - pt \to$ const., $p \to \pm\sqrt{2E}$ as $t \to +\infty$, while in the other two quadrants this happens as $t \to -\infty$. Since at large times Φ_t^0 makes $x > 0$ when $p > 0$ and $x < 0$ when $p < 0$, the wave operators $\Omega_\pm = \lim_{t\to\pm\infty} \Phi_{-t} \circ \Phi_t^0$ exist provided that $p \neq 0$, i.e., $D_+ = D = T^*(\mathbb{R})\backslash\{\mathbb{R} \times 0\}$. The sets $\mathcal{R}_+ = \text{I} \cup \text{III}$ and $\mathcal{R}_- = \text{II} \cup \text{IV}$ are not merely different, but in fact fill up disjoint quadrants (Figure 3.19). Accordingly, the limit

$$S = \lim_{t\to\infty} \Phi_{-t}^0 \circ \Phi_{2t} \circ \Phi_{-t}^0$$

does not exist anywhere, whereas

$$S_H = \lim_{t\to\infty} \Phi_{-t} \circ \Phi_{2t}^0 \circ \Phi_{-t}$$

exists on $\mathcal{R}_-$, which is mapped to $\mathcal{R}_+$.

(3.4.12) **Remarks**

1. The definition of S required asymptotic completeness in the strong form, $\mathcal{R}_+ = \mathcal{R}_-$. This is a valid assumption for harmless potentials after the exclusion of some trajectories from $T^*(M)$. It is not guaranteed in general by the existence

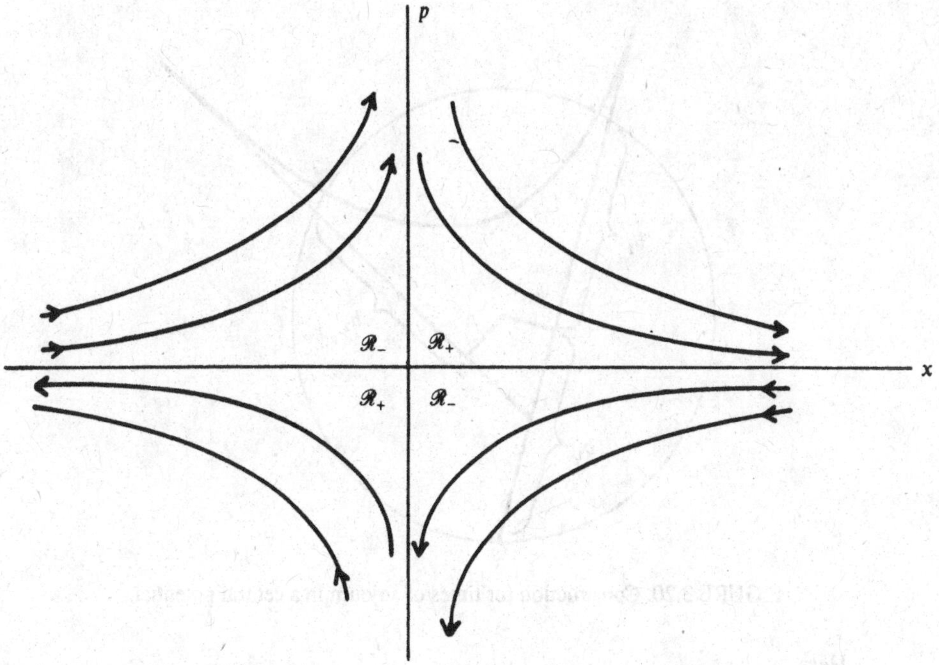

FIGURE 3.19. The domains $\mathcal{R}$ of a system for which S does not exist.

of $\Omega_\pm$, not even for time-reversible Hamiltonian systems. This is shown by Example 3, in which the existence of open domains of definition $D_+ = D_-$ of $\Omega_\pm$ and the existence of S_H together do not ensure the existence of S.

2. In two dimensions τ is the difference between the time taken on the actual trajectory and the time that would be required on pieces b_1 and b_2 of the tangent free trajectories, as in Figure 3.20. Because $|b_1| = |b_2| = |b_3|$, τ may also be interpreted as the difference of the times of sojourn in a large disk about the center of force, when the true and the free trajectories are compared.
3. The formulas for δ from Examples 1 and 2 play a role in quantum theory under the title "semiclassical approximation for the phase-shift." Although this is an uncontrolled approximation in quantum theory, it is an exact expression for the generator in classical scattering theory.
4. The form of the generators in Examples 1 and 2 follows from (3.2.9) and the property of additivity under composition of mappings: With the notation of Example 1 and

$$
\begin{aligned}
(x_s, p_s) &= \Phi^0_{-t/2}(x_+, p_+), \\
p_- \, dx_- &= p \, dx + dW^0_-, \\
p_+ \, dx_+ &= p_- \, dx_- + dW, \\
p_s \, dx_s &= p_+ \, dx_+ + dW^0_+.
\end{aligned}
$$

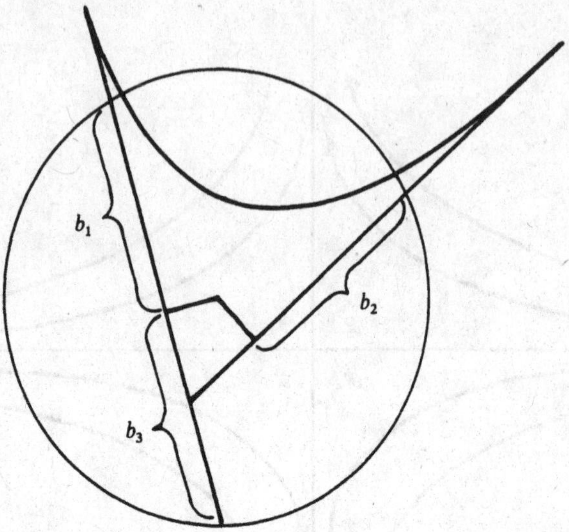

FIGURE 3.20. Construction for times of sojourn in a central potential.

Here,

$$W_-^0 = -\int_{-t/2}^{0} dt'\, \Phi_{t'}^0\left(\frac{p^2}{2}\right) = -\frac{t}{2}\frac{p^2}{2},$$

and

$$\begin{aligned} W &= \int_0^t dt' \left\{\Phi_{t'}\left(\frac{p_-^2}{2}\right) - V(\Phi_{t'}(x_-))\right\} \\ &\to \int_{x_-}^{x_+} dx' \sqrt{p^2 - 2V(x')} - t\,\frac{p^2}{2}, \end{aligned}$$

provided that t is so large that $V(x_-)$ tends to 0 and consequently $\{\dots\} = \Phi_{t'}(p_-^2) - E = p_-^2 + 2V(x_-) - 2V(\Phi_{t'}(x_-)) - E$ tends to $p^2 - 2V(\Phi_{t'}(x_-)) - p^2/2$. Also, for large t,

$$W_+^0 = -\int_{-t/2}^{0} dt'\, \Phi_{t'}^0\left(\frac{p_+^2}{2}\right) = -\frac{t}{2}\frac{p^2}{2}.$$

If the equalities for the 1-forms are added, we see that

$$p_s\, dx_s = p\, dx + d\left(\int_{x_-}^{x_+} dx'\, [p^2 - 2V(x')]^{1/2} - tp^2\right).$$

Now,

$$pt = x_+ - x_- - (x_s - x), \quad \text{so} \quad -tp^2 = -\int_{x_-}^{x_+} dx\, p + p(x_s - x),$$

and therefore

$$p_s\,dx_s = p\,dx + d((x_s - x)p + 2\delta(p)), \qquad \text{yielding} \qquad x_s = x - 2\frac{\partial\delta(p)}{\partial p}.$$

This direct calculation of the generators can be extended to higher dimensions.

As canonical transformations, S and $\Omega_\pm$ preserve volumes in phase space. This seems at first to contradict the fact that $\Omega_\pm$ map almost the whole phase space onto the complement of the trajectories that remain in a finite region, i.e., onto a strictly smaller set. The paradox is easily resolved by noting that both volumes in question are infinite, and this leads to

(3.4.13) **Levinson's Theorem for Classical Scattering**

Let $M = \mathbb{R}$ and $\chi_b\colon T^*(M) \to \mathbb{R}$ be the characteristic function of the set $b = \bigcup_n b_n$ consisting of trajectories that remain finite, and suppose that $V \in C_0^\infty$. Then

$$\int_{-\infty}^{\infty} dx\,dp\,\chi_b(x,p) = -\int_{-\infty}^{\infty} dp\,|p|\tau(p).$$

Proof: Begin by finding the image under Ω_+ of the rectangle $\{(x,p)\colon |x| < R, 0 < p < R'\}$, where R is large enough that $\operatorname{supp}(V) \subset \{x\colon |x| < R\}$. On the side $\{x = R\}$, Ω_+ acts like $\mathbf{1}$, and on the side $\{x = -R\}$ it acts like S^{-1}. The side $\{p = 0\}$ is lifted above the region containing the finite trajectories, and $\{p = R'\}$ is mapped to $\sqrt{R'^2 - 2V(x)}$, and thus remains invariant as $R' \to \infty$ (see Figure 3.21). As R and R' become infinite, the difference between the volume of the image of the rectangle differs from the original volume by the region of trajectories that remain finite and by the integral over τp. This, together with the same reasoning for $p < 0$, proves the theorem. □

(3.4.14) **Remarks**

1. This theorem was first proved in a quantum-mechanical setting, by complex integration, which obscures its geometric meaning.
2. It is easy to generalize Levinson's theorem for larger classes of potentials. In higher dimensions correction terms appear.

One can thus calculate the angle Θ between $\mathbf{p}_- := \tau_-\mathbf{p}$ and $\mathbf{p}_+ := \tau_+\mathbf{p}$ from S: $S_*(\mathbf{p}_-) = \mathbf{p}_+$. Because in experiments the trajectories are only statistically known, it is useful to make

(3.4.15) **Definition**

The angle Θ between $\mathbf{p}_-$ and $\mathbf{p}_+$ is called the **scattering angle**, and the **differential scattering cross-section** is defined as $d\sigma =$ (number of particles scattered into the solid angle $[\Theta, \Theta + d\Theta] \times [\varphi, \varphi + d\varphi]$ per unit time)/(number of incident particles per unit time and per unit of surface area) $= 2\pi\sigma(\Theta)\sin\Theta\,d\Theta\,d\varphi$.

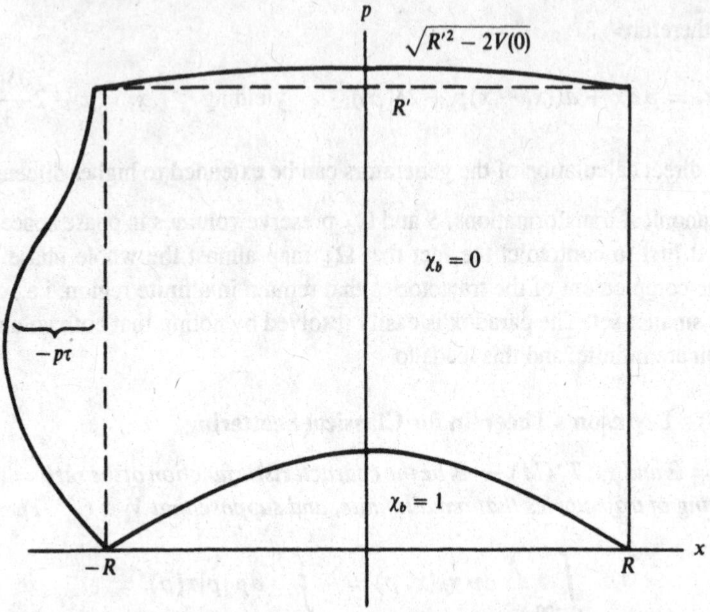

FIGURE 3.21. The geometric meaning of Levinson's theorem.

(3.4.16) **Remarks**

1. If the initial (unnormalized) distribution of particles ρ (cf. (1.3.1)) specifies $\mathbf{p}_-$ precisely but leaves $\mathbf{x}$ completely unrestricted, i.e., $\rho(\mathbf{x}, \mathbf{p}) \cong \delta^3(\mathbf{p}_- - \mathbf{k})$, then the **impact parameter** $b = |\mathbf{L}|/|\mathbf{p}_-|$ has a probability distribution $2\pi b\, db \cdot f$, where f is the number of incident particles per time and per unit area (Figure 3.22). For a central potential in $\mathbb{R}^3$ the number of scattered particles per unit time is

$$f 2\pi b\, db = f\frac{2\pi}{4E} d|\mathbf{L}|^2 = \frac{f\pi}{E} L \frac{dL}{d\Theta}\, d\Theta = 2\pi f \sigma(\Theta) \sin\Theta\, d\Theta,$$

where it is supposed that the relationship between Θ and L is known and is bijective. (If the mass m is not set to 1, E should be replaced with mE.)
2. In $\mathbb{R}^2$ the unit of surface area becomes instead a unit of length. Correspondingly, $2\pi b\, db \to db$ and $\sin\Theta\, d\Theta\, d\varphi \to d\Theta$, and so

$$\sigma(\Theta) = \frac{1}{\sqrt{2E}} \frac{dL}{d\Theta}.$$

(3.4.17) **Examples**

1. The $1/r^2$ potential in $\mathbb{R}^2$. By (3.4.11; 2), $\Theta = \pi((-L/\sqrt{2\alpha + L^2}) + 1)$, which implies

$$L^2 = 2\alpha \frac{(\Theta - \pi)^2}{\Theta(2\pi - \Theta)}$$

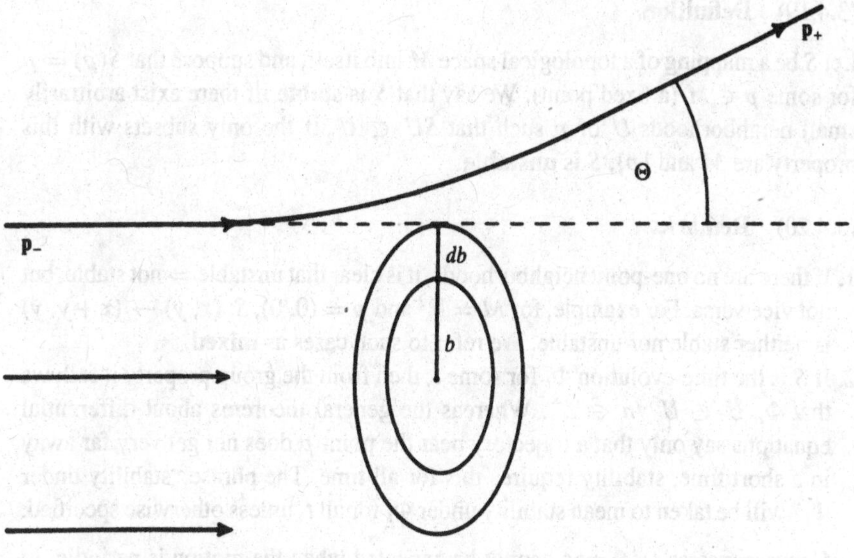

FIGURE 3.22. The scattering angle and scattering cross-section.

and hence

$$\sigma(\Theta) = \sqrt{\frac{\alpha}{E}} \frac{\partial}{\partial \Theta} \frac{\pi - \Theta}{\sqrt{\Theta(2\pi - \Theta)}}.$$

2. The $1/r^2$ potential in $\mathbb{R}^3$. By conservation of momentum this problem reduces to a planar one (see §5.3), so L becomes the same function of Θ as in Example 1, and therefore

$$\sigma(\Theta) = \frac{\alpha}{2E \sin \Theta} \frac{\partial}{\partial \Theta} \frac{(\pi - \Theta)^2}{\Theta(2\pi - \Theta)}.$$

(3.4.18) **Remarks**

1. The angular distribution is strongly bunched up in the forward direction and is not integrable as $\Theta \to 0$. This is because the particles with very large b can still be scattered, although not very much.
2. As $\alpha \to 0$, σ approaches 0, as it must. The cross-section in $\mathbb{R}^3$ is linear in α because only $\sqrt{\alpha/E}$ has the dimension of length.
3. Θ is independent of E because the canonical transformation $x \to \lambda x$, $p \to \lambda^{-1} p$ sends H to $\lambda^{-2} H$, but leaves Θ unchanged.

After having dealt with the trajectories that go off to infinity, we study those that always remain in compact neighborhoods of some equilibrium position. For such trajectories there is hope that the linearized theory (3.3.16; 3) will be useful as a basis of comparison. The intuitive notion of the stability of an equilibrium position is now made into a general

(3.4.19) Definition

Let S be a mapping of a topological space M into itself, and suppose that $S(p) = p$ for some $p \in M$ (a fixed point). We say that S is **stable** iff there exist arbitrarily small neighborhoods U of p such that $SU \subset U$. If the only subsets with this property are M and $\{p\}$, S is **unstable**.

(3.4.20) Remarks

1. If there are no one-point neighborhoods, it is clear that unstable $\Rightarrow$ not stable, but not vice versa. For example, for $M = \mathbb{R}^2$ and $p = (0, 0)$, $S: (x, y) \to (x + y, y)$ is neither stable nor unstable. We refer to such cases as **mixed**.
2. If S is the time-evolution Φ_t for some t, then from the group property it follows that $\Phi_{nt} U \subset U \; \forall n \in \mathbb{Z}^+$. Whereas the general theorems about differential equations say only that a trajectory near the point p does not get very far away in a short time, stability requires this for all time. The phrase "stability under Φ_t" will be taken to mean stability under Φ_t for all t, unless otherwise specified.

Convergence as $t \to \pm\infty$ cannot be expected when the motion is periodic. In order to define an average time-dependence, let us assume that $T^*(M)$ equals $\mathbb{R}^{2m}$. Then it is possible to define the sum of two points of $T^*(M)$, and thereby to define the sum of diffeomorphisms. This enables us to make

(3.4.21) Definition

Given the canonical flows Φ_t and Φ_t^0 on $\mathbb{R}^{2m}$, the linearity of $\mathbb{R}^{2m}$ defines sums $(\Phi_1 + \Phi_2)(x)$ as $\Phi_1(x) + \Phi_2(x)$, and thereby also the integral over diffeomorphisms. Let

$$C_\pm := \lim_{T \to \pm\infty} \frac{1}{T} \int_0^T dt \, \Phi^0_{-t} \circ \Phi_t,$$

assuming that this **Cesàro average** exists and is a local canonical transformation on a neighborhood $\mathcal{D}_\pm \subset \mathbb{R}^{2m}$.

(3.4.22) Remarks

1. If $\Omega_\pm$ exists, then so does $C_\pm$, and $C_\pm = \Omega_\pm^{-1}$ but not necessarily vice versa.
2. If $\bigcup_t \Phi^0_{-t} \circ \Phi_t$ is bounded (i.e., $\bigcup_t \Phi^0_{-t} \circ \Phi_t(x)$ is a bounded subset of $\mathbb{R}^{2m}$ $\forall x \in \mathbb{R}^{2m}$), and Φ_t^0 is linear, then $C_\pm$ has the same effect as $\Omega_\pm$, as it maps both flows canonically onto each other; for

$$\Phi^0_\tau \circ C_\pm = \lim_{T \to \pm\infty} \frac{1}{T} \int_0^T dt \, \Phi^0_{-t+\tau} \circ \Phi_t = \lim_{T \to \pm\infty} \frac{1}{T} \int_{-\tau}^{T-\tau} dt' \, \Phi^0_{-t'} \circ \Phi_{t'} \Phi_\tau$$

$$= C_\pm \circ \Phi_\tau + \lim_{T \to \pm\infty} \frac{1}{T} \int_{-\tau}^0 dt' \, (\Phi^0_{-t'} \circ \Phi_{t'} - \Phi^0_{-T-t'} \circ \Phi_{T+t'}) \circ \Phi_\tau,$$

and the last term approaches zero.
3. Frequently the limit $\lim_{\alpha \to 0} \alpha \int_0^\infty \exp(-\alpha t) \Phi^0_{-t} \circ \Phi_t \, dt$ also equals C_+, and is easier to handle.

By using $C_\pm$ it is possible to reduce the problem of stability for complex analytic systems under Φ_t to the problem under Φ_t^0. Write $(x, p) \in \mathbb{C}^{2m}$ as the single variable z, and suppose that $z = 0$ is a fixed point of Hamilton's equations. Then for a complex analytic X_H we can write Hamilton's equations in the form

$$\dot{z} = f(z) = Az + \cdots, \tag{3.4.23}$$

where A is a constant matrix and ... denotes the terms of higher order. The linearized theory is based on the equation $\dot{z} = Az$, which produces the comparison flow $\Phi_t^0 : z \to \exp(tA)z$.

(3.4.24) Stability Criterion for Analytic Vector Fields

Let Φ_t be the flow of an analytic vector field on $\mathbb{C}^m$ with fixed point 0 and linear part A. Then Φ_t is stable at $z = 0$ (as $t \to \pm\infty$) if and only if:

(a) *A is diagonable and has purely imaginary eigenvalues; and*

(b) *both $C_\pm$ exist on neighborhoods $\mathcal{R}_\pm$ of 0.*

(3.4.25) Remarks

1. The canonical flow Φ_t^0 comes from a series-expansion of H, and thus $\exp(tA)$ is a symplectic matrix. That does not imply that $\exp(tA)$ and A are diagonable (for instance, $\begin{vmatrix} 1 & b \\ 0 & 1 \end{vmatrix}$ is symplectic but not diagonable); instead, it must be explicitly assumed. Distinguish between A diagonable $\Leftrightarrow A = T$ (diagonal matrix) T^{-1}, $T \in GL(m, \mathbb{C}) \Leftrightarrow$ any n-fold degenerate eigenvalue has n linearly independent eigenvectors; and A unitarily diagonable (T is unitary) $\Leftrightarrow AA^t = A^tA$ (i.e., A is normal) $\Leftrightarrow$ any n-fold degenerate eigenvalue has n orthogonal eigenvectors (see also (3.1.13; 4)).
2. Part (a) of the theorem is a necessary condition for stability, which goes back to Liapunov. It is not restricted to Hamiltonian vector fields, and implies more specifically:

   (i) All eigenvalues of A have real parts $< 0 \Rightarrow$ stable as $t \to \infty$.

   (ii) Some eigenvalue of A has real part $> 0 \Rightarrow$ unstable as $t \to \infty$.

   Because of (3.1.8; 2) the only possibility for canonical flows is (ii). Even so, having all eigenvalues purely imaginary is not sufficient by itself, as shown by the flow $\dot{x} = -y + x(x^2 + y^2)$, $\dot{y} = x + y(x^2 + y^2)$ on $\mathbb{R}^2$. The fixed point $(x, y) = (0, 0)$ is unstable, because with polar coordinates the equations become $\dot{r} = r^3$, $\dot{\varphi} = 1$, which are solved by

$$r(t) = r(0)(1 - r(0)^2 2t)^{-1/2}, \qquad \varphi(t) = \varphi(0) + t.$$

If $r(0) \neq 0$, the particle spirals off to infinity in a finite time, whereas

$$A = \begin{vmatrix} 0 & -1 \\ 1 & 0 \end{vmatrix}$$

of the linearized equation is diagonable and has eigenvalues $\pm i$. It is obvious that condition (b) fails.

3. Criterion (b) shows what strong conclusions can be drawn from the assumption of analyticity and stability in the complex plane. Roughly speaking, it states that stable flows are linear flows, possibly written in complicated coordinates. It is not very useful for determining the stability of particular systems, since it is no easier to prove than the existence of $C_\pm$. There are a few other sufficient criteria for stability, but they are not applicable to the Hamiltonian systems that will interest us here. Hence our results about this question are somewhat deficient.

Proof:

(i) *(a) and (b) ⇒ stable.* As with Ω_t, we can suppose that $\mathcal{R}_\pm$ are invariant under Φ_t, and map this flow diffeomorphically by $C_\pm$ to $\exp(tA)$. Since stability is defined purely topologically, it is unaffected by diffeomorphisms, and we need only investigate the stability of $\exp(tA)$. This is guaranteed by (a) (cf. the following examples and Problem 5).

(ii) *(b) but not (a) ⇒ not stable.* If (a) fails, then Φ_t^0 is not stable at (0, 0), and by the same argument as in (i), Φ_t is likewise not stable.

(iii) *stable ⇒ (b).* This part of the proof is somewhat involved, and will not be given here (see [14]). □

(3.4.26) **Examples**

1. Let $T^*(M)$ be $\mathbb{R}^2$. We investigate the form of $\exp(tA)$, which has to leave the canonical 2-form invariant. Taking the derivative of

$$\exp(tA^t)W\exp(tA) = W \quad \text{(see (3.1.8; 2))},$$

results in the requirement that $A^tW + WA = 0$. This is plainly sufficient to make the first equation hold. Since

$$W = \frac{1}{2}\begin{vmatrix} 0 & 1 \\ -1 & 0 \end{vmatrix},$$

A has the form

$$A = \begin{vmatrix} a & b+c \\ b-c & -a \end{vmatrix}.$$

The eigenvalues are $\pm\sqrt{a^2+b^2-c^2}$. Thus we must distinguish among three cases:

(i) $a^2 + b^2 - c^2 > 0$. Φ_t^0 is like a dilatation:

$$A = T\begin{vmatrix} \lambda & 0 \\ 0 & -\lambda \end{vmatrix} T^{-1}, \qquad e^{At} = T\begin{vmatrix} e^{\lambda t} & 0 \\ 0 & e^{-\lambda t} \end{vmatrix} T^{-1};$$

(ii) $a^2 + b^2 - c^2 < 0$. Φ_t^0 is like a rotation:

$$A = T\begin{vmatrix} i\omega & 0 \\ 0 & -i\omega \end{vmatrix} T^{-1}, \qquad e^{At} = T\begin{vmatrix} e^{i\omega t} & 0 \\ 0 & e^{-i\omega t} \end{vmatrix} T^{-1}; \text{ and}$$

(iii) $a^2 + b^2 - c^2 = 0$. Φ_t^0 is like linear motion:

$$A = T\begin{vmatrix} 0 & \lambda \\ 0 & 0 \end{vmatrix} T^{-1}, \qquad e^{At} = \begin{vmatrix} 1 & \lambda t \\ 0 & 1 \end{vmatrix} T^{-1}.$$

Here λ and $\omega \in \mathbb{R}$, and T is a similarity transformation. The fixed point **0** is unstable for (i) (the hyperbolic case—an oscillator with an imaginary frequency); stable for (ii) (the elliptic case—an oscillator with a real frequency); and mixed for (iii) (the linear case—an oscillator with frequency zero). The trajectories in phase space are shown in Figure 3.23. Observe that a canonical flow has neither sources nor sinks, since it preserves volume. If it contracts in one direction, it has to expand in another.

2. With the notation $z = q + ip$ as in (3.4.23), let

$$H_0 = \frac{|z|^2}{2}, \qquad H = \frac{|z - z_0|^2}{2}.$$

Then

$$\Phi_t^0(z) = e^{-it}z, \qquad \Phi_t(z) = e^{-it}(z - z_0) + z_0,$$

and therefore

$$\Phi_{-t}^0 \circ \Phi_t(z) = z - z_0 + e^{it}z_0.$$

Hence $C_\pm(z) = z - z_0$, and C indeed transforms H_0 into H.

3. Again using the notation from above, let

$$H_0 = \frac{|z|^2}{2} \quad \text{and} \quad H = \frac{|z|^2}{2} + \frac{\lambda}{4}|z|^4.$$

Then

$$\Phi_{-t}^0 \circ \Phi_t(z) = e^{i\lambda|z|^2 t}z,$$

so $C_\pm(z) = 0$, and therefore C does not exist. This H does not give rise to an analytic X_H.

(3.4.27) Problems

1. In (3.4.4) we assumed the convergence of $\Phi_{-t} \circ \Phi_t^0$ and $(\Phi_{-t} \circ \Phi_t^0)^{-1}$. Find homeomorphisms Ω_t of $D = \{(x, y) \in \mathbb{R}^2 : x^2 + y^2 \leq 1\}$ such that as $t \to \infty$, $\Omega_t \to \mathbf{1}$, but $\Omega_t^{-1} \not\to \mathbf{1}$. (Here the arrow $\to$ denotes pointwise convergence.)

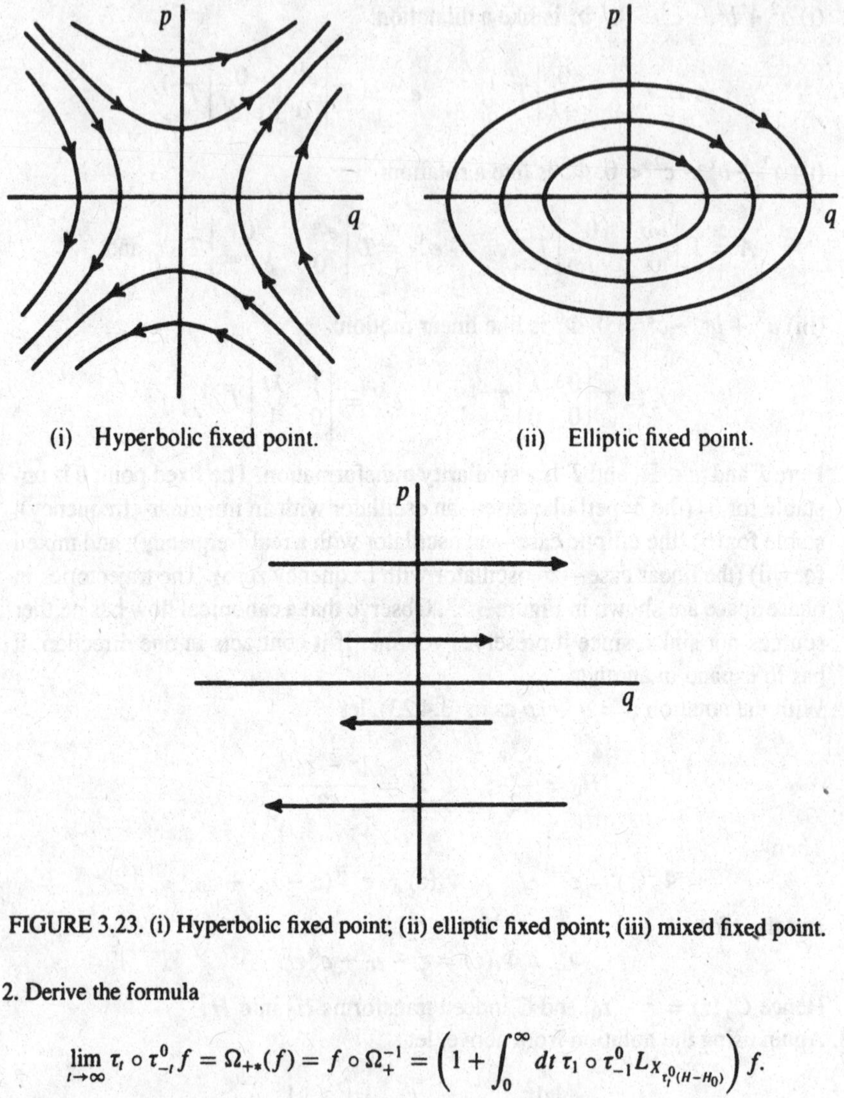

FIGURE 3.23. (i) Hyperbolic fixed point; (ii) elliptic fixed point; (iii) mixed fixed point.

2. Derive the formula

$$\lim_{t\to\infty} \tau_t \circ \tau^0_{-t} f = \Omega_{+*}(f) = f \circ \Omega_+^{-1} = \left(1 + \int_0^\infty dt\, \tau_1 \circ \tau^0_{-1} L_{X_{\tau^0_t(H-H_0)}}\right) f.$$

3. Verify that the Møller transformation $(\bar{r}, p) \to (\bar{,} \bar{p})$ of (3.4.6; 2) is canonical.

4. Show that for a measure-preserving transformation, stability (3.4.19) is equivalent to

$$\forall W\ \exists V \subset W : SV = V$$

(where V and W are neighborhoods).

5. What is $\exp(tA)$ of (3.4.23) for free motion $\dot{q}^i = p_i$, $\dot{p}_i = 0$ on $\mathbb{R}^m$?

6. On $\mathbb{R}^2 \backslash \{\mathbf{0}\} = \mathbb{R}^+ \times T^1$, let the Hamiltonian be

$$H = \frac{p^2}{2} + \frac{2\alpha + L^2}{2r^2} - \frac{\beta}{r}, \qquad \beta > 0,$$

(the relativistic Kepler problem). If $E < 0$, almost every trajectory fills a two-dimensional submanifold densely. Generalize the constants found in (3.4.3; 3) and find out why they are not globally definable for $E < 0$.

(3.4.28) Solutions

1. Let Ω_t be **1** on $\Gamma_t := \{(r, \varphi): 1/t \leq r \leq 1 - 1/t, 0 \leq \varphi \leq 2\pi - 1/t\}$, and define it on $C\Gamma_t$ such that $\Omega_{t|\partial D} = \mathbf{1}$, $\Omega_t(0) = c_t \to 0$, $\Omega_t(1 - 1/2t, 2\pi - 1/2t) = 0$. Since $\bigcup_t \Gamma_t = D \backslash \{0\} \backslash \partial D$, $\lim_{t\to\infty} \Omega_t(x) = x \; \forall x \in D$. But $\Omega_t^{-1} \not\to \mathbf{1}$, because $\Omega_t^{-1}(0) \to \mathbf{1}$.[3]

2. Use $\tau_t = \exp(tL_H)$ and $\tau_t^0 = \exp(tL_{H_0})$. An expansion in a series shows

$$\frac{d}{dt} e^{tL_H} e^{-tL_{H_0}} = e^{tL_H}(L_H - L_{H_0}) e^{-tL_{H_0}} = e^{tL_H} L_{H-H_0} e^{-tL_{H_0}}$$
$$= e^{tL_H} e^{-tL_{H_0}} L_{\tau_t^0 (H-H_0)}$$

(see (2.5.9; 2)), and integration by t gives the result.

3. It is elementary to calculate $dr \wedge dp = d\bar{r} \wedge d\bar{p}$.

4. By (3.4.14), S stable $\Leftrightarrow \forall W \; \exists U \subset W$ with $SU \subset U \Rightarrow V = \bigcup_{n\geq 1} S^n U \subset U \subset W$ and $SV = V$. Conversely, if $\forall W \; \exists V \subset W$ with $SV = V$, then S is stable.

5. $A^2 = 0 \Rightarrow e^{At} = 1 + At$,

$$A = \begin{bmatrix} \begin{matrix} 0 & 1 \\ 0 & 0 \end{matrix} & & & \\ & \begin{matrix} 0 & 1 \\ 0 & 0 \end{matrix} & & \\ & & \ddots & \\ & & & \begin{matrix} 0 & 1 \\ 0 & 0 \end{matrix} \end{bmatrix} \Bigg\} 2m.$$

6. We find

$$\tau_+ p_x = \sqrt{2H} \cos\left(\varphi + \frac{L}{\sqrt{2\alpha + L^2}} \left(\frac{\pi}{2} - \arctan \frac{pr\sqrt{2\alpha + L^2}}{L^2 + 2\alpha - \beta r}\right)\right)$$

(by calculating $\varphi(r)$, for instance). At the equilibrium point $r = (L^2 + 2\alpha)/\beta$ the argument of the arctangent goes through ∞, and it is not possible to continue this function uniquely. For $\beta < 0$, $\tau_+ p_x$ is defined on $T^*(M)$.

3.5 Perturbation Theory: Preliminaries

Continuous changes in H influence the time-evolution for finite times continuously. However, quantities involving infinitely long times, such

---

[3]For this example I am grateful to W. Schmidt, University of Colorado, Boulder.

as constants of the motion, can exhibit behavior that is highly discontinuous.

In this section we look at Hamiltonians of the form $H = H_0 + \lambda H_1$, $\lambda \in \mathbb{R}$, and study how the dynamics depends on λ.

We first make some general observations about the time-evolution of an observable $f \to \exp(tL_H)f$, starting with the most tractable case, in which all functional dependence is analytic. If we differentiate the series for the exponential function by time term by term we get

(3.5.1)
$$\begin{aligned}\frac{d}{dt}e^{-tL_{H_0}}e^{tL_{H_0+\lambda H_1}}f &= e^{-tL_{H_0}}(L_{H_0+\lambda H_1} - L_{H_0})e^{tL_{H_0+\lambda H_1}}f \\ &= \lambda e^{-tL_{H_0}}\{\exp(tL_{H_0+\lambda H_1})f, H_1\}.\end{aligned}$$

Integrating this by t,

(3.5.2)
$$e^{-tL_{H_0}}e^{tL_{H_0+\lambda H_1}}f = f + \lambda\int_0^t dt_1\,\{\exp(-t_1L_{H_0})\exp(t_1L_{H_0+\lambda H_1})f, H_1(-t_1)\},$$

where

(3.5.3)
$$g(t) := e^{tL_{H_0}}g$$

gives the time-evolution according to H_0. Operating on (3.5.2) with $\exp(tL_{H_0})$ yields

(3.5.4) **Estimate**

$$\forall f, H, H_0 \in C_0^\infty\ \exists c, k \in \mathbb{R}^+: \|e^{tL_H}f - e^{tL_{H_0}}f\| \le c\lambda|t|e^{k|t|}\ \forall t,$$

where

$$\|f\| = \sup_{x\in T^*(M)}|f(x)|.$$

(3.5.5) **Remarks**

1. This means that the effect of the perturbation can at first only grow linearly with t, but later it can grow exponentially.
2. Equation (3.5.2) holds at least when H and H_0 are assumed in C_0^∞, which guarantees the existence of a flow.
3. Estimate (3.5.4) can be extended to $H = p^2 + V(x)$, $V \in C_0^\infty$. It does not, however, hold unrestrictedly; it is possible for particles to run off to infinity in finite times.

Proof: From (3.5.2),

$$\begin{aligned}\|e^{tL_H}f - e^{tL_{H_0}}f\| &\le \lambda\int_0^t dt_1\,\|\{\exp(t_1L_H)f, H_1\}\| \\ &\le \lambda t\sup_{0\le t_1\le t}\|\{\exp(t_1L_H)f, H_1\}\|.\end{aligned}$$

Here note that whereas $\exp(tL_H)f = f \circ \Phi_t$ has the same supremum as f, the derivative $d(f \circ \Phi_t) = df \circ T(\Phi_t)$ implied in the brackets $\{\,\}$ may grow with t. This rate of growth, however, is boundable using the group structure of Φ_t: First, notice that

$$\|\{\,\}\| = \|(d(f \circ \Phi_{t_1})|X_{H_1})\| \leq \|T(\Phi_{t_1})\| \cdot \|df\| \cdot \|X_{H_1}\|,$$

where it is understood that the norm $\|\ \|$ of a vectorial or matrix quantity includes taking the supremum over the components. From the inequality $\|A \cdot B\| \leq \|A\| \cdot \|B\|$, it then follows that

$$\|T(\Phi_t)\| \leq c^n \qquad \text{for all} \quad |t| \leq n\tau,$$

where

$$c = \sup_{-\tau \leq t \leq \tau} \|T(\Phi_t)\|.$$

Since no t dependence is hidden in the other factors, this yields (3.5.4). □

Iterating (3.5.2) leads to

(3.5.6)
$$\begin{aligned} e^{-tL_{H_0}} e^{tL_{H_0+\lambda H_1}} f = f + \sum_{n=1}^{r-1} \lambda^n \int_0^t dt_1 \int_0^{t_1} dt_2 \cdots \int_0^{t_{n-1}} dt_n \\ \cdot \{\{\cdots\{f, H_1(-t_n)\}, \ldots, H_1(-t_2)\}, H_1(-t_1)\} \\ + \lambda^r \int_0^t dt_1 \int_0^{t_1} dt_2 \cdots \int_0^{t_{r-1}} dt_r \, \{\{\cdots\{\exp(-t_r L_{H_0}) \\ \times \exp(t_r L_H) f, H_1(-t_r)\}, \ldots, H_1(-t_2)\}, H_1(-t_1)\}. \end{aligned}$$

If the latter term goes to 0 as $r \to \infty$, then we have shown the validity of the

(3.5.7) Perturbation Series

$$\begin{aligned} e^{tL_{H_0+\lambda H_1}} f = f(t) + \sum_{n \geq 1} \lambda^n \int_0^t dt_1 \int_{t_1}^t dt_2 \cdots \int_{t_{n-1}}^t dt_n \\ \cdot \{\{\cdots\{f(t), H_1(t_n)\}, \ldots, H_1(t_2)\}, H_1(t_1)\}. \end{aligned}$$

(3.5.8) Remarks

1. If only the first few terms of (3.5.7) are kept, it can give a completely false picture of the time-dependence. For an oscillator with a changed frequency, $|\sin(1+\lambda)t - \sin t| \leq 2$ for all t, whereas the Taylor series in λ, $\sin(1+\lambda)t = \sin t + \lambda t \cos t + 0(\lambda^2)$ appears to grow with t.
2. Even if the canonical flow can also be expanded in λ, it is not necessarily true that the constants of motion can be expanded. Consider free motion on T^2: $(\varphi_1, \varphi_2) \to (\varphi_1 + t\omega_1, \varphi_2 + t\omega_2)$. If the frequencies have a rational ratio,

$\omega_1/\omega_2 = g_1/g_2$, then $\sin(g_2\varphi_1 - g_1\varphi_2)$ is a constant. The rationality of the ratio can be destroyed by an arbitrary small perturbation in the frequency, leaving no time-independent constants. Accordingly, a series expansion in λ with $g_1 \to g_1(1+\lambda)$ leads to $\varphi_2 g_1 \cos(g_2\varphi_1 - g_1\varphi_2)$, which is not globally defined.

3. The limit as $r \to \infty$ exists under the assumptions of (3.5.4). It can then be shown, in analogy with (3.5.4), that for fixed t the norm of the rth term is bounded by $c^r/r!$. This means that the perturbation series converges as well as the Taylor series for the exponential function.

(3.5.9) **Examples**

1. $\{H_0, H_1\} = 0$, so $H_1(t) = H_1$. Then for the perturbation series (3.5.7),

$$e^{tL_{H_0+\lambda H_1}} = e^{t\lambda L_{H_1}} e^{tL_{H_0}},$$

which can also be shown by making a series expansion in t and using the formulas $L_{H_0+\lambda H_1} = L_{H_0} + \lambda L_{H_1}$ and $L_{H_0}L_{H_1} = L_{H_1}L_{H_0}$.

2. The driven oscillator,

$$H_0 = \tfrac{1}{2}(p^2+q^2), \qquad H_1 = q\colon H_1(t) = q\cos t + p\sin t,$$

$$\left\{H_1(t_1), \begin{pmatrix} q(t) \\ p(t) \end{pmatrix}\right\} = \begin{pmatrix} -\sin(t_1 - t) \\ \cos(t_1 - t) \end{pmatrix} \Rightarrow \left\{H_1(t_1)\left\{H_1(t_2), \begin{pmatrix} q \\ p \end{pmatrix}\right\}\right\} = 0 \Rightarrow$$

$$\begin{aligned} e^{tL_H} q &= q\cos t + p\sin t + \lambda(\cos t - 1), \\ e^{tL_H} p &= -q\sin t + p\cos t - \lambda\sin t, \end{aligned}$$

which is the correct solution of $\dot q = p$, $\dot p = -q - \lambda$.

The most that can be said in complete generality about the influence of H_1 is (3.5.7). But often more precise estimates can be made for perturbations of integrable systems. Let $\bar I_j$ and $\bar\varphi_j$, $j = 1, \ldots, m$, be action and angle variables, and let us study

(3.5.10) $$H(\bar I, \bar\varphi) = H_0(\bar I) + \lambda H_1(\bar I, \bar\varphi).$$

Since H_1 depends periodically on $\bar\varphi$, it can be expanded in a Fourier series:

(3.5.11) $$\begin{aligned} H_1(\bar I, \bar\varphi) &= \sum_k \tilde H_k(\bar I) e^{i(\bar\varphi\cdot k)}, \\ \tilde H_k(\bar I) &= = (2\pi)^{-m} \int d\bar\varphi_1 \cdots d\bar\varphi_m e^{-i(\bar\varphi\cdot k)} H_1(\bar I, \bar\varphi), \end{aligned}$$

$$\begin{aligned} k &= (k_1, \ldots, k_m) \in \mathbb{Z}^m, \\ (\bar\varphi\cdot k) &:= \bar\varphi_1 k_1 + \bar\varphi_2 k_2 + \cdots + \bar\varphi_m k_m. \end{aligned}$$

Next we make a canonical transformation to the variables I, φ with the generator $S(I, \bar{\varphi})$:

(3.5.12) $$\bar{I}_j = I_j + \lambda \frac{\partial S}{\partial \bar{\varphi}_j}, \qquad \varphi_j = \bar{\varphi}_j + \lambda \frac{\partial S}{\partial I_j},$$

so that for some value of I, say $I = 0$, the system remains integrable to order λ. To do this, let

(3.5.13) $$\omega_j \frac{\partial S}{\partial \bar{\varphi}_j} + H_1(I, \bar{\varphi}) = \tilde{H}_{k=0}(I), \qquad \omega_j = \left. \frac{\partial H_0(I)}{\partial I_j} \right|_{I=0},$$

which is formally solved by

(3.5.14) $$S(I, \bar{\varphi}) = - \sum_{k \neq 0} \frac{e^{i(k \cdot \bar{\varphi})}}{i(\omega \cdot k)} \tilde{H}_k(I).$$

Then

(3.5.15) $$H = H_0(I) + \lambda \tilde{H}_{k=0}(I) + \lambda^2 H_2(I, \bar{\varphi}),$$

where

$$H_2 = \lambda^{-2} \left[H_0 \left(I + \lambda \frac{\partial S}{\partial \bar{\varphi}} \right) - H_0(I) - \lambda \omega_j \frac{\partial S}{\partial \bar{\varphi}_j} \right] + \lambda^{-1} \left[H_1 \left(I + \lambda \frac{\partial S}{\partial \bar{\varphi}}, \bar{\varphi} \right) - H_1(I, \bar{\varphi}) \right].$$

(3.5.16) **Remarks**

1. If $|I| < \lambda$, then H_2 remains finite as $\lambda \to 0$.
2. The term with $k = 0$ is obviously to be dropped from the sum (3.5.14). Because we dropped it, we removed $\tilde{H}_{k=0}(I)$, and thereby caused a change in the frequency of order λ.
3. The denominator of (3.5.14) vanishes whenever the frequencies have rational ratios. When that happens we assume that $\det(\partial^2 H / \partial I_i \, \partial I_j) \neq 0$, and instead of $I = 0$ we consider a nearby value of I for which the frequencies do not have rational ratios.
4. Even when the ratios of the frequencies are irrational, the denominators $(\omega \cdot k)$ may become arbitrarily small, and the convergence of the formal series (3.5.14) must be checked (**"the problem of small denominators"**).
5. The connection between the foregoing result and our earlier analysis is as follows: In (3.4.5; 1) we saw that $\Omega_{\pm}$ transforms H into H_0, and (3.5.2) provides an expansion for $\Omega(t)$. Hence it is reasonable to try to apply the same transformation to H, but using a convergence factor in the sense of (3.4.21) for the limit

$t \to \infty$:

$$\begin{aligned}\Omega_+ H &= \lim_{\alpha\to 0}\left(1-\lambda\int_0^\infty dt\, e^{-\alpha t} L_{H_1(t)}\right) H + O(\lambda^2)\\ &= H_0 + \lim_{\alpha\to 0}\lambda\left(\int_0^\infty dt\, e^{-\alpha t} L_{H_0} H_1(t) + H_1\right) + O(\lambda^2)\\ &= H_0 + \lambda \lim_{\alpha\to 0}\int_0^\infty dt\, \alpha e^{-\alpha t} H_1(t) + O(\lambda^2)\\ &= H_0 + \lambda \tilde{H}_{k=0}(I) + O(\lambda^2).\end{aligned}$$

(3.5.17) **Lemma**

Let H_1 be analytic on the domain $|\mathrm{Im}\,\bar{\varphi}_j| < \rho$, $|\bar{I}_j| < r$. If there exist $\tau > m+1$, $c \in \mathbb{R}^+$ such that

$$|(\omega\cdot k)|^{-1} \le c|k|^{\tau-m-1}, \qquad \forall k \ne 0,$$

where

$$|k| \equiv |k_1| + |k_2| + \cdots + |k_m|, \qquad k_i \in \mathbb{Z},$$

then the series (3.5.14) *converges to an analytic function on the same domain, and the estimate*

$$\|S\|_{\rho-\delta,r} \le \bar{c}\delta^{-\tau+1}\|H_1\|_{\rho,r}, \qquad 0 < \delta < \rho < 2,$$

where

$$\|F(I,\varphi)\|_{\rho,r} := \sup_{|I_j|<r}\ \sup_{|\mathrm{Im}\,\varphi_j|<\rho} |F(I,\varphi)|,$$

$$\bar{c} = c2^{3m}\left(\frac{2}{e}(\tau-m-1)\right)^{\tau-m-1},$$

holds.

(3.5.18) **Remarks**

1. The requirement of analyticity is not too severe for us.
2. By Cauchy's theorem, this estimate implies

$$\left\|\frac{\partial S}{\partial\bar{\varphi}_j}\right\|_{\rho-\delta,r} \le \bar{c}\delta^{-\tau}\|H_1\|_{\rho,r},$$

$$\left\|\frac{\partial S}{\partial I_j}\right\|_{\rho-\delta,r-\varepsilon} \le \bar{c}\delta^{-\tau+1}\varepsilon^{-1}\|H_1\|_{\rho,r},$$

$$\left\|\frac{\partial^2 S}{\partial I_i\,\partial\bar{\varphi}_j}\right\|_{\rho-\delta,r-\varepsilon} \le \bar{c}\delta^{-\tau}\varepsilon^{-1}\|H_1\|_{\rho,r}.$$

Therefore the matrix

$$\frac{\partial \varphi_i}{\partial \bar{\varphi}_j} = \delta_{ij} + \lambda \frac{\partial^2 S}{\partial I_i \, \partial \varphi_j}$$

is invertible for small enough λ, as is required for (3.1.6).

3. The condition on the ω's means that they must be sufficiently rationally independent. When they have rational ratios, resonance behavior occurs, which can amplify the effect of the perturbation dramatically. Perturbation methods can founder even when the system is only in the vicinity of a resonance.
4. As $|k|$ increases, the condition of (3.5.17) becomes weaker and thus makes the complement of the set M of ω's that satisfy it smaller. To show that (3.5.14) converges, it suffices to have boundedness by any power of $|k|$. However, if $\tau > 2m$, then the measure of

$$CM \equiv \{\omega \in \mathbb{R}^m : \exists k \in \mathbb{Z}^m \setminus \{0\} \quad \text{such that} \quad (\omega \cdot k) < \frac{|k|^{m+1-\tau}}{C}, \quad \text{and} \quad |\omega| < m \in \mathbb{R}^+\}$$

approaches zero as $1/C$ (Problem 7), although the set, which consists of points violating (3.5.17), contains all rational points. Thus CM is a strange example of an open, dense set of small measure. In this it resembles $\bigcup_{n=1}^{\infty}(\nu_n - \varepsilon 2^{-n}, \nu_n + \varepsilon 2^{-n})$, where $\nu_1, \nu_2, \ldots$ are the rational numbers and $\varepsilon > 0$. The measure of this set is less than $\sum_{n=1}^{\infty} \varepsilon 2^{-n+1} = 2\varepsilon$; that is, it is arbitrarily small, although the set is open and dense.
5. Geometrically, if $m = 2$, CM is the union of the interiors of the strips bounded by

$$\omega_2 = \frac{k_1}{k_2}\omega_1 \pm \frac{1}{C}|k_2|^{-1}(|k_1| + |k_2|)^{m+1-\tau}$$

in the $(\omega_1 - \omega_2)$ plane. If $\tau > 3$, the strips lie more narrowly along the lines through the origin with rational slope k_1/k_2 when k_1 and k_2 become greater, that is, when the ω's are more distantly rationally related. Resonances that are only distantly related have only a small effect, just like widely separated dissonant overtones in music, since they are excited only to a slight extent.
6. Since the rational numbers are dense in $\mathbb{R}$, every $x \in (0, 1)$ may be approximated arbitrarily well by p/q, with $p, q \in \mathbb{N}$. If this is done with decimal numbers, i.e., $q_n = 10^n$, then $|x - p_n/q_n| \leq 1/q_n$. This is certainly not optimal: for a given maximal value of the denominator the best approximation is obtained with a continued fraction. To write down the continued fraction, one would first find the integer part a_1 of $1/x$, then find the integer part a_2 of $1/(1/x - a_1)$, etc. The number x can thus be written

$$x = \cfrac{1}{a_1 + \cfrac{1}{a_2 + \cdots + \cfrac{1}{a_n + r_n}}}, \qquad r_n \in (0, 1).$$

With $r_n = 0$ we obtain a rational number p_n/q_n, for which Liouville was able to show that

$$\left| x - \frac{p_n}{q_n} \right| \le \frac{1}{a_{n+1} \cdot q_n^2}.$$

Furthermore, there is no rational number p'_n/q'_n closer than this to x with $q'_n < q_n$. Of course, many numbers can be approximated with better accuracy than $1/q_n^2$; if x is rational, then for some n, $r_n = 0$. The most irrational number, i.e., the number that is the least well approximated by rational numbers, is the one with the smallest a's, i.e., $a_i = 1$ for all i. It is not a transcendental number; rather,

$$g = \frac{1}{1 + \frac{1}{1+\cdots}}$$

satisfies the quadratic equation $g^2 + g = 1$, viz., $g = (\sqrt{5} - 1)/2$. The number g is a fetish of number worshippers, and appears often in their works, for instance as the golden mean: If the unit interval is divided in two in such a way that the ratio of the greater interval to the whole equals the ratio of the lesser interval to the greater, then $g/1 = (1 - g)/g$. For our purposes we note that the frequency condition is violated when there exist k_1 and $k_2 \in \mathbb{N}$ for which

$$\left(\left| \frac{\omega_1}{\omega_2} \right| - \frac{k_1}{k_2} \right) < \frac{\varepsilon}{k_2} (k_1 + k_2)^{3-\tau}.$$

If $\tau < 3$, then it is not possible for the reversed inequality to hold for all k_1 and $k_2 \in \mathbb{N}$, since even with decimal numbers we eventually have $|\alpha - k_1/k_2| < c/k_2$. With the use of continued fractions, it is even possible, according to the above, to exclude $\tau \le 4$. It follows from Problem 7 that this is the sharpest result possible; in that problem it is shown that if $\tau > 4$, then the measure of all $\alpha \in (0, 1)$ that violate $|\alpha - k_1/k_2| < \varepsilon/k_2^{\tau-2}$ goes to 0 with ε.

Proof: Since H is periodic in φ, we may move the path of integration for φ_j in (3.5.11) so that $\operatorname{Im} \varphi_j = \pm\rho$, with the same sign as k_j. There results

$$\|\tilde{H}_k(I)\|_r = \sup_{|I_j| < r} \left| \int \frac{d\varphi_1 \cdots d\varphi_m}{(2\pi)^m} e^{-i(k\cdot\varphi)} H_1(I, \varphi) \right| \le e^{-|k|\rho} \|H_1\|_{\rho,r}.$$

This means that

$$\begin{aligned} \|S\|_{\rho-\delta,r} &\le \sup_{|\operatorname{Im} \varphi_j| < \rho - \delta} \sum_{k \ne 0} e^{i(k\cdot\varphi) - |k|\rho} c|k|^{\tau-m-1} \|H_1\|_{\rho,r} \\ &\le \sum_{k \ne 0} e^{-|k|\delta} c|k|^{\tau-m-1} \|H_1\|_{\rho,r}, \end{aligned}$$

In order to bound the sum $\sum_k$, use the inequality

$$|k|^\sigma \le \left(\frac{2\sigma}{e\delta} \right)^\sigma e^{|k|\delta/2} \quad \forall \sigma > 0, \delta > 0,$$

and the elementary calculation that

$$\sum_{k_1} e^{-|k_1|\delta/2} = \frac{2}{1-e^{-\delta/2}} - 1 < \frac{8}{\delta}, \quad \forall \delta < 2.$$

Together these produce the inequality of (3.5.17):

$$\|S\|_{\rho-\delta,r} \leq c\|H_1\|_{\rho,r}\left(\frac{2(\tau-m-1)}{e\delta}\right)^{\tau-m-1}\sum_{k\neq 0} e^{-|k|\delta/2}$$

$$< c\|H_1\|_{\rho,r}8^m\delta^{-\tau+1}\left[\frac{2}{e}(\tau-m-1)\right]^{\tau-m-1}.$$

The same bound for the sum also shows the analyticity. □

(3.5.19) **Corollary**

With the assumptions of (3.5.17), *the effect of the perturbation for trajectories with* $|I_j| < \lambda$ *is simply the change in the frequency*

$$\omega_j \to \omega_j + \lambda \left.\frac{\partial \tilde{H}_{k=0}}{\partial I_j}\right|_{I=0},$$

to $O(\lambda^2)$. *Consequently,* $\dot{I}$ *is* $O(\lambda^2)$, *and the condition* $|I_j| < \lambda$ *continues to hold for* $|t| < c\lambda^{-1}$.

(3.5.20) **Remarks**

1. This corollary accords with intuition. For $\lambda = 0$, the I's are constant, and the φ's move with a finite angular velocity. Hence one would guess that for small λ, the average of H over φ is dominant.
2. Corollary (3.5.19) improves on (3.5.4), but with stronger assumptions, in that in (3.5.4) we only learned that $\forall |t| \leq c\lambda^{-1}$, the difference between the flows Φ_t^0 and Φ_t is $O(1)$, whereas now we learn that for this range of times only the frequencies are changed, to accuracy $O(\lambda)$.
3. The independence of the frequencies is necessary in order to exclude some trivial counterexamples to this corollary. For $m = 1$, independence simply means $\omega \neq 0$. If $H_0 = 0$ and $H_1 = \cos\varphi$, and therefore, $\tilde{H}_{k=0}(I) = 0$, then I and φ are constant in perturbation theory to $O(\lambda^2)$. On the other hand, this H generates the flow $\varphi \to \varphi$, $I \to I + \lambda t \sin\varphi$, which means that for $t \sim 1/\lambda$, the change of $O(1)$ rather than $O(\lambda)$.
4. The original action variables $\bar{I}_j$ depend on time to order $O(\lambda)$, while $\partial S/\partial\bar{\varphi}$ is a periodic function with mean zero. As a consequence, when averaged over time, these variables vary only to $O(\lambda^2)$. (This is referred to as the absence of "secular terms.")
5. In the Solar System the characteristic time is a year, and the perturbation due to planetary interactions is about 1/1000 the strength of the Sun. Corollary (3.5.19)

leaves us with the hope that over a millennium the planetary orbits will only vary by tenths of a percent. The verification of actual stability over 10^9 years is too difficult for analytic estimates, and can only be carried out with special-purpose computers [21].

6. The **adiabatic theorem** of classical mechanics follows from (3.5.19), and states that if the perturbation is switched on very slowly, the action variables remain unchanged. More precisely: Suppose that a family $H(q, p; \lambda)$ of Hamiltonians has action variables $I(q, p; \lambda)$ and that the frequency condition of (3.5.17) is satisfied. Now make λ time-dependent, $\lambda = \varepsilon t$, and try to solve the equations

$$\dot{q}(t) = H_{,p}(q(t), p(t); \varepsilon t), \qquad \dot{p}(t) = -H_{,q}(q(t), p(t); \varepsilon t).$$

Then there exist $c_i > 0$ and $\varepsilon_0 > 0$ such that

$$|I(q(t), p(t); \varepsilon t) - I(q(0), p(0), 0)| < \varepsilon c_1 \quad \text{for all} \quad \varepsilon < \varepsilon_0,\ |t| \le \frac{c_2}{\varepsilon_0}.$$

For a proof see Problem 8.

(3.5.21) **Examples**

1. $H_0 = \omega_j \bar{I}_j$ *and* $H_1(\bar{\varphi})$ *is independent of the* $\bar{I}$*'s.* In this case, H_2 vanishes, and perturbation theory actually yields the exact solution:

$$\bar{\varphi}_j = \varphi_j, \qquad \bar{I}_j = I_j - \lambda \sum_{k \neq 0} \frac{e^{i(k\cdot\varphi)} k_j}{(\omega \cdot k)} H_k,$$

where $\dot{\varphi}_j = \omega_j$ and $\dot{I}_j = 0$, is the solution to

$$\dot{\bar{\varphi}}_j = \omega_j, \qquad \dot{\bar{I}}_j = -\lambda \sum_k i k_j e^{i(k\cdot\varphi)} \tilde{H}_k.$$

2. *A driven oscillator.* $H = (p^2 + q^2)/2 + \lambda q$. We must first transform to action and angle variables in such a way that the point $\bar{I} = 0$ can be conveniently chosen:

$$p = \sqrt{2(\bar{I} + \rho) \cos \bar{\varphi}}, \qquad q = \sqrt{2(\bar{I} + \rho) \sin \bar{\varphi}}, \qquad \rho \in \mathbb{R}^+,$$
$$\Rightarrow H = \bar{I} + \rho + \lambda \sqrt{(2(\bar{I} + \rho) \sin \bar{\varphi}}, \quad \tilde{H}_{k=0} = 0.$$

Thus the frequency remains 1, and we calculate

$$S = \sqrt{2(I + \rho)} \cos \bar{\varphi}, \qquad \bar{I} = I - \lambda \sqrt{2(I + \rho)} \sin \bar{\varphi},$$
$$\varphi = \bar{\varphi} + \lambda \cos \bar{\varphi} / \sqrt{2(I + \rho)}.$$

Note that the transformation $(q, p) \to (q - \lambda, p)$ changes H into $(p^2 + q^2)/2 - \lambda^2/2$. If we put

$$\sqrt{2(\bar{I} + \rho)}(\sin \bar{\varphi}, \cos \bar{\varphi}) = (\sqrt{2(I + \rho)} \sin \varphi - \lambda, \sqrt{2(I + \rho)} \cos \varphi),$$

i.e.,

$$\bar{I} = I - \lambda\sqrt{2(I+\rho)}\sin\varphi + \lambda^2/2,$$

$$\cos\bar{\varphi} = \left[1 - \lambda\sqrt{\frac{2}{I+\rho}}\sin\varphi + \frac{\lambda^2}{2(I+\rho)}\right]^{-1/2}\cos\varphi,$$

then H becomes $I + \rho - \lambda^2/2$. Perturbation theory produces the same transformation up to $O(\lambda^2)$. Since $[\cdots]^{-1/2}$ has singularities for nonreal λ, a power series in λ would only have a finite radius of convergence, although the action and angle variables exist $\forall\lambda \in \mathbb{R}$.

3. *An oscillator with a changing frequency.* $H = (p^2+q^2)/2 + \lambda q^2 = \bar{I} + 2\lambda\bar{I}\sin^2\bar{\varphi}$. Now $\bar{H}_{k=0}(\bar{I}) = \bar{I}$, and the frequency changes to $1+\lambda$, the first-order approximation to $\sqrt{1+2\lambda}$. Moreover,

$$S = \frac{I}{2}\sin 2\bar{\varphi} \quad\Rightarrow\quad \bar{I} = I(1+\lambda\cos 2\bar{\varphi}), \qquad \varphi = \bar{\varphi} + \frac{\lambda}{2}\sin 2\bar{\varphi}.$$

Once again this constitutes the first two terms in a series in λ of a transformation, which turns H into $I\sqrt{1+2\lambda}$ (Problem 4). Perturbation theory converges for $|\lambda| < \frac{1}{2}$, while for $\lambda < -\frac{1}{2}$ the behavior becomes exponential rather than periodic. The time-evolution is not singular at $\lambda = -\frac{1}{2}$ (Problem 2); but at that point action and angle variables no longer exist.

(3.5.22) Problems

1. For a linear differential equation, $\exp(tL_H)$ can be written as a matrix $\exp(tA)$. What is the formula analogous to (3.5.2) for $\exp(t(A_0 + \lambda A_1))$?

2. Apply (3.5.7) to $H_0 = p^2/2$, $H_1 = -q^2/2$.

3. Calculate H_2 from (3.5.21; 2), as a cautionary example.

4. For what I and φ does H from (3.5.21; 3) become $I\sqrt{1+2\lambda}$? Compare with perturbation theory.

5. For what frequencies ω is perturbation theory (3.5.19) applicable to an oscillator with a periodic external force, $H = (p^2+q^2)/2 + \lambda q\cos\omega t$ (cf. (3.3.17; 3))?

6. For what values of λ would perturbation theory be expected to be useful for the pendulum $H = (\bar{I}+\omega)^2/2 + \lambda\cos\bar{\varphi}$?

7. For $m = 2$, show that the measure of CM (3.5.18; 4) for $\tau > 4$ tends to 0 as C grows without bound.

8. Prove (3.5.20; 6).

   Step 1: Find the Hamiltonian on the extended phase space that produces the correct equations of motion for φ and I, with $\lambda = \varepsilon t$.

   Step 2: Transform this perturbed Hamiltonian with the analogue of (3.5.14).

(3.5.23) Solutions

1.

$$\frac{d}{dt}e^{-tA_0}e^{t(A_0+\lambda A_1)} = \lambda e^{-tA_0}A_1 e^{t(A_0+\lambda A_1)} \rightarrow$$

$$e^{t(A_0+\lambda A_1)} = e^{tA_0} + \lambda \int_0^t dt_1\, e^{(t-t_1)A_0}A_1 e^{t_1(A_0+\lambda A_1)}.$$

2. $q(t) = q + tp$, $p(t) = p$. $H_1(t_n) = -(q+t_n p)^2/2$. $\{H_1(t_n), q(t)\} = (t_n - t)q(t_n) \Rightarrow$ $\{\{\cdots\{q(t), H_1(t_n)\}, \ldots, H_1(t_2)\}, H_1(t_1)\} = (-1)^n(t_1-t_2)(t_2-t_3)\cdots(t_n-t)q(t_1)$. Using the formula $\int_0^1 dt\, t^p(1-t)^q = p!q!/(p+q+1)!$, integrating

$$\int_0^t dt_1 \int_{t_1}^t dt_2 \cdots \int_{t_{n-1}}^t dt_n$$

produces the series

$$q\frac{t^{2n}}{2n!} + p\frac{t^{2n+1}}{(2n+1)!},$$

which are the Taylor coefficients of $q\cosh\lambda^{1/2}t + p\lambda^{-1/2}\sinh\lambda^{1/2}t$.

3. $H_2 = \lambda^{-1}[\sqrt{2(I+\rho-\lambda\sqrt{2(I+\rho)}\sin\bar{\varphi})} - \sqrt{2(I+\rho)}]\sin\bar{\varphi}$.

4.

$$\sqrt{2\bar{I}}\sin\bar{\varphi} = \sqrt{2I/\sqrt{1+2\lambda}}\sin\varphi,\ \sqrt{2\bar{I}}\cos\bar{\varphi}$$
$$= \sqrt{2I\sqrt{1+2\lambda}}\cos\varphi \rightarrow$$

$$\bar{I} = I\sqrt{1+2\lambda}\{\cos^2\varphi + \sin^2\varphi/(1+2\lambda)\},\ \cos\varphi = \sqrt{\bar{I}/I\sqrt{1+2\lambda}}\cos\bar{\varphi}.$$

Expanding in a power series in λ gives (3.5.21; 3).

5. Introduce action and angle variables for (q, p), and make the substitution $(t, E) \rightarrow (t/\omega, \omega E)$. Then $\mathcal{H} = I - \omega E + \lambda\sqrt{2I}\sin\varphi\cos t$. The irrationality of the ratios of the frequencies is only assured when $\omega \notin \mathbb{Q}$.

6. By (3.3.14), the action variable for H equals

$$I(E) = \frac{1}{2\pi}\oint d\varphi\sqrt{E+\lambda\cos\varphi},$$

which is analytic in λ for $|\lambda| < E$.

7. We may suppose $0 \le \omega_1/\omega_2 \equiv \alpha \le 1$. Consider the measure of

$$B_\varepsilon \equiv \{\alpha \in [0,1]: \exists(k_1,k_2) \in \mathbb{Z}^2\backslash\{(0,0)\}: |\alpha k_1 - k_2| < \varepsilon|\mathbf{k}|^{-n}\},$$

where $\varepsilon = 1/c\omega_2 < 1$ and $n = \tau - 3 > 1$. This is contained in the set of α's with $|\alpha - k_2/k_1| < \varepsilon|k_1|^{-n-1}$. The k's must have the same sign, and $|k_2| < |k_1| + 1$. Thus, for $n > 1$,

$$\mu(B_\varepsilon) \le 2\sum_{\substack{k_1>0\\ k_1+1>k_2>0}} \varepsilon k_1^{-n-1} \le 2\varepsilon\sum_{k_1>0}\frac{k_1+1}{k_1^{n+1}} = \varepsilon\bar{c}, \quad \text{with} \quad \bar{c} < \infty.$$

8. Let $S_1(\lambda)$ be the generator of the transformation $(q, p) \to (\bar{\varphi}, \bar{I})$: $p\,dq = \bar{I}\,d\bar{\varphi} + dS_1$. On the extended phase space, however, with $\lambda = \varepsilon t$, dS_1 gains a contribution $\varepsilon S_{1,\lambda}\,dt$. Hence the canonical 2-form is

$$dq \wedge dp - dt \wedge dE = d\bar{\varphi} \wedge d\bar{I} - dt \wedge (d(E - \varepsilon S_{1,\lambda}));$$

$\bar{\varphi}$ and $\bar{I}$, when supplemented with $\bar{t} = t$ and $\bar{E} = E + \varepsilon S_{1,\lambda}$, become canonical coordinates. Then $\mathcal{H} = H - E = H + \varepsilon S_{1,\lambda} - \bar{E}$, where, even though S_1 is not a globally defined variable, $S_{1,\lambda}$ is. We now consider $\varepsilon S_{1,\lambda}$ as the perturbation, and if

$$S_{1,\lambda} = \sum_k \tilde{H}_k(\bar{I}, \varepsilon\bar{t}) e^{i(k,\bar{\varphi})},$$

then

$$S = -\sum_{k\neq 0} \frac{\tilde{H}_k(I, \varepsilon\bar{t}) e^{i(k,\bar{\varphi})}}{i(\omega \cdot k)}$$

generates a transformation

$$\bar{I} = I + \varepsilon\frac{\partial S}{\partial \bar{\varphi}}, \qquad \bar{\varphi} = \varphi - \varepsilon\frac{\partial S}{\partial I},$$
$$\bar{E} = E + \varepsilon\frac{\partial S}{\partial \bar{t}}, \qquad \bar{t} = t,$$

where $\bar{E}$ differs from E only to $O(\varepsilon^2)$, since $\partial S/\partial \bar{t} = \varepsilon\, \partial S/\partial \lambda$. To $O(\varepsilon^2)$, the effective Hamiltonian thus becomes $H(I) + \varepsilon \tilde{H}_{k=0}(I)$, and $\bar{I}$ and I differ only by ε times a function $p(t) = \partial S/\partial \bar{\varphi}$. Hence $|\bar{I} - I| < \varepsilon|p(t)| + c\varepsilon^2 t$, and p depends on t via $\tilde{H}_k$ and $\bar{\varphi}$. The latter dependence is periodic, while the former is bounded independently of ε for $0 < t < 1/\varepsilon$ by $\sup_{0<\lambda<1} \tilde{H}_k(I, \lambda)$. Therefore both summands satisfy the claims made in (3.5.20; 6).

3.6 Perturbation Theory: The Iteration

If an integrable system is perturbed, many of the invariant tori are completely destroyed, while others are only deformed. If the perturbation is sufficiently small, the ones that are only deformed fill up most of the phase space.

In §3.5 a perturbed integrable system was transformed into another integrable system, up to $O(\lambda^2)$. The question arises of whether this procedure can be repeated to eliminate the perturbation completely. There has long been a wide-spread opinion that with an arbitrarily small perturbation—a **"speck of dust"**—all the constants other than H are destroyed, and the trajectory winds around densely through the energy surface (is ergodic). Thanks to the work of Kolmogoroff, Arnold, and Moser, it is now known that it is not so. Even if there exist no constants other than H, for small λ, enough m-dimensional submanifolds exist so that in most cases the system acts virtually like an integrable system.

Here we shall discuss the simplest nontrivial case

$$(3.6.1)\qquad \begin{aligned} H_0(\bar{I},\bar{\varphi}) &= \bar{I}_j\omega_j + \tfrac{1}{2}C_{ij}(\bar{\varphi})\bar{I}_i\bar{I}_j, \\ H_1(\bar{I},\bar{\varphi}) &= A(\bar{\varphi}) + B_j(\bar{\varphi})\bar{I}_j, \qquad j=1,\ldots,m, \end{aligned}$$

and try to reduce H_1 to zero by a series of transformations. If only H_0 survives, then the torus $I_j = 0$ is time-invariant, and is filled by the trajectory $\varphi_j \to \varphi_j + \omega_j t$.

(3.6.2) **Remarks**

1. We may allow C to depend on $\bar{\varphi}$, because the equations of motion for $H = H_0(I^{(\infty)}, \varphi^{(\infty)})$,

$$\dot{\varphi}_j^{(\infty)} = \omega_j + C_{j\ell}^{(\infty)} I_\ell^{(\infty)}, \qquad \dot{I}_j^{(\infty)} = -\frac{1}{2}\frac{\partial C_{i\ell}^{(\infty)}}{\partial \varphi_j} I_i^{(\infty)} I_\ell^{(\infty)},$$

are satisfied by $I_j^{(\infty)}(t) = 0$, $\varphi_j^{(\infty)}(t) = \varphi_j^{(\infty)}(0) + \omega_j t$. Therefore H_0 fails to be integrable, but has an invariant torus.
2. The analysis made below works just as well when H_0 contains terms of higher orders in I and when H_1 is more complicated.
3. To a first approximation the frequencies are changed by $\tilde{B}_j(k=0)$, which could affect the rational independence required in (3.5.17). If $\det C \neq 0$, however, we can recover the old frequencies ω_j from $\partial H_0^{(1)}/\partial I_{j|I=0}$ by shifting the I coordinate, and are justified in continuing to use (3.5.17) in the succeeding steps.
4. A more effective technique than using a power series is to iterate the procedure of §3.5. In this way, if A and B are $O(\lambda)$, we go to powers $\lambda^2, \lambda^4, \lambda^8, \ldots$, and after the nth iteration H_1 is only $O(\lambda^{2n})$. This idea goes back to Newton, who used it to approximate the zero of a function $f: \mathbb{R} \to \mathbb{R}$ by a sequence $\{x_n\}$ which satisfies the recursion relation $x_{n+1} - x_n = -f(x_n)/f'(x_n) \equiv \varepsilon_n$. If $f \in C^2(\mathbb{R})$, then

$$f(x_n) = f(x_{n-1}) + \varepsilon_{n-1} f'(x_{n-1}) + O(\varepsilon_{n-1}^2) = O(\varepsilon_{n-1}^2)$$

and, if $|f'(x_n)| > c$ for all n, then also $\varepsilon_n = O(\varepsilon_{n-1}^2)$. If x_1 is chosen favorably, then ε_n converges to zero much more rapidly than would be the case for a primitive power series.

The generator S from (3.5.12) is now chosen so as to produce a point transformation combined with a φ-dependent shift in I. This leaves the form of $H_0 + H_1$ invariant. Next we try to make the new A and B quadratic in the old ones, which we consider as small to first order.

(3.6.3) **Step 1**

If the X and Y of the generators (for $\xi \in \mathbb{R}^m$)

$$S(I, \bar{\varphi}) := \xi_j \bar{\varphi}_j + X(\bar{\varphi}) + I_j Y_j(\bar{\varphi}),$$

$$\varphi_j = \bar{\varphi}_j + Y_j(\bar{\varphi}), \qquad \bar{I}_j = I_j + \frac{\partial X}{\partial \bar{\varphi}_j} + I_\ell \frac{\partial Y_\ell}{\partial \bar{\varphi}_j} + \xi_j,$$

satisfy the equations

$$A(\bar{\varphi}) + \omega_j \frac{\partial X}{\partial \bar{\varphi}_j} = \tilde{A}(k=0),$$

$$B_j(\bar{\varphi}) + \omega_i \frac{\partial Y_j}{\partial \bar{\varphi}_i} + C_{ji}\left(\frac{\partial X}{\partial \bar{\varphi}_i} + \xi_i\right) = 0,$$

then

$$H = H_0 + H_1 = \omega_j I_j + \tfrac{1}{2} C_{ij}^{(1)} I_i I_j + A^{(1)} + B_j^{(1)} I_j + \tilde{A}(k=0) + \omega_j \xi_j,$$

$$A^{(1)} = \tfrac{1}{2} C_{ij}\left(\frac{\partial X}{\partial \bar{\varphi}_i} + \xi_i\right)\left(\frac{\partial X}{\partial \bar{\varphi}_j} + \xi_j\right) + B_j\left(\xi_j + \frac{\partial X}{\partial \bar{\varphi}_j}\right),$$

$$B_\ell^{(1)} = C_{ij}\left(\frac{\partial X}{\partial \bar{\varphi}_i} + \xi_i\right)\frac{\partial Y_\ell}{\partial \bar{\varphi}_j} + B_j \frac{\partial Y_\ell}{\partial \bar{\varphi}_j},$$

$$C_{\ell m}^{(1)} = C_{ij}\left(\delta_{i\ell} + \frac{\partial Y_\ell}{\partial \bar{\varphi}_i}\right)\left(\delta_{jm} + \frac{\partial Y_m}{\partial \bar{\varphi}_j}\right).$$

(3.6.4) **Remarks**

1. This follows the notation of (3.5.11) for the Fourier coefficients. It is easy to obtain an expression for H by substitution. ($\tilde{A}(0)$ and $\omega_j \xi_j$ are independent of φ and I, and therefore inessential.)
2. If we want to express the coefficients of H in terms of φ, we must first convince ourselves that the mapping $\bar{\varphi} \to \varphi$ is bijective. Since φ is independent of I, the structure of the Hamiltonian (3.6.1) is preserved.
3. Although S is not periodic in φ, dS is, and hence it can be defined globally.

(3.6.5) **Estimate of S**

In order to solve the equations for X, ξ, and Y, let us suppose as we did in (3.5.17) that A, B, and C are analytic for $\|\text{Im } \bar{\varphi}_j| < \rho$. Such functions f always satisfy the convenient estimate that $\| f, \bar{\varphi}\|_{\rho-h} \le h^{-1} \| f \|_\rho$. The first equation of (3.6.3) is satisfied if we write

(3.6.6) $$X(\bar{\varphi}) = -\sum_{k \neq 0} \frac{e^{i(k\cdot\bar{\varphi})}\tilde{A}(k)}{i(\omega \cdot k)},$$

from which the quantity

(3.6.7) $$E_i(\bar{\varphi}) := C_{ij}(\bar{\varphi}) \frac{\partial X(\bar{\varphi})}{\partial \bar{\varphi}_j}$$

which appears in the second equation can be calculated. The $k = 0$ and $k \neq 0$ parts of this equation become

$$\tilde{C}_{ji}(0)\xi_i = -\tilde{B}_j(0) - \tilde{E}_j(0);$$

(3.6.8)
$$Y_j(\bar{\varphi}) = -\sum_{k\neq 0} \frac{e^{i(k\cdot\bar{\varphi})}}{i(\omega\cdot k)}[\tilde{B}_j(k) + \tilde{C}_{ji}(k)\xi_i + \tilde{E}_j(k)].$$

Lemma (3.5.17) gives bounds for X and Y. The I-dependence is now explicit, and we can forget about the r in the norm $\| \|_{\rho,r}$ (assuming that τ is always ≥ 1):

(3.6.9)
$$\|X\|_{\rho-h} \leq ch^{-\tau+1}\|A\|_\rho,$$
$$\left\| \frac{\partial X}{\partial \bar{\varphi}_j} \right\|_{\rho-h} \leq ch^{-\tau}\|A\|_\rho, \qquad j = 1, \ldots, m.$$

With the resulting inequality,

(3.6.10)
$$\|E_j\|_{\rho-h} \leq cmh^{-\tau}\|C\|_\rho\|A\|_\rho,$$
$$\|C\|_\rho \equiv \max_{i,j} \|C_{ij}\|_\rho,$$

the remaining quantities can be bounded:

(3.6.11)
$$|\xi_i| \leq m\|\tilde{C}(0)^{-1}\|\{\|B\|_\rho + cmh^{-\tau}\|C\|_\rho\|A\|_\rho\},$$
$$\left\| \frac{\partial Y_j}{\partial \bar{\varphi}_i} \right\|_{\rho-2h} \leq \{\|A\|_\rho\|C\|_\rho mc^2h^{-2\tau} + c\|B\|_\rho h^{-\tau}\}$$
$$\cdot \{1 + m\|C\|_\rho\|\tilde{C}(0)^{-1}\|\}.$$

(3.6.12) Remarks

1. Define the norm $\|v\| = \max_i |v_i|$ for a vector (v_i) and $\|M\| = \max_{i,j} |M_{ij}|$ for a matrix (M_{ij}). Then for a product of a matrix and a vector, $\|Mv\| \leq m\|M\| \cdot \|v\|$, and for the product of two matrices, $\|M_1 \cdot M_2\| \leq m\|M\|_1 \cdot \|M_2\|$.
2. The bound for $\partial Y_i/\partial\bar{\varphi}_j$ shows that the matrix $\partial\varphi_i/\partial\bar{\varphi}_j$ is invertible for small enough A and B, and $\bar{\varphi} \to \varphi$ is a diffeomorphism.

These bounds will now be used to determine by how much $H_2 \equiv A^{(1)} + B_j^{(1)} I_j$ is reduced with respect to H_1. (Recall that $\|C\|_\rho \geq \|\tilde{C}(0)\| \geq (m\|\tilde{C}(0)^{-1}\|)^{-1}$.)

(3.6.13) Estimate for H_2

Let

$$\|H_1\|_\rho = \max\{\|A\|_\rho, \|B\|_\rho\},$$
$$\|H_2\|_{\rho-3h} = \sup_{|\mathrm{Im}\,\varphi_j|<\rho-3h} \max\{\|A^{(1)}\|, \|B^{(1)}\|\},$$

and $\Gamma = \max\{1, m\|C\|_\rho, m\|\tilde{C}(0)^{-1}\|\}$. If $\|H_1\|_\rho \leq h^{3\tau}/(16c^3 3^\tau \Gamma^6)$ and $\frac{1}{4}ch^{-\tau} \geq 1$, then $\|H_2\|_{\rho-3h} \leq \|H_1\|_\rho^2 16c^3 h^{-3\tau}\Gamma^6$.

Proof: If we substitute from (3.6.9) and (3.6.11) into (3.6.3), we must bear in mind that the $\sup_{|\text{Im}\,\bar{\varphi}_j|<\rho-2h}$ is taken in (3.6.11) and that it must be rewritten in terms of φ. If the condition on $\|H_1\|_\rho$ holds, then $|\bar{\varphi} - \varphi(\bar{\varphi})| < h\ \forall |\text{Im}\,\bar{\varphi}_j| < \rho - 2h$, and the strip $\{\varphi: |\text{Im}\,\varphi_j| < \rho - 3h\}$ is contained in the image of the strip $\{\bar{\varphi}: |\text{Im}\,\bar{\varphi}_j| < \rho - 2h\}$. The estimate then follows. □

The simple recursive form of (3.6.13) invites us to repeat the procedure n times. However, note that in the norm ρ is reduced by $3h$. In the nth step h_n must be chosen small enough so that $\sum_{n=1}^{\infty} h_n < \rho/3$. The only way that the factor $h_n^{-3\tau}$ in (3.6.13) can be controlled is that $\|H_{n-1}\|$ is squared and is sufficiently small, provided that $\|H_1\|$ is. Moreover, C is changed by the transformation, and we have to check whether Γ continues to be bounded. At any rate, the above analysis can be repeated, producing an

(3.6.14) **Estimate of H_n**

Let $h_n := h3^{-n+1}$ and $\rho_n := \rho - 3\sum_{j=1}^{n=1} 3^{-j+1}h > \rho - 9h/2$. Then, in the notation of (3.6.13) (and writing $\Gamma_{n-1} := \max\{1, m\|C^{(n-1)}\|_{\rho_{n-1}}, m\|\tilde{C}^{(n-1)}(0)^{-1}\|\}$),

$$\|H_n\|_{\rho_n} \leq \|H_{n-1}\|_{\rho_{n-1}}^2 16c^3 h^{-3\tau} 3^{3\tau n} \Gamma_{n-1}^6 3^{-6\tau}.$$

This enables us to let $n \to \infty$:

(3.6.15) **Convergence of the Iteration**

The recursion formula

$$x_n = x_{n-1}^2 \gamma \delta^n$$

is solved by

$$x_n = \frac{(\gamma\delta^3 x_1)^{2^{n-1}}}{\gamma\delta^{n+2}}.$$

Thus

$$\|H_n\|_{\rho_n} \leq (16c^3 h^{-3\tau} 3^{3\tau}\Gamma^6 \|H_1\|_\rho)^{2^{n-1}} (16c^3 h^{-3\tau}\Gamma^6 3^{3\tau n})^{-1},$$

and if $\Gamma > \Gamma_n\ \forall n$ and

$$\|H_1\|_\rho < \frac{h^{3\tau}}{16c^3 3^\tau \Gamma^6},$$

then

$$\|H_n\|_{\rho_n} \geq \|H_n\|_{\rho-9h/2}$$

converges to zero.

(3.6.16) Remarks

1. The generalization of the condition in (3.6.13), i.e., that at the nth step φ is changed by less than h_n, now reads

$$\|H_n\|_{\rho_n} c\Gamma^3 h^{-\tau} 3^{\tau(n-1)} \leq \tfrac{1}{4}$$

   and is guaranteed when the iteration converges.
2. This procedure converges uniformly in the strip $|\mathrm{Im}\,\varphi_j| < \rho - 9h/2$, and thus the limiting function is analytic in that region.

We must next convince ourselves that we can bound Γ at each step, in which case the small denominators are under control.

(3.6.17) Estimate for $C^{(n)}$

Inequality (3.6.11) implies that

$$\|C^{(n)}\|_{\rho_n} \leq \|C^{(n-1)}\|_{\rho_{n-1}}(1 + 4\|H_n\|_{\rho_n} 4c^2 h^{-2\tau} 3^{2\tau(n-1)}\Gamma^3)^2.$$

Estimating $C^{(n)}$ is thus a matter of checking the convergence of

$$\prod_{n=1}^{\infty}(1+x_n)^2 = \exp\left[2\sum_{n=1}^{\infty}\ln(1+x_n)\right] \leq \exp\left[2\sum_{n=1}^{\infty}x_n\right],$$

where

$$x_n \leq \frac{(16c^3 3^{3\tau} h^{-3\tau}\Gamma^6\|H_1\|_\rho)^{2^{n-1}}}{4\Gamma^3 h^{-\tau} 3^{\tau(n+2)}}.$$

But this clearly converges, because for $16c^3 h^{-3\tau} 3^{3\tau}\Gamma^6\|H_1\|_\rho < \frac{3}{4}$, it is easy to see that

$$\|C^{(n)}\|_{\rho_n} < 2\|C\|_\rho \quad \forall n.$$

It is somewhat more troublesome to deal with $\tilde{C}^{(n)}(0)^{-1}$:

(3.6.18) Estimate of $\tilde{C}^{(n)}(0)^{-1}$

We begin with

$$\tilde{C}_{ij}^{(1)}(0) = (2\pi)^{-m}\int d\varphi_1\cdots d\varphi_m\,(\delta_{i\ell} + Y_{i,\ell})C_{\ell k}(\delta_{kj} + Y_{j,k}).$$

This is averaged over the φ's, but Y and C are given as functions of $\bar{\varphi}$. Transforming to barred variables and using matrix notation ($Y' = \partial Y_i/\partial\bar{\varphi}_j$, etc.),

$$\|\tilde{C}(0) - \tilde{C}^{(1)}(0)\| = \|(2\pi)^{-m}\int d\bar{\varphi}_1\cdots d\bar{\varphi}_m \cdot \{C - (1+Y')C(1+Y'^{t})\det(1+Y')\}\|$$

$$= \|(2\pi)^{-m} \int d\bar{\varphi}_1 \cdots d\bar{\varphi}_m \{C(1 - \det(1 + Y'))$$
$$- (CY'' + Y'C + Y'CY'^t) \det(1 + Y')\}\|$$
$$\leq \|C\|_{\bar{\rho}} \frac{3m\|Y'\|_{\bar{\rho}} + m^2\|Y'\|^2_{\bar{\rho}}}{1 - m\|Y'|_{\bar{\rho}}} \quad \forall \bar{\rho} > 0,$$

where we have used $|\det(1 + Y')| \leq (1 - m\|Y'\|)^{-1}$ (Problem 1). From this we get a bound for $\tilde{C}^{(1)}(0)^{-1}$, for

$$\|\tilde{C}^{(1)}(0)^{-1}\| \leq \|\tilde{C}(0)^{-1}\|(1 - m^2\|\tilde{C}(0)^{-1}\| \cdot \|\tilde{C}(0) - \tilde{C}^{(1)}(0)\|)^{-1}$$

(Problem 2). By (3.6.11), $\|Y'\|_{\rho-2h} \leq 4\Gamma^3 c^2 h^{-2\tau}\|H_1\|_\rho/m$, and generalizing this to the nth step yields

$$\frac{\|\tilde{C}^{(n)}(0)^{-1}\|}{\|\tilde{C}^{(n-1)}(0)^{-1}\|} \leq \Bigg(1 - 2m^2\|\tilde{C}^{(n-1)}(0)^{-1}\| \, \|C\|_{\rho_n} \cdot 2$$
$$\cdot \frac{12\Gamma^3 c^2 h^{-2\tau} 3^{2\tau(n-1)}\|H_n\|_{\rho_n} + 16\Gamma^6 c^4 h^{-4\tau} 3^{4\tau(n-1)}\|H_n\|^2_{\rho_n}}{1 - 4\Gamma^3 c^2 h^{-2\tau} 3^{2\tau(n-1)}\|H_n\|_{\rho_n}}\Bigg)^{-1}.$$

Under the same circumstances as in (3.6.17), $\|\tilde{C}^{(n)}(0)^{-1}\|$ is bounded in n (cf. Problem 3), $\|\tilde{C}^{(n)}(0)^{-1}\| < 2\|\tilde{C}(0)^{-1}\| \; \forall n$.

Now that we have convinced ourselves of the boundedness of Γ, we collect the results of Kolmogorov, Arnold, and Moser in the

(3.6.19) K–A–M Theorem

For H as in (3.6.1), suppose that:

(a) $|(\omega \cdot k)|^{-1} < c|k|^{\tau-m-1} \; \forall k \in \mathbb{Z}^m \setminus \{\mathbf{0}\}$, *for some* $c > 0$ *and* $\tau \geq 1$;

(b) A, B, *and* C *are analytic in* $|\text{Im}\, \bar{\varphi}_j| < \rho$; *and*

(c)

$$\sup_{|\text{Im}\, \bar{\varphi}_j| < \rho} \max(|A|, |B_j|) \leq h^{3\tau} \Bigg(16c^3 3^\tau m^6 \times \Bigg(2 \sup_{|\text{Im}\, \bar{\varphi}_j| < \rho} \max\{\|C\|, \|\tilde{C}(0)^{-1}\|\}\Bigg)^6\Bigg)^{-1},$$

and h is less than both $2\rho/9$ and $c^{-1/\tau}$.

Then there exists a canonical transformation to $\varphi^{(\infty)}$, $I^{(\infty)}$, which is analytic in φ for $|\text{Im}\, \varphi_j| < \rho - 9h/2$ and affine[4] in I, such that

$$H = \omega_j I_j^{(\infty)} + \tfrac{1}{2} C_{ij}^{(\infty)}(\varphi^{(\infty)}) I_i^{(\infty)} I_j^{(\infty)},$$

[4]Affine = inhomogeneous linear, that is, of the form $I \to aI + b$.

FIGURE 3.24. The invariant tori.

where $C^{(\infty)}$ *is analytic for* $|\text{Im}\ \varphi_j^{(\infty)}| < \rho - 9h/2$.

(3.6.20) **Remarks**

1. As already mentioned, this theorem can be extended to a wider class of H's, and analyticity can be weakened to sufficiently-often differentiability.
2. Condition (c) is by no means necessary, but as yet no one has been able to improve it in such a way that the strong fall-off with m is essentially better. For systems with many degrees of freedom, the perturbation has to be so ridiculously small for this theorem to apply that it is questionable whether there is any physical relevance in such cases.
3. This result shows that the torus $I = 0$, which is invariant under Φ_t^0, gets deformed only moderately when the perturbation stays small, which happens when there is a bound on the independence of the frequencies. Initially, our transformation tells us nothing about nonzero values of I. The analogous procedure can only be carried out at other values of I if the frequency condition is fulfilled. Hence it is not possible to transform the system into something integrable on a whole neighborhood of $I = 0$.
4. Although the open regions on which the conditions of (3.6.19) are violated are dense, their total measure on compact sets goes to zero for vanishingly small perturbations. The motion on these sets is chaotic. They are divided into disconnected components by the invariant surfaces only when $m = 2$, in which case the trajectory cannot leave the part of phase space bounded by the invariant tori (see Figure 3.24).
5. Measure-theoretical ideas are usually more relevant than topological ones, when one wants to determine whether the invariant submanifolds are mathematically pathological. Thinking topologically, one would call the invariant submanifolds

exceptional as complements of open, dense sets. However, since they have large Liouville measures, they are more the rule than the exception. On the other hand, since the unstable region is dense, and any measurement suffers from some inexactness, one can never be certain of being in a region of stability.

(3.6.21) **Example**

The Double Pendulum. Even in two dimensions this is a nonintegrable system which nicely illustrates the K–A–M theorem. If the two angles are called φ_1 and φ_2, and masses and lengths are set equal to one for simplicity, then the kinetic energy of the system becomes

$$2T = \dot{\varphi}_1^2 + (\dot{\varphi}_1 \sin \varphi_2)^2 + (\dot{\varphi}_1(1 + \cos \varphi_2) + \dot{\varphi}_2)^2.$$

The gravitational potential energy is

$$V = 2 \cos \varphi_1 + \cos(\varphi_1 + \varphi_2) + 3$$

(in units where $g = 1$ and with a convenient added constant). This gives a Hamiltonian flow without any further constants, but in the limits $E \downarrow 0$ and $E \uparrow \infty$ it leads to a simple situation. In the former limit there are oscillations around the minima of V at $\varphi_1 = \pi$ and $\varphi_2 = 0$. In the latter limit V becomes negligible, which makes the angular momentum L a second constant. In between the system is chaotic, and by varying E we can observe the break-up of the K–A–M-tori. Analytically this is impossible, and we have to resort to computer studies [22]. They show how the tori $L =$ const. get deformed for $E = 20$, and for $E = 10$, most of them dissolve into chaotic regions. Near $E = 6$ the most robust torus, the frequency ratio of which is the golden mean $(\sqrt{5} - 1)/2$, disappears. At small energies $E < 1.5$ other invariant tori form, and the usual picture for the harmonic oscillator emerges at $E = 0$. These features are best illustrated in the **Poincaré sections**, which show where the orbit in the (L_1, φ_1)-plane traverses the submanifold $\varphi_2 = 0$, $H = E$. There the periodic orbits are points, quasi-periodic orbits are lines, and chaotic regions become bands. Figure 3.25 shows this for some representative energies.

We conclude by summarizing the general facts learned about Hamiltonian systems and seeing what impression they give us about the global structure of a canonical flow.

In the standard case for physics, where forces decrease with distance, particles with enough energy can escape. Their momenta will approach constants as $t \to \pm\infty$, with which the system will be integrable in the parts of phase space where enough particles escape. All trajectories will be diffeomorphic to $\mathbb{R}$ in these regions, and the canonical flow can be transformed into a linear flow.

Orbits that always return on themselves fill up higher-dimensional manifolds, and it is in this situation that the most complicated things can happen. In the special case of an integrable system, these manifolds are m-dimensional tori, which collapse to lower-dimensional figures in those parts of phase space where the

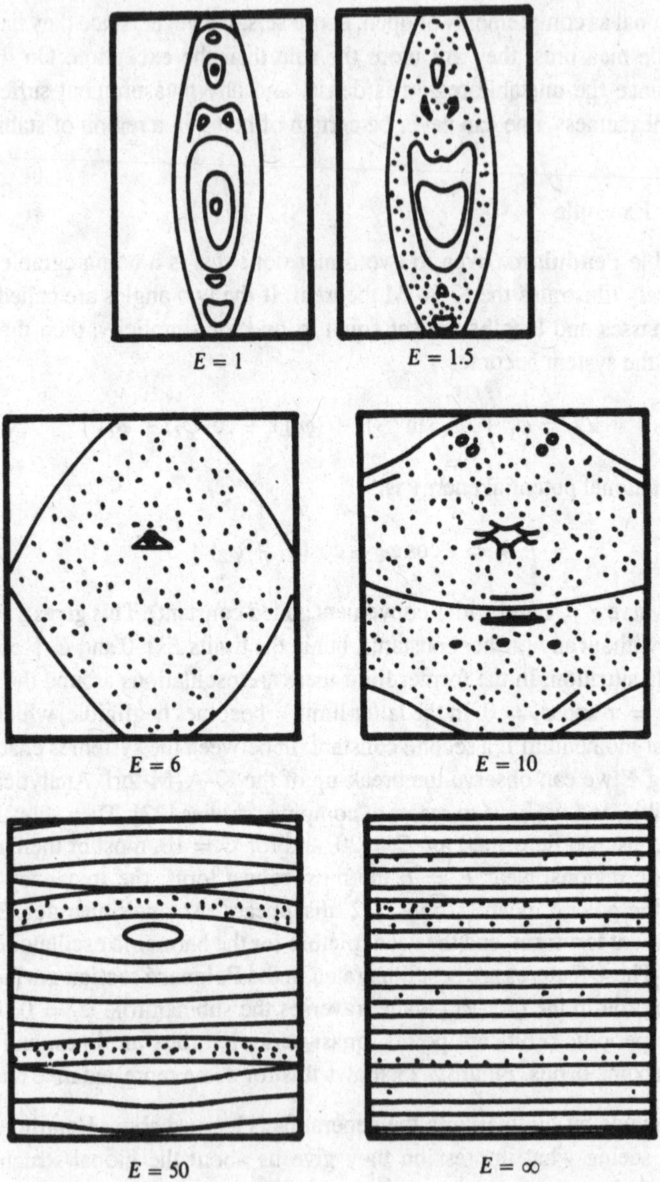

FIGURE 3.25. Poincaré sections of the double pendulum at various energies.

frequencies have rational ratios. Perturbations often destroy many of these tori, making the trajectory cover a $(2m-1)$-dimensional region, while many other tori remain, for which the system acts like an integrable system.

(3.6.22) Problems

1. For a Hermitian matrix M such that $\|M\| < 1/m$, show that $|\det(1 + M)| < 1/(1 - m\|M\|)$.

2. Assuming that $m^2\|M_1\|/\|M_0\| < 1$, show that $\|(M_0 + M_1)^{-1}\| \leq \|M_0^{-1}\|(1 - m^2\|M_1\| \cdot \|M_0\|^{-1})^{-1}$.

3. Complete Estimate (3.6.18).

4. Investigate the convergence of the perturbation series for

$$H = \frac{\omega}{2}(p^2 + q^2) + \frac{c}{8}(p^2 + q^2)^2 + \frac{b}{2}q^2$$

on $T^*(\mathbb{R})$. (With dimensionless action and angle variables, ω, c, and b have the same dimensions as H.)

(3.6.23) Solutions

1. Putting M into diagonal form, we see

$$\det(1 + M) = \exp[\operatorname{tr}\ln(1 + M)] = \exp\left[\sum_{j=1}^{\infty}\frac{(-1)^{j+1}}{j}\operatorname{tr} M^j\right]$$
$$\leq \exp\left[\sum_{j=1}^{\infty}\frac{1}{j}(m\|M\|)^j\right] = \frac{1}{1 - m\|M\|}.$$

2.

$$(M_0 + M_1)^{-1} = M_0^{-1} - M_0^{-1}M_1(M_0 + M_1)^{-1} \Rightarrow \|(M_0 + M_1)^{-1}\|$$
$$\leq \|M_0^{-1}\| + \|(M_0 + M_1)^{-1}\|m^2\|M_0\|^{-1}\|M_1\|.$$

3. Substituting from (3.6.15) into the last formula of (3.6.18), recalling that Γ, τ, and $ch^{-\tau} \geq 1$, and generously conceding factors, the recursion formula simplifies to

$$x_n \leq \frac{x_{n-1}}{1 - 3^{-\tau(n+2)}},$$

where

$$x_n := \|\tilde{C}^{(n)}(0)^{-1}\|.$$

Since $-\ln(1 - x) \leq x 2\ln 2\ \forall 0 \leq x \leq 1/2$,

$$\prod_{n=1}(1 - 3^{-\tau(n+2)})^{-1} \leq \exp\left(2\ln 2\, 3^{-2\tau}\sum_{n=1}3^{-\tau n}\right)$$
$$= \exp\left(2\ln 2\frac{3^{-\tau}}{3^{\tau} - 1}\right) < 2,$$

and so $\|\tilde{C}^{(n)}(0)^{-1}\| < 2\|\tilde{C}(0)^{-1}\|\ \forall n$.

4. In (3.6.19), (a) holds for $\tau = 1$ and $c = 1/\omega$ and (b) holds for all h. If $m = 1$, we can replace $\|C\| \, \|\tilde{C}(0)^{-1}\|$ with 1, so that the term $(\ldots)^6$ of (c) simplifies to $(2c)^2$. The condition is then satisfied if

$$\sup_{|\operatorname{Im}\varphi|<h} |b \sin^2 \varphi| = |b| \frac{e^{2h}}{4} \leq \frac{h^3 c^{-2} \omega^3}{3 \cdot 16 \cdot 2^2},$$

i.e., for $h = \frac{3}{2}$,

$$b \leq \frac{\omega^3}{c^2} \frac{9}{e^3 2^7} = \frac{\omega^3}{c^2} \cdot 0.003.$$

The numerical factors of this calculation could be easily improved.

4

Nonrelativistic Motion

4.1 Free Particles

The study of free particles is the foundation of kinematics, and can be used as a basis of comparison for realistic systems. The canonical flow for free particles is linear.

In this section we apply the mathematical methods that we have developed to the problems posed in (1.1.1) and (1.1.2). We begin with the trivial case of a free particle, in order to illustrate the various concepts. In other words, let $M = \mathbb{R}^3$ and $T^*(M) = \mathbb{R}^6$, and choose a chart $(\mathbb{R}^6, \mathbf{1})$, calling the coordinates x_i and p_i. By (2.3.26) H has the simple form

(4.1.1)
$$H = \frac{|\mathbf{p}|^2}{2m}.$$

(4.1.2) **Theorem**

The point transformations that leave H invariant are the elements of the Euclidean group E_3:

$$x_i \to M_{ij}x_j + \lambda_i, \qquad MM^t = \mathbf{1}, \qquad \lambda_i \in \mathbb{R}^3.$$

On $T^*(\mathbb{R}^3)$ they induce the mapping $p_i \to M_{ij}p_j$.

Proof: See Problem 4. □

(4.1.3) **Corollary**

To each one-parameter subgroup of E_3 there corresponds a canonical flow that leaves H invariant, and hence there is a constant of the motion. As discussed in (3.2.6; 2) *and* 3), *the six generators are*

$$p_i, \qquad L_i = \varepsilon_{ijk} x_j p_k.$$

(4.1.4) **Remarks**

1. Since $T^*(\mathbb{R}^3)$ has only six dimensions, and the trajectory takes up at least one dimension, the six constants cannot be algebraically independent. The relation connecting them is $(\mathbf{p} \cdot \mathbf{L}) = 0$.
2. The important question is now whether there exists a five-parameter subgroup of E_3 that furnishes five independent constants. One might think of the following sort of construction: for fixed (x, p), the mapping E_3: $(x, p) \to (x', p')$ gives a mapping $E_3 \to$ the energy shell, which is surjective but not injective. The stabilizer of $(\mathbf{0}, \mathbf{p})$ is $\{M, \lambda: \lambda_i = 0, M\mathbf{p} = \mathbf{p}\}$, which is a one-parameter subgroup, and so the factor group should be five-dimensional and mapped bijectively onto the energy surface. The flaw in this argument is that E_3 is not Abelian, and thus not every subgroup is an invariant subgroup, and this five-parameter factor group does not exist.

For extended phase space, the transition to a uniformly moving frame of reference,

$$\mathbf{x} \to \mathbf{x} + \mathbf{v}t, \qquad t \to t, \tag{4.1.5}$$

is a good candidate for a three-parameter, Abelian invariance group. Taking (4.1.5) as a point transformation, the momenta transform according to (2.4.34; 3) as

$$\mathbf{p} \to \mathbf{p}, \qquad E \to E + (\mathbf{p} \cdot \mathbf{v}), \tag{4.1.6}$$

but this fails to leave $\mathcal{H} = |\mathbf{p}|^2/2m - E$ invariant. However, suppose that the momenta are transformed as

$$\mathbf{p} \to \mathbf{p} + m\mathbf{v}, \qquad E \to E + (\mathbf{p} \cdot \mathbf{v}) + \frac{m|\mathbf{v}|^2}{2}. \tag{4.1.7}$$

With (4.1.5), this is a three-parameter group of canonical transformations Φ_v that leaves $\mathcal{H}$ invariant:

$$\Phi_v^* \mathcal{H} = \mathcal{H} \circ \Phi_v = \mathcal{H}, \qquad \Phi_{v_1} \circ \Phi_{v_2} = \Phi_{v_1 + v_2}, \tag{4.1.8}$$

although it is not a point transformation.

We combine these transformations with the previous ones in

(4.1.9) **Definition**

The transformations of extended phase space,

$$x_i \to M_{ij}x_j + v_i t + \lambda_i, \qquad v_j, \lambda_j \in \mathbb{R}^3, \qquad MM^t = \mathbf{1},$$
$$t \to t + c, \qquad c \in \mathbb{R},$$
$$p_i \to M_{ij}p_j + mv_i,$$
$$E \to E + p_i v_i + \frac{m}{2}v_i^2,$$

form the **Galilean group**. The ten generators, p_i, L_i, $\mathcal{H}$, and $K_j := p_j t - x_j m$, correspond to the ten parameters, λ_i, M_{ij}, c, and v_j.

(4.1.10) **Remarks**

1. The condition $\mathcal{H} = 0$ defines a seven-dimensional manifold in extended phase space. Hence six constants suffice to determine a trajectory, and there must be four relationships connecting the ten generators. Specifically, we have $H = |\mathbf{p}|^2/2m$, and $[\mathbf{p} \times \mathbf{K}] = m\mathbf{L}$. The relationships $(\mathbf{p} \cdot \mathbf{L}) = (\mathbf{K} \cdot \mathbf{L}) = 0$ follow from these.
2. The Poisson brackets of the generators cannot be expressed in terms of the constants alone. The mass appears on the right side of $\{p_j, K_i\} = \delta_{ij}m$, but as a numerical constant it is not the generator of a transformation; $dm = 0 = X_m$, and the Poisson bracket of the mass with any quantity is zero. Together with m, the ten constants of motion generate an eleven-parameter group, the factor group of which by the center is the Galilean group. "Center" refers to the one-parameter group generated by m, which is trivially realized on phase space.
3. There is a subgroup that furnishes six independent constants, and from which all ten generators of the Galilean group can be constructed. It is generated by $\mathbf{p}$ and $\mathbf{K}$ (see Remark 1). The mass m again appears in the Poisson brackets

$$\{p_i, K_j\} = \delta_{ij}m, \qquad \{p_i, p_j\} = \{K_i, K_j\} = 0,$$

and so the group of transformations,

$$\mathbf{x} \to \mathbf{x} + \boldsymbol{\lambda} + \mathbf{v}t, \qquad t \to t,$$
$$\mathbf{p} \to \mathbf{p} + m\mathbf{v}, \qquad E \to E + (\mathbf{p} \cdot \mathbf{v}) + \frac{m|\mathbf{v}|^2}{2}, \tag{4.1.11}$$

on phase space is isomorphic to the factor group by the center. The subgroup is minimal in the sense that with fewer parameters it is impossible to produce six constants of motion. The Galilean group (4.1.9) is not distinguished by any special property: it is neither the largest invariance group nor the smallest one that produces all the constants of the motion.

(4.1.12) *N* **Free Particles**

In order to generalize the above discussion to cover

$$H = \sum_{i=1}^{N} \frac{|\mathbf{p}_i|^2}{2m_i},$$

use the canonical transformation $\mathbf{x}_i \to \mathbf{x}_i/\sqrt{m_i}$, $\mathbf{p}_i \to \mathbf{p}_i\sqrt{m_i}$ to put it into the form

$$H = \frac{1}{2}\sum_{i=1}^{N} |\mathbf{p}_i|^2.$$

Then on $T^*(M) = T^*(\mathbb{R}^{3N}) = \mathbb{R}^{6N}$ there are the $3N + 3N(3N-1)/2$ constants of motion p_μ and $p_\mu x_\nu - x_\mu p_\nu$, $\mu, \nu = 1, \ldots, 3N$. The system is integrable with the p_μ, although the submanifold $p_\mu = \alpha_\mu =$ const. is not compact, but instead diffeomorphic to $\mathbb{R}^{3N}$. All trajectories are submanifolds of the form $x_\mu \to x_\mu + tp_\mu$, and for $p \neq 0$ they are diffeomorphic to $\mathbb{R}$.

(4.1.13) **Problems**

1. Show that $\bar{x} = x + vt$, $\bar{t} = t$, $\bar{p} = p + mv$, $\bar{E} = E + pv + mv^2/2$ is a canonical transformation both with and without the term $mv^2/2$. What is the reason that $mv^2/2$ appears in (4.1.7)?
2. Calculate the Poisson brackets of the generators of the Galilean group.
3. Discuss the group of canonical transformations generated by $D = (\mathbf{x} \cdot \mathbf{p})$.
4. Show that the Euclidean group is the largest group of point transformations that leaves H invariant.

(4.1.14) **Solutions**

1. We verify that $\{\bar{x}_i, \bar{E}\} = v_i - v_i = 0$, which is the only nontrivial Poisson bracket. Without the $mv^2/2$, the canonical transformations do not form a group:

$$\frac{d}{d\lambda}\exp[\lambda L_{(\mathbf{K}\cdot\mathbf{v})}]E_{|\lambda=0} = \{E, \mathbf{K}\cdot\mathbf{v}\} = (\mathbf{p}\cdot\mathbf{v}),$$
$$\frac{d^2}{d\lambda^2}\exp[\lambda L_{(\mathbf{K}\cdot\mathbf{v})}]E_{|\lambda=0} = \{(\mathbf{p}\cdot\mathbf{v}), (\mathbf{K}\cdot\mathbf{v})\} = m|\mathbf{v}|^2.$$

2.

$$\{p_i, p_j\} = \{p_i, E\} = \{L_i, E\} = \{K_i, K_j\} = 0,$$
$$\{L_i, L_j\} = \varepsilon_{ijk}L_k, \qquad \{p_i, L_j\} = \varepsilon_{ijk}p_k, \qquad \{K_i, L_j\} = \varepsilon_{ijk}K_k,$$
$$\{E, K_j\} = p_j, \qquad \{p_i, K_j\} = m\delta_{ij}.$$

3. D generates the group of dilatations $(\mathbf{x}, \mathbf{p}) \to (e^\lambda\mathbf{x}, e^{-\lambda}\mathbf{p})$. This leaves $\mathbf{L}$ invariant, so the dilatations together with the Euclidean group generate a seven-parameter group of

canonical transformations. A dilatation changes H by $e^{-2\lambda}$, and hence the equation $\ddot{x} = 0$ remains unchanged.

4. The new coordinates $\bar{x}$ have to satisfy $\bar{x}_{m,i}(x)\bar{x}_{m,k}(x) = \delta_{ij}$ for all x, which implies

$$\begin{aligned}
\bar{x}_{m,ij}\bar{x}_{m,k} + \bar{x}_{m,i}\bar{x}_{m,kj} &= 0,\\
\bar{x}_{m,jk}\bar{x}_{m,i} + \bar{x}_{m,j}\bar{x}_{m,ik} &= 0,\\
-\bar{x}_{m,ki}\bar{x}_{m,j} - \bar{x}_{m,k}\bar{x}_{m,ji} &= 0,\\
\hline
2\bar{x}_{m,i}\bar{x}_{m,kj} = 0 \Rightarrow \bar{x}_{m,kj} &= 0.
\end{aligned}$$

Consequently the transformation is inhomogeneous-linear, and thus of the form (4.1.2). Substitution yields $MM^t = \mathbf{1}$.

4.2 The Two-Body Problem

Mathematical physics was born when Newton solved the Kepler problem. This problem has lost none of its attraction over the centuries, especially as it has remained soluble while the theory has become more and more refined (with relativistic electrodynamics and gravitation, and both nonrelativistic and relativistic quantum mechanics).

For equations (2.3.21) through (2.3.26) in the case $N = 2$, we have

(4.2.1)
$$M = \mathbb{R}^3 \times \mathbb{R}^3 \backslash \{(\mathbf{x}_1, \mathbf{x}_2) \colon \mathbf{x}_1 = \mathbf{x}_2\},$$
$$H = \frac{|\mathbf{p}_1|^2}{2m_1} + \frac{|\mathbf{p}_2|^2}{2m_2} + \frac{\alpha}{|\mathbf{x}_1 - \mathbf{x}_2|}, \qquad \alpha = e_1 e_2 - \kappa m_1 m_2.$$

We give the solution in several steps:

(4.2.2) Separation into Center-of-Mass and Relative Coordinates

The point transformation $\mathbb{R}^6 \to \mathbb{R}^6$:

$$\mathbf{x}_{cm} = \frac{m_1\mathbf{x}_1 + m_2\mathbf{x}_2}{m_1 + m_2}, \qquad \mathbf{x} = \mathbf{x}_1 - \mathbf{x}_2,$$

induces the transformation

$$\mathbf{p}_{cm} = \mathbf{p}_1 + \mathbf{p}_2, \qquad \mathbf{p} = \mathbf{p}_1\frac{m_2}{m_1 + m_2} - \mathbf{p}_2\frac{m_1}{m_1 + m_2},$$

on $T^*(\mathbb{R}^6)$. This makes

$$H = \frac{|\mathbf{p}_{cm}|^2}{2(m_1 + m_2)} + \frac{|\mathbf{p}|^2(m_1 + m_2)}{2m_1 m_2} + \frac{\alpha}{|\mathbf{x}|} =: H_s + H_r.$$

(4.2.3) Remarks

1. In these coordinates M has been restricted to $\mathbb{R}^3 \times (\mathbb{R}^3 \backslash \{0\})$. $H \in C^\infty(T^*(M))$, in order to remove the singularity of the potential.
2. Since H_{cm} depends only on $\mathbf{p}_{cm}$, and H_r only on $\mathbf{x}$ and $\mathbf{p}$, the time-evolution is the Cartesian product of the flows determined by H_{cm} and H_r, and $\{H_{cm}, H_r\} = 0$.
3. H_{cm} has the form of the H of §4.1 (with mass $m_1 + m_2$), so we consider this part of the problem solved and only work with the second part.
4. H_r is a limiting case of (4.2.1), in which one particle has infinite mass, and the other has the **reduced mass** $m := m_1 m_2/(m_1 + m_2)$.

(4.2.4) Constants of the Motion

$$\{H_r, \mathbf{L}\} = \{H_r, \mathbf{F}\} = 0,$$

$$\mathbf{L} = [\mathbf{x} \times \mathbf{p}], \qquad \mathbf{F} = [\mathbf{p} \times \mathbf{L}] + m\alpha \frac{\mathbf{x}}{r},$$

when $m = m_1 m_2/(m_1 + m_2)$ and $r = |\mathbf{x}|$. $\mathbf{F}$ is known as the **Lenz vector.**

Proof: The angular momentum $\mathbf{L}$ is constant due to the invariance of H_r under $x_i \to M_{ij} x_j$, $p_i \to M_{ij} p_j$. The constancy of $\mathbf{F}$ can be directly verified by calculating

$$\dot{\mathbf{x}} = \{\mathbf{x}, H_r\} = \frac{\mathbf{p}}{m}, \qquad \dot{\mathbf{p}} = \frac{\alpha \mathbf{x}}{r^3},$$

$$\dot{\mathbf{F}} = \frac{\alpha}{r^3}[\mathbf{x} \times [\mathbf{x} \times \mathbf{p}]] + \alpha \frac{\mathbf{p}}{r} - \alpha \frac{\mathbf{x}}{r^3}(\mathbf{x} \cdot \mathbf{p}) = 0. \qquad \square$$

(4.2.5) Remarks

1. The only elements of E_3 that remain as invariances are the rotations, because H_r is not left unchanged by displacements. If $\alpha \neq 0$, the constants $[\mathbf{p} \times \mathbf{L}]$ generalize to $\mathbf{F}$. See Problem 1 for the transformations that are generated by $\mathbf{F}$ and leave H_r invariant.
2. $\mathbf{L}$ should be thought of as the internal angular momentum (spin), and should be distinguished from the total angular momentum $[\mathbf{x}_{cm} \times \mathbf{p}_{cm}]$. Both of these angular momenta are conserved, but the angular momenta of the individual particles, $[\mathbf{x}_1 \times \mathbf{p}_1]$ and $[\mathbf{x}_2 \times \mathbf{p}_2]$ are not.
3. Only five of the seven constants of motion in the second factor, H_r, $\mathbf{L}$, and $\mathbf{F}$, can be independent. Two relationships among them are

$$(4.2.6) \qquad (\mathbf{L} \cdot \mathbf{F}) = 0 \quad \text{and} \quad |\mathbf{F}|^2 = 2m|\mathbf{L}|^2 H_r + m^2\alpha^2.$$

Thus $\mathbf{F}$ lies in the plane perpendicular to $\mathbf{L}$ (the plane of motion), and its length is fixed by L and H_r.

(4.2.7) The Invariance Group

The invariance group is determined by the Poisson brackets of the constants of motion, which can be calculated as

$$\text{(4.2.8)} \qquad \begin{aligned} \{L_i, L_j\} &= \varepsilon_{ijk} L_k, & \{L_i, F_j\} &= \varepsilon_{ijk} F_k, \\ \{F_i, F_j\} &= -2mH_r \varepsilon_{ijk} L_k, & \{H_r, L_i\} &= \{H_r, F_i\} = 0 \end{aligned}$$

(Problem 2). The flows generated by $\{L_i\}$ and $\{F_j\}$ leave the submanifold $H_r = E \in \mathbb{R}$ invariant, and since on that manifold the Poisson brackets of $\{L_i\}$ and $\{F_j\}$ are themselves linear combinations of $\{L_i\}$ and $\{F_j\}$, according to (3.2.7), these flows form a group, the center of which is the time-evolution. This group is isomorphic, respectively, to $SO(4)$, E_3, or $SO(3, 1)$ on the submanifold $H_r = E$, when $E < 0$, or > 0. In order to see this, suppose that $E < 0$, and define

$$A_i = \frac{1}{2}\left(L_i + \frac{F_i}{\sqrt{-2mE}}\right), \qquad B_i = \frac{1}{2}\left(L_i - \frac{F_i}{\sqrt{-2mE}}\right).$$

Since

$$\{A_i, A_j\} = \varepsilon_{ijk} A_k, \qquad \{B_i, B_j\} = \varepsilon_{ijk} B_k, \qquad \{A_i, B_j\} = 0,$$

it is apparent that the invariance group is isomorphic to $SO(3) \times SO(3) = SO(4)$. For $E = 0$ the claim follows from (4.2.8), and if $E > 0$, A and B can be defined as above, with the appropriate signs.

(4.2.9) Remarks

1. It is not possible to factorize $SO(4)$ into the form (rotations generated by $\mathbf{L}$) $\times$ (some other rotations); the situation is more complicated than that.
2. When $E = 0$, $\mathbf{F}$ plays the same rôle as $\mathbf{p}$ in E_3.
3. Since the flows provide continuous mappings of the group into phase space, the noncompactness of E_3 and $SO(3, 1)$ is equivalent to the existence of unbounded trajectories for $H_r \geq 0$.

(4.2.10) The Shape of the Trajectories

The most convenient way to calculate the projection of a trajectory onto M is with the aid of (4.2.4):

$$\mathbf{F} \cdot \mathbf{x} = |\mathbf{L}|^2 + m\alpha r \quad \Rightarrow \quad r = \frac{|\mathbf{L}|^2}{|\mathbf{F}| \cos\varphi - m\alpha}, \qquad \varphi = \measuredangle(\mathbf{F}, \mathbf{x}).$$

These are conic sections, which lie in the region $\{(\mathbf{L} \cdot \mathbf{x}) = 0$ (the plane of motion)$\}$ $\cap \{(\mathbf{F} \cdot \mathbf{x}) > m\alpha r\}$. There are three cases to be distinguished (cf. Figure 4.1):

(a) $H_r > 0$. According to (4.2.6), $|\mathbf{F}| > |m\alpha|$, and r becomes infinite when $\varphi = \arccos(m\alpha/|\mathbf{F}|)$. The trajectory is hyperbolic (or linear, if $\alpha = 0$).

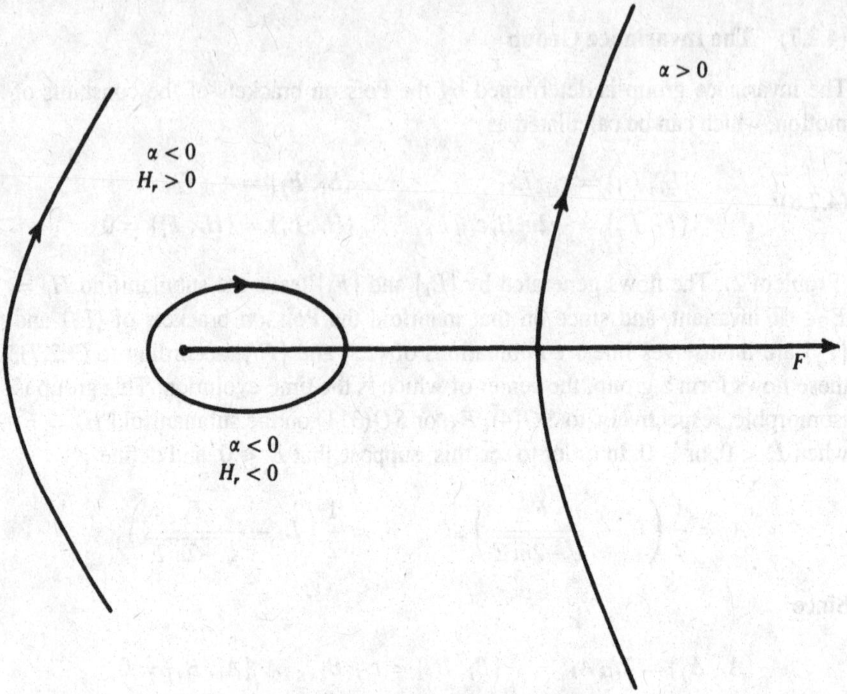

FIGURE 4.1. The Kepler trajectories.

(b) $H_r = 0$. $|\mathbf{F}| = |m\alpha|$, and r becomes infinite when $\varphi = \pi$, if $\alpha = 0$. The trajectory is parabolic.

(c) $H_r < 0$. $|\mathbf{F}| < |m\alpha|$, and r is always finite. The trajectory is elliptic if $\alpha < 0$, or a point if $\alpha = 0$.

(4.2.11) **Remarks**

1. Cases (b) and (c) only occur when $\alpha \leq 0$.
2. Trajectories that pass through the origin have $\mathbf{L} = 0$. The canonical flow exists on the invariant submanifold $T^*(\mathbb{R}^3 \setminus \{0\}) \setminus \{(\mathbf{x}, \mathbf{p}): [\mathbf{x} \times \mathbf{p}] = 0\})$, where H_r generates a complete vector field.
3. The trajectory of $\mathbf{p}(t)$ always lies on some circle (Problem 6).

(4.2.12) **The Elapsed Time**

The momentum canonically conjugate to r is $p_r := (\mathbf{x} \cdot \mathbf{p})/r$, in terms of which

$$H_r = \frac{p_r^2}{2m} + \frac{|\mathbf{L}|^2}{2mr^2} + \frac{\alpha}{r}.$$

Thus the radial motion on the invariant submanifold $\mathbf{L} = \text{const.}$ is like a one-dimensional motion where the original potential gets an additional term from the centrifugal force, $|\mathbf{L}|^2/2mr^2$. Integrating

$$\dot{r} = \frac{p_r}{m} = \sqrt{\frac{2}{m}}\sqrt{E - \frac{\alpha}{r} - \frac{|\mathbf{L}|^2}{2mr^2}}$$

yields (see Problem 7)

$$t - t_0 = \int_{r_0}^{r} \frac{dr'\, r' \sqrt{m/2}}{\sqrt{r'^2 E - \alpha r' - |\mathbf{L}|^2/2m}} = \left| r\sqrt{\frac{m}{2E}}\sqrt{1 - \frac{\alpha}{rE} - \frac{|\mathbf{L}|^2}{2mr^2 E}} \right.$$

(4.2.13)
$$+ \frac{\alpha\sqrt{m/2}}{2|E|^{3/2}} \left(\begin{matrix} \ln\left(Er - \frac{\alpha}{2} + \sqrt{E\left(r^2 E - \alpha r - \frac{|\mathbf{L}|^2}{2m}\right)}\right) \\ \arcsin \frac{2Er - \alpha}{\sqrt{\alpha^2 + 2E|\mathbf{L}|^2/m}} \end{matrix} \right) \Bigg|_{r_0}^{r}$$

$$\text{for} \quad \begin{matrix} E > 0, \\ E < 0. \end{matrix}$$

(4.2.14) Corollary (Kepler's Third Law)

If $E < 0$, then $r = r_0$ when $t = t_0 + \tau$, where $\tau = 2\pi a^{3/2}\sqrt{m/|\alpha|}$, and $a = |\alpha|/2|E|$ is the major semiaxis of the ellipse.

(4.2.15) Remarks

1. Conservation of angular momentum reduces any problem with a central force to a one-dimensional problem. That the orbits are in fact closed, the radial and angular frequencies being degenerate, is a consequence of the richer invariance group of the $1/r$ potential. H depends only on the sum of the action variables.
2. Since τ is proportional to $a^{3/2}$, and thus grows faster than the circumference of the orbit, the speed is less for larger orbits. This is in accordance with the virial theorem, which we shall discuss later, and which says that kinetic energy $\sim$ potential energy, and thus speed $\sim a^{-1/2}$.
3. Corollary (4.2.14) is particularly well illustrated in the Solar System by the extremes, Mercury and Pluto. The radii of their orbits and their periods are roughly 1:100 and 1:1000, respectively.
4. The time-evolution $r(t)$ is the inverse function of the elapsed time (4.2.14), which is not expressible in terms of elementary functions.

When $H_r \geq 0$, the particles escape to infinity, and a number of quantities approach constants. Hence we utilize the concepts introduced in (3.4.1), making the restriction to the invariant submanifold $H_r > 0$.

(4.2.16) Asymptotic Constants of Motion

$$\mathbf{p}, \frac{\mathbf{x}}{r}, \frac{1}{r} \in \mathcal{A}.$$

Proof: To see that the momenta are asymptotically constant when $E > 0$, we use the equation $|\dot{\mathbf{p}}| = |\alpha|/r^2$, and conclude from (4.2.14) that as $t \to \infty$, r becomes proportional to t. It follows that the Cauchy convergence criterion

$$|\mathbf{p}(2T) - \mathbf{p}(T)| = \left| \int_T^{2T} dt\, \dot{\mathbf{p}}(t) \right| \le |\alpha| \int_T^{2T} \frac{dt}{r(t)^2} \le \frac{\text{const.}}{T} \to 0$$

is satisfied for $T_2 > T_1 \to \infty$, and likewise as $t \to -\infty$. Hence $\mathbf{p}_\pm := \tau_\pm \mathbf{p}$ exist for $E > 0$. This implies that the particles escape at definite angles, since the convergence of $p(t)$ implies that of the Abel mean:

$$\tau_\pm \left(\frac{\mathbf{x}}{r} \right) = \lim_{T \to \pm\infty} \frac{\left(\frac{1}{T}\right) \mathbf{x}(0) + \left(\frac{1}{T}\right) \int_0^T dt\, \mathbf{p}(t)/m}{\left| \left(\frac{1}{T}\right) \mathbf{x}(0) + \left(\frac{1}{T}\right) \int_0^T dt\, \mathbf{p}(t)/m \right|} = \pm \frac{\mathbf{p}_\pm}{|\mathbf{p}_\pm|}.$$

Last, $\tau_\pm(1/r) = 0$ when $E > 0$. □

Functions of the form $K(\mathbf{x}/|\mathbf{x}|, \mathbf{p})$ converge trivially as $t \to \pm\infty$, and have the limits $K = K(\pm\mathbf{p}_\pm/|\mathbf{p}_\pm|, \mathbf{p}_\pm)$.

(4.2.17) **Corollaries**

1. $E = |\mathbf{p}_\pm|^2/2m$.
2. $\mathbf{F} = [\mathbf{p}_\pm \times \mathbf{L}] \pm \eta \mathbf{p}_\pm$, *where* $\eta \equiv \alpha\sqrt{m/2E}$.

(4.2.18) **Remarks**

1. The latter equation is easily solved for the limiting momenta:
$$\mathbf{p}_\pm = \frac{[\mathbf{L} \times \mathbf{F}]}{L^2 + \eta^2} \pm \frac{\eta}{L^2 + \eta^2}\mathbf{F}.$$
This implies what is intuitively obvious, that $\mathbf{p}_+$ and $-\mathbf{p}_-$ are related by reflection about $\mathbf{F}$.
2. The Møller transformation (3.4.4) using the free motion for the comparison flow Φ_t^0 simply does not exist in this case. By (4.2.14), for large times
$$r - t\sqrt{\frac{2E}{m}} \sim \frac{|\alpha|\sqrt{m/2}}{2E^{3/2}} \ln t,$$
whereas for Φ_t^0, $r - t\sqrt{2E/m} \sim$ const. If the potential fell off as $r^{-1-\varepsilon}$, there would be no logarithmic term in t, and $\Phi_{-t} \circ \Phi_t^0$ would converge.
3. There exist other simple kinds of time-evolution that the flow approaches asymptotically;
$$H_0 = \frac{|\mathbf{p}|^2}{2m} + \frac{m\alpha}{|\mathbf{p}|t}$$
generates such a flow. However, it depends on t explicitly, and so Φ_t^0 is not a one-parameter group, which causes the consequences of Definition (3.4.4) to lose some of their elegance.

(4.2.19) The Scattering Transformation

Since $\Omega_\pm$ do not exist, we define S by using an algebra $\mathcal{A}_s \subset \mathcal{A}$ on which τ_- is an automorphism and $\mathcal{A}_s \in \{H_r\}'$:

$$\tau_-^{-1} \circ \tau_+ f = f \circ S^{-1}, \qquad \forall f \in \tau_- \mathcal{A}_s = \tau_+ \mathcal{A}_s = \mathcal{A}_s \in \{H_r\}'.$$

Since we are interested in the momenta, a reasonable choice for $\mathcal{A}_s$ is the algebra generated by $\mathbf{p}_-$ and $\mathbf{L}$. Then from (4.2.18; 1) we can calculate

(4.2.20) $$\mathbf{p}_+ = \mathbf{p}_- \circ S^{-1} = \frac{L^2 - \eta^2}{L^2 + \eta^2}\mathbf{p}_- + \frac{2\eta}{L^2 + \eta^2}[\mathbf{p}_- \times \mathbf{L}].$$

(4.2.21) Remarks

1. Schematically, the situation looks like this:

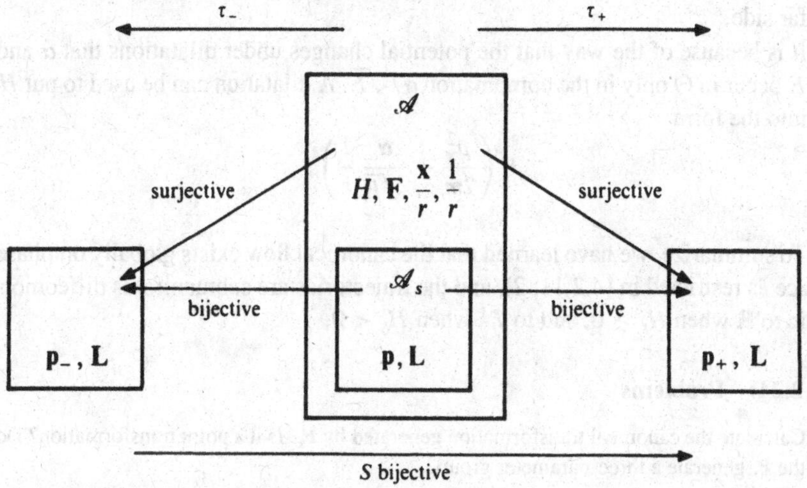

2. S depends on the choice of $\mathcal{A}_s$. If $\mathcal{A}_s$ were chosen as the set of constants of motion, S could be set to $\mathbf{1}$.
3. Its action on p_- and L does not suffice to determine S as a canonical transformation on the manifold $H_r > 0$. We shall not pursue this, since further stipulations all suffer from arbitrariness.
4. As stressed in (3.4.10; 2), S does not leave all the constants of motion invariant. It transforms p_- into p_+, and when applied to $\mathbf{F}$,

$$\mathbf{F} \circ S^{-1} = \mathbf{F} + 2\eta \mathbf{p}_+.$$

(4.2.22) Proposition

By (4.2.20), *the scattering angle is*

$$\Theta = \arccos \frac{\mathbf{p}_+ \cdot \mathbf{p}_-}{2mE} = \arccos \frac{L^2 - \eta^2}{L^2 + \eta^2},$$

so

$$b^2 = \frac{|\mathbf{L}|^2}{|\mathbf{p}_-|^2} = \frac{\eta^2}{|\mathbf{p}_-|^2} \frac{1 + \cos\Theta}{1 - \cos\Theta}$$

(cf. (3.4.16; 1)). *Then the differential scattering cross-section* (3.4.15) *can be calculated as*

$$\sigma(\Theta) = \frac{\alpha^2}{16E^2} \sin^{-4} \frac{\Theta}{2}.$$

(α/E is the turning radius, i.e., the minimum distance from the particle to the scattering center.)

(4.2.23) **Remarks**

1. σ is independent of the sign of α, although if $\alpha > 0$ the particle turns around on the near side of the scattering center, while if $\alpha < 0$ it turns around on the far side.
2. It is because of the way that the potential changes under dilatations that α and E occur in Θ only in the combination $\alpha/\sqrt{E}$. A dilatation can be used to put H into the form

$$E\left(\frac{p^2}{2m} + \frac{\alpha}{\sqrt{E}} \frac{1}{r}\right).$$

To summarize, we have learned that the canonical flow exists globally on phase space as restricted in (4.2.11; 2), and the trajectories are submanifolds diffeomorphic to $\mathbb{R}$ when $H_r \geq 0$, and to T^1 when $H_r < 0$.

(4.2.24) **Problems**

1. Calculate the canonical transformation generated by $\mathbf{F}_i$. Is it a point transformation? Do the $\mathbf{F}_i$ generate a three-parameter group?
2. Verify equations (4.2.8).
3. Show that the trajectories (4.2.10) are conic sections with foci at the origin.
4. Use (4.2.14) to calculate how long a body with r_0 equal to the radius of the Earth's orbit and $v_0 = 0$ takes to fall into the sun.
5. Calculate the scattering angle for (4.2.10; (a)).
6. Calculate the projection of the trajectory onto the second factor of $T^*(\mathbb{R}^3) = \mathbb{R}^3 \times \mathbb{R}^3$ (i.e., $\mathbf{p}(t)$).
7. With the variables $u: r = a(1 + \varepsilon \cos u)$, $a = |\alpha/2E|$ = the major semiaxis, and $\varepsilon = |\mathbf{F}|/m\alpha$ = the eccentricity, equation (4.2.14) is written for $E < 0$, $\alpha < 0$, as

$$\frac{2|E|^{3/2}}{\alpha\sqrt{m/2}}(t - t_0) = u - \varepsilon \sin u$$

(Kepler's equation). Interpret this geometrically.

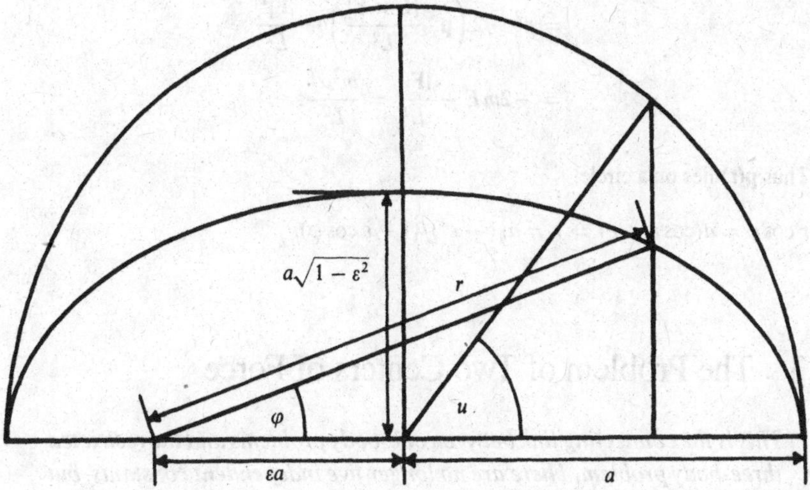

FIGURE 4.2. The variables used in Kepler's equation.

(4.2.25) **Solutions**

1.

$$\{F_i, p_k\} = \delta_{ik}|\mathbf{p}|^2 - p_i p_k + m\alpha\left(\frac{\delta_{ik}}{r} - \frac{x_i x_k}{r^3}\right),$$
$$\{F_i, x_k\} = \delta_{ik}(\mathbf{p}\cdot\mathbf{x}) - 2x_i p_k + x_k p_i.$$

It is not a point transformation, because F is not linear in p. Since $\{F_i, F_j\}$ can not be expressed by the F_k, they do not generate a group.

2. Since L_i generates rotations and **L** and **F** are vectors,

$$\{L_i, L_j\} = \varepsilon_{ijk}L_k, \qquad \{L_i, F_j\} = \varepsilon_{ijk}F_k.$$

For the calculation of $\{F_i, F_j\}$, use: $F_i = x_i|\mathbf{p}|^2 - p_i(\mathbf{x}\cdot\mathbf{p}) + max_i/r$.

3. $r \pm \sqrt{(x-A)^2+z^2} = C \Leftrightarrow (C-r)^2 = A^2 - 2Ax + r^2 \Leftrightarrow r = (A^2 - C^2)/(2A\cos\varphi - 2C)$, where $x = r\cos\varphi$ and $z = r\sin\varphi$.

4. The major semiaxis of the trajectory is half the radius of the Earth's orbit, and it takes half an orbital period to fall into the Sun, so the answer is $2^{-5/2}$ years.

5. $\Theta = \pi - 2\varphi$, where φ is the angle at $r = \infty$. Hence, according to (4.2.6),

$$\cos\Theta = -\cos 2\varphi = 1 - 2\cos^2\varphi = 1 - 2\frac{m^2\alpha^2}{|\mathbf{F}|^2} = 1 - 2\frac{\eta^2}{L^2+\eta^2}.$$

6.

$$[\mathbf{L}\times\mathbf{F}] = \mathbf{p}L^2 + m\alpha\left[\mathbf{L}\times\frac{\mathbf{x}}{r}\right] \Rightarrow (\mathbf{p}\cdot[\mathbf{L}\times\mathbf{F}])$$
$$= L^2\left(p^2 + m\frac{\alpha}{r}\right) \Rightarrow \left|\mathbf{p} - \frac{[\mathbf{L}\times\mathbf{F}]}{L^2}\right|^2$$

$$= p^2 - 2\left(\mathbf{p}\cdot\frac{[\mathbf{L}\times\mathbf{F}]}{L^2}\right) + \frac{|\mathbf{F}|^2}{L^2}$$

$$= -2mE + \frac{|\mathbf{F}|^2}{L^2} = \frac{m^2\alpha^2}{L^2}.$$

Thus $\mathbf{p}(t)$ lies on a circle.

7. $r\cos\varphi = a(\cos u + \varepsilon) \Rightarrow r = a(1-\varepsilon^2)/(1-\varepsilon\cos\varphi)$.

4.3 The Problem of Two Centers of Force

This is the connecting link between one-body problems and the restricted three-body problem. There are no longer five independent constants, but only three. Even so, the system is integrable.

In the last section we saw that the two-body problem can be reduced to the problem of a single particle in the force-field of an infinitely heavy, immovable object. This suggests that the first three-body problem to study is one where one particle is so light that it does not influence the motion of the other two. At this point the nature of the problem depends on whether it involves the electrical or the gravitational force. Let M be the mass of the heavy particles and m the mass of the light one. Then from (4.2.14) we can easily estimate the order of magnitude of the orbital frequencies ω_H of the heavy particles and ω_L of the light one. This is just because the centrifugal and centripetal forces are balanced; thus if R is the orbital radius,

$$MR\omega_H^2 \cong \frac{\kappa M^2 + e^2}{R^2},$$
$$mR\omega_L^2 \cong \frac{\kappa m M + e^2}{R^2}. \tag{4.3.1}$$

If gravitation predominates, that is, $\kappa m M \gg e^2$, then $\omega_H^2 = \omega_L^2 = \kappa M/R^3$, and the motion of the heavy particles cannot be neglected when one studies the motion of the light one. This is a direct consequence of the fact discovered by Galileo, that all masses are accelerated equally strongly in a gravitational field. The case of dominant gravitational forces is known as the restricted three-body problem. It is of obvious interest for space travel, but is rather difficult to attack analytically; we shall study it in the next section. It is somewhat simpler when the electrical force predominates, $\kappa M^2 \ll e^2$, as happens with elementary particles. In that case, $\omega_L^2/\omega_H^2 = M/m$, and the heavy particles move slowly compared with the light one when M/m is large. This would be appropriate for the simplest kinds of molecules, with two nuclei and one electron, except that the important physical properties lie outside the domain of classical physics. We shall return to this problem when we treat the quantum theory.

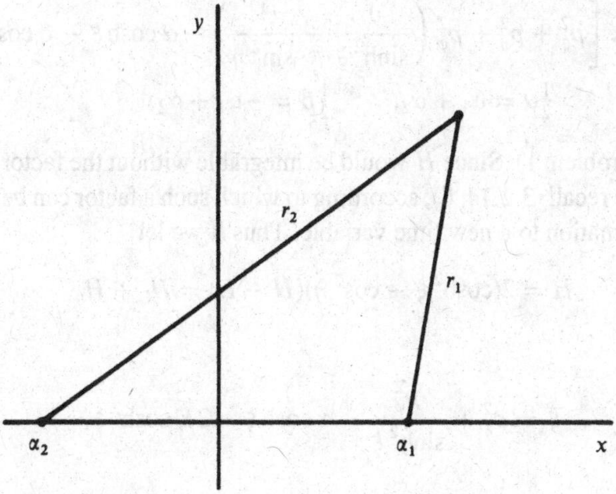

FIGURE 4.3. The centers of force.

(4.3.2) The Hamiltonian

For mathematical convenience we can set the two centers of force at (1, 0, 0) and (−1, 0, 0) without loss of generality, and start off with the manifold

$$M_0 = \mathbb{R}^3 \backslash \{(1, 0, 0), (-1, 0, 0)\}. \tag{4.3.3}$$

It will be necessary to restrict M_0 and $T^*(M_0)$ further in order to avoid some complications. Let r_1 and r_2 be the distances of the light particle from the two centers, of strengths α_1 and α_2 (see Figure 4.3), and suppose that $m = 1$. Then the Hamiltonian of the problem becomes

$$H = \frac{|\mathbf{p}|^2}{2} - \frac{\alpha_1}{r_1} - \frac{\alpha_2}{r_2}. \tag{4.3.4}$$

Since r_1 and r_2 have the rather unwieldy form $\sqrt{(x \pm 1)^2 + y^2 + z^2}$ in Cartesian coordinates, it is convenient to introduce new coordinates that take advantage of the rotational symmetry.

(4.3.5) Elliptic Coordinates

Elliptic coordinates use the diffeomorphism (M_1, Ψ):

$$\begin{aligned} M_1 &= \mathbb{R}^3 \backslash (\mathbb{R}, 0, 0) \subset M_0, \\ \Psi(M_1) &= \mathbb{R}^+ \times (0, \pi) \times T^1 \ni (\xi, \eta, \varphi), \\ \Psi^{-1}: (\xi, \eta, \varphi) &\to (\cosh\xi \cos\eta, \sinh\xi \sin\eta \cos\varphi, \sinh\xi \sin\eta \sin\varphi) \\ &= (x, y, z) \in M_1, \end{aligned}$$

with which H is written as

$$H = \tfrac{1}{2}(\cosh^2\xi - \cos^2\eta)^{-1}$$

$$(4.3.6) \qquad \times \left[p_\xi^2 + p_\eta^2 + p_\varphi^2 \left(\frac{1}{\sinh^2 \xi} + \frac{1}{\sin^2 \eta} \right) - \alpha \cosh \xi - \beta \cos \eta \right],$$

$$\tfrac{1}{2}\alpha = \alpha_1 + \alpha_2, \qquad \tfrac{1}{2}\beta = -\alpha_1 + \alpha_2,$$

on $\Psi(M_1)$ (Problem 1). Since H would be integrable without the factor $(\cosh^2 \xi - \cos^2 \eta)^{-1}$, we recall (3.2.14; 6), according to which such a factor can be eliminated by a transformation to a new time variable. Thus if we let

$$\mathcal{H} = 2(\cosh^2 \xi - \cos^2 \eta)(H - E) = H_\xi + H_\eta,$$

where

$$(4.3.7) \qquad \xi = p_\xi^2 + \frac{p_\varphi^2}{\sinh^2 \xi} - \alpha \cosh \xi - 2E \cosh^2 \xi,$$

and

$$H_\eta = p_\eta^2 + \frac{p_\varphi^2}{\sin^2 \eta} - \beta \cos \eta + 2E \cos^2 \eta,$$

on extended phase space; then on the submanifold $\mathcal{H} = 0$, $\mathcal{H}$ describes the time-evolution with a parameter s such that $dt/ds = 2(\cosh^2 \xi - \cos^2 \eta) > 0$.

(4.3.8) Constants of the Motion

On extended phase space, $\mathcal{H}$, E, H_ξ, and p_φ are all constant and mutually independent.

(4.3.9) Remarks

1. Because $dt/ds > 0$, anything that is constant in s is also constant in t. Restricting ourselves again to $T^*(M_1)$, we can use H, p_φ, and $H_\xi = -H_\eta$ as three independent constants of motion, replacing E with H.
2. The conservation of p_φ comes from the cylindrical symmetry of the problem. The canonical flows coming from H_ξ and H_η are rather complicated.
3. Since the Poisson brackets of any two of the four conserved quantities (or, respectively, H, p_φ, and H_ξ) vanish, the system is integrable on $T^*(M_1)$.
4. No additional constants can be found, so the invariance group of $\mathcal{H}$ (resp. H) is a four-parameter (three-parameter) Abelian group.

(4.3.10) Effective Potentials

Integration of the equations of motion with $\mathcal{H}$ from (4.3.7) leads to two one-dimensional problems with the potentials:

$$(4.3.11) \qquad \begin{aligned} V_\xi &= \frac{p_\varphi^2}{\sinh^2 \xi} - \alpha \cosh \xi - 2E \cosh^2 \xi, \qquad \xi \in \mathbb{R}^+, \\ V_\eta &= \frac{p_\varphi^2}{\sin^2 \eta} - \beta \cos \eta + 2E \cos^2 \eta, \qquad \eta \in (0, \pi), \end{aligned}$$

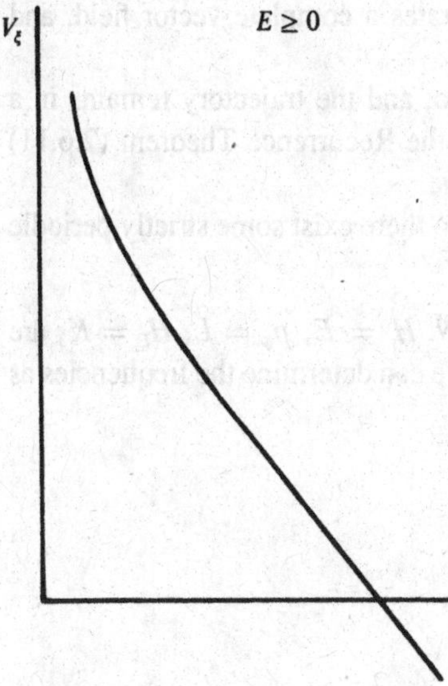

FIGURE 4.4. Effective potential.

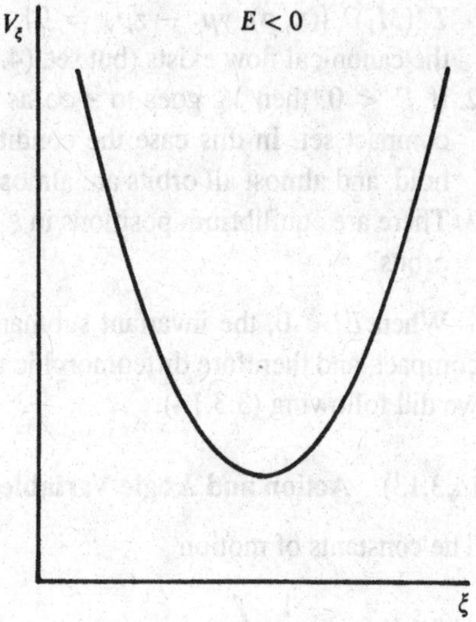

FIGURE 4.5. Effective potential.

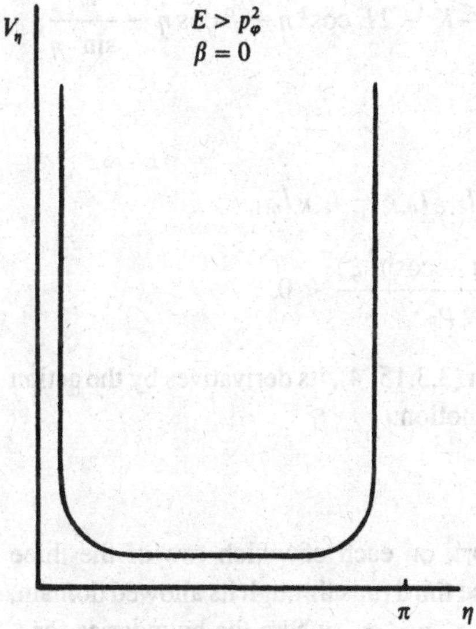

FIGURE 4.6. Effective potential.

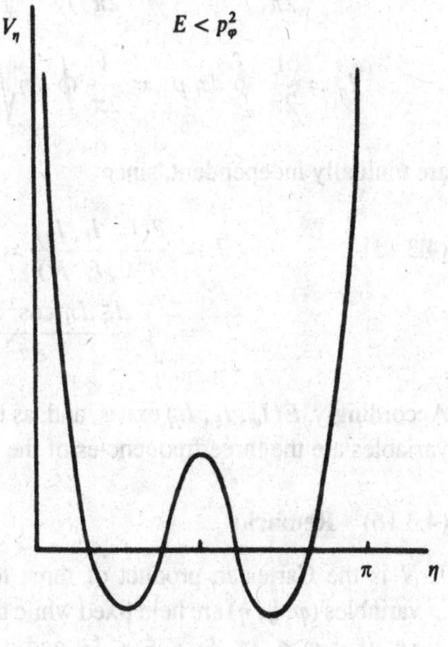

FIGURE 4.7. Effective potential.

(see Figures 4.4–4.7).

(4.3.12) Remarks

1. If $p_\varphi \neq 0$, the effective potential V_ξ becomes infinite as $\xi \to 0$ (as does V_η as $\eta \to 0$ or $\eta \to \pi$), and the trajectory can never leave M_1. On

$T^*(M_1)\setminus\{(\mathbf{x},\mathbf{p}): yp_z - zp_y = 0\}$, H generates a complete vector field, and the canonical flow exists (but see (4.3.17)).
2. If $E < 0$, then V_ξ goes to $+\infty$ as $\xi \to \infty$, and the trajectory remains in a compact set. In this case the conditions of the Recurrence Theorem (2.6.11) hold, and almost all orbits are almost periodic.
3. There are equilibrium positions in ξ and η, so there exist some strictly periodic orbits.

When $E < 0$, the invariant submanifolds $N: H = E$, $p_\varphi = L$, $H_\xi = K$, are compact, and therefore diffeomorphic to T^3. We can determine the frequencies as we did following (3.3.14).

(4.3.13) **Action and Angle Variables**

The constants of motion

(4.3.14)
$$I_\varphi = \frac{1}{2\pi}\oint d\varphi\, p_\varphi = L,$$
$$I_\xi = \frac{1}{2\pi}\oint d\xi\, p_\xi = \frac{1}{2\pi}\oint d\xi\sqrt{K + 2E\cosh^2\xi + \alpha\cosh\xi - \frac{L^2}{\sinh^2\xi}},$$
$$I_\eta = \frac{1}{2\pi}\oint d\eta\, p_\eta = \frac{1}{2\pi}\oint d\eta\sqrt{-K - 2E\cos^2\eta + \beta\cos\eta - \frac{L^2}{\sin^2\eta}},$$

are mutually independent, since

(4.3.15)
$$J := \frac{\partial(I_\varphi, I_\xi, I_\eta)}{\partial(L, E, K)} = I_{\xi,E}I_{\eta,K} - I_{\xi,K}I_{\eta,E}$$
$$= \oint \frac{d\xi\, d\eta(\cos^2\eta - \cosh^2\xi)}{8\pi^2 p_\xi p_\eta} < 0.$$

Accordingly, $E(I_\varphi, I_\xi, I_\eta)$ exists, and, as in (3.3.15; 4), its derivatives by the action variables are the three frequencies of the motion.

(4.3.16) **Remarks**

1. N is the Cartesian product of three tori, on each of which two of the three variables (φ, ξ, η) are held fixed while the third runs through its allowed domain, i.e., $0 < \varphi \le 2\pi$, $\xi_1 \le \xi \le \xi_2$, and $\eta_1 \le \eta \le \eta_2$, where the boundaries for ξ and η are the values for which $p_\xi = 0$ and $p_\eta = 0$.[1] The frequencies correspond to the orbital periods of these variables.
2. The frequencies cannot be written explicitly as functions of the I's or of E, K, and L, because the integrals in (4.3.13) cannot be expressed in terms of ordinary

---

[1]In Figures 4.5 and 4.7 these are given by the intersections with horizontal lines at heights K and $E - K$.

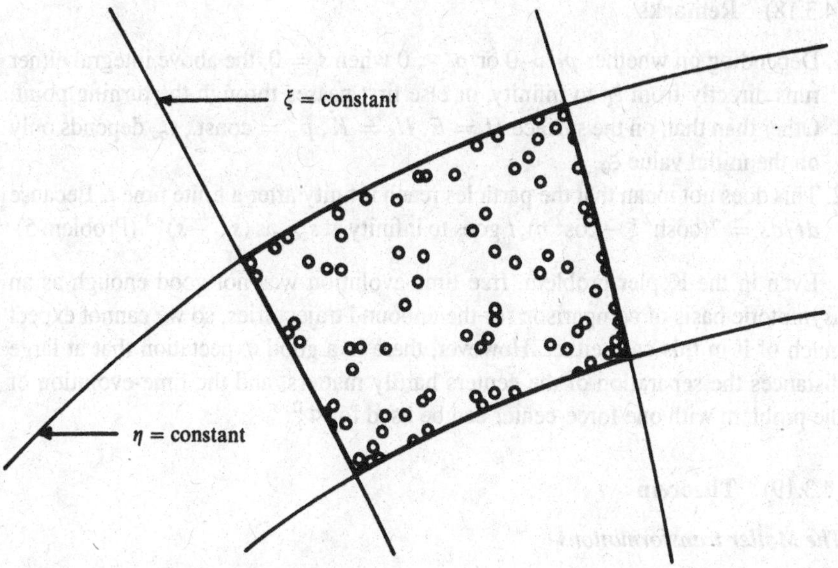

FIGURE 4.8. A computer calculation of the points at which the trajectory intersects the plane $\varphi = 0$. These points fill the projection of an invariant torus to the plane densely.

functions. The frequencies can be written as

$$\omega_\varphi = \frac{I_{\xi,K} I_{\eta,L} - I_{\xi,L} I_{\eta,K}}{J},$$

$$\omega_\xi = \frac{I_{\eta,K}}{J}, \qquad \omega_\eta = \frac{I_{\xi,K}}{J}.$$

Since these functions of E, K, and L vary continuously from one torus to the next one, they are generally not in rational ratios to one another, and the trajectory fills some three-dimensional region densely. Hence there are no additional constants of the motion in that part of phase space.

3. Since the curves $\xi =$ const. and $\eta =$ const. are, respectively, ellipses and hyperbolas in Cartesian coordinates, the projection of N to the plane $\varphi = 0$ is the region with the points shown in Figure 4.8, which should be pictured three-dimensionally as if it were rotated about the x-axis.

(4.3.17) $E > 0$: Unbounded Trajectories

When $E > 0$, the vector field generated by H_ξ is not complete; instead, the particles reach infinity at a finite value of s,

$$s_\infty = \frac{1}{2} \int_{\xi_0}^{\infty} \frac{d\xi}{\sqrt{K - (p_\varphi^2/\sinh^2\xi) + \alpha\cosh\xi + 2E\cosh^2\xi}} < \infty$$

(cf. Figure 4.4 and (4.3.7) with $H_\xi = K$).

(4.3.18) Remarks

1. Depending on whether $p_\xi > 0$ or $p_\xi < 0$ when $s = 0$, the above integral either runs directly from ξ_0 to infinity, or else first passes through the turning point. Other than that, on the surface $H = E$, $H_\xi = K$, $p_\varphi =$ const., s_∞ depends only on the initial value ξ_0.
2. This does not mean that the particles reach infinity after a finite time t. Because $dt/ds = 2(\cosh^2\xi - \cos^2\eta)$, t goes to infinity at s_∞, as $(s_\infty - s)^{-1}$ (Problem 5).

Even in the Kepler problem, free time-evolution was not good enough as an asymptotic basis of comparison for the unbound trajectories, so we cannot expect much of it in this case either. However, there is a good expectation that at large distances the separation of the centers hardly matters, and the time-evolution of the problem with one force-center can be used for Φ_t^0.

(4.3.19) Theorem

The Møller transformations

$$\Omega_\pm = \lim_{t\to\pm\infty} \Phi_{-t} \circ \Phi_t^0$$

exist, where Φ_t^0 is the flow with $\alpha_1' = 0$ and $\alpha_2' = \alpha_1 + \alpha_2$ (i.e., (α, β) becomes (α, α)). The domains are

$$D_\pm = \left\{(\mathbf{x}, \mathbf{p}): \frac{|\mathbf{p}|^2}{2} - \frac{\alpha}{r_2} > 0,\ yp_z - zp_y \neq 0\right\}.$$

(4.3.20) Remarks

1. Φ_t^0 is arbitrary in that the single force-center can be put anywhere at all. The Møller transformations exist as long as the strength of the force center is α.
2. Because $H_0 \circ \Omega_\pm^{-1} = H_{|\mathcal{R}_\pm}$ (3.4.5; 1), we see

$$\mathcal{R}_\pm = \left\{(\mathbf{x}, \mathbf{p}): \frac{|\mathbf{p}|^2}{2} - \frac{\alpha_1}{r_1} - \frac{\alpha_2}{r_2} > 0,\ (y, z) \neq \mathbf{0},\ yp_z - zp_y \neq 0\right\}.$$

Although the proof is not difficult in principle, it requires some involved calculations, and will not be done here.

(4.3.21) Corollaries

1. *$\Omega_\pm$ map the flows $\Phi^0_{t|D_\pm}$ to $\Phi_{t|\mathcal{R}_\pm}$. Since $\Omega_\pm$ are diffeomorphisms, Φ_t must have five independent constants on $\mathcal{R}_\pm$, just like Φ_t^0 on $D_\pm$. From* (3.4.5; 3) *and* (4.2.4), *these constants are $\tau_\pm(\mathbf{L}) = \mathbf{L} \circ \Omega_\pm^{-1} = \lim_{t\to\pm\infty} \mathbf{L}(t)$ and $\tau_\pm(\mathbf{F}) = \mathbf{F} \circ \Omega_\pm^{-1} = \lim_{t\to\pm\infty} \mathbf{F}(t)$, where $(\tau_\pm(\mathbf{L}) \cdot \tau_\pm(\mathbf{F})) = 0$.*

2. *$\Omega_\pm(\beta)$ depend on β continuously, and $\Omega_\pm(\alpha) = \mathbf{1}$. The trajectories of Φ_t with $E > 0$ are mapped by $\Omega_\pm^{-1}$ diffeotopically (see (2.6.15; 5)) onto those of Φ_t^0. Since the one set of trajectories is continuously deformed into the other, no knots can form. With the Møller transformations it is easy to make such global statements, which are otherwise hard to find.*

To summarize what we have learned about how the separation into two centers of force affects the flow: The unbound trajectories ($E > 0$) are only moderately deformed, and the flow can be transformed diffeomorphically to a linear one. The periodic orbits get wound up like balls of string, most of them filling three-dimensional regions densely.

(4.3.22) Problems

1. Derive (4.3.7) by calculating $T^*(\Psi)$.
2. Derive (4.3.7) by introducing elliptic coordinates in $L = |\dot{\mathbf{x}}|^2/2 - V$.
3. Use the equations of motion to verify that $dH_\xi/dt = 0$.
4. Use the Hamilton–Jacobi equation (3.2.16) to separate this problem.
5. Show that if $E > 0$, then t goes to infinity as $1/(2E(s_\infty - s))$.

(4.3.23) Solutions

1. Note that

$$\cosh^2\xi \begin{vmatrix}\cos^2\eta \\ \sin^2\eta\end{vmatrix} + \sinh^2\xi \begin{vmatrix}\sin^2\eta \\ \cos^2\eta\end{vmatrix} = \cosh^2\xi - \begin{vmatrix}\sin^2\eta \\ \cos^2\eta\end{vmatrix} \Rightarrow$$

$$\tfrac{1}{2}(r_1 + r_2) = \cosh\xi, \qquad \tfrac{1}{2}(r_1 - r_2) = \cos\eta \quad \Rightarrow \quad r_1 r_2 = \cosh^2\xi - \cos^2\eta.$$

That takes care of the potential energy. For the kinetic energy, calculate

$$T(\Psi)^t \cdot T(\Psi) = \begin{pmatrix} \cosh^2\xi - \cos^2\eta & 0 & 0 \\ 0 & \cosh^2\xi - \cos^2\eta & 0 \\ 0 & 0 & \sinh^2\xi\sin^2\eta \end{pmatrix} \Rightarrow$$

$$|\mathbf{p}|^2 = (p_\xi, p_\eta, p_\varphi)(T(\Psi)^{-1} \cdot T(\Psi)^{-1t}) \begin{pmatrix} p_\xi \\ p_\eta \\ p_\varphi \end{pmatrix}$$

$$= \frac{p_\xi^2 + p_\eta^2}{\cosh^2\xi - \cos^2\eta} + \frac{p_\varphi^2}{\sinh^2\xi\sin^2\eta}.$$

2.

$$|\dot{\mathbf{x}}|^2 = (\dot{\xi}^2 + \dot{\eta}^2)(\cosh^2\xi - \cos^2\eta) + \dot{\varphi}^2\sinh^2\xi\sin^2\eta \quad \Rightarrow$$

$$(p_\xi, p_\eta, p_\varphi) = (\dot{\xi}(\cosh^2\xi - \cos^2\eta), \dot{\eta}(\cosh^2\xi - \cos^2\eta), \dot{\varphi}\sinh^2\xi\sin^2\eta).$$

3. $$\frac{dH_\xi}{dt} = -2p_\xi H_{,\xi} - \frac{p_\xi}{\cosh^2\xi - \cos^2\eta}\left[\frac{p_\varphi^2 2\cosh\xi}{\sinh^3\xi} + \alpha\sinh\xi + 2H\sinh\xi\cosh\xi\right] = 0.$$

4. Let $f = S - Et$. The equation

$$\left(\frac{\partial S}{\partial \xi}\right)^2 + \left(\frac{\partial S}{\partial \eta}\right)^2 + \left(\frac{\partial S}{\partial \varphi}\right)^2 (\sinh^{-2}\xi + \sin^{-2}\eta)$$
$$- \alpha \cosh \xi - \beta \cos \eta - E(\cosh^2 \xi - \cos^2 \eta) = 0$$

can be solved by supposing that S separates as

$$S = S_1(\xi) + S_2(\eta) + \varphi p_\varphi.$$

5. Asymptotically,

$$\frac{d\xi}{ds} = e^{\xi}\sqrt{2E} \quad \Rightarrow \quad s = s_\infty - \frac{e^{-\xi}}{\sqrt{2E}}, \quad t = 2\int ds \frac{e^{2\xi}}{4} = \frac{1}{2E(s_\infty - s)}.$$

4.4 The Restricted Three-Body Problem

The motion of a particle in the field of two rotating masses is already so complicated—even when all the motion is in a plane—that only rather fragmentary statements can be made about it.

As mentioned above, the restricted three-body problem has to do with the motion of a particle that is light that its influence on the motion of the other two particles is negligible. Such an assumption is justified for, say, the flight of a spacecraft to the Moon. We need a way to take the motion of the centers of force (with masses m and μ and coordinates $\mathbf{x}_m(t)$ and $\mathbf{x}_\mu(t)$) into account. So let us define a time-dependent

(4.4.1) **Hamiltonian**

$$H = \tfrac{1}{2}|\mathbf{p}|^2 - \kappa\left(\frac{m}{|\mathbf{x}_m(t) - \mathbf{x}|} + \frac{\mu}{|\mathbf{x}_\mu(t) - \mathbf{x}|}\right).$$

(4.4.2) **Remarks**

1. We shall only consider the case in which the heavy particles move in circular orbits, and therefore $\mathbf{x}_m$ and $\mathbf{x}_\mu$ describe circles about the center of mass, which we may take as the origin.
2. If both $\mathbf{p}$ and $\mathbf{x}$ are in the orbital plane at any time, then they are in it for all times. This will be the main variant of the problem treated here.
3. In (4.4.1) the mass of the light particle has been set to 1, since it factors out of the equations of motion. For simplicity we shall henceforth use units where the separation of the heavy particles $R = \kappa = m + \mu = 1$. In these units the frequency of the heavy particles $\omega = \sqrt{\kappa(m+\mu)}/R^{3/2}$ equals 1, and the only essential parameter that remains is $\mu/m \leq 1$. The masses m and μ are, respectively, at distances μ and m from the origin.

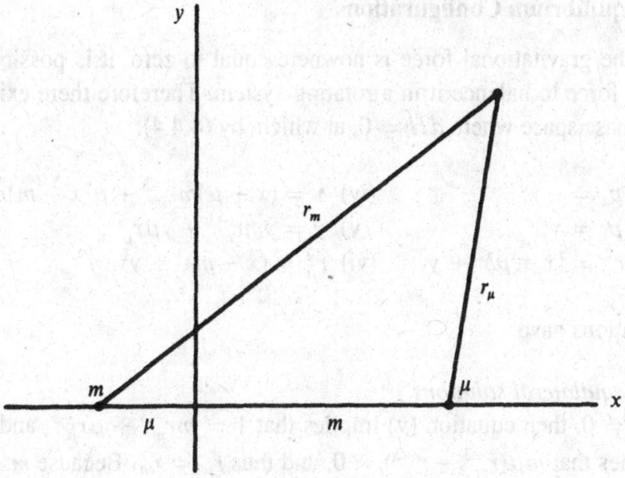

FIGURE 4.9. The coordinates used in the restricted three-body problem.

(4.4.3) Rotating Coordinates

In a coordinate system that rotates with the heavy particles, the centers of force are fixed. As in Example (3.2.15; 2), H gains a term from the centrifugal force, and if it is written out in components it is

(4.4.4)
$$H = \tfrac{1}{2}(p_x^2 + p_y^2) - xp_y + yp_x - \frac{m}{[(x+\mu)^2 + y^2]^{1/2}} - \frac{\mu}{[(x-m)^2 + y^2]^{1/2}}$$

$$\dot{x} = p_x + y, \qquad \dot{y} = p_y - x,$$
$$\dot{p}_x = p_y - V_{,x}, \qquad \dot{p}_y = -p_x - V_{,y},$$
$$V_{,x} = \frac{m(x+\mu)}{[(x+\mu)^2 + y^2]^{3/2}} + \frac{\mu(x-m)}{[(x-m)^2 + y^2]^{3/2}},$$
$$V_{,y} = \frac{ym}{[(x+\mu)^2 + y^2]^{3/2}} + \frac{y\mu}{[(x-m)^2 + y^2]^{3/2}}.$$

(4.4.5) Remarks

1. Changing to a rotating coordinate system is a point transformation in extended phase space, but we can just as well use H on the phase space $T^*(\mathbb{R}^2 \backslash \{(m,0)\} \backslash \{(-\mu,0)\})$.
2. Since H does not depend explicitly on the time in the rotating system, it is a constant, known as **Jacobi's constant**. However, no other constant, which would make the system integrable, is to be found.
3. In order for H to generate a complete vector field, i.e., for collisions to be avoided, we shall have to restrict phase space more than this, but it is not yet clear exactly how this is to be done.

(4.4.6) Equilibrium Configurations

Although the gravitational force is nowhere equal to zero, it is possible for the centrifugal force to balance it in a rotating system. Therefore there exist critical points in phase space where $dH = 0$, at which, by (4.4.4),

$$\begin{aligned} &\text{(i) } p_x = -y; && \text{(iv) } x = (x+\mu)mr_m^{-3} + \mu(x-m)r_\mu^{-3}; \\ &\text{(ii) } p_y = x; && \text{(v) } y = ymr_m^{-3} + y\mu r_\mu^{-3}; \\ &\text{(iii) } r_m^2 = (x+\mu)^2 + y^2; && \text{(vi) } r_\mu^2 = (x-m)^2 + y^2. \end{aligned}$$

These equations have:

(a) *Two equilateral solutions*
If $y \neq 0$, then equation (v) implies that $1 = mr_m^{-3} + \mu r_\mu^{-3}$, and then (iv) implies that $m\mu(r_m^{-3} - r_\mu^{-3}) = 0$, and thus $r_\mu = r_m$. Because $m + \mu = 1$, it follows that $r_m = r_\mu = 1$, independently of m/μ. Consequently, both configurations for which the three particles are at the corners of an equilateral triangle are in equilibrium.

(b) *Three collinear solutions*
If $y = 0$, then there are clearly three solutions, since the curves $f = x$ and

$$f = m\frac{x+\mu}{|x+\mu|^3} + \mu\frac{x-m}{|x-m|^3}$$

have three points of intersection (Figure 4.10).

(4.4.7) Remarks

1. The equilibrium configurations are zero-dimensional trajectories. In the original system, they correspond to circular orbits with frequency 1.
2. The equilateral solutions were known to Lagrange, though he drew no conclusions about their meaning for astronomy. In turns out, however, that there are real bodies in the Solar System approximately in such a configuration. A group of asteroids, the **Trojans**, nearly make an equilateral triangle with the Sun and Jupiter. Since their masses are vanishingly small on this scale, and since all motion in the Solar System is roughly in a plane, the system of equations (4.4.5) is applicable.
3. The collinear solutions do not ever appear to be realized, probably because of their instability. Other large planets have an appreciable influence on the asteroids, and it is important to study whether they deform the asteroids' orbits only moderately, or destroy them altogether. We shall see below that the collinear solutions are always unstable, whereas the equilateral ones are stable so long as the mass of Jupiter is less than 4% that of the Sun, which happens to be the case.

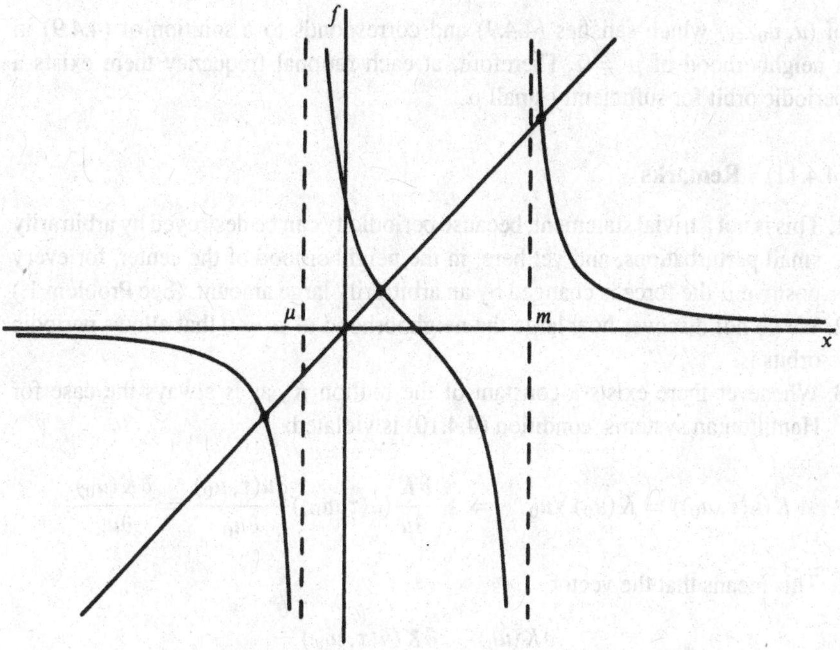

FIGURE 4.10. Determination of the collinear solutions.

(4.4.8) Periodic Orbits

At this point it is of interest to ask whether there are other periodic solutions in addition to these five. In the special case $\mu = 0$ we already know that a great many trajectories are periodic—in the rotating system these are the Kepler orbits of a single force-center and a mass m with a rational frequency. According to the following argument, which goes back to Poincaré, under the right circumstances, for small enough μ, there exist periodic orbits with the same frequencies. Let $u(t, u_0; \mu)$, where u stands for x, y, p_x, or p_y, be the solution of the equation of motion with initial condition $u(0, u_0; \mu) = u_0$, and suppose an orbital period τ is specified. We ask for what initial values u_0 the orbit has the period τ. For these values the equation

(4.4.9) $$u(\tau, u_0, \mu) = u_0$$

should hold, and we can regard it as the equation for $u_0(\mu)$. From our study of the Kepler problem we know that for rational τ equation (4.4.9) has a solution for $\mu = 0$. From the theory of differential equations [1, 10.7], we learn that u is differentiable in μ and u_0, in particular on regions in which the trajectories do not encounter the centers of force. If $u(\tau, u_0; 0)$ satisfies

(4.4.10) $$\det\left(\frac{\partial(u - u_0)}{\partial u_0}\right) \neq 0,$$

then this determinant remains different from zero in a neighborhood of $\mu = 0$. This being so, (2.1.10; 3) guarantees the existence of a five-dimensional submanifold of (u, u_0, μ) which satisfies (4.4.9) and corresponds to a solution of (4.4.9) in a neighborhood of $\mu = 0$. Therefore, at each rational frequency there exists a periodic orbit for sufficiently small μ.

(4.4.11) **Remarks**

1. This is not a trivial statement, because periodicity can be destroyed by arbitrarily small perturbations, and yet here, in the neighborhood of the center, for every positive μ the force is changed by an arbitrarily large amount. (See Problem 1.)
2. We do not discover how large the neighborhood of $\mu = 0$ that allows periodic orbits is.
3. Whenever there exists a constant of the motion K, as is always the case for Hamiltonian systems, condition (4.4.10) is violated:

$$K(u(\tau, u_0)) = K(u_0)\ \forall u_0 \quad \Rightarrow \quad \frac{\partial K}{\partial u}(u(\tau, u_0))\frac{\partial u(\tau, u_0)}{\partial u_0} = \frac{\partial K(u_0)}{\partial u_0}.$$

This means that the vector

$$\frac{\partial K(u_0)}{\partial u_0} = \frac{\partial K(u(\tau, u_0))}{\partial u}$$

is an eigenvector of the transposed matrix $\partial(u(\tau, u_0) - u_0)/\partial u_0$ with the eigenvalue 0, and so the determinant of the matrix vanishes. The problem can be surmounted, however, because if there exists a constant of the motion, then any one of the equations (4.4.9) automatically holds when the other three do.

(4.4.12) **Example**

We investigate the orbits of the unperturbed problem, $\mu = 0$, and $m = 1$. In plane polar coordinates,

$$H = \frac{1}{2}\left(p_r^2 + \frac{p_\varphi^2}{r^2}\right) - p_\varphi - \frac{1}{r},$$

and the equations of motion become

$$(\dot{r}, \dot{\varphi}, \dot{p}_r, \dot{p}_\varphi) = \left(p_r, \frac{p_\varphi}{r^2} - 1, -\frac{1}{r^2} + \frac{p_\varphi^2}{r^3}, 0\right).$$

The solution

$$u := (r, \varphi, p_r, p_\varphi) = ((\omega + 1)^{-2/3}, \omega t, 0, (\omega + 1)^{-1/3})$$

is a circular orbit. The matrix of derivatives $\partial u/\partial u_0$ for any solution is determined by the differential equation that follows from the equations of motion. Defining

$\gamma = (\omega + 1)^{1/3}$ and $\tau = 2\pi/\omega$, we calculate (Problem 2) that

$$\left.\frac{\partial u}{\partial u_0}\right|_{t=\tau, u_0=(\gamma^{-2},0,0,\gamma^{-1})}$$

$$= \begin{bmatrix} \cos\tau\gamma^3 & 0 & \frac{1}{\gamma^3}\sin\tau\gamma^3 & \frac{2}{\gamma}(1-\cos\tau\gamma^3) \\ -2\gamma^2\sin\tau\gamma^3 & 1 & -\frac{2}{\gamma}(1-\cos\tau\gamma^3) & -3\gamma^4\tau + 4\gamma\sin\tau\gamma^3 \\ -\gamma^3\sin\tau\gamma^3 & 0 & \cos\tau\gamma^3 & 2\gamma^2\sin\tau\gamma^3 \\ 0 & 0 & 0 & 1 \end{bmatrix}.$$

According to (4.4.11; 3) the matrix $(\partial u/\partial u_0) - \mathbf{1}$ has determinant 0, because

$$\left.\frac{\partial H}{\partial u}\right|_{u_0} = \left(\frac{1}{r^2} - \frac{p_\varphi^2}{r^3}, 0, p_r, \frac{p_\varphi}{r^2} - 1\right) = (0, 0, 0, \omega),$$

where $(r, \varphi, p_r, p_\varphi) = u_0 = (\gamma^{-2}, 0, 0, \gamma^{-1})$, and $(0, 0, 0, 1)$ is an eigenvector of $(\partial u/\partial u_0)^t$ with eigenvalue 1. Let us look only at the r, φ, and p_r components of u and consider

$$\begin{aligned} u_r(\tau; r, 0, p_r, p_\varphi; \mu) - r &= 0, \\ u_\varphi(\tau; r, 0, p_r, p_\varphi; \mu) - 2\pi &= 0, \\ u_{p_r}(\tau; r, 0, p_r, p_\varphi; \mu) - p_r &= 0, \end{aligned} \tag{4.4.13}$$

as equations for the initial values (r, p_r, p_φ). The Jacobian in this case can be calculated as

$$24\pi \frac{(\omega+1)^{4/3}}{\omega} \sin^2\left(\frac{\pi}{\omega}\right)$$

(Problem 3), which is nonzero for $1/\omega \notin \mathbb{Z}$. It was assumed above that ω was rational, $\omega = p/q$, with $p, q \in \mathbb{Z}$, and the condition on $1/\omega$ is satisfied unless $p = 1$. It is then easy to verify (Problem 4) that the three equations (4.4.13) suffice to prove that $u_{p_\varphi}(\tau)$ returns to its initial value even if $\mu \neq 0$. Thus, unless ω is in resonance with the rotating force-centers, i.e., for $1/\omega \notin \mathbb{Z}$, there are periodic orbits with frequency ω for nonzero μ.

(4.4.14) Remarks

1. When $\mu = 0$, p_φ is also a constant of the motion. The vector $\partial p_\varphi/\partial u = (0, 0, 0, 1)$, which has the same direction as $\partial H/\partial u|_{u_0}$. There is no other eigenvector of $(\partial u/\partial u_0)^t$ with eigenvalue 1, and thus $(\partial u/\partial u_0) - \mathbf{1}$ has a submatrix of rank 3.
2. The mass ratios in the Solar System are so extreme that it is not sheer madness to believe in the relevance of this result for astronomy without knowing exactly how large μ is allowed to be.

(4.4.15) Stability of the Periodic Orbits

The equilibrium configurations are fixed points of the canonical flow Φ_t, and in §3.4 it was explained precisely what stability means in this context. Periodic orbits are invariant under some Φ_τ, and so (3.4.21) defines stability. Then Theorem (3.4.24) gives a necessary criterion for the stability of the orbits: A must be diagonable and have purely imaginary eigenvalues. In this case, the variable z is (x, y, p_x, p_y), and the matrix is

$$A = \begin{bmatrix} 0 & 1 & 1 & 0 \\ -1 & 0 & 0 & 1 \\ -V_{,xx} & -V_{,xy} & 0 & 1 \\ -V_{,xy} & -V_{,yy} & -1 & 0 \end{bmatrix}. \tag{4.4.16}$$

Looking at the equilateral equilibrium configurations,

$$x = \tfrac{1}{2} - \mu, \qquad y = \frac{\sqrt{3}}{2},$$

we see that

$$V_{,xx} = \tfrac{1}{4}, \qquad V_{,xy} = -\frac{3\sqrt{3}}{4}(1 - 2\mu), \qquad V_{,yy} = -\tfrac{5}{4},$$

and the eigenvalues λ of A satisfy the equation

$$\lambda^2 = -\tfrac{1}{2} \pm \tfrac{1}{2}\sqrt{1 - 27\mu(1 - \mu)} \tag{4.4.17}$$

(Problem 5). If $\mu(1 - \mu) < \frac{1}{27}$, i.e., $\mu/m < 0.040$, then all the eigenvalues are imaginary and nondegenerate, and the orbits are possibly stable. Otherwise, they are certainly not stable.

(4.4.18) **Remarks**

1. A similar calculation for the collinear solutions reveals that the eigenvalues of A always have nonzero real parts. Thus those orbits fail to be stable for any value of μ/m.
2. The necessary stability condition of (3.4.24) is too strong for our purposes, since we only need real, and not complex, stability. To decide whether periodic orbits are stable, one can apply the K–A–M Theorem (3.6.19) (μ is the perturbation parameter, as the system is integrable for $\mu = 0$), which gives invariant two-dimensional tori for the perturbed system. On the three-dimensional surface H = const., the orbits that are within the tori can never lead outside them, and so stability can be proved as soon as some frequency condition is fulfilled [6, §34].
3. A beautiful realization of the restricted three-body problem occurs in the asteroids between Mars and Jupiter. Here the Sun and Jupiter are the two heavy bodies. According to the K–A–M theorem the orbits which are disturbed the most are the ones whose frequencies ω are rationally related to Jupiter's, $\omega/\omega_{\text{Jup}} = p/q$, with $p, q \in \mathbb{Z}$. For small p and q the resonance effect should be most noticeable and those asteroids knocked out of orbit. Actually, there are significant gaps in the distribution of asteroids at $\omega/\omega_{\text{Jup}} = 2, \frac{7}{3}, \frac{5}{2}$, and 3. (See Figure 4.11.)

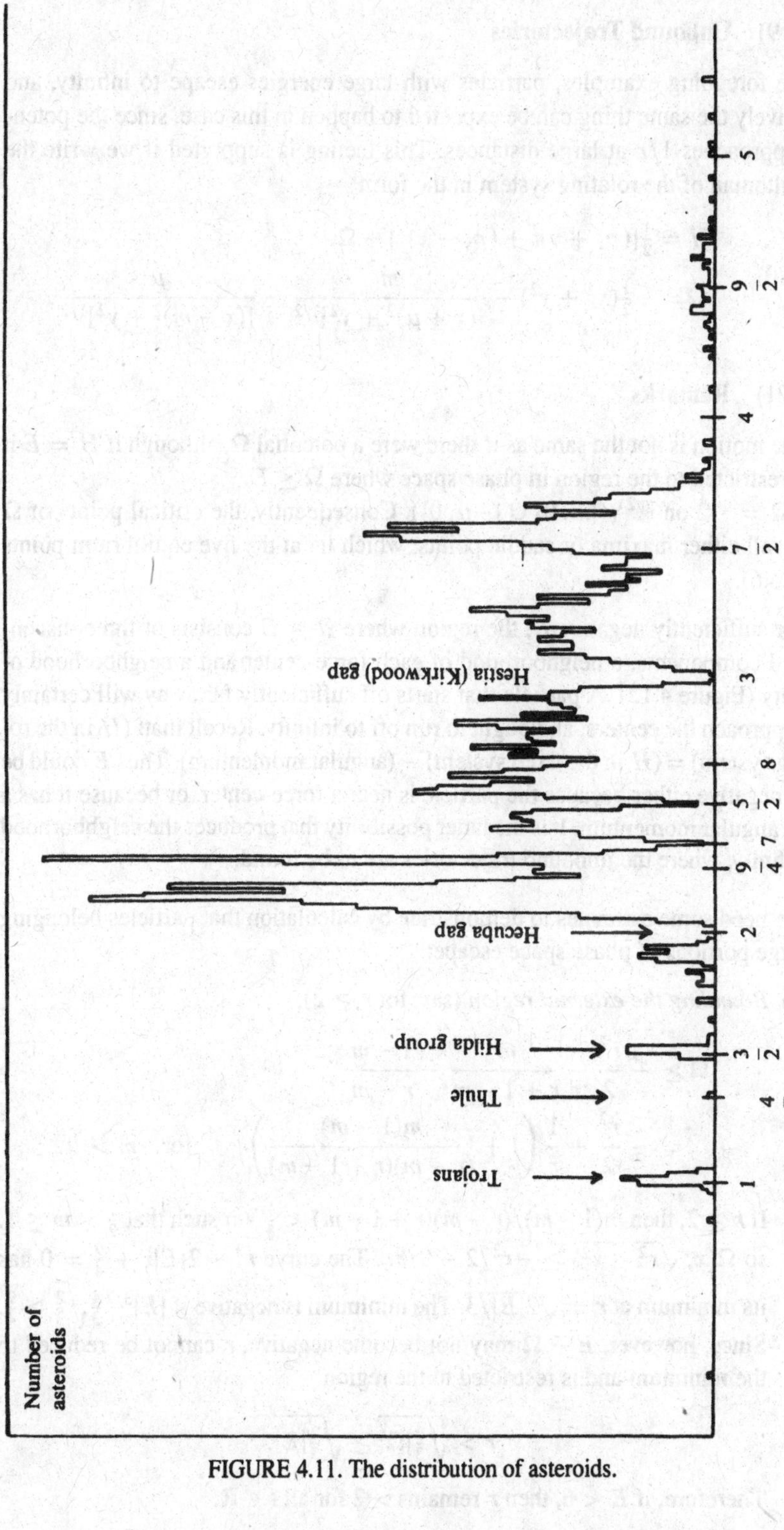

FIGURE 4.11. The distribution of asteroids.

(4.4.19) Unbound Trajectories

In the foregoing examples, particles with large energies escape to infinity, and intuitively the same thing can be expected to happen in this case, since the potential approaches $1/r$ at large distances. This feeling is supported if we write the Hamiltonian of the rotating system in the form

$$\begin{aligned} H &= \tfrac{1}{2}[(p_x + y)^2 + (p_y - x)^2] + \Omega, \\ \Omega &= -\tfrac{1}{2}(x^2 + y^2) - \frac{m}{[(x+\mu)^2 + y^2]^{1/2}} - \frac{\mu}{[(x-m)^2 + y^2]^{1/2}}. \end{aligned} \tag{4.4.20}$$

(4.4.21) Remarks

1. The motion is not the same as if there were a potential Ω, although if $H = E$ it is restricted to the region in phase space where $\Omega \leq E$.
2. $\Delta\Omega = -2$ on $\mathbb{R}^2 \setminus (\{m, 0\} \cup \{-\mu, 0\})$. Consequently, the critical points of Ω are all either maxima or saddle points, which lie at the five equilibrium points (4.4.6).

For sufficiently negative E, the region where $\Omega \leq E$ consists of three disconnected components, a neighborhood of each force-center and a neighborhood of infinity (Figure 4.12). A particle that starts off sufficiently far away will certainly not approach the centers, and ought to run off to infinity. Recall that: (H in the rotating system) = (H in the fixed system) − (angular momentum). Thus E could be very negative either because the particle is near a force-center, or because it has a large angular momentum. It is the latter possibility that produces the neighborhood of infinity, where the unbound trajectories are to be found.

We need some estimates to demonstrate by calculation that particles belonging to large portions of phase space escape:

(a) *Bounding the external region* (say, for $r \geq 2$).

$$\begin{aligned} \Omega &\geq -\frac{r^2}{2} - \frac{m}{r+1-m} - \frac{1-m}{r-m} \\ &= -\frac{r^2}{2} - \frac{1}{r}\left(1 + \frac{m(1-m)}{(r-m)(r+1-m)}\right) \qquad \text{for} \quad m > \mu. \end{aligned}$$

If $r \geq 2$, then $m(1-m)/(r-m)(r+1-m) < \frac{1}{8}\ \forall m$ such that $\frac{1}{2} \leq m \leq 1$, so $\Omega(x, \sqrt{r^2 - x^2}) \geq -r^2/2 - 9/8r$. The curve $r^3 - 2|E|r + \frac{9}{4} = 0$ has its minimum at $r = \sqrt{2|E|/3}$. The minimum is negative if $|E|^{3/2}\frac{4}{3}\sqrt{\frac{2}{3}} \geq \frac{9}{4}$. Since, however, $E - \Omega$ may not become negative, r cannot be reduced to the minimum and is restricted to the region

$$r > \sqrt{\tfrac{2}{3}|\Omega|} \geq \sqrt{\tfrac{2}{3}|E|}.$$

Therefore, if $E < 6$, then r remains > 2 for all $t \in \mathbb{R}$.

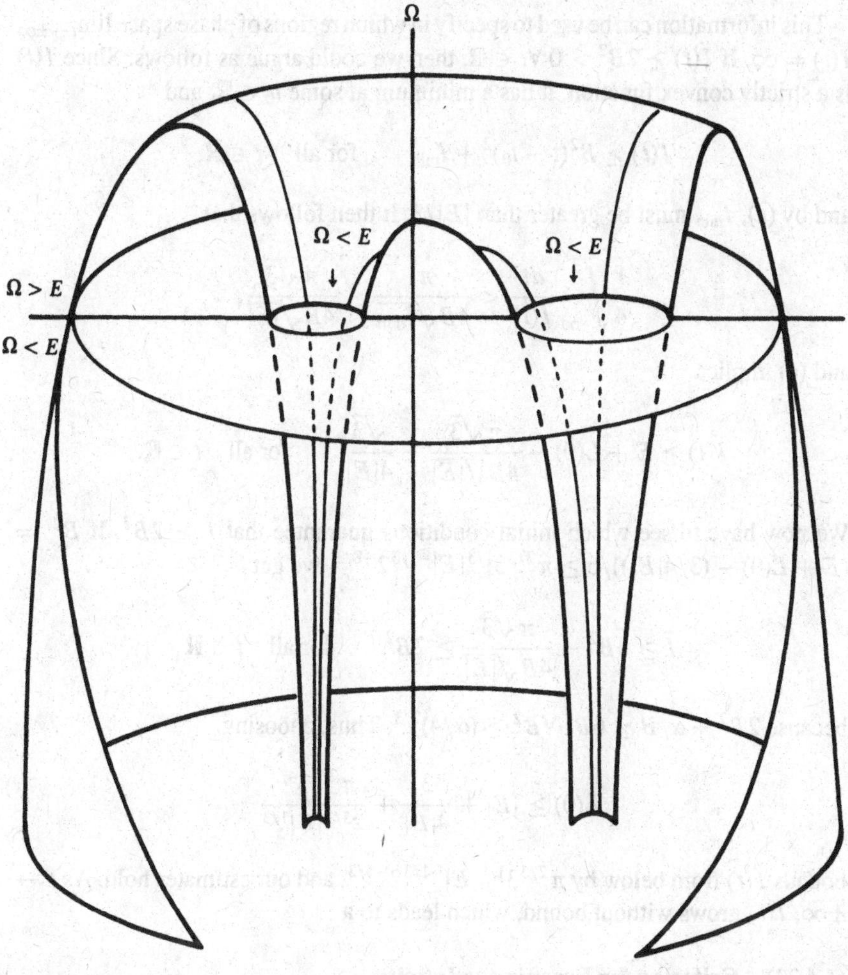

FIGURE 4.12. A cross-section of parts of the surface $\Omega(x, y)$ (somewhat like a volcano with two craters).

(b) *Bounding the angular momentum.* Since the angular momentum L is conserved for the $1/r$ potential, it ought not to vary much in this case for trajectories at large distances. It follows from the equations of motion (Problem 6) that if $r(t') > 2\ \forall t'$ such that $0 \leq t' \leq t$, and $L := xp_y - yp_x$, then

$$|L(t) - L(0)| \leq \int_0^t \frac{dt'}{4I(t')}, \qquad \text{where} \quad I \equiv \frac{r^2}{2}.$$

(c) *Convexity of the moment of inertia.* For free particles, $I(t)$ is a quadratic function, and in certain other situations it is possible to show that it is at least convex. It is easy to discover (Problem 7) that

$$\ddot{I}(t) \geq H + L(0) - \int_{-\infty}^{t} dt' \frac{1}{4I(t')} - \frac{1}{4I(t)}.$$

This information can be used to specify in which regions of phase space $\lim_{t\to\pm\infty} I(t) = \infty$. If $\ddot{I}(t) \geq 2B^2 > 0 \; \forall t \in \mathbb{R}$, then we could argue as follows: Since $I(t)$ is a strictly convex function, it has a minimum at some $t_0 \in \mathbb{R}$, and

$$I(t) \geq B^2(t - t_0)^2 + I_{\min} \qquad \text{for all} \quad t \in \mathbb{R},$$

and by (a), $I_{\min}$ must be greater than $|E|/3$. It then follows that

$$\frac{1}{4}\int_{-\infty}^{\tau} \frac{dt}{I(t)} < \frac{\pi}{4B\sqrt{I_{\min}}} \leq \frac{\pi\sqrt{3}}{4B\sqrt{|E|}},$$

and (c) implies

$$\ddot{I}(t) \geq E + L(0) - \frac{\pi\sqrt{3}}{4B\sqrt{|E|}} - \frac{\sqrt{3}}{4|E|} \qquad \text{for all} \quad t \in \mathbb{R}.$$

We now have to see which initial conditions guarantee that $\ddot{I} \geq 2B^2$. If $B^2 := (E + L(0) - (3/4|E|))/6 \geq \pi^{2/3}3^{1/3}|E|^{-1/3}2^{-8/3}$, we get

$$\ddot{I} \geq 6B^2 - \frac{\pi\sqrt{3}}{4B\sqrt{|E|}} \geq 2B^2 \qquad \text{for all} \quad t \in \mathbb{R}$$

because $2B^2 + \alpha/B \leq 6B^2 \; \forall B^2 \geq (\alpha/4)^{2/3}$. Thus, choosing

$$L(0) \geq |E| + \frac{3}{4|E|} + \frac{\pi^{2/3}3^{4/3}}{2^{5/3}|E|^{1/3}}$$

bounds $\ddot{I}(t)$ from below by $\pi^{2/3}3^{1/3}|E|^{-1/3}2^{-5/3}$, and our estimates hold. As $t \to \pm\infty$, $I(t)$ grows without bound, which leads to a

(4.4.22) **Criterion for Escaping to Infinity**

If for some trajectory

$$E < -6, \qquad L(0) \geq |E| + \frac{3}{4|E|} + \frac{\pi^{2/3}3^{4/3}}{2^{5/3}|E|^{1/3}},$$

then $\lim_{t\to\pm\infty} r(t) = \infty$.

(4.4.23) **Remarks**

1. The strong focus on the assumption that $\ddot{I} \geq 2B^2$ might leave the impression of a circular argument. However, we can see that if the escape criterion is satisfied, then $\ddot{I}$ must indeed remain greater than $2B^2$ on any finite interval: The closed set $\{t \geq 0 : \ddot{I} \geq 2B^2\}$ cannot be of the form $[0, c]$ with $c < \infty$, since the possibility that $\ddot{I}(c) = 2B^2$ has been excluded. This justifies our reasoning a posteriori.
2. As with the problem with two fixed centers of force, the Møller transformation using the flow generated by a $1/r$ potential as Φ_t^0 exists. There are three constants

of the motion (five in the three-dimensional case) in this part of phase space, and the trajectories are homotopic to Kepler hyperbolas.

3. Trajectories that get near the force-centers can become quite complicated. For instance, the following rather surprising statement can be made about trajectories perpendicular to the plane of motion of the heavy particles, if they travel in ellipses [14, III.5]: $\exists m > 0$ such that for every sequence $s_k > m$, there exists a trajectory for which the time between the kth and $(k+1)$th intersection with the plane of the ellipse is exactly s_k.
4. In the situation depicted in Figure 4.10, any trajectory once in the vicinity of one of the force-centers always remains nearby. One might be tempted to apply the Recurrence Theorem (2.6.11) in this case, but it does not work, because collisions cannot be avoided, and no time-invariant region in phase space that is compact in the momentum coordinates as well as the spatial ones can be found.

The flow Φ_t of the restricted three-body problem is, as we see, not known in full detail, and our analysis only gives us the impression that trajectories at a respectable distance from the force-centers evolve smoothly. But if a particle happens to approach too near, it can dance around in a completely crazy way.

(4.4.24) Problems

1. With the two-dimensional harmonic oscillator, it is easy to see that even for arbitrarily small $\mu \neq 0$, the orbits of

$$H(\mu) = \tfrac{1}{2}(p_x^2 + p_y^2 + x^2 + y^2 + \mu(x^2 - y^2))$$

that are periodic when $\mu = 0$ can be destroyed, so that no periodic orbits at all remain. Why doesn't Poincaré's argument work in this case?

2. Calculate $\partial u / \partial u_0$ from (4.4.12).

3. Same problem for (4.4.13). What is the determinant of this matrix?

4. Prove that $u_{p_\varphi}(\tau)$ is in fact equal to p_φ in (4.4.12).

5. Calculate the eigenvalues of A from (4.4.16) for the equilateral equilibrium configurations.

6. Let $p := xp_y - yp_x$ and $I := r^2/2$. Use (4.4.5) to show the following bound for the angular momentum in terms of the moment of inertia:

$$|p_\varphi(0) - p_\varphi(t)| \leq \int_0^t \frac{dt'}{4I(t')},$$

when $I(t') > 2\ \forall t'$ such that $0 \leq t' \leq t$.

7. Use Problem 6 to show that

$$\ddot{I}(t) \geq H + L(0) - \int_{-\infty}^{t} dt' \, \frac{1}{4I(t')} - \frac{1}{4I(t)}$$

if $I(t') > 2\ \forall t'$ such that $0 \leq t' \leq t$.

(4.4.25) Solutions

1. In polar coordinates,

$$H(0) = \frac{1}{2}\left(p_r^2 + \frac{p_\varphi^2}{r^2} + r^2\right).$$

Letting $u := (r, \varphi, p_r, p_\varphi)$ and the solution $u_0 := u(\tau; r, 0, 0, r^2) = (r, \tau, 0, r^2)$, we calculate from the equation for $H(0)$ that

$$A = \left.\frac{\partial X_H}{\partial u}\right|_{u=u_0} = \begin{array}{|c|c|c|c|} \hline 0 & 0 & 1 & 0 \\ \hline -2/r & 0 & 0 & 1/r^2 \\ \hline -4 & 0 & 0 & 2/r \\ \hline \end{array}$$

Then $A \cdot (a, b, c, d) = (c, -2a/r + d/r^2, -4a + 2d/r, 0)$, and thus $\text{rank}(A) = \dim(A \cdot \mathbb{R}^4) = 2$. Consequently, $\text{rank}(e^{tA} - 1) = \sum_{n=1}^{\infty}(tA)^n/n!$ is also equal to 2, and there is no nonsingular (3×3)-submatrix of $\partial(u - u_0)/\partial u_0$. The system of equations has no solution.

2. Because $\dot{u}(t, u_0) = X_H(u(t, u_0))$, the matrix of derivatives satisfies the homogeneous differential equation

$$\frac{d}{dt}\frac{\partial u}{\partial u_0} = \frac{\partial X_H}{\partial u}\frac{\partial u}{\partial u_0},$$

in which $\partial X_H/\partial u$ depends on the solution $u(t, u_0)$, which is assumed known. For circular orbits this matrix is independent of t, and since $u(0, u_0) = u_0$ implies the initial condition $\partial u/\partial u_{0|t=0} = 1$, the solution of the differential equation is simply

$$\frac{\partial u}{\partial u_0} = \exp\left(t\frac{\partial X_H}{\partial u}\right).$$

Hence we have to calculate the matrix

$$\partial X_H(u(t, u_0))/\partial u =: A$$

(u being given by the circular orbit), and then exponentiate it. From (4.4.12) it follows that

$$X_H = \left(p_r, \frac{p_\varphi}{r^2} - 1, -\frac{1}{r^2} + \frac{p_\varphi}{r^3}, 0\right),$$

and

$$\frac{\partial X_H}{\partial u} = \begin{bmatrix} 0 & 0 & 1 & 0 \\ \frac{-2p_\varphi}{r^3} & 0 & 0 & \frac{1}{r^2} \\ \frac{2}{r^3} - \frac{3p_\varphi}{r^4} & 0 & 0 & \frac{2p_\varphi}{r^3} \\ 0 & 0 & 0 & 0 \end{bmatrix},$$

and, in particular, for the circular orbit

$$A = \begin{bmatrix} 0 & 0 & 1 & 0 \\ -2\gamma^5 & 0 & 0 & \gamma^4 \\ -\gamma^6 & 0 & 0 & 2\gamma^5 \\ 0 & 0 & 0 & 0 \end{bmatrix}.$$

In order to calculate $\exp(\tau A)$, we put A into Jordan normal form with a nonsingular matrix C (which is not necessarily unitary, since A is not Hermitian):

$$A = CNC^{-1},$$

where N is a matrix the diagonal elements of which are the eigenvalues of A, and which may have nonvanishing elements immediately above the diagonal, but all other elements are zero. Such a matrix can be easily exponentiated, and $\exp(\tau A) = C\exp(\tau N)C^{-1}$. The eigenvalues of A can be calculated from $\det|A-\lambda| = \lambda^2(\lambda^2+\gamma^3) = 0$ to be $\lambda = 0, 0$, and $\pm i(\omega+1)$. Explicit calculations of the various matrices are:

$$C = \begin{bmatrix} 0 & \frac{2}{3} & 1 & 1 \\ 1 & 0 & 2i\gamma^2 & -2i\gamma^2 \\ 0 & 0 & i\gamma^3 & -i\gamma^3 \\ 0 & \frac{\gamma}{3} & 0 & 0 \end{bmatrix}, \qquad C^{-1} = \begin{bmatrix} 0 & 1 & \frac{-2}{\gamma} & 0 \\ 0 & 0 & 0 & \frac{3}{\gamma} \\ \frac{1}{2} & 0 & \frac{1}{2i\gamma^3} & \frac{-1}{\gamma} \\ \frac{1}{2} & 0 & \frac{-1}{2i\gamma^3} & \frac{-1}{\gamma} \end{bmatrix},$$

$$N = \begin{bmatrix} 0 & -\gamma^5 & 0 & 0 \\ 0 & 0 & 0 & 0 \\ 0 & 0 & i\gamma^3 & 0 \\ 0 & 0 & 0 & -i\gamma^3 \end{bmatrix}, \qquad e^{\tau N} = \begin{bmatrix} 1 & -\gamma^5\tau & 0 & 0 \\ 0 & 1 & 0 & 0 \\ 0 & 0 & e^{i\tau\gamma^3} & 0 \\ 0 & 0 & 0 & e^{-i\tau\gamma^3} \end{bmatrix}.$$

Multiplication of the matrices yields the result given in (4.4.12).

3. We need to calculate the determinant of

$$\begin{bmatrix} u_{r,r}-1 & u_{r,p_r} & u_{r,p_\varphi} \\ u_{\varphi,r} & u_{\varphi,p_r} & u_{\varphi,p_\varphi} \\ u_{p_r,r} & u_{p_r,p_r}-1 & u_{p_r,p_\varphi} \end{bmatrix}.$$

If $\mu = 0$, and

$$S := \sin\tau\gamma^3 \qquad \text{and} \qquad C := \cos\tau\gamma^3,$$

then this matrix equals

$$\begin{bmatrix} C-1 & \frac{S}{\gamma^3} & \frac{2}{\gamma}(1-C) \\ -2\gamma^2 S & -\frac{2}{\gamma}(1-C) & -3\gamma^4\tau + 4\gamma S \\ -\gamma^3 S & C-1 & 2\gamma^2 S \end{bmatrix}.$$

Writing $\tau = 2\pi/\omega$ and $\gamma^3 = \omega + 1$ and taking the determinant yields

$$24\pi\frac{(\omega+1)^{4/3}}{\omega}\sin^2\pi/\omega.$$

4. We know that $H(r, 2\pi, p_r, u_{p_\varphi}(\tau)) = H(r, 0, p_r, p_\varphi)$, and that if $\mu = 0$,

$$\frac{\partial H}{\partial p_\varphi} = \frac{p_\varphi}{r^3} - 1 = \omega \neq 0.$$

Hence in some neighborhood of $\mu = 0$, $\partial H/\partial p_\varphi$ has a definite sign between p_φ and $u_{p_\varphi}(\tau)$; but then $u_{p_\varphi}(\tau)$ must equal p_φ for the first equality to hold.

5. For a block matrix

$$\begin{bmatrix} a & b \\ c & d \end{bmatrix},$$

if $db = bd$, then

$$\det\begin{vmatrix} a & b \\ c & d \end{vmatrix} = \det|ad - bc|.$$

Therefore

$$\det|A-\lambda| = \lambda^4 + \lambda^2 + \tfrac{27}{16}(1-(1-2\mu)^2),$$

and the eigenvalues of A are

$$\lambda_{1,2}^2 = -\tfrac{1}{2} \pm \tfrac{1}{2}\sqrt{1-27\mu(1-\mu)}.$$

6. The amount of nonconservation of the angular momentum ($|\mathbf{L}|^2 \equiv L_z^2 \equiv L^2$) is

$$\dot{L}_z = -xV_{,y} + yV_{,x} = m\mu y\{[(x+\mu)^2+y^2]^{-3/2} - [(x-m)^2+y^2]^{-3/2}\} \quad \Rightarrow$$

$$|\dot{L}_z| \le \frac{m\mu}{(r-m)^2} \le \frac{1}{2r^2}\ \forall r \ge 2 \quad \Rightarrow \quad |L_z| \le \frac{1}{4I}, \quad \text{if } I \ge 2.$$

7.

$$\ddot{I} = \frac{d}{dt}\mathbf{x}\cdot\mathbf{p} = p_x^2 + p_y^2 - xV_{,x} - yV_{,y}$$

$$= H + L + \tfrac{1}{2}(p_x^2+p_y^2) + m\mu\left\{\frac{x+\mu}{[(x+\mu)^2+y^2]^{3/2}} - \frac{x-m}{[(x-m)^2+y^2]^{3/2}}\right\}.$$

As in Problem 6,

$$|\{\ \}| \le \frac{1}{(r-m)^2} \le \frac{1}{2r^2} \quad \text{and} \quad L(t) \ge L(0) - \int_{-\infty}^{t}\frac{dt}{4I(t)}, \quad p_x^2+p_y^2 \ge 0.$$

4.5 The N-Body Problem

Although the system of equations appears hopelessly complicated, it is possible not only to find exact solutions, but even to make some general propositions.

Since time immemorial many of the top minds have applied their mathematical skills to equations (1.1.1) and (1.1.2) with $n \ge 3$, but without great success. We will pick up some of the more amusing pieces from their efforts, and by doing so we hope to illustrate how one typically approaches the problem. We restrict ourselves to the case of pure gravitation; the inclusion of a Coulomb force requires only trivial changes and brings hardly any new insight. As usual we begin with the

(4.5.1) Constants of Motion

As in the two-body problem, the flow factorizes into the motion of the center of mass and relative coordinates. The center-of-mass part has the maximal number of constants, because of the Galilean invariance, while the relative part has only the conserved angular momentum, from invariance under rotations.

(4.5.2) The Case $N = 3$

The overall phase space is 18-dimensional, while the phase space of the relative motion is only 12-dimensional. In the latter there are only four constants, the angular momentum and the energy, which are not enough for the equations to be integrable.

(4.5.3) **Remarks**

1. Whereas it can be proved in the restricted three-body problem [14, VI.8] that there are no additional constants of motion other than Jacobi's constant, in this case a classic **theorem of Bruns** implies that there are no other integrals that are algebraic functions of the Cartesian coordinates $\mathbf{x}$ and $\mathbf{p}$. But since we attribute no special status to any coordinate system, it is not clear that this statement is of much value.
2. The known integrals do not separate off any compact part of phase space on which the recurrence theorem might be applied. It is even possible that collision trajectories are dense in regions of positive measure.
3. Additional constants certainly exist in the parts of phase space where scattering theory operates [19].
4. Computer studies of the restricted three-body problem have found parts of phase space where the trajectories—even trajectories that remain finite—form manifolds of a lower dimension, as if there existed more constants of the motion [16]. This phenomenon could also show up in the n-body problem.

(4.5.4) **Exact Solutions**

If all N particles move in a plane, it can easily happen that gravity and the centrifugal force balance each other. Let us consider the Cartesian coordinates in the plane as complex numbers, and set

$$x_j(t) + iy_j(t) = z(t)z_j,$$
$$z(t)\colon \mathbb{R} \to \mathbb{C} \qquad \text{and} \qquad z_j \in \mathbb{C}.$$

This assumption means that the configuration of the particles in the plane has the same shape at all times. Letting $\kappa = 1$, the equations of motion,

$$\ddot{z}z_i = \sum_{j \neq i} \frac{(z_j - z_i)m_j}{|z_j - z_i|^3} \frac{z}{|z|^3}, \tag{4.5.5}$$

can be decomposed into the Kepler problem in the plane,

$$\ddot{z} = -\omega^2 \frac{z}{|z|^3}, \tag{4.5.6}$$

and the algebraic equation,

$$-\omega^2 z_i = \sum_{j \neq i} \frac{z_j - z_i}{|z_j - z_i|^3} m_j. \tag{4.5.7}$$

Hence, each particle moves in a Kepler trajectory about the collective center of mass (since (4.5.7) $\Rightarrow \sum_i z_i m_i = 0$).

(4.5.8) **Remarks**

1. Since we know that there are solutions to (4.5.6) for all $\omega \in \mathbb{R}$, only (4.5.7) needs to be discussed.
2. The total energy of the motion is

$$E = |\dot{z}|^2 \sum_i |z_i|^2 \frac{m_i}{2} - \frac{1}{2|z|} \sum_{i \neq j} \frac{m_i m_j}{|z_i - z_j|}$$

$$= \left\{ \frac{|\dot{z}|^2}{2} - \frac{\omega^2}{|z|} \right\} \sum_i |z_i|^2 m_i,$$

i.e., the energy of the Kepler trajectory times the moment of inertia. The particles remain in a bounded region iff $E < 0$.

(4.5.9) **The Specialization to $N = 3$**

It is necessary to distinguish two cases:

(a) All $|z_i - z_j| = R$, making an equilateral triangle. Adding the three equations (4.5.7) gives

$$(-\omega^2 R^3 + m_1 + m_2 + m_3) \sum_i z_i = 3 \sum_i z_i m_i = 0,$$

which is solved by

$$\omega^2 R^3 = m_1 + m_2 + m_3.$$

But then all three equations are satisfied.

(b) $|z_1 - z_3| \neq |z_2 - z_3|$. If the coordinate system is chosen so that $z_3 = 0$, then because of (4.5.7),

$$\mathrm{Re}\left(\frac{m_1 z_1}{|z_3 - z_1|^3} + \frac{m_2 z_2}{|z_3 - z_2|^3} \right) = 0.$$

But since also $\mathrm{Re}(m_1 z_1 + m_2 z_2) = -\mathrm{Re}\, m_3 z_3 = 0$, it must be true that $\mathrm{Re}\, z_1 = \mathrm{Re}\, z_2 = 0$, and all three particles are in a line. These are the same as the equilibrium configurations of the special case treated in §4.4.

It is now pertinent to ask whether particles invariably run off to infinity when $E \geq 0$ and the center of mass is fixed. This is in fact so, according to the virial theorem mentioned above, which we now state:

(4.5.10) **The Virial Theorem**

Let $I := \sum_j m_j |\mathbf{x}_j|^2$, $T := \sum_j |\mathbf{p}_j|^2/2m$, *and* $V := -\kappa \sum_{i>j} m_i m_j / |\mathbf{x}_i - \mathbf{x}_j|$. *Then* $\ddot{I} = 2(E + T)$.

Proof: See Problem 1. □

(4.5.11) **Corollaries**

1. *Since $T \geq 0$, $I(t) \geq I(0) + t\dot{I}(0) + t^2 E$. Thus if $E > 0$, then $\lim_{t\to\infty} I(t) = \infty$, unless the system collapses first. Hence at least one particle travels arbitrarily far away.*

2. *If $I(t)$ approaches zero, meaning that the system collapses, it must reach zero in a finite time, and cannot do it asymptotically. This is because $I \to 0$ only if $V \to -\infty$ (Problem 2), and if $\lim_{t\to\infty} V = -\infty$, then, because $\ddot{I} = 4E - 2V$, there exists a time t_0 such that $\ddot{I} > 1$ $\forall t > t_0$. But then $\lim_{t\to\infty} I(t)$ cannot be zero.*

3. *If the motion is almost periodic, then for any $\varepsilon > 0$ there exists a sequence $\tau_i \to \infty$ with $|\dot{I}(0) - \dot{I}(\tau_i)| < \varepsilon$ $\forall i$. Then*

$$\lim_{i\to\infty} \frac{1}{\tau_i} \int_0^{\tau_i} dt(2T + V) = \lim_{i\to\infty} \frac{1}{\tau_i}(\dot{I}(\tau_i) - \dot{I}(0)) = 0.$$

According to this, the average of the potential energy is twice the total energy, which is obviously only possible if $E < 0$.

If $N \geq 3$, it is energetically possible that one of the particles gets catapulted off to infinity. The requisite energy can always be produced if the other particles come close enough together. It might be supposed that whenever the kinetic energy of a particle exceeds its potential energy, the particle flies off, never to be seen again. However, the energy of an individual particle is not conserved, and a closer analysis is needed to see if this is really true.

(4.5.12) **A Lower Bound for the Kinetic Energy**

For simplicity we look at the situation where $N = 3$, $\kappa = m_1 = m_2 = m_3 = 1$. Let $E < 0$, and

(4.5.13)
$$r_m := \min_{i \neq j} |\mathbf{x}_i - \mathbf{x}_j|.$$

Then, because $|V| \geq |E|$, we have the bound

(4.5.14)
$$r_m \leq r^* := \frac{3}{|E|}.$$

To separate off the center-of-mass motion, we introduce the coordinates

(4.5.15)
$$\mathbf{s} = \frac{1}{\sqrt{3}}(\mathbf{x}_1 + \mathbf{x}_2 + \mathbf{x}_3),$$
$$\mathbf{u} = \frac{1}{\sqrt{2}}(\mathbf{x}_1 - \mathbf{x}_2),$$
$$\mathbf{x} = \frac{1}{\sqrt{6}}(\mathbf{x}_1 + \mathbf{x}_2 - 2\mathbf{x}_3)$$

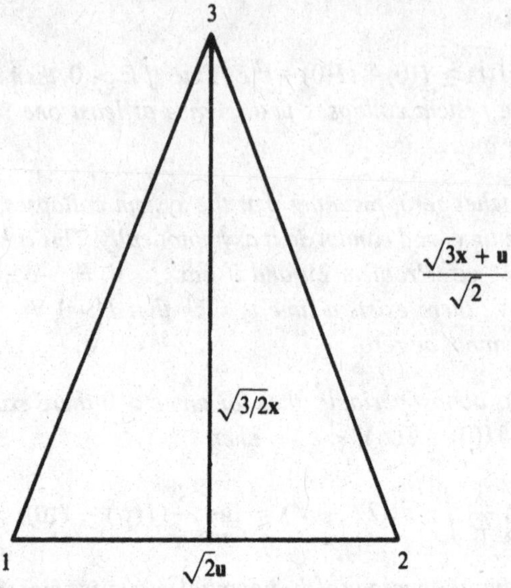

FIGURE 4.13. Center-of-mass and relative coordinates for three bodies.

(see Figure 4.13). Since this transformation is orthogonal on $\mathbb{R}^9$, the kinetic energy is simply

(4.5.16)
$$T = \tfrac{1}{2}(|\dot{\mathbf{s}}|^2 + |\dot{\mathbf{u}}|^2 + |\dot{\mathbf{x}}|^2),$$

and the potential energy is

(4.5.17)
$$V = -\frac{1}{\sqrt{2}|\mathbf{u}|} - \frac{\sqrt{2}}{|\sqrt{3}\mathbf{x}+\mathbf{u}|} - \frac{\sqrt{2}}{|\sqrt{3}\mathbf{x}-\mathbf{u}|}.$$

The total angular momentum is the sum of $[\mathbf{s}\times\dot{\mathbf{s}}]$, the angular momentum of the center of mass, and $[\mathbf{u}\times\dot{\mathbf{u}}]+[\mathbf{x}\times\dot{\mathbf{x}}]$, the internal angular momentum. Both are separately conserved. If particle #3 escapes, then the interesting coordinate is $\mathbf{x}$, the motion of which is governed by

(4.5.18)
$$\ddot{\mathbf{x}} = -\sqrt{\frac{2}{3}}\left(\frac{\mathbf{x}+\mathbf{u}/\sqrt{3}}{|\mathbf{x}+\mathbf{u}/\sqrt{3}|^3} + \frac{\mathbf{x}-\mathbf{u}/\sqrt{3}}{|\mathbf{x}-\mathbf{u}/\sqrt{3}|^3}\right).$$

Of course, $\mathbf{u}$ depends on the time in some unknown way, but because of (4.5.14), if $|\mathbf{u}|$ is initially $< r^*/\sqrt{2}$, then

(4.5.19)
$$\sqrt{\frac{3}{2}}r - \frac{r^*}{2} > r^* \quad\Leftrightarrow\quad r > \sqrt{\frac{3}{2}}r^*,$$

where $r = |\mathbf{x}|$, implies that $|\mathbf{x}_{1,2} - \mathbf{x}_3| > r^*$. Given this, $|\mathbf{u}|$ must always be less than $r^*/\sqrt{2}$. Therefore, $\dot{r}$ is bounded below as

(4.5.20)
$$\frac{\dot{r}(t)^2}{2} \geq \frac{\dot{r}(0)^2}{2} - \sqrt{\frac{2}{3}}\left(\frac{1}{r(0)+r^*/\sqrt{6}} + \frac{1}{r(0)-r^*/\sqrt{6}}\right)$$

$$+\sqrt{\frac{2}{3}}\left(\frac{1}{r(t)+r^*/\sqrt{6}}+\frac{1}{r(t)-r^*/\sqrt{6}}\right)$$

(see Problem 3).

This produces a crude

(4.5.21) Criterion for Escaping to Infinity

If at any time $|\mathbf{u}| < r^*/\sqrt{2}$ and

$$r > \sqrt{\frac{3}{2}}r^*, \qquad \dot{r} > 0, \qquad \frac{\dot{r}^2}{2} > \sqrt{\frac{2}{3}}\left(\frac{1}{r+r^*/\sqrt{6}}+\frac{1}{r-r^*/\sqrt{6}}\right),$$

then $\dot{r}$ is greater than some positive number at all later times, and particle #3 cannot be prevented from escaping.

(4.5.22) Remarks

1. It is possible to relax the conditions so that $\dot{r}$ is not necessarily greater than 0; even an initially incoming particle escapes if its energy and momentum are great enough (Problem 4). The other particles stay in Kepler trajectories, so no collisions take place. This shows that there are open regions in phase space, of infinite measure, in which particles do not collide, and X_H is complete. In regions for which one particle escapes, the number of constants of motion that exist is even maximal. For instance, if particle #3 breaks loose, then the Møller transformation relative to a Hamiltonian in which the potential has been replaced by

$$-\frac{m_1 m_2}{|\mathbf{x}_1-\mathbf{x}_2|}-\frac{m_3(m_1+m_2)}{|\mathbf{x}_3-(m_1\mathbf{x}_1+m_2\mathbf{x}_2)/(m_1+m_2)|},$$

exists. The reason is that the difference between this potential and $-\sum_{i>j} m_i m_j |\mathbf{x}_i-\mathbf{x}_j|^{-1}$ goes as $|\mathbf{x}_3|^{-3}$ for large $\mathbf{x}_3$; this produces a factor $|t|^{-3}$ in the time-derivative of $\Phi_{-t}\circ\Phi_t^0$, which guarantees convergence as $t\to\pm\infty$.

2. To get a feeling for the numbers that come up in (4.5.21), let us rewrite the last condition with $\rho = \sqrt{\frac{3}{2}}r =$ the distance from particle #3 to the center of mass of #1 and #2:

$$\frac{2}{3}\frac{\dot{\rho}^2}{2} \geq \frac{1}{\rho+r^*/2}+\frac{1}{\rho-r^*/2}.$$

Thus the condition means that the potential energy of particle #3 is less than its kinetic energy with a reduced mass $\frac{2}{3}$. The reduced mass

$$\frac{m_3(m_1+m_2)}{m_1+m_2+m_3},$$

in this case of particle #3 and the pair $(1, 2)$, is already familiar from (4.2.3; 4). With this correction, our initial supposition about the energetics is correct.

3. When $N = 4$, there are unbound trajectories for which particles can reach infinity in a finite time [15]. These involve a linear configuration of the particles #1, 2, 3, and 4 (in that order), in which #3 and #4 draw steadily nearer together. The energy thereby released is transmitted to #2, which runs faster and faster between particle #1 and the pair (3, 4), forcing them apart. (The orbit of particle #2 through two reversals is regarded as the limit of a Kepler ellipse with infinite eccentricity. The particles are reflected by the $1/r$ potential, and do not pass through each other.)
4. Computer studies of the three-body problem indicate that sooner or later some particle gains enough energy that (4.5.21) holds, and the system breaks up. This instability, known in atomic physics as the Auger effect, may well be characteristic for all systems with $1/r$ potentials. It is even suspected that, in the equal-mass case, the trajectories for which the system breaks apart may be dense in large parts of phase space. Of course, the physically relevant question is how large the probability of a break-up in a realistic time is. Unfortunately, present analytic methods fail to give an answer, and we must have recourse to the calculating machines, according to which there is a large probability that the system breaks up within 100 natural periods, $\tau \sim R^{3/2}\sqrt{\kappa M}$ [8].

The meagerness of these results makes it clear that the system of equations, (1.1.1) and (1.1.2) for large N, is too difficult for present-day mathematics to handle effectively. Though it is considered the correct expression of the laws of nature, its useful content is slight, because only a very few relevant propositions can be derived from it. Later, in the context of quantum mechanics, we shall return to the same equations and get useful information from them from another point of view. It will not be possible—or even desirable—to calculate the details of all the trajectories; yet one can predict quite a bit about the statistical behavior of the system.

(4.5.23) Problems

1. Derive (4.5.10).
2. Show that $I \to 0 \Rightarrow V \to -\infty$.
3. Prove (4.5.21). *Hint*: Use

$$\ddot{x} = \frac{\mathbf{x}\cdot\ddot{\mathbf{x}}}{r} + \frac{|\dot{\mathbf{x}}|^2}{r} - \frac{(\mathbf{x}\cdot\dot{\mathbf{x}})^2}{r^3} \geq \frac{(\mathbf{x}\cdot\ddot{\mathbf{x}})}{r}.$$

4. Use (4.5.18) to estimate how $\mathbf{L} := [\mathbf{x}\times\dot{\mathbf{x}}]$ varies in time and, with Problem 3, to show that particles with sufficiently large $r_0 := r(0)$, $\mathbf{L}_0 := \mathbf{L}(0)$, and $|\dot{r}_0|$ escape even if $\dot{r}_0 < 0$.

(4.5.24) Solutions

1. $\dot{I} = 2\sum_j(\mathbf{x}_j \cdot \mathbf{p}_j)$ is twice the generator of a dilatation, and is known as the virial. We know that $\{\dot{I}, T\} = 4T$ and $\{\dot{I}, V\} = 2V$, from which (4.5.10) follows.

2. Since the function $1/x$ is convex for $x > 0$, Jensen's inequality implies that

$$|V|\left(\sum_{i\neq j} m_i m_j\right)^{-1} = \frac{\kappa}{2}\sum_{i\neq j}\frac{m_i m_j}{|\mathbf{x}_i - \mathbf{x}_j|}\left(\sum_{i\neq j} m_i m_j\right)^{-1}$$
$$\geq \frac{\kappa}{2}\left(\sum_{i\neq j} m_i m_j |\mathbf{x}_i - \mathbf{x}_j|\right)^{-1}\sum_{i\neq j} m_i m_j.$$

Then with the triangle and Cauchy–Schwarz inequalities we get

$$\sum_{i\neq j} m_i m_j |\mathbf{x}_i - \mathbf{x}_j| \leq \sum_{i\neq j} m_i m_j (|\mathbf{x}_i| + |\mathbf{x}_j|)$$
$$\leq 2M\sum_i m_i|\mathbf{x}_i| \leq 2M\left(\sum_i m_i\right)^{1/2}\left(\sum_i m_i|\mathbf{x}_i|^2\right)^{1/2},$$

where $M = \sum_i m_i$. So finally,

$$V \leq \frac{-\kappa}{4M^{3/2}}\left(\sum_{i\neq j} m_i m_j\right)^2\left(\sum_i m_i|\mathbf{x}_i|^2\right)^{-1/2}.$$

3. The inequality of the hint implies that

$$\ddot{r} \geq -\sup_{|\mathbf{u}|<r^*/\sqrt{2}} \frac{\mathbf{x}}{r}\cdot\sqrt{\frac{2}{3}}\left[\frac{\mathbf{x}+\mathbf{u}/\sqrt{3}}{|\mathbf{x}+\mathbf{u}/\sqrt{3}|^3} + \frac{\mathbf{x}-\mathbf{u}/\sqrt{3}}{|\mathbf{x}-\mathbf{u}/\sqrt{3}|^3}\right]$$
$$\geq -\sup_{|\mathbf{u}|<r^*/\sqrt{2}}\sqrt{\frac{2}{3}}\left[\frac{1}{|\mathbf{x}+\mathbf{u}/\sqrt{3}|^2} + \frac{1}{|\mathbf{x}-\mathbf{u}/\sqrt{3}|^2}\right]$$
$$= -\frac{\sqrt{2/3}}{(r+r^*/\sqrt{6})^2} - \frac{\sqrt{2/3}}{(r-r^*/\sqrt{6})^2} = \frac{\partial}{\partial r}\sqrt{\frac{2}{3}}\left(\frac{1}{r+r^*/\sqrt{6}} + \frac{1}{r-r^*/\sqrt{6}}\right),$$

because $(\mathbf{a}\cdot\mathbf{b}) \geq -|\mathbf{a}|\,|\mathbf{b}|$, and it is clear that the greatest forces occur when some particle approaches as near as possible to particle #3. Multiplication by $\dot{r}$ and integration produce (4.5.21).

4. Let $L_m \in \mathbb{R}^+$ be such that $|L(t)| \geq L_m$ for all t. From (4.5.18),

$$\ddot{r} = \frac{\mathbf{x}\cdot\ddot{\mathbf{x}}}{r} + \frac{L^2}{r^3} \geq -\sqrt{\frac{8}{3}}\frac{1}{(r-r^*/\sqrt{6})^2} + \frac{L_m^2}{r^3} = -\frac{\partial}{\partial r}V_m(r),$$

where

$$V_m := -\sqrt{\frac{8}{3}}\frac{1}{r-r^*/\sqrt{6}} + \frac{L_m^2}{2r^2}.$$

Choose the initial values r_0 and $\dot{r}_0$ so that if v and r_m are defined by

$$\tfrac{1}{2}v^2 = \tfrac{1}{2}\dot{r}_0^2 + V_m(r_0) = V_m(r_m),$$

then the larger of the two solutions for r_m is greater than $2r^*/\sqrt{6}$. Then (because the force from V_m is always less than the actual force) $r > r_m\ \forall t$; and we can calculate that $V_m(r) < v^2 r_m^2/2r^2$; that $|\dot{r}| > v\sqrt{1 - r_m^2/r^2}$; and so, finally, that

$$vt < \left(\int_{r_m}^{r_0} + \int_{r_m}^{r}\right) \frac{dr\, v}{\sqrt{v^2 - 2V_m(r)}} < \left(\int_{r_m}^{r_0} + \int_{r_m}^{r}\right) \frac{dr\, v}{\sqrt{v^2 - v^2 r_m^2/r^2}}$$
$$= \sqrt{r_0^2 - r_m^2} + \sqrt{r^2 - r_m^2} < r_0 + r.$$

We still need to show that L_0 can be chosen consistently with these calculations, in which case we conclude that r gets arbitrarily large. Since (4.5.18) tells us that

$$|\dot{\mathbf{L}}| = |[\mathbf{x} \times \ddot{\mathbf{x}}]| \le \frac{2}{3} \frac{rr^*}{(r - r^*/\sqrt{6})^3},$$

the inequality

$$L > L_0 - \frac{4}{3}\int_{r_m}^{\infty} \frac{dr\, rr^*}{\dot{r}(r - r^*/\sqrt{6})^3} > L_0 - \frac{2}{3}\int_{r_m}^{\infty} \frac{\sqrt{6}\, dr\, r^2 r_m/v}{(r - r_m/2)^3\sqrt{r^2 - r_m^2}}$$
$$= L_0 - \frac{C}{v}, \qquad \text{where} \qquad C := \frac{2}{3}\int_1^{\infty} \frac{dx\sqrt{6}x^2}{(x - \frac{1}{2})^3\sqrt{x^2 - 1}},$$

holds. Using $L_m \le L_0 - C/v$ and observing that $|\mathbf{L}| \le |\dot{r}r|$, we reduce the problem to satisfying the condition

$$r_0^2(v^2 - \dot{r}_0^2) + \sqrt{\frac{2}{3}} \frac{4r_0^2}{r_0 - r^*/\sqrt{6}} \le \left(L_0 - \frac{C}{v}\right)^2 \le \left(\dot{r}_0 r_0 - \frac{C}{v}\right)^2.$$

Since r^* depends on the total energy, and can be chosen independently of r_0, $\dot{r}_0$, and L_0, we discover that there are open regions of phase space, with $v^2 \lesssim \dot{r}_0^2 \gg 1/r_0 \ll 1/r^*$, where this condition is satisfied.

5

Relativistic Motion

5.1 The Hamiltonian Formulation of the Electrodynamic Equations of Motion

The theory of special relativity replaces the Galilean group with the Poincaré group. This makes the equations of motion of a particle in an external field only slightly more complicated. However, physics at high velocities looks quite different from its nonrelativistic limit.

Newton's equations, as we know, are only an approximation, and have to be generalized to (1.1.4) or (1.1.6) when the speed of a particle approaches the speed of light. In order to solve these equations in some physically interesting cases, we first put (1.1.4) into Hamiltonian form. This means that the motion takes place in extended configuration space, which is a particular subset of $\mathbb{R}^4$. We shall just concern ourselves with one-body problems, since even when there are only two bodies only special solutions are known if the interaction is relativistic—and therefore not instantaneous (cf. [12]).

(5.1.1) **Relativistic Notation**

(a) Let $x^0 := t$ and $\mathbf{x} := (x^1, x^2, x^3)$, choosing the units of time so that $c = 1$. The x^α, $\alpha = 0, 1, 2, 3$, are Cartesian coordinates on extended configuration space M_e, and we define a pseudo-Riemannian tensor field (see (2.4.9)),

$$(5.1.2) \qquad \eta = \eta_{\alpha\beta}\, dx^\alpha\, dx^\beta, \qquad \eta_{\alpha\beta} = \begin{pmatrix} & 0 & 1 & 2 & 3 \\ 0 & -1 & & & \\ 1 & & 1 & & \\ 2 & & & 1 & \\ 3 & & & & 1 \end{pmatrix}$$

on this chart. This corresponds to a scalar product on $T(M_e)$,

$$\langle \partial_\alpha | \partial_\beta \rangle = \eta_{\alpha\beta}, \qquad \partial_\alpha = \frac{\partial}{\partial x^\alpha},$$

and hence to a bijection $T(M_e) \to T^*(M_e)$. The inverse mapping is given by the contravariant tensor field $\eta^{\alpha\beta}\partial_\alpha\partial_\beta$, where $\eta_{\alpha\beta}\eta^{\beta\gamma} = \delta^\gamma_\alpha$. Since $\eta_{\alpha\beta}\eta_{\beta\gamma} = \delta_{\alpha\gamma}$, the matrix $\eta^{\alpha\beta}$ is numerically the same as $\eta_{\alpha\beta}$ in (5.1.2).

(b) The motion of a particle is described by writing its coordinates x^α as functions $x^\alpha(s)$ of a parameter s, the **proper time**. We shall denote differentiation by s with a dot: $\dot{x}^\alpha := dx^\alpha/ds$, and normalize s so that $\dot{x}^\alpha\dot{x}^\beta\eta_{\alpha\beta} = -1$; if $|\dot{\mathbf{x}}| \ll 1$, then $\dot{x}^0 \approx 1$, and s becomes equal to t (cf. Problem 2).

(5.1.3) The Equations of Motion

By (1.1.4), the equations

$$m\ddot{x}^\alpha\eta_{\alpha\beta} = e\dot{x}^\alpha F_{\alpha\beta}, \qquad F := \begin{pmatrix} 0 & E_1 & E_2 & E_3 \\ -E_1 & 0 & -B_3 & B_2 \\ -E_2 & B_3 & 0 & -B_1 \\ -E_3 & -B_2 & B_1 & 0 \end{pmatrix}$$

hold for the motion of a particle in an **electric field E** and a **magnetic field B**.

(5.1.4) Remarks

1. Since $F_{\alpha\beta} = -F_{\beta\alpha}$,

$$\ddot{x}^\alpha\dot{x}^\beta\eta_{\alpha\beta} = \frac{1}{2}\frac{d}{ds}\dot{x}^\alpha\dot{x}^\beta\eta_{\alpha\beta} = 0.$$

The normalization of (b) of (5.1.1) is consistent with (5.1.3).

2. If $\gamma = dt/ds$, then the three spatial coordinates of (5.1.3) can be written as

$$\frac{d}{dt}\left(m\gamma\frac{d\mathbf{x}}{dt}\right) = e\left(\mathbf{E} + \left[\frac{d\mathbf{x}}{dt} \times \mathbf{B}\right]\right).$$

In other words, the rate of change of the momentum, using the relativistic mass $m\gamma$, equals the **Lorentz force**, which, as well as causing an acceleration in the direction of $e\mathbf{E}$, causes a positive (i.e., counterclockwise) torque, looking in

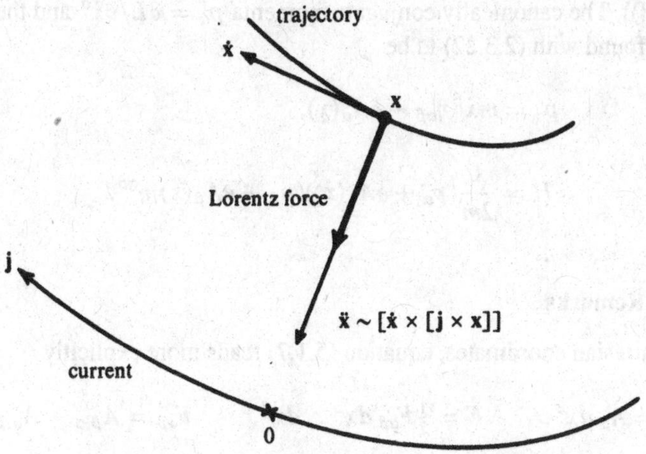

FIGURE 5.1. The direction of the Lorentz force.

the direction of $e\mathbf{B}$. Since a stationary electric current $\mathbf{j}$ at the origin induces a magnetic field

$$\mathbf{B} = \frac{[\mathbf{j} \times \mathbf{x}]}{r^3}$$

at the point $\mathbf{x}$, this means that parallel currents attract (see Figure 5.1).

3. The time component of (5.1.3) expresses conservation of energy:

$$\frac{d}{dt} m\gamma = e\left(\mathbf{E} \cdot \frac{d\mathbf{x}}{dt}\right).$$

4. The nonrelativistic limit, $\gamma \to 1$ and $\mathbf{B} \to \mathbf{0}$, reproduces the earlier equations (1.1.1).

(5.1.5) The Lagrangian and Hamiltonian

The electromagnetic field is a 2-form on M_e, though not an arbitrary one, as it satisfies the homogeneous form of **Maxwell's equations,**

(5.1.6) $$dF = 0.$$

We shall concern ourselves only with manifolds on which

(5.1.7) $$F = dA, \qquad A \in E_1(M_e),$$

follows from (5.1.6) (cf. (2.5.6; 3)). This makes equations (5.1.3) the Euler–Lagrange equations of the Lagrangian

(5.1.8) $$L(x(s), \dot{x}(s)) = \frac{m}{2}\dot{x}^\alpha \dot{x}^\beta \eta_{\alpha\beta} - e\dot{x}^\alpha A_\alpha(x)$$

(cf. (2.3.18)). The canonically conjugate momenta $p_\alpha = \partial L/\partial \dot{x}^\alpha$ and the Hamiltonian are found with (2.3.22) to be

(5.1.9)
$$p_\alpha = m\dot{x}^\beta \eta_{\alpha\beta} - eA_\alpha(x),$$
$$\mathcal{H} = \frac{1}{2m}(p_\alpha + eA_\alpha(x))(p_\beta + eA_\beta(x))\eta^{\alpha\beta}.$$

(5.1.10) **Remarks**

1. With Cartesian coordinates, equation (5.1.7) reads more explicitly

$$A = A_\alpha\, dx^\alpha, \qquad F = \tfrac{1}{2}F_{\alpha\beta}\, dx^\alpha \wedge dx^\beta, \qquad F_{\alpha\beta} = A_{\beta,\alpha} - A_{\alpha,\beta},$$

or, separating the space and time coordinates,

$$A_\alpha = (V, -\mathcal{A}): \mathbf{B} = \nabla \times \mathcal{A}, \qquad \mathbf{E} = -\dot{\mathcal{A}} - \nabla V.$$

2. The canonical form ω on $T^*(M_e)$ is

$$\omega = \sum_\alpha dx^\alpha \wedge dp_\alpha$$

or, written differently,

$$\{x^\alpha, p_\beta\} = \delta^\alpha_\beta.$$

Equation (5.1.7) determines A only up to a **gauge transformation** $A \to A + d\Lambda$, where $\Lambda \in C^\infty(M_e)$, which leaves the equations of motion, but not $\mathcal{H}$, invariant. If the gauge transformation is combined with the canonical transformation $x^\alpha \to x^\alpha$, $p_\beta \to p_\beta - e\Lambda_{,\beta}(x)$ (cf. Problem 4), then $\mathcal{H}$ is left unchanged. The canonical momenta p have no gauge-invariant meaning (and thus neither does the origin of $T^*(M_e)$), although $\dot{x}^\alpha$ is gauge-invariant.

3. Conversely, according to (2.5.6; 3), A's that produce the same F differ on starlike regions at most by a gauge transformation.

4. The Poisson brackets between x^α and $\dot{x}^\beta$ are still

$$\{x^\alpha, \dot{x}^\beta\} = \frac{\eta^{\alpha\beta}}{m},$$

although

$$\{\dot{x}^\alpha, \dot{x}^\beta\} = \frac{-e}{m^2}\eta^{\alpha\gamma}\eta^{\beta\delta}F_{\gamma\delta}(x)$$

is now not zero, but depends only on the gauge-invariant quantity F.

5. Since $\mathcal{H}$ does not depend explicitly on s, it is a constant of the motion. Expressed in terms of $\dot{x}$, it is

$$\mathcal{H} = \frac{m\dot{x}^\alpha \dot{x}^\beta}{2}\eta_{\alpha\beta},$$

so we shall always work with the submanifold $\mathcal{H} = -m/2$, in accordance with the normalization of s in (b) of (5.1.1).

6. Our canonical formulation of the equations of motion makes use of the vector potential A, which not only depends on the gauge, but even fails to exist globally on many manifolds. It is possible to get by without A. The diffeomorphism φ of extended phase space: $\varphi(x, p) = (x, p + eA(x))$, casts $\mathcal{H}$ into the form $\varphi_*(\mathcal{H}) = p^\alpha p^\beta \eta_{\alpha\beta}/2m$. This diffeomorphism is not canonical, as $\varphi_*(\omega) = \omega - eF$, and $\varphi_*(\{A, B\}) = \{\varphi_*(A), \varphi_*(B)\}_{\varphi_*(\omega)}$, where $\{\ \}_{\varphi_*(\omega)}$ is the Poisson bracket calculated with $\omega - eF$ rather than ω. Since

$$\varphi_*(\omega) \wedge \varphi_*(\omega) \wedge \varphi_*(\omega) \wedge \varphi_*(\omega) = \omega \wedge \omega \wedge \omega \wedge \omega,$$

$\varphi_*(\omega)$ is nondegenerate, and provided that $dF = 0$, $\varphi_*(\omega)$ may be used as a symplectic form on $T^*(M_e)$. This means that the influence of F on the motion may be taken into account by changing ω rather than $\mathcal{H}$: $\mathcal{H}$ remains $m\dot{x}^2/2$, but the Poisson brackets of the velocities do not vanish. Only F appears in this formulation; but since A is generally simpler than F, we nevertheless prefer to use (5.1.5).

(5.1.11) Example: Free Particles

We would like to compare the case $e = 0$ with the results of §4.1. This time,

$$\mathcal{H} = \frac{1}{2m} p_\alpha p_\beta \eta^{\alpha\beta} = \frac{1}{2m}(|\mathbf{p}|^2 - p_0^2).$$

And now $E = m\dot{t}$ is $-p_0$, so the additional term E in (3.2.12) is changed by the factor $-p_0/2m$. Consequently, the largest group of point transformations on $T^*(M_e)$ that leave $\mathcal{H}$ invariant is the **Poincaré group** (= displacements + Lorentz transformations):

$$\text{(5.1.12)} \qquad \begin{aligned} x^\alpha &\to \Lambda^\alpha_\beta x^\beta + \lambda^\alpha, \quad \lambda \in \mathbb{R}^4, \qquad \Lambda^t \eta \Lambda = \eta, \\ p_\alpha &\to \eta_{\alpha\gamma} \Lambda^\gamma_\delta \eta^{\delta\beta} p_\beta. \end{aligned}$$

Let ε be the parameter of the one-dimensional subgroup with the infinitesimal elements

$$\text{(5.1.13)} \qquad \lambda^\alpha = \varepsilon e^\alpha, \qquad \Lambda^\alpha_\beta = \delta^\alpha_\beta + \varepsilon L^\alpha_\beta, \qquad (\eta L)^t = L^t \eta = -\eta L;$$

then the generator of the subgroup is

$$\text{(5.1.14)} \qquad p_\alpha e^\alpha + L^\alpha_\beta p_\alpha x^\beta$$

(cf. (3.2.6)). This gives us $4 + 6 = 10$ constants: the 6 generators of the Lorentz transformation corresponding to L^α_β (cf. Problem 5) can also be written as

$$\text{(5.1.15)} \qquad \mathcal{M}^{\alpha\beta} := p^\alpha x^\beta - p^\beta x^\alpha, \text{ where } p^\alpha := \eta^{\alpha\beta} p_\beta,$$

on account of the antisymmetry required in (5.1.13); the 4 generators corresponding to the e^α unite the energy and the momentum as the energy-momentum, in which $-p_0 = p^0 = m\gamma$ is the relativistic energy. The three spatial components of $\mathcal{M}^{\alpha\beta}$ are the angular momentum

(5.1.16) $$\mathbf{L} = [\mathbf{x} \times \mathbf{p}];$$

while the center-of-mass theorem, coming from $\mathcal{M}^{0\beta}$, $\beta = 1, 2, 3$, now reads:

(5.1.17) $$\mathbf{K} := \mathbf{p}t - p^0\mathbf{x} \text{ is constant.}$$

On the seven-dimensional submanifold where $-2m\mathcal{H} = p_0^2 - |\mathbf{p}|^2 = m^2$, there are three relationships of interdependence,

(5.1.18) $$[\mathbf{p} \times \mathbf{K}] = p^0\mathbf{L},$$

as there must be. Thus, concerning the number of independent constants of the motion and the structure of the trajectories the same general facts hold as in the nonrelativistic case of §4.1 despite the change of the invariance group.

(5.1.19) **Problems**

1. Calculate the Poisson brackets of the generators of the Poincaré group, and compare with those of the Galilean group.
2. Show that $\dot{x}^\alpha \dot{x}^\beta \eta_{\alpha\beta} = -1$ implies $|d\mathbf{x}/dt| < 1$. Thus (5.1.4; 1) shows that electromagnetic forces can never cause particles to move faster than light.
3. Suppose that the Lagrangian for relativistic motion in a scalar field $\Phi \in \mathcal{T}_0^0(M_e)$ is
$$L = \tfrac{1}{2}\dot{x}^\alpha \dot{x}^\beta \eta_{\alpha\beta} - \Phi(x).$$
Is it possible for a particle to be accelerated to faster than the speed of light?
4. Show that (5.1.10; 3) is a canonical transformation. (Check the Poisson brackets.)
5. Show that condition (5.1.13) defines a six-dimensional submanifold of the (4×4)-matrices (cf. (2.1.10; 3)).

(5.1.20) **Solutions**

1.
$$\{p_i, p_j\} = \{p_i, p_0\} = \{L_i, p_0\} = 0,$$
$$\{L_i, L_j\} = \varepsilon_{ijk}L_k, \qquad \{p_i, L_j\} = \varepsilon_{ijk}p_k, \qquad \{K_i, L_j\} = \varepsilon_{ijk}K_k,$$
$$\{p^0, K_j\} = p_j, \qquad \{K_i, K_j\} = -\varepsilon_{ijm}L_m, \qquad \{p_i, K_j\} = p^0\delta_{ij}.$$

This differs from the Galilean group in the last two relationships; the Galilean group has m instead of p_0 on the right side (see (4.1.10; 3)). In the nonrelativistic limit, p_0 approaches m, and so in this sense the Poincaré group goes over to the Galilean group;

but note that although the elements of the Poincaré group are point transformations on $T^*(M_e)$, those of the Galilean group are not.

2.

$$dt^2 - |d\mathbf{x}|^2 = ds^2 \quad \Rightarrow \quad \left|\frac{d\mathbf{x}}{dt}\right|^2 = 1 - \left(\frac{ds}{dt}\right)^2 < 1.$$

3. Yes, $\ddot{x}^\alpha \eta_{\alpha\beta} = \Phi_{,\beta}$. If, say, $\Phi = -|\mathbf{x}|^2/2$, then $\ddot{x}^0 = 0$, and one solution is $x^0 = s$, and $\mathbf{x}(t) = \mathbf{x}(0)\cosh t + \dot{\mathbf{x}}(0)\sinh t$. The velocity $\dot{\mathbf{x}}(t)$ gets arbitrarily large.

4. The only nontrivial Poisson bracket is

$$\{p_\alpha + e\Lambda_{,\alpha}, p_\beta + e\Lambda_{,\beta}\} = e(\Lambda_{,\beta\alpha} - \Lambda_{,\alpha\beta}) = 0.$$

5. Let $L_{\alpha\beta}$ be the components of ηL. For the 16 functions $N_{\alpha\beta} := L_{\alpha\beta} + L_{\beta\alpha}$, equation (5.1.13) implies $N_{\alpha\beta} = 0$. Because $dN_{\alpha\beta} = dN_{\beta\alpha}$, only six of the differentials $dN_{\alpha\beta}$ are linearly independent.

5.2 The Constant Field

This is an integrable system, the relativistic generalization of the elementary example of motion in a constant field. It also contains the Larmor precession in a constant magnetic field as a special case.

In this section we discuss the motion in an electromagnetic field the Cartesian components of which are constant. That a field $F_{\alpha\beta}$ has constant strength means that the potentials are linear in x:

(5.2.1)
$$A_\beta = \tfrac{1}{2}x^\alpha F_{\alpha\beta}.$$

Since $A_\beta \in C(\mathbb{R}^4)$, we may set $M_e = \mathbb{R}^4$. The invariance of (5.1.9) under the Poincaré group is broken by the addition of A. Specifically, the p_α are no longer constant, even though the equations of motion (5.1.3) are translation-invariant. A displacement $x^\alpha \to x^\alpha + \lambda^\alpha$ causes $A_\beta \to A_\beta + \frac{1}{2}\lambda^\alpha F_{\alpha\beta}$, which is a gauge transformation with $\Lambda = \frac{1}{2}\lambda^\alpha F_{\alpha\beta}x^\beta$ (cf. (5.1.10; 4)). Adding the gauge function $\frac{1}{2}F_{\alpha\beta}x^\beta$ to the generators of the displacements produces (cf. Problem 4).

(5.2.2) The Constants of the Motion

$$\frac{d}{ds}\left(p_\alpha + \frac{e}{2}F_{\alpha\beta}x^\beta\right) = 0.$$

In the gauge-invariant formulation (5.1.10; 6) these constants generate the translations $x \to x + \lambda$, $p \to p$.

Although the Lorentz transformations do not even leave the equations of motion invariant, they can still be used to put the problem into a convenient form.

(5.2.3) **The Transformation Relations of the Field Tensor**

Using a Lorentz transformation

$$\bar{x} = \Lambda x, \qquad \Lambda^t \eta \Lambda = \eta,$$

the field strength, being a 2-form, must transform as

(5.2.4) $$\bar{F} = \eta \Lambda \eta F \eta \Lambda^t \eta.$$

The 1-form A transforms as p:

(5.2.5) $$\bar{A} = \tfrac{1}{2}\bar{x}\bar{F} = \tfrac{1}{2} x \Lambda^t \eta \Lambda \eta F \eta \Lambda^t \eta = \tfrac{1}{2} x F \eta \Lambda^t \eta = \eta \Lambda \eta A.$$

Under spatial rotations, **E** and **B** change as vectors, but the transformation to a moving reference frame, generated by the **K** of (5.1.17), causes them to mix. For instance, K_i generates the one-parameter group

(5.2.6) $$\Lambda = \begin{pmatrix} \frac{1}{\sqrt{1-v^2}} & \frac{-v}{\sqrt{1-v^2}} & 0 & 0 \\ \frac{-v}{\sqrt{1-v^2}} & \frac{1}{\sqrt{1-v^2}} & 0 & 0 \\ 0 & 0 & 1 & 0 \\ 0 & 0 & 0 & 1 \end{pmatrix}$$

for $v \in (-1, 1)$. Transformation formula (5.2.4), when expressed in terms of **E** and **B**, reads:

(5.2.7) $$\begin{aligned} \bar{E}_1 &= E_1, & \bar{B}_1 &= B_1, \\ \bar{E}_2 &= \frac{E_2 - B_3 v}{\sqrt{1-v^2}}, & \bar{B}_2 &= \frac{B_2 + E_3 v}{\sqrt{1-v^2}}, \\ \bar{E}_3 &= \frac{E_3 + B_2 v}{\sqrt{1-v^2}}, & \bar{B}_3 &= \frac{B_3 - E_2 v}{\sqrt{1-v^2}}. \end{aligned}$$

(5.2.8) **The Normal Form of the Field Tensor**

The first point to recall in the discussion of the possible ways that **E** and **B** can be changed by Λ is that there are two invariants

(5.2.9) $$^*(F \wedge F) = (E \cdot B) \qquad \text{and} \qquad ^*(F \wedge {}^*F) = 2(|\mathbf{B}|^2 - |\mathbf{E}|^2).$$

Hence the statements $|\mathbf{B}| \gtrless |\mathbf{E}|$, $|\mathbf{B}| = |\mathbf{E}|$, and $\mathbf{E} \perp \mathbf{B}$ are Poincaré-invariant. If both of our invariant quantities are zero, then $|\mathbf{E}| = |\mathbf{B}|$ and $\mathbf{E} \perp \mathbf{B}$. Otherwise, it is always possible to make **E** and **B** parallel with a Lorentz transformation: First, the plane of **E** and **B** can be rotated to the (2–3)-plane, and then v can be chosen in (5.2.6) so that $\bar{E}_2/\bar{B}_2 = \bar{E}_3/\bar{B}_3$ (Problem 1).

Because of this, we shall assume that $E_2 = E_3 = B_2 = B_3 = 0$ in what follows, and let $E =: E_1$ and $B =: B_1$. The special case that both invariants vanish will be treated in §5.4. Now F and A have the components

$$(5.2.10)\qquad F = \begin{pmatrix} 0 & E & 0 & 0 \\ -e & 0 & 0 & 0 \\ 0 & 0 & 0 & -B \\ 0 & 0 & B & 0 \end{pmatrix}, \qquad A = \tfrac{1}{2}(-Ex, Et, zB, -yB).$$

Accordingly, $\mathcal{H}$ divides into two parts, one of which depends only on the 0 and 1 coordinates, and the other only on the 2 and 3 coordinates:

$$(5.2.11)\quad \mathcal{H} = \frac{1}{2m}\left[\left(p_y + \frac{ezB}{2}\right)^2 + \left(p_z - \frac{eyB}{2}\right)^2 + \left(p_x + \frac{etE}{2}\right)^2 - \left(p_0 - \frac{exE}{2}\right)^2\right] =: \mathcal{H}_B + \mathcal{H}_E.$$

(5.2.12) The Flow of $\mathcal{H}_B$

$\mathcal{H}_B$ acts on a four-dimensional phase space with coordinates $(y, z; p_y, p_z)$. Hence two additional constants of motion suffice to determine the trajectories completely. If $B = 0$, we simply get the free flow, so we suppose $B \neq 0$. Then the two constants of (5.2.2), with $\alpha = 2$ and 3, are proportional to

$$(5.2.13)\qquad \bar{y} := \frac{y}{2} + \frac{p_z}{eB} \qquad \text{and} \qquad \bar{z} := \frac{z}{2} - \frac{p_y}{eB}.$$

The two constants of motion determine the fixed centers of the **Larmor orbits** in the $(y$–$z)$-plane:

$$(5.2.14)\qquad (\bar{y} - y)^2 + (\bar{z} - z)^2 = \frac{2m}{e^2B^2}\mathcal{H}_B.$$

The trajectories are thus circles of radius $\sqrt{2m\mathcal{H}_B}/eB$ and centers $(\bar{y}, \bar{z})$.

(5.2.15) Remarks

1. If $(\bar{y}, \bar{z}) \neq (0, 0)$, the angular momentum $(y\dot{z} - z\dot{y})m$ is not conserved. However, since $\mathcal{H}$ is invariant under rotations about the x-axis, the generator of those rotations, known as the canonical angular momentum,

$$\bar{L}_1 := yp_z - zp_y,$$

must be a constant. Yet $\bar{L}_1$ is not gauge-invariant, and normally has no physical significance. In the gauge we have chosen for A, $\bar{L}_1$ may be expressed in terms of the gauge-invariant constants we have already found:

$$2eB\bar{L}_1 = -2m\mathcal{H}_B + e^2B^2(\bar{y}^2 + \bar{z}^2).$$

This is as expected, since there is room for only three independent constants in the four-dimensional phase space associated with $\mathcal{H}_B$.

2. There has to be a nonvanishing Poisson bracket between two of the three constants, and, in fact,

$$\{\bar{z}, \bar{y}\} = \frac{1}{eB}.$$

Since $\{\bar{y}, \bar{z}\}$ does not depend on the coordinates and has a vanishing Poisson bracket with any other observable, nothing prevents the invariance group generated by $\bar{y}$ and $\bar{z}$ (Problem 5) from being Abelian. Jacobi's identity guarantees that for all $f \in C^\infty(T^*(M))$, $\{\bar{z}, \{\bar{y}, f\}\} = \{\bar{y}, \{\bar{z}, f\}\}$ (cf. (3.2.8; 2)).

3. Since the Poisson brackets of $\bar{y}$ and $\bar{z}$ with the individual contributions to $\mathcal{H}_B$ vanish,

$$\{\bar{y}, \dot{y}\} = \{\bar{y}, \dot{z}\} = \{\bar{z}, \dot{y}\} = \{\bar{z}, \dot{z}\} = 0,$$

and since from (5.1.10; 4)

$$\{\dot{y}, \dot{z}\} = \frac{eB}{m^2},$$

the transformation

$$(y, z; p_y, p_z) \rightarrow \left(eB\bar{z}, \frac{m^{3/2}}{eB}\dot{y}; \bar{y}, m^{1/2}\dot{z}\right)$$

is canonical. Calling the second pair of these canonical coordinates,

$$\left(\frac{m^{3/2}}{eB}\dot{y}, m^{1/2}\dot{z}\right),$$

q and p makes $\mathcal{H}_B$ simply the Hamiltonian of a harmonic oscillator with the **cyclotron frequency** $\omega := eB/m$:

$$\mathcal{H}_B = \tfrac{1}{2}(p^2 + \omega^2 q^2).$$

This is a degenerate case of the situation described in (3.3.8). One of the two frequencies is necessarily zero and the other eB/m. The physical significance of this is that the circular orbits have a frequency that depends neither on their center nor on their radius.

4. The formulas for the trajectories in terms of the constants, and explicitly using the parameters, are

$$y(s) = \bar{y} + \frac{\sqrt{2m\mathcal{H}_B}}{eB}\cos\omega(s - s_0),$$

$$z(s) = \bar{z} - \frac{\sqrt{2m\mathcal{H}_B}}{eB}\sin\omega(s - s_0).$$

(5.2.16) The Flow of $\mathcal{H}_E$

Up to some differences of signs, for $E \neq 0$ this is handled in much the same way. The remaining two constants from (5.2.2) are proportional to

$$(5.2.17) \qquad \bar{x} = \frac{x}{2} + \frac{p_0}{eE} \quad \text{and} \quad \bar{t} = \frac{t}{2} - \frac{p_x}{eE};$$

so the trajectories are the hyperbolas

$$(5.2.18) \qquad (\bar{x} - x)^2 - (\bar{t} - t)^2 = \frac{2m}{e^2 E^2} \mathcal{H}_E.$$

(5.2.19) Remarks

1. Since E_1 is unchanged by the Lorentz transformation (5.2.6) (cf. (5.2.7)), the generator of the transformation,
$$\bar{K}_1 = xp_0 + tp_x = \frac{m}{eE}\mathcal{H}_E + \frac{eE}{2}(\bar{x}^2 - \bar{t}^2),$$
is a constant of the motion, though it is not independent of the other constants already found.
2. By calculating the Poisson brackets,
$$\{\bar{x}, \bar{t}\} = \frac{-1}{eE}, \qquad \{\dot{x}, \dot{t}\} = +\frac{eE}{m^2},$$
and
$$\{\bar{x}, \dot{x}\} = \{\bar{x}, \dot{t}\} = \{\bar{t}, \dot{x}\} = \{\bar{t}, \dot{t}\} = 0,$$
we see that
$$\left(eE\bar{t}, \frac{m^{3/2}}{eE}\dot{x}; \bar{x}, m^{1/2}\dot{t}\right)$$
are canonical coordinates, and that
$$\mathcal{H}_e = -\tfrac{1}{2}(p^2 - \nu^2 q^2), \qquad q = \frac{m^{3/2}}{eE}\dot{x},$$
$$p = m^{1/2}\dot{t}, \qquad \nu = \frac{eE}{m}.$$
3. With the right changes of signs in the oscillator potential, the coordinates become hyperbolic functions of the proper time,
$$x(s) = \bar{x} + \frac{\sqrt{2m|\mathcal{H}_E|}}{eE} \cosh \nu(s - s_0),$$
$$t(x) = \bar{t} + \frac{\sqrt{2m|\mathcal{H}_E|}}{eE} \sinh \nu(s - s_0).$$

4. As regards the total number of constants of the motion, so far we have found six independent constants, but they are not independent of $\mathcal{H} = \mathcal{H}_E + \mathcal{H}_B$. Therefore there is a constant still to be found. We can find it by noting that the inverse function to the sinh, arcsinh, exists globally, and that $s - \bar{s}_0$ can be expressed in terms of $t - \bar{t}$. Thus

$$y - \frac{\sqrt{2m\mathcal{H}_B}}{eB} \cos\left[\omega \operatorname{arcsinh} \frac{eE}{\sqrt{2m|\mathcal{H}_E|}}(t - \bar{t})\right]$$

is an additional independent constant. Consequently, the trajectories are one-dimensional submanifolds, which are diffeomorphic to $\mathbb{R}$, since in extended phase space there are no closed trajectories.

(5.2.20) Motion as Seen from Another Frame of Reference

If **E** and **B** are parallel, we have seen that the motion parallel to the field is hyperbolic, and the motion perpendicular to it is circular. Unless $|\mathbf{E}|^2 - |\mathbf{B}|^2 = (\mathbf{E} \cdot \mathbf{B}) = 0$, the general case can be obtained from this by making a Lorentz transformation perpendicular to the field. For example, $\mathbf{E} = 0$, $\mathbf{B} = B(1, 0, 0)$ can be turned into

$$\mathbf{E} = \left(0, 0, \frac{Bv}{\sqrt{1 - v^2}}\right), \qquad \mathbf{B} = \left(-\frac{B}{\sqrt{1 - v^2}}, 0, 0\right)$$

by a Lorentz transformation in the 2-direction. The circular orbits around the x-axis then move in the 2-direction, perpendicular to **E** (Figure 5.2). In the new coordinate system this is interpreted by saying that there is an electric field present to which the particle tries to move parallel. At the same time, **B** causes the trajectory to bend, more strongly at lower speeds, and therefore smaller z, producing a drift velocity in the 2-direction, as shown in Figure 5.2.

(5.2.21) Problems

1. Find the v in the Lorentz transformation that makes **E** and **B** parallel, if it does not happen that $|\mathbf{E}|^2 - |\mathbf{B}|^2 = (\mathbf{E} \cdot \mathbf{B}) = 0$.

2. Discuss the equations of motion in a homogeneous magnetic field without using the constants of motion.

3. Express $F \wedge F$ in terms of **E** and **B**.

4. Verify (5.2.2) by using the equations of motion.

5. What group of invariances of $\mathcal{H}_B$ (resp. $\mathcal{H}_B$) generates the $\bar{y}$ and $\bar{z}$ of (5.2.13) (resp. the $\bar{x}$ and $\bar{t}$ of (5.2.17))?

6. What are the shapes of the trajectories of the momenta and the velocities?

7. Write (5.2.9) as the * of a 4-form (cf. (2.4.29; 4)).

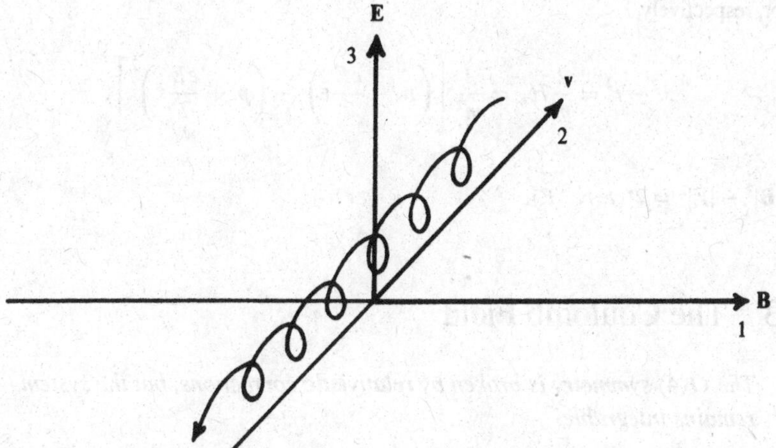

FIGURE 5.2. The influence of an electric field perpendicular to **B**. The trajectory stays in the (2–3)-plane, and slides along perpendicularly to **E**.

(5.2.22) **Solutions**

1. By (5.2.7), the equation $\bar{E}_2/\bar{E}_3 = \bar{B}_2/\bar{B}_3$ implies

$$v^2 - v\left(\frac{E_2^2 + E_3^2 + B_2^2 + B_3^2}{E_2 B_3 - E_3 B_2}\right) + 1 = 0.$$

   The equation $v^2 - 2\alpha v + 1$ has a solution $|v| < 1$ if $|\alpha| > 1$, which is the case, as $|\mathbf{E}|^2 + |\mathbf{B}|^2 > 2|\mathbf{E}| \cdot |\mathbf{B}| > 2|[\mathbf{E} \times \mathbf{B}]|$.

2. The equation $\ddot{x}(s) = M\dot{x}(s)$, $M^i_k = (e/m)\eta^{ij} F_{jk}$, has the solution $\dot{x}(s) = \exp(sM)\dot{x}(0)$. To integrate this once more, we look for the subspace where M is nonsingular, on which $x(s) = M^{-1}\exp(sM)\dot{x}(0) + \text{const}$. On the subspace where M is singular, $x(s) \sim s$.

3. $$F \wedge F = -dx^0 \wedge dx^1 \wedge dx^2 \wedge dx^3 \cdot (\mathbf{E} \cdot \mathbf{B}).$$

4. $$\dot{p}_\alpha + \frac{e}{2} F_{\alpha\beta}\dot{x}^\beta = m\ddot{x}_\alpha + eF_{\alpha\beta}\dot{x}^\beta = 0.$$

5. $$(y, z; p_y, p_z) \to \left(y + \frac{a_3}{eB}, z - \frac{a_2}{eB}, p_y + \frac{a_2}{2}, p_z + \frac{a_3}{2}\right),$$

   or, respectively,

$$(x, t; p_x, p_0) \to \left(x + \frac{a_0}{eE}, t - \frac{a_1}{eE}; p_x + \frac{a_1}{2}, p_0 + \frac{a_0}{2}\right), \qquad a_i \in \mathbb{R}.$$

6. Both are circles (or, respectively, hyperbolas), because

$$\dot{y}^2 + \dot{z}^2 = \frac{2}{m}\mathcal{H}_B = \frac{1}{m^2}\left[\left(p_z - \frac{eB}{2}\bar{y}\right)^2 + \left(p_y + \frac{eB}{2}\bar{z}\right)^2\right]$$

or, respectively,

$$\dot{x}^2 - \dot{t}^2 = \frac{2}{m}\mathcal{H}_E = \frac{1}{m^2}\left[\left(p_x - \frac{eE}{2}\bar{t}\right)^2 - \left(p_t + \frac{eE}{2}\bar{x}\right)^2\right].$$

7. $|\mathbf{B}|^2 - |\mathbf{E}|^2 = 2^*(F \wedge {}^*F)$.

5.3 The Coulomb Field

The $O(4)$ symmetry is broken by relativistic corrections, but the system remains integrable.

The motion described by (5.1.9) with $eA = (\alpha/r, 0, 0, 0)$, where $r = |\mathbf{x}|$, was extremely important in the infancy of atomic physics. Although one needs quantum mechanics to talk about atoms, the classical solution is still of interest—not merely to contrast with the result of quantum mechanics, but also to make the connection to the gravitational case, which we shall come to later.

(5.3.1) **The Hamiltonian**

$$\mathcal{H} = \frac{1}{2m}\left(|\mathbf{p}|^2 - \left(p_0 + \frac{\alpha}{r}\right)^2\right)$$

is $C^\infty(T^*(\mathbb{R} \times (\mathbb{R}^3 \setminus \{0\})))$, and we can use polar coordinates to transform the extended configuration space into $\mathbb{R} \times \mathbb{R}^+ \times S^2$. Since the only spatial coordinate that shows up explicitly in $\mathcal{H}$ is r, we know the

(5.3.2) **Constants of the Motion**

$$\dot{\mathbf{L}} = \frac{d}{ds}[\mathbf{x} \times \mathbf{p}] = 0, \qquad \dot{p}_0 = 0 \qquad \text{and} \qquad \dot{\mathcal{H}} = 0.$$

(5.3.3) **Remarks**

1. From these five constants of the motion it is possible to construct four with vanishing Poisson brackets. Although the Poisson brackets of different components of $\mathbf{L}$ do not vanish, $|\mathbf{L}|^2$, as a scalar, is rotationally invariant, and so

$$\{L_i, |\mathbf{L}|^2\} = 0, \qquad i = 1, 2, 3.$$

Hence $\mathcal{H}$, p_0, $L := |\mathbf{L}|$, and, e.g., L_3 are four independent constants with vanishing Poisson brackets, and the system is integrable.

2. The time-dependence can be determined from $\dot{t} = -(p_0 + \alpha/r)/m$, once $r(s)$ is known.

3. In this section $\mathcal{H}$ can be separated in (x, p) and in (t, E), so we only study the motion in ordinary phase space $T^*(\mathbb{R}^3\setminus\{\mathbf{0}\})$. It is governed by

$$\mathcal{H} + \frac{\mathbf{p}_0^2}{2m} := H = \frac{|\mathbf{p}|^2}{2m} + \frac{\alpha}{r}\frac{p^0}{m} - \frac{\alpha^2}{2mr^2}.$$

The connection between H and the energy $E = p^0$ is the equation

$$p^0 = m\sqrt{1 + \frac{2H}{m}} = \frac{\alpha}{r} + \sqrt{m^2 + |\mathbf{p}|^2},$$

since $\mathcal{H}$ has to equal $-m/2$. The energy p^0 is to be regarded as a constant in H, although its value is not independent of H.

4. Compared with (4.2.2), H has an extra negative contribution, because

$$\sqrt{m^2 + |\mathbf{p}|^2} - m \le \frac{|\mathbf{p}|^2}{2m}.$$

If $H < 0$, the trajectories in $T^*(\mathbb{R}^3\setminus\{\mathbf{0}\})$ remain in compact sets, and Arnold's theorem applies. In order to compute the frequencies, we construct the

(5.3.4) Action and Angle Variables

We have to form combinations of H, L, and L_3, the conjugate variables of which are the three angle variables on T^3. As we have seen, L_3 generates a rotation about the z-axis. To ascertain what angle is conjugate to L, first note that

$$\{|\mathbf{L}|^2, x_i\} = 2L\{L, x_i\} = 2(\mathbf{L}\cdot\{\mathbf{L}, x_i\}) = 2[\mathbf{L}\times\mathbf{x}]_i \tag{5.3.5}$$

(cf. (3.1.12; 4)). This implies that L generates a rotation about $\mathbf{L}$, i.e., in the plane of motion:

$$\{L, \mathbf{x}\} = \left[\frac{\mathbf{L}}{L}\times\mathbf{x}\right], \quad \text{and} \tag{5.3.6}$$

$$\{L, \mathbf{p}\} = \left[\frac{\mathbf{L}}{L}\times\mathbf{p}\right].$$

If we assume that $\mathbf{L}$ does not point in the z-direction, and define the angles φ and χ by

$$\cos\varphi = \frac{L_1}{\sqrt{|\mathbf{L}|^2 - L_3^2}},$$

$$\cos\chi = \frac{[\mathbf{L}\times\mathbf{x}]_3}{r\sqrt{|\mathbf{L}|^2 - L_3^2}} = \frac{[(\mathbf{e}_3\times\mathbf{L})\cdot\mathbf{x}]}{r\sqrt{|\mathbf{L}|^2 - L_3^2}}, \quad \text{where} \quad \mathbf{e}_i = \frac{\mathbf{x}_i}{r},$$

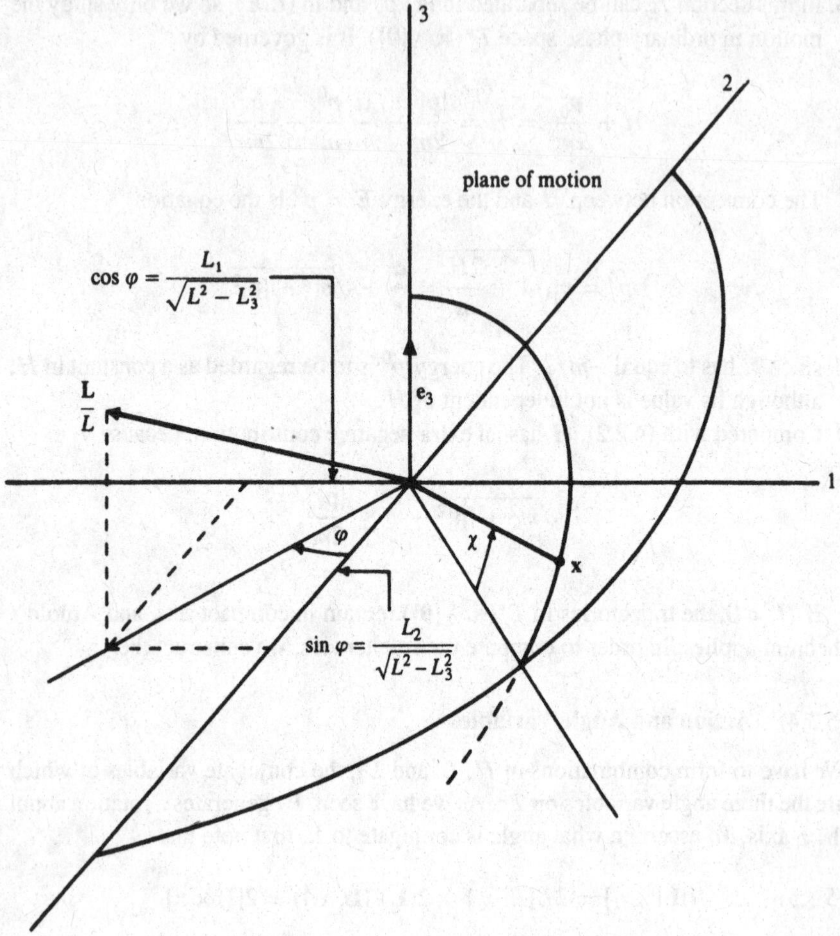

FIGURE 5.3. Action and angle variables.

then, geometrically, $(\mathbf{L} \times \mathbf{e}_3)/\sqrt{|\mathbf{L}|^2 - L_3^2}$ is the unit vector perpendicular to $\mathbf{L}$ in the (1–2)-plane, and thus χ is its angle with $\mathbf{x}$ and φ is its angle with the 1-axis (see Figure 5.3). Thus

$$(5.3.7) \qquad \{L_3, \varphi\} = \{L, \chi\} = -1 \qquad \text{and} \qquad \{L_3, \chi\} = \{L, \varphi\} = 0;$$

because a rotation about the z-axis changes φ while leaving χ unchanged, whereas a rotation about $\mathbf{L}$ leaves φ alone but changes χ (Problem 5).

We may choose r and $p_r := \mathbf{x} \cdot \mathbf{p}/r$ as the third pair of canonical coordinates. Although they are not action and angle variables, their Poisson brackets with the observables in (5.3.7) are all zero: Since both of them are rotationally invariant, their Poisson brackets with all the L's vanish. The only fact that remains to be checked for them to be canonical coordinates is that $\{p_r, \chi\} = 0$. This follows from the observation that $rp_r = (\mathbf{x} \cdot \mathbf{p})$ generates a dilatation (4.1.13; 3), under which **L** and $\mathbf{x}/r$ are invariant. With these variables H can be written

(5.3.8)
$$H = \frac{1}{2m}\left(p_r^2 + \frac{L^2}{r^2}\right) + \frac{\alpha p^0}{rm} - \frac{\alpha^2}{2mr^2}$$

(cf. (4.2.12)). The action variables conjugate to φ, χ, and r, as defined in (3.3.14), exist if $H < 0$ and $L > \alpha$, and are

$$I_\varphi = L_z, \qquad I_\chi = L,$$

and

(5.3.9)
$$I_r = \frac{1}{2\pi}\oint dr\, p_r = -\sqrt{L^2 - \alpha^2} + \frac{\alpha p^0}{2}\sqrt{\frac{2}{m|H|}}$$

(Problem 1). The Hamiltonian H can easily be expressed in terms of the action variables:

(5.3.10)
$$H = -\frac{\alpha^2 p^{02}}{2m(I_r + \sqrt{L^2 - \alpha^2})^2},$$

and an explicit calculation of the frequencies (in s) (3.3.15; 4) gives

(5.3.11)
$$\omega_\varphi = 0, \qquad \omega_\chi = \frac{\alpha^2 p^{02}(L/\sqrt{L^2 - \alpha^2})}{m(I_r + \sqrt{L^2 - \alpha^2})^3},$$
$$\omega_r = \frac{\alpha^2 p^{02}}{m(I_r + \sqrt{L^2 - \alpha^2})^3}.$$

(5.3.12) **Remarks**

1. Since $\omega_\varphi = 0$ (due to conservation of angular momentum the plane of motion is fixed), there is a two-dimensional invariant torus. This torus is densely filled by the orbit, however, unless
$$\frac{\omega_\chi}{\omega_r} = \frac{L}{\sqrt{L^2 - \alpha^2}}$$
is rational, in which case the orbits are closed.
2. In the nonrelativistic limit, $\alpha^2/r^2 \to 0$, we had $L/\sqrt{L^2 - \alpha^2}$ equal to 1, and there were invariant one-dimensional tori, the Kepler orbits. In the present case the projection of the orbit to configuration space is rosette-shaped (Figure 5.4), where the angle of each successive perihelion increases by $2\pi/\sqrt{1 - \alpha^2/L^2}$, and the orbit is in general dense in a ring-shaped region. Because of the relativistic mass increase near the center, r does not return to its initial value at the same time as χ, and the orbit precesses. The existence of a second frequency is also what gives rise to the **fine structure** of spectral lines in atomic physics, which was one of the early experimental confirmations of the theory of relativity.

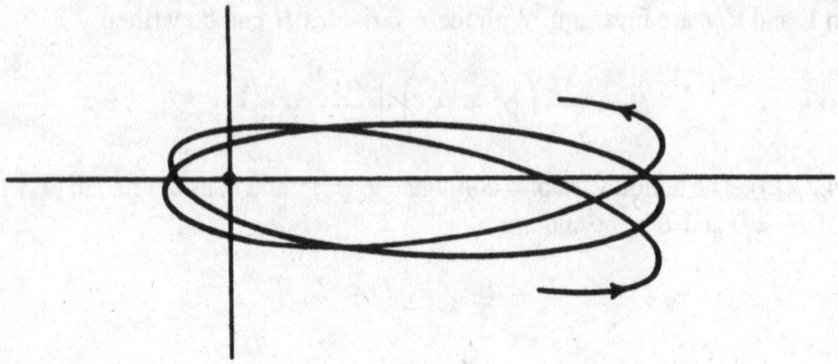

FIGURE 5.4. A rosette-shaped orbit in the relativistic Kepler problem.

3. From (5.3.8) we see that the motion is like one-dimensional motion with an effective potential

$$\frac{\alpha p^0}{rm} + \frac{L^2 - \alpha^2}{2mr^2}.$$

If $\alpha < 0$ and $L \leq |\alpha|$, then this potential is monotonic in r, and the particle spirals inevitably into the singularity, reaching it after a finite time. But if $\alpha > 0$ this cannot happen, as follows from Remark (5.3.3; 3) (see Problem 4). If we are interested in the completeness of $X_{\mathcal{H}}$, we must reduce phase space in the attractive case to

$$T^*(\mathbb{R}^4)\setminus\{(x, p): L \leq |\alpha|\};$$

it is no longer sufficient to remove a submanifold of a lower dimension.

(5.3.13) Unbound Trajectories

In the part of phase space where $H > 0$ and $L > |\alpha|$, the following three facts can be proved:

(a) $\lim_{t\to\pm\infty} r(t) = \infty$;

(b) $\mathbf{p}$ and $\mathbf{x}/r \in \mathcal{A}$ (see (3.4.1)); and

(c) $\Omega_\pm := \lim_{t\to\pm\infty} \Omega_{-t} \circ \Omega_t^0$ exist, where Ω_t^0 is generated by $H + \alpha^2/2mr^2$, the Hamiltonian with a pure $1/r$ potential.

(5.3.14) Remarks

1. The scattering angle $\Theta = \measuredangle(\mathbf{p}_-, \mathbf{p}_+)$ can be written explicitly in terms of the constants of motion (Problem 3), although the function $\sigma(\Theta)$ of (3.4.12) can only be written implicitly in terms of them.
2. Fact (c) implies that this part of phase space has the maximal number of constants of motion, five. For instance, we could take $\mathbf{L}$ and $\mathbf{F} \circ \Omega_\pm^{-1} = \lim_{t\to\pm\infty} \mathbf{F}(t)$, where $\mathbf{F}$ was defined in (4.2.4), with $m \to p^0$. For $H > 0$, the invariance group is still $SO(3, 1)$, as in (4.2.7) (cf. (3.4.5; 3)).

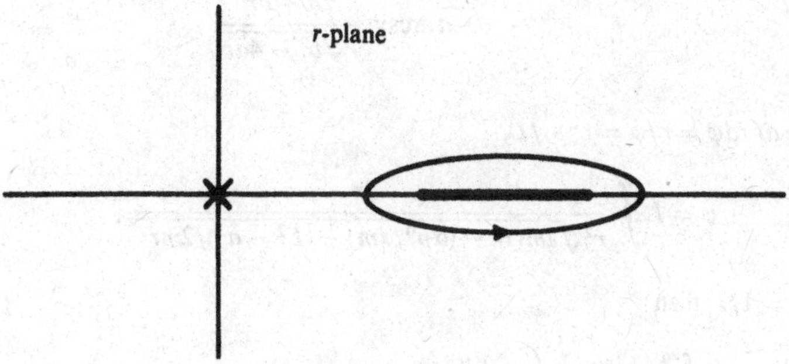

FIGURE 5.5. Integration path in the complex r-plane.

(5.3.15) **Problems**

1. Calculate the integral (5.3.9) for I_r.
2. Write the angle variable φ, conjugate to I_r in terms of r and the constants of motion.
3. Calculate the scattering angle Θ, and show that it approaches zero in the limit of large energies ($p^0 \to \infty$, $L \to \infty$, with $L/p^0 = b$ constant).
4. Calculate the maximum of the effective potential V_{eff} for $\alpha > 0$, and show that it is greater than $p_\infty^2/2m$, where p_∞ is the value of $|\mathbf{p}|$ at $r = \infty$.
5. Verify (5.3.7).

(5.3.16) **Solutions**

1. The integral for I_r can be expressed as an indefinite integral in terms of elementary functions (see Problem 2), but it is simpler to evaluate the definite integral with complex integration. In the complex r-plane the integral has a pole at the origin and a branch cut along an interval on the real axis, which is contained in the integration region (Figure 5.5). By stretching the path of integration, we see that the integral just picks up the residues of the poles at $r = 0$ and $r = \infty$:

$$I_r = -\sqrt{L^2 - \alpha^2} + \frac{\alpha p^0}{2}\sqrt{2/m|H|}.$$

2.

$$\varphi_r = \int dr \sqrt{2m\left(H - \frac{\alpha p^0}{mr} - \frac{L^2 - \alpha^2}{2mr^2}\right)} = \tfrac{1}{2}\sqrt{2mHr^2 - 2\alpha p^0 r - L^2 + \alpha^2}$$

$$+ \frac{\alpha p^0}{\sqrt{-2mH}} \arcsin \frac{4mHr - 2\alpha p^0}{\sqrt{4\alpha^2 p^{02} + 4H(L^2 - \alpha^2)}}$$

$$+ \sqrt{L^2 - \alpha^2} \arcsin \frac{2(L^2 - \alpha^2)/r + 2\alpha p^0}{\sqrt{4\alpha^2 p^{02} + 4H(L^2 - \alpha^2)}},$$

because

$$\int dr \sqrt{\frac{a}{r^2} + \frac{b}{r} + c} = \sqrt{a + br + cr^2} + \frac{b}{2}\frac{1}{\sqrt{-c}} \arcsin \frac{2cr + b}{\sqrt{b^2 - 4ac}}$$

$$- \sqrt{-a} \arcsin \frac{2a + br}{r\sqrt{b^2 - 4ac}}.$$

3. Because $dr/d\varphi = \dot{r}/\dot{\varphi} = r^2 p_r/L$,

$$\varphi = L \int \frac{dr}{r^2 \sqrt{2m(H - (\alpha p^0/rm) - (L^2 - \alpha^2)/2mr^2)}},$$

and if $u = 1/r$, then

$$\pi - \Theta = 2L \int_0^{u_0} \frac{du}{\sqrt{2mH - 2\alpha p^0 u - (L^2 - \alpha^2)u^2}}$$

$$= \frac{2L}{\sqrt{L^2 - \alpha^2}} \arccos \left(1 + \frac{2mH(L^2 - \alpha^2)}{\alpha^2 p^{02}}\right)^{-1/2} \quad \to \quad 2 \arccos 0 = \pi.$$

4. $V_{\text{eff}} = (\alpha/r)(p^0/m) - (\alpha^2/2mr^2)$ has its maximum $(p^0)^2/2m$ at $r = \alpha/p^0$, and according to (5.3.3; 3),

$$\frac{(p^0)^2}{2m} = \frac{m}{2} + \frac{p_\infty^2}{2m}.$$

5. The only one that is not obvious is $\{L, \chi\}$. We calculate $\{L, \cos\chi\}$:

$$\left\{L, \frac{L_y x - L_x y}{r\sqrt{L^2 - L_z^2}}\right\} = \frac{1}{rL\sqrt{L^2 - L_z^2}}(L_y(L_y z - L_z y) - L_x(L_z x - L_x z)).$$

The part in parentheses may be rewritten as $L^2 z - L_z(\mathbf{L} \cdot \mathbf{x}) = L_z^2$, so

$$\{L, \cos\chi\} = \frac{Lz}{r\sqrt{L^2 - L_z^2}} = \sin\chi.$$

5.4 The Betatron

Although this problem is not integrable, it is not only possible to solve for particular trajectories, but for suitably chosen magnetic fields it is even possible to determine the time-evolution for a larger class of initial conditions.

One of the most ingenious applications of the law of induction

$$\nabla \times \mathbf{E} = -\frac{\partial \mathbf{B}}{\partial t} \tag{5.4.1}$$

is the betatron. Its mechanism is based on the following ideas: If a current $\mathbf{j}$ starts to flow in a current loop, it induces a magnetic field $\mathbf{B}$ in the interior of the loop, and $\dot{\mathbf{B}}$ produces an electric field circulating about $\mathbf{B}$ in the same direction as a Larmor orbit, so as to oppose $\mathbf{j}$ (Problem 2). In order to see when the Lorentz force is able to counterbalance the centrifugal force, let us integrate (5.4.1) around a circle of radius a about the z-axis. If $B_z =: B$ depends only on the distance from the z-axis and t, then

(5.4.2)
$$E_\varphi = \frac{1}{2\pi a}\oint \mathbf{E}\cdot d\mathbf{s} = -\int \frac{\partial B}{\partial t}\frac{dS}{2\pi a} = -\frac{a}{2}\frac{d}{dt}\bar{B},$$
$$\bar{B} := \frac{1}{\pi a^2}\int d\,SB.$$

On the other hand, the (nonrelativistic) equation of motion for a particle in this field implies

(5.4.3)
$$\frac{d}{dt}mv = eE_\varphi = -\frac{ea}{2}\frac{d}{dt}\bar{B}$$

or, if the particle was at rest before the current was switched on,

(5.4.4)
$$mv = -\frac{ea}{2}\bar{B}.$$

Hence the centrifugal and centripetal forces will balance at a circular orbit whenever

(5.4.5)
$$\frac{mv^2}{a} = -evB(a) = -\frac{ev}{2}\bar{B}, \quad \text{which implies} \quad B(a) = \frac{\bar{B}}{2};$$

that is, when the field at the orbit is half as strong as its average over the disk enclosed by the orbit. In this section we shall go through the details of the theory sketched above.

(5.4.6) The Hamiltonian

To exploit the symmetry of the problem we use cylindrical coordinates for space, or in other words a chart $\mathbb{R}^3\backslash\{(0,0,\mathbb{R})\} \to \mathbb{R}^+ \times S^1 \times \mathbb{R}$, with coordinates ρ, φ, and z. A field $B(\rho, t)$ in the z-direction comes from a vector potential $e\mathcal{A}$ with a covariant φ-component $-A(\rho, t)$ such that

(5.4.7)
$$eB_z = -\frac{1}{\rho}\frac{\partial}{\partial\rho}A, \qquad eE_\varphi = -\frac{1}{\rho}\frac{\partial A}{\partial t}$$

(cf. Problem 3). This shows that a circulating electric field is created when $\mathbf{B}$ is switched on, according to the law of induction. The motion of a particle in extended phase space is then controlled by

(5.4.8)
$$\mathcal{H} = \frac{1}{2m}\left(p_z^2 + p_\rho^2 + \frac{1}{\rho^2}(p_\varphi - A)^2 - p_0^2\right).$$

Since this depends explicitly only on ρ and t (through A), we obtain the

(5.4.9) Constants of the Motion

$$\dot{p}_\varphi = \dot{p}_z = 0.$$

(5.4.10) Remarks

1. All together, including $\mathcal{H}$ we have three constants, one too few to integrate the problem completely. The complexity is comparable to that of the restricted three-body problem, since only one constant, $\mathcal{H}$, is available for the (ρ, t) motion.
2. We do not need to find the most general solution; the betatron is only operated when the particles are nearly at rest before it is switched on. For that reason we study the

(5.4.11) Orbits with $p_\varphi = p_z = 0$

In this case, $A = 0 \Rightarrow \dot{\varphi} = 0$, as

(5.4.12)
$$\dot{\varphi} = \frac{p_\varphi - A}{m\rho^2}.$$

Therefore $p_\varphi = 0$ means that the particle can have moved at most radially before A was switched on. The ρ coordinate of the motion obeys the equations

(5.4.13)
$$\dot{\rho} = \frac{p_\rho}{m}, \qquad \ddot{\rho} = \frac{\dot{p}_\rho}{m} = -\frac{\partial}{\partial \rho}\frac{A^2}{2m^2\rho^2}.$$

As long as

(5.4.14)
$$\left.\frac{\partial}{\partial \rho}\frac{A^2(\rho, t)}{2m\rho^2}\right|_a = 0$$

for all t, $\rho = a = \text{const.}$ is a solution.

(5.4.15) Remarks

1. Equation (5.4.14) is equivalent to (5.4.5) (Problem 4), showing that the naïve arguments at the beginning of this section are correct, even relativistically (provided Stokes's theorem is applicable).
2. A question of practical importance is whether the orbits are stable, since the initial condition chosen in (5.4.11) will never hold exactly. It is clear that the z-component of the motion is free, and therefore unstable. Some z-dependence must be introduced into A to prevent this (see [10]), but that will not concern us here. Stability in the radial direction depends on the form of A.

Instead of making a thorough investigation of the stability question, we shall rely on the theorist's freedom to leave the construction of suitable fields to the experimenter, and look only at a soluble

(5.4.16) **Example**

$$A = \rho\sqrt{\mu^2(\rho - a)^2 + \nu^2 t^2}.$$

This satisfies condition (5.4.14) and produces the fields

(5.4.17)
$$B_z = \frac{\mu^2(\rho - a)(2\rho - a) + \nu^2 t^2}{\rho\sqrt{\mu^2(\rho - a)^2 + \nu^2 t^2}},$$
$$E_\varphi = \frac{-\nu^2 t}{\sqrt{\mu^2(\rho - a)^2 + \nu^2 t^2}}.$$

At the orbit $\rho = a$, the electric field is constant, and the magnetic field grows linearly in t. The angular velocity (using s) then grows linearly with the ordinary time t according to (5.4.12) (cf. Problem 1):

(5.4.18)
$$\dot{\varphi}\Big|_{\rho=a} = \frac{p_\varphi - a\nu t}{ma^2}.$$

If $p_z = p_\varphi = 0$, then

(5.4.19)
$$\mathcal{H} = \frac{1}{2m}(p_\rho^2 - p_0^2 + \mu^2(\rho - a)^2 + \nu^2 t^2).$$

Except for the sign of p_0^2, this brings us back to the two-dimensional harmonic oscillator. There are two constants of motion; the ρ and t contributions to $\mathcal{H}$ separate, and the equations can be integrated without difficulty (the coefficients being determined by $1 = \dot{t}^2 - \dot{\varphi}^2 - \dot{\rho}^2$):

(5.4.20)
$$\rho(s) = a + c\sin\frac{\mu}{m}(s - s_0), \qquad c = p_\rho(0)/\mu,$$
$$t(s) = \frac{\sqrt{c^2\mu^2 + m^2}}{\nu}\sinh\frac{\nu}{m}s.$$

(5.4.21) **Remarks**

1. As for stability, the motion is stable if we only look at ρ, because the set where $|\rho - a| < c$ is invariant for all $c \in \mathbb{R}^+$. But if we look at both ρ and t (or ρ and φ), then it is not stable, because an arbitrarily small c can bring about arbitrarily large changes in $t(s)$, for s large enough.
2. Since a discussion of stability for arbitrary A is difficult, it is quite common to simply invoke the adiabatic theorem (cf. (3.5.20; 6)), the popular version of which goes roughly as follows: if the field is turned on very slowly, so that it changes by only a tiny fraction during each period, then it is safe to do calculations as if the field were constant. But it is not really possible to formulate the theorem in precisely this way, even though in the special example (5.4.16) it did turn out that the ρ and t dependences separated, and the growth-rate ν did not enter into $\rho(s)$—showing up only in the connection between s and t.

(5.4.22) Problems

1. Calculate the speed $v = |d\mathbf{x}/dt|$ of a particle following the trajectory (5.4.20) with $\rho = a$, and verify that $|v| < 1$.
2. How do the directions of the forces in the betatron square with the fact mentioned in §5.1, that antiparallel currents repel?
3. Calculate the coordinates of $\mathbf{B} = \nabla \times \mathcal{A}$ and $\mathbf{E} = -\dot{\mathcal{A}}$ in cylindrical coordinates, using the covariant components A of $\mathcal{A}$ (cf. (2.5.4)).
4. Show that (5.4.5) and (5.4.14) are equivalent. (Recall that B is an orthogonal component, while A is a covariant component.)

(5.4.23) Solutions

1.

$$v = a\frac{d\varphi}{dt} = \frac{a\dot{\varphi}}{\dot{t}} = -\tanh\left(\frac{\nu}{m}s\right).$$

2.

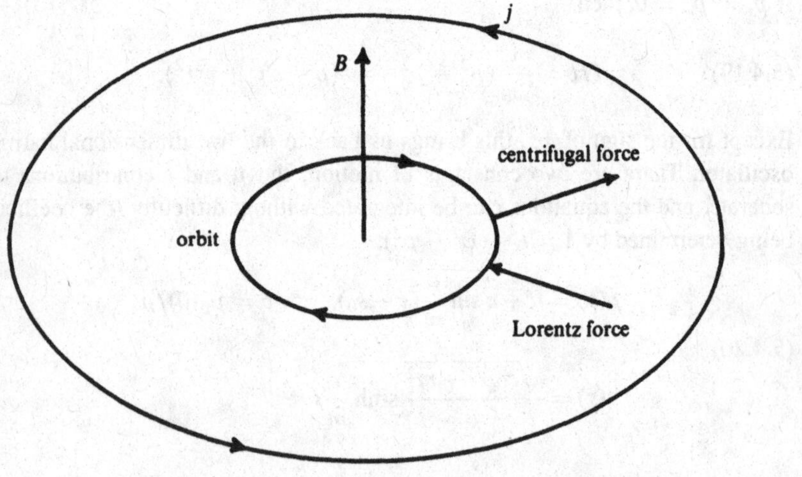

FIGURE 5.6. The directions of the forces in the betatron.

3. In cylindrical coordinates (z, ρ, φ), $g_{ii} = (1, 1, \rho^2)$, so $(\mathcal{A}_z, \mathcal{A}_\rho, \mathcal{A}_\varphi) = (A_z, A_\rho, A_\varphi/\rho)$. To calculate $\nabla \times \mathcal{A} = {}^*(dA)$, we generalize the * operation of (2.4.33; 1) to get ${}^*(dA)_i = \varepsilon_{ikj} g^{k\ell} g^{jm} (dA)_{\ell m} \sqrt{g}$, where $g := \det(g_{ik}) = \rho^2$. Thus

$$A = A_z\,dz + A_\rho\,d\rho + A_\varphi\,d\varphi,$$

$$dA = (A_{z,\rho} - A_{\rho,z})\,d\rho \wedge dz + (A_{\rho,\varphi} - A_{\varphi,\rho})\,d\varphi \wedge d\rho + (A_{\varphi,z} - A_{z,\varphi})\,dz \wedge d\varphi,$$

$${}^*(dA) = \underbrace{\frac{1}{\rho}(A_{\varphi,\rho} - A_{\rho,\varphi})}_{(\nabla\times A)_z}dz + \underbrace{\frac{1}{\rho}(A_{z,\varphi} - A_{\varphi,z})}_{(\nabla\times A)_\rho}d\rho + \underbrace{\rho(A_{\rho,z} - A_{z,\rho})}_{(\nabla\times A)_\varphi}d\varphi,$$

which corresponds to (5.4.7).

4.

$$\bar{B} = \frac{1}{\pi a^2}\int_0^a 2\pi\rho\, d\rho \frac{A_{,\rho}}{\rho} = \frac{2}{a^2}A(a),$$

and

$$0 = \frac{\partial}{\partial\rho}\frac{A}{\rho}\bigg|_a = \frac{A_{,\rho}}{\rho} - \frac{A}{\rho^2}\bigg|_a = B(a) - \frac{\bar{B}}{2}.$$

Since there is still one arbitrary constant in A, we may set $A(0) = 0$.

5.5 The Traveling Plane Disturbance

The rich invariance group of this problem furnishes more constants of motion than are required for integrability. Nevertheless, the trajectory generally covers a two-dimensional submanifold of space.

Classical studies of the scattering of light by a charged particle deal with the motion of the particle in a plane electromagnetic wave. More recently, laser technology has made intense pulses of light available, and the interest in solving the equations of motion in the field of a disturbance moving in some direction at the speed of light has increased.

(5.5.1) The Field and the Hamiltonian

Let x be the direction of propagation, the fields depending only on the combination $t - x$. So that we can use the solutions of Maxwell's equations in a vacuum, $\mathbf{E}$ and $\mathbf{B}$ are taken perpendicular to each other and to the x-axis. Such a situation is described by the vector potential (5.1.7)

$$eA = (0, 0, f(t-x), g(t-x)).$$

This makes the fields

$$e\mathbf{E} = (0, f', g') \quad \text{and} \quad e\mathbf{B} = (0, -g', f');$$

and the motion of a particle in these fields is governed by

$$\mathcal{H} = \frac{1}{2m}[(p_y + f)^2 + (p_z + g)^2 + p_x^2 - p_t^2].$$

(5.5.2) Remarks

1. Both invariants $|\mathbf{E}|^2 - |\mathbf{B}|^2$ and $(\mathbf{E}\cdot\mathbf{B})$ vanish identically. If f' and g' are constant, we get the constant field with vanishing invariants.
2. We can assume that $f, g \in C^\infty(\mathbb{R})$, to be able to work on $M_e = \mathbb{R}^4$.

(5.5.3) **The Invariance Group**

The electromagnetic field tensor F for (5.5.1) is invariant under a five-parameter subgroup of the Poincaré group. Since f and g depend only on $t - x$, it is clearly unchanged by displacements in the spatial directions y and z or in the "**lightlike**" direction $x + t$. It turns out (Problem 1) that the combined Lorentz transformations and rotations generated by $K_z + L_y$ and $K_y - L_z$ also do not affect F. Consequently, F is invariant under the group generated by p_y, p_z, $p_x + p_t$, $K_z + L_y$, and $K_y - L_z$ (Problem 2). Yet not all of these transformations leave A, and thus $\mathcal{H}$, invariant. Since M_e is starlike, Remark (5.1.10; 3) applies, and A can at most be regauged. In fact, it can be calculated that the two kinds of Lorentz transformations change A only by the gauge transformations whose gauge functions are

$$\Lambda_g = \int_0^{t-x} du\, g(u) \quad \text{and} \quad \Lambda_f = \int_0^{t-x} du\, f(u).$$

We have thus accounted for all the

(5.5.4) **Constants of the Motion**

$$p_y, p_z, p_x + p_t,$$

$$K_z + L_y + \Lambda_g = p_z(t - x) + z(p_t + p_x) + \int_0^{t-x} du\, g(u),$$

and

$$K_y - L_z + \Lambda_f = p_y(t - x) + y(p_t + p_x) + \int_0^{t-x} du\, f(u)$$

are constant.

(5.5.5) **Remarks**

1. The group generated by the five constants of the motion is isomorphic to the invariance group of the field (Problem 2).
2. The Poisson brackets of p_y, p_z, and $p_x + p_t$ vanish; so, counting $\mathcal{H}$, we have all four constants necessary for integrability.
3. Counting $\mathcal{H}$ there are altogether six constants, so one more would be needed to determine a trajectory in $T^*(M_e)$ completely. As there is no other such constant, we have to resort to quadrature. Set the origin of s at the point where the trajectory crosses the plane $t = x$. Then from

$$\frac{d}{ds}(t - x) = -\frac{p_t + p_x}{m} =: \alpha > 0,$$

it follows that

$$t(s) - x(s) = \alpha s.$$

Using $\mathcal{H} = -m/2$, we find that

$$\frac{d}{ds}(t+x) = \frac{p_x - p_t}{m} = \alpha^{-1}\left(1 + \frac{(p_y+f)^2 + (p_z+g)^2}{m^2}\right),$$

$$t(s) + x(s) = 2t(0) + \alpha^{-2}\int_0^{\alpha s} du\left[1 + \frac{(p_y + f(u))^2 + (p_z + g(u))^2}{m^2}\right].$$

If we collect all these results, we obtain the

(5.5.6) **Explicit Solution for the Coordinates as Functions of Proper Time**

$$t(s) = t(0) + \frac{s}{2}\left(\frac{1}{\alpha} + \alpha\right) + \frac{1}{2m^2\alpha^2}\int_0^{\alpha s} du[(p_y + f(u))^2 + (p_z + g(u))^2],$$

$$x(s) = x(0) + \frac{s}{2}\left(\frac{1}{\alpha} - \alpha\right) + \frac{1}{2m^2\alpha^2}\int_0^{\alpha s} du[(p_y + f(u))^2 + (p_z + g(u))^2],$$

$$y(s) = y(0) + s\frac{p_y}{m} + \frac{1}{m\alpha}\int_0^{\alpha s} du\, f(u),$$

$$z(s) = z(0) + s\frac{p_z}{m} + \frac{1}{m\alpha}\int_0^{\alpha s} du\, g(u).$$

Because there is one less than the maximal number of constants, the trajectory is generally a Lissajou figure in (x, y).

(5.5.7) **Example**

The superposition of two plane waves:

$$f = A_1 \cos\omega_1 u, \quad g = A_2\cos(\omega_2 u + \delta),$$
$$e\mathbf{E} = (0, -A_1\omega_1 \sin\omega_1 u, -A_2\omega_2\sin(\omega_2 u + \delta)),$$
$$e\mathbf{B} = (0, A_2\omega_2\sin(\omega_2 u + \delta), -A_1\omega_1\sin\omega_1 u).$$

The solution (5.5.6) with $x(0) = t(0)$ is computed as

$$\begin{Bmatrix} t(s) \\ x(s) \end{Bmatrix} = t(0) + \frac{s}{2}\left(\frac{1}{\alpha} \pm \alpha + \frac{p_y^2 + p_z^2}{m^2}\right) + \frac{p_y A_1}{\omega_1\alpha^2 m^2}\sin\omega_1\alpha s$$

$$+ \frac{p_z A_2}{\omega_2\alpha^2 m^2}(\sin(\omega_2\alpha s + \delta) - \sin\delta) + \frac{A_1^2 + A_2^2}{4m^2\alpha}s + \frac{A_1^2}{8\alpha^2 m^2\omega_1}$$

$$\cdot \sin 2\omega_1\alpha s + \frac{A_2^2}{8\alpha^2 m^2\omega_2}(\sin 2(\omega_2\alpha s + \delta) - \sin 2\delta),$$

(5.5.8)

$$y(s) = y(0) + s\frac{p_y}{m} + \frac{A_1}{\omega_1\alpha m}\sin\omega_1\alpha s,$$

$$z(s) = z(0) + s\frac{p_z}{m} + \frac{A_2}{\omega_2\alpha m}(\sin(\omega_2\alpha s + \delta) - \sin\delta).$$

For ω's with irrational ratios, the trajectory fills a two-dimensional region.

(5.5.9) **Special Cases**

1. **Circularly polarized waves.** $A_1 = A_2 = A$, $\omega_1 = \omega_2 = \omega$, and $\delta = \pi/2$. The invariance group of this field has an extra parameter, and there is an extra constant,

$$2L_x + \frac{p_x - p_t}{\omega} \tag{5.5.10}$$

(Problem 3). The solution (for $x^\beta(0) = 0$) accordingly simplifies to

$$\begin{Bmatrix} t(s) \\ x(s) \end{Bmatrix} = \frac{s}{2\alpha}\left(1 \pm \alpha^2 + \frac{A^2 + p_y^2 - p_z^2}{m^2}\right) + \frac{A}{m^2\alpha^2\omega}(p_y \sin \omega\alpha s + p_z(\cos\omega\alpha s - 1)),$$

$$\begin{aligned} y(s) &= \frac{A}{\omega\alpha m}\sin\omega\alpha s + s\frac{p_y}{m}, \\ z(s) &= \frac{A}{\omega\alpha m}(\cos\omega\alpha s - 1) + s\frac{p_z}{m}. \end{aligned} \tag{5.5.11}$$

If $p_y = p_z = 0$, then the particle describes a circular orbit in the plane perpendicular to the wave, with its velocity in the direction of **B** and perpendicular to **E** (Figure 5.7).

2. **Linearly polarized waves.** $A_2 = 0$. If $p_y = p_z = y(0) = z(0) = t(0) = 0$, and $\alpha^2 = 1 + A^2/2m^2$, then the orbit is shaped like a bow tie (Figure 5.8):

$$\begin{aligned} \begin{Bmatrix} t(s) \\ x(s) \end{Bmatrix} &= \frac{s}{2\alpha}\left(1 \pm \alpha^2 + \frac{A^2}{2m^2}\right) + \frac{A^2}{8\alpha^2 m^2\omega}\sin 2\omega\alpha s, \\ y(s) &= \frac{A}{\omega\alpha m}\sin\omega\alpha s, \\ z(s) &= 0. \end{aligned} \tag{5.5.12}$$

(5.5.13) **Remark**

To understand this motion, recall the result of §5.2, where the trajectory was as shown in Figure 5.9 when **E** and **B** were related in the same way but were constant fields. For smaller y, this trajectory has a smaller velocity, and thus a smaller Larmor radius. In a plane wave, the fields start to change direction as soon as the particle goes through the origin. The radius of curvature decreases, and the particle returns to the origin, where it encounters fields of the opposite polarity and follows a mirror-image path.

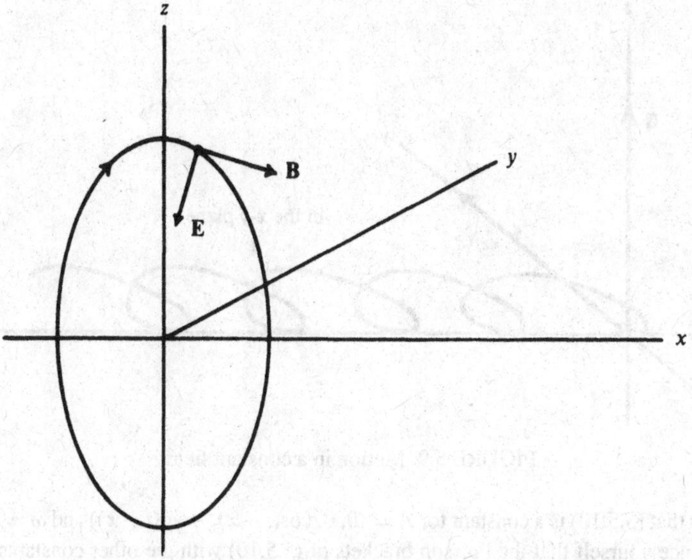

FIGURE 5.7. A circularly polarized wave.

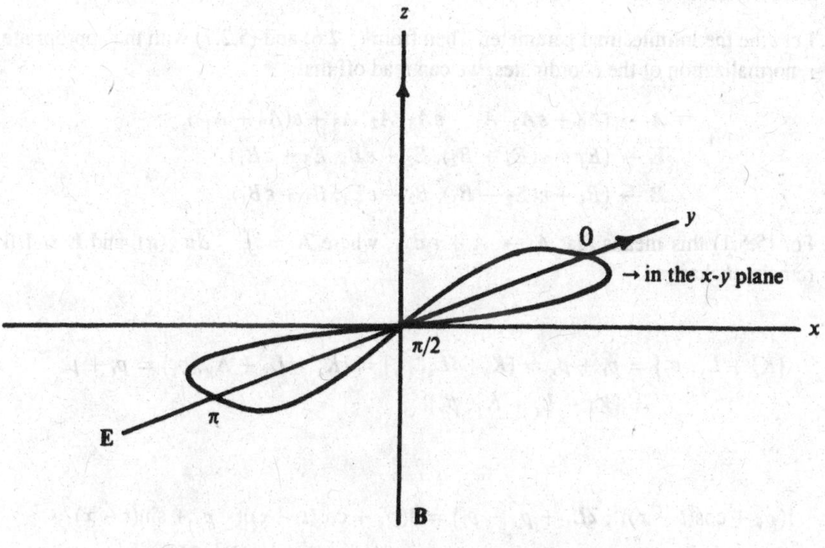

FIGURE 5.8. Motion in a linearly polarized wave.

(5.5.14) **Problems**

1. Calculate the effect of the infinitesimal transformation generated by $K_z + L_y$ on A, $\mathbf{E}$, and $\mathbf{B}$. How do the equations in (5.5.1) change?

2. Calculate the Poisson brackets of the generators of the Poincaré group that leave F invariant, and of the invariance group of $\mathcal{H}$.

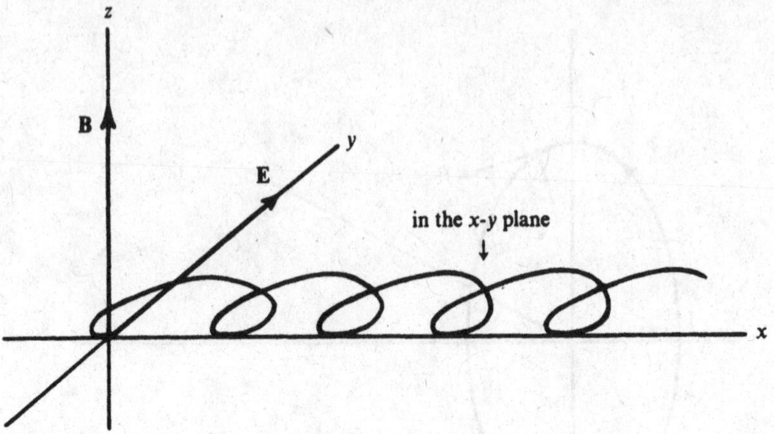

FIGURE 5.9. Motion in a constant field.

3. Show that (5.5.10) is a constant for $A = (0, 0, \cos(t - x), -\sin(t - x))$ and $\omega = 1$. Then convince yourself that the Poisson brackets of (5.5.10) with the other constants can all be written in terms of the other constants.

(5.5.15) **Solutions**

1. Let ε be the infinitesimal parameter. Then from (5.2.6) and (5.2.7) with the appropriate renormalization of the coordinates, we can read off that

$$A \to (A_0 + \varepsilon A_3, A_1 - \varepsilon A_3, A_2, A_3 + \varepsilon(A_0 + A_1)),$$
$$\mathbf{E} \to (E_1 - \varepsilon(E_3 + B_2), E_2 + \varepsilon B_1, E_3 + \varepsilon E_1),$$
$$\mathbf{B} \to (B_1 + \varepsilon(E_2 - B_3), B_2 - \varepsilon E_1, B_3 + \varepsilon B_1).$$

For (5.5.1) this means that $A \to A + \varepsilon\, d\Lambda$, where $\Lambda = \int_0^{t-x} du\, g(u)$, and $\mathbf{E}$ and $\mathbf{B}$ remain unchanged.

2.

$$\{K_z + L_y, p_z\} = p_t + p_x = \{K_y - L_z, p_y\} = \{K_z + L_y + \Lambda_g, p_z\} = p_t + p_x$$
$$= \{K_y - L_z + \Lambda_f, p_y\}.$$

3.

$$\{(p_y + \cos(t - x))^2, 2L_x + p_x - p_t\} = 4(p_y + \cos(t - x))(-p_z + \sin(t - x)$$
$$= -\{(p_z - \sin(t - x))^2, 2L_x + p_x - p_t\}.$$

5.6 Relativistic Motion in a Gravitational Field

In the nonrelativistic limit the equations are very similar to the electrodynamic equations. On the other hand, in their exact form they have a simple geometrical interpretation.

In order to feel comfortable with the complicated-looking system of equations (1.1.6) and (1.1.7) and to see how it compares with its electromagnetic analogue (1.1.4), we start with

(5.6.1) The Nonrelativistic Limit

By this phrase we mean that $|d\mathbf{x}/dt| \ll 1$, but that terms of first order in $d\mathbf{x}/dt$ are to be kept. Moreover, the statement that the gravitational field is weak will mean that $g_{\alpha\beta}$ equals the $\eta_{\alpha\beta}$ of (5.1.2) plus a small quantity. With these approximations, $\overset{-1}{g}_{\alpha\beta}$ equals $\eta_{\alpha\beta}$ minus the same small quantity and s and t can be identified (see (5.6.6; 3)). From field theory we shall learn that a "small" mass M moving with velocity $\mathbf{j}$, $|\mathbf{j}| \ll 1$, at the origin produces a gravitational potential

$$(5.6.2) \qquad g_{\alpha\beta} = \eta_{\alpha\beta} + \frac{4M\kappa}{r}(j_\alpha j_\beta + \tfrac{1}{2}\eta_{\alpha\beta}), \qquad (j_\alpha) = (-1, v_1, v_2, v_3)$$

at the point $\mathbf{x}$, if $|\mathbf{x}| = r \gg M\kappa$. Substituting (5.1.2) into (1.1.6) and (1.1.7) gives the equations of motion

$$(5.6.3) \qquad \frac{d^2\mathbf{x}}{dt^2} = -M\kappa\frac{\mathbf{x}}{r^3} - \frac{4M\kappa}{r^3}\left[\frac{d\mathbf{x}}{dt} \times [\mathbf{j} \times \mathbf{x}]\right] \qquad \text{(Problem 1).}$$

(5.6.4) Remarks

1. The mass m of the particle moving in the gravitational field does not appear in equation (5.6.3); Galileo's discovery that particles of all masses respond identically to a gravitational field is a universal law of nature.
2. A velocity-dependent term of the same form as the Lorentz force (5.1.4; 2) is added to the Newtonian force. Both force terms have the opposite effect to that of their electrodynamic counterparts. Masses of the same sign attract, and mass-currents in the same direction repel. This has been put forth as a confirmation of Mach's principle: if, say, a rotating cylinder encloses another body that rotates along with it, then the forces between the mass-currents act to oppose the centrifugal force in the interior of the cylinder (the H. Thirring effect).[1] If there were nothing else in the Universe, then, according to Mach, there could be no centrifugal force when the angular velocities were equal, since the statement that the two bodies rotate would be meaningless.
3. We shall see in §5.7 how (5.6.3) is altered if $r < M\kappa$ and $|d\mathbf{x}/dt| \sim 1$.

So that we can discuss (1.1.6) and (1.1.7) in the framework of our formalism, we next write down

---

[1] In the electrodynamic case, the centrifugal force on the inner current is counterbalanced when the outer current flows in the opposite direction, as in the betatron.

(5.6.5) The Lagrangian Form of the Equations of Motion

The Lagrangian

$$L = \frac{m}{2}\dot{x}^\alpha \dot{x}^\beta g_{\alpha\beta}(x(s))$$

has equations (1.1.6) and (1.1.7) as its Euler–Lagrange equations ((2.3.20) with s in place of t, and $i = 0, 1, 2, 3$).

Proof: Problem 2. □

(5.6.6) Remarks

1. The factor $m/2$ is, of course, unimportant, and is there only to reproduce (5.1.11) for $g_{\alpha\beta} = \eta_{\alpha\beta}$.
2. We are not able to choose the normalization of (5.1.1; b), $\dot{x}^\alpha \dot{x}^\beta \mu_{\alpha\beta} = -1$, as we did in (5.1.3), since these quantities are not independent of s. Moreover, in the case at hand L itself is a constant; it is quadratic in $\dot{x}$ and in fact equals $\mathcal{H}$. Hence we shall normalize s by requiring $\dot{x}^\alpha \dot{x}^\beta g_{\alpha\beta} = -1$.
3. The argument made in (5.1.20; 2) is no longer valid, and it does not follow from the equations of motion that $|d\mathbf{x}/dt| < 1$ (cf. (5.1.19; 3)). If $\mathbf{j} = \mathbf{0}$ in the gravitational potential (5.6.2), we get

$$\dot{t}^2\left(1 - \frac{2M\kappa}{r}\right) - |\dot{\mathbf{x}}|^2\left(1 + \frac{2M\kappa}{r}\right) = 1,$$

so

$$\left|\frac{d\mathbf{x}}{dt}\right|^2 < \left(1 - \frac{2M\kappa}{r}\right)\left(1 + \frac{2M\kappa}{r}\right)^{-1}.$$

Thus there still exists a maximum velocity, which depends on x through the $g_{\alpha\beta}$, but in other situations might not be less than 1. This could cause some uneasiness, as it sounds as though a gravitational field could accelerate a particle to faster than the speed of light. But note that the maximum velocity is a universal bound for particles of all masses, and is likewise a maximum for photons. As will later be discussed, in this case x and t do not gauge the same lengths and times as one would measure with real yardsticks and clocks. As it would actually be measured, in units where the speed of light is 1, the maximum velocity is also 1.
4. In the nonrelativistic limit as $\dot{x}_0 \to 1$, L tends to $m(g_{00}/2 + \dot{\mathbf{x}}/2)$, so $-g_{00}/2$ plays the role of the gravitational potential.

Using (2.3.22), we can immediately pass from (5.6.5) to

(5.6.7) The Hamiltonian Form of the Equations of Motion

The Hamiltonian

$$\mathcal{H} = \frac{1}{2m} p_\alpha p_\beta g^{\alpha\beta}(x), \qquad g^{\alpha\beta} g_{\beta\gamma} = \delta^\alpha_\gamma,$$

generates a locally canonical flow equivalent to (1.1.6) and (1.1.7).

(5.6.8) **Remarks**

1. The g's ought to be at least C^1: at the points where they are singular, either the chart must be changed, or else extended configuration space must be restricted. One may either have a global flow and be able to extend it over the whole extended phase space or not, depending on the global structure of the extended configuration space as a manifold.
2. The normalization of (5.6.6; 2) is equivalent to $\mathcal{H} = -m/2$.
3. The quantity p/m is only loosely connected with the real velocity:

$$\frac{p_\alpha}{m} = \dot{x}^\beta g_{\beta\alpha}, \qquad \dot{\mathbf{x}} = \frac{d\mathbf{x}}{dt} \cdot \frac{dt}{ds},$$

and $d\mathbf{x}/dt$ is again different from the velocity as measured with real yardsticks and clocks.

According to (2.4.14), $g_{\alpha\beta}$, a symmetric tensor of degree two, gives the extended configuration space a pseudo-Riemannian structure—where we assume that g invariably has one negative and three positive eigenvalues. The **universality** of gravitation gives a real, physical meaning to the spatial and temporal intervals defined formally with g; in Chapter 6 we shall discuss in detail how gravitation influences actual yardsticks and clocks just so that the distances and times they measure are the same as the ones coming locally from g. Put more concretely, the distance between a point (x^0, x^1, x^2, x^3) and a point $(x^0, x^1 + dx^1, x^2, x^3)$ goes as $\sqrt{g_{11}}\, dx^1$ rather than as dx^1 when $dx^1 \to 0$. In equation (5.6.2) with $\mathbf{j} = \mathbf{0}$, $g_{00} = -1 + 2M\kappa/r$ and $g_{11} = 1 + 2M\kappa/r$, and so the times and distances measured with actual clocks at this point are $\sqrt{1 - 2M\kappa/r}\, dt$ and $\sqrt{1 + 2M\kappa/r}\, dx^1$, making the limiting velocity (5.6.6; 3) again 1. However, as $r \to \infty$, dx^1 and dt approach the real length and time elements as measured out there, giving an external observer the impression that yardsticks must contract and clocks run slow if they are at small r in a gravitational potential. But there are no such things as ideal clocks and yardsticks that could directly measure dt and dx at small r, because gravity affects all objects equally. Hence it only makes sense to speak of the metric structure determined by g (not η). If $ds^2 = -dx^\alpha\, dx^\beta\, g_{\alpha\beta} > 0$, then the points x^α and $x^\alpha + dx^\alpha$ have a timelike separation and ds has the significance of a proper time—it is the interval measured by a clock that is itself moving from x^α to $x^\alpha + dx^\alpha$ in such a way that only dx^0 is nonzero in its rest frame. This is the operational meaning of the

(5.6.9) **Geodetic Form of the Equations of Motion**

Equations (1.1.6) and (1.1.7) are the Euler–Lagrange equations of the variational principle

$$W = \int ds \sqrt{-g_{\alpha\beta}\dot{x}^\alpha \dot{x}^\beta}, \qquad DW = 0.$$

Proof: Problem 3. □

(5.6.10) **Remarks**

1. To be more precise, W is determined as follows: Let u and v be two points in the extended configuration space such that there is a trajectory $x(s_0) = u$, $x(s_1) = v$, with $g_{\alpha\beta}(x(s))\dot{x}^\alpha(s)\dot{x}^\beta(s) < 0\ \forall s$ such that $s_0 \leq s \leq s_1$. For all trajectories satisfying these conditions, W is defined as the above integral, i.e., $s_1 - s_0$. The choice of s is immaterial: if $s \to \bar{s}(s)$, where $\bar{s}$ is monotonic and differentiable, then W is unchanged.
2. The previous comment shows that W is precisely the time interval that would be read off a clock that moved along $x(s)$.
3. The condition (5.6.9) that W is stationary actually requires it to be a maximum. To see this, consider the case $g_{\alpha\beta} = \eta_{\alpha\beta}$. Let $u = (0, 0, 0, 0)$, and choose the coordinate system so that $v = (t_1, 0, 0, 0)$. Then

$$W = \int_0^{t_1} dt \sqrt{1 - \left|\frac{d\mathbf{x}}{dt}\right|^2}.$$

Obviously, $0 < W \leq t_1$. The maximum is achieved by the trajectory $x(s) = (s, 0, 0, 0)$, which satisfies the Euler–Lagrange equation $\ddot{x} = 0$. The infimum 0 is not actually achieved, although it is approached arbitrarily nearly by particles moving almost as fast as light, whose proper time $s = t\sqrt{1 - |d\mathbf{x}/dt|^2}$ runs very slowly. At any point it is possible to put g in the form η by choosing the right coordinates (see (5.6.11)), and so the trajectory that satisfies the Euler–Lagrange equations locally maximizes the proper time when the points u and v are sufficiently close together. This is not necessarily so if u and v are far apart (cf. (5.7.17; 1)).
4. In mathematical terminology, the trajectories are characterized as the timelike **geodesics** in extended configuration space, given a pseudo-Riemannian structure by g.

If the coordinate system is changed, $x \to \bar{x}$, then by (2.4.36) g transforms as a tensor of degree two,

$$g_{\alpha\beta}(x) \to g_{\alpha\beta}(\bar{x}) = g_{\gamma\delta}(x)\frac{\partial x^\gamma}{\partial \bar{x}^\alpha}\frac{\partial x^\delta}{\partial \bar{x}^\beta}.$$

Hence if $g_{\alpha\beta} \neq \eta_{\alpha\beta}$, then it is possible that $g_{\alpha\beta}$ is merely $\eta_{\alpha\beta}$ on some different chart, and not a true gravitational field. Then the Γ's in the equations of motion (1.1.6) are merely fictitious forces, like the ones encountered in an accelerating reference frame. In electrodynamics we met with gauge potentials $A_i = \Lambda_{,i}$, which do not produce any fields. Now we see that there are g's that produce the equations of free motion, written in different coordinates; in the volume on field theory we shall learn of criteria for when this happens. Of course, when we talk about such a possibility we are considering g throughout the whole manifold, for at any individual point there is always the

(5.6.11) Principle of Equivalence

For all $x \in M_e$, there are coordinates, the **Riemann normal coordinates**, such that $g_{\alpha\beta}(x) = \eta_{\alpha\beta}$ and $\Gamma^\alpha_{\beta\gamma}(x) = 0$, where Γ is defined as in (1.1.7).

Proof: Problem 4. □

(5.6.12) Remarks

1. Fictitious forces counterbalance gravity in this coordinate system. This is what happens in Einstein's famous free-falling elevator. A passenger does not detect any gravity, because it affects all bodies in the elevator equally, irrespective of their masses or any other such properties.
2. In (3.2.15; 1) we eliminated a constant gravitational field by transforming to an accelerating system. The term in (5.6.3) that resembles the Lorentz force can be counterbalanced by the Coriolis force in some rotating system (3.2.15; 2).
3. Γ does not generally also vanish at nearby points, the principle of equivalence holds only in the infinitely small limit. A gradient in the gravitational field, a **tidal force**, would be detectable.

(5.6.13) Problems

1. Use (1.1.6), (1.1.7), and (5.6.2) to derive (5.6.3).
2. Calculate the Euler–Lagrange equations of L in (5.6.5).
3. Show that (1.1.6) and (1.1.7) satisfy the Euler–Lagrange equations of (5.6.9).
4. Prove (5.6.11).

(5.6.14) Solutions

1. Since the derivatives of g are of first order in κ, we can set $(g^{-1})_{\alpha\beta} = \eta_{\alpha\beta}$ to that order, and since α is a spatial index in (5.6.3),

$$\Gamma^\alpha_{\beta\gamma} = \tfrac{1}{2}(g_{\alpha\beta,\gamma} + g_{\alpha\gamma,\beta} - g_{\beta\gamma,\alpha}) = -\frac{2M\kappa}{r^3}\{(j_\alpha j_\beta + \tfrac{1}{2}\eta_{\alpha\beta})x^\gamma + (j_\alpha j_\gamma + \tfrac{1}{2}\eta_{\alpha\gamma})x^\beta - (j_\beta j_\gamma + \tfrac{1}{2}\eta_{\beta\gamma})x^\alpha\} \quad x^0 = 0.$$

In the nonrelativistic limit, we get

$$\frac{d^2x^\alpha}{dt^2} = -\Gamma^\alpha_{00} - 2\Gamma^\alpha_{0\beta}\cdot\frac{dx^\beta}{dt} = \frac{M\kappa}{r^3}\left\{-x^\alpha - 4\frac{dx^\beta}{dt}(j_\alpha x_\beta - j_\beta x_\alpha)\right\}, \quad \alpha = 1, 2, 3,$$

which is the same as (5.6.3).

2.

$$\frac{d}{ds}\frac{\partial L}{\partial \dot{x}^\alpha} = \frac{d}{ds}(m\dot{x}^\beta g_{\alpha\beta}(x)) = m\{\ddot{x}^\beta g_{\alpha\beta} + \dot{x}^\beta\dot{x}^\gamma \tfrac{1}{2}(g_{\alpha\beta,\gamma} + g_{\alpha\gamma,\beta})\}$$
$$= \frac{\partial L}{\partial x^\alpha} = \frac{m}{2}\dot{x}^\beta\dot{x}^\gamma g_{\beta\gamma,\alpha}.$$

3. In general, a function $f(L)$ gives the Euler–Lagrange equations.

$$\frac{d}{ds}\left(\frac{\partial f}{\partial L}\frac{\partial L}{\partial \dot{x}^\alpha}\right) = \frac{\partial f}{\partial L}\frac{\partial L}{\partial x^\alpha}.$$

By (5.6.8; 2) L, and hence also $\partial f/\partial L$, are independent of s for the solutions of the Euler–Lagrange equations of (5.6.5). Hence (5.6.5) and (5.6.9) are equivalent.

4. Let us first choose the point in question, x, as the origin of the coordinate system. The symmetric matrix $g_{\alpha\beta}(0)$ is orthogonally diagonable, and its eigenvalues can be renormalized to ± 1 by scaling. From the assumptions on the g's we allow, we thus find $g_{\alpha\beta}(0) = \eta_{\alpha\beta}$. The claim will be proved if we can show that $g_{\alpha\beta}(x) = \eta_{\alpha\beta} + O(x^2)$. If we expand g,

$$g_{\alpha\beta}(x) = \eta_{\alpha\beta} + A_{\alpha\beta\gamma}x^\gamma + O(x^2);$$

then a change of charts

$$x^\gamma = \bar{x}^\gamma + \tfrac{1}{2}c^\gamma_{\alpha\beta}\bar{x}^\alpha\bar{x}^\beta$$

produces a transformed matrix

$$\begin{aligned} g_{\alpha\beta} &= (\eta_{\gamma\delta} + A_{\gamma\delta\rho}x^\rho)(\delta^\gamma_\alpha + c^\gamma_{\alpha\sigma}\bar{x}^\sigma)(\delta^\delta_\beta + c^\delta_{\beta\sigma}\bar{x}^\sigma) + O(x^2) \\ &= \eta_{\alpha\beta} + x^\sigma(A_{\alpha\beta\sigma} + c_{\beta\alpha\sigma} + c_{\alpha\beta\sigma}) + O(x^2). \end{aligned}$$

If $c_{\alpha\beta\sigma} := \eta_{\alpha\tau}c^\tau_{\beta\sigma} = -\frac{1}{2}A_{\alpha\beta\sigma}$, then the linear term goes away because $A_{\alpha\beta\sigma} = A_{\beta\alpha\sigma}$. Note that the accelerating system used in (3.2.15) to get rid of gravity nonrelativistically is a special case of the above quadratic transformation.

5.7 Motion in the Schwarzschild Field

Relativistic motion in the gravitational field of a point mass is the same as in the analogous electromagnetic field as regards the structure of the invariance group. However, what goes on at small r is physically quite remarkable.

Soon after Einstein published his field equations, an exact solution describing the field of a point mass was discovered by a young physicist already marked by death.

This simple situation exhibits the essential peculiarities of the relativistic theory of gravitation, because of which it is highly significant both physically and astronomically. It is generally known as

(5.7.1) The Schwarzschild Solution

The fields $g_{\alpha\beta}$ created by a mass M at the origin provide extended configuration space with the pseudometric

$$\begin{aligned} g = dx^\alpha\, dx^\beta\, g_{\alpha\beta} = \left(1 - \frac{r_0}{r}\right)^{-1} dr^2 + r^2(d\vartheta^2 + \sin^2\vartheta\, d\varphi^2) \\ - \left(1 - \frac{r_0}{r}\right) dt^2, \end{aligned}$$

where

$$r_0 := 2M\kappa,$$

in spatial polar coordinates.

(5.7.2) **Remarks**

1. At the radius r_0 the gravitational energy of a mass is of the same order as its rest energy: $Mm\kappa/r_0 \sim mc^2$ (units with $c = 1$ were used in (5.7.1)). According to (1.1.3), $2m_p\kappa/c^2 \sim 10^{-52}$ in cgs units for a proton, and the Earth contains about 10^{51} protons, making r_0 on the order of millimeters for the Earth. The Sun is a million times heavier, and so its $r_0 \sim$ km.
2. If $r \gg r_0$, then we get (5.6.2), $g_{\alpha\beta} \to \eta_{\alpha\beta}$, and the coordinates are the intervals one would actually measure (cf. Problem 4).
3. If $r = r_0$, then $g_{00} = 0$. This does not necessarily mean that anything special happens at such a point. For example, if $r = 0$, then in polar coordinates $g_{\vartheta\vartheta} = g_{\varphi\varphi} = 0$ as well, but all that has happened is that the chart has become unsuitable at that point.
4. A singularity in g_{rr} at $r = r_0$ seems more serious than the one just cited, but it need not be so. For instance, if the coordinate x is used on the circle $x^2+y^2 = 1$, then the line element $ds^2 = dx^2 + dy^2 = dx^2/(1 - x^2)$ is singular at $x = \pm 1$. Yet these points are as good as any other points on the circle, and it is only a question of the chart failing there.
5. If instead of (t, r), the coordinates

$$u = \sqrt{\frac{r}{r_0} - 1}\,\exp\left(\frac{r}{2r_0}\right)\cosh\left(\frac{t}{2r_0}\right),$$
$$v = \sqrt{\frac{r}{r_0} - 1}\,\exp\left(\frac{r}{2r_0}\right)\sinh\left(\frac{t}{2r_0}\right)$$

are introduced, then the metric becomes

$$g = \frac{4r_0^3}{r}e^{-r/r_0}(du^2 - dv^2) + r^2(d\vartheta^2 + \sin^2\vartheta\, d\varphi^2)$$

(Problem 3), and the singularity at $r = r_0$ magically disappears. The region $\{r > r_0, -\infty < t < \infty\}$ in the old chart is mapped to $I = \{|v| < u\}$ (Figure 5.10). This gives us access to new territory, and the solution can be extended to $r = 0$. By inverting the transformation,

$$t = 2r_0 \arctan\left(\frac{v}{u}\right), \qquad \left(\frac{r}{r_0} - 1\right)\exp\left(\frac{r}{r_0}\right) = u^2 - v^2,$$

we see that the territory gained is the region where $u^2 - v^2 > -1$ (Figure 5.11). We shall later return to the physical significance of the new territory we have opened up.

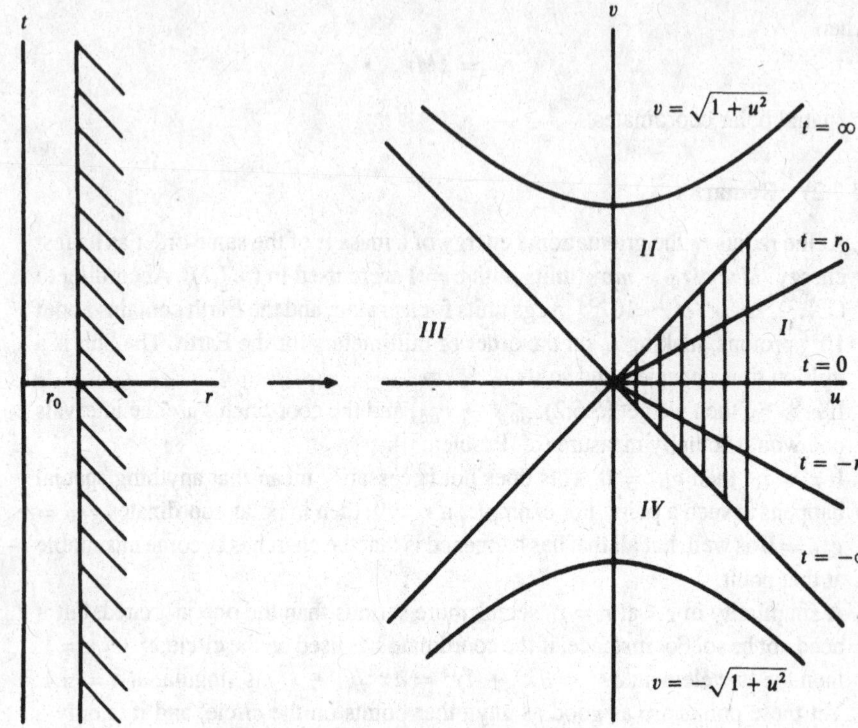

FIGURE 5.10. Extending the Schwarzschild solution.

In order to make the comparison with §4.2 and §5.3 easier, we next solve the equations of motion in coordinates $(t, r, \vartheta, \varphi)$. The coordinate t can only be used in region I of Figure 5.10. But r can also be used in region II until $r = 0$ ($v = \sqrt{u^2+1}$). Accordingly, we turn our attention to the determination of $r(s)$, $\vartheta(s)$, and $\varphi(s)$. Subtitution of (5.7.1) into (5.6.7) produces

(5.7.3) **The Hamiltonian**

$$\mathcal{H} = \frac{1}{2m}\left(|\mathbf{p}|^2 - \frac{r_0}{r}p_r^2 - \left(1 - \frac{r_0}{r}\right)^{-1} p_0^2\right).$$

Since the only coordinate used in $\mathcal{H}$ other than the momenta is r,

(5.7.4) **The Constants of the Motion**

(in s) are

$$\mathbf{L} = [\mathbf{x} \times \mathbf{p}], \qquad p_0 \qquad \text{and} \qquad \mathcal{H}.$$

These are just the same as in the electrical problem of §5.3. The construction of action and angle variables requires only a minor modification: In polar coordinates,

$$\mathcal{H} = \frac{1}{2m}\left(p_r^2\left(1 - \frac{r_0}{r}\right) + \frac{L^2}{r^2} - \frac{p_0^2}{1 - r_0/r}\right) = -\frac{m}{2}. \tag{5.7.5}$$

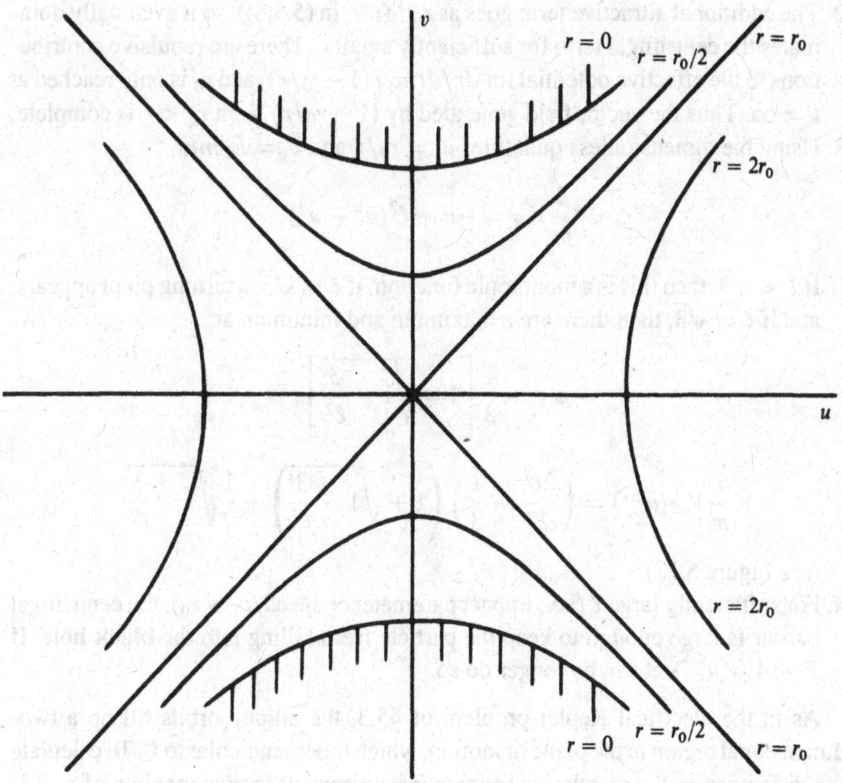

FIGURE 5.11. Regions where the complete Schwarzschild solution is valid.

Hence (cf. (5.6.8; 3))

$$p_r = \frac{m\dot{r}}{1 - r_0/r}.$$

Thus we can write

(5.7.6) $$\frac{m}{2}\dot{r}^2 - \frac{mr_0}{2r} + \frac{L^2}{2mr^2} - \frac{L^2 r_0}{2mr^3} = \frac{m}{2}\left(\frac{p_0^2}{m^2} - 1\right) =: E = \text{const.}$$

This is the equation of energy conservation in a one-dimensional system with an

(5.7.7) Effective Potential

$$V_{\text{eff}}(r) = -\frac{mK\kappa}{r} + \frac{L^2}{2mr^2} - \frac{L^2 r_0}{2mr^3}.$$

(5.7.8) Remarks

1. The first two terms are Newtonian and centrifugal potentials, as in the nonrelativistic theory (4.2.12).

2. The additional attractive term goes as r^{-3} (r^{-2} in (5.3.8)), so it eventually dominates the centrifugal term for sufficiently small r. There are repulsive contributions to the effective potential for $dr/dt = \dot{r}(1 - r_0/r)$, and r_0 is only reached at $t = \infty$. Thus the vector field generated by $(1 - r_0/r)^{-1}$ on $r_0 < r$ is complete.
3. Using the dimensionless quantities $u := r_0/r$ and $\ell := L/mr_0$,

$$\frac{2}{m} V_{\text{eff}} = -u + \ell^2(u^2 - u^3).$$

If $\ell < \sqrt{3}$, then this is a monotonic function; if $\ell = \sqrt{3}$, a turning point appears; and if $\ell > \sqrt{3}$, then there are a maximum and minimum at

$$u_{\mp} = \frac{1}{3}\left[1 \mp \sqrt{1 - \frac{3}{\ell^2}}\right],$$

$$\frac{2}{m} V_{\text{eff}}(u_{-}^{-1}) = \left(\frac{2\ell^2}{27} - \frac{1}{3}\right)\left(1 + \sqrt{1 - \frac{3}{\ell^2}}\right) + \frac{1}{9}\sqrt{1 - \frac{3}{\ell^2}}$$

(see Figure 5.12).
4. For sufficiently large ℓ (i.e., impact parameter × speed/$c \gg r_0$), the centrifugal barrier is large enough to keep the particle from falling into the **black hole**. If $E < V_{\text{eff}}(u_{-}^{-1})$, it can no longer do so.

As in the electrical Kepler problem of §5.3, the elliptic orbits fill up a two-dimensional region in the plane of motion, which is perpendicular to **L**. To calculate the difference in the angular coordinate φ between successive maxima of r, it is most convenient to start with

$$\text{(5.7.9)} \qquad \frac{dr}{d\varphi} = \frac{\dot{r}}{\dot{\varphi}} = \frac{mr^2}{L}\sqrt{\frac{p_0^2}{m^2} - 1 + \frac{r_0}{r} - \frac{L^2}{m^2r^2}\left(1 - \frac{r_0}{r}\right)},$$

from which we obtain an elliptic integral for the

(5.7.10) **Precession Angle**

$$\Delta\varphi := \oint \frac{dr\, L}{mr^2\sqrt{\frac{p_0^2}{m^2} - 1 + \frac{r_0}{r} - \frac{L^2}{m^2r^2}\left(1 - \frac{r_0}{r}\right)}} - 2\pi.$$

Expanding in the $1/r^3$ term to first order gives

$$\Delta\varphi = \frac{3\pi}{2}\left(\frac{r_0 m}{L}\right)^2 + o\left(\left(\frac{r_0 m}{L}\right)^2\right)$$

(Problem 5).

(5.7.11) **Remarks**

1. The radius of a nearly circular orbit is $R = 2L^2/r_0m^2$, and if $R \gg r_0$, then $\Delta\varphi = 3\pi r_0/R$.

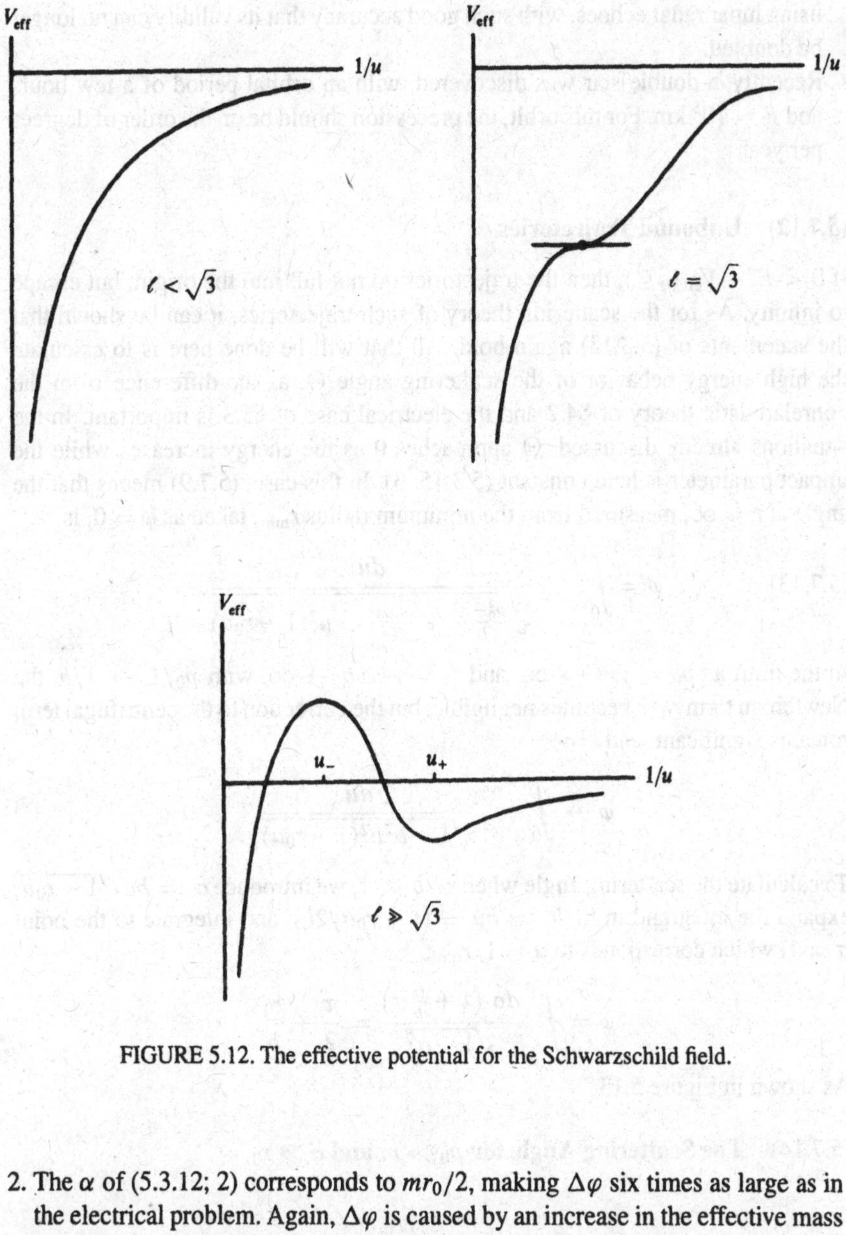

FIGURE 5.12. The effective potential for the Schwarzschild field.

2. The α of (5.3.12; 2) corresponds to $mr_0/2$, making $\Delta\varphi$ six times as large as in the electrical problem. Again, $\Delta\varphi$ is caused by an increase in the effective mass at small r. An explanation for the increase in $\Delta\varphi$ in this case can also be made in the context of Mach's principle, according to which inertia is due to nearby masses, which increase the effective mass of a particle.
3. Since $r_0/R \sim 1$ km/10^8 km for the motion of Earth around the sun, the precession is a tiny effect of a few seconds of arc per century, and much smaller than other perturbations of the orbit. However, the effect seems be to confirmed for the inner planets, after making every imaginable correction, to within one percent accuracy. The predictions of Einstein's theory have also been confirmed

using lunar radar echoes, with such good accuracy that its validity can no longer be doubted.

4. Recently, a double star was discovered, with an orbital period of a few hours and $R \sim 10^5$ km. For this orbit, the precession should be on the order of degrees per year.

(5.7.12) **Unbound Trajectories**

If $0 < E < V_{\text{eff}}(u_-^{-1})$, then the trajectories do not fall into the origin, but escape to infinity. As for the scattering theory of such trajectories, it can be shown that the statements of (5.3.13) again hold. All that will be done here is to calculate the high-energy behavior of the scattering angle Θ, as the difference from the nonrelativistic theory of §4.2 and the electrical case of §5.3 is important. In the situations already discussed, Θ approaches 0 as the energy increases while the impact parameter is held constant (5.3.15; 3). In this case, (5.7.9) means that the angle at $r = \infty$, measured from the minimum radius r_{min}, taken as $\varphi = 0$, is

(5.7.13)
$$\varphi = \int_0^{1/r_{\text{min}}} \frac{du}{\sqrt{\frac{p_0^2 - m^2}{L^2} + \frac{r_0 u m^2}{L^2} - u^2(1 - r_0 u)}}.$$

In the limit as $p_0 = \gamma m \to \infty$, and $L = \gamma m v b \to \infty$, with $p_0/L \to 1/b$, the Newtonian term r_0/r becomes negligible, but the correction to the centrifugal term remains significant, and

$$\varphi \to \int_0^{1/r_{\text{min}}} \frac{b\, du}{\sqrt{1 - b^2 u^2 (1 - r_0 u)}}.$$

To calculate the scattering angle when $r_0/b \ll 1$, we introduce $\sigma := bu\sqrt{1 - r_0 u}$, expand the integrand in r_0/b, set $bu \cong (1 + r_0\sigma/2b)$, and integrate to the point $\sigma = 1$, which corresponds to $u = 1/r_{\text{min}}$:

$$\varphi = \int_0^1 \frac{d\sigma \left(1 + \frac{r_0}{b}\sigma\right)}{\sqrt{1 - \sigma^2}} = \frac{\pi}{2} + \frac{r_0}{b}.$$

As shown in Figure 5.13.

(5.7.14) **The Scattering Angle for $p_0 \gg m$ and $b \gg r_0$**

is

$$\Theta = -\frac{2r_0}{b}.$$

(5.7.15) **Remarks**

1. The negative sign means that gravity is attractive.
2. Light with a wavelength much less that r_0 behaves like a particle with $\gamma \to \infty$ in the Schwarzschild field. Formula (5.7.14) has been verified with good accuracy for the **deflection of light** when it passes near the Sun.

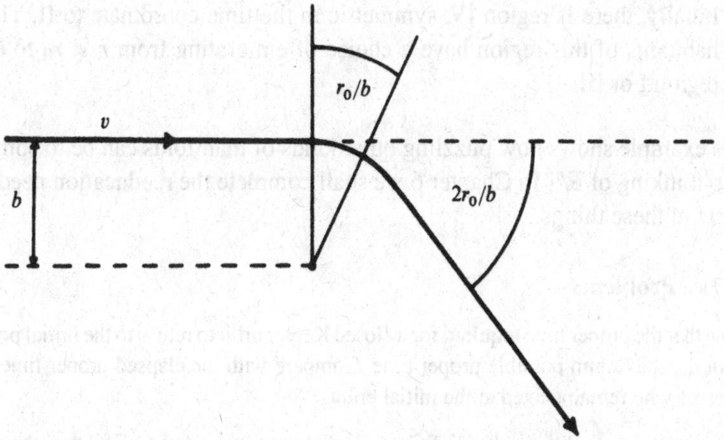

FIGURE 5.13. The impact parameter and the scattering angle.

As we have seen, a large enough angular momentum (and hence impact parameter) can keep a particle from falling into the black hole. But radial trajectories are not restricted to the region $r > r_0$, so we need to change to the variables u and v of Figure 5.11 to discuss them. If a radial line is timelike, then $|dv| > |du|$, and its slope in the diagram is necessarily steeper than 45°. No creature that was once inside r_0 could ever contrive to get to the other side of a pulse of light emitted from one of these lines. Bearing these facts in mind, let us imagine an

(5.7.16) **Expedition to $r < r_0$**

1. In region II, every trajectory, whether subject to gravity alone or in combination with an electromagnetic force, reaches $r = 0$. The electromagnetic forces cannot prevent the trajectories from being timelike lines. Thus ∂r is a timelike direction, and the fall into the center is as inevitable as aging is for us in region I. It is no more possible to stay at a fixed r inside r_0 than it is to make time stand still outside.

2. The line dividing regions I and II, $v = u > 0$, is the same as the curve $t = \infty$; this only means that signals sent by someone falling into the center appear to take an infinitely long time to an observer at $r > r_0$, which corresponds to the fact mentioned in §5.6, that clocks appear to run slow in gravitational fields. As measured by proper time, the fall is of short duration: $\sim r_0 \sim 10^{-5}$ seconds for stars. It is straightforward to figure out that light signals decrease in intensity as $\exp(-t/r_0)$, and so for practical purposes they die out immediately.

3. On the other side of r_0 there is another world symmetric to ours, region III. There is no way to know anything about it, as no trajectory can go from III to I. At best, if someone ventured into region II, he might learn about III just before it was all over.

4. Finally, there is region IV, symmetric in the time coordinate to II. The inhabitants of this region have a choice of emigrating from $r < r_0$ to either region I or III.

This example shows how puzzling other kinds of manifolds can be to someone used to thinking of $\mathbb{R}^n$. In Chapter 6 we shall complete the reeducation needed to understand these things.

(5.7.17) Problems

1. Show that the proper time required for a closed Kepler orbit to return to the initial position is not the maximum possible proper time. Compare with the elapsed proper time of an observer who remains fixed at the initial point.

2. For what $\bar{r}(r)$ does the speed of light in the radial direction equal 1—and thus the metric can be written as $F(\bar{r})(d\bar{r}^2 - dt^2) + G(\bar{r})\,d\Omega^2$?

3. With the coordinates

$$u = h(\bar{r} + t) + g(\bar{r} - t) \quad \text{and} \quad v = h(\bar{r} + t) - g(\bar{r} - t),$$

the metric of Problem 2 becomes

$$f^2(u, v)(du^2 - dv^2) + r^2(u, v)\,d\Omega^2, \qquad \text{where} \qquad f^2 = \frac{1 - r_0/r}{4h'g'}.$$

Find h and g so that the singularity at $r = r_0$ goes away, making sure that the answer is consistent with (5.7.2; 5) in the old variables.

4. Find coordinates $\bar{r}(r)$ such that the metric has the form

$$f(\bar{r})(d\bar{r}^2 + \bar{r}^2\,d\Omega^2) - g(\bar{r})\,dt^2.$$

Verify that $\bar{r} \to \infty$, $g(\bar{r}) \to 1 - r_0/\bar{r}$ and $f \to 1 + r_0/\bar{r}$, in agreement with (5.6.2).

5. Expand (5.7.10) to first order in $L^2 r_0/m^2 r^3$, and calculate the integral using complex integration as in (5.3.16; 1).

(5.7.18) Solutions

1. In the variables of (5.7.8; 3) the proper time s and the time t are related by

$$ds = dt\sqrt{\frac{1 - r_0/r}{1 + r^2\dot{\varphi}^2}} = dt\sqrt{\frac{1 - u}{1 + \ell^2 u^2}}$$

for circular orbits ($dr = 0$); whereas for the proper time s_0 of a stationary observer ($dr = d\varphi = 0$), $ds_0 = dt\sqrt{1 - u}$. If the fixed observer and an orbiting one meet again at the time t, then their proper times are in the ratio

$$\frac{s_0}{s} = \sqrt{1 + \ell^2 u^2} > 1.$$

2. $\bar{r} = \int dr/(1 - r_0/r) = r + r_0 \ln((r/r_0) - 1)$. The metric becomes

$$(1 - r_0/r)(d\bar{r}^2 - dt^2) + r^2\, d\Omega^2.$$

3. Let $\begin{Bmatrix} h \\ g \end{Bmatrix} = \frac{1}{2}\exp((\bar{r} \pm t)/2r_0)$. Then

$$f^2 = (1 - r_0/r)4r_0^2 \exp(-(r + r_0 \ln((r/r_0) - 1))/r_0) = 4r_0^3 \exp(-r/r_0)/r.$$

4. The form of the metric requires that $r^2 = (dr/d\bar{r})^2/(1 - r_0/r)\cdot\bar{r}^2$, which can be integrated to $r = \bar{r}(1 + (r_0/4\bar{r}))^2$. The metric then becomes

$$(1 + (r_0/4\bar{r}))^4(d\bar{r}^2 + \bar{r}^2\, d\Omega^2) - (1 + (r_0/4\bar{r}))^{-2}(1 - (r_0/4\bar{r}))^2\, dt^2.$$

5. It is only necessary to calculate the residue at $r = O$:

$$\oint \frac{dr}{r\sqrt{-Ar^2 + Br - C + D/r}} = \oint \frac{dr}{r\sqrt{-Ar^2 + Br - C}}$$
$$- \frac{1}{2}\oint \frac{dr\, D}{r^2(-Ar^2 + Br - C)^{3/2}} + O(D^2) = \frac{2\pi}{\sqrt{C}} + 2\pi\frac{3}{4}\frac{DB}{C^{5/2}} + O(D^2).$$

If $A = 1 - p_0^2/m^2$, $B = r_0$, $C = L^2/m^2$, and $D = r_0L^2/m^2$, this gives (5.7.10).

5.8 Motion in a Gravitational Plane Wave

The invariance group and methods of solution are the same as in the electromagnetic problem, but there are also some new aspects to consider.

Every reasonable gravitational field theory contains a counterpart to electromagnetic waves. An experiment detecting gravity waves would be one of the basic foundations of any theory of gravitation, but as yet they have not been convincingly detected. In order to observe the effects of these waves, we must first study how particles would behave in them. This chapter closes with a short discussion of the problem and a comparison with the electromagnetic counterpart of §5.5.

Shortly after Einstein wrote down the equations for $g_{\alpha\beta}$, approximate solutions were found exhibiting wave properties. But because of the nonlinearity of the problem it was not obvious that these approximations closely resembled real solutions. Right up to the present day, no-one has succeeded in pushing through the details showing how a source produces gravity waves. Nonetheless, exact plane-wave solutions of the equations can be found fairly easily. They are as follows:

(5.8.1) The Field and the Hamiltonian

The gravitational potential

$$g_{\alpha\beta} = \begin{pmatrix} -1 & 0 & 0 & 0 \\ 0 & 1 & 0 & 0 \\ 0 & 0 & \frac{1}{f(t-x)} & 0 \\ 0 & 0 & 0 & \frac{1}{g(t-x)} \end{pmatrix}$$

satisfies Einstein's equations with no matter, if $f^{1/2}(f^{-1/2})'' + g^{1/2}(g^{-1/2})'' = 0$, and by (5.6.7) it makes the Hamiltonian

$$\mathcal{H} = \frac{1}{2m}[-p_t^2 + p_x^2 + p_y^2 f(t-x) + p_z^2 g(t-x)].$$

(5.8.2) **Remarks**

1. There exist fictitious fields of the form (5.8.1), which are only the fields of free motion in an accelerating coordinate system (5.1.11). In the second volume, *Classical Field Theory*, we shall study Einstein's equations, and when we derive the above solutions we shall see that iff $(f^{-1/2})'' = (g^{-1/2})'' = 0$, then the field is necessarily fictitious (Problem 1).
2. Although f and g were not required to be continuous in §5.5, here we assume the existence of $(f^{-1/2})''$ and $(g^{-1/2})''$, so that the above condition makes sense.

If we compare $\mathcal{H}$ with (5.5.1), we see that there is the same invariance under displacements as before, which ensures that (5.8.1) is integrable. If we slightly modify the two new quantities introduced in (5.5.3), we obtain the same number of

(5.8.3) **Constants of the Motion**

$$p_y, \qquad p_z, \qquad p_x + p_t,$$

$$p_z \int_0^{t-x} du\, g(u) + z(p_t + p_x),$$

and

$$p_y \int_0^{t-x} du\, f(u) + y(p_t + p_x)$$

are constant.

Proof: Follows immediately from

$$\dot{z} = p_z g/m,$$
$$\dot{y} = p_y f/m,$$

and

$$\dot{t} - \dot{x} = -(p_t + p_x)/m. \qquad \square$$

(5.8.4) **Remarks**

1. The invariance group generated by the five constants is isomorphic to the electromagnetic invariance group (5.5.3) (Problem 4).

2. Just as before, one more constant would be needed to determine the trajectory completely. The remaining integration of the equation of motion is accomplished exactly as in (5.5.5; 3). We first deduce that

$$t(s) - x(s) = \alpha s,$$

and

$$t(s) + x(s) = t(0) + x(0) + \alpha^{-2} \int_0^{\alpha s} du \left[1 + \frac{f(u)p_y^2 + g(u)p_z^2}{m^2} \right],$$

where we have again taken $u(0) = 0$; and from this we arrive at

(5.8.5) The Explicit Solution for the Coordinates as Functions of the Proper Time

$$t(s) = t(0) + \frac{s}{2}\left(\frac{1}{\alpha} + \alpha\right) + \frac{1}{2m^2\alpha^2}[p_y^2 F(\alpha s) + p_z^2 G(\alpha s)],$$

$$x(s) = x(0) + \frac{s}{2}\left(\frac{1}{\alpha} - \alpha\right) + \frac{1}{2m^2\alpha^2}[p_y^2 F(\alpha s) + p_z^2 G(\alpha s)],$$

$$y(s) = y(0) + \frac{p_y}{m\alpha} F(s\alpha),$$

$$z(s) = z(0) + \frac{p_z}{m\alpha} G(s\alpha),$$

$$F(u) = \int_0^u du'\, f(u'), \qquad G(u) = \int_0^u du'\, g(u'),$$

$$\alpha = -\frac{p_t + p_x}{m}.$$

(5.8.6) Remark

In particular, the solution with $\alpha = 1$ and $p_y = p_z = 0$ is $(t(s), x(s), y(s), z(s)) = (s + t(0), x(0), y(0), z(0))$. This might lead one to think that gravity waves, unlike light waves, have no effect on particles that are initially at rest, and thus are impossible to detect. But remember that the coordinates (t, x, y, z) are simply not the same as actually measured intervals; on the contrary, they are precisely the coordinates adjusted for the trajectories of initially stationary particles. The displacement of particles in the y (or z) direction is actually $dy\, f^{-1/2}$ (resp. $dz\, g^{-1/2}$), which varies with $t - x$ for fixed dy and dz. Thus a gravity wave causes accelerations perpendicular to its direction of propagation.

(5.8.7) Example

The condition of (5.8.1), that $(f^{-1/2})'' f^{1/2} + (g^{-1/2})'' g^{1/2} = 0$, is fulfilled, for example, by $f^{-1/2}(u) = \cos ku$ and $g^{-1/2}(u) = \cosh ku$. In order to construct a pulse with a length τ, we can piece it together from solutions of $(f^{-1/2})'' = (g^{-1/2})'' = 0$, such that f and g are twice differentiable:

	$f^{-1/2}(u)$	$g^{-1/2}(u)$
$u < 0$	1	1
$0 < u < \tau$	$\cos ku$	$\cosh(ku)$
$\tau < u$	$\cos k\tau + k(\tau - u)\sin k\tau$	$\cosh(k\tau) - k(\tau - u)\sinh(k\tau)$

The "δ-pulse" would be a limiting case, as $\tau \to 0$ and $k = 1/\sqrt{\tau} \to \infty$, which means that:

	$f(u)$	$g(u)$	$F(u)$	$G(u)$
$u < 0$	1	1	u	u
$u > 0$	$\frac{1}{(1-u)^2}$	$\frac{1}{(1+u)^2}$	$\frac{u}{1-u}$	$\frac{u}{1+u}$

(5.8.8) **Remarks**

1. For $u := t - x > 0$, the metric is

$$g = dx^2 + (1-u)^2\,dy^2 + (1+u)^2\,dz^2 - dt^2$$

showing that the wave causes stationary bodies to be compressed in the y-direction and stretched in the z-direction; i.e., it is a quadrupole field.

2. Since g_{yy} is zero when $u = 1$, the chart (t, x, y, z) can only be used when $t - x < 1$. However, the singularity at $t - x = 1$ is only an apparent one, as by (5.8.2; 1) the g of Remark 1 is simple η written in another coordinate system. Specifically, if

$$\begin{aligned} T &= t - (1-u)y^2/2 + (1+u)z^2/2, \\ X &= x - (1-u)y^2/2 + (1+u)z^2/2, \\ &\qquad\qquad\qquad\qquad u = t - x = T - X, \\ Y &= (1-u)y, \\ Z &= (1+u)z, \end{aligned}$$

then

$$g = dX^2 + dY^2 + dZ^2 - dT^2$$

(Problem 1).

3. The curves $(x, y, z) =$ const. are the trajectories of particles initially at rest. These particles get focused in the y-direction when $u = 1$ so that the y-coordinate of their separation is zero. It sounds as if particles must be able to move faster than light, if they can be arbitrarily far apart at $u = 0$ and become focused like this, but if we look at what happens in the chart (T, X, Y, Z), which reproduces the actually measured distances, times, and velocities when $u > 0$, then in the new coordinates the trajectory $(t, x, y, z) = (s, 0, m, 0)$ becomes

$$X = -\frac{m^2/2}{1 + m^2/2}(1 - T), \qquad Y = \frac{m}{1 + m^2/2}(1 - T), \qquad Z = 0.$$

While it is true that a particle reaches the origin $\mathbf{x} = \mathbf{0}$ at time $T = 1$ for all m, it never moves faster than light:

$$\left(\frac{dX}{dT}\right)^2 + \left(\frac{dY}{dT}\right)^2 = \frac{m^2 + m^4/4}{(1 + m^2/2)^2} < 1.$$

Since the gravitational disturbance has reached a given trajectory at the time $T = -m^2/2$, particles that have started off farther away have had more time to approach the origin.

4. Remark 3 only appears to single out the x-axis to focus all particles onto. Actually, they are focused onto all trajectories $(t, x, y, z) = (s, 0, m, 0)$ for all m. Any point can of course be considered the origin, if the coordinates are displaced relative to (T, X, Y, Z).

A gravity wave excites **quadrupole oscillations** perpendicular to its direction of propagation in all objects it passes through. However, the intensities to be expected are so small that one should not be disturbed by the lack of evidence for them as yet.

(5.8.9) Problems

1. Verify (5.8.8; 2).
2. Check that $g_{\alpha\beta}\dot{x}^\alpha \dot{x}^\beta = -1$ for the solution (5.8.5).
3. Integrate $f^{1/2}(f^{-1/2})'' + g^{1/2}(g^{-1/2})'' = 0$ by making the ansatz that $f = L^{-2}\exp(2\beta)$ and $g = L^{-2}\exp(-2\beta)$.
4. Calculate the Poisson brackets of the constants (5.8.3) and compare with (5.5.14; 2).

(5.8.10) Solutions

1. With

$$U = u = T - X = t - x,$$
$$V = X + T = x + t - (1-u)y^2 + (1+u)z^2, \quad x + t = v,$$

we find

$$\begin{aligned} -dU\,dV + dY^2 + dZ^2 &= -du[dv - 2y\,dy(1-u) + 2z\,dz(1+u) + du(y^2 + z^2)] \\ &\quad + (dy(1-u) - du\,y)^2 + (dz(1+u) + du\,z)^2 \\ &= -du\,dv + dy^2(1-u)^2 + dz^2(1+u)^2. \end{aligned}$$

2.

$$\begin{aligned} \dot{x}^2 - \dot{t}^2 + \frac{\dot{y}^2}{f} + \frac{\dot{z}^2}{g} &= \frac{1}{4}\left(\frac{1}{\alpha} - \alpha + \frac{p_y^2 f + p_z^2 g}{\alpha m^2}\right)^2 \\ &\quad - \frac{1}{4}\left(\frac{1}{\alpha} + \alpha + \frac{p_y^2 f + p_z^2 g}{\alpha m^2}\right)^2 + \frac{p_y^2}{m^2} f + \frac{p_z^2}{m^2} g = -1. \end{aligned}$$

3. The ansatz leads to $L'' + (\beta')^2 L = 0$, which is solved by

$$\beta(u) = \int_0^u du' \sqrt{-L''(u')/L(u')}.$$

Then we can take L with $L''/L < 0$ and calculate β.

4. $\{p_z G(u)+z(p_0+p_x), p_z\} = p_0+p_x = \{p_y F(u)+y(p_0+p_x), p_y\}$, and the other Poisson brackets vanish. Therefore the invariance group is isomorphic to the one generated by the constants (5.5.4), and thus to the invariance group of the electromagnetic field tensor. ($G = \int g$ and $F = \int f$.)

(5.8.11) Some Unsolved Problems

1. The stability criterion (3.4.24) is practically useless, since it is difficult to prove that the limits $C_\pm$ exist. There also exist numerous other stability criteria, which, however, do not apply to Hamiltonian systems. Can they be put to any use?
2. The K–A–M Theorem (3.6.19) guarantees that there exist invariant surfaces for systems of several degrees of freedom only when the perturbations are very small. Do bigger perturbations really destroy all invariant surfaces, or is the trouble only that the method of proof is poor? For almost optimal estimates in simple cases, see R.S. MacKay and I.C. Percival, Converse K–A–M: Theory and Practice, *Commun. Math. Phys.* **98** (1985), 469–512; and A. Celletti and L. Chierchia, Construction of Analytic K–A–M Surfaces and Effective Stability Bounds, *Commun. Math. Phys.* **118** (1988), 119–161.
3. If the perturbation is large, so that the K–A–M theorem does not hold, under what circumstances is a system in fact ergodic; in other words, when is a trajectory dense in the surface of constant energy?
4. Suppose that the whole energy surface is not filled densely. When is the trajectory dense in a sufficiently representative part of the energy surface so that the time-average of an observable equals its average over the energy surface?
5. With the same supposition as in Problem 4, how long is it until the time-average is, say, within 1 per mil of its limit for infinitely long times?
6. In the N-body problem there were the following classes of trajectories:

   (a) trajectories with collisions;

   (b) trajectories where one particle escapes; and

   (c) trajectories for which all particles stay in a finite region.

   How large are the parts of phase space comprising each class, or comprising the closure of the trajectories in each class, when $N \geq 3$?
7. One thing that seems to happen in the gravitational N-body problem, $N \geq 3$, is that small clusters of tightly bound particles form. These have a tendency to become more tightly bound and to cause more loosely bound clusters to break up—which means that the specific heat is negative: the hot become hotter and the cold colder. Can such behavior be derived from the equations of motion?

8. In the relativistic case, even the two-body problem is still unsolved, because one has to worry about the infinite number of degrees of freedom of the field. The other choice, eliminating the infinite number of degrees of freedom, means that the force depends not on the position of a particle, but on its whole history. How can one cope with this mathematically? The question is not purely academic, since the relativistic two-body problem may be realistic for double stars with black holes or neutron stars.

6

The Structure of Space and Time

6.1 The Homogeneous Universe

In physics, space and time are defined by the way yardsticks and clocks behave, which in turn is determined by the equations of motion. It is this reasoning that gives a concrete significance to the mathematical structure of our formalism.

The first step toward a theory of relativity was the recognition that space and time are homogeneous. Homogeneity is expressed in the invariance of physical laws under spatial and temporal displacements, and implies that no point of the manifold is special. However, it is compatible with a structure in which certain directions are favored over others.

Mathematically, this is expressed by regarding

(6.1.1) **M_e as a Cartesian Product**

$$M_e = \mathbb{R} \times \mathbb{R} \times \mathbb{R} \times \mathbb{R}$$
$$= \text{time} \times (\text{up–down}) \times (\text{east–west}) \times (\text{north–south}).$$

In other words, canonical projections are specified onto the four coordinates. Such a situation results in a translation-invariant, not necessarily isotropic, Hamiltonian,

(6.1.2) $$H = \sum_{i=1}^{N} \varepsilon_i(\mathbf{p}_i) + \sum_{i<j} V_{ij}(\mathbf{x}_i - \mathbf{x}_j).$$

where ε_i are rotation-invariant functions of the momenta.

(6.1.3) **Remarks**

1. M_e is still understood as the extended configuration space of a single particle, and the potentials are assumed regular enough that M_e can be taken as all of $\mathbb{R}^4$.
2. If V is of the form

$$V(\mathbf{x}) = \tfrac{1}{2}\left(\sqrt{x_1^2\omega_1^2 + x_2^2\omega_2^2 + x_3^2\omega_3^2} - a\right)^2,$$

then molecules made up of such particles would oscillate at different rates in different orientations, so the directions of the coordinate axes could be determined by experiment.
3. In order to break the Galilean invariance, under $\mathbf{x} \to \mathbf{x}+\mathbf{v}t, t \to t$, the functions ε must be more complicated than $|\mathbf{p}|^2$, for example,

$$-\sum_{\alpha=1}^{3} k_\alpha^{-2}\cos(p_\alpha k_\alpha).$$

4. Hamiltonians like this are not merely to be found in science fiction, but are standard for the motion of electrons in anisotropic crystals.

If the H of (6.1.2) is not dilatation-invariant, then a variety of possible definitions of length are compatible with the fundamental laws. For example, if a molecule consisting of particles as in (6.1.3; 2) is brought to rest oriented in the i-direction, it has a natural length a/ω_i. It could be used as both a yardstick and a clock, and because of the homogeneity of M_e it could be used equally well at all places and times. It gives M_e an additional mathematical structure:

(6.1.4) **M_e as a Riemannian Space**

The Hamiltonian of (6.1.2) and Remarks (6.1.3; 2 and 3) provide M_e with the metric

$$g = dt^2\left(\frac{\omega}{2\pi}\right)^2 + \frac{dx_1^2}{a_1^2} + \frac{dx_2^2}{a_2^2} + \frac{dx_3^2}{a_3^2},$$

$$\omega = \text{one of the } \omega_i, \qquad a_i = \frac{a}{\omega_i}.$$

(6.1.5) **Remarks**

1. This g uses the units defined by the molecule of (6.1.3; 2). After a time $\delta t = 2\pi/\omega$, the molecule has completed one cycle; and, once it is in position, it defines the unit lengths $\delta x_i = a_i$ in the various directions.
2. The metric (6.1.4) is an indisputable choice only if all yardsticks and clocks behave identically. And if that is the case, then we would not say that the molecule

has different lengths in different directions, but would prefer to use scaled coordinates x_i/a_i.
3. A pseudo-Riemannian metric, $g = -dt^2(\omega/2\pi)^2 + \cdots$, could also be postulated. As long as the time axis is distinguished, this g contains the same information as the Riemannian one (6.1.4).
4. Even if H is not dilatation-invariant, it can happen that the equations of motion do not naturally define any yardsticks and clocks; invariances of the equations of motion do not necessarily correspond to invariances of H. For example, the nonrelativistic equations of motion (1.1.1) and (1.1.2) are invariant under

$$\mathbf{x} \to \lambda^2 \mathbf{x} \quad \text{and} \quad t \to \lambda^3 t, \qquad \lambda \in \mathbb{R}^+,$$

which is not related to a one-parameter group of invariances of H. That matter actually does define distances and times is a quantum mechanical fact. The universal constants e, m, and $\hbar$ can be combined to form the Bohr radius $\hbar^2/me^2$ and the Rydberg frequency $me^4/2\hbar^3$. It is only in quantum theory that the solutions of the fundamental equations with realistic forces (1.1.2) naturally define the distances and times that make the physical M_e a Riemannian manifold.

6.2 The Isotropic Universe

On the surface of the Earth the vertical direction is apparently more special than the compass directions. The isotropy that space would otherwise possess is destroyed when the direction up–down is singled out. Supposing that some scientists regarded this direction as really fundamental, they might well use the Hamiltonian

$$\mathcal{H} = \sum_i \varepsilon_i(|\mathbf{p}_i|) + \sum_{i>j} V_{ij}(|\mathbf{x}_i - \mathbf{x}_j|) + g\sum_i m_i z_i \tag{6.2.1}$$

to describe the laws of nature.

(6.2.2) **Remarks**

1. Since the kinetic and potential energies depend only on the magnitudes of vectors, there is still a rotational invariance about the z-axis. The invariance group is thus extended from $\mathbb{R} \times \mathbb{R} \times \mathbb{R} \times \mathbb{R}$, as in §6.1, to $\mathbb{R} \times \mathbb{R} \times E_2$.
2. Accordingly, M_e would now be considered as the Cartesian product $\mathbb{R} \times \mathbb{R} \times \mathbb{R}^2$ = times × (up–down) × the Earth's surface. (We draw a distinction between $\mathbb{R} \times \mathbb{R}$ and $\mathbb{R}^2$; by writing the former we presuppose projections onto the two particular axes, while in the latter no direction is preferred.)

Our hypothetical scientists would eventually realize, by the time they invented space travel, anyway, that our position relative to the center of the Earth is not really so important. They would then get rid of the g in their universal formula (6.2.1), and conclude that

$$M_e = \mathbb{R} \times \mathbb{R}^3 = \text{time} \times \text{space}. \tag{6.2.3}$$

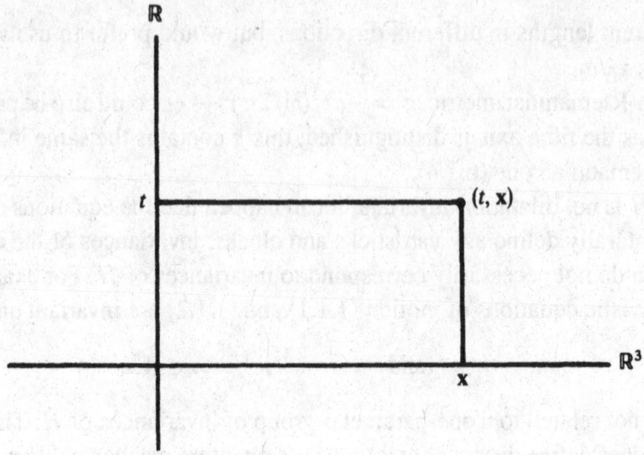

FIGURE 6.1. Canonical projections of M_e.

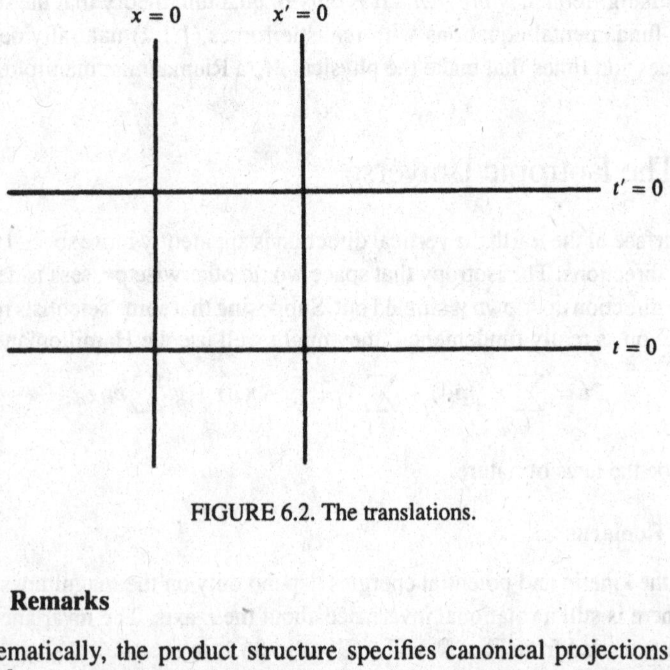

FIGURE 6.2. The translations.

(6.2.4) **Remarks**

1. Mathematically, the product structure specifies canonical projections onto $\mathbb{R}$ and $\mathbb{R}^3$, as shown in Figure 6.1. Given any two points, the question whether they were at different places at the same time, or whether they were at the same place at different times, has a unique answer. All suitable coordinate systems differ from each other only by translations and spatial rotations, which do not alter these facts (Figure 6.2).
2. The existence of the projection onto $\mathbb{R}^3$ distinguishes a system at rest from one that moves at a uniform velocity. One may then ask how bodies that are absolutely at rest can be distinguished from moving ones. For instance, in one dimension the kinetic energy of two bodies could be $-k^{-2}(\cos kp_1 + \cos kp_2)$, distinguishing a particular rest-frame. When rewritten in center-of-mass and

relative coordinates, as in (4.2.2), with $m_1 = m_2$, it becomes

$$-\frac{2}{k^2}\cos(pk)\cos\left(\frac{p_{cm}k}{2}\right).$$

Since V depends only on $x_1 - x_2$, $\mathbf{p}_{cm}$ is a constant, but

$$\dot{\mathbf{x}}_{cm} = \frac{1}{k}\sin\frac{p_{cm}k}{2}\cos pk.$$

The momentum $\mathbf{p}$ is time-dependent, and $\dot{\mathbf{x}}_{cm}$ is independent of time only if $\mathbf{p}_{cm} = \mathbf{0}$. Moving molecules do not have constant center-of-mass velocity. This objective criterion would have to be recognized even by observers moving relative to the molecules.

3. As mentioned in the previous section, a Hamiltonian of the form (6.2.1) can be used to define distances and times (possibly with the help of $\hbar$), and thereby give M_e a Riemannian structure. The sum of the scalar products in $\mathbb{R}$ and $\mathbb{R}^3$ produces a metric

$$g = v^2\,dt^2 + |d\mathbf{x}|^2 \tag{6.2.5}$$

(and the difference produces a pseudometric), where v is a ratio between distances and times defined by (6.1.4).

Now that we have a picture of space and time as they would appear to a creature raised in a crystal, we turn our attention back to the world in which we ourselves live, at first as it appears without closer examination.

6.3 M_e According to Galileo

We now consider the standard laws (1.1.1) and (1.1.2), which are invariant under the Galilean group (4.1.9). With this invariance, the words "rest" and "uniform motion" have no absolute meaning, since the coordinate systems represented in Figure 6.3 are equally valid. There is no chart-independent way to say two events (elements of M_e) happen at the same place at different times. However, simultaneity still has an absolute meaning, unaffected by Galilean transformations.

Put mathematically (see (2.2.15)), we now consider

(6.3.1) **M_e as a Vector Bundle $\mathbb{R}^4$, with $\mathbb{R}$ (Time) as its Basis, and $\mathbb{R}^3$ (Space) as a Fiber**

(6.3.2) **Remarks**

1. There is no canonical projection $\mathbb{R}^4 \to \mathbb{R}^3$, onto space, and thus $M_e \neq \mathbb{R} \times \mathbb{R}^3$. While it is true that on a chart M_e has the form of a product, it is trivializable

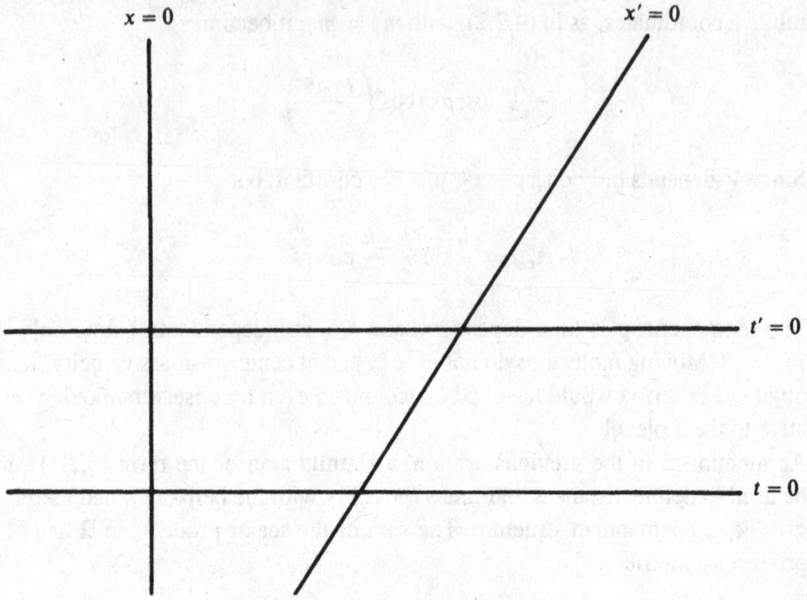

FIGURE 6.3. A Galilean transformation.

without being trivial. The physical meaning of this is that observers whose coordinate systems are in uniform relative motion can make their decomposition of M_e into space and time, but the submanifolds where $\mathbf{x} = \mathbf{0}$ determined in each system differ—though they are equally valid.

2. The fiber $\mathbb{R}^3$ has an affine structure but lacks a distinguished origin $\mathbf{x} = \mathbf{0}$. This is similar to the situation in phase space, $T^*(\mathbb{R}^3)$, where the origin $\mathbf{p} = \mathbf{0}$ of the fibers is undetermined, as it can be freely translated by Galilean transformations $\mathbf{p} \to \mathbf{p} + m\mathbf{v}$.
3. The projection onto the basis is a universal rule for synchronizing clocks. It is not difficult for all observers to agree on the rule in practice, because arbitrarily fast speeds can be attained: the submanifold $t =$ const. is uniquely characterized by the property that no trajectory can pass through two of its points, no matter how great the velocity is, a fact that would be perceived the same in all frames of reference.
4. Distances and times can be defined with the aid of $\hbar$ so as to give space and time a Riemannian structure. Since $M_e \neq \mathbb{R} \times \mathbb{R}^3$, there is no chart-independent way for M_e to inherit this structure. What M_e lacks is an orthogonal coordinate system in which the metric could be written as a sum, as in (6.2.5). Although the time interval between two points can be defined by using the bundle projection, the spatial interval between two points at different times depends on the positions of the reference frames at those times, which are not independent of the chart.

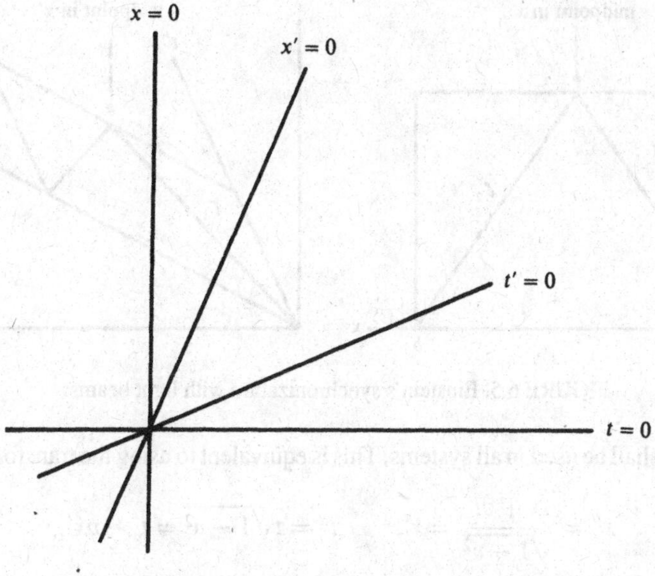

FIGURE 6.4. A Lorentz transformation.

6.4 M_e as Minkowski Space

We next investigate what structure the relativistic equations (1.1.3) give to space and time, united as **Minkowski space**. In addition, we suppose that the complete system of equations, (1.1.3) along with Maxwell's equations for F, is invariant under the Poincaré group (5.1.12). That is, everything looks the same in two reference frames after the coordinates have been transformed as in (5.1.12). The bundle structure of M_e is lost, because, compared with Figure 6.3, the special transformation (5.2.6) can be represented as shown in Figure 6.4.

(6.4.1) **Conclusion**

Simultaneity is defined in Minkowski space only with charts, and it is different with different charts.

(6.4.2) **Remarks**

1. Arbitrarily fast speeds are no longer available, as in (6.3.2; 3), to synchronize clocks with.
2. The synchronization defined in the charts (t, x) and (t', x') is the requirement that beams of light, ℓ_1 and ℓ_2, emitted at the points a and b simultaneously in one chart meet at the midpoint, and is known as **Einstein's synchronization**, as shown in Figure 6.5.
3. It would of course be possible simply to decree that the projection onto the t-axis defined by Einstein's synchronization in a certain system, the "rest frame of the

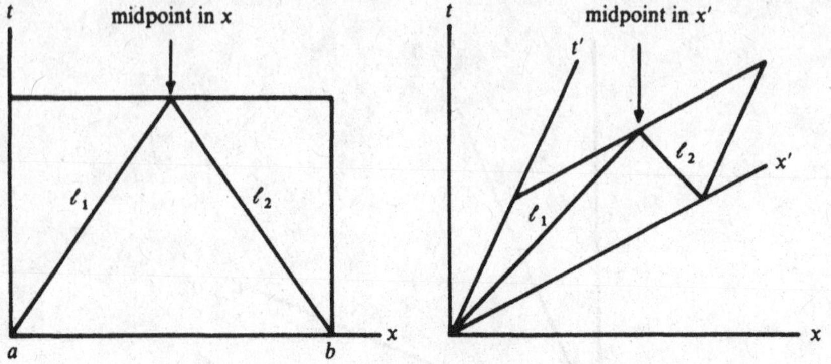

FIGURE 6.5. Einstein's synchronization with light beams.

ether," shall be used in all systems. This is equivalent to using the transformation,

$$x'' = \frac{x + vt}{\sqrt{1 - v^2}} = x', \qquad t'' = t\sqrt{1 - v^2} = t' - vx',$$

in which case $t = 0$ coincides with $t'' = 0$. The advantage of x'' and t'' is that they represent distances and times as measured with real meter sticks and clocks, but the reasons not to use them are:

(a) Nature has no preference for any system.

(b) The users of (t'', x'') would feel put out for having to use such complicated equations.

(c) These transformations do not form a group.

Yet this chart cannot be excluded on purely logical grounds; even the most peculiar kinds of coordinates are allowed by our definition of a manifold. In this way it would be possible to save the notion of absolute simultaneity, but no one is willing to make the necessary sacrifices any more; philosophical principles are not as persuasive as mathematical elegance.

We have remarked in (6.3.2; 1) that if there is no canonical projection to $\mathbb{R}^3$, then spatial intervals are necessarily chart-dependent, unless the points involved are simultaneous. Since now even simultaneity is chart-dependent, so are all spatial and temporal intervals—just as the length of a body in the 1-direction lost its absolute meaning in the transition from §6.1 to §6.2. But what has happened now is more unusual, and requires a more detailed discussion.

The first point to make clear is that the coordinates (t', x') reflect times and distances as actually measured just as much as (t, x) do: A yardstick of unit length, with one end at $(t, x) = (t, 0)$ and the other at $(t, x) = (t, 1)$, corresponds to a certain solution ℓ of the system of equations. The Lorentz-invariance of the equations ensures that there exists a Lorentz-transformed solution ℓ' with ends at $(t', x') = (t', 0)$ and $(t', 1)$. In other words, if the unit yardstick is moving, it is still

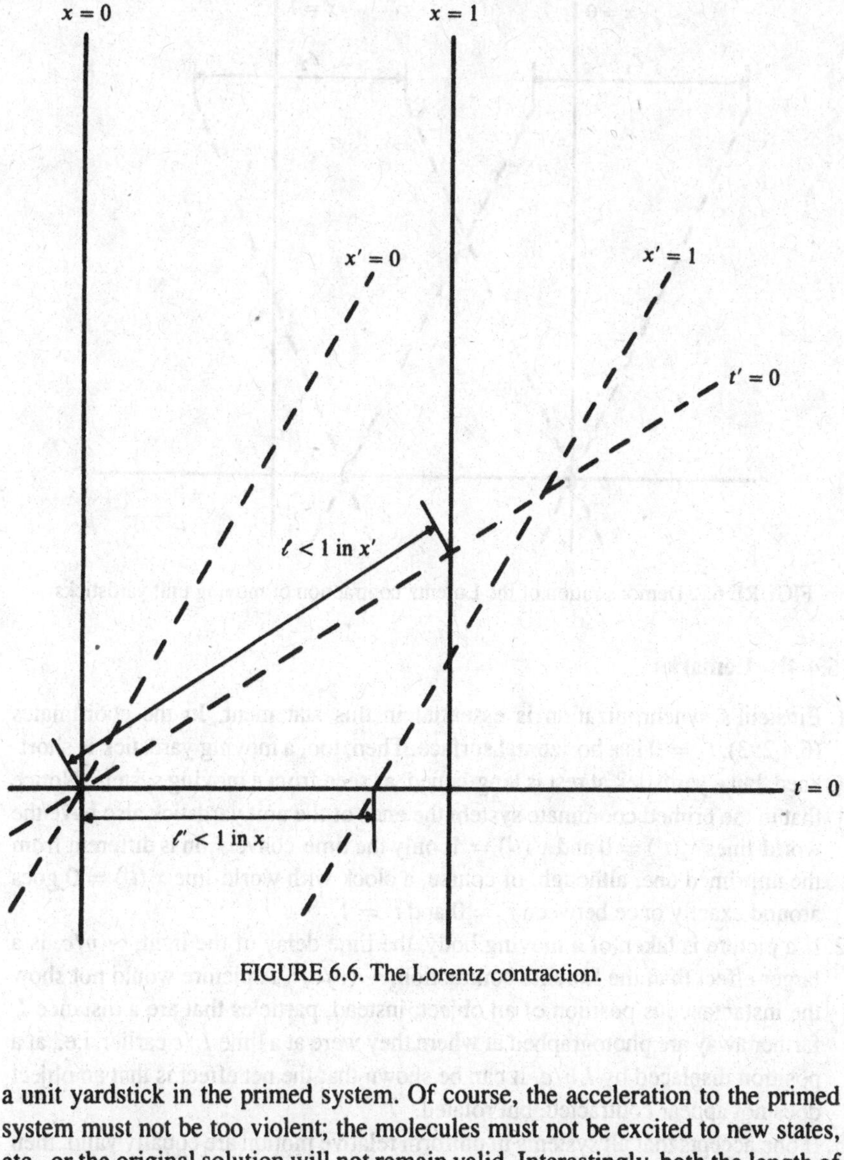

FIGURE 6.6. The Lorentz contraction.

a unit yardstick in the primed system. Of course, the acceleration to the primed system must not be too violent; the molecules must not be excited to new states, etc., or the original solution will not remain valid. Interestingly, both the length of ℓ' in the unprimed system and of ℓ in the primed one are equal to $\sqrt{1 - v^2} < 1$. The apparent contradiction is resolved if one notes that the length in each system is the distance between the ends at the same time, but "at the same time" means something different in the two systems (Figure 6.6).

(6.4.3) Conclusion: The Lorentz Contraction

In Minkowski space, moving bodies are contracted by the factor $\sqrt{1 - v^2}$ in the direction of motion.

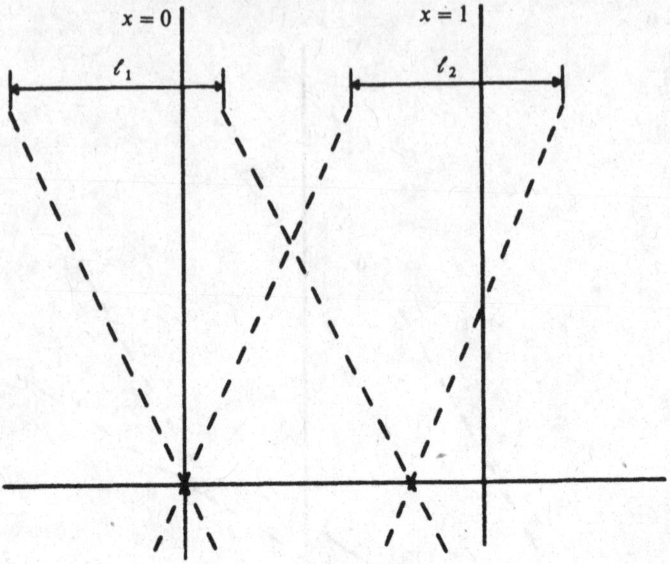

FIGURE 6.7. Demonstration of the Lorentz contraction of moving unit yardsticks.

(6.4.4) **Remarks**

1. Einstein's synchronization is essential in this statement. In the coordinates (6.4.2; 3), $t'' = 0$ is a horizontal surface. Then, too, a moving yardstick is shortened, but a yardstick at rest is lengthened, as seen from a moving system. Notice that in the primed coordinate system the ends of the unit yardstick also have the world-lines $x'(t') = 0$ and $x'(t') = 1$; only the time-convention is different from the unprimed one, although, of course, a clock with world-line $x'(t') = 0$ goes around exactly once between $t' = 0$ and $t' = 1$.
2. If a picture is taken of a moving body, the time-delay of the light, $\sim v/c$, is a larger effect than the Lorentz contraction, $\sim (v/c)^2$. A picture would not show the instantaneous position of an object; instead, particles that are a distance L farther away are photographed at where they were at a time L/c earlier, i.e., at a position displaced by Lv/c. It can be shown that the net effect is that an object does not appear contracted, but rotated.
3. If one accepts that all systems in uniform relative motion are equally valid, then the Lorentz contraction can be demonstrated without reference to the synchronization of clocks. Imagine that two identical yardsticks ℓ_1 and ℓ_2 are sent by each other with equal but opposite velocities, and that the positions of their ends are marked at the instant that they coincide. From symmetry, one concludes that both ends get marked at the same time (Figure 6.7).

The situation concerning the length of time between two events is similar. A periodic solution of the equations of motion can be used as a clock, and the Lorentz-transformed solution would describe a moving clock. If the two clocks are objects with the trajectories $x = 0$ and, respectively, $x' = 0$, and periods τ and τ', then what happens is depicted in Figure 6.8.

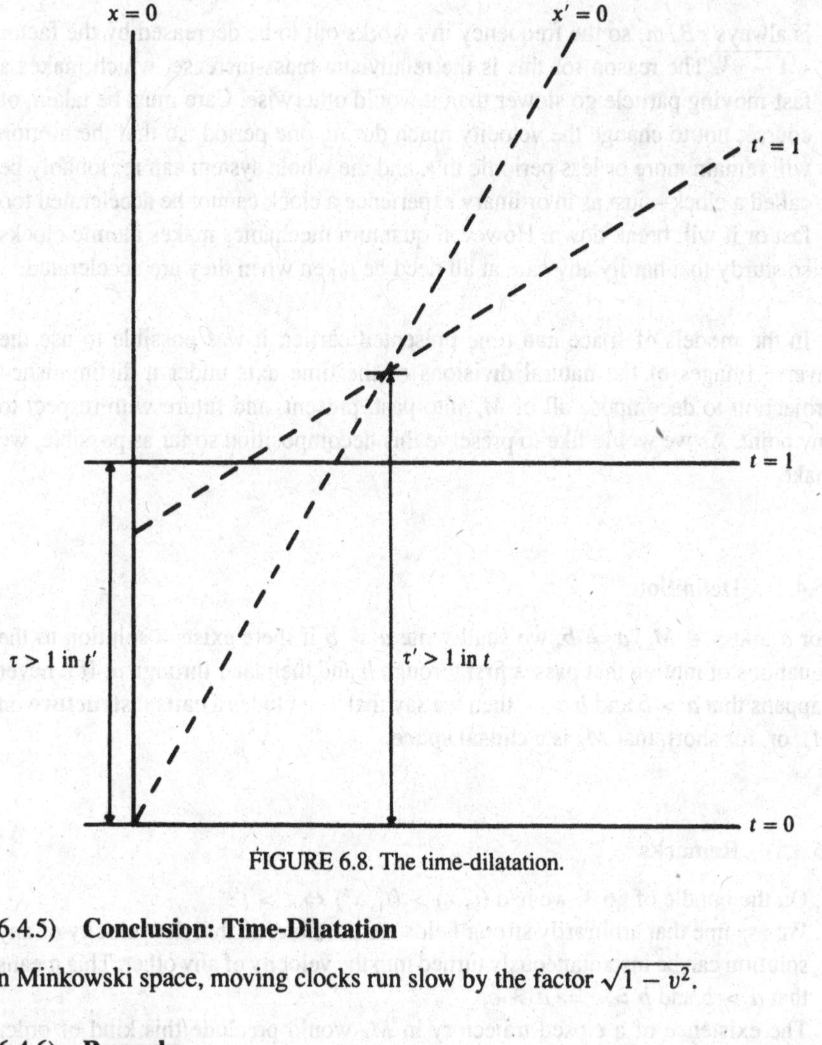

FIGURE 6.8. The time-dilatation.

(6.4.5) Conclusion: Time-Dilatation

In Minkowski space, moving clocks run slow by the factor $\sqrt{1 - v^2}$.

(6.4.6) Remarks

1. In the coordinate system (6.4.2; 3), moving clocks also run slow, but stationary clocks as seen from a moving frame of reference run fast. The apparent contradiction vanishes in the coordinates (t'', x''), too.
2. The slowing of time for moving bodies can be demonstrated without reference to synchronization, for instance for rotational motion. Fast-moving muons in a storage ring live much longer than stationary ones. In fact, time-dilatations by a factor of 100 or more have been observed in storage rings; but it is also possible to measure the miniscule amount of retardation during an airplane's flight around the Earth.
3. One might wonder whether a given clock, accelerated to some velocity, would actually run slow in consequence of the equations of motion. Let us take the Larmor motion as a model of a clock, and accelerate the particle with an electrical field parallel to **B**. We have already calculated in §4.2 that the frequency in s

is always eB/m, so the frequency in t works out to be decreased by the factor $\sqrt{1-v^2}$. The reason for this is the relativistic mass-increase, which makes a fast-moving particle go slower than it would otherwise. Care must be taken, of course, not to change the velocity much during one period, so that the motion will remain more or less periodic in t, and the whole system can reasonably be called a clock—just as in ordinary experience a clock cannot be accelerated too fast or it will break down. However, quantum mechanics makes atomic clocks so sturdy that hardly any care at all need be taken when they are accelerated.

In the models of space and time presented earlier, it was possible to use the inverse images of the natural divisions of the time axis under a distinguished projection to decompose all of M_e into past, present, and future with respect to any point. As we would like to preserve this decomposition so far as possible, we make

(6.4.7) **Definition**

For a and $b \in M_e$, $a \neq b$, we shall write $a > b$ if there exists a solution to the equations of motion that passes first through b and then later through a. If it never happens that $a > b$ and $b > a$, then we say that $>$ includes a **causal structure** on M_e, or, for short, that M_e is a **causal space**.

(6.4.8) **Remarks**

1. On the bundle of §6.3, we had $(t, x) > (t', x') \Leftrightarrow t > t'$.
2. We assume that arbitrarily strong fields are possible, so that the velocity of any solution can be instantaneously turned into the velocity of any other. This means that $a > b$ and $b > c \Rightarrow a > c$.
3. The existence of a closed trajectory in M_e would preclude this kind of order relation. K. Gödel constructed a solution of Einstein's equations for the metric of space–time for which there exist closed, timelike geodesics. Their existence is thus compatible with the laws of nature as we know them. The nature of time in such a manifold is radically different from that in Minkowski space. In the latter an absolute time can be defined by (6.4.2; 3), but this is not possible in the Gödel universe.
4. The past with respect to a is $\{b \in M_e : a > b\}$, the future is $\{b \in M_e : b > a\}$, and the rest of M_e could be referred to as the present. What goes on at a can influence only its future, and be influenced only by its past.

The causal relationships on a more general manifold can be rather strange; there might, for example, be some point in the present for which no trajectory can pass through it and ever reach the future (cf. (6.5.5; 2)). The state of affairs is fortunately more clear-cut in

(6.4.9) **Minkowski Space as a Causal Space**

In Minkowski space,

$$(t, x, y, z) > (t', x', y', z') \quad \Leftrightarrow$$
$$t > t' \quad \text{and} \quad (t - t')^2 > (x - x')^2 + (y - y')^2 + (z - z')^2.$$

(6.4.10) **Remarks**

1. In all the models through §6.3, the present was $\{(t', x', y', z'): t' = t\}$, a submanifold of lower dimensionality. In Minkowski space its interior is a four-dimensional submanifold.
2. Unlike before, it is now possible to have two trajectories such that no point of either trajectory lies in the future (or the past) of any point of the other trajectory. An example would be the two hyperbolic trajectories

$$\mathrm{I} = \{(\sinh s, \cosh s, 0, 0): s \in \mathbb{R}\},$$

and

$$\mathrm{II} = \{(\sinh s, -\cosh s, 0, 0): s \in \mathbb{R}\},$$

of particles responding to the electric field $\mathbf{E} \cong (x_1/|x_1|, 0, 0)$ coming from a surface charge (Figure 6.9). Observers on these trajectories can never see each other directly, although people in between could know about both of them.
3. The causal structure defines a topology, in which the open sets are unions of $U_{ac} := \{b \in M_e: a < b < c\}$, where a and c are any points of M_e. This topology is identical to the ordinary one.
4. Zeeman has proved a surprising theorem: every bijection $f: \mathbb{R}^4 \to \mathbb{R}^4$ that preserves the causal structure of (6.4.9) (i.e., $x > y \Leftrightarrow f(x) > f(y)$) is the product of a Poincaré transformation and a dilatation.

6.5 M_e as a Pseudo-Riemannian Space

Our final task is to take the influence of gravitation on space and time into account. The principle of equivalence (5.6.11) states that suitable coordinates can be used at any point to make $g_{\alpha\beta} = \eta_{\alpha\beta}$ and all the derivatives of $g_{\alpha\beta}$ zero. In other words, gravity is not detectable at a single point, and from the discussion of §6.4 we conclude that in these coordinates

(6.5.1) $$g = dx_1^2 + dx_2^2 + dx_3^2 - dt^2$$

reproduces distances and times as actually measured, supposing that the measurement is confined to an infinitesimal neighborhood of the point of space–time. Other coordinates could obviously be used, in which the actual metric of (6.5.1) is transformed from $\eta_{\alpha\beta}$ to $g_{\alpha\beta}$ at the point in question. At any rate, we can consider

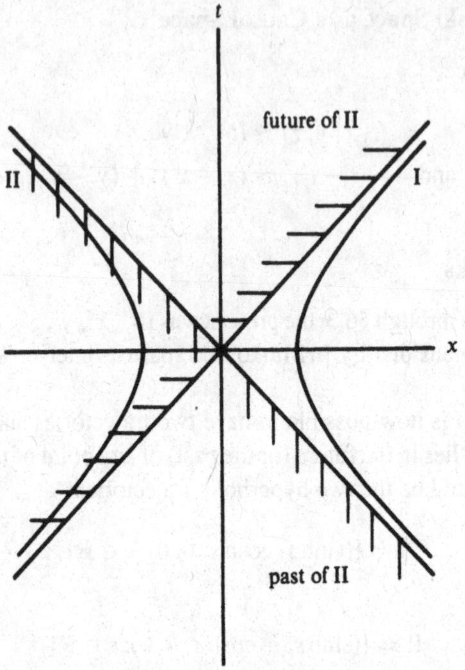

FIGURE 6.9. The future and past for hyperbolic trajectories.

(6.5.2) M_e as a Space with a Pseudo-Riemannian Metric

In a gravitational field, $g_{\alpha\beta}\, dx^\alpha\, dx^\beta$ gives the metric as actually observed.

(6.5.3) Remarks

1. This interpretation depends essentially on the universality of gravitation. If different particles acted differently in a gravitational field, then no single transformation could make the field vanish at a given point for all the particles. Universality also requires that Maxwell's equations in the coordinate system (6.5.1) have the same form as in the absence of gravity. If that is the case, then the forces which are at work on the yardsticks and clocks are also unaffected by gravity, and everything works as if there were no gravity. In mathematical terminology, the tangent space is a Minkowski space.
2. The attempt to interpret other fields geometrically has always failed from the lack of similar universal characteristics. If measuring instruments that have been built differently are affected differently by a field, then a geometrical interpretation is not convincing.
3. From time to time someone comes up with a "theory of gravitation in flat space." The established equations (1.1.6) and (1.1.7) are typically used, but in some coordinate system $\eta_{\alpha\beta}\, dx^\alpha\, dx^\beta$, rather than $g_{\alpha\beta}\, dx^\alpha\, dx^\beta$, is interpreted as the metric. Then it is explained that yardsticks and clocks fail to measure this metric because they are influenced by the gravitational potential. But Nature does

not select a special coordinate system, and real yardsticks and clocks in fact measure $g_{\alpha\beta}\, dx^\alpha\, dx^\beta$. Although logically possible, the attempt to save "absolute Minkowski space" is as artificial as the attempt in (6.4.2; 3) to choose an "absolute rest frame" arbitrarily.

4. Although the $g_{\alpha\beta}$ appear in the equations of motion, they are no more observable than the electrical potentials. Their influence can be regarded as that of a universal scaling transformation, which cannot be perceived locally.
5. The conformal group might turn out to be more fundamental than the Poincaré group, and our units of length would basically be accidental. The Riemannian structure of space would then be a mere fleeting apparition, contingent on the scope of our current understanding. A causal structure, however, would remain.

One occasionally encounters the notion that one would need gravitational theory to be able to use an accelerating coordinate system. Actually, it goes the other way around. An accelerating reference system can always be used to reduce the situation in a gravitational field locally to that of Minkowski space. Differences from §6.4 occur only for nonlocal phenomena.

(6.5.4) **Alterations of the Geometry by Gravitation**

1. Although a universal dilatation could not be perceived infinitesimally, if the $g_{\alpha\beta}$ depend on x, then yardsticks at different places will have different lengths, and if they could be laid next to each other, the difference could be measured. Similarly, any differences between atomic oscillations used as clocks can be detected by a comparison of the frequencies of light emitted at different points. If we consider the Schwarzschild metric (5.7.1), which is time-independent, then electromagnetic waves are solutions of Maxwell's equations proportional to $\exp(i\omega t)$. The frequency in t is a constant throughout space, and consequently the frequency in s is different at different points (this is the origin of the **gravitational red-shift**). Modern experimental techniques can measure the red-shift from the change in the Earth's gravitational potential due to a difference in altitude of only a few meters.
2. Variable $g_{\alpha\beta}$ of course destroy the large-scale Euclidean geometry of space. In particular, Pythagoras's theorem would no longer hold, as the sides of a triangle would be measured by yardsticks contracted by different amounts. These effects, however, are mainly of theoretical interest, since space on the Earth is flat to within the precision of our measuring instruments; the only properties we can measure across greater distances are the direction, frequency, and intensity of light rays.
3. Previously, we always considered M_e as $\mathbb{R}^4$ with some additional structure, and thus $T(M_e)$ was always a Cartesian product—which we described by saying that M_e was parallelizable. That is, given two vectors at different points, it was possible to say whether they were parallel. In Minkowski space this meant that four-velocities at different points could be compared, in practice by using light beams. Two observers have the same velocity if and

only if they do not appear Doppler-shifted to each other. But if the manifold determined by g is not parallelizable, then there is no way to say whether two bodies are at rest with respect to each other unless they are at the same point. Light signals are no longer useful for determining whether vectors are parallel, because they are affected by gravity.

(6.5.5) Alteration of the Causal Structure

1. There exist manifolds that are not causal. For example, let

$$M_e = \{\mathbf{x} \in \mathbb{R}^5 : x_1^2 + x_2^2 + x_3^2 - x_0^2 - x_4^2 = -r^2\},$$

and induce a metric on it from $\mathbb{R}^5$ with $g = dx_1^2 + dx_2^2 + dx_3^2 - dx_0^2 - dx_4^2$. Then $(x_0, x_1, x_2, x_3, x_4) = r(\cos s, 0, 0, 0, \sin s)$ is a closed timelike geodesic, which, according to (6.4.8; 3), precludes a causal structure.

2. If M_e is Minkowski space with some parts removed, then there may exist two particles with disjoint futures. For example, in the Schwarzschild field there can be observers with trajectories

$$\mathrm{I} = \{(u, v, \vartheta, \varphi) = (1 + \cosh s, \sinh s, 0, 0), s \in \mathbb{R}\}$$

and

$$\mathrm{II} = \{(u, v, \vartheta, \varphi) = (-1 - \cosh s, \sinh s, 0, 0), s \in \mathbb{R}\}.$$

Unlike what we saw in (6.4.10; 2), no observer in between can have seen both of them (Figure 6.10).

Our discussion has shown how different laws of nature imprint their structure on the space–time manifold. It is clear that it would be presumptuous to try to state what the true essence of space and time is. The most we can discover is that facet of the essence which is reflected in our present knowledge of the laws of nature. Space–time takes on different features, depending on what we choose to regard as fundamental or as accidental.

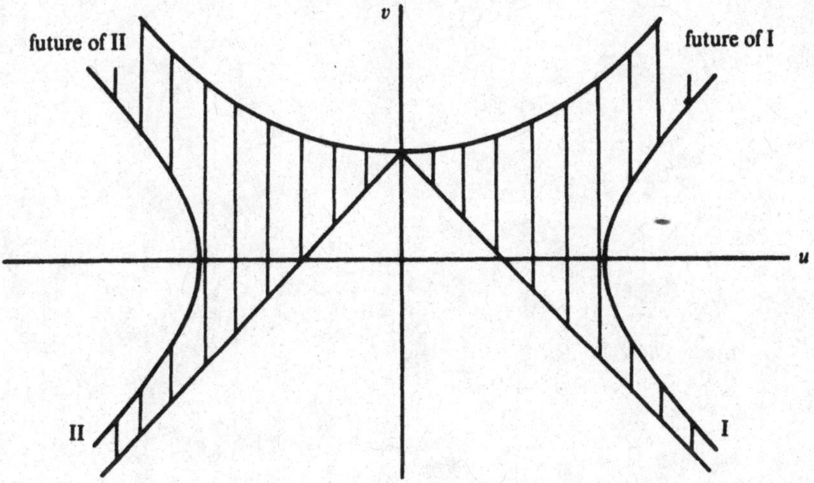

FIGURE 6.10. The futures of two uniformly accelerated observers.

Part II

Classical Field Theory

7

Introduction to Classical Field Theory

7.1 Physical Aspects of Field Dynamics

Electric and magnetic fields are dynamically interconnected in such a way that an electromagnetic disturbance propagates with a universal velocity in empty space. By studying this phenomenon we gain a qualitative understanding of field radiation and are led to expect analogous gravitational behavior.

The unification of the theories of electric and magnetic phenomena was one of the great scientific events of the nineteenth century. Whereas stationary electric fields $\mathbf{E}$ have sources at the positions of the charges but are irrotational ($\boldsymbol{\nabla} \times \mathbf{E} = \mathbf{0}$), changing magnetic fields produce circulating electromotive forces. In contrast, magnetic fields $\mathbf{B}$ are always sourceless and circulate around currents and places where there is a time-dependent electric field. The dynamical interrelation of the two fields is described by Maxwell's equations: If we consider empty space (no sources or currents), then in units where $c = 1$ they require that

$$\text{(7.1.1)} \qquad \oint_{\partial N} d\mathbf{s} \cdot \mathbf{E} = -\int_N d\mathbf{S} \cdot \dot{\mathbf{B}}, \qquad \oint_{\partial N} d\mathbf{s} \cdot \mathbf{B} = \int_N d\mathbf{S} \cdot \dot{\mathbf{E}},$$

for integrals over arbitrary surfaces N with boundaries ∂N; and if the surface is closed, then

$$\text{(7.1.2)} \qquad \oint_N d\mathbf{S} \cdot \mathbf{E} = \oint_N d\mathbf{S} \cdot \mathbf{B} = 0.$$

We shall later recognize these apparently independent relationships as different aspects of a single fact, that the field-strength form and its dual form are closed.

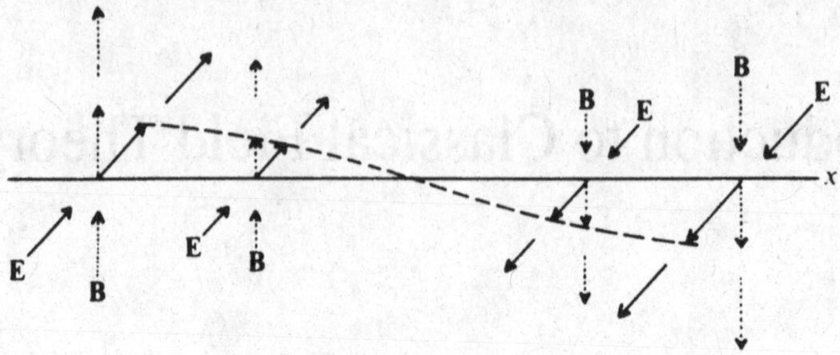

FIGURE 7.1. The fields in a plane wave.

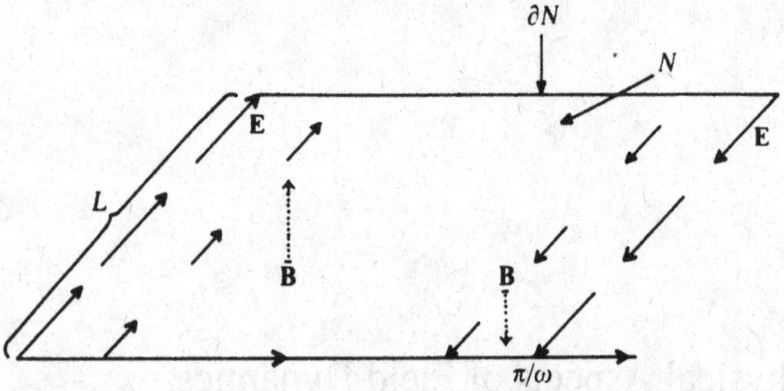

FIGURE 7.2. Illustrating the integral form of Maxwell's equations.

Before going more fully into this geometrical interpretation, let us try to come to an intuitive understanding of the physical consequences of these equations.

(7.1.3) Electromagnetic Waves

At a fixed time, a field

(7.1.4) $$E_y = B_z = \cos(\omega(x - t)), \quad \text{all other components } 0,$$

looks as shown in Figure 7.1.

This is obviously free of sources, and it satisfies (7.1.1), as is easily seen on the surface chosen in Figure 7.2. Since the wave in Figure 7.1 moves to the right, $\dot{E}$ and $\dot{B}$ have the same sign as E and B in that region, and so for $t = 0$,

$$\oint_{\partial N} \mathbf{E} \cdot d\mathbf{s} = -2L = -L\omega \int_0^{\pi/\omega} dx \, \sin \omega x = -\int_N \dot{\mathbf{B}} \cdot dS.$$

We see that unlike in the stationary situation, where the electric field of a point source falls off as $1/r^2$, it is dynamically possible for it to travel through space at the speed of light, normalized here to one, without decaying, that is, without the pulse losing intensity.

It can also be seen from the relationships (7.1.1) and (7.1.2), though less directly, that any change in the fields propagates at the speed of light. We shall later

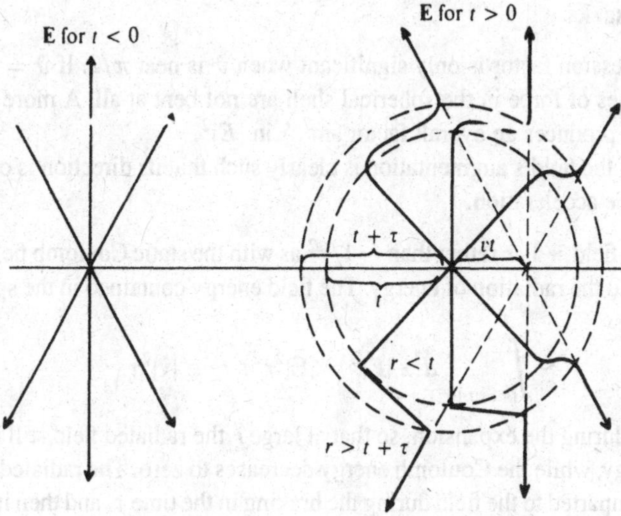

FIGURE 7.3. The field of bremsstrahlung.

study how this follows from the structure of the characteristics of the equivalent differential equation. For the moment let us take this property as given, and use it to investigate how an accelerated charge shakes off some of its Coulomb field and emits it as radiation.

(7.1.5) The Production of Electromagnetic Radiation

A charge e moving with a constant velocity $\mathbf{v}$ in the x-direction emits no radiation (by Lorentz invariance). However, if it is brought to a stop at the origin during the time $-\tau < t < 0$, then its Coulomb field at some time $t > 0$ looks as follows: At distances $r > t+\tau$ from the origin it is equal to the field of the moving charge, since those parts of space have not yet learned of the braking of the particle. The lines of force out there point at the spot $\mathbf{x} \simeq \mathbf{v}t$ rather than at $\mathbf{x} = \mathbf{0}$, where the charge remains as $t > 0$. Thus, the lines of force are displaced by $\mathbf{v}t \sim \dot{\mathbf{v}}\tau t$ compared with the field at $r < t$. At $r < t$ the field is that of a charge at rest at $\mathbf{x} = \mathbf{0}$, as the field has already forgotten that the charge ever moved. In between, in the spherical shell $t < r < t + \tau$, the lines of force progress continuously and without sources. Hence they must bend, and, as shown in Figure 7.3, as r increases they must get folded more closely together in the parts of the spherical shell that are at most at right angles to $\mathbf{v}$. This causes the field strength to increase by a factor of $r\dot{v}$: The increase in the density of the field lines is proportional to (displacement of the field lines)/(thickness of the spherical shell) $\sim \dot{v}\tau t/\tau \sim r\dot{v}$, because at $r = t \gg \tau$ one would see the field of a charge at $x = t\mathbf{v} \sim t\tau\dot{\mathbf{v}}$, and thus the lines of force are displaced by this amount. The field in the spherical shell is consequently not $|\mathbf{E}| = e/r^2$, but

$$(7.1.6) \qquad |\mathbf{E}| \sim \frac{e\dot{v}}{r}, \quad \text{for } t < r < t + \tau \quad \text{and} \quad \vartheta := \measuredangle(x, v) \sim \frac{\pi}{2}.$$

(7.1.7) **Remarks**

1. The compression factor is only significant when ϑ is near $\pi/2$. If $\vartheta = 0$ or π, then the lines of force in the spherical shell are not bent at all. A more careful calculation produces an overall factor $\sin^2\vartheta$ in $|E|^2$.
2. The sign of the field's augmentation is clearly such that its direction is opposite to that of the acceleration.

An electric field $\sim 1/r$ rather than $\sim 1/r^2$ as with the static Coulomb field leads immediately to the radiation of energy. The field energy contained in the spherical shell,

$$\sim \int_{t<r<t+\tau} d^3x\,|\mathbf{E}|^2 \sim |\mathbf{E}|^2 r^2 \tau \sim e^2 |\dot{\mathbf{v}}|^2 \tau$$

is conserved during the expansion, so that at large r the radiated field still has just as much energy, while the Coulomb energy decreases to zero. The radiated energy is evidently imparted to the field during the braking in the time τ, and then it travels off to infinity at the speed of light within the spherical shell $t < r < t + \tau$. In this way we obtain the basic formula of radiation.

(7.1.8) **Larmor's Formula**

$$\text{Radiated energy per unit time} = \frac{2}{3}\frac{e^2}{4\pi}\dot{v}^2.$$

(7.1.9) **Remarks**

1. In order to get a feel for the order of magnitude of the radiated energy, it is most convenient to use the fundamental length of electrodynamics, the classical radius of an electron, i.e., the radius at which the Coulombic potential energy equals the rest energy of an electron: $r_c = e^2/4\pi mc^2 \sim 10^{-13}$ cm. (As we set $c = 1$ this is equivalent to about 10^{-23} s.) If an electron at almost the speed of light is stopped in a time $\tau \sim r_c$ over a distance $b \sim \tau$, then $\dot{v} \sim v/\tau \sim 1/\tau \sim 1/b$, so the braking releases a radiation energy $\sim \tau e^2 \dot{v}^2 \sim e^2/b$, which is roughly the rest energy of an electron ($\sim \frac{1}{2}$ MeV) if $b \sim r_c$. Modern accelerators have made such breakneck occurrences commonplace.
2. Electrons are most easily braked with an electric field **E**, and it is natural to ask what the connection between the radiated energy and the energy of the field **E** is. Because $m\dot{\mathbf{v}} = e\mathbf{E}$, an electron radiates $E^2(e^2/m)^2\tau$, which is the fraction of the energy of the field **E** contained in a volume $\tau \times$ (the classical electron radius)^2. If the electron is not subjected to a single braking, but is moved periodically, as in a light wave, then we are interested in the cross-section, defined as (the radiated energy per unit time)/(the incident energy per unit time and surface area). Since the energy density of a light wave is E^2, the energy incident in a time τ and a unit of surface area is $E^2\tau$, and the scattering cross-section is about r_c^2. This means that the electron is about 10^{-13} cm across, in the sense that it blocks a surface area $r_c^2 \sim 10^{-26}$ cm^2 from a beam of light. However,

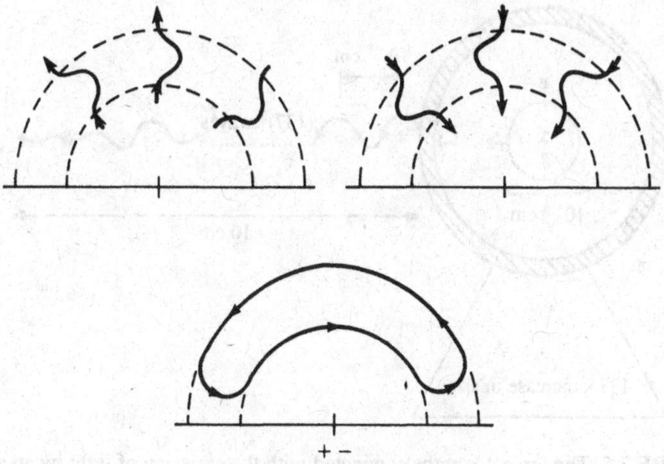

FIGURE 7.4. The electric field of an oscillating dipole.

quantum effects often make an electron act more as if it were 10^{-11} cm across. An electron is best pictured as about 10^{-11} cm across but fairly transparent, so that it scatters light only weakly. The explanation for why matter is so often opaque, though the field is predominantly influenced by the rather transparent electrons it contains, will be discussed later.

3. The precise numerical factor given in (7.1.8) comes about because:

(a) We use units in which the Coulomb field is $e/4\pi r^2$, giving an extra $(4\pi)^{-2}$.

(b) According to (7.1.7; 1) the energy has an angular distribution $\sim \sin^2 \vartheta$. Integrated over a spherical surface, this gives $\frac{2}{3} \cdot 4\pi$.

(c) The energy density is actually $(|\mathbf{E}|^2 + |\mathbf{B}|^2)/2$, but the contribution from **B** equals that from **E**.

The source of light is ordinarily atoms, in which negative charges orbit positive ones, while the total charge is neutral. If a single charge oscillates, then the pattern of field lines produced is that of a repetition of one-time brakings (Figure 7.4, left). If two charges oscillate around each other, then this field must be superposed on one that is oppositely directed and has a phase-lag (Figure 7.4). As we see, oscillating dipoles emit circulating electric fields, in which $|\mathbf{E}|$ is again given by (7.1.6). If ω is the frequency and L the amplitude of oscillation, then $\omega^2 L$ replaces $\dot{v}$ in (7.1.6). This produces the formula for

(7.1.10) **Dipole Radiation**

$$\text{Radiated energy per cycle} \sim e^2 L^2 \omega^3.$$

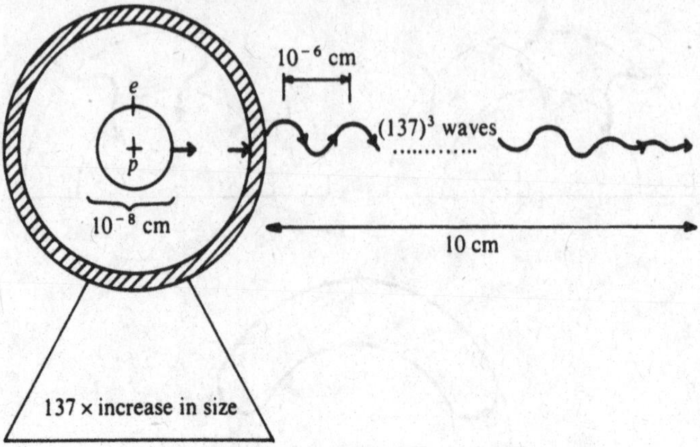

FIGURE 7.5. The typical lengths connected with the emission of light by an atom.

(7.1.11) **Application to Atoms**

For an atom, one would set L equal to the Bohr radius $r_b := \hbar^2/me^2 = (\hbar c/e^2)^2 r_c = (137)^2 r_c \sim 10^{-8}$ cm. The electron velocity ωL is roughly $(137)^{-1}$, so the period is $\sim (137) r_b \sim (137)^3 r_c \sim 10^{-15}$ s. Then from (7.1.10) the energy loss per cycle is $\sim (137)^{-3} e^2/r_b$, and an electron has to orbit $(137)^3$ times to radiate an energy e^2/r_b. The available energies of excitation are on the order of $e^2/r_b \sim 10\,\text{eV}$, so we expect the lifetime of an excited state to be about $(137)^3 \cdot 10^{-15}$ s $\sim 10^{-8}$ s. Of course, a more exact analysis of what goes on requires a quantum-theoretical analysis of the process, but the only quantum-mechanical quantity needed for a preliminary orientation is r_b. The creation of light may be typically outlined in this way: An atom emits $(137)^3$ waves in 10^{-8} s. Since the wavelength is about $1/\omega \sim 137 r_b$, the resultant wave-packet is $(137)^4 r_b \sim 10$ cm long (Figure 7.5). These figures also show up in the widths of the emitted spectral lines and in the coherence length of the radiation.

The original estimate of bremsstrahlung in (7.1.8) used only the electric field's form $\sim \mathbf{x}/r^3$. It is tempting to reason the same way with gravitation, replacing e^2 with the formally analogous quantity κm_p^2. Before spending too much time on the details of this analogy, one would likely note that the numbers involved are discouragingly large. As discussed in (1.1.1), the coupling constants differ by a factor of 10^{36}. Whereas an atom takes 10^{-8} s to emit a photon, it would require $10^{36} \cdot 10^{-8}$ s $\sim 10^{28}$ s to bring a graviton into being. This is 10^{10} times the age of the Universe, and it seems idle to speculate more about such questions. But masses, unlike charges, all have the same sign, and therefore large bodies benefit from tremendous coherence effects. We shall shortly see that collapsing stars ought to release gigantic energies through gravitational radiation.

(7.1.12) **The Production of Gravitational Radiation**

The points to reconsider in the derivation of (7.1.5) are:

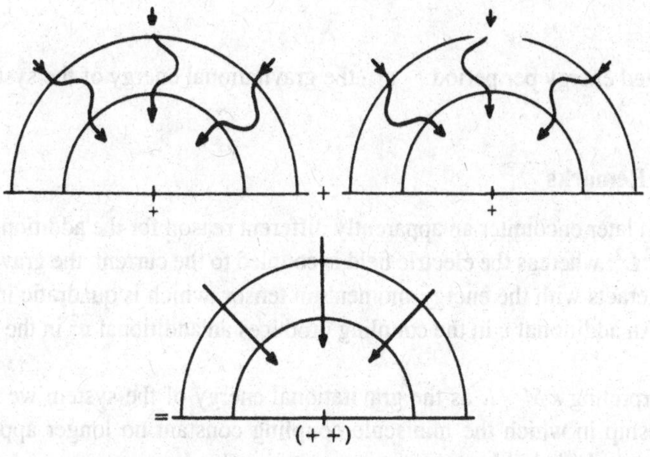

FIGURE 7.6. The gravitational field of two oscillating masses.

(a) One of the facts used was that $\mathbf{E} = e\mathbf{x}/r^3$. The potential energy e^2/r of two charges e corresponds to the gravitational energy $\kappa M^2/r$ of two masses M. Hence the analogue of the Coulomb field is $\sqrt{\kappa}M\mathbf{x}/r^3$, the square of which is the energy density.

(b) The second fact used was that the electric field propagates at the speed of light. We shall later learn that this is also a property of the gravitational field, as a consequence of Einstein's equations. However, the gravitational field in turn influences the speed of light, which complicates the details of the radiation problem, though the orders of magnitude should not be affected.

(c) In the electromagnetic case the center of charge was accelerated, but the analogy fails at this point. Since the gravitational field is coupled to all masses, and the center of mass moves uniformly, this kind of radiation never occurs. This is apparent, for instance, if two masses oscillate around each other, for which two equal fields with a phase shift must be superposed, but this time with the same sign. Then in directions perpendicular to the oscillation, the large component in the direction of $\dot{\mathbf{v}}$ cancels out (see Figure 7.6), leaving a field $\sim 1/r^2$.

(d) There can be electromagnetic quadrupole radiation even with a stationary center of charge; and in the situation of (c), the asymptotic fields do not quite cancel out at 45°. A more careful calculation shows that they are only reduced by a factor $v \sim \omega L$. The same thing is true for gravitational waves, leading to

(7.1.13) **Graviational Radiation of Rotating Masses**

$$\text{Radiated energy per unit time} \sim \kappa M^2 \omega^6 L^4 \sim \frac{\kappa M^2}{L} v^5 \omega,$$

or

Radiated energy per period $\sim v^5 \cdot$ the gravitational energy of the system.

(7.1.14) **Remarks**

1. We shall later encounter an apparently different reason for the additional factor $v^2 = \omega^2 L^2$: whereas the electric field is coupled to the current, the gravitational field interacts with the energy-momentum tensor, which is quadratic in the velocity. An additional v in the coupling produces an additional v^2 in the radiated energy.
2. By interpreting $\kappa M^2/L$ as the gravitational energy of the system we obtain a relationship in which the miniscule coupling constant no longer appears, or, more precisely, is hidden within the gravitational energy.
3. We have not taken the trouble to puzzle out the exact numerical factor in (7.1.13). Moreover, it only applies in the context of the linear approximation to the Einsteinian theory. No exact solutions, showing how gravitational waves are created, have yet been found.

In §1.1 we have discussed the qualitative features of cosmic phenomena. Let us use these rough numbers to calculate

(7.1.15) **The Order of Magnitude of Gravitational Radiation**

(a) Planets: Using Earth as an example, $v \sim 10^{-4}$, giving us a factor 10^{-20} in (7.1.13). That means that in Earth's 10^{10}-year long history it has lost only one 10^{10}th of its potential energy through radiation, causing it to draw closer to the Sun by 10^{-10} times the radius of its orbit, which is $\sim 10^{-10} \cdot 10^8$ km $\sim$ 10 m.

(b) Double stars: For double stars $v \sim 10^{-3}$, and the orbital period is often only 10^{-3} years. Each year they radiate 10^{-12} of their gravitational energy which was too little to be measured until quite recently.

(c) Black holes: If a star collapses under its own weight, then its contents are compressed into nuclear matter, and it turns into a neutron star with a radius of perhaps 10 km (cf. §10.5). If further compressed, it produces such a strong gravitational field that the gravitational energy is comparable to the rest energy, and relativistic effects appear. The gravitational analogue of the classical radius of an electron is the Schwarzschild radius κM, which is on the order of a kilometer for stars. Kepler orbits at this distance become relativistic, as the potential energy approaches the rest energy. The implosion of a neutron star can lead to a situation where $v \sim 1$ and $\kappa M^2/L \sim M$, and then, by (7.1.13), the radiated energy becomes equal to the rest energy of a star. Even if a more exact calculation would show that the true figure was a few percent, the energies involved are certainly huge.

Unfortunately, not much is known about gravitational radiation experimentally. It is predicted by all sensible theories, however, since our derivation has only posited a finite propagation speed for the gravitational field.

In all of the above discussion the motion of the sources of the fields has been assumed given. In fact the fields also affect the sources, and one really ought to solve the equations for the coupled system. We encountered a simple example of such a problem in the scattering of light, where an electron was accelerated by a light wave, which in turn caused the light impinging on a surface $\sim r_c^2$ to be scattered. That derivation, though, was for a free electron, and certainly does not hold for electrons in matter. Such a small cross-section would render normal matter, with interatomic distances $\sim (137)^2 r_c$, highly transparent. The reason this is not the case is that bound electrons exhibit resonance behavior, increasing $\dot{v}$ compared with (7.1.9; 2).

(7.1.16) The Scattering of Light by Bound Electrons

The electrons sit in the electric field $\mathbf{E}$ of the binding force as well as the light wave. Idealizing the binding as a harmonic force with strength $m\omega_0^2$, making the frequency of atomic electron oscillations ω_0, would lead one to expect $m(\ddot{\mathbf{x}} + \omega_0^2 \mathbf{x}) = e\mathbf{E}$ for the equation of motion of the electrons. A periodic electric field then gives rise to an acceleration

$$\dot{\mathbf{v}} = \ddot{\mathbf{x}} \sim \frac{eE}{m}\left(1 - \frac{\omega_0^2}{\omega^2}\right)^{-1},$$

which increases the scattering cross-section by $(1 - \omega_0^2/\omega^2)^{-2}$. Since the frequency ω of visible light, which is emitted by other atoms, may be near ω_0, the resonance denominator can be very small, and light will not penetrate far. Of course, realistic matter has several different resonance frequencies, which is why it is colorful.

Because of the situation just sketched, it is tempting to simplify the coupled system of matter and an electromagnetic field by replacing the matter with the boundary condition that the field does not penetrate it. This is the approach usually taken in optical problems like diffraction. We shall find that even with this simplification such problems can only be solved with difficulty, if any value is placed on precision.

If one wants to undertake a serious analysis of the coupled system of matter plus field, then it must be clearly understood that the dangers of the electric and gravitational cases lie in opposite quarters: Since like charges repel, but all masses attract, the Coulomb force tends to make a system explode, while gravity tends to make it collapse. These tendencies of the static forces are preserved in the relativistic generalizations of the equations, and cannot be abolished by any mathematical sleight of hand. We shall not be able to answer the question of what holds the electronic charge together, and classical electrodynamics will remain an incomplete theory. It is replaced by quantum electrodynamics at short distances, of course, but even that theory cannot explain why the electrical energy of a point charge is finite. On the other hand, the most refined mathematics only confirms the naive expectation that, with gravity, a sufficiently massive object collapses under its own weight. It seems exceedingly difficult to get around these singularity theorems; at the very least, we can say that the prognosis of disaster is likely to be true.

7.2 The Mathematical Formalism

*The central concepts of classical field theory are the algebra $\{E_p\}$ of forms and their exterior derivatives. With the *-mapping $E_p \to E_{m-p}$ defined by the metric, they allow Maxwell's and Einstein's equations to be written in analogous ways.*

The calculus of É. Cartan is especially useful in classical field theory, because it formalizes the mathematical constructions used in the field equations. The most important rules of calculation will be briefly summarized here; the reader is referred to Chapter 2 or to the mathematics books cited in the Bibliography for more precise definitions of the concepts.

The space of tensor fields over a manifold M[1] has a linear structure;[2] every element can be written as a linear combination of certain basis elements, where the coefficients are the tensor components. In the example of the covariant vector fields, any m (= the dimension of M here and henceforth) linearly independent vector fields $e^i(x)$, $i = 1, \ldots, m$, $x \in M$, can be used; any vector field V can be written as

$$V = \sum_{i=1}^{m} V_i(x) e^i(x).$$

(7.2.1) **Remarks**

1. Independence means that the $e^i(x)$ are independent vectors for all x.
2. A basis is occasionally called an m-frame, or tetrad for $m = 4$. For any i, e^i is a vector, and not a component of a vector, or any such thing.
3. There does not generally exist a global basis, as a linearly independent set of e^i cannot normally be continuously extended to all of M. (According to (2.6.17; 6), on the surface of a sphere there does not exist any continuous, nowhere vanishing vector field.) However, on the domain U of a chart there is always a natural basis, as discussed below. Therefore, for local processes like differentiation there is no danger in assuming the existence of a basis.
4. A new basis can always be formed from an old one by $e^i(x) \to \bar{e}^i(x) = A^i{}_j(x) e^j(x)$, where $\det A^i{}_j(x)$ is different from zero for all x. No basis is distinguished from the others in the absence of more structure.

---

[1]As in the first volume, a sufficiently differentiable manifold with a boundary will be referred to simply as a manifold and differentiability will likewise be implicitly assumed for tensor fields.

[2]More precisely, a module structure. The coefficients are functions of M and not simply real numbers.

With the formation of the tensor product $\otimes$, the e^i also provide a basis for the p-fold contravariant tensor fields. On U any tensor field can be written

$$t = \sum_{j_k} t_{j_1 \cdots j_p}(x) e^{j_1} \otimes e^{j_2} \otimes \cdots \otimes e^{j_p}. \tag{7.2.2}$$

Among tensor fields, the totally antisymmetric ones ($\equiv$ p-forms) are especially important, because they define p-dimensional volume elements. An antisymmetric tensor product $\wedge$ (the wedge, or exterior, product) can be introduced on them. Like the (finite) tensor product, it is associative and distributive, but $e^i \wedge e^j = -e^j \wedge e^i := e^i \otimes e^j - e^j \otimes e^i$. Since we shall frequently use the product bases, we introduce the abbreviation

$$e^{i_1 i_2 \cdots i_p} := e^{i_1} \wedge e^{i_2} \wedge \cdots \wedge e^{i_p}. \tag{7.2.3}$$

With (7.2.3) any p-form can be written

$$\omega = \sum_{i_k} \omega_{i_1 \cdots i_p} \frac{e^{i_1 \cdots i_p}}{p!}. \tag{7.2.4}$$

(7.2.5) **Remarks**

1. The basis $e^{i_1 \cdots i_p}$ has only $\binom{m}{p}$ independent elements, due to antisymmetry. If $p = 0$, then p-forms are ordinary functions, and if $p > m$ they are defined as zero, as otherwise antisymmetry would be impossible.
2. The p-forms on a manifold are a linear space (in fact a module), denoted by $E_p(M)$.
3. The wedge product

$$e^{i_1 \cdots i_p} \wedge e^{i_{p+1} \cdots i_{p+q}} = e^{i_1 \cdots i_{p+q}} = (-1)^{pq} e^{i_{p+1} \cdots i_{p+q}} \wedge e^{i_1 \cdots i_p}$$

creates an obviously associative mapping $E_p \times E_q \rightarrow E_{p+q}$, and imparts the structure of a (graded) algebra to the space $\bigcup_p E_p$ of forms.

(7.2.6) **The Exterior Differential**

The elementary differential operations, gradient, divergence, and curl, generalize to a linear mapping $d\colon E_p(M) \to E_{p+1}(M)$, obeying the rules:

(a) $d(\omega_1 + \omega_2) = d\omega_1 + d\omega_2$, $\omega_i \in E_p$;

(b) $d(\omega_1 \wedge \omega_2) = (d\omega_1) \wedge \omega_2 + (-1)^p \omega_1 \wedge d\omega_2$, $\omega_1 \in E_p$, $\omega_2 \in E_q$; and

(c) $d(d\omega) = 0$, $\omega \in E_p$;

for all $p, q = 0, 1, \ldots, m$.

(7.2.7) **Remarks**

1. Rules (a) and (b) describe how d acts with respect to the algebraic operations, by which Leibniz's rule gets an additional $(-1)^p$ from antisymmetry.
2. The coordinates x^i, $i = 1, \dots, m$, of a point of M may be thought of as a mapping $U \to \mathbb{R}$, $M \supset U =$ the domain of the chart. That is, they are an element of $E_0(U)$. The dx^i are thus m 1-forms, known as the **natural basis** of $E_1(U)$. If $p = 0$, then (a) and (b) are the usual rules of differentiation, so the exterior differential of a function $U \to \mathbb{R}: x \to f(x)$ becomes $df = dx^i\, \partial f/\partial x^i$. If $p = 0$, then d is the gradient, and $f_{,i}$ are its components in the natural basis. Because of (c), in the natural basis
$$de^{j_1 \cdots j_p} = 0, \qquad 0 \le p \le m,$$
and the exterior differential of a p-form
$$\omega = \omega_{j_1 \cdots j_p} \frac{e^{j_1 \cdots j_p}}{p!}$$
(with the summation convention) becomes
$$d\omega = (d\omega_{j_1 \cdots j_p}) \wedge \frac{e^{j_1 \cdots j_p}}{p!} = \omega_{j_1 \cdots j_p, k} \frac{e^{k j_1 \cdots j_p}}{p!}.$$
3. Local coordinates can be introduced on any n-dimensional submanifold N so that $x_{n+1} = x_{n+2} = \cdots = x_m = 0$. The **restriction** $\omega_{|N}$ of a form is defined by setting $dx_{n+1|N} = dx_{n+2|N} = \cdots = dx_{m|N} = 0$, and letting the restriction commute with $+$ and $\wedge$. It is then easy to see that $d\omega_{|N} = d(\omega_{|N})$.
4. Forms with vanishing exterior differentials are called **closed**, and forms that can be written as exterior differentials are called **exact**. Rule (c) means that exact $\Rightarrow$ closed. The opposite implication holds on starlike manifolds but not in general. It always holds locally.

(7.2.8) **The Integral**

Under a coordinate transformation
$$x \to \bar{x}(x), \qquad dx^i = \frac{\partial x^i}{\partial \bar{x}^j} d\bar{x}^j,$$
the natural basis of $E_m(U)$ transforms as an m-dimensional volume element:
$$e^{1 \cdots m} = \det\left(\frac{\partial x^i}{\partial \bar{x}^j}\right) \bar{e}^{1 \cdots m},$$
and it is possible to define a coordinate-independent integral over ω by

(7.2.9)
$$\int_U \omega := \int dx^1 \cdots dx^m\, \omega_{1 \cdots m}(x),$$
$$\omega \in E_m^0(U) := \text{the } m\text{-forms with compact support.}$$

As $\omega \in E_m^0(M)$ can be written as a finite sum $\sum_i \omega_i$ of m-forms each supported in the domain of a chart, its integral is defined by (7.2.9) and $\int \sum_i \omega_i = \sum_i \int \omega_i$. More generally, the integral of $\omega \in E_n^0(M)$ over an n-dimensional submanifold N is defined as the integral of the restriction of ω to N:

(7.2.10) $$\int_N \omega := \int_N \omega_{|N}.$$

Integration is the inverse of differentiation d in the sense that Gauss's and Stokes's theorems generalize to

(7.2.11) $$\int_N d\omega = \int_{\partial N} \omega, \qquad \omega \in E_{n-1}^0(M),$$

where ∂N is the boundary of N.[3]

(7.2.12) **Riemannian Structure**

A symmetric, covariant tensor field of the second degree

(7.2.13) $$g = g_{ik} e^i \otimes e^k, \qquad \text{with} \quad \det(g_{ik}(x)) \neq 0 \quad \forall x$$

creates an isomorphism between the covariant and contravariant vector fields and lets the spaces of the two types of vector fields be identified. A scalar product is defined by

(7.2.14) $$\langle e^i(x) | e^k(x) \rangle = g^{ik}(x), \qquad g^{ij} g_{jk} = \delta^i{}_k.$$

If the matrix g_{ik} is positive, then the space is said to be **Riemannian**, and otherwise **pseudo-Riemannian**. The $e_i := g_{ik} e^k$ form a dual basis, with $\langle e^j | e_i \rangle = \delta^j{}_i$; and $v^k := g^{ki} v_i$ are called contravariant components of a vector $v = v_i e^i = v^i e_i$.

The scalar product as map $E_1 \times E_2 \to E_0$ can be generalized in two ways.

(7.2.15) **The Scalar Product Between p-Forms**

The pseudo-Riemannian structure also induces a bilinear map $E_p \times E_p \overset{\langle|\rangle}{\to} E_0$. Expressed in components it reads

$$\langle \omega | \nu \rangle = \frac{1}{p!} \sum_{(i)(j)} \omega_{j_1 \cdots j_p} \nu_{i_1 \cdots i_p} g^{j_1 i_1} \cdots g^{j_p i_p} = \langle \nu | \omega \rangle.$$

In a Riemannian space $\langle \omega | \omega \rangle^{1/2}$ measures the p-dimensional volume defined by ω, if it is the product of 1-forms.

(7.2.16) **The Interior Product of a q-Form and a p-Form**

A bilinear map $E_p \times E_q \to E_{p-q}$, $p \geq q$: $(\omega, \nu) \to i_\nu \omega$ can be defined by the rules:

---

[3]We shall occasionally use this theorem when ∂N is not actually a manifold, but has corners or some other harmless handicap.

(i) $i_\nu \omega = \langle \omega | \nu \rangle$ for $p = q$;

(ii) $i_\nu(\omega_1 \wedge \omega_2) = (i_\nu \omega_1) \wedge \omega_2 + (-1)^{p_1} \omega_1 \wedge i_\nu \omega_2$ for $\nu \in E_1, \omega_i \in E_{p_i}$; and

(iii) $i_{\nu_1 \wedge \nu_2} = i_{\nu_2} \circ i_{\nu_1}$.

By $\langle | \rangle$ the E_p are identified with their dual spaces and the interior product is the transpose of the exterior product:

$$\langle \nu \wedge \omega | \mu \rangle = i_{\nu \wedge \omega} \mu = i_\omega (i_\nu \mu) = \langle \omega | i_\nu \mu \rangle.$$

In an m-dimensional manifold the space E_m is one-dimensional and a pseudo-Riemannian structure distinguishes the $\varepsilon \in E_m$ which are normalized: $\langle \varepsilon | \varepsilon \rangle = (-1)^s = \det g / |\det g|$. If such an ε exists globally the manifold is called **orientable**. We shall assume from now on that there is globally given an orientation ε. In this case E_p can be identified with E_{m-p}.

(7.2.17) **The Hodge Duality Map**

The isomorphism $E_p \xrightarrow{*} E_{m-p}$ is defined by ${}^*\omega = i_\omega \varepsilon$. It has the properties:

(i) $\varepsilon = {}^*1, {}^*\varepsilon = (-1)^s$;

(ii) $* \circ * = (-1)^{p(m-p)+s}$;

(iii) $i_\nu {}^*\omega = {}^*(\omega \wedge \nu) = (-1)^{pq} i_\omega {}^*\nu$ for $\omega \in E_p, \nu \in E_q$; and

(iv) for $\nu, \omega \in E_p$,

$$\nu \wedge {}^*\omega = \varepsilon i_\nu \omega = \omega \wedge {}^*\nu = \varepsilon (-1)^s i_{*\nu} {}^*\omega i^* \nu^* \omega.$$

One should notice:

ad (i). E_0 and E_m are one-dimensional and, at a point, isomorphic to R. ε is the $*$-image of the number 1.

ad (ii). Except for a sign the map $*$ is its own inverse. Unfortunately, one cannot get rid of these signs by using another ε.

ad (iii). Duality changes the interior product into the exterior product.

ad (iv). The word duality has its root in the fact that E_{m-p} is the dual space for E_p if we define a scalar product $\{\ ,\ \}$ by $\nu \wedge \omega = \varepsilon\{\nu, \omega\}$. It is related to i by

$$\{\nu, \omega\} = (-1)^{p(m-p)+s} i_\nu {}^*\omega.$$

So far the scalar product and other algebraic operations have been understood pointwise, that is $\langle \omega | \nu \rangle \in E_0 := C(M)$ assigns a number to each point at a manifold. In an orientable Riemannian manifold $\int {}^*\langle \omega | \nu \rangle$ maps $E_p \times E_p$ into R and is a scalar product in the sense of a Hilbert space.

In terms of a basis the definitions imply the

(7.2.18) **Rules**

(a) $^*e^{i_1\cdots i_p} = g^{i_1 j_1}\cdots g^{i_p j_p} e^{j_{p+1}\cdots j_m}\varepsilon_{j_1\cdots j_m}\frac{\sqrt{|\det g|}}{(m-p)!}$;

(b) $e_j \wedge {}^*e^{j_1\cdots j_p} = \sum_{r=1}^{p}(-1)^{r+p}\delta^{j_r}{}_j {}^*e^{j_1\cdots j_{r-1}j_{r+1}\cdots j_p} = (-1)^{p+1}{}^*i_{e_j}e^{j_1\cdots j_p}$; and

(c) $i_{e_j}{}^*e^{j_1\cdots j_p} = {}^*e^{j_1\cdots j_p j}$.

The **codifferential**

(7.2.19) $$\delta := {}^*d^*(-1)^{m(p+1)+s}, \qquad d = {}^*\delta^*(-1)^{m(p+1)+1+s},$$

is a generalization of the divergence. It is a linear mapping $E_p \to E_{p-1}$, and can be coupled with d to construct the generalization Δ of the Laplace operator for $E_p \to E_p$:

(7.2.20) $$\Delta = \delta d + d\delta \quad \text{(the **Laplace–Beltrami operator**).}$$

With a natural basis $e^{j_1\cdots j_p}$ of $E_p(\mathbb{R}^m)$, where $g = dx^i \otimes dx^k\,\eta_{ik}$, $\eta_{ik} = \pm 1$ if $i = k$ and otherwise 0,

(7.2.21)
$$\delta(fe^{i_1\cdots i_p}) = \sum_{j=1}^{p} f^{i_j}_{,} e^{i_1\cdots i_{j-1}i_{j+1}\cdots i_p}(-1)^{j-1}, \qquad f \in E_0(\mathbb{R}^m),$$
$$\Delta(fe^{i_1\cdots i_p}) = f_{,k}{}^k e^{i_1\cdots i_p}, \qquad f^k_{,}\eta_{ki} =: f_{,i},$$

(Problem 6). If, in particular, $p = 1$ and f_k are the components of a vector field, then δ is the ordinary divergence $f^k{}_{,k}$ and Δ is the operator $(\eta^{-1})^{ki} = \partial^2/\partial x^i\,\partial x^k$ as applied to the individual components. (If $m = 3$ and $p = 1$, then $\Delta = -\nabla\times\nabla\times + \nabla\nabla\cdot$).

Problem 7 is to derive the following set of

(7.2.22) **Rules**

(a) $dd = \delta\delta = 0$, $d\Delta = \Delta d$, $\delta\Delta = \Delta\delta$;

(b) $\delta^* = (-1)^p{}^*d$, ${}^*\delta = (-1)^{p+1}d^*$; and

(c) $d\delta^* = {}^*\delta d$, ${}^*d\delta = \delta d^*$, ${}^*\Delta = \Delta^*$.

In Chapter 2 we learned moreover that if a vector field v is specified, then

(7.2.23) **The Lie Derivative**

$$L_v = i_v \circ d + d \circ i_v \colon E_p \to E_p:$$
$$L_v(\omega_1 + \omega_2) = L_v\omega_1 + L_v\omega_2,$$
$$L_v(\omega_1 \wedge \omega_2) = (L_v\omega_1) \wedge \omega_2 + \omega_1 \wedge L_v\omega_2,$$

gives the rate of change of a form under the action of the flow generated by the contravariant components of v. (If $v = v_i\, dx^i$ and $f \in E_0(\mathbb{R}^n)$, then $L_v f = v^i\, \partial f/\partial x^i$.)

To calculate the derivative in a different coordinate system (Problems 1 and 2) it is convenient to use a so-called orthogonal basis. Since g_{ik} is a symmetric matrix, it can be transformed into a diagonal matrix η_{ik} with only ± 1 as eigenvalues, by some change of basis. This determines the e^i up to a local Lorentz transformation:

(7.2.24)
$$e^i(x) \to \Lambda^i{}_k(x) e^k(x),$$
$$\eta_{k\ell}\Lambda^k{}_m(x)\Lambda^\ell{}_n(x) = \eta_{mn} \quad \forall x.$$

An orthogonal basis is not necessarily natural, and the exterior differentials of the orthogonal e^i may not vanish. To handle the general case we define

(7.2.25) **The Affine Connections $\omega^i{}_k \in E_1(M)$**

$$de^i = -\omega^i{}_k \wedge e^k, \qquad dg_{ik} = \omega_{ik} + \omega_{ki},$$
$$\omega_{ik} = g_{ij}\omega^j{}_k.$$

(7.2.26) **Remarks**

1. In §10.1 we shall show that these properties determine the $\omega^i{}_k$ uniquely.
2. It follows from (7.2.25) for the differentials of the bases in E_p that

$$de^{j_1\cdots j_p} = -\omega^{j_1}{}_j \wedge e^{j j_2\cdots j_p} - \omega^{j_2}{}_j \wedge e^{j_1 j\cdots j_p} - \cdots$$
$$\cdots - \omega^{j_p}{}_j \wedge e^{j_1\cdots j_{p-1} j},$$

and likewise

$$d^*e^{j_1\cdots j_p} = -\omega^{j_1}{}_j \wedge {}^*e^{j j_2\cdots j_p} - \omega^{j_2}{}_j \wedge {}^*e^{j_1 j\cdots j_p} - \cdots$$
$$\cdots - \omega^{j_p}{}_j \wedge {}^*e^{j_1\cdots j_{p-1} j}$$

(Problem 8).

3. The ω's generalize the Γ's of (1.1.7), and hence play the role of the field strength in gravitational theory. They transform inhomogeneously under a change of basis (Problem 9):

$$\omega^k{}_r \to A^k{}_s \omega^s{}_j (A^{-1})^j{}_r - (A^{-1})^j{}_r\, dA^k{}_j.$$

This means in particular that there is always a basis for which all the $\omega^k{}_r$ vanish at a point (see (5.6.11)).

(7.2.27) **Partial Differential Equations**

All of the equations that will interest us are generally covariant (i.e., chart-independent), and hence the only kind of differentiation they contain is the exterior

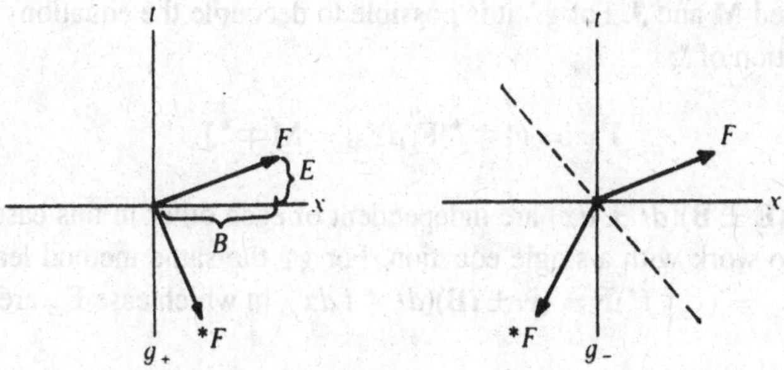

FIGURE 7.7. The action of $*$ on $\mathbb{R}^2$.

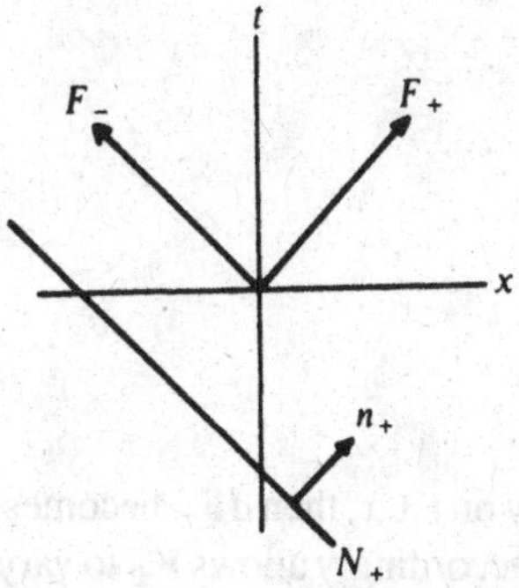

FIGURE 7.8. The hypersurface N_+ and its normals n_+.

differential. Orientability and the pseudo-Riemannian structure of the world enter only through the $*$-mapping. The prototype of this kind of equation for a vector field is the specification of the divergence and curl of the field, or more generally of δF and dF for $F \in E_p$.

Partial differential equations determine the derivatives in certain directions, while in others they may allow the fields to vary freely, and even have discontinuities. These directions, known as the **characteristics**, are the key to an understanding of the implications of the equations. To exhibit this fact in its germinal form, we discuss the simplest nontrivial

(7.2.28) **Example**

$M = \mathbb{R}^2$ with the metrics $g_\pm \equiv dx^2 \pm dt^2$. According to (7.2.17; 2), if $p = 1$, then $^*\circ^* = \mp\mathbf{1}$, and so * is a linear transformation of vectors, and has eigenvalues $\pm i$ for g_+ and ± 1 for g_-. More specifically, if $\mathbf{F} = \mathbf{E}\,dt + \mathbf{B}\,dx$, then $^*\mathbf{F} = -\mathbf{B}\,dt \pm \mathbf{E}\,dx$; for g_+, * is a rotation through 90°, and for g_- it is a reflection about the axis $x = -t$ (Figure 7.7).

Let us therefore consider the system of equations for $\mathbf{F}$,

$$d\mathbf{F} = \mathbf{M}, \qquad \delta\mathbf{F} = \mathbf{J},$$

with specified $\mathbf{M}$ and $\mathbf{J}$. For g_- it is possible to decouple the equations using the diagonalization of *:

$$\mathbf{F}_\pm := (1 \mp {}^*)\mathbf{F}: d\mathbf{F}_\mp = \mathbf{M} \mp {}^*\mathbf{J}.$$

The $\mathbf{F}_\pm = (\mathbf{E} \pm \mathbf{B})(dt \pm dx)$ are independent of each other in this case, and we need only to work with a single equation. For g_+ the same method leads to the equations $\mathbf{F}_\pm = (1 \mp i^*)\mathbf{F} = (\mathbf{E} \pm i\mathbf{B})(dt \mp i\, dx)$, in which case $\mathbf{F}_\pm$ are complex conjugates.

If $n_\pm = dx \pm dt$ and $N_\pm$ are lines parallel to $x = \mp t$, so that $n_{\pm|N_\pm} = 0$, then for g_- we get $i_{n_+}n_+ = i_{n_-}n_- = 0$ and therefore

(a) $n_\pm \wedge \mathbf{F}_\pm = \mathbf{0}$;

(b) $\mathbf{F}_{\pm|N_\pm} = 0$; and

(c) $i_{n_\pm}\mathbf{F}_\pm = \mathbf{0}$.

(7.2.29) **Conclusions**

(a) If $\mathbf{E}\pm\mathbf{B}$ depends only on $t \pm x$, then $d\mathbf{F}_\pm$ becomes $(\mathbf{E}\pm\mathbf{B})' n_\pm \wedge n_\pm = \mathbf{0}$. The system of equations accordingly allows $\mathbf{F}_\pm$ to vary freely in those directions; it is even possible for $\mathbf{E}+\mathbf{B}$ to be discontinuous in the direction of n_+. There is no other direction $n \in E_1$ such that $n \wedge \mathbf{F}_+ = \mathbf{0}$ or $n \wedge \mathbf{F}_- = \mathbf{0}$; the equations dictate how $\mathbf{F}_\pm$ varies in all the remaining directions. This state of affairs does not carry over to g_+, for which there are no real directions where $n \wedge \mathbf{F}_+ = \mathbf{0}$ or $n \wedge \mathbf{F}_- = \mathbf{0}$.

(b) The initial-value problem would be formulated by specifying as the initial data the restriction of $\mathbf{F}_\pm$ to some $(m-1)$-dimensional submanifold N, in the hope that this might determine $\mathbf{F}$. If $n_{\pm|N} = 0$ at any point of N, then $\mathbf{F}_{\pm|N}$ also vanishes at that point, making it impossible to choose the initial data at will. It is possible that they even leave $\mathbf{F}$ indeterminate. By fact (b) there exist $\mathbf{F}_+ \neq \mathbf{0}$ for which $\mathbf{F}_{+|N} = \mathbf{0}$ and $d\mathbf{F}_+ = \mathbf{0}$. If $n_{\pm|N} \neq 0$, the only condition on the initial data is $d(\mathbf{F}_{+|N}) = d\mathbf{F}_{+|N} = (\mathbf{M} - {}^*\mathbf{J})_{|N}$, which automatically vanishes (because there exists no 2-form on a one-dimensional N).

(c) By (7.2.23) the Lie derivative L_{e_k} in the direction x^k is given by $i_{e_k} \circ d + d \circ i_{e_k}$ with $e_k = g_{kj}\, dx^j$. Thus $L_{e_\pm}$ with $e_\pm \equiv gn_\pm = n_\mp$ is the Lie derivative in the direction $n_\pm$ (parallel to $x = \pm t$), and because $L_{e_-}\mathbf{F}_+ = i_{e_-}d\mathbf{F}_+ + di_{e_-}\mathbf{F}_+ = i_{e_-}d\mathbf{F}_+ = i_{e_-}(\mathbf{M} - {}^*\mathbf{J})$, the rate of change of $\mathbf{F}_+$ in the direction n_- is determined. As we shall see later from the explicit solution, F is determined by an arbitrary set of initial data on N when:

(i) $n_{\pm|N}$ is different from zero at all points of N; and

(ii) every line $t \pm x =$ const. intersects N at exactly one point.

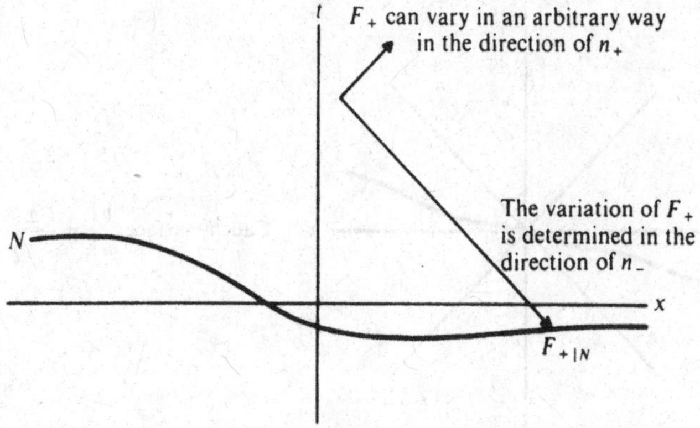

FIGURE 7.9. Determination of F_+ from its restrictions to the Cauchy surface N.

An $(m-1)$-dimensional hypersurface N (in other words, in two dimensions, a curve) that satisfies (i) and (ii) is called a **Cauchy surface**. It is nowhere tangent to the characteristics $n_\pm$, and it must be large enough to determine $\mathbf{E}\pm\mathbf{B}$ everywhere. (See Figure 7.9.)

(7.2.30) **Remarks**

1. Not all surfaces that are everywhere spacelike and extend to infinity are Cauchy surfaces. (See Figure 7.10.)
2. Observations like the foregoing can normally answer only local questions. Global questions are more difficult to treat effectively. Some seemingly innocuous manifolds, which are dear to the hearts of cosmologists, have no Cauchy surfaces at all.
3. Since $\mathbf{E}+\mathbf{B}$ can be an arbitrary function of $t+x$, one might want to risk using discontinuous functions, for which $M_\pm\colon n_\pm\big|_{M_\pm}=0$ are surfaces where $\mathbf{E}\pm\mathbf{B}$ has jump discontinuities. If such functions are admitted, the classical notion of differentiation is fraught with unnecessary difficulties. For instance, the equation $(\partial/\partial u)(\partial/\partial v)\Phi=0$ has the solution $\Phi=\varphi(u)+\psi(v)$, where φ is arbitrary and ψ is differentiable, but this solution does not satisfy $(\partial/\partial v)(\partial/\partial u)\Phi=0$.

To avoid these inconveniences it is natural to try to use distributions φ, defined by their integrals against suitable test-functions. It is then always permissible to differentiate, as defined through integration by parts: $\int f\varphi^{(n)}:=(-1)^n\int f^{(n)}\varphi$.

A strict formulation of the problem requires the theory of locally convex topological vector spaces. As we shall not need the more profound theorems, we content ourselves with making the

(7.2.31) **Definition**

$$\Theta(x)=\begin{cases}1 & \text{if } x\geq 0\\ 0 & \text{if } x<0\end{cases}\qquad \text{(Heaviside's step function)},$$

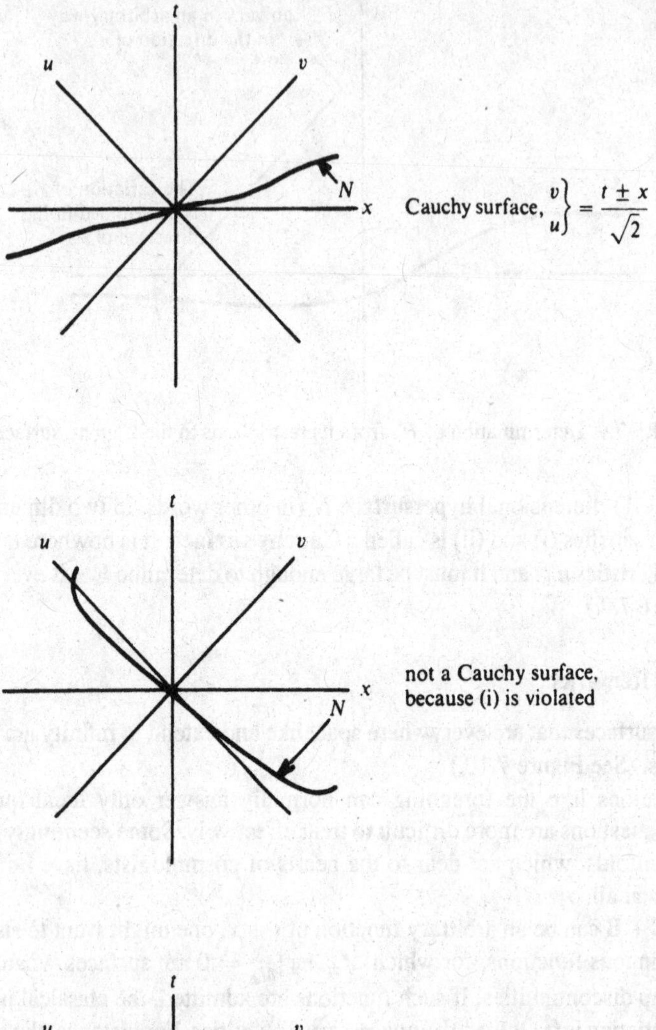

t
u
v
N
x
Cauchy surface, $\left.\begin{matrix} v \\ u \end{matrix}\right\} = \frac{t \pm x}{\sqrt{2}}$
t
u
v
x
N
not a Cauchy surface,
because (i) is violated
t
u
v
x
N
not a Cauchy surface,
because (ii) is violated

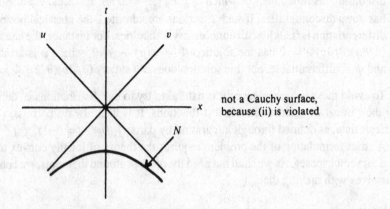

FIGURE 7.10.

$$\delta(x) = \frac{d\theta(x)}{dx} \qquad \text{(Dirac's delta function)},$$

and the

(7.2.32) **Warning**

Distributions form a vector space but not an algebra; they can be added but not always multiplied [41], [56].

A special distribution, the Green function,[4] is particularly useful for solving the equations we have to deal with. This lets one exploit the linearity of the equations, the consequence of which is that the sum of the solutions for several individual sources is a solution of the equation with the sum of the sources. If the solution for a point source is known, then it can be used to construct the solution for an arbitrary inhomogeneity by superposition. Putting these ideas into practice as applied to E_p requires a bit more algebra and replacing the sums with integrals.

The first thing to do is to generalize the delta function so that when it is multiplied by a p-form and integrated it reproduces the value of the p-form at a point $\bar{x} \in M$ (the set of all such values is denoted by $E_{p|\bar{x}}$). Since we are only able to integrate m-forms, and the result ought to be in $E_{p|\bar{x}}$, we need a $\delta_{\bar{x}} \in E_{p|\bar{x}} \otimes E_{m-p}(M)$ such that[5]

$$\int \delta_{\bar{x}} \wedge F = F(\bar{x}) \qquad \text{for all} \qquad F \in E_p(M). \tag{7.2.33}$$

The distribution $\delta_{\bar{x}}$ vanishes away from $\bar{x}$ and has the form (Problem 10)

$$\delta_{\bar{x}} = \bar{e}_{i_1 \cdots i_p} \otimes {}^*e^{i_1 \cdots i_p} \frac{(-1)^{p(m-p)}}{p!} \delta(x^1 - \bar{x}^1)\delta(x^2 - \bar{x}^2) \cdots \delta(x^m - \bar{x}^m), \tag{7.2.34}$$

in a natural basis $e^{i_1 \cdots i_p}$ of a neighborhood of $\bar{x}$, and where $\bar{e}_{i_1 \cdots i_p}$ is the basis in $E_{p|\bar{x}}$ and x^i are the coordinates to be integrated over. The Green function $G_{\bar{x}} \in E_{p|\bar{x}} \otimes E_{m-p}(M)$ satisfies the equation

$$\Delta G_{\bar{x}} = -\delta_{\bar{x}}, \tag{7.2.35}$$

where $\Delta = d\delta + \delta d$ acts on x^i with $\bar{x}$ fixed. In the cases we are interested in, we shall verify the existence of a Green function by explicit construction. Equation (7.2.35) determines $G_{\bar{x}}$ only up to a solution of the homogeneous equation. Using

[4]We follow Jackson's [20] lead and abandon the ungrammatical phrase "the Green's function."

[5]In de Rham's terminology, $\delta_{\bar{x}}$ would be called a p-form-valued current. In the older literature one often encounters the term Green's dyadic for the three-dimensional case. We shall write $E_p(M)$ even when the components are distributions rather than differentiable functions.

$G_{\bar{x}}$, Green's formula generalizes as

$$F(\bar{x}) = (-1)^{p+m} \int_N [dG_{\bar{x}} \wedge \delta F - \delta G_{\bar{x}} \wedge dF]$$

(7.2.36)

$$- \int_{\partial N} [\delta G_{\bar{x}} \wedge F + (-1)^{mp+p+s*} dG_{\bar{x}} \wedge {}^*F],$$

which expresses $F \in E_p(M)$ at the point $\bar{x}$ in terms of dF and δF on some m-dimensional submanifold $N \ni \bar{x}$ and the values of the restrictions of F and ${}^*F$ to ∂N. (For the derivation of (7.2.36), see Problem 11.)

(7.2.37) **Example**

We return to (7.2.28). $M = \mathbb{R}^2$, $p = 1$, $g = g_-$, and $(-1)^s = -1$. Using the "lightlike" coordinates that diagonalize $*$,

$$\begin{matrix} v \\ u \end{matrix} = \frac{t \pm x}{\sqrt{2}}, \quad \text{for which} \quad g_- = \begin{vmatrix} 0 & -1 \\ -1 & 0 \end{vmatrix} \quad \text{and} \quad (-1)^s = -1,$$

$${}^*du = du, \qquad {}^*dv = -dv, \qquad {}^*(du \wedge dv) = -\mathbf{0},$$

and (omitting the $\otimes$ sign)

$$\begin{aligned} \delta_{\bar{x}} &= (d\bar{x}\, dt - d\bar{t}\, dx)\delta(x - \bar{x})\delta(t - \bar{t}) \\ &= (d\bar{v}\, du - d\bar{u}\, dv)\delta(u - \bar{u})\delta(v - \bar{v}). \end{aligned}$$

The Green function can be written with the aid of the step function (7.2.31) as

$$G_{\bar{x}} = (d\bar{v}\, du - d\bar{u}\, dv)\tfrac{1}{2}\Theta(\bar{u} - u)\Theta(\bar{v} - v),$$

because

$$\begin{aligned} dG_{\bar{x}} &= \tfrac{1}{2}[\delta(v - \bar{v})\Theta(\bar{u} - u)\, d\bar{v} + \Theta(\bar{v} - v)\delta(\bar{u} - u)\, d\bar{u}]\, du \wedge dv \\ \delta G_{\bar{x}} &= \tfrac{1}{2}[\delta(v - \bar{v})\Theta(\bar{u} - u)\, d\bar{v} - \Theta(\bar{v} - v)\delta(\bar{u} - u)\, d\bar{u}], \end{aligned}$$

with which it is easy to verify that

$$\Delta G_{\bar{x}} = (d\bar{u}\, dv - d\bar{v}\, du)\delta(u - \bar{u})\delta(v - \bar{v}) = -\delta_{\bar{x}}.$$

If $F = \varphi\, du + \psi\, dv$, then the field equations $dF = M$, $\delta F = J$, explicitly read:

$$\begin{aligned} dF &= du \wedge dv(\psi_{,u} - \varphi_{,v}) = M := du \wedge dv\, m(u, v), \\ \delta F &= \qquad - \psi_{,u} - \varphi_{,v} \quad = J(u, v). \end{aligned}$$

Now the integral over N of (7.2.36) contributes

$$-\int dGJ + \int \delta GM = -\frac{1}{2} \int du \wedge dv(\delta(v - \bar{v})\Theta(\bar{u} - u)\, d\bar{v}$$

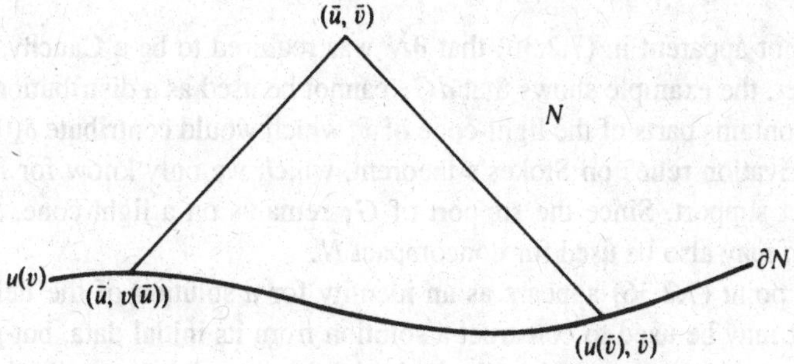

FIGURE 7.11. Variables for the fields at a boundary.

$$
\begin{aligned}
&+ \Theta(\bar{v} - v)\delta(\bar{u} - u)\, d\bar{u})J(u, v) \\
&+ \frac{1}{2}\int du \wedge dv(\delta(v - \bar{v})\Theta(\bar{u} - u)\, d\bar{v} \\
&- \Theta(\bar{v} - v)\delta(\bar{u} - u)\, d\bar{u})m(u, v) \\
= &\int_{u(\bar{v})}^{\bar{u}} du\, \tfrac{1}{2}(m(u, \bar{v}) - J(u, \bar{v}))\, d\bar{v} \\
&- \int_{v(\bar{u})}^{\bar{v}} dv\, \tfrac{1}{2}(m(\bar{u}, v) + J(\bar{u}, v))\, d\bar{u},
\end{aligned}
$$

if the submanifold ∂N is defined as $\{u = u(v)\} = \{v = v(u)\}$.

This contribution clearly satisfies $2\psi_{,u} = m - J$ and $2\varphi_{,v} = -m - J$, and approaches zero as $(\bar{u}, \bar{v})$ approaches ∂N. To calculate the boundary term $F_{\partial N} := \int_{\partial N} \cdots$ note that ∂N has the orientation of $-dx$ for this boundary integral (see (2.6.6; 1)):

$$
\begin{aligned}
\int_{\partial N} \delta G \wedge F &= \frac{d\bar{v}}{2}\int_{\partial N} (\varphi\, du + \psi\, dv)\delta(v - \bar{v})\Theta(\bar{u} - u) \\
&- \frac{d\bar{u}}{2}\int_{\partial N} (\varphi\, du + \psi\, dv)\delta(u - \bar{u})\Theta(\bar{v} - v), \\
\int_{\partial N} {}^*dG \wedge {}^*F &= \frac{d\bar{v}}{2}\int_{\partial N} (-\varphi\, du + \psi\, dv)\delta(v - \bar{v})\Theta(\bar{u} - u) \\
&+ \frac{d\bar{u}}{2}\int_{\partial N} (-\varphi\, du + \psi\, dv)\delta(u - \bar{u})\Theta(\bar{v} - v);
\end{aligned}
$$

since dv contributes positively and du negatively, the net result is that

$$
F_{\partial N} = d\bar{v}\, \psi(u(\bar{v}), \bar{v})\Theta(\bar{u} - u(\bar{v})) + d\bar{u}\, \varphi(\bar{u}, v(\bar{u}))\Theta(\bar{v} - v(\bar{u})).
$$

As long as $\bar{x} \in N$, so that $\Theta(\bar{u} - u(\bar{v}))\Theta(\bar{v} - v(\bar{u})) = 1$, this solves the homogeneous equation, because ψ is independent of $\bar{u}$ and φ is independent of $\bar{v}$. It also clearly satisfies the boundary conditions.

(7.2.38) **Remarks**

1. It was not apparent in (7.2.36) that ∂N was required to be a Cauchy surface. However, the example shows that $dG_{\bar{x}}$ cannot be used as a distribution on ∂N if ∂N contains parts of the light-cone of $\bar{x}$, which would contribute $\delta(0)$.
2. The derivation relies on Stokes's theorem, which we only know for forms of compact support. Since the support of $G_{\bar{x}}$ remains on a light-cone, Stokes's theorem may also be used for noncompact N.
3. At this point (7.2.36) appears as an identity for a solution of the field equations. It may be used to construct a solution from its initial data, but gives no information about how arbitrarily the initial data may be chosen. In the example, if ∂N is a Cauchy surface, then the solution meets all requirements with arbitrary $\psi(u(v), v)$ and $\varphi(u, v(u))$ on ∂N. The initial values can be specified independently, since ∂N never contains two points of the same light-cone, while information about an initial condition only propagates along its characteristics. If the metric were g_+, then the situation would be quite different. The solutions would be analytic functions of $x + it$, and would be determined everywhere by their values on an arbitrarily small curve segment.
4. The equations of constraint mentioned in (7.2.29(b)) arise in four dimensions.

In particular, formula (7.2.36) solves Maxwell's equations, and electrodynamics consists largely in working out special cases of it. The normal detour via the introduction of potentials is inconvenient as well as unnecessary, as one really wants to express F in terms of the boundary values on F and ${}^*F$, and not of the potentials. Unfortunately, the canonical formulation of the equations of motion, and hence also quantum mechanics, use potentials. Yet even the potentials can be obtained from (7.2.36). On the other hand, the currents analogous to J and M appearing in Einstein's equations depend nonlinearly on the fields, so in that case (7.2.36) is useful only in the linear approximation.

(7.2.39) **Problems**

1. The metric $g = d\rho^2 + \rho^2\, d\varphi^2 + dz^2$ in cylindrical coordinates for $\mathbb{R}^3 \backslash \{\mathbf{0} \times \mathbb{R}\}$. Calculate ω_{ik} for the orthogonal basis $e^1 = dz, e^2 = d\rho, e^3 = \rho\, d\varphi$. Write $\nabla \times A$ and $\nabla \cdot A$ in the components of this basis and of the natural basis. What is the connection between them?

2. The same question for spherical coordinates and $\mathbb{R}^3 \backslash \{\mathbf{0} \times \mathbb{R}\}$, for which $g = dr^2 + r^2\, d\vartheta^2 + r^2 \sin^2 \vartheta\, d\varphi^2$.

3. Calculate Δ from (7.2.20) for $\rho = 0$ in the natural basis, and specialize to the cases of cylindrical and spherical coordinates on $\mathbb{R}^3 \backslash \{\mathbf{0} \times \mathbb{R}\}$.

4. Prove the normalization ${}^* \circ {}^* = (-1)^{p(m-p)+s}1$. Using (7.2.16).

5. Derive the rules (7.2.18):

$$
{}^*e^{i_1 \cdots i_p} = g^{i_1 j_1} \cdots g^{i_p j_p} e^{j_{p+1} \cdots j_m} \varepsilon_{j_1 \cdots j_m} \frac{\sqrt{|g|}}{(m-p)!},
$$

$$
e_j \wedge {}^*e^{j_1 \cdots j_p} = \sum_r (-1)^{r+p} \delta^{j_r}{}_j\, {}^*e^{j_1 \cdots j_{r-1} j_{r+1} \cdots j_p},
$$

$$i_{e_j}{}^*e^{j_1\cdots j_p} = {}^*e^{j_1\cdots j_p j}.$$

6. Verify (7.2.21).
7. Verify (7.2.22).
8. Check (7.2.26; 2).
9. Derive the transformation law for the ω's (7.2.26; 3).
10. Show that $\delta_{\bar{x}}$ from (7.2.34) has the property (7.2.33).
11. Prove (7.2.36).
12. Find a manifold M and a $J \in E_p(M)$ for which the equation $dF = J$ has no (global) solution, although $dJ = 0$.

(7.2.40) Solutions

1. Orthogonal basis: $\eta_{ik} = \delta_{ik}$, $0 = de^1 = de^2$, $de^3 = e^{23}/\rho$. Hence only $\omega_{32} = -\omega_{23} = e^3/\rho \neq 0$, and $de^{12} = de^{32} = 0$, $de^{31} = -\omega^{32} \wedge e^{21}$. Letting $A = a_i e^i$,

$$dA = (a_{1,2} - a_{2,1})e^{21} + (a_{2,3}/\rho - a_{3,2} - a_3/\rho)e^{32} + (a_{3,1} - a_{1,3}/\rho)e^{13},$$
$${}^*dA = (a_{3,2} + a_3/\rho - a_{2,3}/\rho)e^1 + (a_{1,3}/\rho - a_{3,1})e^2 + (a_{2,1} - a_{1,2})e^3,$$
$${}^*A = a_1e^{23} + a_2e^{31} + a_3e^{12}, \qquad {}^*d^*A = a_{1,1} + a_{2,2} + a_{3,3}/\rho + a_2/\rho.$$

Natural basis: $\sqrt{|g|} = \rho$. Let $A = A_z\, dz + A_\rho\, d\rho + A_\varphi\, d\varphi$. Then

$${}^*A = \rho[A_z\, d\rho \wedge d\varphi + A_\rho\, d\varphi \wedge dz + \rho^{-2}A_\varphi\, dz \wedge d\rho],$$
$${}^*d^*A = \rho^{-1}[(\rho A_z)_{,z} + (\rho A_\rho)_{,\rho} + (\rho^{-1}A_\varphi)_{,\varphi}],$$
$${}^*dA = \rho^{-1}[(A_{\varphi,\rho} - A_{\rho,\varphi})dz + (A_{z,\varphi} - A_{\varphi,z})d\rho + \rho^2(A_{\rho,z} - A_{z,\rho})d\varphi].$$

The connection is that $(A_z, A_\rho, A_\varphi) = (a_1, a_2, \rho a_3)$.

2. Orthogonal basis: $e^1 = dr$, $e^2 = r\, d\vartheta$, $e^3 = r\sin\vartheta\, d\varphi$, $de^1 = 0$, $de^2 = dr \wedge d\vartheta$, $de^3 = r\cos\vartheta\, d\vartheta \wedge d\varphi + \sin\vartheta\, dr \wedge d\varphi \Rightarrow \omega_{21} = e^2/r$, $\omega_{31} = \sin\vartheta\, d\varphi$, $\omega_{32} = \cos\vartheta\, d\varphi \Rightarrow de^{12} = 0$, $de^{13} = -e^{23}/r\tan\vartheta$, $de^{23} = 2e^{23}/r$. Let $A = a_ie^i$. Then

$${}^*d^*A = a_{1,1} + \frac{2}{r}a_1 + \frac{1}{r}a_{2,2} + \frac{a_2}{r\tan\vartheta} + \frac{a_{3,3}}{r\sin\vartheta},$$
$${}^*dA = \left[\frac{a_{3,2}}{r} + \frac{a_3}{r\tan\vartheta} - \frac{a_{2,3}}{r\sin\vartheta}\right]e^1 + \left[\frac{a_{1,3}}{r\sin\vartheta} - \frac{a_3}{r} - a_{3,1}\right]e^2$$
$$+ \left(a_{2,1} + \frac{a_2}{r} - \frac{a_{1,2}}{r}\right)e^3.$$

Natural basis: $\sqrt{|g|} = r^2\sin\vartheta$. Let $A = A_r\, dr + A_\vartheta\, d\vartheta + A_\varphi\, d\varphi$. Then

$${}^*A = r^2\sin\vartheta[A_r\, d\vartheta \wedge d\varphi + r^{-2}A_\vartheta\, d\varphi \wedge dr + r^{-2}\sin^{-2}\vartheta A_\varphi\, dr \wedge d\vartheta],$$
$${}^*dA = \frac{dr}{r^2\sin\vartheta}(A_{\varphi,\vartheta} - A_{\vartheta,\varphi}) + \frac{d\vartheta}{\sin\vartheta}(A_{r,\varphi} - A_{\varphi,r}) + d\varphi\, \sin\vartheta(A_{\vartheta,r} - A_{r,\vartheta}),$$
$${}^*d^*A = \frac{1}{r^2\sin\vartheta}\left[(r^2\sin\vartheta A_r)_{,r} + (\sin\vartheta A_\vartheta)_{,\vartheta} + \left(\frac{A_\varphi}{\sin\vartheta}\right)_{,\varphi}\right].$$

The connection is that $(A_r, A_\vartheta, A_\varphi) = (a_1, ra_2, r\sin\vartheta a_3)$.

3.
$$\Delta f = \frac{1}{\sqrt{g}}\frac{\partial}{\partial x^\alpha}(f_{,\beta}g^{\alpha\beta}\sqrt{g}).$$

Cylindrical coordinates:

$$\Delta = \frac{1}{\rho}\left[\rho\frac{\partial^2}{\partial z^2} + \frac{\partial}{\partial\rho}\rho\frac{\partial}{\partial\rho} + \frac{1}{\rho}\frac{\partial^2}{\partial\varphi^2}\right].$$

Spherical coordinates:

$$\Delta = \frac{1}{r^2\sin\vartheta}\left[\frac{\partial}{\partial r}r^2\sin\vartheta\frac{\partial}{\partial r} + \frac{\partial}{\partial\vartheta}\sin\vartheta\frac{\partial}{\partial\vartheta} + \frac{\partial}{\partial\varphi}\frac{1}{\sin\vartheta}\frac{\partial}{\partial\varphi}\right].$$

4. From (7.2.16) we infer

$$\begin{aligned}\langle{}^*\omega|{}^*\omega\rangle &= \langle{}^*\omega|i_\omega\varepsilon\rangle = \langle\omega\wedge{}^*\omega|\varepsilon\rangle\\ &= (-1)^{p(m-p)}\langle{}^*\omega\wedge\omega|\varepsilon\rangle = (-1)^{p(m-p)}\omega|i_{*\omega}\varepsilon\rangle = (-1)^{p(m-p)}\langle\omega|{}^{**}\omega\rangle.\end{aligned}$$

The fact that ${}^{**}\omega$ can differ from ω only by a sign is a general property of duality. Therefore ${}^{**}\omega = (-1)^{p(m-p)+s}\omega$ is equivalent to $\langle\omega|\omega\rangle = (-1)^s\langle{}^*\omega|{}^*\omega\rangle$. It is obviously enough to show this for a basis

$$e^{i_1\cdots i_p} =: \varepsilon_p \in E_p$$

such that $\varepsilon = \varepsilon_p\wedge\varepsilon_{m-p}$ with $i_{\varepsilon_p}\varepsilon_{m-p} = 0$. Then

$$\begin{aligned}\langle{}^*\varepsilon_p|{}^*\varepsilon_p\rangle &= \langle\varepsilon_{m-p}|\varepsilon_{m-p}\rangle\langle\varepsilon_p|\varepsilon_p\rangle\langle\varepsilon_p|\varepsilon_p\rangle\\ &= \langle\varepsilon|\varepsilon\rangle\langle\varepsilon_p|\varepsilon_p\rangle = (-1)^s\langle\varepsilon_p|\varepsilon_p\rangle.\end{aligned}$$

5. In a basis we have

$$\varepsilon = \frac{e^{i_1\cdots i_m}}{m!}\varepsilon_{i_1\cdots i_m}\sqrt{|g|},$$

where

$$g = \det g_{ik} = (\det\langle e^i|e^k\rangle)^{-1}.$$

Then (a) follows from the definition ${}^*\omega = i_\omega\varepsilon$ and the rules (7.2.16) for i. Parts (b) and (c) are a consequence of (7.2.17(iii)).

6.

$$\begin{aligned}\delta(fe^{i_1\cdots i_p}) &= {}^*[(-1)^{m(p+1)+s}f_{,k}e^k\wedge{}^*e^{i_1\cdots i_p}]\\ &= {}^*\left[(-1)^{(m+1)(p+1)+s}\sum_{j=1}^{p} f^{,i_j}\,{}^*e^{i_1\cdots i_{j-1}i_{j+1}\cdots i_p}(-1)^{j+1}\right]\\ &= \sum_{j=1}^{p} f^{,i_j}e^{i_1\cdots i_{j-1}i_{j+1}\cdots i_p}(-1)^{j+1},\end{aligned}$$

$$d\delta(fe^{i_1\cdots i_p}) = \sum_{j=1}^{p} f^{,i_j}{}_{,i_0}e^{i_0 i_1\cdots i_{j-1}i_{j+1}\cdots i_p}(-1)^{j-1},$$

$$\delta d(fe^{i_1\cdots i_p}) = \sum_{j=1}^{p} f^{i_j}_{,\,i_0} e^{i_0 i_1\cdots i_{j-1} i_{j+1}\cdots i_p}(-1)^j,$$

$$\Rightarrow \Delta(fe^{i_1\cdots i_p}) = \sum_{k=1}^{m} f^{k}_{,\,k} e^{i_1\cdots i_p}.$$

7. (a) $d\Delta = d\delta d = \Delta d, \delta\Delta = \delta d\delta = \Delta\delta$;
(b) $\delta^* = {}^*d^{**}(-1)^{m(m-p+1)+s} = (-1)^{p*}d, d^* = {}^*\delta^{**}(-1)^{m(m-p)+1+s} = (-1)^{p+1*}\delta$; and
(c) $d\delta^* = (-1)^p d^* d = {}^*\delta d, \delta d^* = (-1)^{p+1}\delta^*\delta = {}^*d\delta$.

8. Letting $\omega_k{}^i = \eta^{ij}\omega_{kj}$, we start with the identity

$$\omega_{k_1}{}^i \varepsilon_{ik_2\cdots k_m} + \omega_{k2}{}^i \varepsilon_{k_1 i\cdots k_m} + \cdots + \omega_{k_m}{}^i \varepsilon_{k_1\cdots k_{m-1}i} = 0$$

$\forall k_1 = 1, \dots, m; \dots; k_m = 1, \dots, m$. To verify this, consider the three cases:

(i) All k_i are different. Then i must equal the k that is missing in ε, and $\omega_{kj} = -\omega_{jk}$; since η is diagonal, $\omega_k{}^i = 0$ if $i = k$.

(ii) Two of the k_i are equal, say $k_1 = k_2$. There remains

$$\omega_{k_1}{}^i \varepsilon_{ik_1\cdots} + \omega_{k_1}{}^i \varepsilon_{k_1 i\cdots} = 0.$$

(iii) Three k_i are equal. Then all the ε's vanish.

If the identity is multiplied by $e^{k_{p+1}\cdots k_m}$ and the indices are relabeled, then

$$\omega_{k_{p+1}}{}^i \varepsilon_{k_1\cdots k_p i k_{p+2}\cdots m} e^{k_{p+1}\cdots k_m} = -\omega^{k_{p+1}}{}_i \varepsilon_{k_1\cdots k_m} e^{ik_{p+2}\cdots k_m},$$

in the orthogonal basis. In general, because of (7.2.18(a)),

$$\begin{aligned}
(m-p)!\, d^* e^{j_1\cdots j_p} &= \eta^{j_1k_1}\cdots\eta^{j_pk_p}\varepsilon_{k_1\cdots k_m} de^{k_{p+1}\cdots k_m}\\
&= -\eta^{j_1k_1}\cdots\eta^{j_pk_p}\varepsilon_{k_1\cdots k_m}\{\omega^{k_{p+1}}{}_i e^{ik_{p+2}\cdots k_m} + \cdots + \omega^{k_m}{}_i e^{k_{p+1}\cdots k_{m-1}i}\}\\
&= \eta^{j_1k_1}\cdots\eta^{j_pk_p}\{\omega_{k_{p+1}}{}^i \varepsilon_{k_1\cdots k_p i k_{p+1}\cdots k_m} + \omega_{k_{p+2}}{}^i \varepsilon_{k_1\cdots k_{p+1} i k_{p+3}\cdots k_m} + \cdots\\
&\quad + \omega_{k_m}{}^i \varepsilon_{k_1\cdots k_{m-1}i}\} e^{k_{p+1}k_{p+2}\cdots k_m}\\
&= -\eta^{j_1k_1}\cdots\eta^{j_pk_p}\{\omega_{k_1}{}^i \varepsilon_{ik_2\cdots k_m} + \cdots + \omega_{k_p}{}^i \varepsilon_{k_1\cdots k_{p-1} i k_{p+1}\cdots k_m}\} e^{k_{p+1}\cdots k_m}\\
&= -\omega^{j_1}{}_i {}^*e^{ij_2\cdots j_p} - \cdots - \omega^{j_p}{}_i {}^*e^{j_1 j_2\cdots j_{p-1}i}(m-p)!.
\end{aligned}$$

9. Let

$$\begin{aligned}
\bar{e}^j = A^j{}_k e^k : d\bar{e}^j &= dA^j{}_r (A^{-1})^r{}_k \bar{e}^k - A^j \omega^k{}_r (A^{-1})^r{}_s \bar{e}^s = -\bar{\omega}^j{}_k \bar{e}^k\\
\Rightarrow \quad \bar{\omega}^j{}_k &= A^j{}_s \omega^s{}_r (A^{-1})^r{}_k - (A^{-1})^s{}_k\, dA^j{}_s,
\end{aligned}$$

because this also satisfies the second defining equation:

$$\begin{aligned}
\bar{g} = A^{-1\ell} g A^{-1} \quad &\Rightarrow \quad d\bar{g} = A^{-1\ell}(dg)A^{-1} - A^{-1\ell} g A^{-1}(dA)A^{-1}\\
&- A^{-1\ell}(dA^\ell)A^{-1\ell} g A^{-1} = \bar{g}\bar{\omega} + (\bar{g}\bar{\omega})^\ell\\
&= A^{-1\ell} g\omega A^{-1} - A^{-1\ell} g A^{-1}(dA)A^{-1} + A^{-1\ell}(g\omega)^\ell A^{-1} - A^{-1\ell}(gA^{-1}\, dA)^\ell A^{-1}.
\end{aligned}$$

10. According to (7.2.18),

$$\frac{1}{p!}e_{j_1\cdots j_p}\wedge {}^*e^{i_1\cdots j_p}\omega^{j_1\cdots j_p}=\omega^{i_1\cdots i_p}{}^*\mathbf{1}.$$

Thus

$$\bar{e}_{i_1\cdots i_p}\otimes {}^*e^{i_1\cdots i_p}\wedge e_{j_1\cdots j_p}\omega^{j_1\cdots j_p}\frac{(-1)^{p(m-p)}}{p!}=\bar{e}^{i_1\cdots i_p}\omega_{i_1\cdots i_p}{}^*\mathbf{1}.$$

Integrating this $\int dx_1\cdots dx_m$ yields

$$\int\delta_{\bar{x}}\wedge e^{i_1\cdots i_p}\omega_{i_1\cdots i_p}=\bar{e}^{i_1\cdots i_p}\omega_{i_1\cdots i_p}(\bar{x}).$$

11.

$$\begin{aligned}
\delta G_{\bar{x}}\wedge dF &= (-1)^{p+m+1}[d(\delta G_{\bar{x}}\wedge F)-(d\delta G_{\bar{x}})\wedge F],\\
dG_{\bar{x}}\wedge\delta F &= (-1)^{mp+m+s}d^*F\wedge{}^*dG_{\bar{x}}\\
&= (-1)^{mp+m+s}[d({}^*F\wedge{}^*dG_{\bar{x}}-(-1)^{m-p}{}^*F\wedge d^*\,dG_{\bar{x}}]\\
&= (-1)^{mp+m+s}d({}^*F\wedge{}^*dG_{\bar{x}})+(-1)^{mp+p+s+1}{}^*d^*\,dG_{\bar{x}}\wedge F\\
&= (-1)^{mp+m+s}d({}^*F\wedge{}^*dG_{\bar{x}})+(-1)^{m+1+p}\delta\,dG_{\bar{x}}\wedge F\\
&\Rightarrow (-1)^{p+m}[dG_{\bar{x}}\wedge\delta F-\delta G_{\bar{x}}\wedge dF]\\
&= \underbrace{-(\delta d+d\delta)G_{\bar{x}}}_{=\delta_{\bar{x}}}\wedge F+d(\delta G_{\bar{x}}\wedge F+(-1)^{mp+p+s}{}^*F\wedge{}^*dG_{\bar{x}}),
\end{aligned}$$

and when this is integrated over N, we get (7.2.36).

12. $M-T^1$ and $J=d\varphi\in E_1(T^1)$, where φ is the angle on the torus. If $F\in E_0(T^1)$, then it would be $\varphi+$ constant, which cannot be defined continuously over all of T^1, even though $d\varphi$ exists on all of T^1.

7.3 Maxwell's and Einstein's Equations

Field strengths are described by 2-forms, the exterior differentials of which in the electric (resp. gravitational) case are the 3-forms of charge (resp. energy and momentum). These 3-forms, which function as the sources of the fields, are exact, a fact that implies differential and integral conservation theorems.

Maxwell wrote Faraday's discoveries in the form of the equations that now bear his name (Table 7.1), and which describe all electromagnetic phenomena. The way they fuse space and time, and the electric and magnetic fields, is so awe-inspiring that Boltzmann, quoting Goethe, exclaimed, "Was it a god who traced these signs?"[6] It is regrettable that Cartan's calculus did not exist at that time,

[6] Faust, Part I, Dr. Faust's first soliloquy.

as it has perfected the notation in which all the facts of electromagnetism are so concisely formulated.

As discussed in detail in (§5.1), the electric and magnetic fields **E** and **B** are combined in a single 2-form F, which is written in the natural basis of $\mathbb{R}^4$, with coordinates $x^0 = t$, $\mathbf{x} = (x^1, x^2, x^3)$ and in vector notation, as

(7.3.1)
$$F = \tfrac{1}{2} F_{\alpha\beta}\, dx^\alpha \wedge dx^\beta = dt \wedge (d\mathbf{x} \cdot \mathbf{E}) - dx^1 \wedge dx^2\, B_3 - dx^3 \wedge dx^1\, B_2 - dx^2 \wedge dx^2\, B_1.$$

TABLE 7.1. Maxwell's Equations in the Course of History
The constants c, μ_0, and ε_0 are set to 1, and modern notation is used for the components.

The Homogeneous Equation	**The Inhomogeneous Equation**
Earliest Form	Earliest Form
$\frac{\partial B_x}{\partial x} + \frac{\partial B_y}{\partial y} + \frac{\partial B_z}{\partial z} = 0$	$\frac{\partial E_x}{\partial x} + \frac{\partial E_y}{\partial y} + \frac{\partial E_z}{\partial z} = \rho$
$\frac{\partial E_z}{\partial y} - \frac{\partial E_y}{\partial z} = -\dot{B}_x$	$\frac{\partial B_z}{\partial y} - \frac{\partial B_y}{\partial z} = j_x + \dot{E}_x$
$\frac{\partial E_x}{\partial z} - \frac{\partial E_z}{\partial x} = -\dot{B}_y$	$\frac{\partial B_x}{\partial z} - \frac{\partial B_z}{\partial x} = j_y + \dot{E}_y$
$\frac{\partial E_y}{\partial x} - \frac{\partial E_x}{\partial y} = -\dot{B}_z$	$\frac{\partial B_y}{\partial x} - \frac{\partial B_x}{\partial y} = j_z + \dot{E}_z$
At the End of Last Century	At the End of Last Century
$\nabla \cdot \mathbf{B} = 0$	$\nabla \cdot \mathbf{E} = \rho$
$\nabla \times \mathbf{E} = -\dot{\mathbf{B}}$	$\nabla \times \mathbf{B} = \mathbf{j} + \dot{\mathbf{E}}$
At the Beginning of This Century	At the Beginning of This Century
${}^*F^{\beta\alpha}{}_{,\alpha} = 0$	$F^{\beta\alpha}{}_{,\alpha} = j^\beta$
Mid-Twentieth Century	Mid-Twentieth Century
$dF = 0$	$\delta F = J$

(7.3.2) **The Homogeneous Maxwell Equations**

These are the basis-free statement that F is closed:

(7.3.3)
$$dF = 0.$$

By (7.2.11) this is equivalent to

(7.3.4)
$$0 = \int_{\partial N_3} F,$$

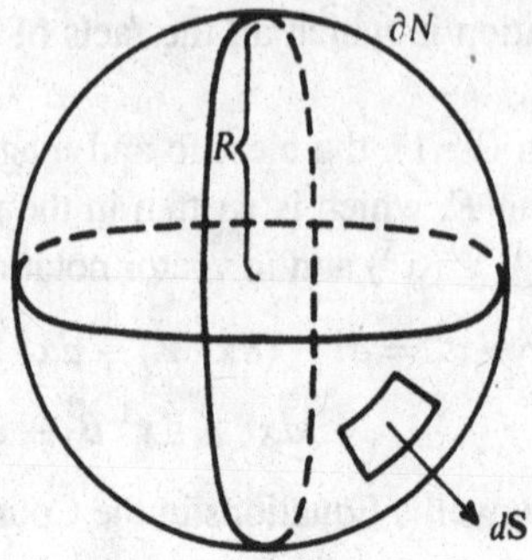

FIGURE 7.12. N_3 of (7.3.5; 1).

if $F_{|N_3} \in E_2^0(N_3)$, and N_3 is three-dimensional. To illustrate this concentrated notation, let us examine some

(7.3.5) Special Cases

1. $N_3 = \{(t, \mathbf{x}) \in \mathbb{R}^4 : t = t_0 \in \mathbb{R}, |\mathbf{x}| \leq \mathbb{R}\}$ (a spherical ball).
   $\partial N = \{(t, \mathbf{x}) \in \mathbb{R}^4 : t = t_0, |\mathbf{x}| = R\}$ (a spherical surface). (See Figure 7.12.)
   Then (7.3.4) says that there are no magnetic charges:

$$\int_{|\mathbf{x}|=R} \mathbf{B} \cdot d\mathbf{S} = 0.$$

   The differential version of this statement, $\nabla \cdot \mathbf{B} = 0$, amounts to $d(F_{|N_3}) = 0$, which follows from (7.3.3) because d commutes with restrictions.

2. $N_3 = \{(t, \mathbf{x}) \in \mathbb{R}^4 : t_0 \leq t \leq t_1, x^3 = 0, |\mathbf{x}| \leq R\}$ (a cylinder).
   $\partial N = \{(t, \mathbf{x}) \in \mathbb{R}^4 : t_0 \leq t \leq t_1, x^3 = 0, |\mathbf{x}| = R\} \cup \{(t, \mathbf{x}) \in \mathbb{R}^4 : t = t_0, x^3 = 0, |\mathbf{x}| \leq R\} \cup \{(t, \mathbf{x}) : t = t_1, x^3 = 0, |\mathbf{x}| \leq R\}$ (the outer cylindrical surface plus top and bottom). (See Figure 7.13.)
   Then (7.3.4) is the integral form of the law of induction:

$$\int_{t_0}^{t_1} dt \int_{\substack{x^3=0\\|\mathbf{x}|=R}} d\mathbf{s} \cdot \mathbf{E} = \int_{\substack{t=t_0, x^3=0\\|\mathbf{x}|\leq R}} d\mathbf{S} \cdot \mathbf{B} - \int_{\substack{t=t_1, x^3=0\\|\mathbf{x}|\leq R}} d\mathbf{S} \cdot \mathbf{B}.$$

   The vanishing of the differential of $F_{|N_3}$ means that

$$0 = d(F_{|N_3}) \quad \Leftrightarrow \quad \dot{B}_3 + (\nabla + \mathbf{E})_3 = 0.$$

As already mentioned, closure is equivalent to exactness on starlike manifolds, and specifically on $\mathbb{R}^4$. This implies that there exists a

(7.3.6) Vector Potential $A \in E_1(\mathbb{R}^4)$

(7.3.7)
$$F = dA.$$

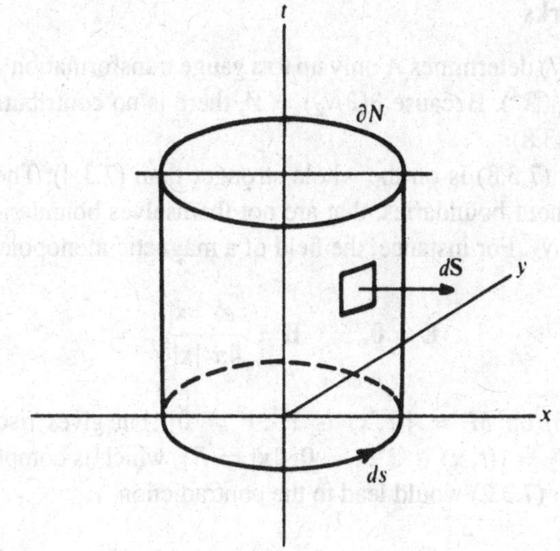

FIGURE 7.13. N_3 of (7.3.5; 2).

In integrated form,

(7.3.8)
$$\int_{N_2} F = \int_{\partial N_2} A,$$

if $F \in E_2^0(N_2)$.

(7.3.9) **Special Cases**

1. $N_2 = \{(t, \mathbf{x}) \in \mathbb{R}^4 : t = x^3 = 0, |\mathbf{x}| \le R\}$ (a disk).
$\partial N_2 = \{(t, \mathbf{x}) \in \mathbb{R}^4 : t = x^3 = 0, |\mathbf{x}| = R\}$ (a circle).
Then (7.3.8) expresses the magnetic flux as a line integral over the spatial part of the vector potential. Writing $A = A\,dt - \mathcal{A}\,d\mathbf{x}$,
$$\int_{\substack{|\mathbf{x}|\le R \\ x^3=t=0}} dx^1 \wedge dx^2\, B_3 = \int_{\substack{|\mathbf{x}|=R \\ x^3=t=0}} d\mathbf{s} \cdot \mathcal{A}.$$

2. $N_2 = \{(t, \mathbf{x}) \in \mathbb{R}^4 : t_0 \le t \le t_1, x^1 = x^2 = 0\}$.
$\partial N = \{(t, \mathbf{x}) \in \mathbb{R}^4 : t = t_0, x^1 = x^2 = 0\} \cup \{(t, \mathbf{x}) \in \mathbb{R}^4 : t = t_1, x^1 = x^2 = 0\}$.
We have been looking only at compact submanifolds, thereby trivially fulfilling the requirement of compact support. If F vanishes outside some compact region, or at least goes to zero fast enough at infinity, then (7.3.8) still holds, and implies
$$\int_{t_1}^{t_2} dt \wedge dx^3\, E_3 = \int_{t=t_0} dx^3\, \mathcal{A}_3 - \int_{t=t_1} dx^3\, \mathcal{A}_3.$$

(7.3.10) **Remarks**

1. Equation (7.3.7) determines A only up to a gauge transformation $A \to A + d\Lambda$, where $\Lambda \in E_0(\mathbb{R}^4)$. Because $\partial(\partial N_2) = \emptyset$, there is no contribution from Λ to the integral (7.3.8).
2. The statement (7.3.8) is on the whole stronger than (7.3.4). There exist manifolds N_2, without boundaries, that are not themselves boundaries of compact submanifolds N_3. For instance, the field of a magnetic monopole,

$$\mathbf{E} = \mathbf{0}, \qquad \mathbf{B} = \frac{e'}{4\pi} \frac{\mathbf{x}}{|\mathbf{x}|^3},$$

satisfies (7.3.3) on $M = \{(t, \mathbf{x}) \in \mathbb{R}^4 : x \neq 0\}$, but gives rise to no vector potential: If $N_2 = \{(t, \mathbf{x}) \in \mathbb{R}^4 : t = 0, \|\mathbf{x}| = R\}$, which is compact and has no boundary, then (7.3.2) would lead to the contradiction

$$0 = \int_{N_2} F = -\int_{|\mathbf{x}|=R} \mathbf{B} \cdot d\mathbf{S} = -e'.$$

N_2 is in fact the boundary of $N_3 = \{(t, \mathbf{x}) \in \mathbb{R}^4 : t = 0, 0 < |\mathbf{x}| < R\}$, but since N_3 is not compact, one cannot conclude that $0 = \int_{N_3} dF = \int_{N_2} F$: for (7.3.4) to be applicable, F must have compact support on N_3. The physical significance of this is that any magnetic sources have been removed from M, whereas the field in M is that of a magnetic monopole. There are theories that also claim there are no electric charges either, but that some complicated structure of M makes it appear that there are electrical sources.
3. $M = \mathbb{R}^4 \backslash \{(t, \mathbf{x}) \in \mathbb{R}^4 : x^1 = x^2 = 0, x^3 \leq 0\}$ is starlike with respect to every point of the positive z-axis, and is contained in the M of Remark 2. Therefore the field of a magnetic monopole satisfies $dF = 0$ on M, and there ought to be a vector potential. In fact,

$$A = \frac{e'}{4\pi} \left(1 - \frac{z}{\sqrt{x^2 + y^2 + z^2}}\right) \frac{x\,dy - y\,dx}{x^2 + y^2}$$

reproduces the field of Remark 2 and is differentiable on M (Problem 6). Equation (7.3.8) does not lead to a contradiction if $N_2 = \{(t, \mathbf{x}) \in \mathbb{R}^4 : t = 0, |\mathbf{x}| = R\} \cap M$, since N_2 is not compact. Physically, A is the potential of an infinitely thin solenoid along the negative z-axis, which has been removed from M. We conclude that the empirical absence of magnetic monopoles indicates that the manifold in which we live has none of the pathologies described here, at least on the small scale.

The inhomogeneous equation describes how F is produced by the sources, which are the charges or, more correctly, the charge densities. Since the charge density ought to yield the charge present in some three-dimensional region, when it is integrated over that region, we describe it as a 3-form ${}^*J$. The corresponding

1-form J consists of a temporal part, the charge density ρ, and a spatial part, the current density $\mathbf{J}$: $J = -\rho\, dt + \mathbf{J}\, d\mathbf{x}$. In terms of J we may write down the

(7.3.11) Inhomogeneous Maxwell Equations

(7.3.12) $$\delta F = J \qquad \text{or} \qquad d^*F = -{}^*J;$$

and in integrated form,

(7.3.13) $$-\int_{\partial N_3} {}^*F = \int_{N_3} {}^*J, \qquad F \in E_2^0(N_3).$$

(7.3.14) Remarks

1. The formulation of these equations requires a manifold with some additional structure, which defines the *-mapping. We do not require the concept of a covariant derivative, which will come up later.
2. Since none of these equations make reference to a chart, we are not restricted to the use of Cartesian coordinates. We shall even postulate the same form of Maxwell's equations for use on more general manifolds than $\mathbb{R}^4$. This will cause the interaction with the gravitational field, and will make the equations describe such things as the bending of light by the Sun.
3. The Poincaré transformations of $\mathbb{R}^4$ are the ones that leave the form of the metric η_{ik}, and consequently also * and Maxwell's equations, invariant.

The equations of (7.3.11) do not have solutions for an arbitrary 1-form J. On the contrary, because of (7.2.6(c)), their first corollary is

(7.3.15) Conservation of Charge

(7.3.16) $$\delta J = 0 \qquad \text{or} \qquad d^*J = 0.$$

When integrated, this is

(7.3.17) $$\int_{\partial N_4} {}^*J = 0 \qquad \text{if} \qquad {}^*J \in E_3^0(N_4).$$

(7.3.18) Special Cases

1. $N_4 = \{(t, \mathbf{x}) : t_0 \le t \le t_1, |\mathbf{x}| \le R\}$. Then (7.3.17) implies that the charge that flows out through the surface equals the change of the charge contained in the sphere between the time t_0 and t_1:

$$\int\limits_{\substack{|\mathbf{x}|<R \\ t=t_0}} d^3x\, \rho - \int\limits_{\substack{|\mathbf{x}|<R \\ t=t_1}} d^3x\, \rho = \int_{t_0}^{t_1} dt \int_{|\mathbf{x}|=R} d\mathbf{S} \cdot \mathbf{J}.$$

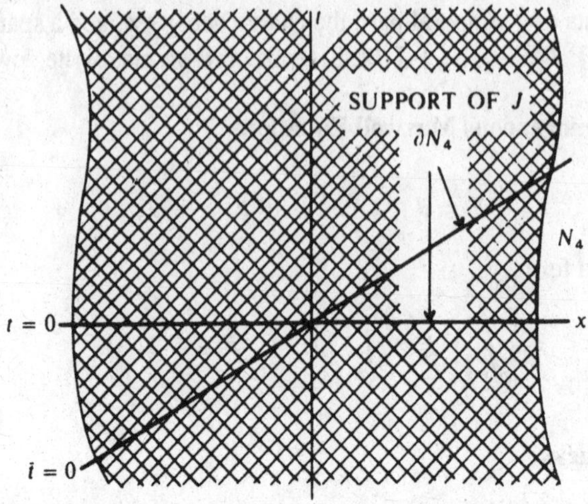

FIGURE 7.14. The Lorentz invariance of the total charge.

2. Let $\partial N_4 = \{(t, \mathbf{x}) : t = 0\} \cup \{(t, \mathbf{x}) : \bar{t} := (t - vx^1)/\sqrt{1 - v^2} = 0\} =: A \cup B$. This fails to be the boundary of a submanifold, because it has sharp bends (see Figure 7.14), although, as already mentioned, the bends could be smoothed out without changing the value of the integral much. It also fails to be compact, so let us assume that J has spatially compact support. Then (recalling that ${}^*dx^0 = -dx^1 \wedge dx^2 \wedge dx^3$, so that $\int_{t=\text{const.}} {}^*J = \int d^3x\, J^0 = \int d^3x\, \rho$),

$$\int_A {}^*J = \int_B {}^*J =: Q.$$

If we calculate the integral over B in Lorentz-transformed coordinates,

$$\bar{x}^1 = \frac{x^1 - vt}{\sqrt{1 - v^2}}, \qquad \bar{x}^2 = x^2, \qquad \bar{x}^3 = x^3, \qquad \bar{t} = \frac{t - vx^1}{\sqrt{1 - v^2}},$$

then we find that

$$Q := \int_{t=0} d^3x\, J^0 = \int_{\bar{t}=0} d^3\bar{x}\, \bar{J}^0, \qquad \bar{J}^0 = \frac{J^0 - vJ^1}{\sqrt{1 - v^2}},$$

since (J^0, J^1), the components of J, transform as (x^0, x^1). The total charge Q is therefore a Lorentz invariant; the increase in the charge density by $\sqrt{1 - v^2}$ is compensated for by the Lorentz contraction of the volume element. One must bear in mind, however, that such concepts as scalar, vector, etc., are not defined for integrated quantities, because the transformation coefficients in general depend on the position.

(7.3.19) **Remarks**

1. The equations (7.3.11) imply more than (7.3.16); conservation of charge necessarily implies the existence of a 2-form ${}^*F$ such that ${}^*J = -d{}^*F$ only if

the manifold is starlike, or has some similar property (cf. (7.2.39; 12)). If t is defined globally on M, and the submanifold $N_3 = \{t = 0\}$ is compact and has no boundary, then it follows from (7.3.11) that

$$Q = \int_{N_3} {}^*J = -\int_{\partial N_3} {}^*F = 0,$$

since $\partial N_3 = \emptyset$. Hence a closed universe must have zero total charge, a conclusion that goes beyond mere charge conservation. The underlying reason is that in this case the lines of force emanating from a charge cannot go to infinity, but must terminate in some opposite charge. The fact is that matter appears to be neutral to a remarkable degree. If the charge on the proton were increased relative to the electron charge, then the $1/r^2$ gravitational force would in effect be weakened for protons and strengthened for electrons. Since it is empirically known that the two forces are proportional to the masses and are weaker than the Coulomb force by a factor 10^{-36}, matter is neutral to at least this accuracy.

2. If F is a periodic 2-form and N_3 the periodicity volume then the contributions from the opposite boundaries to $\int_{\partial N_3} {}^*F$ cancel and $Q = 0$. We shall meet periodic F's in plane wave solutions and thus the total charge in the periodicity volume has to vanish. The usual plane wave solutions actually solve Maxwell's equations for $J = 0$. But there are many wave-type solutions of a plasma coupled to the electromagnetic field and we see that one can get strict periodicity in this case only if particles of both signs of charge are present.

The physical manifestation of the field tensor F is the electromagnetic force, i.e., the exchange of energy and momentum between the field and the charge-carriers. In field theory even the energy and momentum are introduced as local observables, and described by vector fields T^α, where T^0 is the energy-current and T^j, $j = 1, 2, 3$, are the momentum-currents. The components in a basis define the energy-momentum tensor $T^\alpha{}_\beta$:

$$T^\alpha = T^\alpha{}_\beta e^\beta. \tag{7.3.20}$$

If the corresponding dual 3-forms are integrated over a spacelike[7] submanifold N_3, the result is the energy P^0 and the momentum P^j contained in N_3:

$$P^\alpha = \int_{N_3} {}^*T^\alpha. \tag{7.3.21}$$

As in classical mechanics, the Lagrangian formulation of the field equations makes it possible to construct a field energy.

In fact, there will be some close analogies the expressions we met with for classical dynamics, except that field theory is richer in components. The coordinate q is replaced by the vector potential A and $\dot{q}$ by $dA = F$. The Lagrangian $\frac{1}{2}\dot{q}^2$

---

[7] A submanifold is called **spacelike (timelike, lightlike)** if g is positive (negative, zero) when restricted to it.

becomes the 4-form $-\frac{1}{2}F \wedge {}^*F$ and the dual $-{}^*F = \partial\mathcal{L}/\partial\, dA$ plays the role of $p = \partial L/\partial\dot{q}$. The sign is dictated by the requirement that the energy density T_{00} be positive. The counterpart to the energy $H = \dot{q}p - L$ is

$$\begin{aligned} {}^*T^\alpha &= -dA \wedge i^{\alpha *}F - i^\alpha \mathcal{L}, \\ T^\alpha &= \tfrac{1}{2}{}^*((i^\alpha F) \wedge {}^*F - (i^{\alpha *}F) \wedge F) \\ &= -(F^\alpha{}_\tau F^\tau{}_\beta + \tfrac{1}{4}\delta^\alpha{}_\beta F_{\rho\sigma}F^{\rho\sigma})\,dx^\beta, \end{aligned} \tag{7.3.22}$$

where i^α stands for the interior product (7.2.15) with dx^α: $i^\alpha F = F^\alpha{}_\beta\, dx^\beta$. It has the job to pick out the components. We shall derive (7.3.22) in two ways: First by a consideration similar to classical mechanics where the invariance of the Lagrangian under translations produces a conservation law; later, we shall see that it also equals $\partial\mathcal{L}/\partial e^\alpha$. This shows that change $e^\alpha \to L^\alpha{}_\beta e^\beta$ of the basis produces the homogeneous transformation law $T^\alpha \to L^\alpha{}_\beta T^\beta$ for the energy-momentum currents of matter and of the electromagnetic field. Only the Lagrangian of gravity contains de, which produces an inhomogeneous transformation.

As with charge-conservation, the change in the energy-momentum is determined by the codifferential of the currents, δT^α. The rate of change in time of P^α equals a four-dimensional integral over d^*T^α and a surface integral over the spatial parts of the energy-momentum currents. For example, if $N_3 = \{(t, \mathbf{x})\colon t = \pm T, |\mathbf{x}| \le R\}$, then

$$P^\alpha(T) - P^\alpha(-T) = \int\limits_{\substack{-T\le t\le T\\ |\mathbf{x}|\le R}} d^*T^\alpha - \int\limits_{\substack{-T\le t\le T\\ |\mathbf{x}|= R}} {}^*T^\alpha. \tag{7.3.23}$$

If F is coupled to a current, then energy and momentum are exchanged between the field and the current in amounts determined by the codifferential of the energy-momentum currents of the field. This codifferential is determined by Maxwell's equations and is known as the

(7.3.24) **Lorentz Force**

$$\delta T^\alpha = \langle i^\alpha F | J\rangle$$

or, in components,

$$T^{\alpha\beta}{}_{,\beta} = F^{\alpha\beta} J_\beta.$$

Derivation of the Equation for the Lorentz Force

In the natural basis of $\mathbb{R}^n$,

$$L_{e^k}\omega := (i^k \circ d + d \circ i^k)\omega_{j_1\dots j_p} e^{j_1\dots j_p} = \omega_{j_1\dots j_p}{}^{,k} e^{j_1\dots j_p},$$

and hence $L_{e^k}{}^*\omega = {}^*L_{e^k}\omega$. Defining $L^\alpha := L_{e^\alpha}$, this means that

$$\begin{aligned} L^\alpha(F \wedge {}^*F) &= (L^\alpha F) \wedge {}^*F + F \wedge L^{\alpha *}F \\ &= (L^\alpha F) \wedge {}^*F + F \wedge {}^*L^\alpha F = 2(L^\alpha F) \wedge {}^*F \end{aligned}$$

(cf. (7.2.18)(a) and (7.2.23)). Therefore, using (7.2.6) and (7.3.23),

$$\begin{aligned} d^*T^\alpha &= -d[\tfrac{1}{2}i^\alpha(F\wedge {}^*F)-(i^\alpha F)\wedge {}^*F] \\ &= -\tfrac{1}{2}L^\alpha(F\wedge {}^*F)+(L^\alpha F)\wedge {}^*F - i^\alpha F\wedge d^*F = i^\alpha F\wedge {}^*J. \end{aligned}$$

According to (7.2.18)(b) and (c), $\nu\wedge {}^*\omega = {}^*i_\nu\omega(-1)^{p+1}$, for $\nu\in E_1$ and $\omega\in E_p$. Hence

$$\delta T^\alpha = -{}^*d^*T^\alpha = -{}^*(i^\alpha F\wedge {}^*J) = \langle i^\alpha F|J\rangle.$$

(7.3.25) **Remarks**

1. It is important that the i^α in (7.3.23) be an interior product with a natural basis e^α. The $*$-mapping does not commute with L^α in an arbitrary basis, which spoils the above proof. It is also not possible for (7.3.24) to hold in every basis, because if $e^\alpha \to A^\alpha{}_\beta(x)e^\beta$, then the right side would transform linearly with A, while the left side would gain a term containing dA. A physical consequence of this is that fictitious forces must be added to the Lorentz force in an accelerated reference system.
2. If $z(s)$ is the world-line of a particle (see §5.1), then its current is

$$J_\beta(x) = e\int_{-\infty}^{\infty} ds\, \delta^4(x-z(s))\dot z_\beta(s),$$

where $\delta^4(x) := \delta(x^0)\delta(x^1)\delta(x^2)\delta(x^3)$ (cf. Problem 7). The Lorentz force with this J,

$$e\int_{-\infty}^{\infty} ds\, \delta^4(x-z(s))F^{\alpha\beta}(z(s))\dot z_\beta(s),$$

equals $-\delta t^\alpha$, if we choose the energy-momentum currents of a particle so that the energy becomes $\int {}^*t^0 = m\dot z^0 > 0$:

$$\begin{aligned} t^\alpha &= +m\int_{-\infty}^{\infty} ds\, \dot z^\alpha \dot z_\beta(s)\delta^4(x-z)\,dx^\beta: \\ \delta t^\alpha &= +m\int_{-\infty}^{\infty} ds\, \dot z^\alpha \dot z^\beta \frac{\partial}{\partial x^\beta}\delta^4(x-z(s)) \\ &= -m\int_{-\infty}^{\infty} ds\, \dot z^\alpha \frac{d}{ds}\delta^4(x-z(s)) = m\int_{-\infty}^{\infty} ds\, \ddot z^\alpha \delta^4(x-z(s)) \\ &= e\int_{-\infty}^{\infty} ds\, \dot z_\beta F^{\beta\alpha}(z(s))\delta^4(x-z(s)). \end{aligned}$$

Thus the total energy and momentum are formally conserved. Yet the singularity of the field of a point-particle at $z(s)$ results in inconsistencies (cf. §8.4), which are resolved only for continuous matter in §9.1. Note that the Lorentz force is concentrated on the world-line of the particle, and that the equation $\delta(T^\alpha + t^\alpha) = 0$, with the replacement of $\delta^4(x - z(s))$ by $\rho(x - z(s))$ for some continuous function ρ, is no longer true. Hence there is no local energy-momentum conservation for an extended charged particle, unless other forces hold it together.

If T^α is defined in an arbitrary basis by (7.3.23), then it transforms as e^α under a change of basis:

(7.3.26) $$e^\alpha \to \bar{e}^\alpha = A^\alpha{}_\beta e^\beta, \qquad \bar{T}^\alpha = A^\alpha{}_\beta T^\beta.$$

The energy and momentum are combined into a vector-valued 1-form. Thus, if $\delta(T^\alpha + t^\alpha) = 0$, then a global Lorentz transformation treats P^α as a vector (Problem 5). If the $A^\alpha{}_\beta$ are allowed to depend on x, then this statement becomes meaningless, however, and the conservation equation $\delta(\bar{T}^\alpha + \bar{t}^\alpha) = 0$ is false. In fact, in this case

$$d^*(\bar{T}^\alpha + \bar{t}^\alpha) = dA^\alpha{}_\beta \wedge A^{-1\beta}{}_\gamma{}^*(\bar{T}^\gamma + \bar{t}^\gamma).$$

Choosing e^α as the Cartesian dx^α of $\mathbb{R}^4$, i.e., setting the $\omega^\alpha{}_\beta$ of (7.2.25) to zero, makes

(7.3.27) $$\bar{\omega}^\alpha{}_\gamma = -(dA^\alpha{}_\beta)(A^{-1})^\beta{}_\gamma,$$

according to (7.2.26; 3), and therefore

(7.3.28) $$\delta(\bar{T}^\alpha + \bar{t}^\alpha) = -\langle \bar{\omega}^\alpha{}_\gamma | \bar{T}^\gamma + \bar{t}^\gamma \rangle.$$

(7.3.29) **Remarks**

1. It seems odd at first that with general bases there are only nonconservation theorems; they are due to fictitious forces in accelerated reference frames and hence are to be expected on physical grounds. They provide a clue to an understanding of the structure of Einstein's equations.
2. If (7.3.28) is compared with (7.3.24), one sees that ω plays the same role for fictitious forces as F plays for the Lorentz force. ω acts on $T + t$ as F acts on J.

(7.3.30) **Example**

A rotating basis. Let

$$\begin{aligned}
\bar{e}^0 &= dt, \\
\bar{e}^1 &= dx \cos \nu t + dy \sin \nu t, \\
\bar{e}^2 &= -dx \sin \nu t + dy \cos \nu t, \\
\bar{e}^3 &= dz.
\end{aligned}$$

This basis is orthogonal but not natural:

$$d\bar{e}^1 = \nu\, dt \wedge \bar{e}^2, \qquad d\bar{e}^2 = -\nu\, dt \wedge \bar{e}^1 : \bar{\omega}^1{}_2 = -\nu\, dt = -\bar{\omega}^2{}_1.$$

It makes

$$d({}^*\mathcal{T}_1 + {}^*t_1) = \nu\, dt \wedge ({}^*\mathcal{T}_2 + {}^*t_2),$$
$$d({}^*\mathcal{T}_2 + {}^*t_2) = -\nu\, dt \wedge ({}^*\mathcal{T}_1 + {}^*t_1);$$

or, if $P_n(t) \equiv \int_{x^0=t} ({}^*\mathcal{T}_n + {}^*t_n)$, then it follows from Stokes's theorem that

$$P_1(t_1) - P_1(t_0) = \nu \int_{t_0}^{t_1} dt\, P_2(t),$$

$$P_2(t_1) - P_2(t_0) = -\nu \int_{t_0}^{t_1} dt\, P_1(t).$$

These are just like the equations $\dot{P}_1 = \nu P_2$, $\dot{P}_2 = -\nu P_1$ of the mechanics of point particles (cf. (3.2.15; 2)) in a rotating system.

Einstein's theory introduces gravity in (7.3.28) via the principle of equivalence. The gravitational potential is represented by the metric g. In other words, if we use orthogonal bases,[8] so that $g = e^\alpha \otimes e^\beta \eta_{\alpha\beta}$, then the 1-forms e^α are analogous to the A of (7.3.7). Formula (7.3.28) replaces the Lorentz force (7.3.24), and is assumed to hold not only in Minkowski space, but also in the pseudo-Riemannian space determined by g.

(7.3.31) **Remarks**

1. As there are four 1-forms e^α representing gravitational potentials, there is not just one current, but four entering into the force.
2. The ω's are the new field strength, although they do not transform linearly under a change of basis, as F does, but inhomogeneously according to (7.2.26; 3). As a result, they can always be transformed to zero at a point. A physical realization of such a system would be a freely falling elevator, in which there is no gravitational force.
3. Theories have recently been proposed [35] in which the ω's are analogous to A; we shall come back to that in §10.2.

If ω plays the role of F, then

$$de^\alpha = -\omega^\alpha{}_\beta \wedge e^\beta, \qquad \omega_{\alpha\beta} = -\omega_{\beta\alpha},$$

can be viewed as the counterpart of the relation $F = dA$. To construct the inhomogeneous equations, one might try to equate the codifferentials of 2-forms linear in ω to the energy and momentum currents.

Since now there are four currents they ought to be codifferentials of the same number of 2-forms representing the field strength. They should be linear in the

[8] As will be done throughout the rest of this chapter. Then indices are raised and lowered by means of η: $\omega_{\alpha\beta} = \eta_{\alpha\gamma}\omega^\gamma{}_\beta$, etc., which at most changes some signs.

ω's but cannot be identical because the ω's are six independent 1-forms. The right combination turns out to be

(7.3.32) $$F^\gamma = -\tfrac{1}{2} i_{\omega_{\alpha\beta}} e^{\alpha\beta\gamma}.$$

Einstein's equations are the counterpart to inhomogeneous Maxwell's equations and assert that δF^γ equals the energy-momentum currents. However, there is the problem that according to (7.3.28) the currents of matter and electromagnetism alone are not conserved in an arbitrary basis and $\delta F^\gamma = \text{const.}(T^\gamma + t^\gamma)$ would be mathematically inconsistent. This nonconservation is due to fictitious gravitational forces and can be compensated by adding to the contributions from matter and electromagnetism the energy-momentum currents of gravity t^α. It turns out that they are constructed following the pattern of (7.3.23) where de^α replaces dA and F^α replaces F:

$$^*t^\alpha = \frac{1}{16\pi\kappa}[de^\beta \wedge i^{\alpha *} F_\alpha - i^\alpha \mathcal{L}_g].$$

Here κ is the gravitational constant which relates the geometric to the physical quantities. The gravitational Lagrangian

$$\mathcal{L}_g = \frac{1}{16\pi\kappa} {}^*e^{\beta\gamma} \wedge \omega_\gamma{}^\alpha \wedge \omega_{\alpha\beta}$$

is such that

$$^*F^\alpha = \frac{\partial \mathcal{L}_g}{\partial de_\alpha}.$$

These calculations will be done in detail in §10.2. For the moment we just want to show that in analogy to the inhomogeneous Maxwell equation we have

(7.3.33) **Einstein's Equations**

$$\delta F^\gamma = \tfrac{1}{2}\delta(i_{\omega_{\alpha\beta}} e^{\alpha\beta\gamma}) = 8\pi\kappa(T^\gamma + t^\gamma + t^\gamma).$$

(7.3.34) **Remarks**

1. The analogy with (7.3.12) is not quite perfect. Einstein's equations do not contain the codifferentials of $de^\alpha = -{}^*i_{\omega^\alpha{}_\beta} {}^*e^\beta$, but, instead, the 3-form ${}^*e^\beta$ is replaced with $e^{\alpha\beta\gamma}$. The left side of (7.3.33) equals Δe^γ only in special bases.
2. In §10.2 (7.3.33) is compared with the form used by Einstein where only the curvature appears.
3. Equation (7.3.28) follows from (7.3.33) in the same way as the Lorentz force follows from Maxwell's equations.
4. Since t^γ contains ω, it does not transform under a change of basis according to the linear law (7.3.26). Such a law would be inconsistent with the continuity equation

$$\delta(T^\alpha + t^\alpha + t^\alpha) = 0$$

arising from (7.3.33). In particular, in a basis where all the ω's vanish at some point, $\not{t}^\alpha$ also vanishes at that point. Physically speaking, this is a consequence of the principle of equivalence: Since gravity can be transformed away at any point, it can have neither energy nor momentum definitely localized at a point. On the other hand, $\not{t}^\alpha$ may be nonzero even in the absence of gravity (flat space), if an unnatural basis is used. This is to say that fictitious forces produce fictitious energies and momenta.

5. Under a change of basis, equations (7.3.33) transform linearly because the inhomogeneous contribution to $\not{t}^\gamma$ is compensated for in the transformation of the left sides of the equations.
6. There is, of course, not just one $\not{t}^\gamma$ for which the continuity equation (7.3.28) holds; it is always possible to add a closed form to it. With an orthogonal basis, our $\not{t}^\gamma$ agrees with the pseudotensor used by Landau and Lifshitz. At this stage, our definition of $\not{t}^\gamma$ as the source of the term $i_{\omega_{\alpha\beta}} e^{\alpha\beta\gamma}$ may not seem convincing; we shall see later that in some familiar cases it reproduces the gravitational energy of the elementary theory, in a basis that is as Cartesian as possible.

While nonlinearity makes the equations harder to solve, there is no difficulty in drawing general conclusions by using Stokes's theorem and restricting to submanifolds. As with the inhomogeneous Maxwell equations, we obtain some

(7.3.35) **Corollaries**

1.
$$\int_{\partial N_4} ({}^*T^\gamma + {}^*t^\gamma + {}^*\not{t}^\gamma) = 0.$$

If a timelike coordinate t is specified globally on M, and the energy and momentum fall off sufficiently fast at infinity on the submanifold $t = \text{const.}$, then the total energy and momentum are conserved:

$$P^\gamma(t_1) = P^\gamma(t_2),$$
$$P^\gamma(t_2) = \int_{t=t_1} ({}^*T^\gamma + {}^*t^\gamma + {}^*\not{t}^\gamma).$$

2. Equations (7.3.33) allow us to make an even stronger statement,

$$\int_{N_3} ({}^*T^\gamma + {}^*t^\gamma + {}^*\not{t}^\gamma) = -\frac{1}{16\pi\kappa} \int_{\partial N_3} \omega_{\alpha\beta} \wedge {}^*e^{\alpha\beta\gamma}.$$

More specifically, if the submanifold $t = t_0$ is compact, spacelike, and has no boundary, then

$$P^\gamma(t_0) = 0.$$

In analogy with (7.3.19; 1), this means that the total energy and momenta of a closed universe are zero.

(7.3.36) Remarks

1. A rough, order-of-magnitude estimate shows that the negative gravitational energy may well balance the rest energy of the matter in the Universe. The average distance between galaxies is about 6×10^6 light-years. Since the Universe is some 10^{10} years old and about as many light-years across, there are $\sim 10^{11}$ galaxies in all. If each one has about 10^{11} stars at 10^{33} grams apiece, then the Universe has a mass of perhaps 10^{55} grams. Hence the gravitational energy (in cgs units) is on the order of
$$-M\frac{M\kappa}{R} \sim -M \cdot 10^{55} \cdot 10^{-7} \cdot 10^{-28} \sim -M \cdot 10^{20},$$
where the radius of the Universe $R \sim 10^{10}$ light-years $\sim 10^{28}$ cm. This falls short of the rest-energy $Mc^2 \sim M \times 10^{21}$ by about a factor of 10, but there are large uncertainties in our calculation, owing to our ignorance about how much of the energy of the Universe is detectable. Later we shall again have occasion to convince ourselves that t^γ indeed contributes a gravitational energy $\sim -M^2\kappa/R$.
2. Because of the complicated dependence on the basis, there is no intuitively appealing observable corresponding to t^γ. The total energy and momentum P^γ are less sensitive to the choice of basis, as they may be expressed as integrals over ∂N_3 according to (7.3.35; 2). Changes of basis leave them invariant as long as nothing is changed on the boundary. If the space is asymptotically a Minkowski space, then these observables transform as vectors under any transformation that asymptotically approaches a Lorentz transformation.
3. The question of when the energy-momentum forms can be defined globally even if it requires more than one chart to describe the manifold will be discussed later (10.2.13; 8).

So much for the structure of Maxwell's and Einstein's equations. In the remainder of the book, following this preview, we study them in greater detail, and will not only guess a few particular solutions, but also see what general statements can be made about the solutions of these equations.

(7.3.37) Problems

1. What are the consequences of (7.3.13) for $N_3 = \{(t, \mathbf{x}): t = 0, |\mathbf{x}| \le R\}$, and of (7.3.12) restricted to N_3?

2. The same question for $N_3 = \{(t, \mathbf{x}): t_0 \le t \le t_1, x^3 = 0, |\mathbf{x}| \le R\}$.

3. According to (7.3.18; 3), there is a Lorentz-transformed system with a nonzero charge-density $\bar{J}_0$, if $J_0 = 0$ but $J_1 \neq 0$. Explain physically how a neutral system with a current appears charged when seen from a moving reference frame.

4. Show that on $\mathbb{R}^n$ the Lie derivative with respect to the natural basis, $L_k = i_{dx_k} \circ d + d \circ i_{dx_k}$, of $e^{j_1 \cdots j_p}$ simply yields
$$L_k \omega_{j_1 \cdots j_p} e^{j_1 \cdots j_p} = \omega_{j_1 \cdots j_p, k} e^{j_1 \cdots j_p},$$

in the natural basis.

5. Discuss how $P^\alpha := \int_{N_3} {}^*T^\alpha$ is affected by Lorentz transformations.

6. Calculate F with the potential given in (7.3.10; 3).

7. Verify that $\delta J = 0$ for a current of the form

$$J_\alpha(x) = \int_{-\infty}^{\infty} ds\, \dot{z}_\alpha(s)\rho(x - z(s)), \qquad \rho \in E_0^0(\mathbb{R}^4).$$

(7.3.38) **Solutions**

1.

$$-{}^*F = \begin{pmatrix} 0 & B_1 & B_2 & B_3 \\ -B_1 & 0 & E_3 & -E_2 \\ -B_2 & -E_3 & 0 & E_1 \\ -B_3 & E_2 & -E_1 & 0 \end{pmatrix}.$$

Thus (7.3.13) implies that

$$\int_{|\mathbf{x}|=R, t=0} d\mathbf{S}\cdot\mathbf{E} = \int_{|\mathbf{x}|\le R, t=0} d^3x\, \rho(x); \qquad \text{and} \qquad d({}^*F_{|N_3}) = -{}^*J_{|N_3}$$

means that $\nabla\cdot\mathbf{E} = \rho$.

2.

$$\int_{t_0}^{t_1} dt \int_{\substack{x^3=0\\|\mathbf{x}|=R}} ds\cdot\mathbf{B} = \int_{\substack{t=t_1, x^3=0\\|\mathbf{x}|\le R}} d\mathbf{S}\cdot\mathbf{E} - \int_{\substack{t=t_0, x^3=0\\|\mathbf{x}|\le R}} d\mathbf{S}\cdot\mathbf{E} + \int_{t_0}^{t_1} dt \int_{\substack{x^3=0\\|\mathbf{x}|\le R}} d\mathbf{S}\cdot\mathbf{j},$$

or $\nabla\times\mathbf{B} = \mathbf{j} + \dot{\mathbf{E}}$.

3. A neutral current consists of oppositely charged current flowing past each other. Because of the differing definitions of simultaneity, the two currents seem to have different charges in the moving frame (see Figure 7.15).

4.

$$i^k\, d\omega = \omega_{j_1\cdots j_p, j}\, i^k e^{jj_1\cdots j_p} = \omega_{j_1\cdots j_p,}{}^k e^{j_1\cdots j_p} - p\omega^k{}_{j_2\cdots j_p, j} e^{jj_2\cdots j_p},$$
$$di^k\omega = pd(\omega^k{}_{j_2\cdots j_p} e^{j_2\cdots j_p}) = p\omega^k{}_{j_2\cdots j_p, j} e^{jj_2\cdots j_p}.$$

5. Since the natural basis transforms according to $d\bar{x}^\alpha = L^\alpha{}_\beta\, dx^\beta$, it is also true that $\bar{T}^\alpha = L^\alpha{}_\beta T^\beta$. Now note that N_3 looks different in the new coordinates. For instance, if N_3 is the hyperplane $t = 0$, then in the new system it is not $\bar{t} = 0$, but $\bar{t} = v\bar{x}$. It is only if $\delta T^\alpha = 0$ and T^α vanishes fast enough at infinity that

$$\int_{t=0} {}^*T^\alpha = \int_{\bar{t}=0} \bar{T}^\alpha$$

(cf. (7.3.18; 2)).

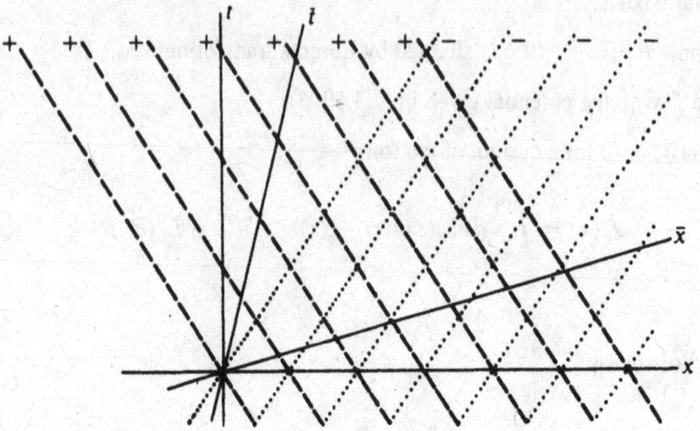

FIGURE 7.15. At any point of time $t = \text{const.}$, x sees 7 −'s to 7 +'s. At any point of time $\bar{t} = \text{const.}$, $\bar{x}$ sees 5 −'s to 7 +'s.

6. $r^2 := x^2 + y^2 + z^2$; then

$$B_x = -\frac{\partial \mathcal{A}_y}{\partial z} = \frac{e^1}{4\pi}\frac{x}{r}\frac{1}{x^2+y^2}\left(1 - \frac{z^2}{r^2}\right) = \frac{e^1}{4\pi}\frac{x}{r^3},$$

$$B_y = \frac{\partial \mathcal{A}_x}{nz} = \frac{e^1}{4\pi}\frac{y}{r^3},$$

$$B_z = \frac{\partial \mathcal{A}_y}{\partial x} - \frac{\partial \mathcal{A}_x}{\partial y} = \frac{e^1}{4\pi}\left\{\frac{z(x^2+y^2)}{r^3(x^2+y^2)} + \frac{1}{x^2+y^2} - \frac{2x^2}{x^2+y^2} + \frac{1}{x^2+y^2} - \frac{2y^2}{x^2+y^2}\right\} = \frac{e^1}{4\pi}\frac{z}{r^3}.$$

7.

$$\frac{\partial}{\partial x^\beta}\int_{-\infty}^{\infty} ds\, \rho(x - z(s))\dot{z}^\beta(s) = -\int_{-\infty}^{\infty} ds\, \dot{z}^\beta(s),$$

$$\frac{\partial}{\partial z^\beta}\rho(x - z(s)) = -\int_{-\infty}^{\infty} ds\, \frac{\partial}{\partial s}\rho(x - z(s)) = 0.$$

No boundary terms appear from the partial integration, if the world-lines are infinite and $\rho \in E_0^0$. The normalization $\langle \dot{z}|\dot{z}\rangle = -1$ is irrelevant: $\delta J = 0$ even if the particles move faster than light.

8

The Electromagnetic Field of a Known Charge Distribution

8.1 The Stationary–Action Principle and Conservation Theorems

If the field equations originate from a stationary–action principle, then a conserved current can be constructed for each parameter of an invariance group.

Field theory may be regarded as a generalization of the mechanics of point-particles, in which the dynamical variables $q_i(t)$ are replaced with fields $\Phi(x, t)$, such as $\mathbf{E}(x, t)$ and $\mathbf{B}(x, t)$. The discrete index i goes over to the continuous variable x, and, accordingly, the sum $\sum_i$ is replaced with an integral $\int d^3x$. A direct transcription of the formalism of Chapter 3, leads to infinite-dimensional manifolds, which we would prefer to avoid. Instead, we merely generalize the stationary–action principle (2.3.20) in order to find the analogues of the constants arising from the invariance properties. It is clear that in field theory the action $\int dt\, L(q, \dot{q})$ involves an integral over a four-dimensional submanifold N_4, and thus requires a 4-form, which allows the construction of a chart-independent integral.

(8.1.1) The Lagrangian Formulation of Field Theory

The action is given by

$$W = \int_{N_4} \mathcal{L}(\Phi, d\Phi),$$

where $\mathcal{L} \in E_4$ is the **Lagrangian.** The field equations result from the requirement that $\delta W = 0 \; \forall N_4$ compact and $\forall \delta\Phi$ such that $\delta\Phi_{|\partial N_4} = 0$.[1]

If we strengthen the homogeneous Maxwell equations to $F = dA$, then in pseudo-Riemannian space, the appropriate

(8.1.2) **Electromagnetic Lagrangian**

is

$$\mathcal{L} = -\tfrac{1}{2} dA \wedge {}^*dA - A \wedge {}^*J.$$

Proof: Making a variation $A \to A + \delta A$ and using (7.2.18)(a), one finds

$$-\delta W = \int_{N_4} \delta A \wedge [{}^*J + d^*dA] + \int_{\partial N_4} \delta A \wedge {}^*dA,$$

which vanishes if $\delta A_{|\partial N_4} = 0$ and $d^*F = -{}^*J$. □

(8.1.3) **Remarks**

1. The variational formulation offers no guarantee of existence or uniqueness of the solutions of the field equations. Nowhere has it been assumed that $d^*J = 0$, though without this condition it is not possible to satisfy $\delta W = 0 \; \forall \delta A$ such that $\delta A_{|\partial N_4} = 0$. The reason is easy to discover. With the gauge transformation $A \to A + d\Lambda$, where $\Lambda_{|\partial N_4} = 0$, W changes by $\int_{N_4} \Lambda \, d^*J$, and is linear in Λ not only for infinitesimal Λ. As a linear functional, either W has no stationary points, or else, if $d^*J = 0$, it has a plateau. Accordingly, either there are no solutions at all, or else the solution is not uniquely fixed by any boundary condition whatsoever, because there is always the possibility of a gauge transformation.
2. According to (5.2.8),

$$-\tfrac{1}{2} F \wedge {}^*F = -\tfrac{1}{4} F_{\sigma\rho} F^{\sigma\rho} {}^*\mathbf{1} = \tfrac{1}{2}(|\mathbf{E}|^2 - |\mathbf{B}|^2)^*\mathbf{1}.$$

The sign of $\mathcal{L}$ has been chosen so that the interaction

$$-A \wedge {}^*J = -{}^*i_J A = -J^\alpha A_\alpha {}^*\mathbf{1}$$

of a point particle moving along the world-line $z(s)$ (cf. (7.3.25; 2)) has the same sign

$$-e \int_{-\infty}^{\infty} ds \, \dot{z}^\alpha(s) A_\alpha(z(s)) \delta^4(x - z(s))^*\mathbf{1}$$

as in (5.1.8). If a term $\frac{1}{2} m \int ds \, \dot{z}^\alpha(s) \dot{z}_\alpha(s)$ were added to the action, then both the field equations and the equation of motion could be derived from the same stationary–action principle by varying $A(x)$ and $z(s)$. The coupled system of equations suffers from difficulties due to the reaction of a particle on itself, as will be discussed more fully in §8.4.

---

[1] We use the symbol δ in §8.1 for variations, rather than for codifferentials.

The advantage of the Lagrangian formulation is that every one-parameter invariance group of $\mathcal{L}$ furnishes a conservation theorem. In field theory the argument goes as follows: If the 4-form $\mathcal{L}$ depends on the p-form Φ_j, then a variation of $\mathcal{L}$ (taking $\partial\mathcal{L}/\partial\Phi_j$ with $\delta\Phi$ commuted through to the left) produces

$$\text{(8.1.4)}\quad \delta\mathcal{L} = \sum_j \delta\Phi_j \wedge \left[\frac{\partial\mathcal{L}}{\partial\Phi_j} - (-1)^p d\frac{\partial\mathcal{L}}{\partial(d\Phi_j)}\right] + d\left(\sum_j \delta\Phi_j \wedge \frac{\partial\mathcal{L}}{\partial(d\Phi_j)}\right),$$

and the field equations require that the term in the square brackets [ ] be zero. If the variation δ is that of a Lie derivative L_v, $v \in E_1$, then by (7.2.23), $\delta\mathcal{L} = L_v\mathcal{L} = di_v\mathcal{L}$, because the exterior differential of a 4-form vanishes on a four-dimensional manifold. (We think of δ as a kind of derivative, and not as an infinitesimal quantity.) Hence (8.1.4) says that a certain 3-form is closed and thus, by use of the *-mapping, that there is a conserved current.

(8.1.5) Noether's Theorem

Suppose that $v \in E_1$, $\delta\Phi = L_v\Phi$ and $\delta\mathcal{L} = L_v\mathcal{L}$. Then

$$d\left[\sum_j L_v\Phi_j \wedge \frac{\partial\mathcal{L}}{\partial(d\Phi_j)} - i_v\mathcal{L}\right] =: +d^*T_v = 0.$$

(8.1.6) Remarks

1. The validity of (8.1.5) is premised on the variation δ including everything that is affected by L_v. For instance, even if $J = 0$, the $\mathcal{L}$ of (8.1.2) involves not only dA but also the metric g, through the *-mapping, and g was not allowed to vary below (8.1.2). Since we supposed that $\delta^*F = {}^*\delta F$, Noether's theorem (8.1.5) applies to (8.1.2) only if $J = 0$ and $L_v{}^* = {}^*L_v$.
2. If L_v is a translation, then (8.1.5) defines what is known as the **canonical energy-momentum tensor.**
3. The field equations remain unchanged when an exact 4-form is added to $\mathcal{L}$, $\mathcal{L} \to \mathcal{L} + dG$, where $G \in E_3$, since the addition only contributes a boundary integral to δW. However, the 3-form in the square brackets in (8.1.5) gets a contribution, which, if G depends only on the fields Φ and not on their derivatives, is di_vG (Problem 1). As an exact 3-form, it does not contribute to integrals over submanifolds without boundaries, but it can affect the conserved observables locally. This difficulty is not encountered in classical mechanics, which is formally a one-dimensional field theory, and where G would be in E_0, and hence $i_vG = 0$. Indeed, an additional dG such that

$$\frac{d}{dt}G(q) = \dot{q}_i\frac{\partial G}{\partial q_i}$$

does not change the Hamiltonian

$$H = \dot{q}_i \frac{\partial \mathcal{L}}{\partial \dot{q}_i} - \mathcal{L}$$

at all.

(8.1.7) Application to the Electromagnetic Field

If $J = 0$ and $L_v{}^* = {}^*L_v$, then

$$d[-(L_v A) \wedge {}^*dA + \tfrac{1}{2} i_v(dA \wedge {}^*dA)]$$
$$= d[-\tfrac{1}{2}(i_v F) \wedge {}^*F + \tfrac{1}{2} F \wedge i_v{}^*F - (di_v A) \wedge {}^*F] = 0.$$

(8.1.8) Remarks

1. As can be seen above, if $v = e^\alpha$, then in addition to the electromagnetic energy-momentum forms ${}^*\mathcal{T}^\alpha$ given in (7.3.23), there is a new term $(di_v A) \wedge {}^*F$. If $J = 0$, then this is an exact 3-form, $(di_v A) \wedge {}^*F = d(i_v A \wedge {}^*F) + i_v A \wedge {}^*J$, so that (8.1.7) is a special case of (7.3.24). Unfortunately, the new term contains A as well as F, and thus depends on the gauge. This does not contradict the gauge-invariance of $\mathcal{L}$ (assuming $J = 0$): Even in the mechanics of point particles the angular momentum $m[\mathbf{x} \times \dot{\mathbf{x}}]$ fails to be translation-invariant, although $\mathcal{L} = m|\dot{\mathbf{x}}|^2/2$ is.
2. The nonuniqueness mentioned in (8.1.6; 3) consists of an additional $d(F \wedge A) = F \wedge F$ in $\mathcal{L}$. Since the corresponding G depends on dA as well as A, the extra term in the conserved observable is changed to $2d((i_v A) \wedge F)$ (Problem 5). This is conserved independently of whether ${}^*L_v = L_v{}^*$, because $F \wedge F$ makes no reference to the metric structure of space–time. The expression $d((i_v A) \wedge F)$ not only depends on the gauge, but it also has the wrong reflection property. For example, the energy density would have a term $\sim \mathbf{B} \cdot \nabla V$, which changes sign if $(t, \mathbf{x}) \rightarrow (t, -\mathbf{x})$, as $(V, \boldsymbol{A}) \rightarrow (V, -\boldsymbol{A})$ and $(\mathbf{E}, \mathbf{B}) \rightarrow (-\mathbf{E}, \mathbf{B})$. In the so-called gauge theories, similar expressions determine what are known as topological charges $\int F \wedge F$, which characterize the topological structure of the bundle.
3. The local energy is defined by the coupling with gravity, and the formula for it is derived by letting the metric vary in $\mathcal{L}$. This variation has no contribution from $F \wedge F$, and $F \wedge {}^*F$ gives rise only to the T^α of (7.3.22), which is gauge-invariant (see (10.2.8)).
4. If $J \neq 0$, then adding $L_v{}^*J$ to the variations in (8.1.2) results in the Lorentz force. If $d^*F = -{}^*J$, then from (8.1.2),

$$L_v\mathcal{L} = -d[(L_v A) \wedge {}^*dA] - A \wedge L_v{}^*J$$
$$= -d[(i_v F) \wedge {}^*F] - (di_v A) \wedge {}^*J - A \wedge L_v{}^*J.$$

On the other hand, due to the rules governing L_v,

$$L_v\mathcal{L} = -\tfrac{1}{2} di_v(F \wedge {}^*F) - L_v(A \wedge {}^*J).$$

Together, these facts imply the formula

$$-d^*T_v := \tfrac{1}{2} d[F \wedge i_v{}^*F - (i_v F) \wedge {}^*F] = -(i_v F) \wedge {}^*J$$

for any vector field such that ${}^*L_v = L_v{}^*$.

5. If $d^*J = 0$ and $\mathcal{L}$ is as in (8.1.2), then $\int_{N_4} \mathcal{L}$ is invariant under $A \to A + d\Lambda$, $\Lambda_{|\partial N_4} = 0$. The gauge group is then a huge (Abelian) invariance group, and is not even locally compact, since it contains arbitrary functions. This leads one to think that there must be an infinite number of conservation theorems. However, they always reduce to trivialities, or rather to identities that hold independently of the field equations. By the way, this is not a characteristic only of field theories, but also occurs in point-particle mechanics (Problem 3). In (8.1.2) the variation $\delta A = d\Lambda$ (and $d^*J = 0$) produces

$$0 = \delta W = -\int_{N_4} \Lambda d(d^*F) + \int_{\partial N_4} \Lambda({}^*J + d^*F).$$

Since $\Lambda_{|\partial N_4} = 0$, this means that $d(d^*F) = 0$, which is true regardless of whether $d^*F = -{}^*J$.

As we have seen, there is a conservation theorem for each v whose Lie derivative does not destroy the structure of $\mathcal{L}$ determined by the metric. Such vector fields are important enough to merit a

(8.1.9) Definition

A vector field v satisfying $L_v g = 0$ on a pseudo-Riemannian manifold with the metric g is known as a **Killing vector field.**

(8.1.10) Remarks

1. Because $L_x L_y - L_y L_x = L_{[x,y]}$, where [,] is the Lie bracket of vector fields (see (2.5.9; 6)), the Killing vector fields form a Lie algebra with [,]. (But not a module: fv, with $f \in E_0$ is not necessarily a Killing vector field if v is.)
2. If an orthogonal basis e^i ($g = e^i \otimes e^j \eta_{ij}$) is used to decompose the Lie derivative of the e^i as $L_v e^i = A^i{}_j e^j$, $A^i{}_j \in E_0$, then v is a Killing vector field iff $A_{ij} = -A_{ji}$, where $A_{ij} := \eta_{ik} A^k{}_j$.
3. Problem 2 is to show that ${}^*L_v = L_v{}^*$ for Killing vector fields v.
4. It is possible for ${}^*L_v\omega$ to equal $L_v{}^*\omega$, $\omega \in E_p$ (for particular values of p), even if v is not a Killing vector field. For example, $L_v e^j = f e^j$, $f \in E_0$, makes $L_v e^{j_1 \cdots j_p} = pf e^{j_1 \cdots j_p}$ and $L_v{}^* e^{j_1 \cdots j_p} = (m-p) f^* e^{j_1 \cdots j_p}$, and so ${}^*L_v\omega = L_v{}^*\omega$ for all $\omega \in E_{m/2}$. Yet L_v generates the conformal transformation $L_v g = 2fg$, and v is not a Killing vector field.

(8.1.11) Examples

M is taken as $\mathbb{R}^4$ with $e^\alpha = dx^\alpha$ and $g = e^\alpha \otimes e^\beta \eta_{\alpha\beta}$ in these examples.

1. The rigid displacement $v = e^\alpha$ leaves g invariant:

$$L_v e^\gamma = di_{e^\alpha} e^\gamma + i_{e^\alpha} de^\gamma = 0.$$

The T_v of (8.1.8; 4) becomes the T^α of (7.3.23), and (8.1.8; 4) reduces to the Lorentz force (7.3.24).

2. $v = x^\beta e^\alpha - x^\alpha e^\beta$ generates a Lorentz transformation

$$L_v e^\gamma = x^\beta i_{e^\alpha} de^\gamma + d(x^\beta i_{e^\alpha} e^\gamma) - (\alpha \leftrightarrow \beta) = -\eta^{\beta\gamma} dx^\alpha + \eta^{\alpha\gamma} dx^\beta.$$

The A of (8.1.10; 2) becomes $A^{\gamma\alpha} = \eta^{\alpha\gamma}\eta^{\beta\sigma} - \eta^{\beta\gamma}\eta^{\alpha\sigma}$, and it satisfies the condition of antisymmetry that characterizes Killing vector fields. Because the interior product is linear, the T_v of (8.1.8; 4) is simply $x^{\beta *}T^\alpha - x^{\alpha *}T^\beta$. Remark (8.1.8; 4) means that

$$d(x^{\beta *}T^\alpha - x^{\alpha *}T^\beta) = x^\beta (i_{e^\alpha} F) \wedge {}^*J - x^\alpha (i_{e^\beta} F) \wedge {}^*J,$$

which implies with (7.3.24) that $dx^\beta \wedge {}^*T^\alpha - dx^\alpha \wedge {}^*T^\beta = 0$. Because $dx_\beta \wedge {}^*dx^\gamma = \delta^\gamma{}_\beta {}^*\mathbf{1}$, the energy-momentum tensor must therefore be symmetric; i.e., $T_{\alpha\beta} = T_{\beta\alpha}$. Although (7.3.23) is symmetric, the canonical energy-momentum tensor (8.1.7) is not; to be sure, the canonical energy-momentum tensor is also conserved if $v = x^\beta e^\alpha - x^\alpha e^\beta$, but x^α cannot simply be factored out of the gauge-dependent term $di_v A$, which contains dv, and the conserved quantity is not $x^\alpha \mathcal{T}^\beta - x^\beta \mathcal{T}^\alpha$.

3. $v = x_\alpha e^\alpha$ generates a dilatation, $L_v e^\gamma = e^\gamma$; and if $F \in E_2(\mathbb{R}^4)$, then according to (8.1.10; 4) $L_v {}^*F = {}^*L_v F$. Then T^α can be used to formulate a new conservation theorem.

$$d(x_\alpha {}^*T^\alpha) = x_\alpha i_{e^\alpha} F \wedge {}^*J.$$

From this and (7.3.24) we conclude that $dx \wedge {}^*T^\alpha = T^{\alpha}_{\alpha}{}^*\mathbf{1} = 0$. We again observe that (7.3.23) satisfies this equation, whereas the canonical energy-momentum tensor (8.1.7) does not.

4. The conformal transformation $L_v g = 2x^\beta g$ is generated by $v = x^\beta x_\alpha e^\alpha - \frac{1}{2} x^\alpha x_\alpha e^\beta$:

$$L_v e^\gamma = x^\beta e^\gamma + x^\gamma e^\beta - \eta^{\gamma\beta} x_\alpha e^\alpha,$$

and the last two terms cancel out in the expression for $L_v g$. According to Remark (8.1.10; 4), it is again true that $L_v {}^*F = {}^*L_v F$ for all $F \in E_2(\mathbb{R}^4)$, and d^*T_v is

$$d(x^\beta x_\alpha {}^*T^\alpha - \tfrac{1}{2} x^{2*} T^\beta) = ((x^\beta x_\alpha i_{e^\alpha} - \tfrac{1}{2} x^2 i_{e^\beta}) F) \wedge {}^*J.$$

The resultant equation

$$0 = x_\alpha\, dx^\beta \wedge {}^*T^\alpha - x_\alpha\, dx^\alpha \wedge {}^*T^\beta + x^\beta\, dx_\alpha \wedge {}^*T^\alpha$$

contains no new information, because the final term vanishes as in Example 3 and the first two vanish as in Example 2.

(8.1.12) **Remarks**

1. The vector fields generating conformal transformations (8.1.11; 4) are not complete (Problem 6); these transformations are not diffeomorphisms of $\mathbb{R}^4$, as they have singularities. This is not serious for the uses we make of them, as we need only the infinitesimal transformations. If a group of diffeomorphisms is desired, then $\mathbb{R}^4$ must be compactified by the addition of points at infinity [36].
2. The Lagrangian $d\Phi \wedge {}^*d\Phi$ for a massless scalar field $\Phi \in E_0$ has no intrinsic length, but even so it fails to be invariant under the dilatation (8.1.11; 3). As a consequence $T^\alpha{}_\alpha \neq 0$ for the energy-momentum tensor of a scalar field even if $m = 0$ (Problem 4).

(8.1.13) **The Properties of the Energy-Momentum Tensor**

Let e^α be an orthogonal basis of a pseudo-Riemannian space ($g = -e^0 \otimes e^0 + \sum_{j=1}^3 e^j \otimes e^j$), and let ${}^*T^\alpha = \frac{1}{2}(i_{e^\alpha} F \wedge {}^*F - F \wedge i_{e^\alpha}{}^*F) = T^\alpha{}_\beta{}^*e^\beta$. Because of the component representation

$$T_{00} = \tfrac{1}{2}(|\mathbf{E}|^2 + |\mathbf{B}|^2), \qquad T^j{}_0 =: S^j, \qquad j = 1, 2, 3;$$

where

$$\mathcal{S} = [\mathbf{E} \times \mathbf{B}] \text{ is } \textbf{Poynting's vector},$$

we find that for all x:

(a) $T_{00}(x) \geq 0$, and $= 0$ only if $F(x) = 0$; and

(b) $\|\mathcal{S}(x)\| \leq T_{00}(x)$.

(8.1.14) **Remarks**

1. If (a) holds in every Lorentz system, then (b) follows. A Lorentz transformation treats $P^\alpha = \int {}^*T^\alpha$ as a vector (7.3.38; 5), and the equation $P^0 \geq 0$ would be violated by the transformation $P^0 \to (P^0 - \mathbf{v} \cdot \mathbf{P})/\sqrt{1 - v^2}$ unless $\|P\| \leq P^0$.
2. In the orthogonal basis, $S^j = T^j{}_0 = T_0{}^j$ doubles as a momentum density and as the rate of energy flow: The change in the total energy can be written as
$$\delta T_0 = \frac{\partial}{\partial t} T_0{}^0 + \frac{\partial}{\partial x^j} T_0{}^j.$$
The physical interpretation of (b) is that electromagnetic energy can never be transmitted faster than light.
3. The positivity of the energy is expressed mathematically as follows: Let N_3 be a spacelike submanifold, and define $*$ with the restriction of g to N_3. Then $*$ converts a 3-form into a numerical function, and the positivity of the energy means that ${}^*(T^0{}_{|N_3}) \geq 0$.
4. The signs in (8.1.13) arise from the signature of the metric, and thus depend on the relationship to the standard basis. However, the equations $T^{\alpha\beta} = T^{\beta\alpha}$ and

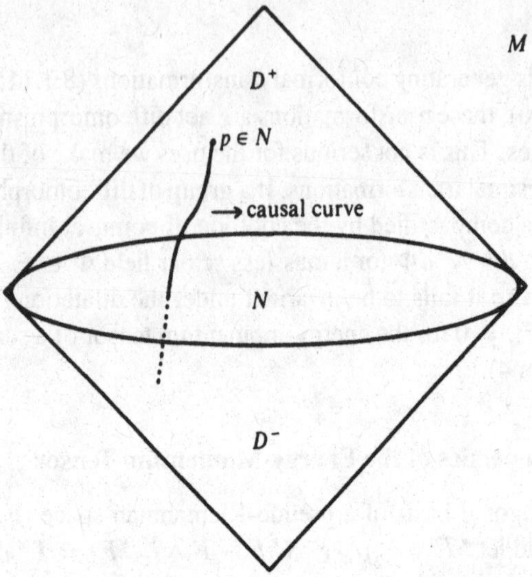

FIGURE 8.1. The domain of influence of a hypersurface $N \subset M$.

$T^\alpha{}_\alpha = 0$, which follow from (8.1.11; 2) and (8.1.11; 3), hold in any basis on account of the transformation law (7.3.26).

Since the existence of a limiting speed of energy transport follows solely from the structure of the energy-momentum tensor, it is possible to prove the uniqueness of the solution of the Cauchy problem without further analysis of the field equations. Nonuniqueness would contradict the causal propagation of the field at the speed of light. To make this impression mathematically precise, we make some intuitively reasonable.

(8.1.15) **Definitions**

(a) A continuous mapping of an interval $I \to M$ is called a **causal curve** iff no two of its points can be connected by a spacelike curve. It is said to be **nonextensible** iff it is not a proper subset of a larger causal curve.

(b) Let M be orientable in time, that is, the forward and backward time directions can be defined smoothly over all of M. The **future** (resp. **past**) **domain of influence** $D^+(N)$ (resp. $D^-(N)$) of a spacelike hypersurface N is the set of all points p of M for which all nonextensible, causal curves through p oriented toward the past (resp. future) intersect N. $D(N) := D^+(N) \cup D^-(N)$ is called the **domain of influence** of N, and if $D(N) = M$ then N is a **Cauchy surface.**

The uniqueness of the Cauchy problem is the statement that F is uniquely determined on $D^\pm(N_3)$ by $F_{|N_3}$, $^*F_{|N_3}$, and J. If $F_{|D^\pm(N_3)} \neq 0$ but $J = F_{|N_3} = {}^*F_{|N_3} = 0$, then F would have to propagate from somewhere outside $D^\pm(N_3)$ to within

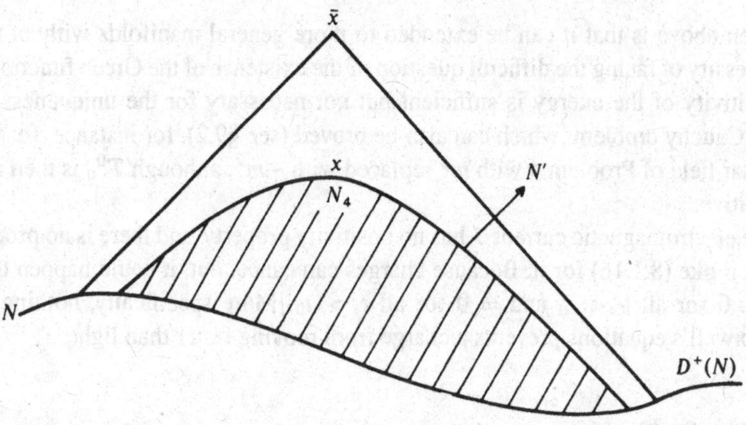

FIGURE 8.2. The region of integration used to prove the uniqueness of the Cauchy problem.

$D^{\pm}(N_3)$ without encountering N_3, which would be possible only if the propagation speed were greater than the speed of light. Since there exist complicated manifolds M for which the shape of $D^{\pm}$ is confusingly tortuous, let us be contented with the

(8.1.16) Uniqueness of the Cauchy Problem in Minkowski Space

Let N be a three-dimensional, compact, spacelike submanifold of $(\mathbb{R}^4, \eta)$, and suppose that F_1 and F_2 are two continuous solutions of Maxwell's equations, $dF_1 = dF_2 = 0$, and $\delta F_1 = \delta F_2 = J$. If $F_{1|N} = F_{2|N}$ and ${}^*F_{1|N} = {}^*F_{2|N}$, then F_1 and F_2 are also equal throughout the interiors of $D^{\pm}(N)$: $F_{1|\,\mathrm{Int}\,D^{\pm}(N)} = F_{2|\,\mathrm{Int}\,D^{\pm}(N)}$.

Proof: Let $x \in \mathrm{Int}\,D^+(N)$. Then there exist $\varepsilon > 0$ and $\bar{x} \in D^+(N)$ such that

$$x \in N' := \{y \in \mathbb{R}^4 : |\bar{\mathbf{x}} - \mathbf{y}|^2 - (\bar{x}^0 - y^0)^2 = -\varepsilon^2,\ y^0 \le \bar{x}^0\}$$
$$\subset \{y \in \mathbb{R}^4 : |\bar{\mathbf{x}} - \mathbf{y}|^2 \le (\bar{x}^0 - y^0)^2\}.$$

Hence

$$(N' \cap D^+(N)) \cup \{y \in N : |\bar{\mathbf{x}} - \mathbf{y}|^2 - (\bar{x}^0 - y^0)^2 \le -\varepsilon^2\} = \partial N_4,$$

where N_4 is a compact, four-dimensional submanifold (see Figure 8.2). If $F = F_1 - F_2$, then $dF = \delta F = 0$ and $F_{|N} = {}^*F_{|N} = 0$. The T^α formed from F therefore satisfies $\delta T^\alpha = 0$ and $T^\alpha{}_{|N} = 0$, and so $0 = \int_{N_4} d^*T^0 = \int_{N' \cap D^+(N)} {}^*T^0$. Since N' is spacelike, ${}^*T^0$ is a nonnegative measure on N', and the vanishing of the integral implies that ${}^*T^0{}_{|N' \cap D^+(N)}$ is zero almost everywhere. Since it is a continuous function, ${}^*T^0{}_{|N' \cap D^+(N)} = 0$; and because of (8.1.13)(b), $F_{|N' \cap D^+(N)} = 0$, and in particular $F_{|x} = 0$. The proof for $D^-(N)$ is similar. □

(8.1.17) Remarks

1. Formula (7.2.36) proves uniqueness by explicit construction, but it assumes the existence of the distributions $G_{\bar{x}}$, $dG_{\bar{x}|N}$, and ${}^*dG_{\bar{x}|N}$. The advantage of the proof

given above is that it can be extended to more general manifolds without the necessity of facing the difficult question of the existence of the Green functions.
2. Positivity of the energy is sufficient but not necessary for the uniqueness of the Cauchy problem, which can also be proved (see §9.2), for instance, for the scalar field of Problem 4 with m^2 replaced with $-m^2$, although $T^0{}_0$ is then not positive.
3. The electromagnetic current J has no positivity property, and there is no proposition like (8.1.16) for it. Because charges can cancel out, it could happen that $J \equiv 0$ for all $t < t_0$ and $\neq 0$ for all $t > t_0$. More specifically, nothing in Maxwell's equations prevents a charge from moving faster than light.

(8.1.18) **Problems**

1. Show that $(L_v\Phi_j) \wedge (\partial\mathcal{L}/\partial\, d\Phi_j) - i_v\mathcal{L} = di_vG$, if $\mathcal{L} = dG(\Phi)$.

2. Show that if v is a Killing vector, then $L_v{}^* = {}^*L_v$. (Use (7.2.26; 2) and (8.1.10; 2).

3. The action $W = \int ds[-\dot{z}^\alpha(s)\dot{z}_\alpha(s)]^{1/2}$ is left invariant not only by $s \to s + \text{const.}$, but even by $s \to f(s)$, for f such that $0 < df/ds < \infty$. What is the resultant conserved quantity?

4. Calculate T^α for $\mathcal{L} = -\frac{1}{2}[d\Phi \wedge {}^*d\Phi + m^2\Phi^*\Phi]$, $\Phi \in E_0$, and investigate which of the properties (8.1.11) and (8.1.13) this T^α has.

5. Calculate the conserved quantity (8.1.5) for $\mathcal{L} = dA \wedge dA$, $A \in E_1$.

6. Integrate the equations $\partial x(t)/\partial t = -a\langle x(t)|x(t)\rangle + 2x(t)\langle a|x(t)\rangle$, which generate the (local) flow of a conformal transformation.

7. Suppose that $\delta(T^\alpha + t^\alpha) = 0$ and that $T^\alpha + t^\alpha$ is independent of time in some system and falls off sufficiently in space. Show that in this reference system

$$\int_{0\le t\le T} dx^j \wedge {}^*(T^\alpha + t^\alpha) = 0,$$

for $j = 1, 2, 3$, and $\alpha = 0, 1, 2, 3$. (In particular, the "self-stress" terms $\int d^3x\, T_{jj}$ vanish.) Conclude that $\delta(T^\alpha + t^\alpha)$ cannot be zero for the point charge (7.3.25; 2).

(8.1.19) **Solutions**

1.
$$\mathcal{L} = d\Phi_j \wedge \frac{\partial G}{\partial \Phi_j}, \qquad (L_v\Phi_j) \wedge \frac{\partial G}{\partial \Phi_j} - i_v\, dG = L_vG - i_v\, dG = di_vG.$$

2. First note that $L_ve^{k_1\cdots k_p} = \sum_j A^{k_j}{}_k e^{k_1\cdots k_{j-1}kk_{j+1}\cdots k_p}(-1)^{j+1}$. The identity of (7.2.40; 8) continues to hold when $A^k{}_j$ is substituted for $\omega^k{}_j$, because $A^k{}_j$ has the same antisymmetry. Hence

$$L_v{}^*e^{k_1\cdots k_p} = \sum_j A^{k_j}{}_k{}^*e^{k_1\cdots k_{j-1}kk_{j+1}\cdots k_p}(-1)^{j+1},$$

and consequently

$$^*L_v(\omega_{k_1\cdots k_p}e^{k_1\cdots k_p}) = (L_v\omega_{k_1\cdots k_p})^* e^{k_1\cdots k_p} + \omega_{k_1\cdots k_p}{}^* L_v e^{k_1\cdots k_p} = L_v(\omega_{k_1\cdots k_p}{}^* e^{k_1\cdots k_p}).$$

3. If $s = \bar{s} + f(\bar{s})$, then as $f \to 0$ with $f = 0$ on the boundary,

$$\begin{aligned} W &= \int d\bar{s}(1+f')\sqrt{-\dot{z}^\alpha(\bar{s})\dot{z}_\alpha(\bar{s}) - 2f\dot{z}^\alpha(s)\ddot{z}_\alpha(\bar{s})} \\ &= W + \int d\bar{s}\left(f'\sqrt{-\dot{z}^\alpha(\bar{s})\dot{z}_\alpha(\bar{s})} - f\frac{\dot{z}^\alpha(\bar{s})\ddot{z}_\alpha(\bar{s})}{\sqrt{-\dot{z}^\alpha(\bar{s})\dot{z}^\alpha(\bar{s})}}\right) + O(f^2). \end{aligned}$$

Integrating this by parts leads to the identity

$$\frac{d}{ds}\sqrt{-\dot{z}^\alpha(s)\dot{z}_\alpha(s)} = \frac{-\dot{z}^\alpha(s)\ddot{z}_\alpha(s)}{\sqrt{-\dot{z}^\alpha(s)\dot{z}_\alpha(s)}},$$

which also holds if $\ddot{z} \neq 0$.

4. $T^\alpha = (i_\alpha\, d\Phi)\wedge{}^*d\Phi - \frac{1}{2}i_\alpha[d\Phi\wedge{}^*d\Phi + m^2\Phi^*\Phi]$. Then $T_{\alpha\beta} = T_{\beta\alpha}$ (Lorentz invariance), but not $T^\alpha{}_\alpha = 0$, even if $m = 0$ (no dilatation invariance). $T_{00} \geq |\dot{\Phi}\nabla\Phi|$ and $T^0{}_j = \dot{\Phi}\nabla_j\Phi$, iff $m^2 \geq 0$.

5. $(L_v A)\wedge\dfrac{\partial\mathcal{L}}{\partial\, dA} - i_v\mathcal{L} = 2(L_v A)\wedge F - i_v(F\wedge F) = 2(di_v A)\wedge F = 2d(i_v A\wedge F).$

6. Letting $x := x(0)$, $x(t) = (x - at\langle x|x\rangle)/(1 - 2t\langle x|a\rangle + \langle x|x\rangle\langle a|a\rangle t^2)$, because

$$\begin{aligned} \frac{\partial x}{\partial t} &= \frac{-a\langle x|x\rangle}{1 - 2t\langle x|a\rangle + \langle x|x\rangle\langle a|a\rangle t^2} + 2\frac{(x - at\langle x|x\rangle)(\langle x|a\rangle - \langle x|x\rangle\langle a|a\rangle t)}{(1 - 2t\langle x|a\rangle + \langle x|x\rangle\langle a|a\rangle t^2)^2} \\ &= -a\langle x(t)|x(t)\rangle + 2x(t)\langle a|x(t)\rangle. \end{aligned}$$

For any $t > 0$ there exists $x \in \mathbb{R}^4$, namely the x for which

$$1 - 2t\langle a|x\rangle + t^2\langle a|a\rangle\langle x|x\rangle = 0,$$

which gets sent off to infinity.

7. Integrate $x^j\, d^*(T^\alpha + t^\alpha) = dx^j \wedge {}^*(T^\alpha + t^\alpha) + d(x^{j*}(T^\alpha + t^\alpha))$ over

$$N_4 := \{(t, \mathbf{x}): 0 \leq t \leq T\}.$$

In the rest-frame of the charge nothing depends on time, and only $t^0 \neq 0$. Since $dx_\alpha \wedge {}^*T^\alpha = 0$ (8.1.11; 3), we obtain the contradiction

$$0 = \sum_{j=1}^{3}\int_{0\leq t\leq T} dx^j \wedge {}^*T^j = \int_{0\leq t\leq T} dx^0 \wedge {}^*T^0 > 0.$$

8.2 The General Solution

The characteristics of Maxwell's equations are hypersurfaces with lightlike normals. The Green function is easy to construct in Minkowski space ($\mathbb{R}^4, \eta$)*, and solves the initial-value problem explicitly.*

If the argument of (7.2.28) is applied to the physically interesting case $m = 4$, $p = 2$, one finds that the Minkowski metric

$$\eta = \begin{vmatrix} -1 & & & \\ & 1 & & \\ & & 1 & \\ & & & 1 \end{vmatrix}$$

contains the g_+ of (7.2.28) as its spatial part and also g_- as a spatiotemporal part. Since in the present case $^*\circ^* = (-1)^{p(m-p)+s} = -1$, the combinations of the fields that diagonalize * are $F \pm i^*F$, and when these are written explicitly, both real and complex characteristic directions are found, so there is a combination of both of the cases of (7.2.28) (Problem 1). We have remarked that the characteristics may in general be surfaces of discontinuity for the solutions. If such a hypersurface is specified by $u(x) = 0$, for $u \in E_0(V)$, $V \subset M$, then there should exist solutions that behave locally like $\Theta(u)$. The exterior differential $d\Theta(u) = \delta(u)\, du$ is then singular at $u = 0$, and if J is a regular function, then the singular contributions to the left sides of the equations $dF = 0$ and $\delta F = J$ must cancel out. These contributions are proportional to du, so we are interested in finding solutions for $J = 0$ that depend only on u: $F = c_{ij}(u)e^{ij}$. For such solutions the equations say that the exterior and interior products of F' with du must vanish, because $dF = du\, c_{ij}{}' \wedge e^{ij} \equiv du \wedge F'$, $d^*F = du \wedge {}^*F' = -{}^*(i_{du}F')$. This argument leads directly to a

(8.2.1) Condition for the Characteristics

Let F be discontinuous where $u = 0$, but suppose that J is continuous there. Then the equations $dF = 0$ and $\delta F = J$ imply that at $u = 0$

$$du \wedge F' = 0 \qquad \textit{and} \qquad i_{du}F' = 0;$$

which are satisfied only if $\langle du|du\rangle = 0$ or $F' = 0$.

Proof: If $du \wedge F' = 0$, then in a local basis using du, F' contains du as a factor, that is, $F' = du \wedge f$, where $f \in E_1$ is independent of du. The second equation then requires that $i_{du}\, du \wedge f = \langle du|du\rangle f - \langle du|f\rangle du = 0$. Because f and du are independent, we conclude that $\langle du|du\rangle = \langle du|f\rangle = 0$. □

(8.2.2) Remarks

1. In other words, either the normal to the surface is lightlike, or else there are no discontinuities in F.

2. This is only a local statement. Whether u can be defined as a global coordinate depends on the large-scale structure of space–time.
3. A by-product of (8.2.1) is the statement that fields with discontinuities must have a special structure; they are the exterior products of two 1-forms, of which one (du) is a null field, and the other (f) is orthogonal to it (also in the sense of the metric η). Both invariants vanish for such fields:

$$^*(F \wedge F) = {}^*(du \wedge f \wedge du \wedge f) = 0,$$

and $^*(F \wedge {}^*F) = i_{du} i_f F = (\langle du|f\rangle)^2 - \langle du|du\rangle\langle f|f\rangle = 0$. If the space and time parts are separated, then $du = dt + \mathbf{n} \cdot d\mathbf{x}$ and $f = dt + \mathbf{f} \cdot d\mathbf{x}$, which requires that $|\mathbf{n}|^2 = 1 = (\mathbf{n} \cdot \mathbf{f})$. In terms of the field strengths, $\mathbf{E} = \mathbf{n} - \mathbf{f}$ and $\mathbf{B} = [\mathbf{n} \times \mathbf{f}]$, so this means that $(\mathbf{n} \cdot \mathbf{E}) = (\mathbf{n} \cdot \mathbf{B}) = (\mathbf{E} \cdot \mathbf{B}) = 0 = |\mathbf{E}|^2 - |\mathbf{B}|^2$. The field (7.1.4) was of this form with $u = x - t$.
4. We also note that if $\langle du|du\rangle = 0$, then it is not sufficient to specify F and $^*F$ at $u = 0$ in order to solve the Cauchy problem: It is possible to choose F not identically zero, such that $f(0) = 0$.
5. Whereas a field satisfying $du \wedge F = i_{du} F = 0$ can vary arbitrarily from one hyperplane $u =$ const. to the next, the Lie derivative L_{du} in the direction $g\,du$ tangential to $u =$ const. is determined by Maxwell's equations:

$$L_{du} F = d i_{du} F + i_{du}\, dF = 0,$$
$$L_{du}{}^* F = d i_{du}{}^* F + i_{du}\, d^* F = -i_{du}{}^* J.$$

The higher Lie derivatives can similarly be calculated, which determines F everywhere, if F and $^*F$ are specified on some surface that can be translated with $e^{L_{du}}$ so as to cover all of M.

After studying these local questions, we will evaluate formula (7.2.36) for Maxwell's equations. This requires the explicit form of the Green function, which can be written down only for the simplest manifolds.

(8.2.3) The Construction of $G_{\bar{x}}$ in Minkowski Space $(\mathbb{R}^4, \eta)$

If e^α is the natural and orthogonal basis, then according to equation (7.2.34), $G_{\bar{x}}$ has the form

$$G_{\bar{x}} = \tfrac{1}{2} \bar{e}_{\alpha\beta} \otimes {}^* e^{\alpha\beta} D_{\bar{x}}(x),$$

for $p = 2$, and where $D_{\bar{x}} \in E_0(\mathbb{R}^4)$ satisfies the equation

$$-D_{\bar{x},\alpha}{}^{\alpha} = \delta^4(x - \bar{x}).$$

The translation-invariance of Minkowski space allows the partial differential equation to be reduced to an ordinary differential equation, by expanding it in a series in the eigenfunctions of the translation operator, i.e., a Fourier series. For the delta-function this expansion is

$$\delta^4(x - \bar{x}) = (2\pi)^{-4} \int_{-\infty}^{\infty} d^4k\, e^{i\langle k|\bar{x}-x\rangle}, \tag{8.2.4}$$

where k^i and x^i are regarded as the components of vectors, and $\langle | \rangle$ stands for the Lorentz scalar product (7.2.14) with $g = \eta$. In Fourier-transformed space the Laplacian (i.e.: $_{,\alpha}{}^{\alpha}$) produces a factor $ik_\alpha ik^\alpha = -\langle k|k\rangle =: -k^2$, so that, finally, $D_{\bar{x}}$ has the Fourier integral representation

$$D_{\bar{x}}(x) = (2\pi)^{-4} \int_{-\infty}^{\infty} d^4k \, e^{i\langle k|\bar{x}-x\rangle}/k^2. \tag{8.2.5}$$

(8.2.6) **Remarks**

1. The k integrals do not converge in the classical sense, but as distributions—this is the content of Fourier's theorem. We have unscrupulously interchanged integrals by k with derivatives by x. Fortunately, distributions are so agreeable that they put up with manipulations that classical analysis considers criminal.
2. Because of the translation-invariance of Minkowski space, $D_{\bar{x}}$ depends only on $x - \bar{x}$.
3. Since $k^2 = (|\mathbf{k}| - k^0)(|\mathbf{k}| + k^0)$, the integrand of (8.2.5) has poles, and it must be decided what to do about them when they are in the integration path. We are not restricted to integrals along the real axis; the analyticity of the integrand of (8.2.4) allows the integration path to be distorted into the complex plane. We use the path denoted as

$$\int_{\frown\frown} d^4k,$$

which passes above the poles in the complex k-plane. Other choices of the integration path would produce integrals differing by the contributions of the residues. This nonuniqueness should not be surprising, because the equation in (8.2.3) determines $D_{\bar{x}}$ only up to a solution of the homogeneous equation, and

$$\eta^{\alpha\beta} \frac{\partial^2}{\partial x^\alpha \, \partial x^\beta} e^{i(\mathbf{k}\cdot\mathbf{x} - k^0 t)} = 0,$$

if $k^0 = \pm|\mathbf{k}|$.

The path shown in Figure 8.3 is chosen on physical grounds, since the Green function it produces corresponds to physically realistic initial conditions. It is denoted $D^{\text{ret}}(\bar{x} - x)$, and an elementary integration (Problem 2) produces the

(8.2.7) **Retarded Green Function**

$$D^{\text{ret}}(x) = \frac{\delta(r - 1)}{4\pi r} = \frac{\delta(x^2)}{2\pi}\Theta(t),$$

$$G_{\bar{x}}^{\text{ret}} = \frac{1}{4\pi} \bar{e}_{\alpha\beta} \otimes {}^*e^{\alpha\beta} \delta((\bar{x} - x)^2)\Theta(\bar{t} - t),$$

$$r := |\mathbf{x}|, \qquad x^2 := \langle x|x\rangle.$$

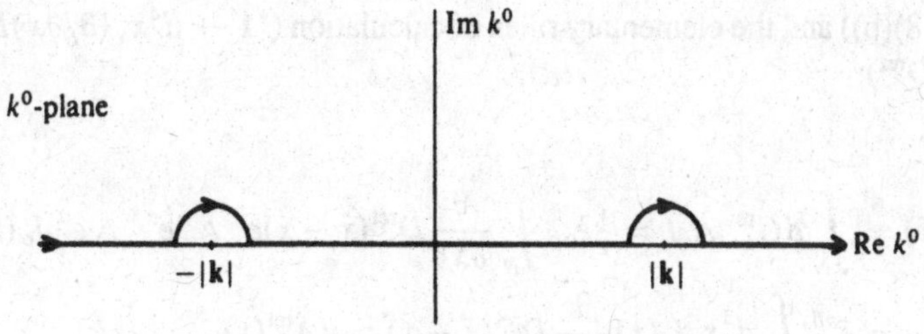

FIGURE 8.3. The path of integration for D^{ret}.

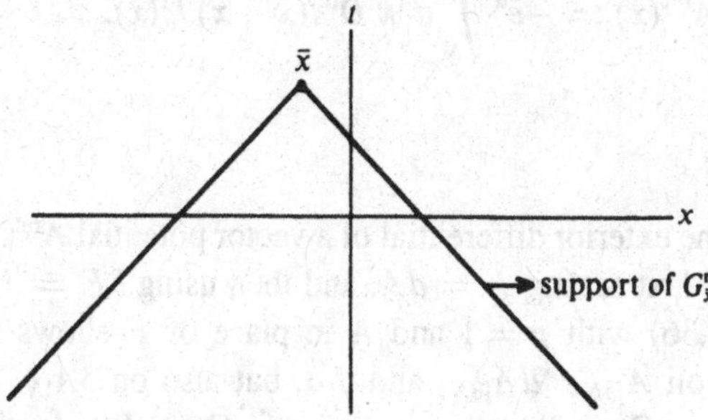

FIGURE 8.4. The support property of the retarded Green function.

(8.2.8) Remarks

1. For the second form of D^{ret} we made use of the formula

$$\delta(f(x)) = \sum_i \frac{1}{|f'(x_i)|}\delta(x - x_i), \qquad f(x_i) = 0,$$

to write

$$\delta(x^2)\Theta(t) = \delta((r+t)(r-t))\Theta(t) = \frac{\Theta(t)}{2r}(\delta(r+t) + \delta(r-t))$$
$$= \frac{\delta(r-t)}{2r}\Theta(t).$$

2. The integral (8.2.5) singles out no particular Lorentz system, and thus D^{ret} depends only on $(x-\bar{x})^2$ and $\Theta(t-\bar{t})$. The preference for one time-direction enters through the choice of the integration path; the path ⌣⌣ would result in $D^{\text{adv}}(\bar{x} - x) := D^{\text{ret}}(x - \bar{x})$.
3. Because $G_{\bar{x}}^{\text{ret}}$ is supported wholly on the negative light-cone of $\bar{x}$ (see Figure 8.4), the integration over infinite space–time regions is justified painlessly.

The means by which the explicit forms of the field strengths can be calculated from (7.2.36) are now ready. The field is the sum of two integrals, one over N and one over ∂N, which we call F^{ret} and F^{boundary}. F^{ret} depends only on J, and F^{boundary} depends only on the boundary values of the field.

We first consider F^{ret}, which can be calculated with the aid of the formula

$$e^\gamma \wedge {}^*e^{\alpha\beta} \wedge e^\sigma = e^\gamma \wedge e^\sigma \wedge {}^*e^{\alpha\beta} = (\eta^{\sigma\beta}\eta^{\gamma\alpha} - \eta^{\sigma\alpha}\eta^{\gamma\beta})^*1$$

(cf. (7.2.18)(b)) and the elementary rules of calculation ($^*\mathbf{1} \to d^4x$, $(\partial/\partial x)D^{\text{ret}} = -(\partial/\partial\bar{x})D^{\text{ret}}$):

(8.2.9)

$$\begin{aligned} F^{\text{ret}}(\bar{x}) &= \int_N dG_{\bar{x}}^{\text{ret}} \wedge J = \tfrac{1}{2}\bar{e}_{\alpha\beta}\int_N \frac{\partial}{\partial x^\gamma} D^{\text{ret}}(\bar{x}-x)e^\gamma \wedge {}^*e^{\alpha\beta} \wedge e^\sigma J_\sigma(x) \\ &= \bar{e}^{\alpha\beta}\int_N d^4x\, J_\alpha(x)\frac{\partial}{\partial\bar{x}^\beta} D^{\text{ret}}(\bar{x}-x) = dA^{\text{ret}}(\bar{x}), \\ A^{\text{ret}}(\bar{x}) &:= -\bar{e}^\alpha \int_N d^4x\, D^{\text{ret}}(\bar{x}-x)J_\alpha(x). \end{aligned}$$

(8.2.10) **Remarks**

1. F^{ret} is precisely the exterior differential of a vector potential A^{ret}. It is common to solve for F by first setting $F = dA$, and then using $\delta F = J$ to determine A. Equation (7.2.36) with $p = 1$ and A in place of F shows that $A_{|\bar{x}}$ does not depend only on $A_{|\partial N}$, $^*dA_{|\partial N}$, and $J_{|N}$, but also on $\delta A_{|N}$. If we rewrite $\int_N \delta G_{\bar{x}} \wedge dA$ as $\int_N G_{\bar{x}} \wedge \delta dA + \int_{\partial N} \cdots = \int_N G_{\bar{x}} \wedge J + \int_{\partial N} \ldots$, then there are three contributions to $A(\bar{x})$, viz., $\int_N G_{\bar{x}} \wedge J$ as in (8.2.9), $\int_N dG_{\bar{x}} \wedge \delta A$, and $\int_{\partial N} \ldots$. Gauge invariance makes it impossible to fix A in terms of J and the boundary values of A and dA; the equations leave open the possibility of a contribution from $d\Lambda$. The solution of the boundary-value problem is unique if we impose the additional condition that $\delta A = 0$ (the **Lorentz gauge**).
2. The A^{ret} of (8.2.9) satisfies

$$\delta A^{\text{ret}}(\bar{x}) = -\int_{\partial N} D^{\text{ret}}(\bar{x}-x)^*J(x) =: j(\bar{x}), \qquad j \in E_0(N)$$

(Problem 4), so F^{ret} satisfies the equations

$$\begin{aligned} dF^{\text{ret}} &= 0, \\ \delta F^{\text{ret}} &= \delta\, dA^{\text{ret}} = \Delta A^{\text{ret}} - d\delta\, A^{\text{ret}} = -\int \Delta G^{\text{ret}} \wedge J - dj = J - dj. \end{aligned}$$

3. Since the support of $G_{\bar{x}}$ is on the negative light-cone of $\bar{x}$, we are not required to choose N compact. If N is taken as $\{(t, x): t_0 \le t \le t_1\}$, then, by (8.2.7), the integral for A^{ret} reads more explicitly

$$A_\alpha^{\text{ret}}(\bar{x}) = -\int_{|\mathbf{x}-\bar{\mathbf{x}}|<\bar{t}-t_0} \frac{d^3x}{4\pi\,|\mathbf{x}-\bar{\mathbf{x}}|} J_\alpha(\bar{t} - |\mathbf{x}-\bar{\mathbf{x}}|, \mathbf{x}).$$

This integral always converges for bounded J_α, even if J does not have spatially compact support (Figure 8.5).
4. Although the use of a different Green function, say D^{adv}, does not change $F(\bar{x})$, it does change the integrals $\int_N$ and $\int_{\partial N}$. For example, with the N of Remark 3, F^{boundary} has contributions only from $t = t_0$, while only the boundary values at

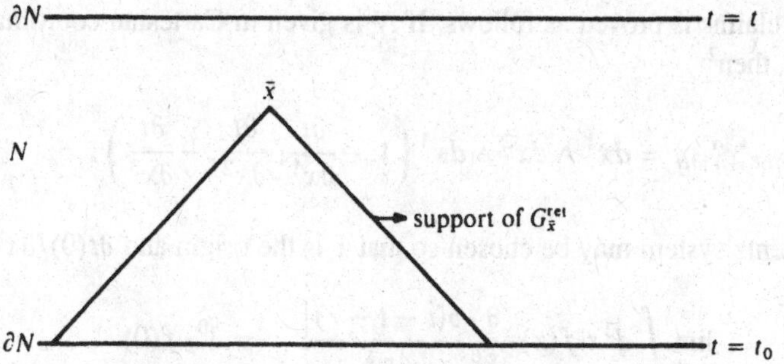

FIGURE 8.5. The region of integration for $F^{ret}(\bar{x})$.

$t = t_1$ would show up in D^{adv}. If we want to know F in the half-space later than $t = t_0$, then D^{ret} is more useful than D^{adv}, because we can let the upper (later) part of N go to infinity. The limit $t_1 \to \infty$ would not necessarily exist for D^{adv}: If we insisted that $F_{|t_0} = 0$, then without the upper boundary term all that would remain of F would be $\int_{t>t_0} dG^{adv} \wedge J$, which does not vanish at $t = t_0$, and thus requires that there be a contribution to $\int_{\partial N}$ at $t = t_1$. Since the equations are invariant under time-reflection $t \to -t$, the appropriate Green function for the time-reversed question would be D^{adv}.

5. With the aid of Green functions with $p = 3$, we can write (8.2.9) in coordinate-free notation as

$$A^{ret} = \int_N G_{\bar{x}}^{ret} \wedge J.$$

In order to study $F^{boundary}$, we use (8.2.7) to write it more explicitly as

(8.2.11)

$$\begin{aligned} F^{boundary}(\bar{x}) &= -\int_{\partial N} (\delta G_{\bar{x}} \wedge F - {}^*dG_{\bar{x}} \wedge {}^*F) \\ &= \bar{e}^{\alpha\beta} \int_{\partial N} \left[\frac{\partial}{\partial \bar{x}^\alpha} D^{ret}(\bar{x} - x) F_{\gamma\beta}(x)^* e^\gamma - \frac{\partial}{\partial \bar{x}^\gamma} D^{ret}(\bar{x} - x) F_\alpha{}^\gamma(x)^* e_\beta \right. \\ &\quad \left. + \frac{1}{2} \frac{\partial}{\partial \bar{x}^\alpha} D^{ret}(\bar{x} - x) F_{\alpha\beta}(x)^* e^\gamma \right] \end{aligned}$$

(Problem 5).

(8.2.12) **Remarks**

1. As in (7.2.38; 1), we see that the boundary ∂N is not arbitrary. If it contained part of the light-cone of a point $\bar{x} \in N$, i.e., the set specified by the equation $t(x) = \bar{t} - |x - \bar{x}|$, then the integral over ∂N would diverge, as

$$\int_{\partial N} D^{ret}(\bar{x} - x)^* e^0 \cong \int \frac{d^3x}{|\mathbf{x} - \bar{\mathbf{x}}|} \delta(t(\mathbf{x}) - \bar{t} - |\mathbf{x} - \bar{\mathbf{x}}|) \sim \int \delta(0).$$

2. If N is bounded by two Cauchy surfaces (7.2.29)(c), then on the earlier surface $F^{boundary}$ takes on the boundary values of F, and F^{ret} goes to zero. The first

of these claims is proved as follows: If N is given in Cartesian coordinates as $t \geq t(x)$, then[2]

$$^*e^\alpha{}_{|\partial N} = dx^1 \wedge dx^2 \wedge dx^3 \left(1, -\frac{\partial t}{\partial x^1}, \frac{\partial t}{\partial x^2}, -\frac{\partial t}{\partial x^3}\right).$$

The Lorentz system may be chosen so that $\bar{x}$ is the origin and $\partial t(0)/\partial x^j = 0$. Because

$$\lim_{\bar{t}\downarrow 0} \int d^3x\, f(x) \frac{\partial}{\partial \bar{x}^\beta} \frac{\delta(\bar{t} - t - r)}{4\pi r}\bigg|_{t=0} = \delta^0{}_\beta f(0)$$

(see Problem 6),

$$\lim_{\bar{x}\downarrow 0} \int_{\partial N} {}^*e^\alpha \frac{\partial}{\partial \bar{x}^\beta} D^{\text{ret}}(\bar{x} - x) f(x) = \delta^\alpha{}_0 \delta^0{}_\beta f(0).$$

If this is substituted into (8.2.11), the two first terms cancel out as $\bar{x} \downarrow 0$, leaving only $F(0)$. On the other hand,

$$\lim_{\bar{x}\downarrow 0} F^{\text{ret}}(\bar{x}) = 0,$$

because the integration region shrinks to zero (see (8.2.12; 3)), so that if J is bounded, nothing is left over.

3. In Problem 7 it is shown, using the explicit form (8.2.12), that in fact $dF^{\text{boundary}} = 0$ and $\delta F^{\text{boundary}} = dj$, because $d(F_{|\partial N}) = 0$ and $d(^*F_{|\partial N}) = -^*J$ (cf. (8.2.10; 2)).
4. Although the boundary values normally propagate by D^{ret} along whole light-cones, particular $F_{|\partial N}$ and $^*F_{|\partial N}$ can be found for which the propagation is approximately along a light ray. In this somewhat vague approximation, the problem of how light propagates reduces to the easier problem of the motion of a massless particle. We made use of a similar simplification in (5.7.15) to calculate the bending of light by a gravitational field.
5. By using D^{ret}, one can take N as the whole half-space later than some Cauchy surface, without changing F^{ret} or F^{boundary}, if the upper boundary is displaced to $t \to +\infty$ (Figure 8.6).

In our verification that $d(F^{\text{ret}} + F^{\text{boundary}}) = 0$ and $d(^*F^{\text{ret}} + {}^*F^{\text{boundary}}) = -^*J$, i.e.,

$$\lim_{\bar{x}\downarrow x} F^{\text{ret}}(\bar{x}) = \lim_{\bar{x}\downarrow x} {}^*F^{\text{ret}}(\bar{x}) = 0, \qquad \lim_{\bar{x}\downarrow x} F^{\text{boundary}}(\bar{x}) = F(x),$$
$$\lim_{\bar{x}\downarrow x} {}^*F^{\text{boundary}}(\bar{x}) = {}^*F(x) \quad \forall x \in \partial N,$$

we needed to know only that $d(F_{|\partial N}) = 0$ and $d(^*F_{|\partial N}) = -^*J_{|\partial N}$. Whereas to use (7.2.36) it is a priori necessary to assume the existence of a solution F, the explicit construction demonstrates the

[2] $^*e^0 = -e^{123}$, but ∂N, as the lower side of N, has a negative orientation.

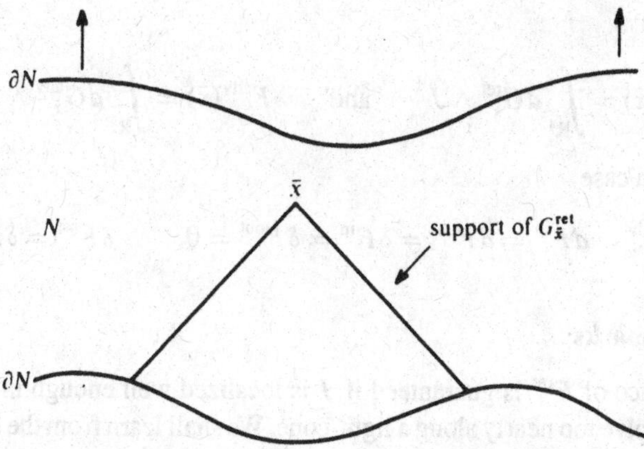

FIGURE 8.6. The displacement of the later boundary of N.

(8.2.13) Existence of the Solution to $dF = 0, \delta F = J$, for Given Initial Values

Let N be the half-space later than a Cauchy surface ∂N in $(\mathbb{R}^4, \eta)$, and suppose that $H \in E_2(\partial N)$, $K \in E_2(\partial N)$, and $J \in E_1(N)$ satisfy the equations

$$dH = 0, \qquad dK = -{}^*J_{|\partial N}.$$

Then

$$F(\bar{x}) = \int_N dG_{\bar{x}} \wedge J - \int_{\partial N} [\delta G_{\bar{x}} \wedge H - {}^*dG_{\bar{x}} \wedge K]$$

is a solution of $dF = 0$ and $\delta F = J$ on N, for which F approaches H and ${}^*F$ approaches K as x approaches ∂N.

(8.2.14) Remarks

1. On a Cauchy surface ∂N, $F_{|\partial N}$ and ${}^*F_{|\partial N}$ are linearly independent, so the boundary values H and K do not depend on each other. If N had characteristic directions, then the above construction would not work, because $\delta G_{\bar{x}}$ and ${}^*dG_{\bar{x}}$ are not distributions on lightlike surfaces.
2. H and K satisfy the restriction of Maxwell's equations to a spacelike hypersurface; $\nabla \cdot \mathbf{H} = 0$ and $\nabla \cdot \mathbf{K} = \rho$. Thus they may be taken as $\mathbf{H} = \nabla \times \mathbf{v}_1$ and $\mathbf{K} = \nabla \times \mathbf{v}_2 - \nabla \int \rho(x)/|\mathbf{x} - \bar{\mathbf{x}}|$, for arbitrary $\mathbf{v}_1$ and $\mathbf{v}_2$.

Often J is sufficiently localized that even $\int_{\mathbb{R}^4} dG_{\bar{x}} \wedge J$ converges. Then the remaining boundary in (7.2.36) can be taken to $t_0 = -\infty$. If $F^{\rm ret}$ is to converge in this case, $F^{\rm boundary}$ must also converge, and in fact it approaches a solution of the free equation ($J = 0$), because $F^{\rm ret}$ then solves Maxwell's equations. This incoming field will be denoted by $F^{\rm in}$, and in the time-reversed situation, using $D^{\rm adv}$ from (8.2.8; 2), it will be called $F^{\rm out}$.

(8.2.15) Definition of the Asymptotic Fields

$$F = F^{\rm in} + F^{\rm ret} = F^{\rm out} + F^{\rm adv},$$

if the integrals

$$F^{\mathrm{ret}}(\bar{x}) = \int_{\mathbb{R}^4} dG_{\bar{x}}^{\mathrm{ret}} \wedge J \qquad \text{and} \qquad F^{\mathrm{adv}}(\bar{x}) = \int_{\mathbb{R}^4} dG_{\bar{x}}^{\mathrm{adv}} \wedge J$$

exist; in which case

$$dF^{\mathrm{ret}} = dF^{\mathrm{adv}} = dF^{\mathrm{in}} = dF^{\mathrm{out}} = \delta F^{\mathrm{in}} = \delta F^{\mathrm{out}} = 0, \qquad \delta F^{\mathrm{ret}} = \delta F^{\mathrm{adv}} = J.$$

(8.2.16) **Remarks**

1. The existence of F^{ret} is guaranteed if J is localized well enough in space and does not evolve too nearly along a light-cone. We shall learn from the hyperbolic motion in §8.3 that it is possible for F^{ret} to converge pointwise while $\delta F^{\mathrm{ret}} \neq J$, because the limit as $t_0 \to -\infty$ does not exist in the appropriate topology.
2. If there exists a T such that $J = 0 \; \forall t < T$, which is, of course, possible only if the total charge $Q = 0$, then $F = F^{\mathrm{in}} \; \forall t < T$. Roughly speaking, F^{in} is the field that existed before J was switched on, and F^{out} is the field that remains after J has been switched off.

   If $N = \mathbb{R}^4$, then F^{ret} is the differential of the

(8.2.17) **Liénard–Wiechert Potentials**

$$F^{\mathrm{ret}} = dA^{\mathrm{ret}},$$

where

$$A_\alpha^{\mathrm{ret}}(\bar{x}) = -\int \frac{d^3x \, J_\alpha(\bar{t} - |\bar{\mathbf{x}} - \mathbf{x}|, x)}{4\pi |\bar{\mathbf{x}} - \mathbf{x}|}.$$

(8.2.18) **The Static Case**

If J is independent of time, then

$$A_\alpha^{\mathrm{ret}}(\bar{\mathbf{x}}) = -\int \frac{d^3x \, J_\alpha(x)}{4\pi |\bar{\mathbf{x}} - \mathbf{x}|},$$

and, in particular, the Coulomb potential

$$V(\bar{\mathbf{x}}) = A_0^{\mathrm{ret}}(\bar{\mathbf{x}}) = \int \frac{d^3x \, \rho(x)}{4\pi |\bar{\mathbf{x}} - \mathbf{x}|}, \qquad \rho(x) := J^0(x) = -J_0(x).$$

A necessary condition for the existence of this potential is that ρ fall off faster than $1/r^2$.

We close this section with a different application of (8.2.15), to rewrite equation (7.3.23) for the work expended. To this end we assume that J is sufficiently well localized in space so that N may be set to $\mathbb{R}^4$. Then

$$P^\alpha(T) - P^\alpha(-T) = \int_{-T \le t \le T} (i^\alpha F^{\mathrm{in}} + i^\alpha F^{\mathrm{ret}}) \wedge {}^*J. \tag{8.2.19}$$

The energy and momentum are thus affected partly by the incoming field and partly by the field caused by J alone. The latter contribution is explicitly

(8.2.20)

$$\int_{-T\leq t\leq T} i^\alpha F^{\text{ret}} \wedge {}^*J = \int_{-T}^{T} d^4\bar{x} \int_{-\infty}^{\infty} d^4x\, J^\beta(\bar{x}) \times \left(J^\alpha(x) \frac{\partial}{\partial \bar{x}^\beta} D^{\text{det}}(\bar{x}-x) - J_\beta(x) \frac{\partial}{\partial \bar{x}_\alpha} D^{\text{ret}}(\bar{x}-x) \right).$$

If J has compact support ($Q = 0$), then we can take the limit $T \to \infty$. By integration by parts, the first term on the right side of (8.2.21) contributes zero, because $\delta J = 0$. By symmetrization $x \leftrightarrow \bar{x}$, in the second term

$$\frac{\partial}{\partial \bar{x}^\alpha} D^{\text{ret}}(\bar{x}-x)$$

is changed to

$$\frac{1}{2}\frac{\partial}{\partial \bar{x}^\alpha}(D^{\text{ret}}(\bar{\mathbf{x}}-x) - D^{\text{adv}}(\bar{x}-x)) =: \frac{1}{2}\frac{\partial}{\partial \bar{x}^\alpha} D(\bar{x}-x).$$

Introducing the **radiation field**

$$F^{\text{rad}} = F^{\text{ret}} - F^{\text{adv}} = F^{\text{out}} - F^{\text{in}}$$

allows one to write

(8.2.21)

$$P_\alpha(\infty) - P_\alpha(-\infty) = \int_{\mathbb{R}^4} (i_\alpha F^{\text{in}} + i_\alpha \tfrac{1}{2} F^{\text{rad}} \wedge {}^*J,$$

$$-\int i_\alpha F^{\text{rad}} \wedge {}^*J = \int d^4\bar{x}\, d^4x\, J^\beta(\bar{x}) J_\beta(x) \frac{\partial}{\partial \bar{x}^\alpha} D(\bar{x}-x).$$

The convolution (8.2.21) becomes a product when Fourier-transformed:[3]

$$\tilde{J}_\beta(k) = \int d^4x\, e^{-i(k|x)} J_\beta(x) = \tilde{J}_\beta{}^*(-k).$$

Since D has the Fourier integral representation

(8.2.22) $$D(x) = i(2\pi)^{-3} \int d^4k\, \delta(k^2)(\Theta(k^0) - \Theta(-k^0))e^{i(k|x)}$$

(Problem 8), there results an expression for the

[3]We shall write * on the right side to denote the complex conjugate.

(8.2.23) Energy and Momentum Lost by the Radiative Reaction of the Field

If $F^{\text{in}} = 0$, $J \in E_1^0(\mathbb{R}^4)$, then

$$P^\alpha(\infty) - P^\alpha(-\infty) = (2\pi)^{-3} \int d^4k\, \Theta(k^0)\delta(k^2)(|\tilde{\mathbf{J}}(k)|^2 - |\tilde{\mathbf{J}}^0(k)|^2)k^\alpha.$$

(8.2.24) Remarks

1. Because $\tilde{F}^{\text{ret}}_{\alpha\beta}(k) = (ik_\alpha \tilde{J}_\beta(k) - ik_\beta \tilde{J}_\alpha(k))/k^2$, the field reflects the frequency distribution of the source. This allows the integrand to be interpreted as the energy and momentum lost to the field with wave-vector k. In particular, as the time becomes infinite, the factor $\delta(k^2)$ makes the loss go mainly into the free field, characterized by $|\mathbf{k}|^2 = (k^0)^2$.
2. If the sources are strictly periodic, the assumption that $J \in E_1^0$ is violated. This shows up as a delta-function in $\tilde{J}$ and a δ^2 in (8.2.23). The physical significance is that a periodic process radiates an infinite amount of energy in an infinite time (cf. (8.3.29; 7)).
3. In Fourier-transformed space, $\delta J = 0$ reads

$$\tilde{\mathbf{J}} \cdot \mathbf{k} = \tilde{J}^0 k^0.$$

If $|\mathbf{k}| = |k^0|$, then $|\mathbf{J}| \geq |J^0|$. Since with $\alpha = 0$ the rest of the integrand is nonnegative, $P^0(\infty) \geq P^0(-\infty)$. This is a consequence of the positivity of the energy. If $F^{\text{in}} = 0$, then $P^0(-\infty) = 0$; thus energy can only be released from the current to the field.

(8.2.25) Problems

1. Write $F \pm i^*F$ out explicitly, and show that $E_1 = B_2 =$ an arbitrary function of $t + z$, and $E_2 = B_1 = E_3 = B_3 = 0$ solves the equations $dF = \delta F = 0$. Show that the field can be written as $du \wedge f$, where $\langle du|du\rangle = \langle du|f\rangle = 0$.
2. Calculate the integral (8.2.5) for D^{ret}, using the integration path (8.2.6; 3).
3. Use D^{ret} to find the Green function for Laplace's equation in three dimensions.
4. Calculate δA for the A of (8.2.9).
5. Calculate $\delta G_{\bar{x}} \wedge F - {}^*dG_{\bar{x}} \wedge {}^*F$ with the $G_{\bar{x}}$ of (8.2.7) and $F = \frac{1}{2}e^{\alpha\beta}F_{\alpha\beta}$.
6. Show that

$$\lim_{t\to 0} \int d^3x\, f(x) \frac{\partial}{\partial x^\alpha} \frac{\delta(t-r)}{4\pi r} = \delta^0{}_\alpha f(0).$$

7. Write (8.2.12) for $\partial N = (0, \mathbf{x})$ explicitly in terms of the components $\mathbf{E}$ and $\mathbf{B}$ of F^{boundary}, and calculate $E_{k,k}$ and $B_{k,k}$.
8. Find the Fourier integral decomposition of the D of (8.2.22).

(8.2.26) **Solutions**

1.

$$F + i^*F = \tfrac{1}{2}(E_1 + B_2)(dt + dx_3)(dx_1 - i\,dx_2) + \tfrac{1}{2}(E_1 - B_2)(dt - dx_3)(dx_1 + i\,dx_2)$$
$$+ \tfrac{1}{2}(E_2 + B_3)(dt + dx_1)(dx_2 - i\,dx_3) + \tfrac{1}{2}(E_2 - B_3)(dt - dx_1)(dx_2 + i\,dx_3)$$
$$+ \tfrac{1}{2}(E_3 + B_1)(dt + dx_2)(dx_3 - i\,dx_1) + \tfrac{1}{2}(E_3 - B_1)(dt - dx_2)(dx_3 + i\,dx_1).$$

The individual terms of this sum are of the required form.

2. If $t \le 0$ (resp. ≥ 0), then the path of integration can be closed in the upper (resp. lower) complex k^0-plane, so that

$$-\frac{1}{2\pi}\int dk^0 \frac{e^{-ik^0 t}}{(k^0 - k)(k^0 + k)} = \frac{\Theta(t)}{k}\sin kt.$$

There remains

$$(2\pi)^{-3}\int d^3k\, e^{i\mathbf{k}\cdot\mathbf{x}}\frac{\Theta(t)}{k}\sin kt = \frac{\Theta(t)}{2\pi^2 r}\int_0^x dk \sin kr \sin kt = \frac{\delta(r-t)}{4\pi r}\Theta(t).$$

3. Because $(\partial^2/\partial t^2) - \Delta)D^{\mathrm{ret}}(x) = \delta^4(x)$, the function $\int_{-\infty}^{\infty} dt\, D^{\mathrm{ret}}(x) = 1/4\pi r$ satisfies the equation $-\Delta(1/4\pi r) = \delta^3(x)$.

4.

$$-\delta A^{\mathrm{ret}}(\bar{x}) = \int_N d^4x\, J_\alpha(x)\frac{\partial}{\partial x^\alpha}D^{\mathrm{ret}}(\bar{x} - x) =: \int_N d({}^*J D^{\mathrm{ret}}_{\bar{x}}) = \int_{\partial N} {}^*J D^{\mathrm{ret}}_{\bar{x}}.$$

5. Using the abbreviation $D_{,\alpha} = (\partial/\partial x^\alpha)D^{\mathrm{ret}}(\bar{x} - x) = -(\partial/\partial \bar{x}^\alpha)D^{\mathrm{ret}}(\bar{x} - x)$ and Rule (7.2.18)(b),

$$dG_{\bar{x}} = \tfrac{1}{2}\bar{e}_{\alpha\beta}e^\gamma \wedge {}^*e^{\alpha\beta}D_{,\gamma} = -\bar{e}^{\alpha\beta}D_{,\alpha}{}^*e_\beta,$$
$${}^*dG_{\bar{x}} \wedge {}^*F = -\tfrac{1}{2}\bar{e}^{\alpha\beta}D_{,\alpha}e_\beta \wedge {}^*e^{\sigma\tau}F_{\sigma\tau} = \bar{e}^{\alpha\beta}D_{,\alpha}F_{\beta\sigma}{}^*e^\sigma,$$
$$\delta G_{\bar{x}} = \tfrac{1}{2}\bar{e}^{\alpha\beta}{}^*(dD \wedge e_{\alpha\beta}) = \tfrac{1}{2}\bar{e}^{\alpha\beta}D_{,}{}^{\gamma}{}^*e_{\gamma\alpha\beta},$$
$$\delta G_{\bar{x}} \wedge F = \tfrac{1}{4}\bar{e}_{\alpha\beta}D_{,\gamma}F_{\sigma\tau}e^{\sigma\tau} \wedge {}^*e^{\gamma\alpha\beta} = -\bar{e}_{\alpha\beta}D_{,\gamma}[F^{\alpha\gamma}{}^*e^\beta - \tfrac{1}{2}F^{\alpha\beta}{}^*e^\gamma].$$

If these equations are combined, then

$$-\delta G_{\bar{x}} \wedge F + {}^*dG_{\bar{x}} \wedge {}^*F = \bar{e}^{\alpha\beta}\left[\frac{\partial}{\partial \bar{x}^\alpha}D(\bar{x} - x)F_{\gamma\beta}(x){}^*e^\gamma \right.$$
$$\left. - \frac{\partial}{\partial \bar{x}_\gamma}D(\bar{x} - x)F_{\alpha\gamma}(x){}^*e_\beta + \frac{1}{2}\frac{\partial}{\partial \bar{x}_\gamma}D(\bar{x} - x)F_{\alpha\beta}(x){}^*e^\gamma\right].$$

6.

$$\alpha = j = 1, 2, 3: \int_0^\infty dr\, d\Omega(f(0)r^2 + f_{,k}(0)r^2x_k + \cdots)\frac{-\partial}{\partial x_j}\frac{\delta(t-r)}{4\pi r}$$
$$= \int_0^\infty dr\, d\Omega(f(0)2x_j + f_{,k}(0)(2x_jx_k + r^2\delta^j{}_k) + \cdots)\frac{\delta(t-r)}{4\pi r} \to 0 \quad \text{as} \quad t \to 0,$$
$$\alpha = 0: \int_0^\infty dr\, d\Omega(f(0)r^2 + f_{,k}(0)r^2x_k + \cdots)\frac{1}{4\pi r}\frac{-\partial}{\partial r}\delta(t-r)$$
$$= f(0) + O(t) \to f(0) \quad \text{as} \quad t \to 0.$$

7.

$$E_k(\bar{x}) = \int d^3x\,[\dot{D}^{\rm ret}(\bar{x}-x)E_k(x) + \varepsilon_{klm} D^{\rm ret}_{,l}(\bar{x}-x)B_m(x)],$$

$$B_k(\bar{x}) = \int d^3x\,[\dot{D}^{\rm ret}(\bar{x}-x)B_k(x) - \varepsilon_{klm} D^{\rm ret}_{,l}(\bar{x}-x)E_m(x)],$$

$$E_{k,k}(\bar{x}) = \int d^3x\,\dot{D}^{\rm ret}(\bar{x}-x)E_{k,k}(x) = \frac{\partial}{\partial \bar{t}}\int d^3x\,D^{\rm ret}(\bar{x}-x)J^0(x),$$

$$B_{k,k}(\bar{x}) = \int d^3x\,\dot{D}^{\rm ret}(\bar{x}-x)B_{k,k}(x) = 0,$$

since **E** and **B** must satisfy the restriction of Maxwell's equations to ∂N.

8.

$$\begin{aligned} D(x) &= (D^{\rm ret}(x) - D^{\rm adv}(x)) \\ &= (2\pi)^{-4}\int d^4k\, e^{i\langle k|x\rangle} \lim_{\varepsilon\downarrow 0}\left[\frac{1}{|\mathbf{k}|^2-(k^0+i\varepsilon)^2} - \frac{1}{|\mathbf{k}|^2-(k^0-i\varepsilon)^2}\right] \\ &= (2\pi)^{-3} i \int d^4k\, e^{i\langle k|x\rangle}\delta(k^2)[\Theta(k^0)-\Theta(-k^0)]. \end{aligned}$$

8.3 The Field of a Point Charge

There is an expression in closed form for the field of a point charge undergoing any given motion. It contains all information about the radiation emitted by an accelerating charge.

Our first application of the formulas derived in the preceding section will be to calculate $A^{\rm ret}$ and $F^{\rm ret}$ for the current (7.3.25; 2); J and $D^{\rm ret}$ will be, respectively, a four-dimensional and a one-dimensional delta-function, which allows the integrals over d^4x and ds to be done. Using the rule $\delta(f(x)) = \sum_i \delta(x-x_i)|f'(x_i)|^{-1}$, where the x_i are the zeros of f, we obtain

(8.3.1)

$$\begin{aligned} A^{\rm ret}_\alpha(\bar{x}) &= -\int d^4x\, D^{\rm ret}(\bar{x}-x)J_\alpha(x) = -e\int_{-\infty}^{\infty} ds\,\dot{z}_\alpha(s)D^{\rm ret}(\bar{x}-z(s)) \\ &= \frac{-e}{2\pi}\int_{-\infty}^{z^0(s)<\bar{x}^0} ds\,\dot{z}_\alpha(s)\delta((\bar{x}-z(s))^2) \\ &= \frac{e}{4\pi}\frac{\dot{z}_\alpha(s_0)}{\langle\dot{z}(s_0)|\bar{x}-z(s_0)\rangle}, \quad (\bar{x}-z(s_0))^2 = 0, \quad \bar{x}^0 > z^0(s_0). \end{aligned}$$

(8.3.2) **Remarks**

1. The negative sign makes (8.3.2) consistent with $\langle\dot{z}(s_0)|\bar{x}-z(s_0)\rangle < 0$.

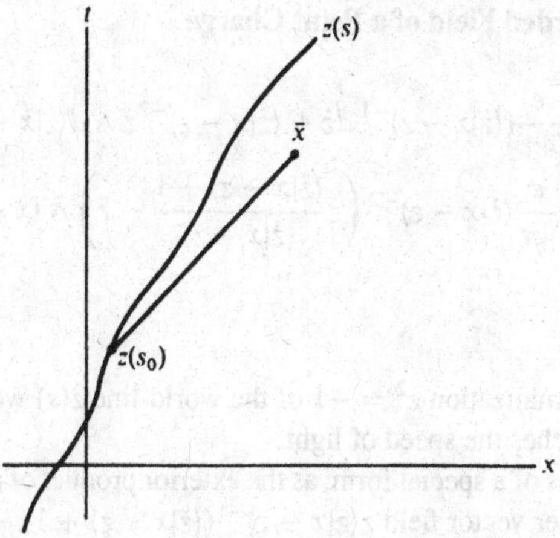

FIGURE 8.7. Determination of $z(s_0)$ on the world-line of a charge.

2. s_0 is a function of $\bar{x}$ (see Figure 8.7).
3. If we consider $\bar{x}_\alpha$, $z_\alpha(s_0(\bar{x}))$, and $\dot{z}_\alpha(s_0(\bar{x}))$ as the components of vector fields x, z, and $\dot{z}$, then (8.3.2) may be written without indices as

$$A^{\text{ret}} = \frac{e}{4\pi}\frac{\dot{z}}{\langle \dot{z}|x-z\rangle}.$$

After the x-integration the coordinate will be called x, rather than $\bar{x}$.

The way that s_0 depends on x must first be known before F^{ret} can be calculated from A^{ret}. This dependence can be determined by noting that $x - z \in E_1(\mathbb{R}^4)$ is a null field:

(8.3.3)

$$0 = \frac{1}{2}\frac{\partial}{\partial x^\beta}(x^\alpha - z^\alpha(s_0(x)))(x_\alpha - z_\alpha(s_0(x))) = x_\beta - z_\beta - \dot{z}^\alpha \frac{\partial s_0}{\partial x^\beta}(x_\alpha - z_\alpha)$$
$$\Rightarrow \quad ds_0 = (x-z)\langle \dot{z}|x-z\rangle^{-1}.$$

The exterior differential of the constituents of A can be calculated from (8.3.4) as

(8.3.4)
$$d\dot{z} = ds_0 \wedge \ddot{z} = (x-z)\wedge \ddot{z}\langle \dot{z}|x-z\rangle^{-1},$$
$$\frac{\partial}{\partial x^\beta}\dot{z}^\alpha(x_\alpha - z_\alpha) = \ddot{z}^\alpha \frac{\partial s_0}{\partial x^\beta}(x_\alpha - z_\alpha) + \dot{z}_\beta - \dot{z}^\alpha \dot{z}_\alpha \frac{\partial s_0}{\partial x^\beta},$$
$$\Rightarrow \quad d\langle \dot{z}|x-z\rangle = z + (x-z)\frac{\langle \ddot{z}|x-z\rangle + 1}{\langle \dot{z}|x-z\rangle},$$

where we have used the normalization $\dot{z}^\alpha \dot{z}_\alpha =: \dot{z}^2 =: \langle \dot{z}|\dot{z}\rangle = -1$ and have also considered $\ddot{z}_\alpha(s_0(x))$ as the components of the vector field $\ddot{z}$. Combining the above formula produces

(8.3.5) The Retarded Field of a Point Charge

$$F^{\text{ret}} = \frac{e}{4\pi}(\langle \dot{z}|x - z\rangle^{-1}\, d\dot{z} + \langle \dot{z}|x - z\rangle^{-2}\dot{z} \wedge d\langle \dot{z}|x - z\rangle)$$

$$= \frac{e}{4\pi}\langle \dot{z}|x - z\rangle^{-2}\left(\dot{z}\frac{\langle \ddot{z}|x - z\rangle + 1}{\langle \dot{z}|x - z\rangle} - \ddot{z}\right) \wedge (x - z).$$

(8.3.6) Remarks

1. By using the normalization $\dot{z}^2 = -1$ of the world-line $z(s)$ we assume that the charge never reaches the speed of light.
2. The 2-form F^{ret} is of a special form, as the exterior product of a null vector field $x - z$ with another vector field $\dot{z}\langle \dot{z}|x - z\rangle^{-1}(\langle \ddot{z}|x - z\rangle + 1) - \ddot{z}$. In contrast to the field in (8.2.2; 3) with discontinuities, the interior product of these fields is not 0 but 1.
3. F^{ret} is the sum of two fields, one of which, $F^{(\ddot{z})}$, contains the terms proportional to $\ddot{z}$, and the other, $F^{(\dot{z})}$, contains only $\dot{z}$. The two fields have different asymptotic behavior, as expected on dimensional grounds; $F^{(\dot{z})}$, the field of the near zone, falls off as a Coulomb field $(1/r^2)$, whereas $F^{(\ddot{z})}$, the field of the far zone, falls off only as $1/r$.

We note some of the special

(8.3.7) Properties of F^{ret} for a Point Charge

(a) $F^{\text{ret}} \wedge F^{\text{ret}} = 0 = F^{(\dot{z})} \wedge F^{(\dot{z})} = F^{(\ddot{z})} \wedge F^{(\ddot{z})}$.

(b) $F^{\text{ret}} \wedge (x - z) = 0 = F^{(\dot{z})} \wedge (x - z) = F^{(\ddot{z})} \wedge (x - z)$.

(c) $i_{x-z}F^{(\ddot{z})} = {}^*((x - z) \wedge {}^*F^{(\ddot{z})}) = 0$.

(d) $f^{(\ddot{z})} \wedge {}^*F^{(\ddot{z})} = 0 = F^{(\dot{z})} \wedge {}^*F^{(\ddot{z})} = F^{\text{ret}} \wedge {}^*F^{(\ddot{z})}$.

Proof:

(a) This holds for any element of E_2 that is an exterior product of two vectors.

(b) F^{ret}, $F^{(\dot{z})}$, and $F^{(\ddot{z})}$ contains $x - z$ as a factor.

(c) The interior product of the two factors of $F^{(\ddot{z})}$ vanishes, and $x - z$ is a null field.

(d) This follows from (c) because of the factor $x - z$. □

(8.3.8) Remarks

1. Since $F \wedge F \sim \mathbf{E} \cdot \mathbf{B}$, Property (a) implies that the electric and magnetic fields are always perpendicular to each other, and this is also a property of $F^{(\dot{z})}$ and $F^{(\ddot{z})}$

considered separately. Therefore it requires more than one charge to produce a magnetic field parallel to an electric field.

2. Because $^*((x-z)\wedge F) = i_{x-z}{}^*F$, Property (b) implies that

$$(x-z)^\beta ({}^*F^{\text{ret}})_{\alpha\beta} = 0.$$

For $\alpha = 0$, this means that $\mathbf{B}$ is also perpendicular to the 3-vector connecting the reference point (where the field is measured) to the position of the charge at the retarded time s_0.

3. If $F^{(\ddot{z})}$ is considered separately, then Remark 2 still applies, and because of Property (c) it applies even if F and $^*F$, that is, $\mathbf{B}$ and $\mathbf{E}$, are interchanged. Therefore the electric field stemming from $F^{(\ddot{z})}$, which dominates in the far zone, is also perpendicular to the 3-vector from the reference point to the position of the charge at the retarded time s_0.

4. According to Property (d), because $F \wedge {}^*F \sim |\mathbf{E}|^2 - |\mathbf{B}|^2$, in the far zone $|\mathbf{E}^{(\ddot{z})}| = |\mathbf{B}^{(\ddot{z})}|$, and if $z(s_0) = 0$, then

$$\mathbf{E}^{(\ddot{z})} = \left[\mathbf{B}^{(\ddot{z})} \times \frac{\mathbf{x}}{r}\right] \qquad \text{and} \qquad \mathbf{B}^{(\ddot{z})} = \left[\frac{\mathbf{x}}{r} \times \mathbf{E}^{(\ddot{z})}\right].$$

The Lagrangian $F^{\text{ret}} \times {}^*F^{\text{ret}} = F^{(\dot{z})} \wedge {}^*F^{(\dot{z})}$ has contributions only from the near zone.

(8.3.9) The Field in the Rest-Frame of the Particle

Let us choose the Lorentz system in which $z(s_0) = 0$, $\dot{z}(s_0) = (1, 0, 0, 0)$, and $\ddot{z}(s_0) = (0, \ddot{\mathbf{z}})$. Then for x in the positive light-cone of $z(s_0)$, the field (8.3.5), written in components, is expressed as

$$\mathbf{E} = \frac{e}{4\pi r}\left\{-\ddot{\mathbf{z}} + \frac{\mathbf{x}}{r^2}(1 + (\mathbf{x}\cdot\ddot{\mathbf{z}}))\right\},$$

$$\mathbf{B} = \frac{e}{4\pi r}\left[\ddot{\mathbf{z}} \times \frac{\mathbf{x}}{r}\right].$$

The magnetic lines of force circle the charge, and the electric lines of force leave the charge initially in the radial direction, and later bend to become perpendicular to $\mathbf{B}$ and to $\mathbf{x}$ in the far zone (Figure 8.8).

(8.3.10) Warning

The description in (8.3.9) is not an instantaneous picture of the field, but specifies the field at a given position $\mathbf{x}$ at the time $t = |\mathbf{x}|$. The field in the spacelike section at $t =$ const. does not depend on $\dot{z}$ and $\ddot{z}$ at a single value of s, but on a whole segment of the particle's trajectory. It can be written down explicitly only for special kinds of trajectories.

(8.3.11) Examples

1. **Uniform motion.** Let

$$z(s) = \frac{s}{\sqrt{1-v^2}}(1, v, 0, 0) = \dot{z}s.$$

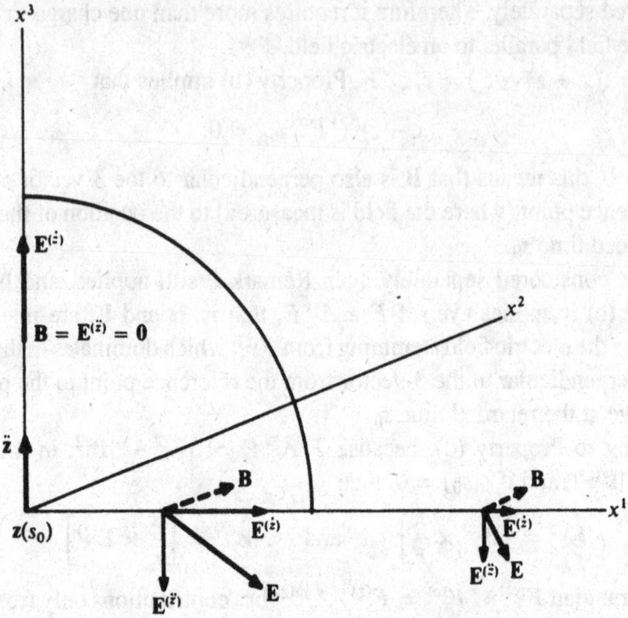

FIGURE 8.8. The fields in the near and far zones.

Then

$$(x - z(s_0))^2 = x^2 - s_0^2 - 2\langle \dot{z}|x\rangle s_0,$$

and thus

$$s_0 = -\langle \dot{z}|z(s_0)\rangle = -\langle \dot{z}|x\rangle - \sqrt{\langle \dot{z}|x\rangle^2 + x^2}.$$

This means that

$$A^{\text{ret}} = \frac{-e}{4\pi} \frac{\dot{z}}{\sqrt{\langle \dot{z}|x\rangle^2 + x^2}}$$

$$= \frac{-e}{4\pi} \frac{(-1, v, 0, 0)}{\sqrt{1 - v^2}} \left[\left(\frac{x_1 - vt}{\sqrt{1 - v^2}} \right)^2 + x_2^2 + x_3^2 \right]^{-1/2},$$

$$F^{\text{ret}} = \frac{e}{4\pi} \frac{x \wedge \dot{z}}{(\langle \dot{z}|x\rangle^2 + x^2)^{3/2}}$$

$$= \frac{e}{4\pi} \begin{vmatrix} 0 & x_1 - vt & x_2 & x_3 \\ -x_1 + vt & 0 & -vx_2 & -vx_3 \\ -x_2 & vx_2 & 0 & 0 \\ -x_3 & vx_3 & 0 & 0 \end{vmatrix}$$

$$\times \frac{1 - v^2}{[(x_1 - vt)^2 + (1 - v^2)(x_2^2 + x_3^2)]^{3/2}}.$$

Note that:

(i) If $v = 0$, the above expression reduces to the usual Coulomb field, $A_0 = e/4\pi r$, $\mathbf{A} = \mathbf{0}$; $\mathbf{E} = e\mathbf{x}/4\pi r^3$, $\mathbf{B} = \mathbf{0}$. When v is not zero, the Coulomb field is simply transformed according to the transformation law for $E_2(\mathbb{R}^4)$ (cf. (5.2.7)), which is automatically taken into account by the covariant notation.

(ii) The electric field points at the simultaneous position of the charge, not to its retarded position. This fact was used in (7.1.5).

(iii) The denominator contains the spatial distance to $z(s_0)$ in the rest-frame of the charge, viz.,

$$|\langle \dot{z}|x - z\rangle| = (\langle \dot{z}|x\rangle^2 + x^2)^{1/2} = \left(\frac{(x_1 - vt)^2}{1 - v^2} + x_2^2 + x_3^2\right)^{1/2}.$$

This is equal neither to the distance $((x_1 - vt)^2 + x_2^2 + x_3^2)^{1/2}$ from the simultaneous position of the charge, nor to the distance from $z(s_0)$, which would be $\langle dt|x - z\rangle$. On the spacelike section $t = 0$, $\langle \dot{z}|x - z\rangle$ equals $r = |\mathbf{x}|$ for the points perpendicular to the direction of motion (i.e., $x_1 = 0$), and is otherwise greater than r. The increase of the denominator is compensated for by the factor $1/\sqrt{1 - v^2}$ of the Lorentz transformation, and the net effect is that the static Coulomb potential is altered as shown in Figure 8.9; A^0 is increased perpendicular to the motion and left unchanged in the x_1-direction. The oblately squashed potential produces an electric field increased with respect to $\mathbf{x}/4\pi r^3$ if $\mathbf{x} \perp \mathbf{v}$ and decreased if $\mathbf{x} \parallel \mathbf{v}$. The increased range of the Coulomb field makes a charged particle at nearly the speed of light cause greater ionization.

(iv) Of course, the decrease in the field in the forward direction changes into an increase when the charge gets near the reference point. At the time $t = r$, the denominator $\langle \dot{z}|x - z\rangle^2$ becomes

$$\frac{(x_1 - vr)^2}{1 - v^2} + x_2^2 + x_3^2 = \frac{(r - x_1 v)^2}{1 - v^2}$$

and for $x_2 = x_3 = 0$ is decreased compared with r^2 by the factor $(1 - v)/(1 + v)$ (Figure 8.10). Incidentally, this factor causes the radiation to bunch strongly in the forward direction (Problem 3). The expression $\langle \dot{z}|x - z\rangle$ must be evaluated at the retarded time, and radiation, once emitted, is not affected if the particle flies off afterward in some different direction, never to come near the reference point.

2. **Uniform acceleration.** Hyperbolic motion, like that of a charged particle in a constant electric field (cf. §5.2), is characterized by $z(s)^2 = \text{const}$. Let

$$z(s) = (\sinh s, \cosh s, 0, 0) = \ddot{z}, \qquad \dot{z} = (\cosh s, \sinh s, 0, 0) = \dddot{z}.$$

Then s_0 can be calculated from $(x - z(s_0))^2 = x^2 + 1 - 2\langle x|z\rangle = 0$. Although the equation for s_0 is transcendental, it is easy to calculate that

$$\dot{z}(s_0) = \left(\frac{t\xi - x_1(1 + x^2)}{2(t^2 - x_1^2)}, \frac{x_1\xi - t(1 + x^2)}{2(t^2 - x_1^2)}, 0, 0\right),$$

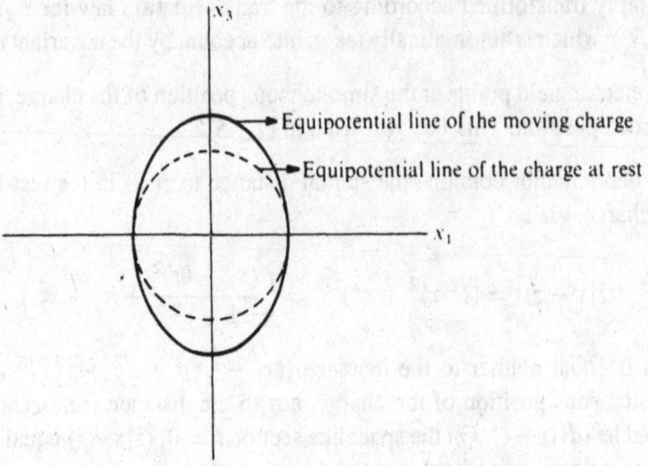

FIGURE 8.9. Lorentz contraction of the Coulomb potential.

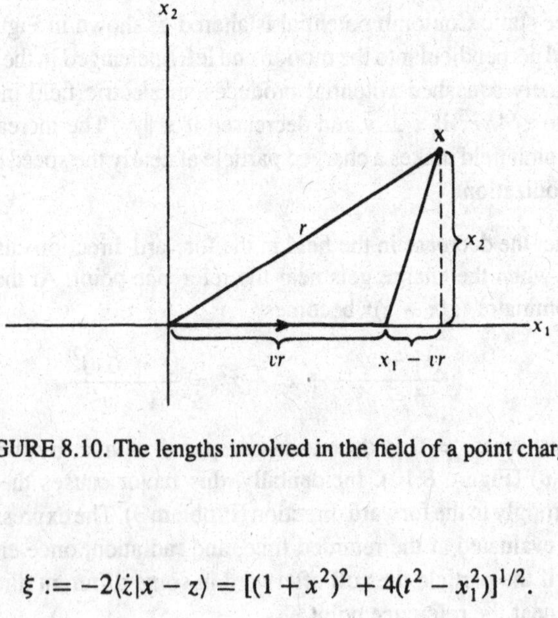

FIGURE 8.10. The lengths involved in the field of a point charge.

where

$$\xi := -2\langle \dot{z}|x - z\rangle = [(1 + x^2)^2 + 4(t^2 - x_1^2)]^{1/2}. \tag{8.3.12}$$

Then

$$A^{\text{ret}} = \frac{-e}{4\pi}\frac{1}{\xi(t^2 - x_1^2)}[t\xi - x_1(1 + x^2), x_1\xi - t(1 + x^2), 0, 0]$$

if $t > -x_1$, and otherwise 0. The general expression (8.3.5) simplifies due to the special properties of the motion $\ddot{z} = z$, $\langle \dot{z}|z\rangle = 0$, and $\langle x|z(s_0)\rangle = (x^2 + 1)/2$:

$$F^{\text{ret}} = \frac{e}{4\pi}\langle \dot{z}|x\rangle^{-2}\left(\dot{z}\frac{x^2 + 1}{2\langle \dot{z}|x\rangle} - z\right) \wedge (x - z).$$

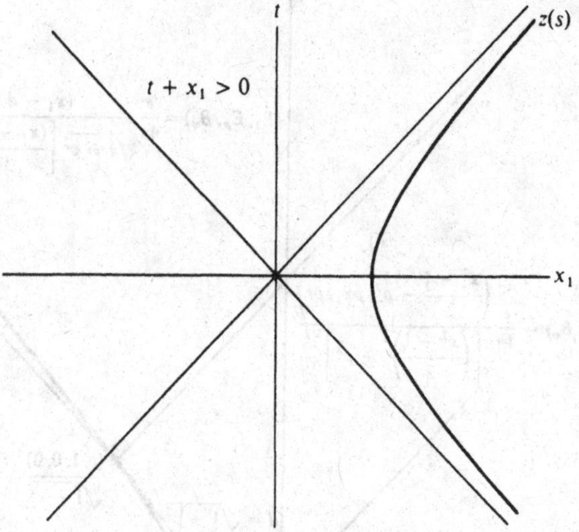

FIGURE 8.11. Hyperbolic motion.

Substitution from (8.3.12) reveals (Problem 1) that, using cylindrical coordinates about the x_1-axis, $\rho^2 = x_2^2 + x_3^2$, the only nonzero components are

$$(8.3.13)\quad (E_1, E_\rho, B_\varphi) = \frac{e}{4\pi}\left[\left(\frac{x^2-1}{2}\right)^2 + \rho^2\right]^{-3/2}\left(\frac{x^2-1}{2} - \rho^2, \rho x_1, \rho t\right)$$

if $t > -x_1$, and otherwise these too are zero. Note that:

(i) The part of space where $t + x_1 \leq 0$ has no field, becaust it cannot be connected to the world-line by any light-cone (Figure 8.11).

(ii) As a practical matter, motion is never strictly hyperbolic. If hyperbolic motion is combined with uniform motion at a velocity $-v$ (resp. v) at the point $(-v, 1, 0, 0,)/\sqrt{1-v^2}$ (resp. $(v, 1, 0, 0)/\sqrt{1-v^2}$), then the results of Examples 1 and 2 in the appropriate regions are simply combined (see Figure 8.12). Interestingly, this field does not converge to (8.3.13) as $v \to 1$, as there remains a contribution from the initial uniform motion:

$$\lim_{v\to 1} \frac{e}{4\pi}\frac{(x_1 - \sqrt{1-v^2} + vt, \rho, -v\rho)(1-v^2)}{[(x_1 - d + vt)^2 + (1-v^2)\rho^2]^{3/2}} = \frac{e}{2\pi}\frac{\delta(x_1+t)}{1+\rho^2}(0, \rho, -\rho)$$

(see Problem 5). This field, which has accumulated on the surface $x_1 = -t$, must be added to (8.3.13), and it also shows up in other physically meaningful limiting processes, such as when the hyperbolic motion is that of one of the particles of a pair-production [57].

(iii) Equation (8.3.13) is not acceptable mathematically, because

$$\delta F^{\mathrm{ret}} - J = \frac{e}{\pi}\frac{\delta(x_1+t)}{(1+\rho^2)^2}(1, 1, 0, 0).$$

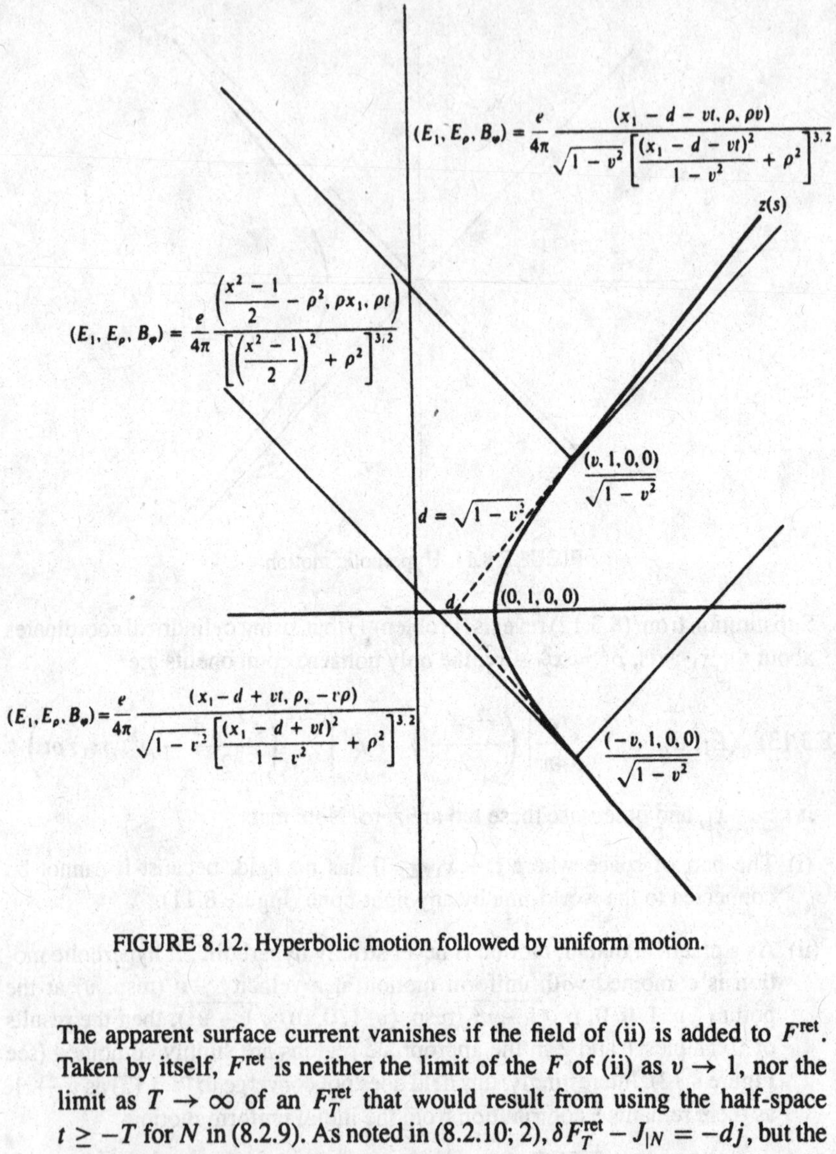

FIGURE 8.12. Hyperbolic motion followed by uniform motion.

The apparent surface current vanishes if the field of (ii) is added to F^{ret}. Taken by itself, F^{ret} is neither the limit of the F of (ii) as $v \to 1$, nor the limit as $T \to \infty$ of an F_T^{ret} that would result from using the half-space $t \geq -T$ for N in (8.2.9). As noted in (8.2.10; 2), $\delta F_T^{\text{ret}} - J_{|N} = -dj$, but the right side of this equation approaches zero as $T \to \infty$ (Problem 6), while $\delta F^{\text{ret}} - J \sim \delta(x_1 + t)$. Its integral over x_1 does not tend to zero, although the integral of dj does.

(iv) If we write the field of a uniformly moving charge in the notation of (8.3.13),

$$(E_1, E_\rho, B_\varphi) = \frac{e}{4\pi}\left[\frac{(x_1 - vt)^2}{1-v^2} + \rho^2\right]^{-3/2} \times \left(\frac{x_1 - vt}{\sqrt{1-v^2}}, \frac{\rho}{\sqrt{1-v^2}}, \frac{v\rho}{\sqrt{1-v^2}}\right),$$

then we see that the term $(x_1 - vt)/\sqrt{1-v^2}$ is replaced in (8.3.13) with $(x^2-1)/2$ and that E_1 contains an additional term $-\rho^2[\ldots]^{-3/2}$. This is the compression factor of the intuitive picture (7.1.5).

(v) At the time when $x^2 = 1$, the fields become

$$(E_1, E_\rho, B_\varphi) = \frac{e}{4\pi\rho}\left(-1, \frac{x_1}{\rho}, \sqrt{1+(1+x_1^2)/\rho^2}\right).$$

The maximum field strength over time consequently falls off only as $1/\rho$ as a function of the distance ρ to the line of flight of the charge.

(vi) Except for the condition $t + x_1 > 0$, E_1 and B_φ are even in x_1, while E_ρ is odd. The fields are just as large where $x_1 < 0$, where the charge never goes, as where $x_1 > 0$. The total radiation field over the line $t = -x_1$ is therefore just as large as the total Coulomb field over the line $t = x_1$, which the world-line of the particle asymptotically approaches.

(vii) Again disregarding the condition that $t + x_1 > 0$, **E** is even in t and **B** is odd. In particular, the magnetic field is zero throughout space at the time $t = 0$.

3. **Rotating charges.** If

$$z(s) = \left(\frac{s}{\sqrt{1-v^2}}, R\cos\frac{vs/R}{\sqrt{1-v^2}}, R\sin\frac{vs/R}{\sqrt{1-v^2}}, 0\right), \tag{8.3.14}$$

the determination of $s_0(x)$ is more difficult than in Example 2, and therefore one usually looks only at the limit $v \to 0$, $R \to 0$, $e \to \infty$, such that $v/R \to \omega$ and $e = 1/R\omega^2$. In order not to be encumbered with the infinite Coulomb field that results, one considers two opposite charges in mirror-image paths about the origin (Hertz's dipole). In this limit, $s_0 = t - r$, $z - x = (r, \mathbf{x})$, $\dot{z} = (1, 0, 0, 0)$, $e\ddot{z} = -(0, \cos\omega(r-t), \sin\omega(r-t), 0)$, and the fields become

$$\mathbf{B} = \frac{e[\ddot{\mathbf{z}} \times \mathbf{x}]}{4\pi r^2}, \quad \mathbf{E} = \frac{[\mathbf{B} \times \mathbf{x}]}{r}, \qquad e\ddot{\mathbf{z}} = (\cos\omega(r-t), \sin\omega(r-t), 0). \tag{8.3.15}$$

These examples illustrate F in three representative cases, for free motion and linear and circular acceleration. Often of greater practical interest than the field strengths are the energy and momentum forms created by the charge. These will be sums (7.3.23) of two terms quadratic in F. Since each component of F is itself a sum of six fractions, it seems that blind substitution would produce 72 fractions. Fortunately, the special structure of the F of a point charge can be used to reduce the algebraic complexity.

(8.3.16) **The Energy-Momentum Forms of the Field of a Point Charge**

With the rules (7.2.18), $\mathcal{T}_\alpha$ can be rewritten

$$\mathcal{T}_\alpha = {}^*((i_\alpha F) \wedge {}^*F - \tfrac{1}{2} i_\alpha(F \wedge {}^*F)) = \frac{e_\alpha}{2}{}^*(F \wedge {}^*F) - i_{i_\alpha F}F.$$

Our F is of the form

$$F = \frac{e}{4\pi}\langle \dot{z}|x-z\rangle^{-2} v \wedge n, \qquad v = \dot{z}\frac{1+\langle \ddot{z}|x-z\rangle}{\langle \dot{z}|x-z\rangle} - \ddot{z}, \qquad n = (x-z),$$

and so the invariants are

$$n^2 = 0, \qquad \langle n|v\rangle = 1, \qquad v^2 = \ddot{z}^2 - \left(\frac{1+\langle \ddot{z}|x-z\rangle}{\langle \dot{z}|x-z\rangle}\right)^2$$

(use $\dot{z}^2 = -1$ and $\langle \dot{z}|z\rangle = 0$). To calculate $\mathcal{T}_\alpha$ we need the equations

$$^*(v \wedge n \wedge {}^*(v \wedge n)) = i_v i_n v \wedge n = 1, \qquad i_\alpha v \wedge n = v_\alpha n - n_\alpha v,$$

and

$$i_{v_\alpha n - n_\alpha v} v \wedge n = v_\alpha n \langle n|v\rangle - n_\alpha (n v^2 - v\langle n|v\rangle).$$

By substituting for the scalar product, one finds that

$$\begin{aligned}\mathcal{T}_\alpha = \left(\frac{e}{4\pi}\right)^2 \langle \dot{z}|x-z\rangle^{-4}\Bigg\{&(x-z)(x-z)_\alpha\left(\ddot{z}^2 - \left(\frac{1+\langle \ddot{z}|x-z\rangle}{\langle \dot{z}|x-z\rangle}\right)^2\right)\\ &+ \ddot{z}(x-z)_\alpha + (x-z)\ddot{z}_\alpha - (\dot{z}(x-z)_\alpha + (x-z)\dot{z}_\alpha)\\ &\times \frac{1+\langle \ddot{z}|x-z\rangle}{\langle \dot{z}|x-z\rangle} + \frac{e_\alpha}{2}\Bigg\}.\end{aligned}$$

(8.3.17) Remarks

1. We have only used F^{ret}, which corresponds to the initial condition $F^{\text{in}} = 0$.
2. The terms that contain $\ddot{z}$ quadratically are recognizable as the $\mathcal{T}_\alpha$ of $F^{(\ddot{z})}$. At large distances they would dominate, as they decrease as $1/r^2$. The contribution from $F^{(\dot{z})}$ goes as $1/r^4$, and the mixed term as $1/r^3$.
3. The structure of $\mathcal{T}_\alpha$ shows $T_{\alpha\beta} = T_{\beta\alpha}$ and $T_\alpha{}^\alpha = 0$.

Poynting's vector $\mathcal{S}_j = [\mathbf{E} \times \mathbf{B}]_j$ is useful for visualizing how the field energy flows. This can best be understood by returning to the representative

(8.3.18) Examples

1. Uniform motion. In cylindrical coordinates, F of the form given in (8.3.11; 2), paragraph (iv), makes Poynting's vector

$$(\mathcal{S}_1, \mathcal{S}_\rho, \mathcal{S}_\varphi) = \left(\frac{e}{4\pi}\right)^2 \frac{\rho v(1-v^2)^2}{[(x_1 - vt)^2 + (1-v^2)\rho^2]^3}(\rho, -x_1 + vt, 0).$$

The streamlines of energy are circles $\rho^2 + (x_1 - vt)^2 = R^2$ around the (simultaneous, not retarded) position of the charge vt. The field energy flows toward the future positions of the charge, in other words, along with the charge. (See Figure 8.13.)

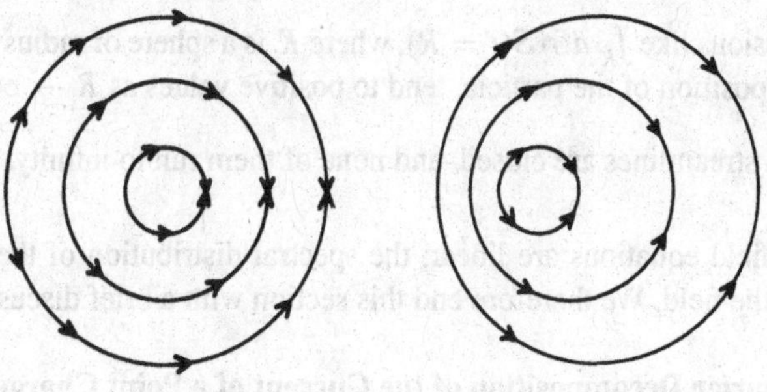

FIGURE 8.13. Streamlines of the energy.

2. Uniform acceleration. From (8.3.11; 2),

$$(S_1, S_\rho, S_\varphi) = \left(\frac{e}{4\pi}\right)^2 \frac{\rho t}{\left[\left(\frac{x^2-1}{2}\right)^2 + \rho^2\right]^3} \left(x_1\rho, \rho^2 + \frac{1-x^2}{2}, 0\right),$$

in the notation used above. The streamlines are again circles, $\rho^2 + (x_1 - \sqrt{R^2+t^2+1})^2 = R^2$, but as the radius R increases, the center of the circle moves ahead to the position of the charge. This makes the flow of energy point more and more outward as ρ increases with fixed x_1 (Figure 8.13); this occurs because $\mathbf{E}$ has a stronger component in the direction of the motion.

3. A rotating charge in the dipole limit. With (8.3.15), we get

$$\mathcal{S} = \frac{\mathbf{x}}{(4\pi)^2 r^3}\left(|\ddot{\mathbf{z}}|^2 - \left(\ddot{\mathbf{z}} \cdot \frac{\mathbf{x}}{r}\right)^2\right) = \frac{\mathbf{x}}{(4\pi)^2 r^3}(1 - \sin^2\vartheta \cos^2(\omega(r-t) - \varphi))$$

in polar coordinates. The energy flows radially outward, and the flow is strongest perpendicular to the direction of the acceleration at the retarded time.

(8.3.19) **Remarks**

1. If we return to the case of uniform motion at the velocity $\pm v$ joined to hyperbolic motion, then we have to add the streamlines of Examples 1 and 2. In the spherical shell

$$\left(t - \frac{v}{\sqrt{1-v^2}}\right)^2 < \left(x - \frac{1}{\sqrt{1-v^2}}\right)^2 + \rho^2 < \left(t + \frac{v}{\sqrt{1-v^2}}\right)^2$$

we would obtain an increased flow of energy outward.

2. Opinions differ as to whether a charge in hyperbolic motion emits radiation, owing to the different possible definitions made when posing global questions like that of radiation to infinity. In §8.4 a local definition will be introduced, using the reaction of the radiation on the charge. At this point we summarize the facts supporting the various opinions, and leave the reader to make up his or her own mind:

(a) $\text{Max}_t\,|S|$ falls off only as $1/\rho^2$.

(b) Expressions like $\int_K d\mathbf{S}\cdot S(t=R)$, where K is a sphere of radius R centered on the position of the particle, tend to positive values as $R \to \infty$.

(c) All the streamlines are closed, and none of them run to infinity.

Since the field equations are linear, the spectral distribution of the sources is inherited by the field. We therefore end this section with a brief discussion of the

(8.3.20) **Fourier Decomposition of the Current of a Point Charge**

$$\tilde{J}_\alpha(k) = \int d^4x\, e^{-i\langle k|x\rangle} J_\alpha(x) = e\int_{-\infty}^{\infty} ds\, \dot{z}_\alpha(s) e^{-i\langle k|z(s)\rangle}.$$

(8.3.21) **Examples**

1. The sudden acceleration of a charge,

$$z(s) = \begin{cases} \dot{\bar{z}}s & \text{if } s > 0, \quad \dot{\bar{z}} = \dfrac{1}{\sqrt{1-\bar{v}^2}}(1, \bar{\mathbf{v}}), \\ \dot{z}s & \text{if } s \le 0, \quad \dot{z} = \dfrac{1}{\sqrt{1-v^2}}(1, \mathbf{v}). \end{cases}$$

The distribution $\tilde{J}_\alpha(k)$ turns out to be

$$\tilde{J}_\alpha(k) = ie \lim_{\varepsilon\to 0}\left(\frac{\dot{z}_\alpha}{\langle k|\dot{z}\rangle + i\varepsilon} - \frac{\dot{\bar{z}}_\alpha}{\langle k|\dot{\bar{z}}\rangle - i\varepsilon}\right).$$

Consequently,

(i) if $\dot{z} = \bar{z}$, then $\tilde{J}_\alpha(k) = e2\pi\dot{z}_\alpha(\langle k|\dot{z}\rangle)$, and if $v = \bar{v} = 0$, $\tilde{J}_\alpha(k)$ is thus $\sim \delta(k^0)$.

(ii) The use of this $\tilde{J}(k)$ in (8.2.23), although it does not satisfy the assumptions needed to justify that formula, produces

$$P^\alpha(\infty) - P^\alpha(-\infty) = e^2\int \frac{d^4k}{(2\pi)^3}\delta(k^2)k^\alpha\Theta(k^0)\left|\frac{\dot{z}}{\langle\dot{z}|k\rangle} - \frac{\dot{\bar{z}}}{\langle\dot{\bar{z}}|k\rangle}\right|^2.$$

It might at first be thought that (8.2.19) could not be used for a point charge, since the field at the position of the particle, and hence also $F^{\text{ret}} \wedge {}^*J$, is infinite. As we shall learn in §8.4, however, $F^{\text{rad}} \wedge {}^*J$ remains finite, and so under the right circumstances (8.2.23) is also applicable to point-particles. It should furthermore be no surprise that the integral diverges for large $|\mathbf{k}|$, as $\ddot{z}$, and consequently also $F^{(\ddot{z})}$, become infinite for this motion. It can be hoped that if the acceleration were made gentler during a time τ, the integrand could be suppressed for $|\mathbf{k}| > 1/\tau$, allowing the integral to converge, while not essentially changing $\tilde{J}$ for $|\mathbf{k}| < 1/\tau$. With this motive, let us leave the integral as it is, but rewrite it by using the continuity equation for current,

$$k^\alpha \tilde{J}_\alpha = k^0\left(\tilde{J}^0 - \left(\frac{\mathbf{k}}{k^0}\cdot\tilde{\mathbf{J}}\right)\right) = 0.$$

This brings the spatial components $\tilde{\mathbf{J}}_\perp$ of $\tilde{J}$ that are perpendicular to $\mathbf{k}$ into play: If $k^2 = 0$, then

$$|\tilde{J}^\alpha \tilde{J}_\alpha| = |\tilde{J}_\perp|^2, \qquad \tilde{\mathbf{J}}_\perp := \tilde{\mathbf{J}} - \frac{\mathbf{k}(\mathbf{k}\cdot\tilde{\mathbf{J}})}{|\mathbf{k}|^2}.$$

Defining $\mathbf{w} := \mathbf{v}/(1 - |\mathbf{v}|\cos\vartheta)$, $\vartheta = \measuredangle(\mathbf{k}, \mathbf{v})$, and defining $\bar{\mathbf{w}}$ analogously, makes the energy loss

$$P^0(\infty) - P^0(-\infty) = e^2 \int \frac{d^3k}{(2\pi)^3}\frac{|\mathbf{w} - \bar{\mathbf{w}}|_\perp^2}{2|\mathbf{k}|^2}.$$

If $|\mathbf{v}|$ and $|\bar{\mathbf{v}}| \ll 1$, then $|\mathbf{w} - \bar{\mathbf{w}}|_\perp^2 = |\mathbf{v} - \bar{\mathbf{v}}|^2 - \langle \mathbf{k}|\mathbf{v} - \bar{\mathbf{v}}\rangle^2/|\mathbf{k}|^2$, and thus the maximum occurs for $\mathbf{k} \perp \mathbf{v} - \bar{\mathbf{v}}$, for wave-vectors perpendicular to the acceleration there is enhancement. For relativistic motion, the denominators $1 - v\cos\vartheta$ and $1 - \bar{v}\cos\vartheta$ strongly favor the directions of $\mathbf{v}$ and $\bar{\mathbf{v}}$ (see Problem 3). The frequency distribution has the characteristic spectrum $d^3k/|\mathbf{k}|^2 \sim dk$ of bremsstrahlung; the same energy is radiated in every frequency interval.

2. A rotating charge. Whereas the field of a Hertz dipole (8.3.15) is purely harmonic in time, the fields of charges rotating in circles with finite radii also exhibit higher harmonic frequencies. Thus it comes about that highly energetic electrons in a magnetic field emit a characteristic x-ray spectrum. Let us calculate the energy emitted at a given harmonic frequency.

A formula analogous to (8.2.23) can be derived (Problem 7) for the energy lost per period ω by a current with periodic time-dependence. The only essential change is the definition of $\tilde{J}$, as we now use a Fourier series in time:

$$\tilde{J}^\alpha(n\omega, \mathbf{k}) := \frac{\omega}{2\pi}\int_0^{2\pi/\omega} dt \int_{\mathbb{R}^3} d^3x\, e^{-i(\mathbf{k}\cdot\mathbf{x} - n\omega t)} J^\alpha(x), \qquad n \in \mathbb{Z}. \tag{8.3.22}$$

There results

$$P^0\left(\frac{2\pi}{\omega}\right) - P^0(0) = \sum_{n\geq 1}\int \frac{d^3k}{(2\pi)^3} 2\pi\delta(|\mathbf{k}|^2 - (n\omega)^2)n\omega \cdot \tilde{J}^\beta(n\omega, \mathbf{k})\tilde{J}_\beta(-n\omega, -\mathbf{k}). \tag{8.3.23}$$

If the current (8.3.14) is written as

$$\begin{aligned} J^\alpha(x) &= e\int_{-\infty}^{\infty} ds\, \dot{z}^\alpha(s)\delta^4(x - z(s)) \\ &= e(1, -v\sin\omega t, v\cos\omega t, 0)\delta^3(\mathbf{x} - \mathbf{z}(t)), \end{aligned}$$

then

$$\tilde{J}(n\omega, \mathbf{k}) = e\omega \int_0^{2\pi/\omega} \frac{dt}{2\pi} e^{i(n\omega t - \mathbf{k}\cdot\mathbf{z}(t))}(1, -v\sin\omega t, v\cos\omega t, 0). \tag{8.3.24}$$

Since the radiation is concentrated in the plane of motion within a sector of opening angle $\sim \sqrt{1/v - 1}$ (Problem 3), it is interesting to calculate the frequency distribution for **k** in the plane of motion, let us say in the x_2-direction. According to the argument in the preceding example, only J^1 contributes to the energy loss. The integral (8.3.24) becomes a Bessel function

(8.3.25)

$$\tilde{J}^1(n\omega, 0, n\omega, 0) = -e\omega v \int_0^{2\pi/\omega} \frac{dt}{2\pi} \sin\omega t e^{in(\omega t - v\sin\omega t)} = -ieRJ_n'(nv).$$

To discuss this result, it is necessary to distinguish the cases $nv \ll 1$ and $nv \gg 1$. In the former case, one can use the well-known expansion ([39, cf. (3.4.3) and (3.4.4)]):

$$J_n'(nv) = \frac{1}{2(n-1)!}\left(\frac{nv}{2}\right)^{n-1}(1 + O((nv)^2)). \tag{8.3.26}$$

This shows that the frequency distribution of (8.3.24) is $\sim (nv)^{2n}/[(n-1)!]^2$, which has a strong maximum at $n = 1$ if $v \ll 1$. If $v = 1$, the simple formula

$$J_n'(n) \to \frac{2^{2/3}}{3^{1/3}\Gamma(\frac{1}{3})} n^{-2/3}$$

(see [39, 9.4.43]) could be used, and the spectrum would be $n^2|\tilde{J}|^2 \sim n^{2/3}$. Since $v < 1$, however, this formula is not valid for high n, and the better asymptotic formula,

$$J_n'(nv) \to -\frac{2}{v}\left(\frac{1-v^2}{4\xi}\right)^{1/4} \frac{\mathrm{Ai}'(n^{2/3}\xi)}{n^{2/3}},$$

$$\tfrac{2}{3}\xi^{3/2} = \ln\frac{1+\sqrt{1-v^2}}{v} - \sqrt{1-v^2} \underset{v\to 1}{\longrightarrow} \frac{(1-v^2)^{3/2}}{3}, \tag{8.3.27}$$

$$Ai'(z) \to \begin{cases} -\dfrac{1}{3^{1/3}\Gamma(\frac{1}{3})} & \text{as } z \to 0, \\[2ex] \dfrac{1}{2\sqrt{\pi}} z^{1/4} e^{-(2/3)z^{3/2}} & \text{as } z \to \infty, \end{cases}$$

must be resorted to. This modifies the spectrum to

$$n \exp\left(-\tfrac{2}{3}n(1-v^2)^{3/2}\right)$$

for $n^{2/3}(1-v^2) \gg 1$. Therefore the spectrum has its maximum at $n \sim (1-v^2)^{-3/2} =: \gamma^3$, after which it decreases exponentially.

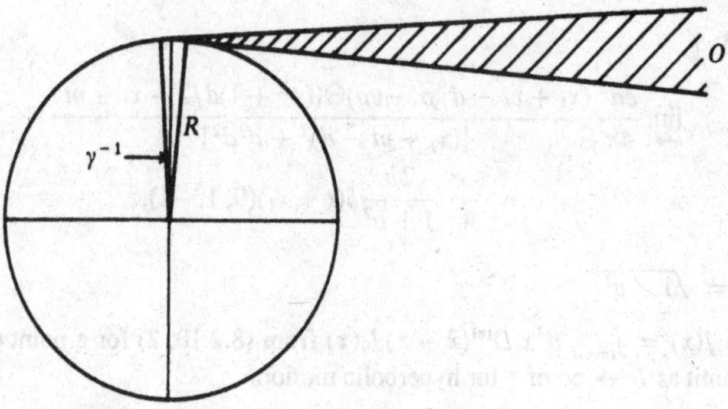

FIGURE 8.14. Radiation from relativistic motion in a circle.

(8.3.28) Remark

The reason that the maximum is at $\sim \gamma^3$ can be understood as follows: An observer O first sees the light that is emitted in an angular interval $\sim \gamma^{-1}$. (See Figure 8.14.) Since the beam of radiation sweeps just barely ahead of the charge, the light arrives at O within a time interval

$$\Delta t = (1 - v)\gamma^{-1} R \sim \gamma^{-3} R$$

(if $v \to 1$), and hence

$$\omega_{\max} \sim \frac{1}{\Delta t} \sim \frac{\gamma^3}{R} \sim \omega\gamma^3.$$

(8.3.29) Problems

1. Substitute from (8.3.12) into the expression for F^{ret}.
2. Calculate the field and Poynting's vector S for a charge moving in a straight line at the speed of light.
3. Calculate the energy distribution T_{00} of the radiation of a fast-moving charge $[\dot{z}(s_0) = 1/\sqrt{1-v^2}(1, 0, 0, v), z(s_0) = 0]$ in the far zone at $x^2 = 0$ and:

   (a) longitudinal acceleration ($\ddot{z}(s_0) \sim (v, 0, 0, 1)$); and

   (b) transverse acceleration ($\ddot{z}(s_0) \sim (0, 0, 1, 0)$).

   At what angle, in the usual polar coordinates, is the maximum as $v \to 1$?
4. Calculate $\int_{t=0} T^\alpha$ for the field of a uniformly moving point charge (8.3.11; 1). Use $\bar{x} = (x_1/\sqrt{1-v^2}, x_2, x_3)$ as the integration variables, and define the classical radius of the electron by the divergent integral

$$\frac{1}{r_c} = \frac{1}{2} \int \frac{d^3\bar{x}}{|\bar{\mathbf{x}}|^4}.$$

   Show that this destroys the vector transformation property of (7.3.37; 5). How come?

5. Show that

$$\lim_{v\to 1}\frac{ed^2}{4\pi}\frac{(x_1+vt-d,\rho,-v\rho)\Theta((\rho^2+1)d/2-x_1-vt)}{[(x_1+vt-d)^2+\rho^2d^2]^{3/2}} = \frac{e}{4\pi}\frac{2\rho}{1+\rho^2}\delta(t+x_1)(0,1,-1),$$

where $d := \sqrt{1-v^2}$.

6. Calculate $j(x) = \int_{t=-T} d^3x\, D^{\mathrm{ret}}(\bar{x}-x)J_0(x)$ from (8.2.10; 2) for a point charge, and take the limit as $T\to\infty$ of j for hyperbolic motion.

7. Calculate $P^0(2\pi/\omega) - P^0(0)$ for a J that depends on time periodically, by using the Fourier series (8.3.22).

(8.3.30) Solutions

1. $((x^2+1)/2\langle\dot{z}|x\rangle)\dot{z} - z = -(2/\xi)(x_1,t,0,0)$.

$$F^{\mathrm{ret}} = \frac{-8}{\xi^3}(x_1,t,0,0) \wedge\left(\frac{t(t^2-x_1^2+\rho^2+1)-x_1\xi}{2(t^2-x_1^2)}, \frac{x_1(t^2-x_1^2+\rho^2+1)-t\xi}{2(t^2-x_1^2)}, x_2, x_3\right)\frac{e}{4\pi}.$$

Using $\mathbf{x}\cdot d\mathbf{x} = x_1\,dx_1 + \rho\,d\rho$, we find (8.3.13).

2. The limit as $v\to 1$ of the fields of (8.3.11; 1) is

$$(E_1, E_\rho, E_\varphi) = \frac{e}{2\pi}\delta(x-t)\frac{1}{\rho}(0,1,1);$$

and this satisfies $\delta F = J$ with

$$J = e\delta(x-t)\delta(y)\delta(z)(1,1,0,0) = \lim_{v\to 1} e\delta(x-vt)\delta(y)\delta(z)(1,v,0,0).$$

S points in the direction x_1, but because of the factor δ^2, this infinitely Lorentz-contracted field has an infinite energy-momentum density.

3.

(a) $T_{00} \sim \dfrac{(1-v^2)^4\sin^2\vartheta}{r^2(1-v\cos\vartheta)^6}$, $\quad \vartheta_{\max} \sim \sqrt{\dfrac{1}{v}-1} \sim \dfrac{m}{E}$.

(b) $T_{00} \sim \dfrac{(1-v^2)^2}{r^2(1-v\cos\vartheta)^6}[(1-v\cos\vartheta)^2 - (1-v^2)\sin^2\vartheta\sin^2\varphi]$.

4. $\displaystyle\int T^0 = \frac{1+v^3/3}{1-v^2}\frac{1}{r_c}$, $\quad \displaystyle\int T^j = v^j\frac{4/3}{1-v^2}\frac{1}{r_c} \neq v^j\int T^0(v=0)/\sqrt{1-v^2}$.

The condition that $\delta T^\alpha = 0$, used in the derivation of the transformation law, is violated at the origin. Even adding a t^α for the particle will not avoid the problem, as P^α equals $(m/\sqrt{1-v^2}, m\mathbf{v}/\sqrt{1-v^2})$. Thus $\delta(T^\alpha + t^\alpha)$ is only formally zero (cf. (7.3.25; 2) and (8.1.18; 7)).

5. If $x_1 \neq t$, then the three components approach zero; on the other hand, with $\alpha = (x + vt)/d - 1$,

$$\int_{-\infty}^{\infty} \frac{dt\, d^2 \Theta(\,)}{[(x_1 + vt - d)^2 + \rho^2 d^2]^{3/2}} \to \int_{-\infty}^{d(1+\rho^2)/2 - x_1} \frac{dt\, d^2}{[(x_1 + vt - d)^2 + \rho^2 d^2]^{3/2}}$$

$$\to \frac{1}{v} \int_{-\infty}^{(\rho^2-1)/2} \frac{d\alpha}{[\alpha^2 + \rho^2]^{3/2}}$$

$$= \frac{1}{v\rho^2} \frac{\alpha}{(\alpha^2 + \rho^2)^{1/2}} \Bigg|_{-\infty}^{(\rho^2-1)/2}$$

$$= \frac{1}{v\rho^2} \left(1 + \frac{\rho^2 - 1}{\rho^2 + 1} \right) \to \frac{2}{1 + \rho^2}.$$

6. If z is the point at which the world-line $z(s)$ crosses the hypersurface at $t = -T$, then

$$j(x) = \frac{e}{2\pi} \delta((x - z)^2) \Theta(x^0 - z^0).$$

If the motion is hyperbolic,

$$z = \left(\frac{-v}{\sqrt{1 - v^2}}, \frac{1}{\sqrt{1 - v^2}}, 0, 0 \right),$$

and as $v \to 1$,

$$\delta(x^2 + 1 - 2\langle x | z \rangle) = \sqrt{1 - v^2}\, \delta(2(x + vt) - (x^2 + 1)\sqrt{1 - v^2}) \to 0.$$

7. Substituting the Fourier series

$$J(x) = \sum_{n \in \mathbf{Z}} \int \frac{d^3k}{(2\pi)^3} \tilde{J}(n\omega, \mathbf{k}) e^{i(\mathbf{k} \cdot \mathbf{x} - n\omega t)}$$

into the integral

$$P^0(2\pi/\omega) - P^0(0) = \int_{0 \le \bar{t} \le 2\pi/\omega} i_0 F^{\text{ret}} \wedge {}^*J$$

$$= \int_{0 \le \bar{t} \le 2\pi/\omega} d^4\bar{x}\, J^\beta(\bar{x}) \int_{\mathbf{R}^4} d^4x \left(J^0(x) \frac{\partial}{\partial \bar{x}^\beta} D^{\text{ret}}(\bar{x} - x) - J_\beta(x) \frac{\partial}{\partial \bar{x}^0} D^{\text{ret}}(\bar{x} - x) \right)$$

and using (8.2.4) and (8.2.5) yields

$$\int d^4x\, D^{\text{ret}}(\bar{x} - x) J^\alpha(x) = \sum_{n \in \mathbf{Z}} \int \frac{d^3k}{(2\pi)^3} \tilde{J}(n\omega, \mathbf{k}) \frac{e^{i(\mathbf{k} \cdot \bar{\mathbf{x}} - n\omega t)}}{|\mathbf{k}|^2 - (n\omega + i\varepsilon)^2}.$$

Then, finally,

$$P^0 \left(\frac{2\pi}{\omega} \right) - P^0(0) = \sum_{n \ge 1} \int \frac{d^3k}{(2\pi)^3} 2\pi \delta(|\mathbf{k}|^2 - (n\omega)^2) n\omega \tilde{J}^\beta(n\omega, \mathbf{k}) \tilde{J}_\beta(-n\omega, -\mathbf{k}).$$

8.4 Radiative Reaction

The radiation of electromagnetic energy causes a reactive force on a charge. The calculation of this force for point particles is tricky, as it involves divergent integrals.

The product of J and F appears in the Lorentz force (7.3.24), but it is not well defined for a point particle, since the field is singular at the particle's position. Let us first look into the less problematical matter of the energy-momentum of the emitted radiation, as we slowly work up to the infinities in the equation for the total energy and momentum.

The starting point is Stokes's theorem for the electromagnetic energy-momentum forms:

$$\int_N d^*T^\alpha = \int_{\partial N} {}^*T^\alpha. \tag{8.4.1}$$

We choose N as the four-dimensional region bounded by the light-cones

$$L_1 = \{x \in \mathbb{R}^4\colon (x - z(s_1))^2 = 0, x^0 > z^0(s_1)\},$$
$$L_2 = \{x \in \mathbb{R}^4\colon (x - z(s_2))^2 = 0, x^0 > z^0(s_2)\},$$

and the cylinder

$$K = \{x \in \mathbb{R}^4\colon (x - z(s_1))^2 + \langle \dot{z}(s_1)|x - z(s_1)\rangle^2 = R^2\},$$

in order to be able to follow the radiation as it goes to infinity (Figure 8.15). We first calculate the part of (8.4.1) coming from the piece of ∂N contained in K. For this purpose R may be increased to ∞, and $ds := s_2 - s_1 \to 0$; the result has the interpretation of the amount of energy-momentum that is lost by the charge between s_1 and $s_1 + ds$ and escapes to infinity. In this limit $\int_{\partial N \cap K} {}^*T_\alpha$ consists only of the contribution of $F^{(\ddot{z})}$ to T_α (cf. (8.3.16)

$$T_\alpha^{(\ddot{z})} = \left(\frac{e}{4\pi}\right)^2 \langle \dot{z}|x - z\rangle^{-4}(x - z)(x - z)_\alpha \left\{\ddot{z}^2 - \left(\frac{\langle \ddot{z}|x - z\rangle}{\langle \dot{z}|x - z\rangle}\right)^2\right\};$$

the integral of this is asymptotically independent of R, while the other terms all have higher powers of R in the denominator. The external surface $\partial N \cap K$ has a height ds above $K \cap L_1$, and in the limit as $ds \to 0$, we need to know ${}^*T_\alpha$ only on $K \cap L_1$. If we write $x - z = R(\dot{z} + n)$, $n \in E_1$, on that surface (see Figure 8.16), then from $x \in K \cap L_1$:

$$0 = R^{-2}(x - z)^2 = -1 + 2\langle \dot{z}|n\rangle + n^2$$

and

$$1 = R^{-2}\{(x - z)^2 + \langle \dot{z}|x - z\rangle^2\} = (-1 + \langle \dot{z}|n\rangle)^2,$$

it follows that $n^2 = 1$ and $\langle \dot{z}|n\rangle = 0$, and hence that

$$-\langle \dot{z}|x - z\rangle = R \qquad \text{and} \qquad \langle \ddot{z}|x - z\rangle = R\langle \ddot{z}|n\rangle.$$

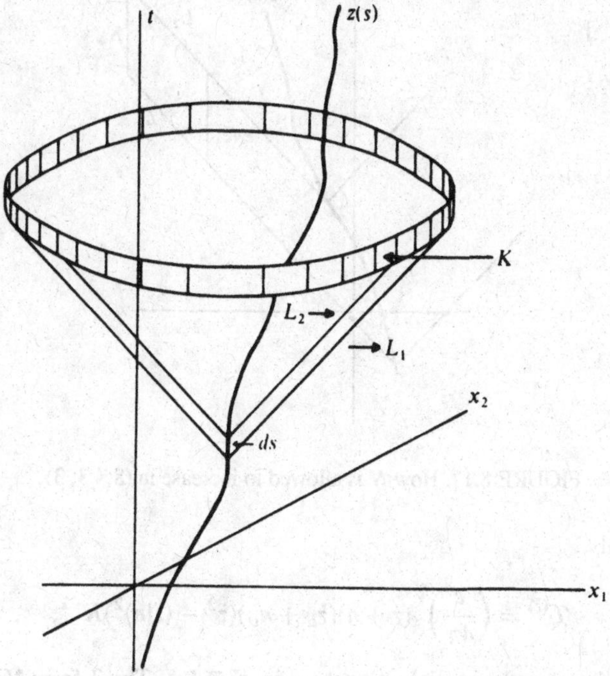

FIGURE 8.15. The hypersurface used for ∂N in (8.4.1).

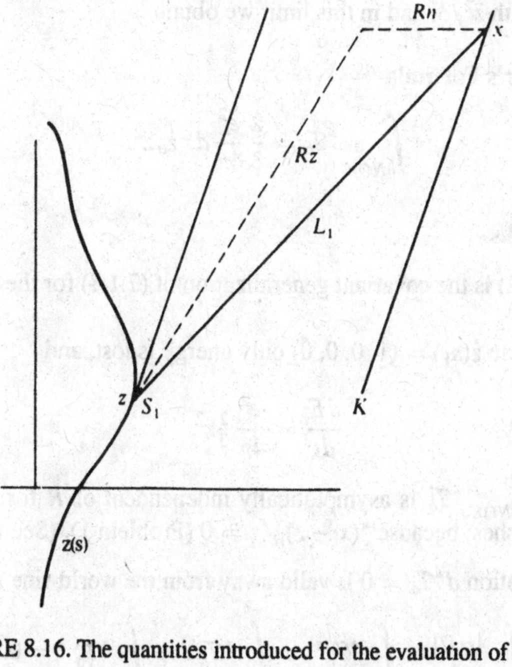

FIGURE 8.16. The quantities introduced for the evaluation of (8.4.1).

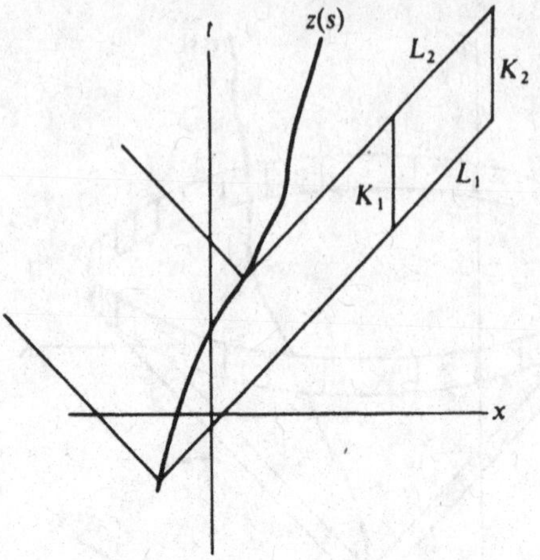

FIGURE 8.17. How N is allowed to increase in (8.4.3; 3).

Consequently,

$$\mathcal{T}_\alpha^{(\ddot{z})} = \left(\frac{e}{4\pi}\right)^2 (\dot{z}+n)(\dot{z}_\alpha + n_\alpha)(\ddot{z}^2 - \langle \ddot{z}|n\rangle^2)R^{-2},$$

where n varies over the spatial unit sphere in $K \cap L_{10}$. The 3-form $^*(\dot{z}+n)$ acts in the integral $\int {}^*\mathcal{T}_\alpha$ as $R^2\, ds\, d\Omega_n$ ($d\Omega$ is the element of solid angle on the unit sphere). By symmetry, all odd powers of n, and hence the part containing n_α, drop out of the integral $\int {}^*\mathcal{T}_\alpha$. By taking the average over the unit sphere we simply replace $\langle \ddot{z}|n\rangle^2$ with $\ddot{z}^2/3$, and in this limit we obtain

(8.4.2) **Larmor's Formula**

$$\int_{\partial N \cap K} {}^*\mathcal{T}_\alpha = \frac{2}{3}\frac{e^2}{4\pi} ds\, \dot{z}_\alpha \ddot{z}^2.$$

(8.4.3) **Remarks**

1. Formula (8.4.2) is the covariant generalization of (7.1.8) for the loss of energy-momentum.
2. In the rest frame $\dot{z}(s_1) = (1, 0, 0, 0)$ only energy is lost, and

$$\frac{dE}{dt} = \frac{e^2}{4\pi}\tfrac{2}{3}\ddot{z}^2.$$

3. The reason $\int_{\partial N \cap K} {}^*\mathcal{T}_\alpha$ is asymptotically independent of R for large R is that $\int_{L_1} {}^*\mathcal{T}_\alpha^{(\ddot{z})}$ vanishes, because $^*(x-z)_{|L_{1,2}} = 0$ (Problem 1). (See Figure 8.17.)

Since the equation $d^*\mathcal{T}_\alpha = 0$ is valid away from the world-line $z(s)$,

$$\int_{K_1} {}^*\mathcal{T}_\alpha^{(\ddot{z})} - \int_{K_2} {}^*\mathcal{T}_\alpha^{(\ddot{z})} = \int_{L_1} {}^*\mathcal{T}_\alpha^{(\ddot{z})} - \int_{L_2} {}^*\mathcal{T}_\alpha^{(\ddot{z})} = 0.$$

(8.4.4) **Examples**

1. **Hyperbolic motion.** If

$$z(s) = a^{-1}(\sinh as, 0, 0, \cosh as) \qquad \text{and} \qquad \dot{z}(s) = (\cosh as, 0, 0, \sinh as),$$

then

$$\frac{dP^\alpha}{ds} = \frac{3}{2}\frac{e^2}{4\pi}\dot{z}^\alpha a^2.$$

The energy contained in the radiation field always increases, while momentum is transferred to it in the negative z-direction when $s < 0$, and in the positive z-direction when $s > 0$. Therefore the charge always radiates forward along its direction of motion.

2. **Synchrotron radiation.** With the current of a rotating charge (8.3.14),

$$\dot{z} = \frac{1}{\sqrt{1-v^2}}(1, -v\sin\omega t, v\cos\omega t, 0),$$

$$\ddot{z} = \frac{-v^2}{R(1-v^2)}(0, \cos\omega t, \sin\omega t, 0),$$

where $\omega = v/R$. As a consequence, δE, the energy loss per period $\cdot\, 2\pi$, is

$$\delta E = \frac{2}{3\omega}\frac{e^2}{4\pi}\ddot{z}^2 = \omega\frac{e^2}{6\pi}\frac{v^2}{(1-v^2)^2}.$$

If we use the value $e^2/4\pi\hbar = \frac{1}{137}$, this is $\sim \omega\hbar(v^2/200)(E/m)^4$. So long as $E/m \sim 1$, the charge clearly needs to undergo more than 200 revolutions to give off a quantum at the ground-state frequency. If $v \to 1$, the rate of energy loss increases rapidly; for, e.g., 5 GeV electrons, $(E/m)^4 \sim 10^{16}$. Accordingly, fast electrons moving in a circle lose quite a bit of energy in the form of synchrotron radiation.

Although $\mathcal{T}_\alpha^{(\ddot{z})}$ does not contribute to the integral over the light-cones L_1 and L_2 in the calculation of the right side of (8.4.1), the other terms in $\mathcal{T}_\alpha$ have infinite integrals! They decrease as r^{-3} and r^{-4}, which diverge when integrated over all space. In order to isolate the causes of the problem, we write

$$\text{(8.4.5)} \qquad d^*\mathcal{T}_\alpha = i_\alpha\left(F^{\text{in}} + \tfrac{1}{2}F^{\text{rad}} + \tfrac{1}{2}(F^{\text{ret}} + F^{\text{adv}})\right) \wedge {}^*J$$

(cf. (8.2.21)). The first term is the Lorentz force from the incoming field, and causes no trouble. We discovered that if the current in (8.2.21) had compact support, then aside from F^{in} only the radiation field F^{rad} contributes to the loss of energy-momentum in the limit of infinite times. For this reason, we next evaluate the contribution of F^{rad} to the left side of (8.4.1) for the point charge. It will turn out that this term is finite, and all the difficulties stem from the last term.

Since J is supported on $z(s)$, we must evaluate F^{rad} on the world-line of the charge. After some simple algebra (Problem 2), we can write the radiation field as

$$\text{(8.4.6)} \qquad F^{\text{rad}}_{\alpha\beta}(x) = e\int_{-\infty}^{\infty} ds\, D(x - z(s))\frac{d}{ds}\left[\frac{\dot{z}_\alpha(x-z)_\beta}{\langle\dot{z}|x-z\rangle} - (\alpha \leftrightarrow \beta)\right].$$

We wish now to let x carefully approach some point of the world-line, say $z(0)$. To do this, we expand the integrand of (8.4.6) about $s = 0$:

$$z(s) - z(0) = s\dot{z} + \frac{s^2}{2}\ddot{z} + \frac{s^3}{6}\dddot{z} + \cdots, \qquad \dot{z} := \dot{z}(0), \quad \text{etc.}$$

(8.4.7)

$$\dot{z}(s) = \dot{z} + s\ddot{z} + \frac{s^2}{2}\dddot{z} + \cdots,$$

and call $x - z(0) = \lambda$. In order that x stay between the retarded time and the advanced time (Figure 8.18), let $\langle \dot{z}|\lambda\rangle = 0$ thus λ is spacelike as it approaches 0. Since

$$D(x) = \frac{\delta(x^2)}{2\pi}(\Theta(x^0) - \Theta(-x^0))$$

and

$$(x - z(s))^2 = \left(\lambda - s\dot{z} - \frac{s^2}{2}\ddot{z}\right)^2 + O(s^3) = \lambda^2 - s^2(1 + \langle\lambda|\ddot{z}\rangle) + O(s^3),$$

if $\lambda \to 0$, then

$$D(x - z(s)) = \frac{\delta(\lambda^2 - s^2)}{2\pi}(\Theta(-s) - \Theta(s)) = \frac{1}{4\pi\lambda}(\delta(s + \lambda) - \delta(s - \lambda)),$$

(8.4.8) $\qquad \lambda := \langle\lambda|\lambda\rangle^{1/2} > 0.$

If the rest of the integrand of (8.4.6) is also expanded about $s = 0$,

$$[\ ] := \frac{N(\lambda)}{s} + A(\lambda) + sB(\lambda) + \frac{s^2}{2}C(\lambda),$$

(8.4.9)

$$\frac{d}{ds}[\ ] = -\frac{N(\lambda)}{s^2} + B(\lambda) + sC(\lambda),$$

then as $\lambda \to 0$, (8.4.6) becomes equals to

$$\frac{e}{4\pi}\lim_{\lambda\to 0}\frac{1}{\lambda}\left(\frac{d}{ds}[\ ]_{|s=-\lambda} - \frac{d}{ds}[\ ]_{|s=\lambda}\right) = -\frac{e}{4\pi}2C(0).$$

Substitution of the series (8.4.7) reveals that

$$-C(0) = \tfrac{2}{3}(\dddot{z}_\alpha\dot{z}_\beta - \dddot{z}_\beta\dot{z}_\alpha),$$

with which we obtain a formula for the

(8.4.10) **Radiation Field on the World-Line**

$$F^{\text{rad}}(z(s)) = \frac{e}{3\pi}\dddot{z}(s) \wedge \dot{z}(s).$$

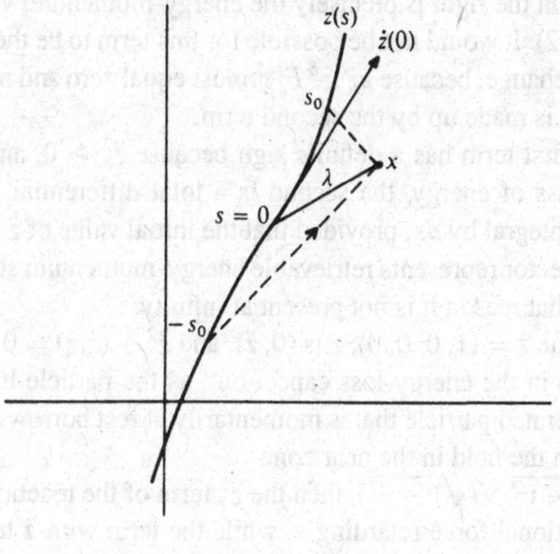

FIGURE 8.18. The limit $\lambda \to 0$.

(8.4.11) **Remarks**

1. The radiation field is purely electric in the rest-frame of the particle:

$$\tfrac{1}{2}(\mathbf{E}^{\text{ret}} - \mathbf{E}^{\text{adv}}) = \frac{e}{4\pi}\tfrac{2}{3}\dddot{\mathbf{z}}, \qquad \tfrac{1}{2}(\mathbf{B}^{\text{ret}} - \mathbf{B}^{\text{adv}}) = 0.$$

2. In the result (8.4.10) it is understood that the term $-\ddot{\mathbf{z}}/r$ in $\mathbf{E}$ of (8.3.9) contributes the limit

$$\lim_{r \to 0} \frac{-\ddot{\mathbf{z}}(s-r) + \ddot{\mathbf{z}}(s+r)}{2r} = \dddot{\mathbf{z}}$$

to F^{rad}.
When averaged over space, $(\mathbf{x}(\mathbf{x}\cdot\ddot{\mathbf{z}}))/r^3$ in like manner contributes $-\ddot{\mathbf{z}}/3$, and the Coulomb field of the near zone and the magnetic field disappear from (8.3.9). Formula (8.4.10) should a priori be averaged over the different possible directions from which $z(s)$ is approached, but we have noted above that it is independent of the direction of λ provided that $\langle \dot{z}|\lambda\rangle = 0$.

If we use (8.4.10) in equation (8.4.5), and conclude from $\langle \dot{z}|\ddot{z}\rangle = 0$ that $\langle \dot{z}|\dddot{z}\rangle = -\ddot{z}^2$, then we get a formula for the

(8.4.12) **Energy and Momentum Lost to the Radiation Field**

$$\frac{dP^{\text{rad}}_\alpha}{ds} := \frac{-d}{2ds}\int_{N_4} {}^*i_J i_\alpha F^{\text{rad}} = \frac{e}{2}\dot{z}^\beta(s) F^{\text{rad}}_{\beta\alpha}(z(s)) = \frac{e^2}{4\pi}\tfrac{2}{3}(\dot{z}_\alpha \ddot{z}^2 - \dddot{z}_\alpha).$$

(8.4.13) **Remarks**

1. The limit $ds \to 0$ is understood in (8.4.12), and N_4 contains the part of $z(s)$ between s and $s + ds$; thus on the right side the derivative of z is taken at the proper time s.

2. The first term on the right is precisely the energy-momentum vector that flows to infinity (8.4.2). It would not be possible for this term to be the whole energy-momentum exchange, because $e\dot{z}^\alpha \ddot{z}^\beta F^{\mathrm{rad}}_{\alpha\beta}$ must equal zero and not $-(e^2/6\pi)\ddot{z}^2$. The deficiency is made up by the second term.
3. Although the first term has a definite sign because $\ddot{z}^2 \geq 0$, and represents an irretrievable loss of energy, the second is a total differential and contributes nothing to an integral by ds, provided that the initial value of $\ddot{z}$ is returned to at the end. This vector represents retrievable energy-momentum stored in the near zone, and for that reason it is not present at infinity.
4. In the rest-frame $\dot{z} = (1, 0, 0, 0)$, $\ddot{z}$ is $(0, \ddot{\mathbf{z}})$, and $\ddot{z}^2 + (\dot{z}|\dddot{z}) = 0 \Rightarrow \dddot{z} = (\ddot{z}^2, \dddot{\mathbf{z}})$. The two terms in the energy loss cancel out, as the particle has no energy to lose. An accelerated particle that is momentarily at rest borrows the energy that it radiates from the field in the near zone.
5. If $\dot{z} = (1/\sqrt{1-v^2}, \mathbf{v}/\sqrt{1-v^2})$, then the $\ddot{z}^2$ term of the reaction on the charge acts like a frictional force retarding $\mathbf{v}$, while the term with $\dddot{\mathbf{z}}$ tends to increase the acceleration of the particle. This leads to all sorts of paradoxical consequences, which we shall return to after having discussed the remaining contribution $\sim F^{\mathrm{ret}} + F^{\mathrm{adv}}$.

(8.4.14) Examples

1. By taking more derivatives in Example (8.4.4; 1),

$$\ddot{z}(s) = a(\sinh as, 0, 0, \cosh as),$$
$$\dddot{z}(s) = a^2(\cosh as, 0, 0, \sinh as),$$

we see that the two terms of (8.4.12) exactly cancel, and $dP^{\mathrm{rad}}/ds = 0$. A charge in hyperbolic motion radiates on credit; the energy is not supplied by the particle, but comes from the near zone. Of course, the debt must be repaid later, once the acceleration stops. For example, if the charge is accelerated from rest to the velocity $v = \operatorname{Tanh} as_0$,

$$\dot{z}(s) = \Theta(-s)(1, 0, 0, 0) + \Theta(s)\Theta(s_0 - s)(\cosh as, 0, 0, \sinh as) + \Theta(s - s_0)\left(\frac{1}{\sqrt{1-v^2}}, 0, 0, \frac{v}{\sqrt{1-v^2}}\right);$$

and then

$$\ddot{z}(s) = \Theta(s)\Theta(s_0 - s)a(\sinh as, 0, 0, \cosh as),$$
$$\dddot{z}(s) = \Theta(s)\Theta(s_0 - s)a^2(\cosh as, 0, 0, \sinh as) + \delta(s)(0, 0, 0, 1) - \delta(s - s_0)\left(\frac{v}{\sqrt{1-v^2}}, 0, 0, \frac{1}{\sqrt{1-v^2}}\right).$$

Hence

$$\dot{z}\ddot{z}^2 - \dddot{z} = -\delta(s)(0, 0, 0, 1) + \delta(s - s_0)\left(\frac{v}{\sqrt{1-v^2}}, 0, 0, \frac{1}{\sqrt{1-v^2}}\right).$$

The force on the particle is the negative of the rate of change of the energy-momentum of the field. At first the particle feels a jolt in the direction of the acceleration from F^{rad}, and later the radiation force operates in the opposite direction, and F^{in} has to pay the energy bill.

2. For the rotating charge of (8.4.4; 2),

$$\dddot{z} = \frac{v^3}{R^2(1-v^2)^{3/2}}(0, \sin\omega t, -\cos\omega t, 0).$$

The second term of (8.4.12) causes no additional loss of energy in this case, but it does intensify the braking action of the first term, opposing the velocity of the charge:

$$\dot{z}\ddot{z}^2 - \dddot{z} = \frac{v^3}{R^2(1-v^2)^{5/3}}(v, -\sin\omega t, \cos\omega t, 0).$$

The calculation of the last term of (8.4.5) is simply a matter of replacing the difference appearing in (8.4.8) with a sum:

(8.4.15) $$D^{\text{ret}}(x - z(s)) + D^{\text{adv}}(x - z(s)) = \frac{1}{4\pi\lambda}(\delta(s-\lambda) + \delta(s+\lambda)).$$

Hence the expansion (8.4.9) results in a contribution $-N(\lambda)/\lambda^3 + B(\lambda)/\lambda$, where

$$N(\lambda) = \frac{\dot{z}_\alpha\lambda_\beta - \dot{z}_\beta\lambda_\alpha}{1 + \langle\ddot{z}|\lambda\rangle}.$$

Not only is $N(\lambda)/\lambda^3$ divergent as $\lambda \to 0$, but it also depends on the direction of λ: If no direction has been singled out in Minkowski space, then the net result must be $\sim \dot{z}_\alpha\ddot{z}_\beta - \dot{z}_\beta\ddot{z}_\alpha$ and the coefficient goes as $1/\lambda$ for dimensional reasons. Actually the next term $B(0)$ is $(\dot{z}_\alpha\dddot{z}_\beta - \dot{z}_\beta\dddot{z}_\alpha)/2$ independently of the direction of λ. Normally only the latter term is retained while N is swept under the rug by some averaging procedure; by this hocus-pocus,

(8.4.16) $$\tfrac{1}{2}(F^{\text{ret}} + F^{\text{adv}})(z(s)) = \frac{e}{4\pi}\ddot{z} \wedge \dot{z} \lim_{\lambda\to 0}\frac{c}{\lambda}.$$

Equation (8.4.16) is not very well defined, but at any rate the numerical factor c is positive when calculated in this way. Then attempts are made to argue away the resulting indeterminacy in the radiative reaction of the field along the following lines: Suppose that the three contributions of (8.4.5) are combined and set equal to minus the rate of change of the energy-momentum of a particle of mass m_0 during the time ds. Then there results

(8.4.17) $$m_0\ddot{z}_\beta = e\dot{z}^\alpha F^{\text{in}}_{\alpha\beta} - \frac{e^2}{4\pi}\tfrac{2}{3}(\dot{z}_\beta\ddot{z}^2 - \dddot{z}_\beta) - \delta m\ddot{z}_\beta,$$
$$\delta m = \lim_{\lambda\to 0}\frac{ce^2}{4\pi\lambda}.$$

Next one calls $m = m_0 + \delta m$ ("mass-renormalization") and smugly solves the

(8.4.18) Renormalized Equation of Motion

$$m\ddot{z}_\beta = e\dot{z}^\alpha F^{\text{in}}_{\alpha\beta} - \frac{e^2}{4\pi}\frac{2}{3}(\dot{z}_\beta \ddot{z}^2 - \dddot{z}_\beta).$$

(8.4.19) Remarks

1. The mass m is clearly what would be measured by the inertia of the particle in an external field. In (8.4.18) there are no apparent infinities.
2. It may seem peculiar that in a theory that is invariant under reversal of the motion, something so obviously not invariant under motion reversal as the radiative reaction force should occur. It arises because of the use of F^{in} and D^{ret}; it gets its sign from the initial conditions, and would have the opposite sign if F^{out} and D^{adv} were used. By using $(D^{\text{ret}} + D^{\text{adv}})/2$, one can even find stationary solutions of the relativistic two-body problem [42] for which nothing at all is radiated. If the system has a finite energy one finds radiation damping for $t \to \pm\infty$.
3. When discussing (8.2.21) we learned that for currents with compact support the total energy-momentum vector transmitted to the field originates in the Lorentz force with F^{rad}. This result is carried over to point-particles in (8.4.16), and it can be seen that $(F^{\text{ret}} + F^{\text{adv}})/2$ simply contributes $\delta m(\dot{z}(\infty) - \dot{z}(-\infty))$, which is the change of the energy-momentum of the self-field attached to the particle.
4. The self-field is eliminated in (8.4.18), and only the particle's coordinates appear. To solve the initial-value problem of the total system, z, $\dot{z}$, F, and $^*F$ must all be known at some time. One might therefore expect that the elimination of F would render it necessary to take the whole previous history, $z(s)$ for $s < 0$, into account. In fact, the only extra quantity that shows up in the limit of a point-particle is $\ddot{z}$, and the solution manifold of the Cauchy problem is only increased by dependence on the three parameters $\ddot{\mathbf{z}}$. Even so, the physically acceptable solution manifold will turn out to be of a lower dimension.
5. The question arises of why we did not circumvent the difficulties connected with point-particles by using charges spread over some positive volume. Unfortunately, it is not easy to obtain a theory in this way that has local conservation of energy and momentum (cf. (7.3.25; 2)).
6. The result (8.4.18) can be explained as follows: The field of a point charge has an infinite energy,
$$\frac{1}{2}\int (|\mathbf{E}|^2 + |\mathbf{B}|^2) \to \infty,$$
and since the particle carries this energy along with it, it is subjected to an infinite mass increase δm by Einstein's principle that energy is mass. In order that m remain finite, it is necessary to start with an infinite negative "bare" mass m_0, obviously a dangerous undertaking. The field $E^{(\ddot{z})} = -e\ddot{z}/r$ causes a reverse acceleration, that is, a braking. This quantity was decomposed above into an infinite term $-\delta m\ddot{z}$ and $+\dddot{z}$. The positive sign of the latter part comes

from the use of D^{ret}: The particle feels the field that was produced a short time beforehand, and if the contribution $-\delta m\ddot{z}$ from $(D^{\text{ret}} + D^{\text{adv}})/2$ is subtracted, the net force is in the direction of the positive rate of change of the acceleration. This tends to make the particle fly on ahead, if the braking term is compensated for to a large extent by an infinite negative inertia. If a particle is accelerated so quickly that the "acceleration force" $\sim \dddot{z}$ is as large as the braking terms $\dot{z}\ddot{z}^2$ and $\ddot{z}$, then the particle takes off under its own steam.

Unfortunately, mathematics cannot be fooled by such simple tricks; the difficulties that were swept under the rug show up later as all sorts of paradoxical consequences of equation (8.4.18).

(8.4.20) **Examples**

1. The run-away solution:

$$\dot{z} = (\cosh[\tau_0 a e^{s/\tau_0}], \sinh[\tau_0 a e^{s/\tau_0}], 0, 0),$$
$$\tau_0 = \frac{e^2}{6\pi m}, \quad a \text{ arbitrary},$$

solves (8.4.18) with $F^{\text{in}} = 0$ (Problem 3). The charge suddenly begins to run away ($\tau_0 \sim 10^{-23}$ s for electrons). Since $\ddot{z}^2 = a^2 e^{2s/\tau_0}$, it radiates a tremendous amount of energy; this is consistent with conservation of energy because on the one hand an acceleration takes energy away from a particle with a negative mass m_0, and on the other hand there is always energy to be tapped from the infinite reservoir of self-energy and pumped into the far zone.

2. A well-aimed shot can bring the flight of the particle to a stop. If F^{in} is such that

$$eE_1(z) = ma\tau_0 \cosh[\tau_0 a]\delta(z^0 - z^3),$$

then (8.4.18) has the solution

$$\dot{z}(s) = \Theta(-s)(\cosh[\tau_0 a e^{s/\tau_0}], \sinh[\tau_0 a e^{s/\tau_0}], 0, 0)$$
$$+ \Theta(s)(\cosh[\tau_0 a], \sinh[\tau_0 a], 0, 0)$$

if $z_3(0) = z_0(0)$, $0 = -\dot{z}^3(0) = \dot{z}^2(0)$ (Problem 3). Such behavior is often felt to be acausal, because the particle starts to accelerate before it is brought to its senses by the pulse from F^{in} (Figure 8.19).

(8.4.21) **Remark**

Not all the solutions of (8.4.18) are crazy (see Problem 4). Attempts have been made to separate sense from nonsense by imposing special initial conditions (cf. [43]). It is to be hoped that some day the real solution of the problem of the charge-field interaction will look different, and the equations describing nature will not be so highly unstable that the balancing act can only succeed by having the system correctly prepared ahead of time by a convenient coincidence.

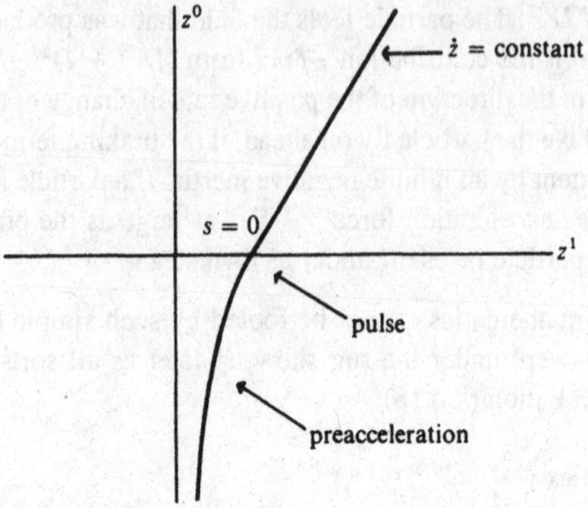

FIGURE 8.19. Motion with preacceleration.

(8.4.22) Problems

1. Verify the claim made in (8.4.3; 3) that $^*(x-z)_{|L_1} = 0$.

2. Derive (8.4.6).

3. Show that the $\dot{z}$'s of Examples (8.4.20; 1) and (8.4.20; 2) solve (8.4.18).

4. Solve (8.4.18) for F^{in} a constant electric field in the x_1-direction with no magnetic field. Use the ansatz $\dot{z}(s) = (\cosh\omega(s), \sinh\omega(s), 0, 0)$. Compare with (5.2.19; 3) and (8.4.20; 1).

(8.4.23) Solutions

1. Let $z(s) = 0$: $r_{|L_1} = t_{|L_1} \Rightarrow dr_{|L_1} = dt_{|L_1}$. Hence

$$^*x = \mathbf{x}\cdot {}^*d\mathbf{x} - t^*\,dt = r^*\,dr - t^*\,dt = r\,d\Omega\wedge dt - t\,d\Omega\wedge dr\colon {}^*x_{|L_1} = 0.$$

2. Let $y := (\bar{x} - z(s))^2$. Then

$$A^{\text{rad}}_{\alpha,\beta}(\bar{x}) = -e\int ds\,\dot{z}_\alpha \frac{\partial y}{\partial \bar{x}^\beta}\frac{ds}{dy}\frac{d}{ds}D(y) = -e\int ds\,D(\bar{x}-z(s))\frac{d}{ds}\frac{\dot{z}_\alpha(s)(\bar{x}-z(s))_\beta}{\langle \dot{z}(s)|\bar{x}-z(s)\rangle}.$$

3. We shall verify Example (8.4.21; 2), and the solution of (8.4.20; 1) will appear as a by-product.

$$\begin{aligned}
\ddot{z} &= \Theta(-s)ae^{s/\tau_0}(\sinh[\;], \cosh[\;], 0, 0),\\
\dddot{z} &= \Theta(-s)a^2e^{2s/\tau_0}(\cosh[\;], \sinh[\;], 0, 0) + \ddot{z}/\tau_0\\
&\quad - \delta(s)a(\sinh\tau_0 a, \cosh\tau_0 a, 0, 0),\\
\dot{z}\ddot{z}^2 - \dddot{z} &= -\ddot{z}/\tau_0 + \delta(s)a(\sinh\tau_0 a, \cosh\tau_0 a, 0, 0).
\end{aligned}$$

Now,

$$\delta(z^0(s) - z^3(s)) = \frac{1}{\dot{z}^0(0)}\delta(s) = \frac{\delta(s)}{\cosh\tau_0 a},$$

and hence $e\dot{z}^\alpha F^{\text{in}}_{\alpha\beta} = \delta(s) ma\tau_0(-\sinh \tau_0 a, \cosh \tau_0 a, 0, 0)$ and $m\ddot{z}_\beta = e\dot{z}^\alpha F^{\text{in}}_{\alpha\beta} - m\tau_0(\dot{z}_\beta \ddot{z}^2 - \dddot{z}_\beta)$.

4.

$$\begin{aligned}
\ddot{z} &= \dot{\omega}(\sinh \omega, \cosh \omega, 0, 0), \qquad \ddot{z}^2 = \dot{\omega}^2, \\
\dddot{z} &= \dot{\omega}^2(\cosh \omega, \sinh \omega, 0, 0) + \ddot{\omega}(\sinh \omega, \cosh \omega, 0, 0), \\
-\dot{z}\ddot{z}^2 + \dddot{z} &= \ddot{\omega}(\sinh \omega, \cosh \omega, 0, 0).
\end{aligned}$$

Equation (8.4.18) requires that $\dot{\omega} = E/m + \tau_0 \ddot{\omega}$, which implies that

$$\omega(s) = a + \frac{E}{m}s + c\tau_0 e^{s/\tau_0},$$

where a and c are constants of integration. Only if $c = 0$, that is, for the special initial condition $\ddot{z}(0)^2 = E^2/m^2$, is there no self-acceleration.

9

The Field in the Presence of Conductors

9.1 The Superconductor

The superconductor is a simple model of a coupled system of equations for charged matter and an electromagnetic field. As a perfect conductor and diamagnet it excludes all electric and magnetic fields from its interior.

Realistic situations do not very closely resemble the idealization discussed in the preceding chapter, where the charge distribution is prescribed. The field in turn influences the motion of the charges, so it would be more correct to analyze the coupled system. For a point-particle the analysis is subject to the difficulties encountered in §8.4. Moreover, the charge-carriers in matter, electrons, and atomic nuclei, are governed by the laws of quantum mechanics, and their motion is a very complicated many-body problem. Every phenomenological description of matter is of necessity either highly idealized or else so general as to contain little information. Notwithstanding that objection, in order to formulate the ideas of this chapter mathematically, we shall single out one of the many models for a superconductor, which can be cast in a simple mathematical form. It is good enough for our purposes, as we shall always consider an extreme case in the examples, for which the charge-carriers in matter are numerous and move about freely. By responding instantaneously to any applied field, they cause the net field within the material to disappear entirely. Later, when we treat the gravitational interaction, this model will serve as our prototype of charged matter.

(9.1.1) London's Equations

Consider the hydrodynamic equations of an incompressible, charged, frictionless fluid in an electromagnetic field, calling the velocity field $\mathbf{v}(\mathbf{x}, t)$:

(9.1.2) $$\frac{d\mathbf{v}}{dt} := \frac{\partial \mathbf{v}}{\partial t} + \boldsymbol{\nabla}\frac{v^2}{2} - [\mathbf{v} \times \boldsymbol{\nabla} \times \mathbf{v}] = \frac{e}{m}(\mathbf{E} + [\mathbf{v} \times \mathbf{B}]).$$

Then from the equation $\boldsymbol{\nabla} \times \mathbf{E} = -\dot{\mathbf{B}}$ results the generalization of Helmholtz's circulation theorem,

(9.1.3) $$\dot{\mathbf{w}} = \boldsymbol{\nabla} \times [\mathbf{v} \times \mathbf{w}], \qquad \text{where} \quad \mathbf{w} := \boldsymbol{\nabla} \times \mathbf{v} + \frac{e}{m}\mathbf{B}.$$

Therefore, if $\mathbf{w}$ is zero at any time, it is always zero. This means that the curl of $\mathbf{v}$ arises only from the vortices created when $\mathbf{B}$ is switched on (cf. §5.4), and the equations simplify to

(9.1.4) $$\frac{\partial \mathbf{v}}{\partial t} + \boldsymbol{\nabla}\frac{v^2}{2} = \frac{e}{m}\mathbf{E}, \qquad \boldsymbol{\nabla} \times \mathbf{v} = -\frac{e}{m}\mathbf{B}.$$

If one now writes

(9.1.5) $$\mathbf{J} = \frac{e\rho\mathbf{v}}{\sqrt{1 - v^2}}, \qquad J^0 = \frac{e\rho}{\sqrt{1 - v^2}},$$

and if ρ is constant, then

(9.1.6) $$J_{\beta,\alpha} - J_{\alpha,\beta} = \frac{\rho e^2}{m} F_{\alpha\beta}$$

to an accuracy of order $v^2 \ll 1$. This equation together with Maxwell's equations will be the foundation of our model. It admits a coordinate-independent formulation,

(9.1.7) $$F = \frac{dJm}{\rho e^2}, \qquad \delta F = J + j.$$

(9.1.8) Remarks

1. The current j consists of charges not participating in the superconductive current J. We shall take j as given, and assume that $\delta j = 0$; then (9.1.7) implies that $dF = \delta J = 0$.
2. We shall ignore the heuristic derivation to the point of not requiring that $\langle J | J \rangle = -e^2\rho^2$, which follows from (9.1.5).
3. For the present, ρ is regarded as a constant, known as the density of the superconducting electrons. There will later be a discussion of the variable ρ.
4. Equation (9.1.7) shows that the manifold is not provided with any additional structure; for instance, there is no distinguished rest-frame.

(9.1.9) The Integral Form of the Equations

If $F = dA$, then (9.1.6) is equivalent to

$$\int_{\partial N_2} \left(J - \frac{\rho e^2}{m} A \right) = 0, \qquad \dim N_2 = 2.$$

Two important special cases are:

(i) $N = \{t = z = 0, x^2 + y^2 \le R^2\}$:

$$\int_{x^2+y^2=R^2} d\mathbf{s} \cdot \mathbf{J} = \frac{\rho e^2}{m} \int_{x^2+y^2=R^2} ds_j A_j = -\frac{\rho e^2}{m} \int_{x^2+y^2<R^2} d\mathbf{S} \cdot \mathbf{B},$$

i.e., the current circulates in proportion to the magnetic flux (cf. Remark (5.1.10; 1), with A_j equal to minus the vector potential).

(ii) $N = \{y = z = 0, t_1 \le t \le t_2\}$:

$$\int_{t=t_2} dx\, J_x - \int_{t=t_1} dx\, J_x = \frac{\rho e^2}{m} \int_{t_1}^{t_2} dt\, dx\, E_x.$$

The rate of change of the superconductive current is given by the integral over the electric field.

(9.1.10) The Elimination of the Superconductive Current

From (9.1.7) we get a second-order equation for F,

(9.1.11) $$\left(-\Delta + \frac{\rho e^2}{m} \right) F = -dj.$$

The solution of (9.1.11) requires a Green function satisfying

(9.1.12) $$\left(-\Delta + \frac{\rho e^2}{m} \right) G_{\bar{x}}^{\mathrm{ret}} = \delta_{\bar{x}},$$

with which

(9.1.13) $$F(\bar{x}) = F^{\mathrm{in}}(\bar{x}) - \int G_{\bar{x}}^{\mathrm{ret}} \wedge dj,$$
$$\left(-\Delta + \frac{\rho e^2}{m} \right) F^{\mathrm{in}} = 0.$$

We have assumed that j decreases sufficiently fast at infinity, so that the integrals can extend over the whole manifold as in (8.2.15), without any boundary terms.

In Minkowski space $(\mathbb{R}^4, \eta)$ it is easy to construct a Green function satisfying (9.1.12). As in (8.2.3) it has the form

(9.1.14) $$G^{\text{ret}}_{\bar{x}} = \tfrac{1}{2}\bar{e}_{\alpha\beta} \otimes {}^{*}e^{\alpha\beta}\,\Delta^{\text{ret}}(\bar{x} - x),$$

where

(9.1.15) $$\Delta^{\text{ret}}(x) = (2\pi)^{-4} \int d^4k\, e^{i\langle k|x\rangle} \left(k^2 + \frac{e^2\rho}{m}\right)^{-1}.$$

The integration path for k^0 again passes above the poles at $\pm\sqrt{|\mathbf{k}|^2 + e^2\rho/m}$, as in Figure 8.3, in order that $\Delta^{\text{ret}}(x) = 0$ for $x^0 < |\mathbf{x}|$. The integral (9.1.15) can be expressed in terms of Hankel functions [41], and if Δ^{ret} is integrated over time, the result is a Yukawa potential,

(9.1.16) $$\int_{-\infty}^{\infty} dt\,\Delta^{\text{ret}}(x) = \frac{e^{-r(\rho e^2/m)^{1/2}}}{4\pi r}$$

(Problem 1). If (9.1.13) is written as

(9.1.17) $$F = F^{\text{in}} + dA^{\text{ret}},$$
$$A^{\text{ret}}_\alpha(\bar{x}) = -\int d^4x\,\Delta^{\text{ret}}(\bar{x} - x)\,j_\alpha(x),$$

then in the static limit,

(9.1.18) $$A^{\text{ret}}_\alpha(\bar{\mathbf{x}}) = -\int \frac{d^3x}{4\pi\,|\mathbf{x} - \bar{\mathbf{x}}|} \exp\left(-|\mathbf{x} - \mathbf{x}|\left(\frac{\rho e^2}{m}\right)^{1/2}\right) j_\alpha(x).$$

(9.1.19) **Remarks**

1. A bounded solution $\sim \exp(i\langle k|x\rangle)$ for F^{in} exists only for $(k^0)^2 = |\mathbf{k}|^2 + \rho e^2/m \geq \rho e^2/m$. The significance of the **plasma frequency** $\sqrt{\rho e^2/m}$ is evident in the following electrostatic situation: Suppose charges e are arrayed along a line at the points nL, $n \in \mathbb{Z}$. If one charge is displaced slightly from equilibrium by $x \ll L$ (see Figure 9.1), then it feels a force $e^2((L + x)^{-2} - (L - x)^{-2}) \sim -e^2xL^{-3}$ from its two nearest neighbors. If this is set equal to $m\ddot{x}$, the equation is oscillatory at the frequency $(e^2\rho/m)^{1/2}$, where we have identified the density ρ as L^{-3}. The oscillations are associated with solutions having $\mathbf{k} = \mathbf{0}$ and $k^0 = (e^2\rho/m)^{1/2}$.
2. There are no nontrivial static solutions ($k^0 = 0$) for F^{in}. According to equation (9.1.18), the field of a static charge does not penetrate a superconductor, but decreases exponentially within a skin-depth also given by $(e^2\rho/m)^{1/2}$. The cause is the induced current J, which can be calculated from (9.1.7) with the boundary condition $F^{\text{in}} = 0$ and (9.1.18) as

$$J_\alpha(\bar{\mathbf{x}}) = -\int \frac{d^3x\, e^{-|\mathbf{x}-\bar{\mathbf{x}}|(\rho e^2/m)^{1/2}}}{4\pi\,|\mathbf{x} - \bar{\mathbf{x}}|} j_\alpha(\mathbf{x}) \frac{\rho e^2}{m}$$

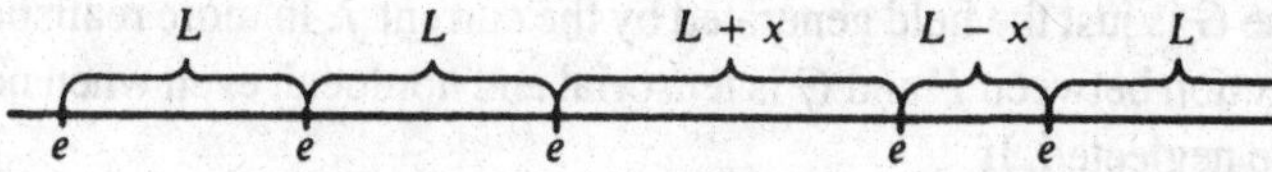

FIGURE 9.1. A chain of oscillating charges.

in the static case; it opposes the original current and completely cancels j at large distances:

$$\int d^3x\, j_\alpha(\mathbf{x}) = -\int d^3x\, J_\alpha(\mathbf{x}).$$

3. The connection between the Fourier transforms of the external and induced currents with the appropriate definition of $(\ldots)^{-1}$ is

$$\tilde{J} = -\left(1 + \frac{mk^2}{\rho e^2}\right)^{-1} \tilde{j}.$$

If $k^2 = 0$, they are equal and opposite. The 2-form G with Fourier transform

$$\tilde{G} = \left(1 + \frac{\rho e^2}{mk^2}\right) \tilde{F}$$

satisfies Maxwell's equations with no superconducting current,

$$dG = 0, \qquad \delta G = j.$$

Since ρ is constant, the fields $\mathbf{D}$ and $\mathbf{H}$ of phenomenological electrodynamics satisfy these same equations, where the factor $1 + \rho e^2/mk^2$ corresponds on the one hand to a dielectric constant

$$\varepsilon(k) = 1 + \frac{\rho e^2}{mk^2}, \qquad \tilde{\mathbf{D}} = \varepsilon \tilde{\mathbf{E}};$$

and on the other to a magnetic susceptibility

$$\kappa(k) = \frac{-1}{1 + \frac{k^2 m}{\rho e^2}}, \qquad \tilde{\mathbf{B}} = (1 + \kappa)\tilde{\mathbf{H}}.$$

Both $\varepsilon(k)$ and $\kappa(k)$ are Lorentz-invariant and commute as convolution operators with translations, but they depend on the frequency, because the relationship between F and G is not local. If $k = 0$, then $\varepsilon = \infty$ and $\kappa = -1$; $\mathbf{E}$ and $\mathbf{B}$ become zero within the material, as they are shielded in a perfect conductor and diamagnet. The theory with the field J does not distinguish a rest frame and the velocity of light is unchanged. Thus $c = 1 = \varepsilon(1 + \kappa)$ and $\kappa \to -1 \Rightarrow \varepsilon \to \infty$. Thus in a Lorentz-invariant situation charge screening and expulsion of the magnetic field go together.

4. In our case G is just the field generated by the current j. In more realistic modes the connection between F and G is tensorial and nonlocal, even when nonlinear effects are neglected. If

$$G_{ik}(x) = \int d^4x'\, K_{iklm}(x, x')F^{lm}(x'),$$

then $dG \neq 0$ and G no longer satisfies Maxwell's equations with the external current.

Now that we have a useful system of equations for the field plus the charges, let us take the opportunity to study the conservation theorems for the total system, by using the

(9.1.20) **Lagrangian Formulation**

If we particularize (9.1.7) *by setting*

$$F = dA, \qquad J = \frac{e\rho}{m}(dS + eA), \qquad S \in E^0,$$

then the Lagrangian

$$\mathcal{L} = -\frac{1}{2}\frac{m}{\rho e^2} J \wedge {}^*J - \tfrac{1}{2} F \wedge {}^*F$$

reproduces equation (9.1.7) *without* j.

Proof: Making a variation of $\mathcal{L}$ as in (8.1.2) yields

$$\delta\mathcal{L} = \frac{\delta S\, d^*J}{e} - d\left(\frac{\delta S^*J}{e}\right) - \delta A \wedge [d^*F + {}^*J] - d[\delta A \wedge {}^*F]$$

(where δ is the variation, not the codifferential). Hence the Euler–Lagrange equations are

$$d^*J = 0, \qquad d^*F = -{}^*J. \qquad \square$$

(9.1.21) **Remarks**

1. Even the "superpotential" S must be changed if a gauge transformation is made; if F and J, and consequently $\mathcal{L}$, are to be invariant under $A \to A + d\Lambda$, $S \to S - e\Lambda$.
2. It is possible to express $J \wedge {}^*J$ as $\overline{(d + ieA)\varphi} \wedge {}^*(d + ieA)\varphi$ by use of a complex field $\varphi = \exp(iS)$. The effect of A in this scheme is to make the exterior differential invariant under $\varphi \to \exp(ie\Lambda(x))\varphi$. (See §10.1).
3. The scalar model discussed so far turned out to be of more importance in elementary particle physics than in condensed matter physics. It showed that in a gauge-invariant theory a "mass term" can appear in Maxwell's equations. Exponentially decaying Green functions correspond in quantum theory to particles

with a mass, and one stumbling block in the unification of weak interactions with electromagnetism was that the particles mediating the weak interaction did not seem to be massless. In fact, they were found to be exceedingly heavy, which is believed to be caused by their interaction with a scalar field.

(9.1.22) The Energy-Momentum Forms

According to Noether's theorem (8.1.5), the 3-forms

$$\frac{(L_v S)^* J}{e} + (L_v A) \wedge {}^*F + i_v \mathcal{L} = \frac{m}{2\rho e^2}(J \wedge i_v{}^*J + (i_v J)^* J)$$
$$+ \tfrac{1}{2}((i_v F) \wedge {}^*F - F \wedge i_v{}^*F) + d((i_v A)^* F)$$

are closed for all Killing vector fields v (cf. Definition (8.1.9)).

(9.1.23) Remarks

1. The gauge-dependent term, $d((i_v A)^* F)$, has again shown up. Since it is exact, the rest of the right side of (9.1.22) must be closed. As will be seen in (10.2.9), only that part interacts with the gravitational field, and will be used as the energy-momentum tensor.
2. All the generators of the Poincaré group can function as v. However, since J is a 1-form, the presence of matter breaks the conformal invariance (see (8.1.10; 4)); the skin-depth is a distinguished length.
3. If v is the generator ∂_α of a translation, then the 3-form (9.1.22) gets a contribution $t_{\alpha\beta}\, dx^\beta$ from the matter, where

$$t_{\alpha\beta} = \frac{m}{\rho e^2}\left[J_\alpha J_\beta - \tfrac{1}{2}\eta_{\alpha\beta} J_\gamma J^\gamma\right].$$

(9.1.24) The Properties of the Energy-Momentum Tensor of Matter

(a) $t_{\alpha\beta} = t_{\beta\alpha}$;

(b) $t_{00} = \dfrac{m}{2\rho e^2}[J_0^2 + |\mathbf{J}|^2] \geq 0$, and $= 0$ only if $J = 0$; and

(c) $t_{0i} = \dfrac{m}{\rho e^2} J_0 J_i, \qquad {t_{00}}^2 \geq \sum_{i=1}^{3} {t_{0i}}^2.$

(9.1.25) Remarks

1. Property (a) follows from Lorentz invariance; but $t_\alpha{}^\alpha \neq 0$, because of the lack of conformal invariance (cf. (8.1.11)).

2. Since there is the same positivity property as in (8.1.13) for the electromagnetic energy-momentum tensor, the argument of (8.1.16) about the uniqueness of the Cauchy problem is again applicable.
3. A relativistic fluid is described phenomenologically by the field $u \in E_1$ of its four-velocity ($\langle u|u \rangle = -1$), its mass-density $\bar{\rho}$, and its pressure p. The energy-momentum tensor $T_{\alpha\beta} = (\bar{\rho} + p)u_\alpha u_\beta + p g_{\alpha\beta}$ is by construction diagonal in the rest system ($u = (1, 0, 0, 0)$) and in no other. Its eigenvalues are $(\bar{\rho}, p, p, p)$. By a comparison with (9.1.23; 3) one arrives at the identification

$$u_\alpha = J_\alpha (J_0^2 - |\mathbf{J}|^2)^{-1/2}, \qquad \bar{\rho} = p = \frac{1}{2}\frac{m}{\rho e^2}(J_0^2 - |\mathbf{J}|^2).$$

In normal matter, $p \sim$ the density of kinetic energy $\sim 10^{-10}\bar{\rho}$, while for radiation $p = \bar{\rho}/3$. Thus the pressure is unrealistically high in this model.
4. If $e = 0$, then S satisfies d'Alembert's equation $\Delta S = 0$. The compressional waves, which may be thought of as sound, therefore propagate at the speed of light, as is to be expected from $\partial p/\partial \bar{\rho} = 1$. Thus S describes a fluid that is as incompressible as possible without allowing sound to travel faster than light.
5. The reader may be wondering what happens to these sound waves if $e \neq 0$. In that case S loses its physical significance, and can be made to disappear by a gauge transformation. It turns out that the sound waves then reappear as a longitudinal oscillatory mode of A, and if $e = 0$, then A can oscillate only transversally (Problem 3).

In practically important problems, one part of space usually contains the metal and all the rest is a vacuum. In that case ρ is not a constant in the model, but instead it changes discontinuously to zero at the metallic surface. A slowly varying field with frequency $k^0 \ll (\rho e^2/m)^{1/2}$ would consequently not penetrate the metal, but would decrease exponentially at the surface within a depth $(\rho e^2/m - (k^0)^2)^{1/2}$. For simplicity we consider the limit $\rho e^2/m \to \infty$, thus disengaging ourselves from the details of the model. The only essential feature that is preserved is that the field F is excluded from within the metal. It is screened by a surface current J, which has a delta-function singularity at the surface, in the limit $\rho e^2/m \to \infty$ (cf. (9.1.19; 2)). The equations (9.1.7) of the model are replaced with the

(9.1.26) **Metallic Boundary Conditions**

Suppose that the four-dimensional submanifold CN is filled with metal, so that $F = 0$ on it, and let the surface ∂N be given locally by the equation $u = 0$. Then the restriction (7.2.7; 3) *$F_{|\partial N}$ of the field to ∂N must vanish, and F is screened by the surface current $\delta(u)(i_{du}F)_{|\partial N}$.*

Proof: It is only necessary to extend the arguments of (8.2.1) to allow the current to have a delta-function singularity. If $F = \Theta(u)F'$, where F' is continuous, then $0 = dF = \delta(u)\, du \wedge F'$ and $-{}^*J = d^*F = \delta(u)\, du \wedge {}^*F'$, where we consider only the singular parts. Because of the $\delta(u)$, it only matters what F' is at $u = 0$, and all the terms containing a factor du drop out of the exterior product. What is left over is simply the restriction of F' according to (7.2.7; 3), and so we conclude from the first of the two equations above that $F'_{|\partial N} = 0$. If we make use of $du \wedge {}^*F = -{}^*i_{du}F$ in the second equation, the claim made about the surface current follows. □

(9.1.27) **Remarks**

1. If $u = x_1$, making $N = \{x_i \in \mathbb{R}: x_1 > 0\}$, then $dx_{1|\partial N} = 0$ and $F_{|\partial N} = (E_2\, dt \wedge dx_2 + E_3\, dt \wedge dx_3 - B_1\, dx_2 \wedge dx_3)_{|\partial N}$. Therefore E_2, E_3, and B_1 must vanish. The interpretation of this is that surface charges do not produce any discontinuous tangential components of the electric field, and surface currents do not produce discontinuous normal components of the magnetic field.
2. The situation is drawn schematically below:

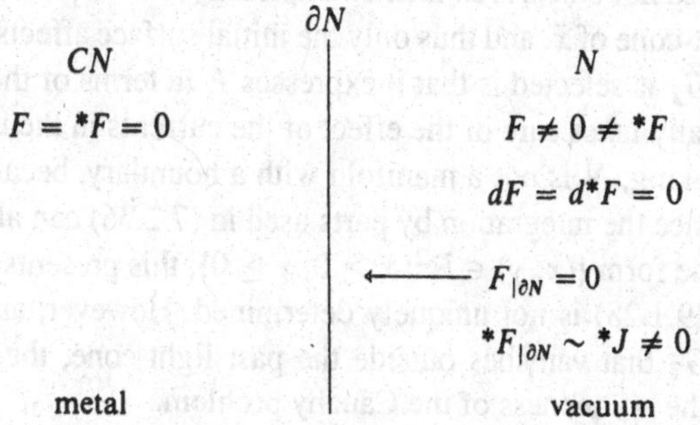

Since the surface current is not specified beforehand, it may be asked how the initial-value problem is to be solved. While the general solution (7.2.36) is always valid, it is not immediately useful, since ${}^*F$ occurs in the surface integral as well as F. It would seem to be necessary to know the restrictions of both to the surface, and we only know that $F_{|\partial N} = 0$. If we manage to find a $G_{\bar{x}}$ such that ${}^*dG_{\bar{x}|\partial N} = 0$, however, then there are no unknown surface contributions, and the solution works as in Chapter 2. In other words, the key to the problem is

(9.1.28) **The Green Function for Metallic Boundary Conditions**

Let $M \subset \mathbb{R}^4$ be a part of Minkowski space bounded by spacelike hypersurfaces ∂M. Suppose that metal fills $CN \cap M$ and that the current j is known in $N \subset M$ and $\partial_v N := \partial N \setminus \partial N \cap \partial M$ is the vertical boundary of N. If $G_{\bar{x}}$ satisfies the equations $-\Delta G_{\bar{x}} = \delta_{\bar{x}}$ and ${}^*dG_{\bar{x}|\partial_v N} = 0\ \forall \bar{x} \in N \setminus \partial N$, then for all $x \in N \setminus \partial N$ the field strength is given by

$$F(\bar{x}) = \int_N dG_{\bar{x}} \wedge j - \int_{\partial M \cap N} [\delta G_{\bar{x}} \wedge F - {}^*dG_{\bar{x}} \wedge {}^*F].$$

(9.1.29) Remarks

1. The situation looks as follows:

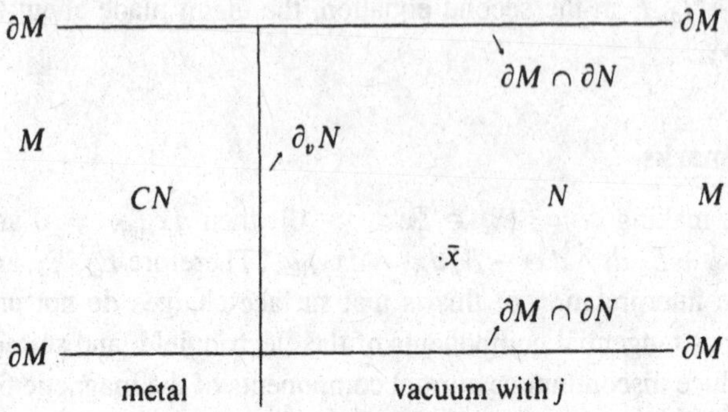

In (9.1.28) it is only relevant to know F and $^*F$ on the initial and final surfaces $\partial M \cap \partial N$, and not on $\partial_v N$. In later examples G_x, like G_x^{ret}, will be zero outside the past light-cone of $\bar{x}$, and thus only the initial surface affects the integral. The reason this $G_{\bar{x}}$ is selected is that it expresses F in terms of the initial data, and it automatically takes care of the effect of the currents in the upper surface.

2. Strictly speaking, N is not a manifold with a boundary, because it has a sharp edge. But since the integration by parts used in (7.2.36) can also be justified on regions of the form $\{(x, y) \in \mathbb{R}^2 : x \geq 0, y \geq 0\}$, this presents no real obstacle.
3. The $G_{\bar{x}}$ of (9.1.28) is not uniquely determined. However, as long as we possess some $G_{\bar{x}}$ that vanishes outside the past light-cone, the formula (9.1.28) guarantees the uniqueness of the Cauchy problem.

In the following sections we shall prove the existence of the $G_{\bar{x}}$ used in (9.1.28) by explicit construction, making use of the well-known method of images from electrostatics. This static method is generalized by means of $G_{\bar{x}}$ for charges in arbitrary motion.

(9.1.30) Problems

1. Calculate the integral (9.1.16).
2. Show that $t_{\alpha\beta,}{}^{\beta} = J_\beta F^\beta{}_\alpha$, with $t_{\alpha\beta}$ as in (9.1.23; 3).
3. Show that the equations

$$dJ = \frac{e^2\rho}{m} F, \qquad \delta F = J, \qquad \frac{e^2\rho}{m} > 0,$$

have three linearly independent solutions $\sim \exp(i\langle k|x\rangle)$ if

$$k^2 = -\frac{e^2\rho}{m},$$

and otherwise have only the trivial solution $F = 0$.

(9.1.31) Solutions

1. If $\mu = \sqrt{e^2\rho/m}$, then

$$\int_{-\infty}^{\infty} dt(2\pi)^{-4} \int d^4k\, e^{i(k|x)}(k^2 + \mu^2)^{-1} = (2\pi)^{-3} \int d^3k\, e^{i\mathbf{k}\cdot\mathbf{x}}(|\mathbf{k}|^2 + \mu^2)^{-1}$$
$$= (2\pi)^{-2} \int_0^{\infty} \frac{k^2\, dk}{k^2 + \mu^2} \int_{-1}^{1} d\eta\, e^{ikr\eta}$$
$$= (2\pi)^{-2} \int_{-\infty}^{\infty} \frac{k\, dk}{k^2 + \mu^2} \frac{e^{ikr}}{ir} = \frac{e^{-\mu r}}{4\pi r}.$$

2. Because of (9.1.6) and $J_{\beta.}{}^{\beta} = 0$,

$$t_{\alpha\beta}{}^{.\beta} = \frac{m}{\rho e^2}[J_{\alpha}{}^{.\beta} J_{\beta} - J^{\beta}{}_{,\alpha} J_{\beta}] = F_{\beta\alpha} J^{\beta}.$$

3. In the Fourier-transformed space the equations become

$$k_\mu k^\rho \tilde{F}_{\rho\nu} - k_\nu k^\rho \tilde{F}_{\rho\mu} = \frac{e^2\rho}{m} \tilde{F}_{\nu\mu}.$$

Multiplying this by $\cdot\, k^\nu$,

$$-k^2 k^\rho \tilde{F}_{\rho\mu} = \frac{e^2\rho}{m} k^\nu \tilde{F}_{\nu\mu}.$$

Since $\tilde{A}_\mu - k_\mu \tilde{S} := -(m/e^2\rho)k^\rho \tilde{F}_{\rho\mu}$ vanishes iff $F \equiv 0$, either $\tilde{F}_{\nu\mu} = 0$, or else $k^2 = -e^2\rho/m$. Then $\tilde{F}$ is of the form $\tilde{F}_{\nu\mu} = k_\nu \tilde{A}_\mu - k_\mu \tilde{A}_\nu$, which vanishes only if $\tilde{A}_\nu \sim k_\nu$. Thus for the three directions other than k there are nonvanishing solutions.

9.2 The Half-Space, the Wave-Guide, and the Resonant Cavity

The general solution of Maxwell's equations with metallic boundary conditions is easy to construct for simple geometric arrangements of the conductors.

Classically, the electromagnetic problem in the presence of metallic surfaces is usually conceived as the quest for particular solutions. Here we shall proceed directly to the more general problem, and solve the Cauchy problem by specifying the $G_{\bar{x}}$ of (9.1.28). The interesting question will be what the causal structure of the Green function is. Like the $G_{\bar{x}}^{\text{ret}}$ of (8.2.7), its support will be restricted to the full past light-cone of $\bar{x}$, but unlike G_x^{ret} not to its surface. This fact is due to waves that reflect from the metallic surfaces and return at some later time. Such echoes may apparently violate causality, as when they give rise to phase velocities greater than the speed of light. In all the problems discussed below we replace the conducting material with a metallic boundary condition. The currents induced on the conductors do not appear, so we may use the symbol J for the externally prescribed current.

We begin with a trivial warm-up exercise, the problem of a plane metallic mirror. The method of solution introduced points the way to the procedure for more complicated problems.

(9.2.1) The Half-Space

In the notation of (9.1.28), *let*

$$M = \{x^i \in \mathbb{R}^4 : x^0 \geq t^0\},$$
$$N = \{x^i \in M : x^1 \geq 0\}.$$

The symbol R will stand for the reflection $(x^0, x^1, x^2, x^3) \to (x^0, -x^1, x^2, x^3)$ in M and at the same for the induced mapping on the space of tensors (2.4.19). *Then*

$$G_{\bar{x}} = (1 + R)G_{\bar{x}}^{\text{ret}},$$

with $G_{\bar{x}}^{\text{ret}}$ from (8.2.7), *is the Green function needed for this problem.*

Proof: The rules for manipulating the diffeomorphism R of the space of vectors are such that it can be interchanged with sums, products, and exterior differentiation:

$$R(\omega + \nu) = R\omega + R\nu, \qquad R(\omega \wedge \nu) = R\omega \wedge R\nu, \qquad \omega, \nu \in E,$$
$$R(e^0, e^1, e^2, e^3) = (e^0, -e^1, e^2, e^3), \qquad e^i = dx^i.$$

However, R reverses the orientation so that $R^*\omega = -{}^*R\omega \, \forall \omega \in E_p$ (cf. (7.2.17; 1)).

We now make use of a

Lemma *If $\omega \in E_p$ and $\omega = -R\omega$, then $\omega_{|\partial_\nu N} = 0$.*

Proof of Lemma: Let

$$\omega = \sum_{(i)} \omega_{i_1 \cdots i_p}(x) e^{i_1 \cdots i_p}, \qquad e^i = dx^i, \qquad \omega_{i_1 \cdots i_p} \in E_0.$$

If $i_1 \cdots i_p$ contains the index 1, then $e^{i_1 \cdots i_p}{}_{|x^1=0} = 0$, because $dx_{1|x_1=0} = 0$. If $i_1 \cdots i_p$ does not contain the index 1, then $\omega_{i_1 \cdots i_p}(x^0, -x^1, x^2, x^3) = -\omega_{i_1 \cdots i_p}(x^0, x^1, x^2, x^3)$, which must vanish when $x^1 = 0$.

This lemma implies the property ${}^*dG_{\bar{x}|\partial_\nu N} = 0$ required in (9.1.28), because

$$R^*d(1 + R)G_{\bar{x}}^{\text{ret}} = -{}^*dR(1 + R)G_{\bar{x}}^{\text{ret}} = -{}^*d(1 + R)G_{\bar{x}}^{\text{ret}}.$$

Observe that none of the operations of this equation affect $\bar{x}$, and that $R^2 = 1$. The property $-\Delta G_{\bar{x}} = \delta_{\bar{x}} \, \forall \bar{x} \in N$ is proved by noting that the factor $\delta((\bar{x} - x)^2)$ has been replaced with $\delta((\bar{x}^0 - x^0)^2 - (\bar{x}^1 + x^1)^2 - (\bar{x}^2 - x^2)^2 - (\bar{x}^3 - x^3)^2)$ in

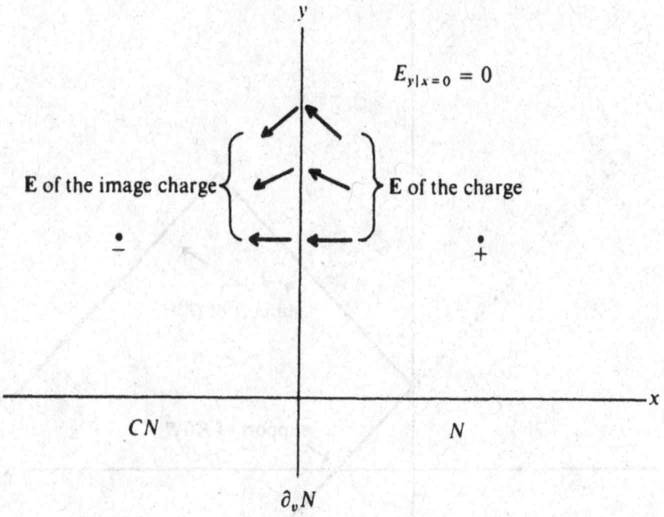

FIGURE 9.2. **E** in the presence of a reflecting half-plane.

$RG^{\text{ret}}_{\bar{x}}$. When Δ acts on this it yields zero unless $x = (\bar{x}^0, -\bar{x}^1, \bar{x}^2, \bar{x}^3)$. However, this fails to be in $N \backslash \partial N$ if $\bar{x} \in N \backslash \partial N$. □

(9.2.2) **Remarks**

1. Since R reverses the orientation, $\int_M R\omega = -\int_M \omega\ \forall \omega \in E_p$. If supp $j \subset N$ the integral $\int_N (1+R)\, dG^{\text{ret}} \wedge J$ can be taken as $\int_M \ldots$, so R does not send points out of the integration region; and because
$$\int_M R\, dG^{\text{ret}} \wedge J = \int_M R(dG^{\text{ret}} \wedge RJ) = -\int_M dG^{\text{ret}} \wedge RJ$$
the integral can be written as
$$F(\bar{x}) = \int_M dG^{\text{ret}}_{\bar{x}} \wedge (1-R)J + \int_{\partial M \cap N} \cdots.$$
The components of $-RJ$ are $(-J_0(Rx), J_1(Rx), -J_2(Rx), -J_3(Rx))$, and thus the field produced by J is as if there were a mirror-image charge of reversed sign at $Rx \in CN$, undergoing the reflected motion. It is easy to see that its field taken together with the field directly produced by J satisfies the metallic boundary conditions (9.1.26) on $\partial_v N$ (Figure 9.2). In actuality there is an induced charge in the metal, not in the interior of CN, but rather, according to (9.1.26), on $\partial_v N$. The surface current $\delta(x^1)(i_{dx^1} F)_{|\partial_v N}$ generates the field in N that would come from $-RJ$.
2. It is likewise possible to take $\int_{\partial M \cap N}$ as $\int_{\partial M}$, which involves only $G^{\text{ret}}_{\bar{x}}$, and the appropriate reflected initial data are to be used on $\partial M \cap CN$. If F and ${}^*F_{|\partial M \cap N}$ originate in an incoming wave, then their values in N are the same as if there were no metal present, and the reflected initial values had been specified on $\partial M \cap CN$.

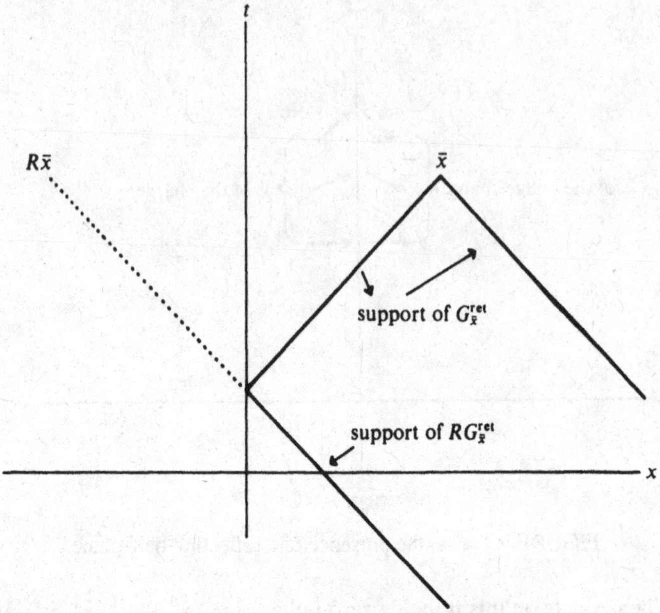

FIGURE 9.3. The support of $G_{\bar{x}}$ with a reflecting half-plane.

3. The support of $G_{\bar{x}}$ in N is contained in the full past light-cone of $\bar{x}$ (Figure 9.3). Metallic boundary conditions produce echo effects, but never really violate causality.

Electromagnetic waves can be transmitted through metallic pipes by a continuous process of reflection. Wave-guides, as these pipes are called, are quite important in communications technology. They are not always made of superconductors, and the limit $\rho e^2/m \to \infty$ is not completely realistic. The field seeps into the walls of the conductor, energy is consumed for Joule heating, and the waves are damped. Nevertheless, the model exhibits the basic mechanism whereby electromagnetic fields can be transmitted like water in a hose.

(9.2.3) The Rectangular Wave-Guide

In the notation of (9.1.28), *let*

$$M = \{(t, x, y, z) \in \mathbb{R}^4 : t \geq t_0\},$$
$$N = \{x^\alpha \in M : 0 \leq x \leq a, 0 \leq y \leq b\},$$

and let R_1, R_2, T_1, and T_2 be the transformations on the space of tensors induced by the diffeomorphisms

$$R_1 : (t, x, y, z) \to (t, -x, y, z),$$
$$R_2 : (t, x, y, z) \to (t, x, -y, z),$$
$$T_1 : (t, x, y, z) \to (t, x + 2a, y, z),$$
$$T_2 : (t, x, y, z) \to (t, x, y + 2b, z),$$

of M. Then

$$G_{\bar{x}} = (1 + R_1)(1 + R_2) \sum_{n=-\infty}^{\infty} \sum_{m=-\infty}^{\infty} T_1^n T_2^m G_{\bar{x}}^{\text{ret}}$$

is the Green function of this problem.

Proof: The diffeomorphisms satisfy the commutation relations

$$\begin{aligned} &T_1 T_2 = T_2 T_1, \qquad T_1 R_2 = R_2 T_1, \qquad T_2 R_1 = R_1 T_2, \\ &R_1 R_2 = R_2 R_1, \\ &T_1 R_1 = R_1 T_1^{-1}, \qquad T_2 R_2 = R_2 T_2^{-1}. \end{aligned}$$

Note that $T_1 R_1$ is the transformation $x \to -x + 2a$, or $(x - a) \to -(x - a)$; in words, a reflection about $x = a$. Similarly, $T_2 R_2$ is a reflection about $y = b$. Since

$$G_{\bar{x}} = R_1 G_{\bar{x}} = R_2 G_{\bar{x}} = T_1 R_1 G_{\bar{x}} = T_2 R_2 G_{\bar{x}}$$

by construction,

$$^*dG_{\bar{x}|x=0} = {}^*dG_{\bar{x}|y=0} = {}^*dG_{\bar{x}|x=a} = {}^*dG_{\bar{x}|y=b} = 0,$$

as in the proof of (9.2.1). Moreover, the image of $(0, a)$ under T_1^n equals $(2na, (2n + 1)a)$, and its image under $R_1 T_1^n$ equals $(-2na, -(2n + 1)a)$, and similar statements hold for T_2 and R_2. As a result, the only part of the sum $\sum_{n,m} (1 + R_1)(1 + R_2)T_1^n T_2^m$ that sends a point of $N \backslash \partial N$ back into $N \backslash \partial N$ is the 1 from the term with $n = m = 0$. Hence that is the only term contributing to $\Delta G_{\bar{x}}$ if x and $\bar{x} \in N$, and therefore $-\Delta G_{\bar{x}} = \delta_x$. □

(9.2.4) **Remarks**

1. Although N is not a manifold, because of its sharp edges, its structure is harmless enough that the integration by parts needed in order to use $G_{\bar{x}}$ is easy to justify.
2. Once again, the support of $G_{\bar{x}}$ in N is contained in the full past light-cone (Figure 9.4). This is why we were able to choose M as a region extending to $t = +\infty$ as in (8.2.9).

The structure of $G_{\bar{x}}$ results from the infinite number of reflections of the field back and forth between the walls. Consequently, there are infinitely many image charges, as depicted in cross-section in Figure 9.5. Their periodic configuration gives rise to characteristic normal modes of oscillation.

(9.2.5) **The Decomposition of $G_{\bar{x}}$ into Normal Modes**

If we write

$$G_{\bar{x}}^{\text{ret}} = \tfrac{1}{2}\bar{e}_{\alpha\beta} \otimes {}^*e^{\alpha\beta} \int \frac{d^4k}{(2\pi)^4} \frac{e^{i(k|\bar{x}-x)}}{k^2} \tag{9.2.6}$$

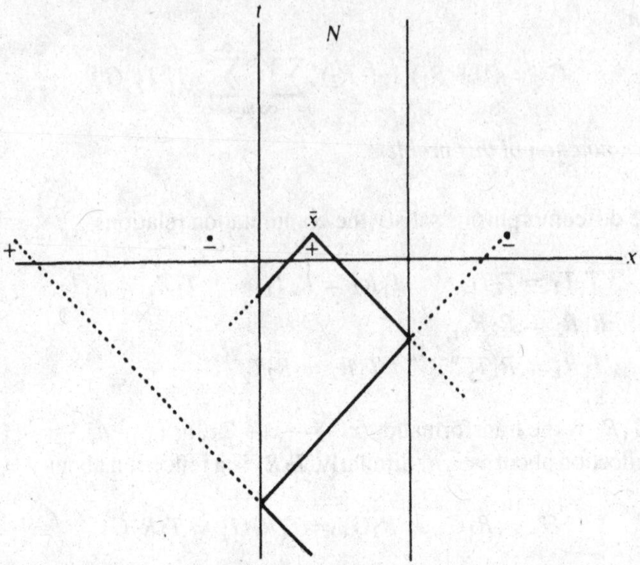

FIGURE 9.4. The support of $G_{\bar{x}}$ in a wave-guide.

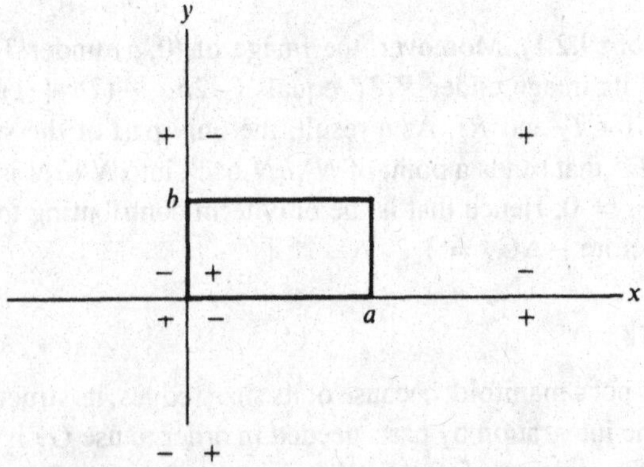

FIGURE 9.5. Image charges in a wave-guide.

and use the invariance of e^{α} under the diffeomorphism

$$T_1^n T_2^m : (t, x, y, z) \to (t, x + 2na, y + 2mb, z),$$

then we find ourselves presented with sums of the form

$$\sum_{n=-\infty}^{\infty} \exp(ik(x + 2na)) = \exp(ikx) \sum_{g \in \mathbb{Z}} \delta\left(k - \frac{\pi g}{a}\right) \frac{\pi}{a}. \tag{9.2.7}$$

Substitution of this into $G_{\bar{x}}^{\text{ret}}$ and summing $\sum_m$ allows the integration $\int dk_x\, dk_y$ to be accomplished with the aid of delta-functions:

$$\text{(9.2.8)} \quad \sum_{n=-\infty}^{\infty} \sum_{m=-\infty}^{\infty} \int_{-\infty}^{\infty} dk_x\, dk_y\, e^{i(k_x(x+2na)+k_y(y+2mb))} = \frac{\pi^2}{ab} \sum_{g_1=-\infty}^{\infty} \sum_{g_2=-\infty}^{\infty} e^{i\pi(g_1 x/a + g_2 y/b)}.$$

The $G_{\bar{x}}$ of (9.2.3) becomes

$$\text{(9.2.9)} \quad G_{\bar{x}} = \tfrac{1}{2}\bar{e}_{\alpha\beta}(1+R_1)(1+R_2)^* e^{\alpha\beta} \sum_{g_1,g_2} \int \frac{d\omega\, dk}{ab(4\pi)^2} \frac{e^{i((\bar{x}-x)\pi g_1/a + (\bar{y}-y)\pi g_2/b + (\bar{z}-z)k - (\bar{t}-t)\omega)}}{(\pi g_1/a)^2 + (\pi g_2/b)^2 + k^2 - \omega^2}.$$

(9.2.10) The Field Produced by a Current J

For the sake of simplicity we consider $M = \mathbb{R}^4$ with no incoming field. We also suppose that the charges are concentrated within N, so that $J_{|\partial N} = 0$. Then, after an integration by parts as in (9.1.13),

$$F(\bar{x}) = -\int G_{\bar{x}} \wedge dJ.$$

The Green function $G_{\bar{x}}$ contains the reflections $(1+R_1)(1+R_2)$; if we always combine a reflected term with one having a g_i of opposite sign, then, for example for $(1+R_1)$, we get

$$\int dx (e^{i(\bar{x}-x)\pi g_1/a} \pm e^{-i(\bar{x}+x)\pi g_1/a})\, dJ(x) = \begin{matrix}\cos \\ i\sin\end{matrix}\, \bar{x}\pi g_1/a \cdot 2 \int dx\, e^{-ix\pi g_1/a}\, dJ(x).$$

The integration over ω can be done as in (8.2.25; 2), leaving the explicit formula

$$\text{(9.2.11)} \quad F = \bar{e}_{\alpha\beta} \sum_{g_1,g_2} f^{\alpha\beta} \int \frac{dk}{2\pi ab} \int d^3x\, dt\, \Theta(\bar{t}-t) \frac{\sin\omega(\bar{t}-t)}{-\omega} \cdot e^{ik(\bar{z}-z)} e^{-i(x\pi g_1/a + y\pi g_2/b)}\, dJ^{\alpha\beta},$$

where $\omega = ((\pi g_1/a)^2 + (\pi g_2/b)^2 + k^2)^{1/2}$ and $f^{\alpha\beta}$ are the normal modes

$$\text{(9.2.12)} \quad f = \begin{pmatrix} 0 & ic_1s_2 & is_1c_2 & -s_1s_2 \\ -ic_1s_2 & 0 & c_1c_2 & ic_1s_2 \\ -is_1c_2 & -c_1c_2 & 0 & is_1c_2 \\ s_1s_2 & -ic_1s_2 & -is_1c_2 & 0 \end{pmatrix},$$

$$c_1 := \cos\bar{x}g_1\pi/a, \qquad s_1 := \sin\bar{x}g_1\pi/a,$$
$$c_2 := \cos\bar{y}g_2\pi/b, \qquad s_2 := \sin\bar{y}g_2\pi/b.$$

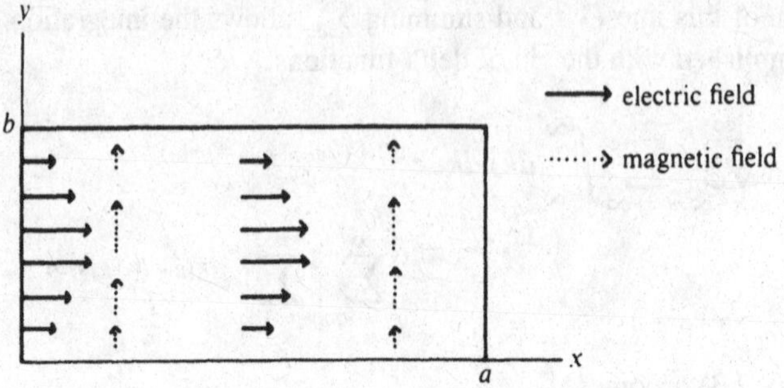

FIGURE 9.6. The progression of the fields in a wave-guide.

(9.2.13) **Remarks**

1. The boundary conditions (9.1.26) are satisfied because $f_{|\partial N} = 0$. If, say, $\bar{x} = 0$, then they can be checked by deleting the x^1 row and column and noting that what is left is $\sim s_1$, which vanishes when $\bar{x} = 0$. It is natural to ask what is meant by the tangential component of **E** or the normal component of **B** at the edges (§7, [21]). It turns out that all the components in question are zero at the corners since $s_i = 0$, and only an axial magnetic field B_3 is left over.
2. The term with $g_1 = g_2 = 0$ is an axial magnetic field. Electric fields first occur when $g_1 = 0$ and $g_2 = 1$ or vice versa. In the former case the field evolves as shown in Figure 9.6. The oscillation of the corresponding solution with $J = 0$ (see (9.2.18; 1)) has a minimal frequency π/b, and the fields depend on y. It is clear that a constant electric field cannot exist within a metallic pipe, because the boundary conditions would require it to vanish identically. On the other hand, we know that waves of a sufficiently high frequency are able to travel through metal pipes, because we can see through them.
3. If $g_{1,2} \neq 0$ are fixed, then equation (9.2.11) describes waves moving in the z-direction at a phase velocity $((\pi g_1/ka)^2 + (\pi g_2/kb)^2 + 1)^{1/2} > 1$. It is a purely geometrical effect that this velocity is greater than the speed of light. It comes about because the waves do not pass directly through the pipe, but are reflected back and forth at the walls. A wave moving directly in the z-direction could not fulfill the boundary conditions; they demand interference from other waves at a certain angle to the z-axis. Such a wave $\sim \exp(i(\mathbf{k} \cdot \mathbf{x} - \omega t))$ with $|\mathbf{k}| = \omega$ necessarily has $\omega/k_z > 1$, as the intersection with a plane of constant phase moves along the z-axis faster than light, see Figure 9.7.
4. The group velocity $\partial\omega/\partial k = k/\omega$ is less than 1. Problem 2 establishes its significance as the flow of energy in the z-direction per energy.
5. The question of whether the signal velocity is ≤ 1 is more pertinent. To answer it, consider the wave-packet

$$g(z, t) = \int_{-\infty}^{\infty} dk\, e^{ikz}[\tilde{g}(k) \cos \omega(k)t + \tilde{\dot{g}}(k)\omega(k)^{-1} \sin \omega(k)t], \tag{9.2.14}$$

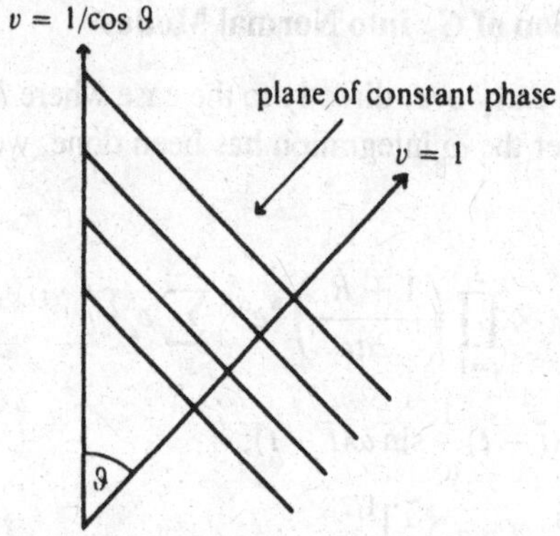

FIGURE 9.7. Speeds faster than light in a wave-guide.

where $\tilde{g}$ and $\tilde{\dot{g}}$ are the Fourier transforms of g and $\partial g/\partial t$ at $t = 0$, and

$$\omega(k) = (k^2 + m^2)^{1/2}, \qquad m^2 = \left(\frac{\pi g_1}{a}\right)^2 + \left(\frac{\pi g_2}{b}\right)^2.$$

It is supposed that g is a reasonable function, meaning that it and its first derivatives are integrable, so that all the relevant integrals converge. If g represents a signal, then it must have a well-defined beginning; for instance, g and $\dot{g}$ may be 0 at $t = 0$, $z > 0$. Then the question becomes whether this "wave-front" moves at the speed of light. Indeed it does; the above assumptions imply that $g = 0$ for all $z > |t|$ (Problem 5). This does not depend at all on $m^2 > 0$, but holds also if $m^2 < 0$, i.e., for phase velocities less than 1 and group velocities greater than 1 (tachyons) the signal velocity is still 1.

In a wave-guide there are waves with a continuous frequency-spectrum $\omega(k) \geq \omega_{\text{min}}$. If the wave-guide is sealed off in the z-direction, then the electromagnetic oscillations have only a discrete spectrum.

(9.2.15) The Resonant Cavity

Let the sets occurring in (9.1.28) *be*

$$M = \{x^\alpha \in \mathbb{R}^4 : x^0 \geq t_0\},$$
$$N = \{x^\alpha \in M : 0 \leq x^i \leq a_i, i = 1, 2, 3\}.$$

Let R_i denote the reflections $x^i \to -x^i$, and T_i the translations $x^i \to x^i + 2a^i$, as well as the induced transformations of the tensor spaces. Then the Green function for this problem is

$$G_{\bar{x}} = \prod_{i=1}^{3} \sum_{n_i=-\infty}^{\infty} (1 + R_i) T_i^{n_i} G_{\bar{x}}^{\text{ret}}.$$

The proof proceeds exactly as for (9.2.3), and will not be repeated here. Since the earlier remarks about the causal structure are still valid, let us immediately make the

(9.2.16) **Decomposition of $G_{\bar{x}}$ into Normal Modes**

The formulas of (9.2.5) carry over directly to the case where N is also bounded in the z-direction, and after the ω integration has been done, we find, in analogy to (9.2.9), that

$$G_{\bar{x}} = \tfrac{1}{2}\bar{e}_{\alpha\beta} \otimes \prod_{i=1}^{3}\left(\frac{1+R_i}{2a_i}\right) {}^{*}e^{\alpha\beta} \sum_{g_i} e^{i\sum_{j=1}^{3}(\bar{x}^j - x^j)\pi g_j/a_j}$$

$$\cdot\, \Theta(\bar{t}-t)\frac{1}{\omega}\sin\omega(\bar{t}-t),$$

$$\omega = \left[\sum_{i=1}^{3}\left(\frac{\pi g_i}{a_i}\right)^2\right]^{1/2}.$$

By making a change of sign in the g_i to replace the reflection of the x_i with a reflection of the $\bar{x}_i$, as in (9.2.10), we can solve for the normal modes:

$$G_{\bar{x}} = \tfrac{1}{2}\bar{e}_{\alpha\beta} \otimes \sum_{g_i} f^{\alpha\beta} {}^{*}e^{\alpha\beta} e^{-i\sum_{j=1}^{3} x^j \pi g_j/a_j}\,\Theta(\bar{t}-t)\frac{\sin\omega(\bar{t}-t)}{a_1 a_2 a_3 \omega}.$$

This time,

$$f = \begin{pmatrix} 0 & -c_1 s_2 s_3 & -s_1 c_2 s_3 & -s_1 s_2 c_3 \\ c_1 s_2 s_3 & 0 & i c_1 c_2 s_3 & i c_1 s_2 c_3 \\ s_1 c_2 s_3 & -i c_1 c_2 s_3 & 0 & i s_1 c_2 c_3 \\ s_1 s_2 c_3 & -i c_1 s_2 c_3 & -i s_1 c_2 c_3 & 0 \end{pmatrix},$$

with the abbreviations

$$\left.\begin{matrix} c_i \\ s_i \end{matrix}\right\} = \left\{\begin{matrix} \cos \\ \sin \end{matrix}\right\} \frac{\bar{x}_i g_i \pi}{a_i}.$$

(9.2.17) **Remarks**

1. The boundary conditions are again satisfied at the edges by the disappearance of the relevant components, and at the corners f is identically zero.
2. There are no static fields with $J = 0$ in a resonant cavity; if $g_i = 0, i = 1, 2, 3$, then $f \equiv 0$. Consequently there is a minimal frequency

$$\omega_{\min} = \left(\sum\left(\frac{\pi}{a_i}\right)^2\right)^{1/2}.$$

(9.2.18) **Problems**

1. Provide the $f^{\alpha\beta}$ of (9.2.12) with coefficients for which there exists a solution of the homogeneous Maxwell equations ($J = 0$).

2. For the f of Problem 1, calculate the averages of T^{00} and T^{03}, and verify that $\overline{T^{03}}/\overline{T^{00}} = k/\omega$.

3. Find the solutions corresponding to Problem 1 for the cylindrical geometry

$$N = \{x^\alpha \in \mathbb{R}^4 : x^2 + y^2 \leq a^2\}.$$

4. What type of oscillation in the cylinder has the lowest $\omega_{\min}$, and how does $\omega_{\min}^2$ times the cross-sectional area compare with the rectangular case?

5. Suppose that the norms

$$\int_{-\infty}^{\infty} dz\, |g(z,0)| =: \|g\|, \qquad \int_{-\infty}^{\infty} dz \left|\frac{\partial}{\partial t} g(z,0)\right| =: \|\dot{g}\|$$

and

$$\int_{-\infty}^{\infty} dz \left|\frac{\partial}{\partial z} g(z,0)\right| =: \|g'\|$$

are finite for the g of (9.2.14). Show that if $g = \dot{g} = 0$ at $t = 0$, $z > 0$, then $g(z,t) = 0$ for all $z > |t|$.

(9.2.19) Solutions

1.

$$f^{\alpha\beta} = \left|\begin{array}{c|c|c|c} & i\dfrac{kg_1\pi}{a}c_1s_2 & i\dfrac{kg_2\pi}{b}s_1c_2 & (\omega^2-k^2)s_1s_2 \\ \hline & & 0 & -i\dfrac{\omega g_1\pi}{a}c_1s_2 \\ \hline & & & -i\dfrac{\omega g_2\pi}{b}s_1c_2 \\ \hline & & & \end{array}\right| \cdot \exp(i\omega t) \quad \text{or} \quad f \to {}^*f.$$

For this f, $B_3 = 0$, and it is known as a *TM* (transverse magnetic) solution, and hence for ${}^*f$, $E_3 = 0$, which is a *TE* (transverse electric) solution.

2. Because $\overline{c_i^2} = \overline{s_i^2} = \frac{1}{2}$, etc., we find that

$$\overline{T^{00}} = \frac{\omega^2}{4}(\omega^2 - k^2), \qquad \overline{T^{03}} = \frac{k\omega}{4}(\omega^2 - k^2).$$

3. If J_n denotes the nth Bessel function, then

$$f^{\mu\nu} = \frac{1}{\rho}\left|\begin{array}{cccc} t & z & \varphi & \rho \\ & (\omega^2-k^2)\rho & -\frac{kn}{\rho} & ik\rho\frac{\partial}{\partial\rho} \\ (-\omega^2+k^2)\rho & 0 & -\frac{n\omega}{\rho} & i\omega\rho\frac{\partial}{\partial\rho} \\ \frac{kn}{\rho} & \frac{n\omega}{\rho} & 0 & 0 \\ -ik\rho\frac{\partial}{\partial\rho} & -i\omega\rho\frac{\partial}{\partial\rho} & 0 & 0 \end{array}\right| \cdot e^{i(kz+n\varphi-\omega t)} J_n(\rho\sqrt{\omega^2-k^2}),$$

$$f^{*\mu\nu} = \frac{1}{\rho}\begin{vmatrix} 0 & 0 & i\omega\frac{\partial}{\partial\rho} & n\omega \\ 0 & 0 & ik\frac{\partial}{\partial\rho} & kn \\ -i\omega\frac{\partial}{\partial\rho} & -ik\frac{\partial}{\partial\rho} & 0 & \omega^2 - k^4 \\ -n\omega & -kn & k^2-\omega^2 & 0 \end{vmatrix} \cdot e^{i(kz+n\varphi-\omega t)} \mathrm{J}_n(\rho\sqrt{\omega^2-k^2}).$$

Solving Maxwell's equations for $f_{|\partial N} = 0$ means that the parts enclosed in the dotted lines must vanish when $\rho = a$. This implies that

$$\mathrm{J}_n(a\sqrt{\omega^2-k^2}) = 0, \qquad \text{i.e.,} \quad \omega = \pm\sqrt{\frac{k^2+j_{ni}^2}{a^2}},$$

where j_{ni} is the ith zero of J_n. Interchanging f and $^*f$ produces a *TE* solution, for which it is required that

$$\mathrm{J}'_n(a\sqrt{\omega^2-k^2}) = 0, \qquad \text{where} \quad \mathrm{J}'_n(\rho) := d\, J_n(\rho)\, d\rho,$$

i.e., $\omega_{\min} = j'_{ni}(a)$, where j'_{ni} is the ith zero of J'_n.

4. Because $j_{01} = 2.40$, $j'_{01} = 3.83$, and $j'_{11} = 1.84$, the *TE* solution with $n = 1$ has the lowest $\omega_{\min}$, and its $\omega^2_{\min} \cdot a^2\pi = \pi(1.84)^2$. This is always somewhat larger than the analogous product for the $g_1 = 1$, $g_2 = 0$ oscillation with a square cross-section:

$$\left(\frac{\pi}{a}\right)^2 \cdot a^2 = \pi \cdot 3.14 \le \pi \cdot (1.84)^2 = \pi \cdot 3.38.$$

5. One can analytically continue g into the upper half-plane, where it goes to zero as $|k|^{-1}$, because

$$|\tilde{g}(u+iv)| = \left|\int_{-\infty}^{\infty} dz\, e^{-iz(u+iv)} g(z,0)\right| = \left|\int_{-\infty}^{0} dz\, e^{zv} e^{-izu} g'(z,0)\frac{i}{u+iv}\right|$$
$$\le \frac{\|g'\|}{\sqrt{u^2+v^2}} \qquad \forall u \in \mathbb{R},\, v \in \mathbb{R}^+.$$

Similarly, $\tilde{\dot{g}}$ is analytic and bounded by $\|\dot{g}\|$ where $v > 0$. Since $\cos[\omega(k)t]$ and $\sin[\omega(k)t]/\omega(k)$ are entire functions in k, the k integral of (9.2.14) can be deformed into the upper half-plane. If we decompose

$$\int_{-\infty}^{\infty} dk \qquad \text{into} \qquad \int_{-\infty}^{-R} dk + \int_{\pi}^{0} i\, d\varphi\, R\exp(i\varphi) + \int_{R}^{\infty} dk,$$

where $k = R\exp(i\varphi)$ is on the semicircle, then $\int_{-\infty}^{-R}$ and $\int_{R}^{\infty}$ go to zero as $R \to \infty$. The remaining integrals are of the form

$$\int_0^{\pi} d\varphi \exp[(z \pm t\sqrt{1+m^2e^{-2i\varphi}/R^2})R(i\cos\varphi - \sin\varphi)]\left[\frac{e^{i\varphi}R\tilde{g}(Re^{i\varphi})}{(1+m^2e^{-2i\varphi}/R^2)\tilde{\dot{g}}(Re^{i\varphi})}\right].$$

Since $|R\tilde{g}(R\exp(i\varphi))|$ and $|\tilde{\dot{g}}(R\exp(i\varphi))|$ remain bounded as $R \to \infty$, $0 \le \varphi \le \pi$, such integrals go to zero if $z > |t|$, by the Riemann–Lebesgue lemma.

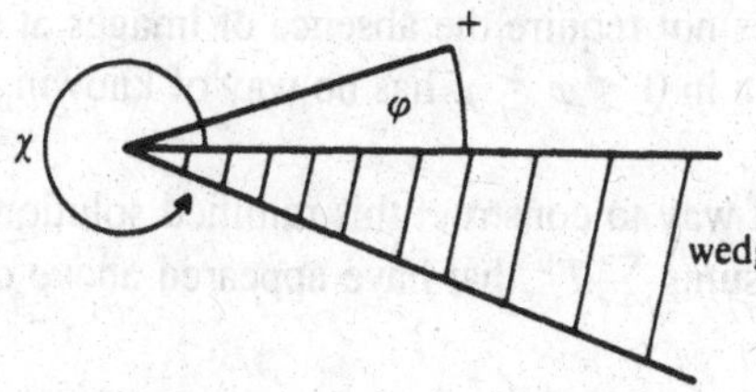

image charges at $-\varphi + 2\chi - 2\pi$

FIGURE 9.8. The image charges for a wedge.

9.3 Diffraction at a Wedge

This was the first diffraction problem to succumb to a rigorous treatment. Not only does it confirm the general outlines of one's naive expectations, but it also displays the wave nature of light with a wealth of complex detail.

The solution of boundary-value problems in somewhat complicated geometrical settings is fraught with difficulties even in the two-dimensional case. For a long time people had to settle for an imprecise theory as handed down primarily from Kirchhoff. The procedure was to substitute into certain exact integral equations the field that would be present if the conductors were removed. By so doing, one obtains a solution of the field equations but violates the boundary conditions. Since the solution involves oscillatory integrals, i.e., the small differences between large positive and negative terms, it is nearly impossible to estimate the errors incurred if one makes a small change in the integrand. Yet the result exhibits the features of experimentally known diffraction patterns, and thus the approximations have remained popular down to the present day among young and old alike. It must nonetheless be counted as a great step forward, that A. Sommerfeld succeeded in solving a nontrivial diffraction problem in 1895; it was only then possible to determine when the approximate theory of diffraction was good, and when it failed.

That problem concerned the diffraction at a wedge, where the metallic surface consisted of the planes where $\varphi = 0$ and where $\varphi = \chi > \pi$. A plausible attempt to solve this problem would proceed by mimicking the solution of the problem of two metallic mirrors at $x = 0$ and $x = a$. Following §9.2, image charges at $x + 2na$ would produce a solution with period $2a$, and then by a reflection at $x = 0$ the boundary conditions on both mirrors could be satisfied. The hitch in trying to transcribe this procedure to the wedge is that if x is the variable, the image charges always remain outside of $0 \le x \le a$ when reflected by R, but if φ is the variable on the torus T^1, then because it is periodic with period 2π, the image charges at $-\varphi + 2n\chi$ would at some point enter the region $0 \le \varphi \le \chi$ (mod 2π). Sommerfeld's brilliant stroke was to forget about the periodicity with period 2π, and to seek a solution with period 2χ. His solution has branches, in the sense that by continuation through the metal $\chi \le \varphi \le 2\pi$ the variable would not return to its initial value in $(0, \chi)$, and consequently image charges could appear there. The problem does not require the absence of images at such values of the coordinates; the solution in $0 \le \varphi \le \chi$ has no way of knowing that φ is actually a variable on T^1.

The most convenient way to construct this ramified solution is to write it as a complex integral. The sums $\sum T^n$ that have appeared above can be represented for analytic f as

$$(9.3.1) \quad \sum_n T^n f(x) = \sum_n f(x+2na) = \int_C \frac{dx'}{2a} \frac{-f(x')}{1-\exp(2\pi i(x'-x)/2a)},$$

where the contour C goes around the poles of the integrand at $x' = x + 2na$.

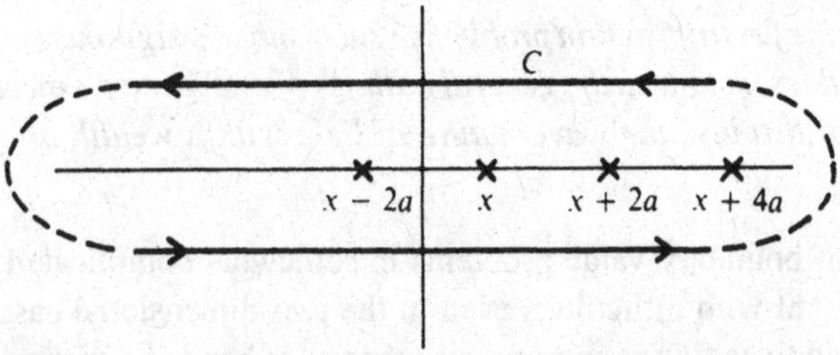

The path of integration in the integral representation of a sum.

The expression (9.3.1) always has a period $2a$ in x, as long as the contour C is flexible enough to avoid the poles as x increases to $x + 2a$. The best choice of C depends on the analytic properties of the function f.

If we try to apply this procedure to the Green function, we soon run up against the difficulty that D^{ret} is not an analytic function but only a distribution, $\delta(r-t)/4\pi r$. However, it can be approximated by analytic functions, for example by

$$(9.3.2) \qquad D^{\text{ret}}(x) = \lim_{\varepsilon \downarrow 0} \operatorname{Re} \int_0^\infty d\omega \frac{e^{i(r-t+i\varepsilon)\omega}}{4\pi^2 r},$$

where the limit is taken in the sense of distributions. Since the expressions that we require are of the form

$$\int D^{\text{ret}}(\bar{x}-x) f(x)\, dx,$$

the integrand will be multiplied by the Fourier transform of f, and for suitable f the integral $\int d\omega$ will be convergent. The integration becomes rather insensitive to the limit $\varepsilon \downarrow 0$, and therefore we may interchange $\varepsilon \downarrow 0$ with other limiting processes without worry. If we use the above expression and the notation of (9.1.28), we arrive at

(9.3.3) The Green Function for the Wedge

In cylindrical coordinates

$$x^\alpha = (t, \rho\cos\varphi, \rho\sin\varphi, z),$$

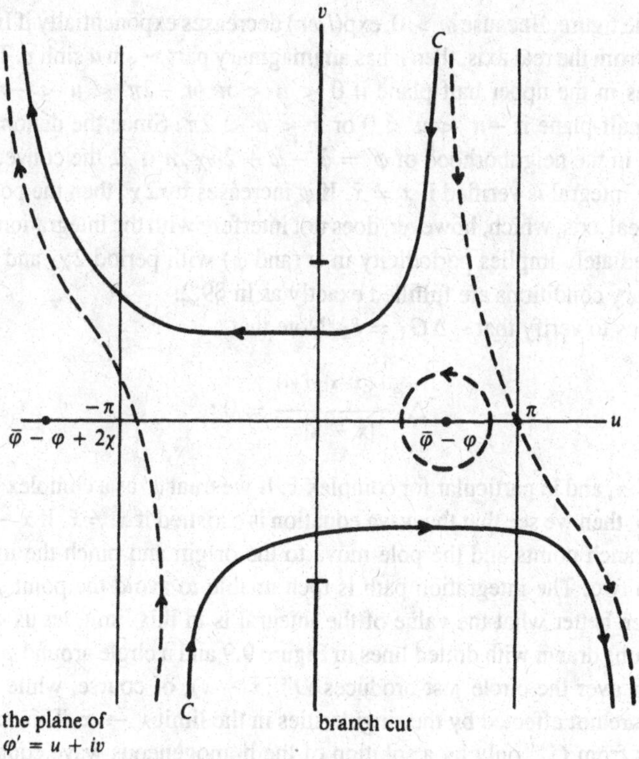

FIGURE 9.9. Paths of integration in (9.3.3).

let

$$M = \{x^\alpha \in \mathbb{R}^4 : t \geq t_0\}, \qquad N = \{x^\alpha \in M : 0 \leq \varphi \leq \chi\}.$$

If R_φ is the transformation of forms induced by $\varphi \to -\varphi$, then the Green function of this problem is

$$G_{\bar{x}} = -\tfrac{1}{2}\bar{e}_{\alpha\beta} \otimes (1 + R_\varphi)^* e^{\alpha\beta} \lim_{\varepsilon \uparrow 0} \operatorname{Re} \int_0^\infty d\omega\, e^{-i\omega(\bar{t} - t + i\varepsilon)}$$
$$\cdot \int_C \frac{d\varphi'/2\chi}{1 - e^{i\pi(\varphi' - \bar{\varphi} + \varphi)/\chi}}\, g([(\bar{z} - z)^2 + \bar{\rho}^2 + \rho^2 - 2\bar{\rho}\rho\cos\varphi']^{1/2}),$$

where

$$g(r) = \frac{e^{i\omega r}}{4\pi^2 r}.$$

The integration contour C consists of two curves, one in the upper and one in the lower half-plane, which come in from infinity, enclose the zeros of the term in brackets [ ], avoid the zeros of the denominator, and then return to infinity (Figure 9.9).

Proof: Let $\varphi' = u + iv$, so $\cos\varphi' = \cos u \cosh v - i \sin u \sinh v$. The zeros of [ ] =: r where $u = 0$ and $\cosh v = ((\bar{z} - z)^2 + \bar{\rho}^2 + \rho^2)/2\bar{\rho}\rho$ require the branch cuts drawn in the figure. Because $\omega > 0$, $\exp(i\omega r)$ decreases exponentially if $\mathrm{Im}\, r > 0$. If φ' is far from the real axis, then r has an imaginary part $\sim \sin u \sinh v$. Therefore g decreases in the upper half-plane if $0 < u < \pi$ or $-2\pi < u < -\pi$, and in the lower half-plane if $-\pi < u < 0$ or $\pi < u < 2\pi$. Since the denominator is small only in the neighborhood of $\varphi' = \bar{\varphi} - \varphi + 2n\chi$, $n \in \mathbb{Z}$, the convergence of the infinite integral is verified if $x \neq \bar{x}$. If φ increases by 2χ, then the poles move along the real axis, which, however, does not interfere with the integration contour. This immediately implies periodicity in φ (and $\bar{\varphi}$) with period 2χ, and thus that the boundary conditions are fulfilled exactly as in §9.2.

It remains to verify that $-\Delta G_{\bar{x}} = \delta_{\bar{x}}$. Note that

$$\Delta \frac{e^{i\omega(|\bar{\mathbf{x}}-\mathbf{x}|-\bar{t}+t)}}{|\bar{\mathbf{x}}-\mathbf{x}|} = 0$$

for all $\bar{x} \neq x$, and in particular for complex $\bar{x}$. If we treat φ' as a complex part to be added to $\bar{\varphi}$, then we see that the wave equation is satisfied if $x \neq \bar{x}$. If $x \to \bar{x}$, then the two branch points and the pole move to the origin and pinch the integration contour in two. The integration path is then unable to avoid the point $\bar{\varphi} - \varphi$. In order to see better what the value of the integral is in this limit, let us deform C into the paths drawn with dotted lines in Figure 9.9 and a circle around $\bar{\varphi} - \varphi$. The integration over the circle just produces $D^{\mathrm{ret}}(\bar{x} - x)$, of course, while the other two parts are not affected by the singularities in the limit $x \to \bar{x}$. This means that $G_{\bar{x}}$ differs from $G^{\mathrm{ret}}_{\bar{x}}$ only by a solution of the homogeneous wave equation, and consequently it satisfies everything required of it. □

(9.3.4) **Remark**

The edge still keeps N from being a manifold. At the edge ($\bar{\rho} \to 0$), $G_{\bar{x}}$ behaves like $(\bar{\rho})^{\pi/\chi}$. Although this approaches zero, if $\chi > \pi$ the derivative, and consequently the nonvanishing components of the field strength, diverge. However, because they approach infinity more slowly than ρ^{-1}, the square-integrability is not in danger, and thus the total energy remains finite.

As the first application of (9.3.3) we find out how it accords with the naive idea that the wedge reflects some of the light and casts a shadow. To this end, we assume in the following that $\chi > \pi$.

(9.3.5) **Geometric Optics**

According to geometric optics, if $|\bar{\varphi} - \varphi| < \pi$, then one can see directly from $\bar{x}$ to x; if $|\bar{\varphi} + \varphi| < \pi$, then the reflected image from the upper surface is also visible; and if $|\bar{\varphi} - \chi + \varphi - \chi| < \pi$, then the image from the lower surface is visible (Figure 9.10). These effects can be separated off and the corrections to geometric optics calculated if we make a decomposition of C into the following parts. Two curves, C_1 and C_2, join $-i\infty - \pi/2$ to $i\infty - 3\pi/2$ and, respectively, $i\infty + \pi/2$ to $-i\infty + 3\pi/2$, and intersect the real axis at $-\pi$ and, respectively, π; and if there is a pole in $(-\pi, \pi)$, then there is also a loop around it (Figure 9.11).

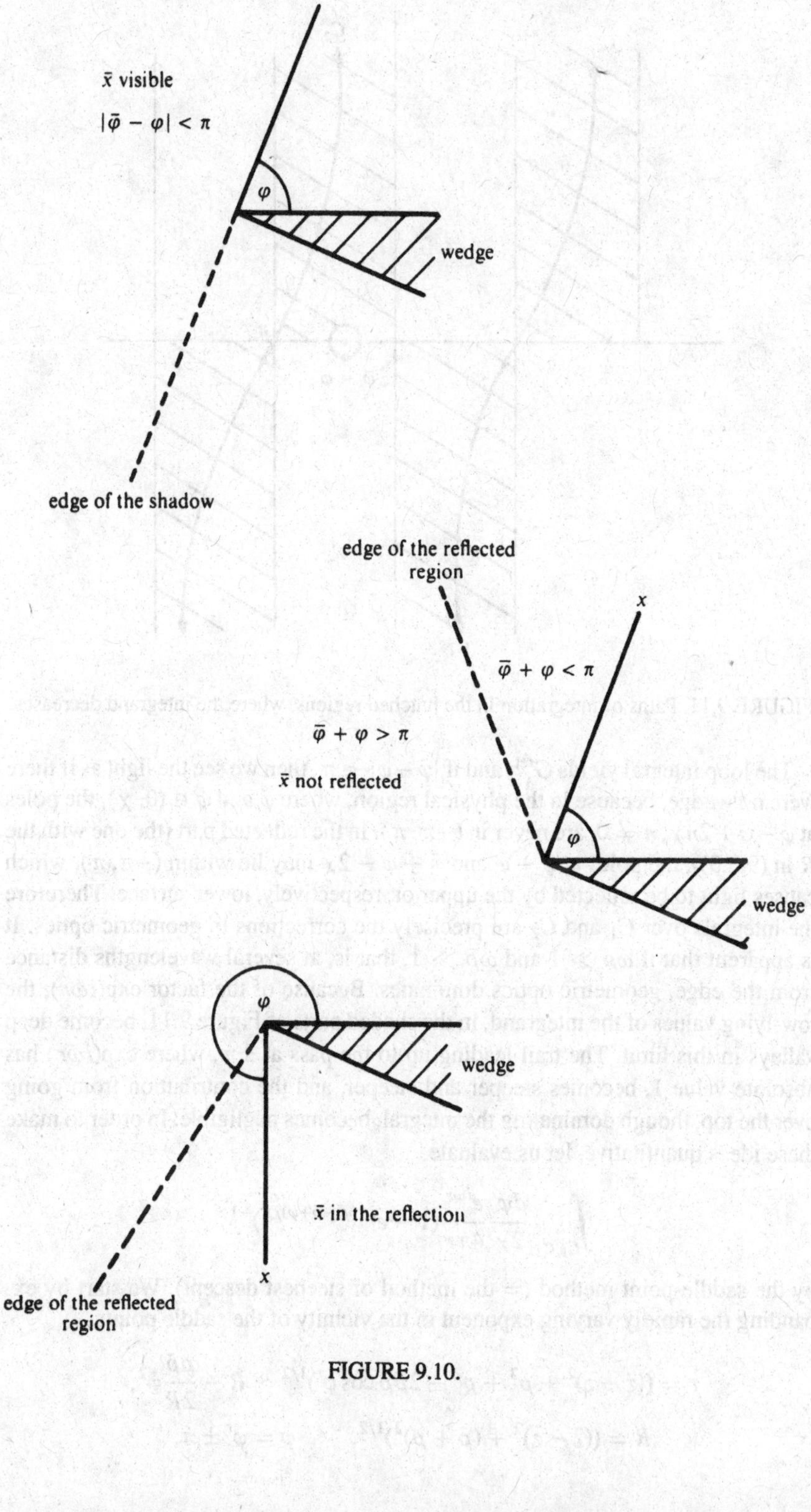

$\bar{x}$ visible
$|\bar{\varphi} - \varphi| < \pi$
φ
wedge
edge of the shadow
edge of the reflected region
x
$\bar{\varphi} + \varphi < \pi$
$\bar{\varphi} + \varphi > \pi$
$\bar{x}$ not reflected
φ
wedge
φ
wedge
$\bar{x}$ in the reflection
x
edge of the reflected region

FIGURE 9.10.

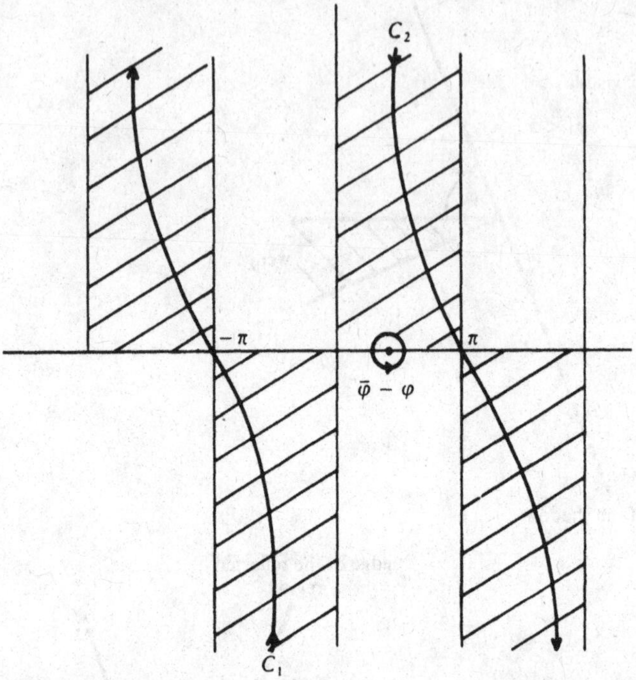

FIGURE 9.11. Paths of integration in the hatched regions, where the integrand decreases.

The loop integral yields $G^{\text{ret}}_{\bar{x}}$, and if $|\bar{\varphi}-\varphi| < \pi$, then we see the light as if there were no wedge, because in the physical region, where $\bar{\varphi}$ and $\varphi \in (0, \chi)$, the poles at $\bar{\varphi}-\varphi+2n\chi, n \neq 0$, are never in $(-\pi, \pi)$. In the reflected part (the one with the R in (9.3.3)), the poles at $\bar{\varphi}+\varphi$ and $\bar{\varphi}+\varphi-2\chi$ may lie within $(-\pi, \pi)$, which causes light to be reflected by the upper or, respectively, lower surface. Therefore the integrals over C_1 and C_2 are precisely the corrections to geometric optics. It is apparent that if $\omega\rho \gg 1$ and $\omega\bar{\rho} \gg 1$, that is, at several wavelengths distance from the edge, geometric optics dominates. Because of the factor $\exp(i\omega r)$, the low-lying values of the integrand, in the shaded parts of Figure 9.11, become deep valleys in this limit. The trail leading up to the pass at $\pm\pi$, where $\exp(i\omega r)$ has absolute value 1, becomes steeper and steeper, and the contribution from going over the top, though dominating the integral, becomes negligible. In order to make these ideas quantitative, let us evaluate

$$\int_{C_1,C_2} \frac{d\varphi'}{2\chi}\frac{e^{i\omega r}}{4\pi r}(1-e^{i\pi(\varphi'-\bar{\varphi}+\varphi)/\chi})^{-1}$$

by the saddle-point method (= the method of steepest descent). We start by expanding the rapidly varying exponent in the vicinity of the saddle points,

$$r = ((\bar{z}-z)^2+\bar{\rho}^2+\rho^2-2\bar{\rho}\rho\cos\varphi')^{1/2} \sim R - \frac{\rho\bar{\rho}}{2R}\tilde{\varphi}^2,$$

$$R = ((\bar{z}-z)^2+(\bar{\rho}+\rho)^2)^{1/2}, \qquad \tilde{\varphi} = \varphi' \pm \pi,$$

and replacing the rest of the integrand by its value at the saddle point. There remains a Gaussian integral over $\bar{\varphi}$, from which we obtain the asymptotically exact formula:

(9.3.6)

$$\int_{C_1 \cup C_2} \frac{d\varphi'}{2\chi} \frac{e^{i\omega r}}{4\pi r} (1 - e^{i\pi(\varphi' - \bar{\varphi} + \varphi)/\chi})^{-1}$$

$$\cong \frac{1}{4\chi} \left(\frac{1}{1 - e^{i\pi(\varphi - \bar{\varphi} - \pi)/\chi}} - \frac{1}{1 - e^{i\pi(\varphi - \bar{\varphi} + \pi)/\chi}} \right) \cdot \frac{e^{i\omega R - i\pi/4}}{\sqrt{2\pi\omega\bar{\rho}\rho R}}$$

$$= -\frac{1}{4} \frac{\sin(\pi^2/\chi)}{\cos\pi(\bar{\varphi} - \varphi)/\chi - \cos\pi^2/\chi} \frac{e^{i\omega R + i\pi/4}}{\sqrt{2\pi\omega\bar{\rho}\rho R}}$$

(cf. Problem 1).

(9.3.7) **Remarks**

1. The correction to geometric optics vanishes as $\bar{\rho} \to \infty$. If $\rho \gg \bar{\rho}$, then it goes as $\bar{\rho}^{-1/2}$, which corresponds to a cylindrical wave emanating from the edge.
2. The cylindrical wave has an angular factor, which diverges at the boundary of the shadow, $|\bar{\varphi} - \varphi| = \pi$. However, the asymptotic expansion (9.3.6) does not apply in that limit, because the pole coincides with the saddle point.
3. If $\chi = \pi$, then the cylindrical wave disappears, as the wedge becomes the plane mirror of (9.2.1). This fact is therefore strictly true, and not merely asymptotic (Problem 2).

If $\chi = 2\pi$, then the expression (9.3.3) simplies so greatly that many additional questions can be answered for which the asymptotic expression (9.3.6) is inadequate.

(9.3.8) **Diffraction at a Half-Plane**

If $\chi = 2\pi$, $\rho \to \infty$, and $\bar{\rho}$ and $\bar{\varphi}$ are arbitrary, then $G_{\bar{x}}$ is asymptotically

$$G_{\bar{x}} = (1 + R_\varphi) G_{\bar{x}}^{\text{ret}} \frac{e^{-i\pi/4}}{\sqrt{2}} \int_{-\infty}^{v} dv' \, e^{i\pi v'^2/2},$$

$$v = \sqrt{\frac{\bar{\rho}\rho\omega}{\pi R}} 2\cos\left(\frac{\bar{\varphi} - \varphi}{2}\right), \qquad R = ((\bar{z} - z)^2 + \rho^2)^{1/2}.$$

(9.3.9) **Remarks**

1. The ω in v should be chosen so that $G_{\bar{x}}^{\text{ret}}$ is represented as the integral of (9.3.2).
2. For (9.3.6) it was necessary to assume that $\omega\bar{\rho}\rho((\bar{z} - z)^2 + \rho^2 + \bar{\rho}^2)^{-1/2} \gg 1$. This is not the case if $\rho \to \infty$ and $\bar{\rho} < 1/\omega$; in this sense (9.3.8) is more general. The boundary of the shadow is also described by (9.3.8), as long as the source of the light is sufficiently far from the edge of the wedge.

Proof of (9.3.8): First observe that

$$G_{\bar{x}}(\varphi) + G_{\bar{x}}(\varphi + 2\pi) = G_{\bar{x}}^{\text{ret}}(\varphi),$$

because

$$\frac{1}{1 - e^{i(\varphi' - \bar{\varphi} + \varphi)/2}} + \frac{1}{1 - e^{i(\varphi' + 2\pi - \bar{\varphi} + \varphi)/2}} = \frac{2}{1 - e^{i(\varphi' - \bar{\varphi} + \varphi)}}.$$

Since this sum has a periodicity 2π, the contributions from the paths C_1 and C_2 of Figure 9.11 cancel out, and only the loop around $\bar{\varphi} - \varphi$ remains. On the other hand, the difference of the same terms is

$$\frac{1}{1 - e^{i(\varphi' - \bar{\varphi} + \varphi)/2}} - \frac{1}{1 + e^{i(\varphi' - \bar{\varphi} + \varphi)/2}} = \frac{i}{\sin(\varphi' - \bar{\varphi} + \varphi)/2},$$

and we shall find a similar factor in the exponent in the limit $\rho \to \infty$. In (9.3.3) in this limit, let us write

$$r \cong R - \frac{\bar{\rho}\rho}{R} \cos\varphi', \qquad R = ((\bar{z} - z)^2 + \rho^2)^{1/2},$$

and compare it with the quantity

$$|\bar{\mathbf{x}} - \mathbf{x}| \cong R - \frac{\bar{\rho}\rho}{R} \cos(\bar{\varphi} - \varphi)$$

occurring in $G_{\bar{x}}^{\text{ret}}$ in place of r; then in the difference there occurs

$$\begin{aligned}\cos(\bar{\varphi} - \varphi) - \cos\varphi' &= 2\cos^2\frac{(\bar{\varphi} - \varphi)}{2} - 2\cos^2\frac{\varphi'^2}{2} \\ &= 2\sin\frac{(\varphi' + \bar{\varphi} - \varphi)}{2} \sin\frac{(\varphi' - \bar{\varphi} + \varphi)}{2}.\end{aligned}$$

If the factor $\exp(i\omega|\bar{\mathbf{x}} - \mathbf{x}|)$ is removed, then in the limit $\rho \to \infty$ we are left with the integral

$$I := \frac{1}{8\pi i} \int_C \frac{d\varphi' e^{2i\omega(\bar{\rho}\rho/R)\sin(\varphi' - \bar{\varphi} + \varphi)/2 \sin(\varphi' + \bar{\varphi} - \varphi)/2}}{\sin(\varphi' - \bar{\varphi} + \varphi)/2}$$

in $G_{\bar{x}}(\varphi) - G_{\bar{x}}(\varphi + 2\pi)$. The denominator can be disposed of by differentiating by $\bar{\rho}$;

$$\frac{\partial I}{\partial \bar{\rho}} = \frac{\omega\rho}{4\pi R} \int_C d\varphi' \sin\frac{(\varphi' + \bar{\varphi} - \varphi)}{2} e^{(2i\omega\bar{\rho}\rho/R)(\cos^2(\bar{\varphi} - \varphi)/2 - \cos^2\varphi'/2)}.$$

Using

$$\sin\frac{(\varphi' + \bar{\varphi} - \varphi)}{2} = \sin\frac{\varphi'}{2}\cos\frac{\bar{\varphi} - \varphi}{2} + \cos\frac{\varphi'}{2}\sin\frac{\bar{\varphi} - \varphi}{2},$$

we see that the second term contributes nothing, because its part of the integrand is even in φ', and the integral $\int_C d\varphi'$ is taken in the opposite direction when $\varphi' \to -\varphi'$. We may choose $\cos\varphi'/2$ as a new integration variable in the first term. Then there is no pole, and the two curves C_1 and C_2 have the same contribution, since the integral changes its sign under $\varphi' \to \varphi' + 2\pi$:

$$\frac{\partial I}{\partial \bar{\rho}} = \frac{\omega\rho}{\pi R}\int_{-i\infty}^{i\infty} e^{-2i(\omega\bar{\rho}\rho/R)(t^2-\cos^2(\bar{\varphi}-\varphi)/2)}dt \cos\frac{\bar{\varphi}-\varphi}{2}$$

$$= \frac{\omega\rho}{\pi R} e^{-i\pi/4} e^{2i(\omega\bar{\rho}\rho/R)\cos^2(\bar{\varphi}-\varphi)/2}\sqrt{\frac{2\pi R}{,\omega\bar{\rho}\rho}}\cos\frac{\bar{\varphi}-\varphi}{2}.$$

In order to do the $\bar{\rho}$ integration, we introduce the variable

$$v = \sqrt{\frac{\bar{\rho}\rho\omega}{\pi R}}2\cos\frac{(\bar{\varphi}-\varphi)}{2},$$

$$dv = d\bar{\rho}\sqrt{\frac{\rho\omega}{\pi\bar{\rho}R}}\cos\frac{(\bar{\varphi}-\varphi)}{2}.$$

Since $I = 0$ for $\bar{\rho} = 0$ where C can be shifted away

$$I = \sqrt{2}e^{-i\pi/4}\int_0^v dv'\, e^{i\pi v'^2/2}.$$

Combining this with the sum $G_{\bar{x}}(\varphi) + G_{\bar{x}}(\varphi + 2\pi)$ and using

$$\sqrt{2}e^{-i\pi/4}\int_{-\infty}^0 dv'\, e^{i\pi v'^2/2} = 1,$$

we obtain (9.3.8). □

To discuss the problem in more detail, we need some

(9.3.10) Properties of Fresnel's Integral

Let

$$F(z) = \int_{-\infty}^z dv\, e^{i\pi v^2/2}.$$

Then $F(\infty) = 2F(0) = 1 + i$, and F behaves asymptotically as

$$z \to 0\colon F(z) = F(0) + z + i\frac{\pi}{6}z^3 + O(z^5),$$

(9.3.11) $$z \to \infty\colon F(z) = F(\infty) + \frac{e^{i\pi z^2/2}}{i\pi z} + O(z^{-3}),$$

$$z \to -\infty\colon F(z) = \frac{e^{i\pi z^2/2}}{i\pi z} + O(z^{-3}).$$

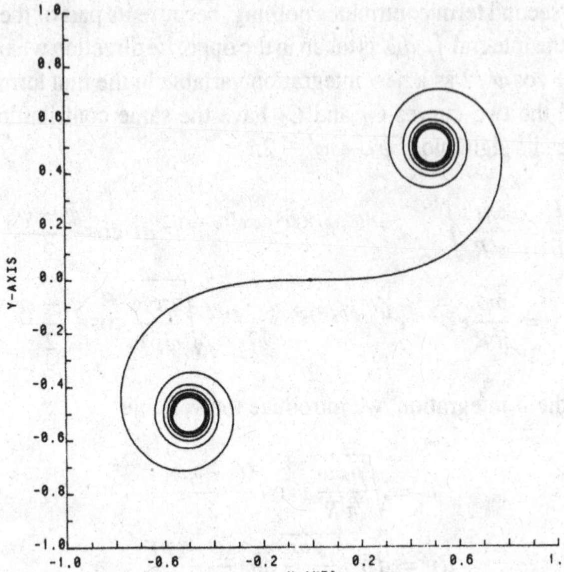

FIGURE 9.12. Cornu's spiral.

If $F(x)$ is graphed as a curve in the complex plane depending on $x \in \mathbb{R}$ as a parameter, it has the form of Cornu's spiral. When $x \leq x_0 > 0$, $|F(x)|$ increases monotonically, and after that point it oscillates about the limiting value $\sqrt{2}$.

We next investigate the regions not covered by (9.3.6):

(9.3.12) The Boundary of the Shadow

Let $\bar{\varphi} - \varphi = \pi + \delta$, $\delta \ll 1$. The time average of the intensity equals the absolute square of the field, and its ratio to the unscreened intensity i^{ret} becomes

$$\frac{i}{i^{\text{ret}}} = \frac{1}{2}\left|F\left(\delta\sqrt{\frac{\bar{\rho}\omega}{\pi}}\right)\right|^2$$

as $\rho \to \infty$ with $\rho/R \to 1$. We have disregarded the reflected light; the interference with the term containing $R\varphi$ produces a correction $O(\bar{\rho}^{-1/2})$. During the transition from light to shadow, the intensity thus first oscillates about the value i^{ret} and then decreases monotonically. The width $\sim \sqrt{\bar{\rho}/\omega}$ of the transition zone, where the shadow is hazy, increases with $\bar{\rho}$, while its angle $1/\sqrt{\bar{\rho}\omega}$ from the edge of the wedge approaches zero as $\bar{\rho}$ increases.

(9.3.13) The Edge

If $\bar{\rho} \to 0$ and $\rho/R \to 1$, and we consider such a small neighborhood of the edge that the contribution from $G_{\bar{x}}^{\text{ret}}$ is effectively constant and hence unaffected by a

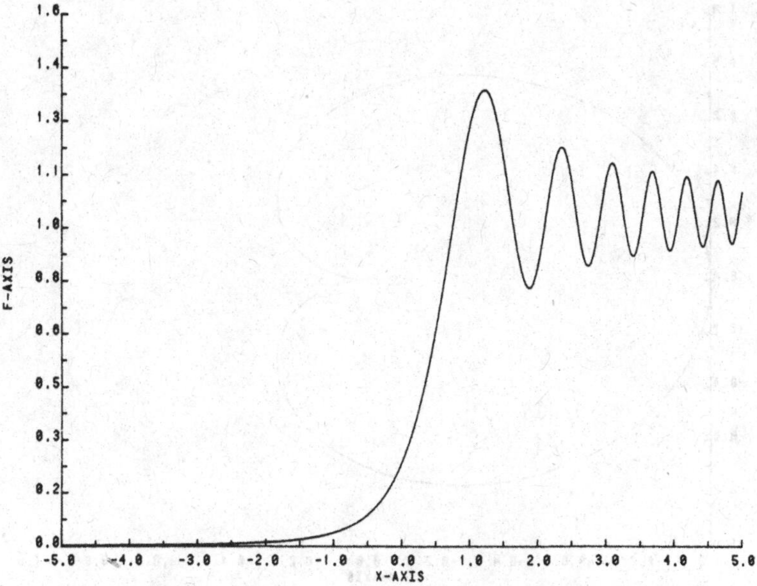

FIGURE 9.13. The intensity at the edge of a shadow.

reflection, then the components that go to zero at the edge are

$$F\left(\sqrt{\frac{\bar{\rho}\omega}{\pi}}2\cos\left(\frac{\bar{\varphi}-\varphi}{2}\right)\right)-F\left(\sqrt{\frac{\bar{\rho}\omega}{\pi}}2\cos\left(\frac{\bar{\varphi}+\varphi}{2}\right)\right)\sim\sqrt{\frac{\bar{\rho}\omega}{2}}2\sin\frac{\bar{\varphi}}{2}\cdot\sin\frac{\varphi}{2}.$$

Thus the intensity dies down in directions along the half-plane, and the angular distribution in $\bar{\varphi}$ with φ fixed has no preference for the incident direction.

(9.3.14) The Plane of the Screen

If $\bar{\varphi}=\pi$, and the factor $G_{\bar{x}}^{\mathrm{ret}}$ is still effectively constant, then the components that vanish at the screen are diminished by

$$F\left(2\sqrt{\frac{\bar{\rho}\omega}{\pi}}\sin\frac{\varphi}{2}\right)-F\left(-2\sqrt{\frac{\bar{\rho}\omega}{\pi}}\sin\frac{\varphi}{2}\right)$$

compared with $G_{\bar{x}}^{\mathrm{ret}}$. It takes a few wavelengths for the field to increase from zero at $\bar{\rho}=0$ to the value it would have were it not for the screening, and it takes longer to do so if the light is incident at a smaller angle φ.

(9.3.15) Remark

In the integral, Kirchhoff's theory of diffraction replaces the field within the screening plane with the value it would have without screening. This is a very good approximation at many wavelengths away from the screen, but it is poor in the vicinity of the screen. It has the unsettling consequence that the field strength

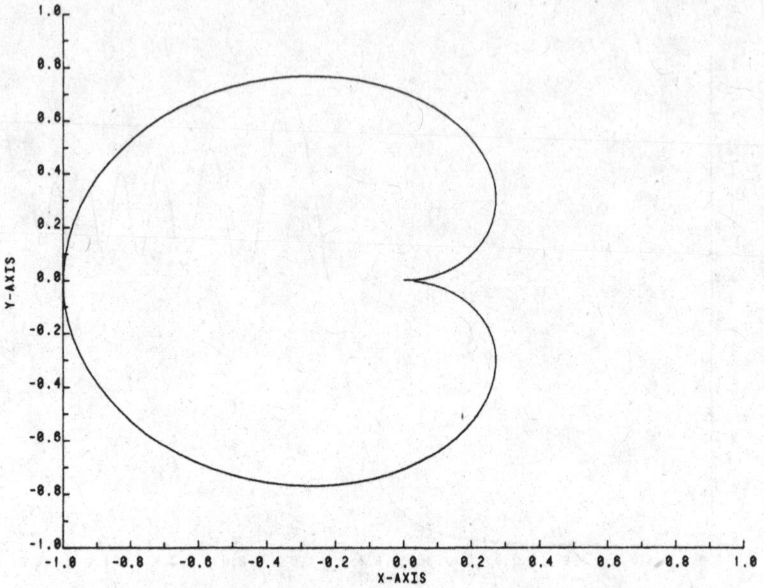

FIGURE 9.14. Polar graph of $i(\varphi)$.

calculated in this manner becomes singular at the edge as $1/\rho$, and is thus not square-integrable, so the field energy is infinite (cf. §25, [21]).

In conclusion, geometric optics gives the right answers if one can take a bird's-eye view, from which the wavelength is not perceptible. But up close there is such a complicated pattern of interference that any simple approximation is doomed to failure.

(9.3.16) Problems

1. Show how and where (9.3.6) is an asymptotic expansion.
2. Check that $G_{\bar{x}} = (1 + R)G_{\bar{x}}^{\text{ret}}$ if $\chi = \pi$.
3. See whether (9.3.6) agrees with (9.3.8).
4. How large is the error in (9.3.11)?

(9.3.17) Solutions

1. Use the saddle-point method discussed, e.g., in Dieudonné, *Infinitesimal Calculus* IX, §1.
2. If $\chi = \pi$, then the function of period 2χ reduces to the function $G_{\bar{x}}^{\text{ret}}$ of period 2π.
3. Applying (9.3.11) to (9.3.8), we get

$$\frac{e^{i\omega|\bar{x}-x|}}{4\pi|\bar{x}-x|}\,\frac{e^{-i\pi/4}}{\sqrt{2}}\,\frac{e^{i\pi v^2/2}}{i\pi v} \cong -\frac{e^{i\omega R}ie^{-i\pi/4}e^{i\omega\bar{\rho}\rho/R}}{8\pi\sqrt{2\pi\omega\bar{\rho}\rho R}\cos(\bar{\varphi}-\varphi)/2}$$

for the correction to geometric optics, in agreement with (9.3.6) (where the R is $\sim R + \bar{\rho}\rho/R$).

4. By integration by parts with $z \in \mathbb{R}^+$,

$$-\int_z^\infty d\nu\, e^{i\pi\nu^2/2} = -\int_z^\infty \frac{d\nu}{i\pi\nu}\frac{\partial}{\partial\nu}e^{i\pi\nu^2/2} = \frac{e^{i\pi\nu^2/2}}{i\pi\nu} - \int_z^\infty \frac{d\nu}{i\pi\nu^2}e^{i\pi\nu^2/2},$$

$$\left|\int_z^\infty \frac{d\nu}{i\pi\nu^2}e^{i\pi\nu^2/2}\right| = \left|\int_z^\infty \frac{d\nu}{i\pi^2\nu^3}\frac{\partial}{\partial\nu}e^{i\pi\nu^2}\right| \le \frac{1}{\pi^2 z^3} + \frac{1}{\pi^2 z^3}.$$

Therefore

$$\left|F(z) - F(\infty) - \frac{e^{i\pi z^2/2}}{i\pi z}\right| \le \frac{2}{\pi^2 z^3} \quad \forall z \in \mathbb{R}^+.$$

9.4 Diffraction at a Cylinder

This problem is so tractable mathematically that all the various phenomena of geometric and wave optics can be worked out.

Despite there being complementary problems to the wave-guide and the resonant cavity, it is an elaborate job to analyze diffraction at a cylinder or at a sphere. Even if it is not required to write down the complete Green function, but is enough to solve for the individual waves that $G_{\bar{x}}$ is constructed from, there are still infinite sums and complex integrals to cope with. These expressions reduce to elementary functions only in particular limits, in which the relevant properties are well displayed. Depending on the ratio of the wavelength to the size of the metallic body, it may be possible for the wave to pass around the obstruction. This phenomenon is somewhat simpler for a cylinder than for a sphere, for which reason we restrict our discussion to cylinders. We also consider only scalar solutions u to the wave equation $-\Box u = 0$, with either Dirichlet or Neumann boundary conditions, $u_{|\partial N} = 0$ or $du_{|\partial N} = 0$. Although it is possible to construct $G_{\bar{x}}$ with the aid of these solutions, the expressions that result are too involved to be illuminating. We are again primarily interested in the limit where the source of the waves is at infinity; therefore we utilize u's incident in a plane and scattering into cylindrical waves. In cylindrical coordinates they would have the form

$$\left(e^{-iqx} + \frac{e^{iq\rho}}{\sqrt{\rho}}f(\varphi)\right)e^{i(kz-\omega t)}$$

as $\rho \to \infty$. If the first term is expanded in a Fourier series,

$$(9.4.1)\quad e^{iqx} = \sum_{n=-\infty}^{\infty} e^{in\varphi}\frac{1}{2\pi}\int_0^{2\pi} d\varphi'\, e^{-i(q\rho\cos\varphi' + n\varphi')} = \sum_{n=-\infty}^{\infty} e^{in(\varphi-\pi/2)}J_n(q\rho),$$

then we can represent the solution $u(\rho, \varphi)$ that satisfies the boundary conditions at $\rho = a$ and $\rho \to \infty$ as a

(9.4.2) Fourier Series

$$u=\sum_{n=-\infty}^{\infty}e^{in(\varphi-\pi/2)}\left[\mathrm{J}_n(q\rho)-\frac{\mathrm{J}_n(qa)}{\mathrm{H}_n^{(1)}(qa)}\mathrm{H}_n^{(1)}(q\rho)\right].$$

In order to evaluate the sum, recall the

(9.4.3) Asymptotic Behavior of $\mathrm{H}_\nu^{(1)}(x)$

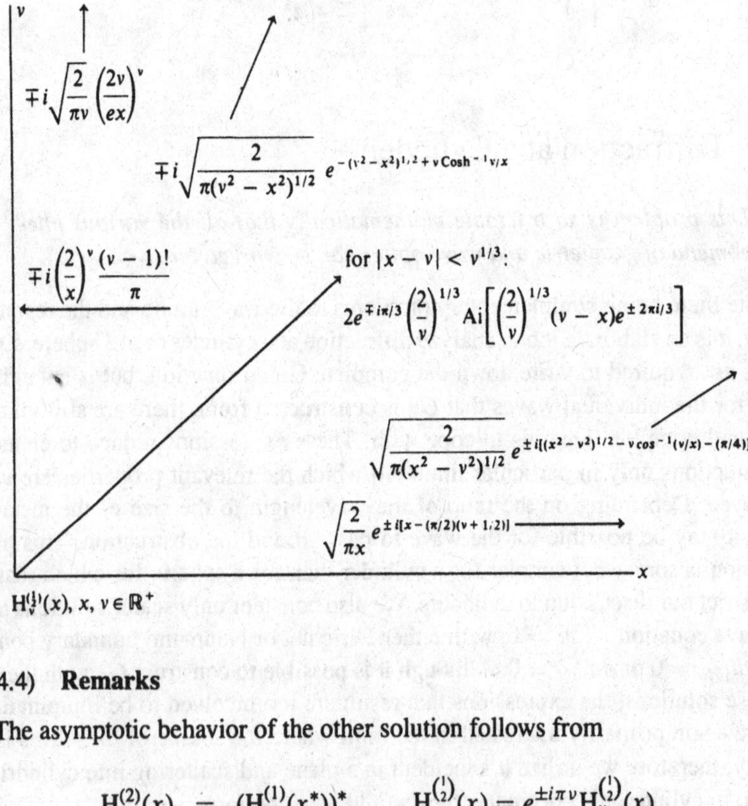

$\mathrm{H}_\nu^{(^1_2)}(x)$, x, $\nu\in\mathbb{R}^+$

(9.4.4) Remarks

1. The asymptotic behavior of the other solution follows from

$$\begin{aligned}\mathrm{H}_\nu^{(2)}(x)&=(\mathrm{H}_{\nu^*}^{(1)}(x^*))^*,\qquad \mathrm{H}_{-\nu}^{(^1_2)}(x)=e^{\pm i\pi\nu}\mathrm{H}_\nu^{(^1_2)}(x),\\ \mathrm{J}_\nu&=\tfrac{1}{2}(\mathrm{H}_\nu^{(1)}+\mathrm{H}_\nu^{(2)}),\\ \mathrm{J}_\nu&\overset{\nu\to\infty}{\sim}\frac{1}{\sqrt{2\pi\nu}}\left(\frac{ex}{2\nu}\right)^{\nu},\quad \overset{x\to 0}{\sim}\left(\frac{x}{2}\right)^{\nu}\frac{1}{\nu!}.\end{aligned}$$

2. Consequently, if n is large, then the nth term in (9.4.2) approaches

$$\frac{e^{in(\varphi-\pi/2)}}{\sqrt{2\pi n}}\left(\frac{e}{2n}\right)^{n}\left[(q\rho)^n-\frac{(qa)^{2n}}{(q\rho)^n}\right],$$

and the sum converges uniformly on compact subsets of the $(x$–$y)$-plane. Even so, the first term in the sum dominates the scattered wave only if $qa\ll 1$. In

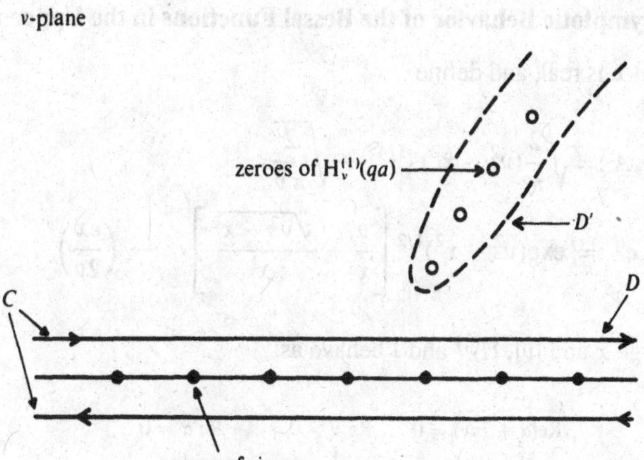

FIGURE 9.15. Paths of integration for (9.4.6).

that case,

$$\frac{J_n(qa)}{H_n^{(1)}(qa)} \to \pi i \left(\left(\frac{qa}{2} \right)^{|n|} \frac{1}{|n|!} \right)^2 n \qquad \text{for } |n| \geq 1,$$

and

$$u = e^{-iqx} + \frac{i\pi}{2 \ln qa} H_0^{(1)}(q\rho) + O((qa)^2).$$

3. If $qa \gg 1$, then we make use of the fact that $H_\nu^{(1)}_{(2)}$ are analytic functions in ν. This allows the sum in (9.4.2) to be turned into an integral over $d\nu$,

(9.4.5)
$$\sum_{n=-\infty}^{\infty} = \frac{1}{2i} \int_C \frac{d\nu\, e^{-i\pi\nu}}{\sin \pi\nu},$$

where C is a contour passing around the integers in the complex plane and excluding the zeros of $H_\nu^{(1)}$ (Figure 9.15), which we shall investigate later.

This integral can be further transformed into an integral along the path D strictly above the real axis, by taking account of (9.4.4; 1). The result of this is the

(9.4.6) **Representation as a Fourier Integral**

$$u = \frac{1}{4i} \int_D \frac{d\nu\, e^{-i\nu\pi/2}}{\sin \pi\nu} 2 \cos \nu(\varphi - \pi) \cdot \frac{H_\nu^{(1)}(qa) H_\nu^{(2)}(q\rho) - H_\nu^{(2)}(qa) H_\nu^{(1)}(q\rho)}{H_\nu^{(1)}(qa)}.$$

To evaluate the above integral by means of the residue theorem, we need to also know the

(9.4.7) Asymptotic Behavior of the Bessel Functions in the Upper ν-Plane

Suppose that x is real, and define

$$A := \sqrt{\frac{2}{\pi}}(\nu^2 - x^2)^{-1/4} \to \sqrt{\frac{2}{\pi\nu}},$$
$$e^\alpha := \exp(\nu^2 - x^2)^{1/2}\left[\frac{\nu}{x} + \frac{\sqrt{\nu^2 - x^2}}{x}\right]^{-\nu} \to \left(\frac{ex}{2\nu}\right)^\nu.$$

Then for large x and $|\nu|$, $\mathrm{H}^{(1)}_{(2)}$ and J behave as:

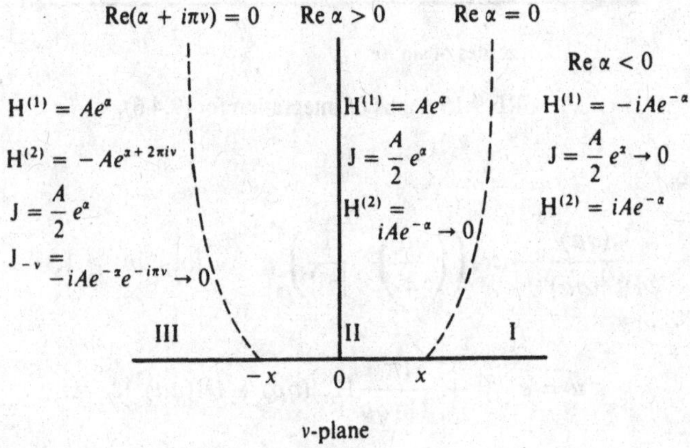

(9.4.8) Remarks

1. The function J is always of the form $(A/2)e^\alpha$, whereas $\mathrm{H}^{(1)}$ (resp. $\mathrm{H}^{(2)}$) changes its asymptotic behavior along the curve Re $\alpha = 0$ (resp. $\mathrm{Re}(\alpha + i\pi\nu) = 0$), on which its zeros are situated. These curves intersect the real axis at $\pm x$ at an angle of $\pm\pi/3$. Far away from the curve Re $\alpha = 0$, $\mathrm{H}^{(1)}$ increases rapidly in the upper half-plane, going as $|\nu|^{\mathrm{Re}\,\nu}$.
2. The zeros ν_m, $m = 1, 2, 3$, of $\mathrm{H}^{(1)}_\nu(qa)$ closest to the real axis are near $\nu = qa$, where the asymptotic representation in terms of Airy functions is applicable; if $qa \gg 1$, then

$$\mathrm{H}^{(1)}_{\nu_m}(qa) = 0, \qquad \nu_m \cong qa + c_m\left(\frac{qa}{2}\right)^{1/3} e^{i\pi/3},$$
$$\mathrm{Ai}(-c_m) = 0, \qquad c_m = (2.3, 4.1, 5.5, 6.7, 7.9, \ldots).$$

3. Farther from the real axis along the curve Re $\alpha = 0$, $\mathrm{H}^{(1)}$ is given simply as the sum of the asymptotic forms of regions I and II, and becomes zero if

$$\sinh\left(\alpha - \frac{i\pi}{4}\right) = 0.$$

This implies that

$$\nu_m \sim \frac{i\pi m}{\ln \dfrac{2\pi m}{eqa}} e^{-(i\pi/2)\ln(2\pi m/eqa)}, \qquad m \gg 1.$$

These facts allow us to discover expressions for u in different regions of the $(\rho\text{–}\varphi)$-plane when $qa \gg 1$.

(9.4.9) The Shadow

The integral in (9.4.6) can be turned into an integral over D' around the zeros of $H^{(1)}_\nu(qa)$, because the integrand goes to zero exponentially fast in the upper half-plane, except along the curve $\operatorname{Re}\alpha = 0$. To see the reason for this, consider the numerator

(9.4.10) $$g_\nu := H^{(1)}_\nu(qa)H^{(2)}_\nu(q\rho) - H^{(2)}_\nu(qa)H^{(1)}_\nu(q\rho) = g_{-\nu}.$$

In the region I ∪ II of (9.4.7), within which the asymptotic form of $H^{(1)}_\nu$ changes, we express $H^{(1)}_\nu$ in terms of J and $H^{(2)}_\nu$. Meanwhile, for the same reason we write $H^{(2)}_\nu$ in terms of $J_{-\nu}$ in II ∪ III:

(9.4.11) $$\begin{aligned} g_\nu &= 2[J_\nu(qa)H^{(2)}_\nu(q\rho) - H^{(2)}_\nu(qa)J_\nu(q\rho)] \\ &= 2e^{i\pi\nu}[H^{(1)}_\nu(qa)J_{-\nu}(q\rho) - J_{-\nu}(qa)H^{(1)}_\nu(q\rho)]. \end{aligned}$$

In either case (9.4.7) shows us the asymptotic expression

(9.4.12) $$g_\nu \sim \frac{2i}{\pi\nu}\left[\left(\frac{a}{\rho}\right)^\nu - \left(\frac{\rho}{a}\right)^\nu\right],$$

which thus holds uniformly in the upper half-plane. Hence, g_ν is essentially $c^{\pm\nu}$, whereas the denominator $H^{(1)}_\nu(qa)$ is asymptotically dominated by $\nu^{\pm\nu}$, which wins out over $c^{\pm\nu}$, as can be seen if we let $\nu = t\exp(i\psi)$, t and $\psi \in \mathbb{R}^+$; then for $t \to \infty$,

(9.4.13) $$\begin{aligned} |\nu^{\pm\nu}| &= e^{t|\cos\psi\ln t - \psi\sin\psi|}, \\ |c^{\pm\nu}| &= e^{t|\cos\psi\ln c|}. \end{aligned}$$

Because all the other factors in (9.4.6) grow at the fastest as $c^{\pm\nu}$, the integrand decreases exponentially in the upper half-plane for all ρ and φ. An exception is made, of course, for the curve $\operatorname{Re}\alpha = 0$ (asymptotically, $\ln t = \psi\tan\psi$), because the denominator can vanish, and the integral $\int_{D'}$ can be rewritten as the sum of the residues:

(9.4.14) $$u = \pi\sum_{m=1}^{\infty}\frac{e^{-i\nu_m\pi/2}}{\sin\pi\nu_m}\cos(\varphi-\pi)\nu_m\frac{H^{(2)}_{\nu_m}(qa)}{\dfrac{\partial}{\partial\nu_m}H^{(1)}_{\nu_m}(qa)}H^{(1)}_{\nu_m}(q\rho).$$

This equation is only useful if the term with $m = 1$ is so much larger than the rest of the sum that it is essentially equal to u. To discover when this happens, we refer again to (9.4.3) to calculate

$$r_m := \frac{H^{(2)}_{\nu_m}(qa)}{\frac{\partial}{\partial \nu_m} H^{(1)}_{\nu_m}(qa)} \tag{9.4.15}$$

for small m, with ν_m as in (9.4.8; 2). We learn that the order of magnitude of this factor is the same for the first few m's.

If $\rho - a \gg a(qa)^{-2/3}$, then the other asymptotic form can be used for $H^{(1)}_{\nu_m}(q\rho)$,

$$\begin{aligned} H^{(1)}_{\nu_m}(q\rho) &= \sqrt{\frac{2}{\pi}}(q^2\rho^2 - \nu_m^2)^{-1/4} \\ &\cdot \exp\left\{ i \left[(q^2\rho^2 - \nu_m^2)^{1/2} - \nu_m \arccos \frac{\nu_m}{q\rho} - \frac{\pi}{4} \right] \right\}. \end{aligned} \tag{9.4.16}$$

Whether the summands of (9.4.14) decrease rapidly with m depends mainly on whether the coefficient of $i\nu_m$ in the exponent is positive. Since

$$\operatorname{Im} \nu_m = \left(\frac{qa}{2}\right)^{1/3} c_m \sqrt{\frac{2}{3}} \gg 1,$$

we see that

$$\sin \pi \nu_m \cong \frac{e^{-i\pi\nu_m}}{2i};$$

and if $|\varphi - \pi| \gg (qa)^{-1/3}$, then also

$$\cos(\varphi - \pi)\nu_m \cong \tfrac{1}{2} e^{-i\nu_m|\varphi - \pi|}.$$

All together, there is an overall factor (up to terms $O((qa)^{-1/2})$) of

$$\exp\left\{ i\nu_m \left[\frac{\pi}{2} - |\varphi - \pi| - \arccos \frac{a}{\rho} \right] \right\},$$

making the condition for the first term of (9.4.14) to dominate that

$$|\varphi - \pi| + \arccos \frac{a}{\rho} - \frac{\pi}{2} < -(qa)^{-1/3}. \tag{9.4.17}$$

This amounts to the geometric condition that there is a shadow with a boundary fuzzy within an angle $\sim (qa)^{-1/3}$ (Figure 9.16). With the numerical value of c_1, the resulting expression for u, up to corrections $\sim (qa)^{-1/3}$, is

$$\begin{aligned} u &\cong \sqrt{\frac{\pi}{2q}} \exp\left(\frac{i\pi}{4}\right) r_1 \frac{\exp(iq\sqrt{\rho^2 - a^2})}{(\rho^2 - a^2)^{1/4}} \\ &\cdot \left[\exp(iqa - 1.5(qa)^{1/3}) \left(\varphi - \frac{\pi}{2} - \arccos \frac{a}{\rho} \right) \right. \\ &\left. + \exp(iqa - 1.5(qa)^{1/3}) \left(-\varphi + \frac{3\pi}{2} - \arccos \frac{a}{\rho} \right) \right]. \end{aligned} \tag{9.4.18}$$

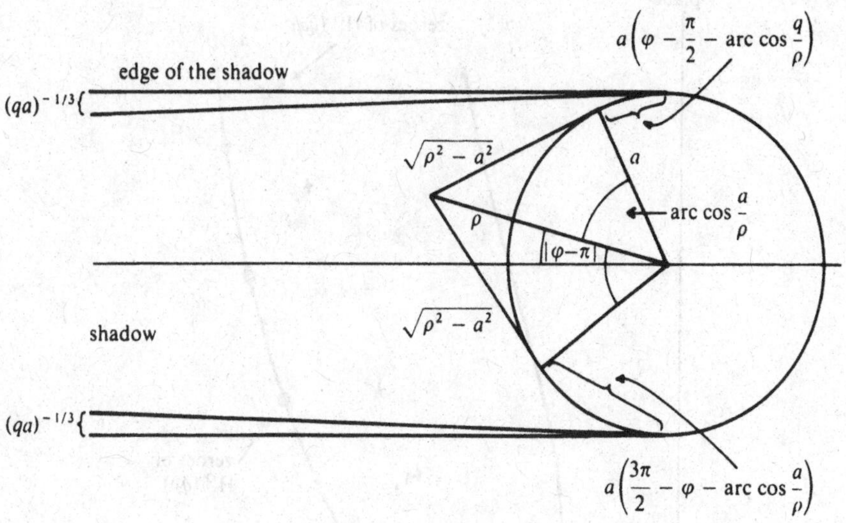

FIGURE 9.16. The shadow according to geometric optics.

The interpretation of this is that waves impinging on $(x, y) = (0, \pm a)$ are initially attenuated as they propagate along the surface to the points from which they can emanate in a straight line to the reference position. The damping factor of such "creeping waves" is correlated with the fuzziness of the shadow. During the linear propagation through the distance $\sqrt{\rho^2 - a^2}$ there is the usual amplitude factor $(\rho^2 - a^2)^{-1/4}$ for cylindrical waves, along with the phase factor $\exp(iq\sqrt{\rho^2 - a^2})$.

(9.4.19) **The Illuminated Region at $|\varphi| > \pi/2$**

Since (9.4.14) is no good outside the shadow, we rewrite the $\cos \nu(\varphi - \pi)$ of (9.4.6) as $\exp(i\nu\pi)\cos\nu\varphi - i\exp(i\nu\varphi)\sin\nu\pi$. Taking the first term of this expression changes (9.4.17) to

(9.4.20) $$|\varphi| + \arccos\frac{a}{\rho} \le \frac{3\pi}{2} - (qa)^{-1/3},$$

and this is always true except exactly in the forward direction. Hence this part produces only the damped waves that manage to slip by the cylinder, and can be neglected in comparison with the remainder, u_2. The $\sin \nu\pi$ cancels out of u_2, so the integration contour can be deformed at will across the real axis, and the integral can be calculated by the saddle-point method. We start by extending the integral

(9.4.21) $$u_2 := -\frac{1}{2}\int d\nu\, e^{i\nu(\varphi-\pi/2)} \frac{H_\nu^{(2)}(q\rho)H_\nu^{(1)}(qa) - H_\nu^{(2)}(qa)H_\nu^{(1)}(q\rho)}{H_\nu^{(1)}(qa)}$$

along a contour D' that passes around the curve $\text{Re}\,\alpha = 0$ so as to stay to the right of $\text{Re}(\alpha + i\pi\nu) = 0$ ($\alpha = \sqrt{\nu^2 - (qa)^2} - \nu\ln(\nu + \sqrt{\nu^2 - (qa)^2})/qa$), and which

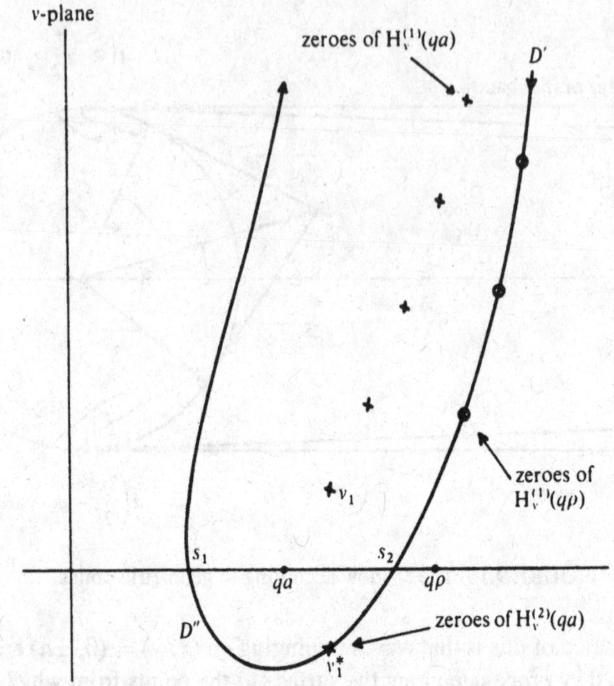

FIGURE 9.17. Path of integration for calculating geometric optics.

then follows a curve containing the zeros of $H_\nu^{(1)}(q\rho)$:

$$\text{(9.4.22)} \qquad \operatorname{Re}\left[\sqrt{\nu^2-(q\rho)^2}-\nu\ln\nu+\frac{\sqrt{\nu^2-(q\rho)^2}}{q\rho}\right]=0.$$

According to (9.4.7), in this region $H_\nu^{(2)}(q\rho)$ decreases exponentially, and the first term in the numerator of (9.4.21) contributes nothing, because it is an entire function in ν. As for the second term, on the left-side portion of D' the factor $H_\nu^{(2)}(qa)$ is exponentially decreasing, whereas on the right-side portion the variable can follow a path through the zeros of $H_\nu^{(1)}(q\rho)$, which are on (9.4.22). On this portion of the contour the factor $\exp(i\nu(\varphi-\pi/2))$ takes care of the exponential decrease, and this part of the contour can be connected to a path D'' that crosses the real axis at a saddle point $s_1 < qa$, proceeds to the first zero ν^* of $H_\nu^{(2)}(qa)$, and then climbs over another saddle point $s_2 > qa$ to reach the valley containing the zeros of $H_\nu^{(1)}(q\rho)$. We next determine s_i in the case $|s_i - qa| > (qa)^{1/3}$; the other case will be considered in (9.4.34). The integrand of u_2 on the left-side portion of D'' looks like

$$\text{(9.4.23)} \qquad \frac{1}{\sqrt{2\pi}}e^{i\nu(\varphi-\pi/2)}((q\rho)^2-\nu^2)^{-1/4}e^{i\pi/4}$$

$$\cdot \exp\left\{i\left[-2(q^2a^2-\nu^2)^{1/2}+(q^2\rho^2-\nu^2)^{1/2}\right.\right.$$

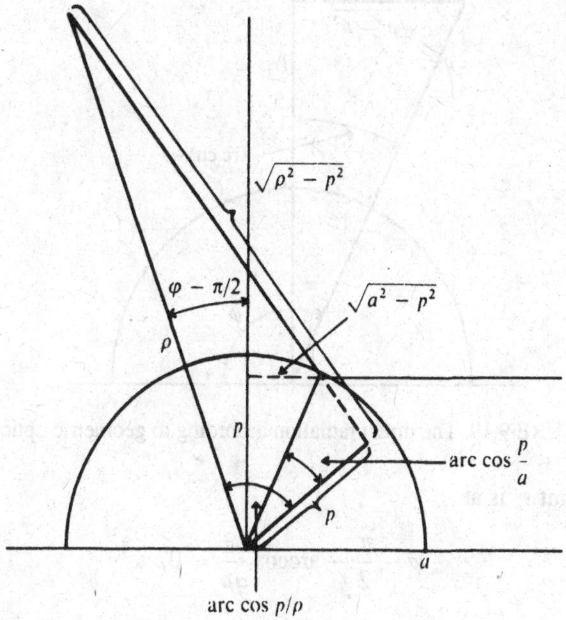

FIGURE 9.18. The reflected light according to geometric optics.

$$+ 2\nu \arccos \frac{\nu}{aq} - \nu \arccos \frac{\nu}{q\rho} \Big]\Big\},$$

because of the asymptotic properties of the Hankel functions. The ν dependence is dominated by the exponent, and the saddle point s_1 is at the position where the derivative by ν vanishes,

$$\varphi - \frac{\pi}{2} + 2 \arccos \frac{s_1}{qa} - \arccos \frac{s_1}{q\rho} = 0. \tag{9.4.24}$$

With the notation $s_1 := qp$, the integrand becomes

$$\frac{\exp(-iq\sqrt{a^2 - p^2})}{\sqrt{q}(\rho^2 - p^2)^{1/4}} \exp\left\{iq\left[\sqrt{\rho^2 - p^2} - \sqrt{a^2 - p^2}\right]\right\} \frac{e^{i\pi/4}}{\sqrt{2\pi}} \tag{9.4.25}$$

at the saddle point. We can now recognize p as the impact parameter of the ray that is reflected to the observation point according to the laws of geometric optics, and that the phase of (9.4.25) is exactly that of the optical path having this reflection (Figure 9.18). Therefore this is the contribution from reflected light.

On the right side portion of D'', $H_\nu^{(2)}(qa)/H_\nu^{(1)}(qa) \sim -1$, and the asymptotic form of the integrand simplifies to

$$-\frac{e^{-i\pi/4}}{\sqrt{2\pi}} \frac{\exp\left\{i\left(\nu\left(\varphi - \frac{\pi}{2} - \arccos \frac{\nu}{q\rho}\right) + \sqrt{\rho^2 q^2 - \nu^2}\right)\right\}}{\sqrt[4]{(q\rho)^2 - \nu^2}}. \tag{9.4.26}$$

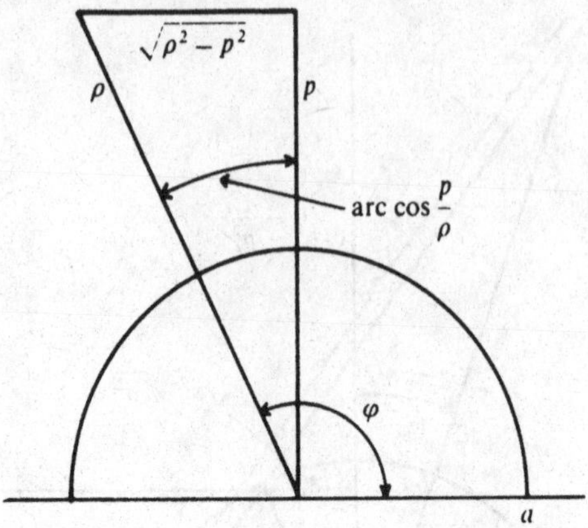

FIGURE 9.19. The direct radiation according to geometric optics.

The saddle point s_2 is at

$$\varphi - \frac{\pi}{2} - \arccos\frac{s_2}{q\rho} = 0, \tag{9.4.27}$$

and, with $p = \nu/q$, the integrand has the value

$$-\frac{e^{-i\pi/4}}{\sqrt{2\pi}}\frac{\exp(iq\sqrt{\rho^2 - p^2})}{\sqrt[4]{(q\rho)^2 - \nu^2}} \tag{9.4.28}$$

at the saddle point. Consequently p is the impact parameter of the incident ray, and the phase of (8.4.28) is simply $\exp(-iqx)$ (Figure 9.19).

To calculate the incoming wave precisely, it is necessary to go through the usual manipulations of the method of steepest descent:

$$\int d\nu\, A(\nu)e^{i\alpha(\nu)} \cong A(s_2)e^{i\alpha(s_2)}\int_{-\infty}^{\infty} d\nu \exp\left\{i\alpha''(s_2)\frac{(\nu - s_2)^2}{2}\right\} \tag{9.4.29}$$
$$= A(s_2)\sqrt{\frac{-2\pi}{i\alpha''(s_2)}}e^{i\alpha(s_2)};$$

since

$$A(s_2) = \frac{-e^{-i\pi/4}}{\sqrt{2\pi}((q\rho)^2 - s_2)^{1/4}}, \tag{9.4.30}$$
$$\alpha''(s_2) = ((q\rho)^2 - s_2)^{-1/2},$$

we find exactly $\exp(iq\sqrt{\rho^2 - p^2}) = \exp(-iqx)$. The corrections to the asymptotic expression can be calculated systematically, and are $O(\rho^{-2})$. Similarly, the contribution u_{2r} of the reflected light can be calculated accurately, to reveal that there is an amplitude factor that controls the spreading of the reflected light. In (9.4.30) the only changes in A are signs, but, because of (9.4.24),

$$\alpha''(\nu) = 2(a^2q^2 - \nu^2)^{-1/2} - (\rho^2q^2 - \nu^2)^{-1/2}. \tag{9.4.31}$$

Together with $u_2 \cong \exp(-iqx) + u_{2r}$, this makes

(9.4.32)
$$u_{2r} = \left(\frac{a^2 - p^2}{\rho^2 - p^2}\right)^{1/4} \left(2 - \left(\frac{a^2 - p^2}{\rho^2 - p^2}\right)^{1/2}\right)^{-1/2} \cdot \exp\{iq[\sqrt{\rho^2 - p^2} - 2\sqrt{a^2 - p^2}]\}.$$

If $\rho \gg a$, then $\sqrt{a^2 - p^2} = a \cos \varphi/2 =: a \sin \vartheta/2$, where $\vartheta = \pi - \varphi$ is the scattering angle. Then

(9.4.33)
$$u_{2r} \cong \sqrt{\frac{a \sin \vartheta/2}{2\rho}} \exp\left\{i\left(q\rho - 2qa \sin\frac{\vartheta}{2}\right)\right\}.$$

The angular distribution $\rho|u_{2r}|^2\, d\vartheta = dp$ is exactly what the geometric law of reflection predicts. Although our derivation works only if $\varphi > \pi/2$, the same result can be obtained for $\varphi < \pi/2$ by making minor changes, and by symmetry $u(\varphi) = u(-\varphi)$.

(9.4.34) The Boundary of the Shadow

The validity of the result of the previous paragraph relied on:

(a) $\varphi > \pi/2$, so that the exponential factor decreases in the upper half-plane;

(b) staying away from the region where $\varphi \to \pi$ and $\rho \to \infty$, in which the creeping waves are insufficiently attenuated; and

(c) $|p - a| > a(qa)^{-2/3}$, so that the saddle point is sufficiently far from qa.

At the boundary of the shadow, assumption (c) fails, and we would have to assume that ρ is not too large, in order to preserve (b). Granting that, equation (9.4.21) could still be used, although both saddle points now move toward qa. The integration contour is divided into D_1, reaching from $aq + i\infty$ to aq; D_2, from aq to $q\rho_1 < q\rho$; and D_3, from $q\rho_1$ to infinity along (9.4.22). The integral decreases exponentially fast along D_1 and $D_3(\varphi > \pi/2)$, and on D_2 the ratio $\mathrm{H}^{(2)}/\mathrm{H}^{(1)} = -1 + 2\mathrm{J}/\mathrm{H}^{(1)}$ can be replaced with -1. The other contributions are straightforward to estimate as being $O((qa)^{1/3}(\varphi - \varphi_0))$, where φ_0 is the angle such that (ρ, φ_0) is on the geometric boundary of the shadow, and we are interested in the values $|\varphi - \varphi_0| \sim (qa)^{-1/2}$.

Under these circumstances we are thus led back to

(9.4.35)
$$u_2 = \frac{e^{-i\pi/4}}{\sqrt{2\pi}} \int_{qa}^{q\rho_1} d\nu \frac{\exp\left\{i\nu\left(\varphi - \frac{\pi}{2} - \arccos\frac{\nu}{q\rho}\right) + i\sqrt{\rho^2 q^2 - \nu^2}\right\}}{[(q\rho)^2 - \nu^2]^{1/4}}$$

(cf. (9.4.26)), since the dominant part of the integral ought to come from the vicinity of qa, and we have used the proper asymptotic form of $\mathrm{H}_\nu^{(1)}(q\rho)$ for that region.

By the substitution

$$\nu := q\rho \sin\omega, \qquad \varphi_0 > \omega > \varphi_1 > \frac{\pi}{2} \tag{9.4.36}$$

($a = \rho \sin\varphi_0$, $\rho_1 = \rho \sin\varphi_1$), (9.4.35) becomes

$$u_2 = \frac{e^{-i\pi/4}}{\sqrt{2\pi}} \int_{\varphi_1}^{\varphi_0} d\omega (|q\rho \cos\omega|)^{1/2} e^{iq\rho[(\varphi-\omega)\sin\omega - \cos\omega]}. \tag{9.4.37}$$

The function in the brackets [ ] in the exponent has its minimum at $\omega = \varphi$, and its leading term as $q\rho \to \infty$ can be approximated by a parabolic function,

$$[\ ] \cong -\cos\varphi \left(1 + \frac{(\omega-\varphi)^2}{2}\right).$$

If we write

$$\rho\cos\varphi = x, \qquad -\frac{iqx(\varphi-\omega)^2}{2} = \frac{i\pi\tau^2}{2}, \tag{9.4.38}$$

then we discover that

$$u_2 = e^{-iqx} \frac{e^{-i\pi/4}}{\sqrt{2|x|q}} \int_{\sqrt{q|x|/\pi}(\varphi-\varphi_0)}^{\sqrt{q|x|/\pi}(\varphi-\varphi_1)} d\tau\, e^{i\pi\tau^2/2} \cdot \left[q|x|\cos\tau \sqrt{\frac{\pi}{q|x|}} + q\rho\sin\varphi \sin\tau \sqrt{\frac{\pi}{q|x|}}\right]^{1/2}. \tag{9.4.39}$$

As $q|x| \to \infty$, the upper limit of integration goes to infinity and $[\ ]^{1/2}$ approaches $\sqrt{q|x|}$; but unlike what we had earlier, the lower limit of integration remains finite if we are in the region $|\varphi - \varphi_0| \sim (q|x|)^{-1/2}$. With Fresnel's integral

$$F(x) = \int_{-\infty}^{x} d\tau\, e^{i\pi\tau^2/2} \tag{9.4.40}$$

we can simply write

$$u_2 = e^{-iqx} \frac{e^{-i\pi/4}}{\sqrt{2}} \left(F(\infty) - F\left((\varphi-\varphi_0)\sqrt{\frac{q|x|}{\pi}}\right)\right). \tag{9.4.41}$$

In other words, F specifies how the incident wave $\exp(-iqx)$ dies down as it passes from light ($\varphi - \varphi_0 < 0$, $F(\infty) = \sqrt{2}\exp(i\pi/4)$) to shadow ($\varphi - \varphi_0 > 0$, $u_2 \to 0$). At the geometric edge of the shadow ($\varphi = \varphi_0$) it has only half the original amplitude, because $F(0) = F(\infty)/2$, and as it passes from that point into the light, it oscillates according to Cornu's spiral around the original amplitude (cf. (9.3.10)).

(9.4.42) The Frauenhofer Region

Finally, we undertake a study of the region $\rho > (qa)a$, where the shadow has already faded out. If $|\varphi - \pi| > (qa)^{-1/3}$, then we know from (9.4.19) that the wave is simply composed of the incident wave plus a wave reflected according to geometric optics. If $|\varphi - \pi| < (qa)^{-1/3}$, then the facts discussed in (9.4.19) do not suffice to cause attenuation of the creeping waves, and we are forced to take drastic steps. We shall rely on the uniformly convergent expansion in the upper ν-plane,

(9.4.43) $$\frac{e^{-i\nu\pi/2}}{\sin \nu\pi} = -2ie^{i\nu\pi/2} \sum_{m=0}^{\infty} e^{2i\pi m\nu}.$$

Note that the condition for the contributions from the residues to decrease sufficiently fast is now

(9.4.44) $$|\varphi - \pi| + \arccos \frac{a}{\rho} < \pi \left(2m + \tfrac{1}{2}\right) - (aq)^{-1/3}$$

(cf. (9.4.17)), which is always true for $m \geq 1$, so we only need to worry about the contribution u_0 with $m = 0$ in the sum. To take care of that, we again make the partition $D' = D_1 \cup D_2 \cup D_3$ of the earlier paragraph, and see that the main contribution again comes from D_2 with $\mathrm{H}^{(2)}/\mathrm{H}^{(1)} \to -1$. We shall only calculate the main contribution, as an estimate of the others shows that they become negligible in comparison with it as $\rho \to \infty$. Choosing $\rho_1 = \rho/2$, we can write the $m = 0$ term of the integral over D_2 of the asymptotic form of the integrand (9.4.6) as

(9.4.45)
$$\begin{aligned} u_{20} = &-\frac{e^{-i\pi/4}}{\sqrt{2\pi}} \int_{qa}^{q\rho/2} \frac{d\nu}{((q\rho)^2 - \nu^2)^{1/4}} \\ &\cdot \exp\left\{i\nu\left(\vartheta + \frac{\pi}{2} - \arccos\frac{\nu}{q\rho}\right) + i\sqrt{(q\rho)^2 - \nu^2}\right\} + (\vartheta \to -\vartheta) \\ = &-e^{i(q\rho - \pi/4)} \sqrt{\frac{q\rho}{2\pi}} \int_{a/\rho}^{1/2} d\tau (1 - \tau^2)^{-1/4} e^{iq\rho\alpha(\tau)} + (\vartheta \to -\vartheta), \end{aligned}$$
$$\alpha(\tau) = \vartheta\tau + \tau \arcsin\tau + \sqrt{1 - \tau^2} - 1.$$

We have introduced the scattering angle $\vartheta = \pi - \varphi > 0$ and substituted from $\tau := \nu/q\rho$. Since the derivative of the exponent $\alpha'(\tau) = \vartheta + \arcsin\tau$ never vanishes, we may make the usual integration by parts,

(9.4.46) $$\begin{aligned} \int_{a/\rho}^{1/2} d\tau\, A(\tau) e^{iq\rho\alpha(\tau)} &= \int_{a/\rho}^{1/2} d\tau \frac{A(\tau)}{iq\rho\alpha'(\tau)} \frac{\partial}{\partial\tau} e^{iq\rho\alpha(\tau)} \\ &= \left.\frac{A(\tau) e^{iq\rho\alpha(\tau)}}{iq\rho\alpha'(\tau)}\right|_{a/\rho}^{1/2} + O((q\rho)^{-2}) \end{aligned}$$

to determine the asymptotic properties. For values $\vartheta \sim (qa)^{-1}$ but $\vartheta \gg a/\rho$, i.e., $\rho \gg aaq$, we thereby find that

(9.4.47) $$u_{20} \overset{\rightarrow}{\sim} \frac{e^{iq\rho}}{\sqrt{\rho}} e^{-i\pi/4} \sqrt{\frac{2}{\pi q} \frac{\sin qa\vartheta}{\vartheta}}.$$

This is the same as the diffraction pattern at a slit or cylinder as calculated with the older theory, in which Formula (7.2.36) is used with u in the lighted regions replaced with the incident wave. However, the older procedure gives no indication of the errors incurred, whereas the above theory is able to specify corrections to the asymptotic formulas.

If $\rho \gg aqa$, then u is of the form

(9.4.48) $$u \cong e^{-iqx} + \frac{e^{iq\rho}}{\sqrt{\rho}} f(\vartheta) =: u_{\text{in}} + u_{\text{scat}}.$$

If the wave is normally incident ($k = 0$ in (9.2.19; 3) and $q = \omega$), f determines the

(9.4.49) Scattering Cross-Section

$$\sigma(\vartheta) := \lim_{\rho\to\infty} \rho \frac{(T^{0\rho})_{\text{scat}}}{(T^{00})_{\text{in}}} = |f(\vartheta)|^2,$$

where the energy-momentum tensor is to be calculated respectively with u_{scat} or u_{in}.

(9.4.50) Remarks

1. The cross-section σ is the radiated energy per unit angle divided by the incident energy per unit length. It has the dimensions of a length. We are using the energy current from (9.1.24) for the scalar field u (see Problem 1); the electromagnetic problem has a somewhat more complicated formulation.
2. In the limit $qa \gg 1$ and $\rho \to \infty$, we find that

$$|f(\vartheta)|^2 = \tfrac{a}{2} \sin \tfrac{\vartheta}{2} \qquad \text{if } \vartheta > (qa)^{-1/3},$$

$$|f(\vartheta)|^2 = \tfrac{2}{\pi q} \left| \tfrac{\sin qa\,\vartheta}{\vartheta} \right| \qquad \text{if } \vartheta \sim (qa)^{-1};$$

in words, in addition to the geometrically scattered light there is an extra forward maximum, originating in the waves that permeate the shadow. Interestingly enough, both components have the geometric cross-section $2a$; but there is hardly any contribution from $(qa)^{-1} \ll q < (qa)^{-1/3}$:

$$2\int_{(qa)^{-1/3}}^{\pi} d\vartheta \frac{a}{2} \sin \frac{\vartheta}{2} = 2a(1 + O((qa)^{-2/3})),$$

$$2\int_0^{(qa)^{-1/3}} d\vartheta \frac{2}{\pi q} \left| \frac{\sin qa\vartheta}{\vartheta} \right|^2 = 2a(1 + O((qa)^{-1/3})).$$

Hence the total cross-section is twice the geometric cross-section. The significance is that not only is the directly incident light reflected, but also the light passing by at a distance a is somewhat deflected.

3. In the limit $qa \ll 1$, the scattering cross-section is strongly dependent on the polarization. Until now we have always assumed that $u_{|\partial N} = 0$, for which (9.4.4; 2) makes the scattering cross-section

$$\sigma(\vartheta) = \frac{\pi a}{2qa|\ln qa|^2}.$$

This can exceed the geometric cross-section by an arbitrary amount, even though it obviously vanishes if q is fixed and $a \to 0$.

4. If $du_{|\partial N} = 0$, then the Fourier coefficients in (9.4.2) are replaced with $J'_n(qa)/H^{(1)\prime}_n(qa)$, and if $qa \to 0$, then the terms with $n = 0, \pm 1$ contribute:

$$\sigma = \frac{\pi(qa)^3}{8} a(1 - 2\cos\vartheta)^2.$$

The total cross-section $a(3\pi^2/4)(qa)^3$ is much smaller than the geometric cross-section.

5. In electromagnetism Remark 3 applies to a field **E** parallel to the z-axis, and Remark 4 to **E** perpendicular to the z-axis. The physical reason for the greater cross-section of Remark 3 is that it is easier to move charges along the axis of the cylinder than transversely.

The example we have discussed has shown how complicated diffraction phenomena can be. Even Cartan's formalism, which makes the algebraic complications trivial, is not very effective for these analytical problems. However, the integral representation (9.4.6) is versatile enough that the whole range of diffraction phenomena can be derived from it, like rabbits pulled from a magician's hat. In general interference of waves renders the solution chaotic. Only in some limits the simple laws of geometrical optics emerge.

(9.4.51) Problems

1. Derive (9.4.49) for the energy-momentum forms (9.1.24) with

$$J_\alpha = u_{,\alpha}, \qquad u_{\text{in}} = e^{-iq(x+t)}, \qquad u_{\text{scat}} \cong \frac{e^{iq(\rho - t)}}{\sqrt{\rho}} f(\varphi).$$

2. Derive the optimal theorem

$$\sigma_{\text{tot}} = \int_{-\pi}^{\pi} d\vartheta\, \sigma(\vartheta) = -2\sqrt{\frac{2\pi}{q}} Re(e^{i\pi/4} f(0)),$$

where f, as in (9.4.48) is defined by the asymptotic form of

$$u = \sum_{n=-\infty}^{\infty} e^{in((\pi/2)-\vartheta)} \left[J_n(q\rho) - \frac{J_n(qa)}{H^{(1)}_n(qa)} H^{(1)}_n(q\rho) \right] \overset{\rho\to\infty}{\longrightarrow} e^{-iqx} + f(\vartheta)\frac{e^{iq\rho}}{\sqrt{\rho}}.$$

Use the formula

$$\frac{J_n}{H_n^{(1)}} = \frac{1}{2}\frac{H_n^{(1)} + H_n^{(2)}}{H_n^{(1)}} =: \tfrac{1}{2}(1 - e^{2i\delta_n}), \qquad \delta_n \text{ real.}$$

3. Calculate the scattering cross-sections given in (9.4.50; 3) and (9.4.50; 4).

(9.4.52) Solutions

1.

$$\frac{(T^{0\rho})}{(T^{00})} = \frac{\dot{u}u_{,\rho}}{\frac{1}{2}(\dot{u}^2 + |\nabla u|^2)}, \qquad \frac{(T^{0\rho})_{\text{scat}}}{(T^{00})_{\text{in}}} = \rho\frac{q^2|f(\varphi)|^2/\rho}{q^2} = |f(\varphi)|^2.$$

Since $|u(t)|^2 = (\text{Re}\, u(t))^2 + (\text{Im}\, u(t))^2 = (\text{Re}\, u(t))^2 + (\text{Re}\, u(t + \pi/2\omega))^2$, the use of the absolute value is equivalent to a time-average.

2.

$$\frac{f(\vartheta)}{\sqrt{\rho}}e^{iq\rho} = \frac{1}{2}\sum_{n=-\infty}^{\infty} e^{in((\pi/2)-\vartheta)}(e^{2i\delta_n} - 1)H_n^{(1)}(q\rho).$$

From the asymptotic expression

$$H_n^{(1)}(q\rho) \cong \sqrt{\frac{2}{\pi q\rho}}e^{i(q\rho-(n\pi/2)-(\pi/4))}$$

one finds that

$$f(\vartheta) = \frac{e^{-i\pi/4}}{\sqrt{2\pi q}}\sum_{n=-\infty}^{\infty} e^{-in\vartheta}(e^{2i\delta_n} - 1), \qquad \sigma = \int_{-\pi}^{\pi}|f(\vartheta)|d\vartheta = \frac{4}{q}\sum_{n=-\infty}^{\infty}\sin^2\delta_n,$$

$$\text{Re}(e^{i\pi/4}f(0)) = \frac{-1}{\sqrt{2\pi q}}\sum_{n=-\infty}^{\infty}(1 - \cos^2\delta_n + \sin^2\delta_n).$$

3. The cross-section of Remark 3 follows from (9.4.4; 2). As for Remark 4,

$$u = \sum_{n=-\infty}^{\infty} e^{in((\pi/2)-\vartheta)}\left(J_n(q\rho) - \frac{J_n'(qa)}{H_n^{(1)\prime}(qa)}H_n^{(1)}(q\rho)\right),$$

$$f(\vartheta) = \sqrt{\frac{2}{\pi q}}e^{-i\pi/4}\left(\frac{J_0'(qa)}{H_0^{(1)\prime}(qa)} + 2\cos\vartheta\frac{J_1'(qa)}{H_1^{(1)\prime}(qa)} + \cdots\right)$$

$$\cong \sqrt{\frac{2}{\pi q}}e^{-i\pi/4}\left(\frac{qa}{2}\right)^2(1 - 2\cos\vartheta)i\pi.$$

10

Gravitation

10.1 Covariant Differentiation and the Curvature of Space

The covariant derivative defines the rate of change of a tensor field in the direction of a vector. Covariant derivatives in two different directions do not in general commute; their commutator determines the curvature of space.

In field theory one has to deal with derivatives of vector fields, and in modern theories there appear quantities which are vectors not in space–time but in an internal space. In both cases one deals with vector bundles where vectors at different points are not canonically oriented toward each other. A chart-independent notion of a derivative requires an additional structure, the so-called connection. It will be the subject of this chapter. As one hopes that eventually space, time, and internal space will turn out to be only different directions in a unifying entity we start with some definitions which allow us to treat both cases in the same way.

(10.1.1) Definition

A **section** of a vector bundle V with basis B and projection Π is a map $\Phi: B \to V$ with $\Pi \circ \Phi = 1_B$. The set of sections is denoted by $S_0(V)$.

(10.1.2) Examples

1. The charged scalar field (9.1.21; 2) Φ has two components φ^1, φ^2 and can be considered as a section of the bundle $B \times \mathbb{R}^2$ where B is space–time. It associates

to $x \in B$ the point $(x; b_1\varphi^1(x) + b_2\varphi^2(x)) \in V$ if $b_{1,2}$ are a basis of $\mathbb{R}^2$, which is the fiber in this case.

2. Besides the charged pions $\pi^\pm$ there exists the neutral pion π^0 with about the same mass. Their description requires a 3-component pseudoscalar field φ^i, $i = 1, 2, 3$, where $\varphi^1 \pm i\varphi^2$ describe $\pi^\pm$ and φ^3 describes π^0. Thus the pion field is a section in a vector bundle with fiber $\mathbb{R}^3$, the so-called isospin space. There are many examples of similar internal degrees of freedom in particle physics.
3. Vector fields (resp. 1-forms) are sections in the bundles $T(B)$ (resp. $T^*(B)$), $S_0(T^*(B)) = T_1^0(B)$, etc. In our case B will be space–time, $\dim B = 4$. Thus the fibers in $T(B)$ will always be $\mathbb{R}^4$. Similarly tensor fields are sections of the tensor bundles.

We shall encounter only situations where, in addition to their vector-space structure, the fibers have another important property which deserves a name.

(10.1.3) **Definition**

A nondegenerate bilinear map (or sequilinear form for complex bundles) $S_0(V) \times S_0(V) \to C(B)$: $(\Phi, \Psi) \to \langle\Phi|\Psi\rangle$ is called a **fiber metric**.

(10.1.4) **Remark**

$S_0(V)$ is a module over $C(B)$ and bilinearity implies $\langle f\Phi|g\Psi\rangle = fg\langle\Phi|\Psi\rangle$ $\forall f, g \in C(B)$. By nondegenerate we mean $\langle\Phi|\Psi\rangle = 0 \; \forall\Phi \in S_0 \Rightarrow \Psi = 0$ at every point of B.

(10.1.5) **Examples**

1. In the Lagrangian for the charged scalar field or the pion field the combination $\sum_i \varphi^{i2}(x)$ appears. It corresponds to the usual metric in $\mathbb{R}^n$ and will be taken as the fiber metric for this bundle.
2. On a Riemannian space M the metric defines a fiber metric in the bundle $T(M)$. We have seen how it can be extended to $T^*(M)$ and the tensor bundles.
3. On any manifold M the canonical 2-form on $T^*(M)$ defines a fiber metric in the bundle $T(T^*(M))$. Note that the fiber metric need not be positive.

Every element of a vector space can be expanded in terms of a basis $\{b_i\}$, but for a vector bundle a basis for the sections may exist only locally, and not globally. In any case the metric is locally determined by its action on a basis, and for a symmetric metric it is expedient to use a basis where the metric assumes the normal form

$$\langle b_i|b_j\rangle = \eta_{ij} \qquad \text{with} \quad \eta_{ij} = \begin{cases} \pm 1 & \text{for } i = j, \\ 0 & \text{for } i \neq j \end{cases}$$

("orthonormal" or simply "orthogonal basis"). This requirement does not fix the b_i uniquely: $b_i = b_k L^k{}_i$ is also an orthogonal basis if $L^t\eta L = \eta$. Even though the η_{ik} do not depend on $x \in B$, the L may do so.

(10.1.6) **Definition**

Let $\{b_i\}$ be a local basis of sections in a vector bundle with fiber metric $\langle b_i | b_j \rangle = \eta_{ij}$. The transformations $b_i \to b_k L^k{}_i$ with $L^t \eta L = \eta$ are called **gauge transformations**. The $L^k{}_i(x)$ form the same group for all $x \in B$, the so-called **gauge group** G of the bundle. If the bundle is trivializable and $L: B \to G$ is a constant map one speaks of global gauge transformations.

(10.1.7) **Examples**

1. For the fields $\Phi = \sum b_i \varphi^i$ with fiber metric $\langle \Phi | \Phi \rangle = \sum_{i=1}^n \varphi^{i^2}$ the gauge group G is $O(n)$.
2. In a pseudo-Riemannian space with signature (n, m) (η has n positive and m negative eigenvalues) the gauge group is $O(n, m)$.
3. In $T(T^*(M))$ with the canonical 2-form as fiber metric the gauge group is the symplectic group.

(10.1.8) **Remark**

There are groups which cannot be characterized by the invariance of a quadratic form. To treat them as a gauge group one has to ascend to the next level of abstraction and define principal fiber bundles where the fibers are the group manifold itself. In our cases the gauge group will always come from the invariance of a fiber metric and we shall not need this construction.

For an exterior derivative of a section we still need a notion which combines sections and p-forms.

(10.1.9) **Definition**

The sections over the vector bundle $B \times V \otimes_\pi \wedge_p T^*(B)$ are called **p-form valued sections** (or vector-valued p-forms), their set is denoted by S_p.[1] They are a linear space, and the exterior product $\wedge$ maps $E_p \times S_q$ into S_{p+q}. Thus $\Phi \in S_p$ can be written $\Phi = \sum_i b_i \nu^i$ with $\nu^i \in E_p$ and b_i the basis in S_0. Given a vector field $u \in T_0^1(B)$ we define an inner product $i_u: S_p \to S_{p-1}$ by

$$i_u \Phi = \sum_j b_j i_u \nu^j.$$

So far there is no chart-independent notion of parallelism between fibers at different points. To define the derivative of a section Φ we need a prescription how to parallel transport $\Phi(x)$ to $x + dx$ such that $\Phi(x + dx) - \Phi(x)$ can be defined. This prescription will be expressed on the infinitesimal level by the connection. We shall wonder later whether and how this parallelism can be extended to the

[1] $\otimes_\pi$ denotes a product of bundles with the same basis. In it the fibers are the tensor product of individual fibers and the basis is the common basis. $\wedge_p\, T^*(B)$ is the p-fold antisymmetric product of the $T^*(B)$.

local or global levels. First we list some formal desiderata for the derivative and then successively narrow it down by requiring that it conserve further structures. To start with we want a derivative D_u of a section in the direction of a vector field u to be linear in u so that one ought to be able to write it $i_u D$, $D: S_0 \to S_1$; hence we concentrate on D.

(10.1.10) **Definition**

A **covariant exterior derivative** D is a map $S_p \to S_{p+1}$ with the properties:

(i) $D(\Phi_1 + \Phi_2) = D\Phi_1 + D\Phi_2$, $\Phi_i \in S_p$; and

(ii) $D(\Phi \wedge \nu) = (D\Phi) \wedge \nu + (-1)^p \Phi \wedge d\nu$, $\Phi \in S_p$, $\nu \in E_q$.

The 1-forms $\omega^k{}_i$ appearing in the expansion of the covariant exterior derivative of a local basis $b_i \in S_0$,

$$Db_i = b_k \omega^k{}_i, \qquad \omega^k{}_i \in E_1,$$

are called the **linear connection** or **gauge potential**.

(10.1.11) **Remarks**

1. Through the rules (i) and (ii) D is completely determined by the ω's. For instance, with the natural basis dx^α in E_1 we can write

$$\Phi = \frac{1}{p!} \sum_{i,(\alpha)} b_i \varphi^{i(\alpha)} dx^{\alpha_1} \wedge \cdots \wedge dx^{\alpha_p}$$

and with the rules (10.1.10) $D\Phi$ becomes

$$D\Phi = \frac{1}{p!} \sum_{i,(\alpha)} b_i (d\varphi^{i(\alpha)} + \omega^i{}_k \varphi^{k(\alpha)}) \wedge dx^{\alpha_1} \wedge \cdots \wedge dx^{\alpha_p}.$$

2. Since the b_i need not be defined globally, the ω will not be either, even though D exists globally.
3. ω can be considered as an element of $S_1(L(F))$, a 1-form valued section of the bundle over the domain of the b_i with the linear transformations of the fibers F as fibers. In this sense one might use the shorthand $D = d + \omega \wedge$. However, one has to keep in mind that under a change of the basis in F the ω's do not transform the way elements of $S_1(L(F))$ do.

(10.1.12) **Transformation of the Gauge Potential Under Gauge Transformations**

Under a change of the local basis $b \to bL$, $L \in S_0(L(F))$ the ω transform as

$$\omega \to L^{-1}\omega L + L^{-1}\, dL.$$

(10.1.13) **Remarks**

1. $dL \in S_1(L(F))$ is to be understood as the matrix with elements $dL^i{}_k$ if the change of basis is $b_k \to \bar{b}_k = b_i L^i{}_k$. Thus, with index notation

$$\bar{\omega}^i{}_k = (L^{-1})^i{}_j \omega^j{}_l L^l{}_k + (L^{-1})^i{}_j \, dL^j{}_k.$$

2. The transformation law follows from the rules (10.1.10):

$$\bar{b} = bL \quad \Rightarrow \quad D\bar{b} = (Db)L + b\,dL = b\omega L + b\,dL = \bar{b}(L^{-1}\omega L + L^{-1}\,dL).$$

3. Looked at from the point of view of the components φ^i of $\Phi = \sum b_i \varphi^i \in S_0$ one can say that the components $d\varphi^i + \omega^i{}_k \varphi^k$ of $D\Phi$ transform like φ^i under the gauge transformation $\varphi^i \to \bar{\varphi}^i = (L^{-1})^i{}_k \varphi^k$. The inhomogeneous term $(dL^{-1})\Phi$ is cancelled by the corresponding term in the transformed ω since $L^{-1}L = 1 \Rightarrow dL^{-1}L = -L^{-1}\,dL$.

(10.1.14) **Examples**

1. It may happen that some basis b_i is covariantly constant, $\omega^i{}_k = 0$. In another basis $\bar{b}_k = b_i L^i{}_k$ we have then

$$D\bar{b}_k = \bar{b}_j (L^{-1})^j{}_i \, dL^i{}_k.$$

If there is no global basis it is therefore not clear whether there is a D with $\omega = 0$ on all of B.

2. The model (9.1.21; 2) of a charged field can be written in suitable terms of two real fields $\varphi^1 = \sqrt{\rho}\cos S$, $\varphi^2 = \sqrt{\rho}\sin S$ or one complex field $\Phi = \varphi^1 + i\varphi^2 = \sqrt{\rho}e^{iS}$. To get a vector bundle we have to allow ρ to vary over $\mathbb{R}^+$. In this case we have the fibers F equal to $\mathbb{R}^2$ or C and the bundle $V = \mathbb{R}^4 \times F$. In the Lagrangian (9.1.20) occurred the combination $d\varphi + iA\varphi$ (or $d\varphi^i + \omega^i{}_k \varphi^k$, $\omega^i{}_k = \begin{pmatrix} 0 & 1 \\ -1 & 0 \end{pmatrix} A(x)$), which is a covariant exterior derivative in this bundle. $S \to S + \Lambda$ corresponds to a change of basis in which case A undergoes the gauge transformations $A \to A + d\Lambda$. Here the gauge transformation and the connection involve only the matrix $\begin{pmatrix} 0 & 1 \\ -1 & 0 \end{pmatrix}$ and therefore the L cancel out in $L^{-1}\omega L$.

So far our motivation for D was more formal. Its geometrical significance becomes more apparent when one defines the covariant derivative in the direction of a vector field over B.

(10.1.15) **Definition**

Given a vector field $u \in \mathcal{T}^1_0(B)$ the **covariant Lie derivative** is defined by

$$\mathcal{L}_u = i_u \circ D + D \circ i_u.$$

It maps S_p into S_p. The particular case $p = 0$ is called **covariant derivative** and is denoted by $D_u := i_u \circ D$.

(10.1.16) **Remarks**

1. Recall how the inner product i_u was defined. The quantity can be expressed more explicitly as $\varphi = \sum_i b_i \varphi^i \in S_0$

$$\mathcal{L}_u \Phi = D_u \Phi = \sum_i b_i((d\varphi^i|u) + (\omega^i{}_k|u)\varphi^k).$$

2. For $\mathcal{L}_u$, as for L_u, the Leibniz rule $\mathcal{L}_u(\nu \wedge \Phi) = (\mathcal{L}_u \nu) \wedge \Phi + \nu \wedge \mathcal{L}_u \Phi$ holds. Generally $\mathcal{L}_{fu} \neq f\mathcal{L}_u$, except that for D_u we have $\forall f \in E_0$: $D_{fu} = f D_u$.
3. Following the construction of the tensor bundles $T_s^r(M)$ out of $T(M)$ one can associate to V its dual bundle V^* and the tensor products

$$V_s^r = \overbrace{V \underset{\pi}{\otimes} V \underset{\pi}{\otimes} \cdots}^{r} \overbrace{V^* \underset{\pi}{\otimes} V^* \underset{\pi}{\otimes} \cdots V^*}^{s}.$$

The differential processes D, and thus $\mathcal{L}_u$, can be carried over to the sections $S_p(V_s^r)$ by postulating the Leibniz rule for $p = 0$,

$$D(\Phi_1 \otimes \Phi_2) = (D\Phi_1) \otimes \Phi_2 + \Phi_1 \otimes D\Phi_2 \qquad \text{for} \quad \Phi_i \in S_0(V),$$

and the invariance of the map

$$(S_0(V^*) \times S_0(V)) \to E_0(B):$$
$$D(\Psi|\Phi) = (D\Psi|\Phi) + (\Psi|D\Phi), \qquad \Psi \in S_0(V^*), \qquad \Phi \in S_0(V).$$

$D(\Psi|\Phi) = d(\Psi|\Phi)$, because $D = d$ on the scalars. In particular, for the dual basis b^{*j}, $(b^{*j}|b_k) = \delta^j{}_k$ we infer from $d\delta^j{}_k = 0$ that $Db^{*j} = -\omega^j{}_k b^{*k}$. One typical example will be $S(L(F))$ where $L(F)$ is identified with $F \otimes F^*$: If $L(F) \ni L = \sum v_i w_i$, $v_i \in F$, $w_i \in F^*$, then $L \cdot u = \sum_i v_i(w_i|u)$. Again $\mathcal{L}_u$ depends only on the direction of u and not on its derivative when applied to $S_0(V_s^r)$, where it is denoted by D_u.

(10.1.17) **Geometrical Signficance of the Covariant Derivative**

As a map $B \to V$, a section $\Phi \in S_0(V)$ also maps $T(B) \xrightarrow{T(\Phi)} T(V)$. With reference to a basis this can be written: $x \to (x, \varphi^i(x))$,

$$T(\Phi): (x, v) \to \left(x, \varphi^i(x); v, v^\alpha \frac{\partial \varphi^i}{\partial x^\alpha}\right).$$

If we identify F and $T(F)$ this looks schematically as follows:

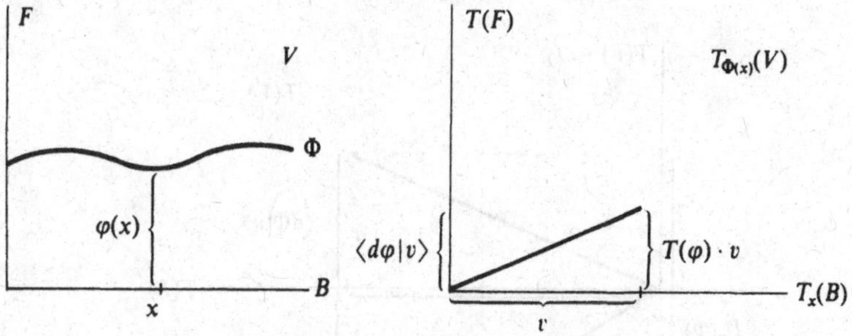

What we now want for $D_v(\Phi)$ is the change of the vertical part of Φ (in the direction of F) as one goes in the direction v of B. However, there is no preferred horizontal direction in $T(V)$—only the vertical direction is defined by $T(\Pi)$. For the components in the vertical direction we need a preferred horizontal direction. The latter is defined by the 1-forms $-\omega^i{}_k$ which can be considered as map[2]

$$(x, v) \to (x, \varphi^i(x); v, -(\omega^i{}_k|v)\varphi^k).$$

As the difference between the two maps $D_v(\Phi)$ measures the change $T(\Phi)v$ relative to the horizontal direction given by ω (see the following illustration). Thus $D_v\Phi$ should be looked at as the change of $\Phi_x \in F$ as one proceeds in the direction of v. It comes about because the components change by $(d\varphi^i|v)$ and the basis is rotated by $(\omega^i{}_k|v)$.

A section whose vertical component does not change will be one with $D_v\Phi = 0$ $\forall v$, that is, $D\Phi = 0$. Such a section is termed covariantly constant; equivalently, $\Phi_x \in F$ is parallel-transported as one moves in B. If one has a covariantly constant basis the ω's are zero and the covariantly constant vectors are the ones with constant components. In this case the notion of parallelism given infinitesimally by D can be extended to the domain of this basis. One can see on $T(S^2)$ that in general the extension even to the local level might be impossible. There the standard prescription for parallel transport along great circles is to keep the angle with them fixed. If one now takes a vector at the north pole and transports it to the south pole one arrives at vectors pointing in different directions if different great circles are followed. Even for points which are a small but nonzero distance apart the above description does not lead to a unique result. Obviously, a necessary condition for $D\Phi = 0$ is $DD\Phi = 0$. Unlike d the square of D is not identically zero. It turns out that $D \circ D$ does not depend on the derivatives of the section one applies it to; rather, our rules (10.1.8) say

(10.1.18)

$$DD(f \wedge \Phi) = D(df \wedge \Phi + (-1)^q f \wedge D\Phi)$$

---

[2] Analogously to $\times_\pi$ the sum $+_\pi$ of vector bundles is defined such that the fibers are the sums of the constituent fibers and the basis is the common basis.

$$= ddf \wedge \Phi + (-1)^{q+1}\, df \wedge D\Phi + (-1)^q\, df \wedge D\Phi$$
$$+ f \wedge DD\Phi = f \wedge DD\Phi \quad \forall f \in E_q, \quad \Phi \in S_p.$$

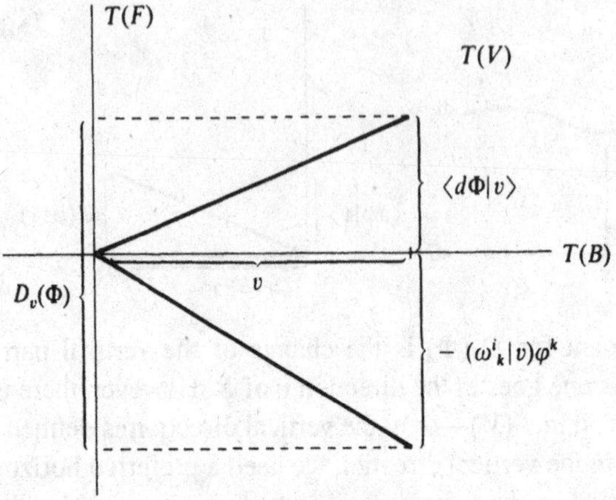

The geometrical meaning of D_v.

Thus $DD\Phi \in S_{p+2}$ can be linearly expressed by Φ and thus defines a 2-form with values in $L(F)$.

(10.1.19) Definition

The **curvature form** $\Omega \in S_2(L(F))$ is defined by

$$DD\Phi = \Omega \wedge \Phi, \qquad \Phi \in S_p.$$

(10.1.20) Remarks

1. Ω is determined by its action on a basis $\Omega b_i =: b_k \Omega^k{}_i$ since (10.1.19) shows $DDb_i\varphi^i = b_k \Omega^k{}_i \varphi^i$. Now

$$DDb_i = D(b_k\omega^k{}_i) = b_k\, d\omega^k{}_i + b_k\omega^k{}_j \wedge \omega^j{}_i$$

and we arrive at "Cartan's second structure equation"

$$\Omega^i{}_k = d\omega^i{}_k + \omega^i{}_j \wedge \omega^j{}_k.$$

(In the tangent bundle of space–time we shall use the letter R for Ω.) Considering $\omega \in S_1(L(F))$ one may write this more compactly as $\Omega = d\omega + \omega \wedge \omega$. Since matrix multiplication is not commutative, $\omega \wedge \nu \neq -\nu \wedge \omega$ for matrix-valued 1-forms, and $\omega \wedge \omega$ need not vanish.

2. For the connection (10.1.13) in $V = \mathbb{R}^4 \times \mathbb{R}^2$ with the Abelian gauge group $\mathcal{A}(2)$ the term $(\omega \wedge \omega)^i{}_k = A \wedge A \varepsilon^i{}_j \varepsilon^j{}_k$ vanishes since $A \wedge A = 0$ and $\Omega^i{}_k = dA\varepsilon^i{}_k = F\varepsilon^i{}_k \left(\varepsilon^i{}_k = \begin{pmatrix} 0 & 1 \\ -1 & 0 \end{pmatrix}\right)$. Thus the electromagnetic field $F = dA$ is the curvature of this connection.

3. Because of (10.1.19), with a change of basis $b_i \to \bar{b}_i = b_k L^k{}_i$,

$$DD\bar{b}_i = b_j \Omega^j{}_k L^k{}_i = \bar{b}_k \Omega^k{}_i,$$

which shows that

$$\bar{\Omega}^i{}_k = (L^{-1})^i{}_j \Omega^j{}_m L^m{}_k.$$

Thus Ω, in contradistinction to ω, transforms under a change of basis as an element of $S_2(L(F))$ ought to transform. One can verify (Problem 4) that the inhomogeneous terms in the transformation law (10.1.12) do cancel out in the combination $d\omega + \omega \wedge \omega$.

4. To see the significance of Ω for the directional derivative we have to use the identity

(10.1.21) $$(i_v i_u DD)\Phi = i_u D i_v D\Phi - i_v D i_u D\Phi - D_{[u,v]}\Phi$$

for $\Phi \in S_0$ (Problem 7). It tells us that if we have two vector fields with $[u, v] = 0$, so that they form two-dimensional surfaces, then $(D_u D_v - D_v D_u)\Phi = (i_v i_u \Omega)\Phi$. We see that Ω measures the difference between the covariant changes of Φ by first proceeding in the direction of u and then in the direction of v and secondly by coming to the same (infinitesimally adjacent) point by proceeding in the opposite order.

We return to the question of covariantly constant sections. $D\Phi = 0$ means that for the components of Φ,

(10.1.22) $$d\varphi^i + \omega^i{}_k \varphi^k = 0.$$

This implies

(10.1.23) $$0 = d(\omega^i{}_k \varphi^k) = (d\omega^i{}_k + \omega^i{}_j \wedge \omega^i{}_k)\varphi^k = \Omega^i{}_k \varphi^k.$$

Thus a covariantly constant section has to be an eigenvector of the matrix $\Omega^i{}_k(x)$ with eigenvalue zero at each point $x \in B$. If we have a covariantly constant basis, then Ω has to be zero since its action on a basis gives zero. This also follows from the fact that in such a basis $\omega = 0$ and therefore Ω is zero, and Ω transforms homogeneously under a change of basis. Conversely, $\Omega = 0$ implies $d\omega^i{}_k + \omega^i{}_j \wedge \omega^j{}_k = 0$ which are the integrability conditions for the system (10.1.22). Frobenius's theorem [41, Vol. I] tells us that they are also sufficient for the local existence of solutions of (10.1.22) with arbitrary initial conditions. This situation deserves a name.

(10.1.24) **Definition**

A connection of a vector bundle is called locally **flat** if one of the two equivalent conditions hold:

(i) $\Omega = 0$; and

(ii) there exists (locally) a covariantly constant basis.

(10.1.25) Remarks

1. If the bundle is parallelizable, that is, there is a global basis, then we get a flat connection by declaring this basis as covariantly constant. However, flatness is a local property, and a flat bundle like the Möbius strip need not be a product. Even if the tangent space of a manifold is a flat product bundle the manifold need not be an open set of $\mathbb{R}^n$ but can have different global properties, as is the case for T^n.
2. Flatness is an intrinsic property of the manifold: Ω does not reproduce the curvature of a surface as imbedded in, say, $\mathbb{R}^3$. Ω vanishes for a cylinder because it could be rolled out flat on a plane without affecting its metric and thereby its intrinsic structure.

The covariant exterior derivative D can be extended in a natural way to $S_p(L(F))$ (see (10.1.16; 3) and when applied to Ω it yields

(10.1.26) Bianchi's Identity

$$D\Omega = 0.$$

(10.1.27) Remarks

1. For the components (10.1.20) of $\Omega = b_k \Omega^k{}_i b^{*i}$ we see from (10.1.16; 3) that

$$0 = D\Omega = b_r(\omega^r{}_k \Omega^k{}_j + d\Omega^r{}_j - \Omega^r{}_i \omega^i{}_j) b^{*j},$$

or

$$d\Omega^i{}_j = -\omega^i{}_k \wedge \Omega^k{}_j + \Omega^i{}_k \wedge \omega^k{}_j.$$

This relation is proved in Problem 6.
2. One might think that $\Omega = D\omega$ and thus $D^2 = 0$ on ω. However, it has to be noted that although Ω is a covariant exterior derivative of ω, in the sense that Ω transforms like $L(F)$ under a change of basis, this D is not the one used in (10.1.26) since ω does not transform like $S_1(L(F))$. Written in components as above we would get

$$(D\omega)^i{}_k = d\omega^i{}_k + 2\omega^i{}_l \wedge \omega^l{}_k,$$

which differs from $\Omega^i{}_k$ by the factor 2 in the last term.

Our general considerations so far do not constrain ω any further. We shall now impose restrictions on the covariant derivative which finally determine it uniquely from the fiber metric.

The fiber metric defines a scalar product in F and thus maps $S_p(V) \times S_0(V)$ into $E_p(B)$. We shall require that the connection conserves this structure in the sense that the derivative of the scalar product contains just the derivatives of the sections involved.

(10.1.28) **Compatibility of the Connection with the Fiber Metric**

In a vector bundle V with fiber metric $\langle | \rangle$ we shall require that

$$D\langle\Psi|\Phi\rangle = d\langle\Psi|\Phi\rangle = \langle D\Psi|\Phi\rangle + (-1)^p\langle\Psi|D\Phi\rangle, \qquad \Psi \in S_p(V), \quad \Phi \in S_0(V).$$

Such connections will be called **metric connections.**

(10.1.29) **Remark**

For the ω's this imposes the condition

$$\begin{aligned} 0 = d\eta_{ik} &= d\langle e_i|e_k\rangle = \langle De_i|e_k\rangle + \langle e_i|De_k\rangle \\ &= \omega^j{}_i\langle e_j \mid e_k\rangle + \langle e_i|e_j\rangle\omega^j{}_k = \omega^j{}_i\eta_{jk} + \omega^j{}_k\eta_{ij}. \end{aligned}$$

Thus if the η are symmetric, $\omega_{ik} := \omega^j{}_k\eta_{ij}$ has to be antisymmetric:

(10.1.30) $$\omega_{ik} = -\omega_{ki}.$$

This means that $\omega_{ik}(x)$ cannot be an arbitrary element of $L(F)$, but must belong to the Lie algebra of the gauge group. In our cases, for the inner gauge symmetries G will be an $O(n)$ or $U(n)$ group, while for the space–time symmetry we have $O(3, 1)$. For an arbitrary basis with $\langle e_i|e_k\rangle = g_{ik}$ (10.1.30) generalizes to $dg_{ik} = \omega_{ik} + \omega_{ki}$.

(10.1.31) **Examples**

1. The Lie algebra of the one-dimensional group $O(2)$ consists of the multiples of $\begin{pmatrix} 0 & 1 \\ -1 & 0 \end{pmatrix}$, and the connection (10.1.14; 2) conserves the $O(2)$-invariant fiber metric $\langle\Phi|\Phi\rangle = (\varphi^1)^2 + (\varphi^2)^2$ since $\omega^i{}_k = \omega_{ik} = -\omega_{ki}$.
2. In the tangent bundle of a Riemannian space $\eta_{ik} = \delta_{ik}$ and thus $\omega^i{}_k = -\omega^k{}_i$. For the pseudo-Riemannian space–time we have

$$\eta_{ik} = \begin{cases} -1 & \text{if } i = k = 0, \\ 1 & \text{if } i = k = 1, 2, 3, \\ 0 & \text{otherwise.} \end{cases}$$

Hence $\omega^i{}_k = -\omega^k{}_i$ for $i, k = 1, 2, 3$, $\omega^i{}_0 = \omega^0{}_i$, and $\omega^i{}_i = 0$.

Finally we shall study special properties of the tangent bundle of a Riemannian space. To indicate that we are considering this special case we shall denote the curvature in $T(M)$ by the traditional letter R. First, we use the fact that in this case the sections $S_0(T(M))$ can be naturally identified with the vector fields $\mathcal{T}_0^1(M)$. Since vector fields $\mathcal{T}_0^1(M)$ map E_1 into E_0 they also map $S_1(T(M))$ into $S_0(T(M)) \equiv \mathcal{T}_0^1(M)$. The vector-valued 1-form which produces the identity map in this way has a special name.

(10.1.32) **Definition**

If the b_i form a local basis in $\mathcal{T}_0^1(M)$ and e^i the dual basis, $(e^i|b_j) = \delta^i{}_j$, then

$$\theta := \sum_i b_i \otimes e^i \in S_1(T(M))$$

is called the **soldering form** and $T := D\theta$ the **torsion**. Metric connections with $T = 0$ are called torsion-free or Levi-Civita connections.

(10.1.33) **Remarks**

1. $D\theta = \sum b_i(de^i + \omega^i{}_k \wedge e^k)$, and for a torsion-free connection we have $de^i = -\omega^i{}_k \wedge e^k$. Unless specified otherwise we shall only consider torsion-free metric connections.
2. In Problem 11 it is verified that $T = 0$ is equivalent to $D_X Y - D_Y X = [X, Y]$, $X, Y \in \mathcal{T}_0^1$, or to $(D_X u|Y) - (D_Y u|X) = (du|X \otimes Y)$, $u \in \mathcal{T}_1^0$.

(10.1.34) **Examples**

1. In a flat Riemannian space we have $\omega^i{}_k = 0$ in the covariantly constant basis. If, in addition, $T = 0$ we see from (10.1.33; 1) that $de^i = 0$, or locally $e^i = dx^i$. Therefore $R = T = 0$ imply the local existence of a natural covariantly constant basis in E_1.
2. Consider polar coordinates in $\mathbb{R}^2 \setminus \{0\}$. The metric is $g = dr^2 + r^2\,d\varphi^2 = e^1 \otimes e^1 + e^2 \otimes e^2$ in the orthogonal basis $e^1 = dr$, $e^2 = r\,d\varphi$. From $de^1 = 0$, $de^2 = dr \wedge d\varphi$ we conclude that for a torsion-free connection which preserves g, $\omega_{12} = -\omega_{21} = -d\varphi = -e^2/r$, $\omega_{11} = \omega_{22} = 0$. Thus $R_{ik} = d\omega_{ik} + \omega_i{}^l \wedge \omega_{lk} = 0$. With a connection for which the orthogonal basis is covariantly constant we get the torsion $b_2\,de^2 = (1/r)\partial\varphi \otimes dr \wedge d\varphi$, since the dual basis is $b_1 = \partial r$, $b_2 = (1/r)\partial\varphi$. It can be verified that in this case the relation $D_X Y - D_Y X = [X, Y]$ from (10.1.31; 2) is violated. If we take $X = b_1$, $Y = b_2$ the left-hand side vanishes since the basis is covariantly constant, but $[b_1, b_2] = [\partial r, 1/r]\partial\varphi = -(1/r^2)\partial\varphi \neq 0$. This definition of the connection could be made concrete by bending a rectangular crystal into a ring so that the atoms sit at the points $r = nr_0$, $\varphi = m\varphi_0$ for $n, m \in \mathcal{N}$ (Figure 10.1). If we lived in such a crystal and defined parallel transport by the lattice of atoms in the usual way, then our space would have torsion.
3. The sphere S^2 with the metric $g = R^2(d\theta^2 + \sin^2\theta\,d\varphi^2)$. In the orthogonal basis $e^1 = R\,d\theta$, $e^2 = R\sin\theta\,d\varphi$ we have $de^1 = 0$, $de^2 = R\cos\theta\,d\theta \wedge d\varphi$, and therefore $\omega_{12} = -\omega_{21} = -\cos\theta\,d\varphi$. This gives rise to a curvature
$$R^1{}_2 = d\omega^1{}_2 = \sin\theta\,d\theta \wedge d\varphi = \frac{1}{R^2} e^1 \wedge e^2.$$
4. The Lorentz hyperboloid $H = \{(x, y, z) \in \mathbb{R}^3 : x^2 + y^2 - z^2 = R^2\}$ and g = restriction of $dx^2 + dy^2 - dz^2$ to H. In cylindrical coordinates $(x, y) =$

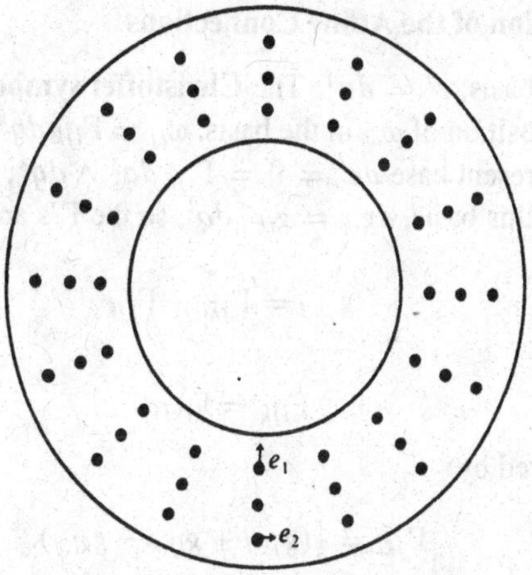

FIGURE 10.1. The bent crystal, a model of a space with torsion.

$\rho(\cos\varphi, \sin\varphi)$ we find that $e^1 = R\,d\rho/\sqrt{\rho^2 - R^2}$ and $e^2 = \rho\,d\varphi$ are an orthogonal basis on H: $g_{|H} = -e^1 \times e^1 + e^2 \times e^2$. $de^1 = 0, de^2 = d\rho \wedge d\varphi \Rightarrow \omega^1{}_2 = -\omega^2{}_1 = -\sqrt{\rho^2/R^2 - 1}\,d\varphi \Rightarrow R^1{}_2 = d\omega^1{}_2 = -\rho R^{-2}/\sqrt{\rho^2/R^2 - 1}\,d\rho \wedge d\varphi = -R^{-2} e^1 \wedge e^2$.

(10.1.35) Remark

In Example 2, those ω's imply for the covariant derivative that

$$D_X e^2 = -(e^2|X)\frac{e^1}{r}, \qquad D_X e^1 = (e^2|X)\frac{e^2}{r}.$$

If X has only the 1-component, then the $D_X e^i$ vanish. The covariant derivative of e^i in the 2-direction is a rotation, which is faster for smaller r. This accords exactly with the following intuitive picture of the rotation of the local basis in polar coordinates.

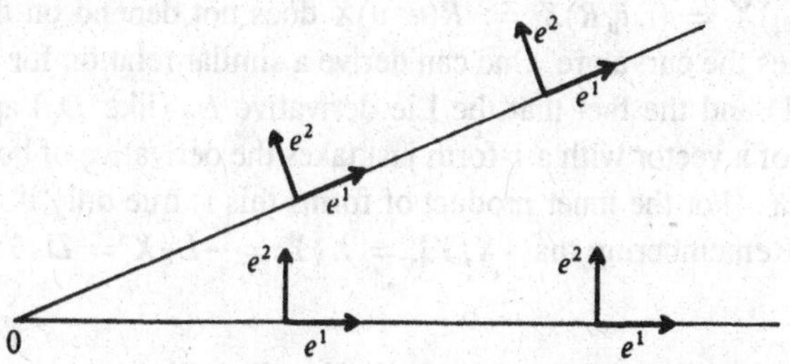

The rotating basis of polar coordinates.

We have seen that the two conditions of invariance of g under parallel transport and absence of torsion imply the relations (7.2.25). There it was stated that they determined the connection ω. We shall now verify this claim by an explicit expression for ω in terms of the derivatives of g.

(10.1.36) Calculation of the Affine Connections

(a) In the natural basis, $e^j = dq^j$: The **Christoffel symbols** $\Gamma_{ijk} \in \mathcal{T}^0_0$ occur in the decomposition of ω_{ij} in the basis, $\omega_{ij} = \Gamma_{ijk}\, dq^k$.
Since in the present case $de^i = 0 = \Gamma_{ijk}\, dq^j \wedge dq^k$, they are symmetric in j and k. In this basis, $dg_{ik} = g_{ik,l}\, dq^l$, so the Γ's are determined by the facts that

$$g_{jk,l} = \Gamma_{jkl} + \Gamma_{kjl}$$

and

$$\Gamma_{ijk} = \Gamma_{ikj},$$

which are solved by

(10.1.37) $$\Gamma_{jkl} = \tfrac{1}{2}(g_{jk,l} + g_{lj,k} - g_{kl,j}).$$

Since this solution is unique, we see that our axioms fix D uniquely (Problem 5).

(b) In an orthogonal basis: From $dg_{ik} = d\eta_{ik} = 0 = \omega_{ik} + \omega_{ki}$, together with

$$(de^j | e_k \otimes e_i) = -(\omega^j{}_l \wedge e^l | e_k \otimes e_i) = (\omega^j{}_k | e_i) - (\omega^j{}_i | e_k),$$

one can derive

(10.1.38) $$(\omega_{kj} | e_i) = \tfrac{1}{2}[(de_j | e_i \otimes e_k) + (de_k | e_j \otimes e_i) - (de_i | e_k \otimes e_j)]$$

(Problem 5).

Next we study the special features which arise because a flow in the base manifold induces in a natural way a flow in all the tensor bundles. Thus the Lie derivative of a section with respect to a vector field in B makes sense, and we can study the

(10.1.39) Commutator of the Covariant Derivative and the Lie Derivative

We noted in (10.1.21) that for vector fields u, v, X the commutator $(D_u D_v - D_v D_u - D_{[u,v]})X = (i_v i_u R)X =: R(u, v)X$ does not depend on the derivative of X but defines the curvature. One can derive a similar relation for $DL_v - L_v D$ using (10.1.21) and the fact that the Lie derivative L_v (like D_v) applied to the inner product of a vector with a 1-form just takes the derivative of both: $L_v i_u X = i_{L_v u} X + i_u L_v X$. (For the inner product of forms this is true only if v is a Killing vector field.) Remembering that $[X, Y] := L_X Y = -L_Y X = D_X Y - D_Y X$, we find

$$\begin{aligned}
\langle (DL_v - L_v D)X | u \rangle &= DL_v X - L_v D_u X + D_{[v,u]} X \\
&= D_u D_v X - D_u D_X v - D_{[u,v]} X - D_v D_u X + D_{D_u X} v \\
&= R(u, v)X + D_{D_u X} v - (R(u, X)v + D_X D_u v + D_{[u,X]} v) \\
&= R(u, v)X + R(X, u)v + D_{D_u X} v - D_X D_u v - D_{D_u X} v + D_{D_X u} v \\
&= R(X, v)u - \langle D_X Dv | u \rangle.
\end{aligned}$$

In the last step we used a cyclic property of R derived in (10.1.44; 2(c)) and $D_X D_u v = D_X i_u Dv = i_{D_X u} Dv + i_u D_X Dv = D_{D_X u} v + \langle D_X Dv | u \rangle$. Thus we see that the commutator does not contain any derivatives of X but second derivatives of v.

(10.1.40) **Commutativity of D and ***

In (7.2.16) we extended the scalar product $\langle | \rangle$ as the interior product i_v. The generalization of (10.1.28) for i_v is $D_X i_v = i_{D_X v} + i_v D_X$, as used above. In particular, we see that $D_X \varepsilon = 0$, since $0 = D_X 1 = D_X \langle \varepsilon | \varepsilon \rangle = 2 \langle D_X \varepsilon | \varepsilon \rangle$ and ε is not degenerate. Therefore $D_X i_v \varepsilon = i_{D_X v} \varepsilon$ and thus $D_X {}^* v = {}^* D_X v$. Here E_p is identified with $S_0(\wedge_p T^*(B))$ and not with $S_p(T^*(B))$. Thus D_X on E_p cannot be written as $i_X D$ where i_X is the map $E_p \to E_{p-1}$ in which sense it is to be understood here.

In a pseudo-Riemannian space vector fields define directions in the vector bundles. Thus the covariant derivative D_X distinguishes the vector fields that are consistent with it in the sense that a vector translated parallelly in its own direction from some point equals the vector at the neighboring point:

(10.1.41) **Definition**

A vector field X such that $D_X X = 0$ is called a **geodesic vector field.**

(10.1.42) **Remarks**

1. The connection with the geodesic lines: Let $z(s)$ be a streamline of a geodesic vector field X, that is, $\dot{z}(s) = X(z(s))$. Then the components of z in the natural basis satisfy the equation $\ddot{z}^i = -\Gamma^i{}_{jk} \dot{z}^j \dot{z}^k$, because with Remark (10.1.16; 3),
$$0 = D_X(X^i \partial_i) = (X^i{}_{,k} X^k + (\omega^i{}_j | \partial_k) X^j X^k) \partial_i,$$
and
$$\ddot{z}^i(s) = \frac{d}{ds} X^i(z(s)) = X^i{}_{,k} \dot{z}^k.$$
From (10.1.36(a)), $(\omega^i{}_j | \partial_k)$ is just $\Gamma^i{}_{jk}$, and we obtain exactly the geodesic equations of motion (1.1.6). The shortest lines are thus the straightest, in the sense of the tangent vector to a curve being transformed into itself under parallel transport along them.
2. The connection with Killing vector fields: Remember that for a Killing vector field X we had $L_X g = 0$, whereas $L_X X = 0$ for any vector field. In contrast, for the covariant derivative $D_X X = 0$ holds only for geodesic vector fields whereas $D_X g = 0$ holds for any vector field. In general "Killing" does not imply "geodesic" or vice versa. There is, however, the following relation: Let v be a Killing vector field (8.1.9), so that $L_v \langle X | X \rangle = 2 \langle L_v X | X \rangle$, and let X be a geodesic vector field. Then
$$L_X \langle v | X \rangle = D_X \langle v | X \rangle = \langle D_X v | X \rangle = \langle D_v X | X \rangle - \langle L_v X | X \rangle$$
$$= \tfrac{1}{2} (D_v \langle X | X \rangle - L_v \langle X | X \rangle) = 0.$$

Consequently, the component of a geodesic vector field in the direction of a Killing vector field is constant along a geodesic line. This reveals a new significance of a Killing vector field v, that it provides a constant of the motion of a particle in a gravitational field, namely the component of the four-momentum $\dot{z}$ of the particle in the direction of v.

3. If one vector field in an orthogonal basis is also natural it is geodesic. To see this, let X be this vector field and v be either X (then $\langle v|X\rangle = \pm 1$) or a member of a natural basis orthogonal to X. In any case $\langle v|X\rangle$ is constant and with (10.1.33; 2)

$$0 = L_X\langle v|X\rangle = D_X\langle v|X\rangle = \langle D_X v|X\rangle + \langle v|D_X X\rangle$$
$$= D_v \tfrac{1}{2}\langle X|X\rangle - \langle L_v X|X\rangle + \langle v|D_X X\rangle.$$

Now $\langle X|X\rangle = \pm 1$ so the first term vanishes and $L_v X = 0$ since the Lie derivatives between members of a natural basis vanish. Therefore all components of $D_X X$ are zero.

In a pseudo-Riemannian space the curvature forms are not arbitrary, but are subject to certain algebraic conditions which we now enumerate:

(10.1.43) Algebraic Identities Satisfied by the Curvature Forms

(a) $R_{ij} = -R_{ji}$; and

(b) $R_{ij} \wedge e^i = 0$.

Proof:

(a) This fact is independent of the basis. If the basis is changed by $\bar{e} = Ae$ (Problem 4), then $R \to ARA^{-1}$ and $g \to A^{t-1}gA^{-1}$, and thus $R_{ij} = g_{ik}R^k{}_j \equiv (gR)_{ij} \to (A^{t-1}gRA^{-1})_{ij}$, which preserves the antisymmetry. Fact (a) is obviously true in an orthogonal basis, since $\omega_{ij} = -\omega_{ji} \Rightarrow \omega_{ik} \wedge \omega^k{}_j = -\omega_{kj} \wedge \omega_i{}^k = -\omega_{jk} \wedge \omega^k{}_i$.

(b) $0 = d\,de^i = -d(\omega^i{}_k \wedge e^k) = -A^i k \wedge e^k$. □

(10.1.44) Consequences

1. The m 3-forms (b) we get from the $\binom{m}{2}$ 2-forms vanish. In the absence of other algebraic conditions there are consequently

$$\binom{m}{2}^2 - m\binom{m}{3} = \frac{m^2(m^2-1)}{12}$$

independent components of the curvature. In one dimension, there is no curvature 2-form, in two dimensions there is one, and in four dimensions there are twenty. However, many of the components can be made to vanish by suitable choices of bases and coordinate systems. Let us take an orthogonal basis, to fix the g_{ik} as η_{ik}, and try to find how many invariants can be constructed from the $m^2(m^2-1)/12$ components of R_{ik} and the m^2 components of the e^i. There are m^2 functions $\partial\bar{x}^i/\partial x^j$ available from a possible change of coordinate system and $m(m-1)/2$ components of A (since $A\eta A^t = \eta$), with which to make $m^2+m(m-1)/2$ of the total of $m^2+m^2(m^2-1)/12$ components vanish. There remain $m^2(m^2-1)/12-m(m-1)/2 = m(m-1)(m-2)(m+3)/12$ invariants. This simple argument does not work in two dimensions, however. Moreover, we have only imposed the algebraic conditions, and there may be some differential dependence among the invariants as well.

2. If $R^i{}_j$ are decomposed in a basis $R^i{}_j = \frac{1}{2}R^i{}_{jkm}e^{km}$, then the **Riemann–Christoffel tensor** $R_{ijkm} = g_{in}R^n{}_{jkm}$ obeys the equations (Problem 8):

   (a) $R_{ijkm} = -R_{ijmk}$;

   (b) $R_{ijkm} = -R_{jikm}$;

   (c) $R_{ijkm} + R_{ikmj} + R_{imjk} = 0$; and

   (d) $R_{ijkm} = R_{kmij}$.

3. If R_{jk} is written in terms of the contractions $i_{e_j}R^j{}_k = R^j{}_{kjm}e^m =: R_k \in E_1$ and $i_{e_k}R^k = R^{jk}{}_{jk} =: R \in E_0$ in such a way that the contractions of the remainders vanish, then those remainders define the **Weyl forms** C_{jk}:

$$R_{jk} := \frac{-R}{(m-2)(m-1)}e_j \wedge e_k + \frac{i}{m-2}(e_j \wedge R_k - e_k \wedge R_j) + C_{jk}.$$

It is clear that the C_{ij} also satisfy (10.1.23), and by construction $i_{e_j}C^j{}_k =: C_k = 0$ (Problem 9). Because $i_{e_j}R_k = i_{e_k}R_j$, the equations $C_k = 0$ pose only $m(m+1)/2$ independent conditions, which leaves $m^2(m^2-1)/12-m(m+1)/2 = m(m+1)(m+2)(m-3)/12$ components for the C_{ij}. In three dimensions all C_{ij} are zero; they first occur in four dimensions, with ten components. The $C^i{}_j$ are important because they remain invariant under conformal transformations $g \to fg$, $f \in E_0$ (Problem 10), and in particular they vanish on all conformally flat spaces. Conversely, if $C = 0$, then $g = f\eta$ [44].

We conclude the section by collecting the formulas that connect the metric and the curvature, to lay the foundation for later calculations:

(10.1.45)
$$\begin{aligned} g &= g_{ik}e^i \otimes e^k, \\ de^i &= -\omega^i{}_j \wedge e^j, \\ dg_{ik} &= g_{ij}\omega^j{}_k + \omega_k{}^j g_{ji}, \\ R^i{}_k &= d\omega^i{}_k + \omega^i{}_j \wedge \omega^j{}_k. \end{aligned}$$

(10.1.46) **Problems**

1. Find a u such that $(du|X\otimes Y) = (D_Xu|Y)-(D_Yu|X)$ is violated in Example (10.1.34; 2) with torsion.

2. Use (7.2.26; 3) to calculate the ω's for the plane with polar coordinates. (The ω's vanish for $dx^{1,2}$, so set $\bar{e}^i = A^i{}_k\, dx^k$, and compare with (10.1.34; 2).)

3. With the help of (7.2.26; 3), find a necessary and sufficient condition to make the ω's vanish by a change of basis.

4. Show that for $\bar{b} = bL$, we have $\Omega = L^{-1}\Omega L$.

5. Derive (10.1.37) and (10.1.38).

6. Prove (10.1.27; 1).

7. Show (10.1.21) by verifying the more general relation for $\omega \in S_1$:

$$i_v i_u D\omega = i_u D(\omega|v) - i_v D(\omega|u) - i_{[u,v]}\omega.$$

Hint: ω can be written $\sum_i b_i v^i$, $b_i \in S_0$, $v^i \in E_1$ and because of linearity it suffices to show the relation for one term. Use

$$i_v i_u\, dv = i_u d(v|v) - i_v d(v|u) - i_{[u,v]}v \qquad \text{for} \quad v \in E_1.$$

8. Derive the equations (10.1.44; 2).

9. Show that $i_{e_j} C^j{}_k =: C_k = 0$.

10. Suppose that $\bar{e}^i = f e^i$ but $\bar{g}_{ik} = g_{ik}$ such that $\bar{g} = f^2 g$. Show that $\bar{C}_{ik} = C_{ik}$.

11. Show $de^j = -\omega^j{}_k e^k \overset{b}{\Leftrightarrow} (D_X u|Y) - (D_Y u|X) = (du|X \otimes Y) \overset{b}{\Leftrightarrow} D_X Y - D_Y X = [X, Y]$, $X, Y \in T_0^1$, $u \in T_1^0$ and e^j a basis in T_1^0.

(10.1.47) Solutions

1. $u = e^2$; $de^2 = dr \wedge d\varphi \neq 0$, but $D_X e^2$ ought to vanish for all $X \in T_0^1$.

2. $A = \begin{vmatrix} \cos\varphi & \sin\varphi \\ -\sin\varphi & \cos\varphi \end{vmatrix}$, $\quad dA = d\varphi \begin{vmatrix} -\sin\varphi & \cos\varphi \\ -\cos\varphi & -\sin\varphi \end{vmatrix}$, $\quad -dA \cdot A^{-1} = d\varphi \begin{vmatrix} 0 & -1 \\ 1 & 0 \end{vmatrix}$.

3. Choose A so that $dA^i{}_k = \omega^i{}_l A^l{}_k$. As remarked in (10.1.20; 3), it is both necessary and sufficient for solubility that $d(\omega^i{}_l A^l{}_k) = R^i{}_l A^l{}_k = 0$, and therefore $R^i{}_k = 0 \Leftrightarrow \omega^i{}_k = 0$.

4. $\bar{\omega} = L^{-1}\omega L + L^{-1}\, dL$,

$$\begin{aligned} d\bar{\omega} + \bar{\omega} \wedge \bar{\omega} &= (dL^{-1}) \wedge \omega L + L^{-1}\, d\omega L - L^{-1}\omega\, dL + (dL^{-1}) \wedge dL \\ &\quad + (L^{-1}\omega L + L^{-1}\, dL) \wedge (L^{-1}\omega L + L^{-1}\, dL) \\ &= L^{-1}(d\omega + \omega \wedge \omega)L. \end{aligned}$$

5.

$$\begin{aligned} \Gamma_{kjl} + \Gamma_{jkl} &= g_{jk,l}, \\ \Gamma_{jlk} + \Gamma_{ljk} &= g_{lj,k}, \\ -\Gamma_{lkj} - \Gamma_{klj} &= -g_{kl,j}, \\ \hline 2\Gamma_{jkl} &= g_{jk,l} + g_{lj,k} - g_{kl,j}. \end{aligned}$$

Let $(de_j|e_i \otimes e_k) = (de_j)_{ik}$, etc. Then

$$\begin{aligned}(de_j)_{ik} &= (\omega_{ji})_k - (\omega_{jk})_i,\\ (de_k)_{ji} &= (\omega_{kj})_i - (\omega_{ki})_j,\\ -(de_i)_{kj} &= -(\omega_{ik})_j + (\omega_{ij})_k,\\ \hline (de_j)_{ik} + (de_k)_{ji} - (de_i)_{kj} &= 2(\omega_{kj})_i.\end{aligned}$$

6.

$$\begin{aligned}d(d\omega^i{}_j + \omega^i{}_s \wedge \omega^s{}_j) &= d\omega^i{}_s \wedge \omega^s{}_j - \omega^i{}_s \wedge d\omega^s{}_j\\ &= \Omega^i{}_s \wedge \omega^s{}_j - \omega^i{}_s \wedge \Omega^s{}_j - \omega^i{}_k \wedge \omega^k{}_s \wedge \omega^s{}_j + \omega^i{}_s \wedge \omega^s{}_k \wedge \omega^k{}_j\\ &= \Omega^i{}_s \wedge \omega^s{}_j - \omega^i{}_s \wedge \Omega^s{}_j.\end{aligned}$$

7. Insert $\omega = b\nu$, $b \in S_0$, $\nu \in E_1$:

$$\begin{aligned}i_v i_u D\omega = (i_u Db)(i_v \nu) - (i_v Db)(i_u \nu)\\ + b(i_u d(i_v \nu) - i_v d(i_u \nu) - i_{[u,v]}\nu).\end{aligned}$$

This equals

$$i_u D(\omega|v) - i_v D(\omega|u) - i_{[u,v]}\omega = i_u D(bi_v \nu) - i_v D(bi_u \nu) - bi_{[u,v]}\nu.$$

8.

(a) holds because both sides are components of a 2-form;

(b) follows from (10.1.43(a));

(c) $0 = R^i{}_j \wedge e^j = R^i{}_{jkm} e^{jkm}$. Because of (a), only the cyclic permutation remains in the sum; and

(d) follows from (a)–(c):

$$\begin{aligned}R_{ijkm} &= -R_{ikmj} - R_{imjk} = R_{kimj} + R_{mijk}\\ &= -R_{kmji} - R_{kjim} - R_{mjki} - R_{mkij}\\ &= 2R_{kmij} + R_{kjim} + R_{jmki} = 2R_{kmij} - R_{jimk} \quad \Rightarrow \quad 2R_{ijkm} = 2R_{kmij}.\end{aligned}$$

9. Because $i_{e_j} e^j = m$ and $e^j i_{e_j} \omega = p\omega$ for all $\omega \in E_p$, we find

$$R_k = i_{e_j} R^j{}_k = \frac{-R(m-1)e_k}{(m-2)(m-1)} + \frac{1}{m-2}[Re_k - R_k - R_k + mR_k] + C_k \quad \Rightarrow \quad C_k = 0.$$

10. It suffices to show that $\bar{R}_{ik} = R_{ik} + \nu_i \wedge e_k - \nu_k \wedge e_i$ for some $\nu_i \in E_1$. Because $d\bar{e}^i = df \wedge e^i + f\, de^i = -\bar{\omega}^i{}_k \wedge \bar{e}^k$ and $\bar{\omega}_{ik} + \bar{\omega}_{ki} = \omega_{ik} + \omega_{ki}$,

$$\bar{\omega}_{ik} = \omega_{ik} + (df|e_i)e_k - (df|e_k)e_i.$$

In a natural basis, $e_s \wedge \omega^s{}_n = 0$, and therefore

$$((df|e_i)e_s - (df|e_s)e_i) \wedge \omega^s{}_k = \nu_k \wedge e_i, \qquad \nu_k = (df|e_s)\omega^s{}_k,$$

and

$$d\bar{\omega}_{ik} - d\omega_{ik} = \nu_i \wedge e_k - \nu_k \wedge e_i, \qquad \nu_i = d(df|e_i).$$

11.

(a) $\Leftarrow$: $(de^j|X \otimes Y)$
$= (D_X e^j|Y) - (D_Y e^j|X) = -(\omega^j{}_k|X)(e^k|Y) + (\omega^j{}_k|Y)(e^k|X) = -(\omega^j{}_k \wedge e^k|X \otimes Y)$.
$\Rightarrow$ If $u = fe$, $f \in E_0$, then $du = df \wedge e + f\, de$ and $D_X u = i_X(df \wedge e + f\, De)$. Thus the first term also satisfies the equation.

(b) Remember that $(du|X \otimes Y) = L_X(u|Y) - L_Y(u|X) - (u|[X, Y])$ and $L_X(u|Y) = D_X(u|Y) = (D_X u|Y) + (u|D_X Y)$.

10.2 Gauge Theories and Gravitation

Energy and momentum imprint a structure on space and time through Einstein's equations, which equate the energy and momentum forms to quantities constructed from $R^\alpha{}_\beta$.

To derive the field equations with the Lagrangian formalism one needs a 4-form as Lagrangian density. However, to get fields equations which are consistent one has to make sure that the action functional has a stationary point. We have seen in §8.1 that in Maxwell's theory this is the case only if the current is conserved. In §9.1 the current was derived from a scalar field S and the field equations for S were just the requirement that the current be conserved. We shall now study this phenomenon more generally with a scalar field $\Phi = \sqrt{\rho}e^{iS}$ where the density ρ is no longer constant. Instead of the complex field φ we shall use two real scalar fields φ^1, φ^2 with $\Phi = \varphi^1 + i\varphi^2$. Expressed in terms of these fields the Lagrangian $(d + ieA)\varphi \wedge {}^*(d + ieA)\varphi$ of (9.1.21; 2) becomes

$$(10.2.1) \quad -\frac{1}{2}\sum_{i=1}^{2}(D\varphi)^i \wedge {}^*(D\varphi)^i, \quad (D\varphi)^i = d\varphi^i + e\varepsilon^i{}_k A\varphi^k, \quad \varepsilon = \begin{pmatrix} 0 & -1 \\ 1 & 0 \end{pmatrix}$$

in the notation (10.1.14; 2). Another term invariant under the local gauge transformation would be a "mass-term" $-\frac{1}{2}m^2 \sum_i \varphi^i \wedge {}^*\varphi^i$. The general total Lagrangian

$$(10.2.2) \qquad \mathcal{L} = -\tfrac{1}{2}dA \wedge {}^*dA - \tfrac{1}{2}\sum_i((D\varphi)^i \wedge {}^*(D\varphi)^i + m^2\varphi^i \wedge {}^*\varphi^i)$$

is invariant under $A \to A + d\Lambda$, $\varphi \to e^{-e\varepsilon\Lambda}\varphi$, $\Lambda \in C(\mathbb{R}^4)$, and the linear dependence of the action $W = \int \mathcal{L}$ on Λ which prevented W in (8.1.3; 1) from having stationary points is absent. Thus there is hope that the Euler equations which result from (10.2.2),

(10.2.3) $$d^*F = -^*J, \qquad D^*D\Phi = m^{2*}\Phi,$$

with

$$^*J = -\frac{\partial \mathcal{L}}{\partial A} = e\varepsilon^j{}_k\varphi^k \wedge {}^*(D\varphi)^j,$$

are consistent. Indeed, $d^*J = 0$ is a consequence of the equations for Φ. To see this most directly consider the variation of $\mathcal{L}$ with Φ:

(10.2.4) $$\delta\mathcal{L} = \sum_i \delta\varphi^i \left[\frac{\partial\mathcal{L}}{\partial\varphi^i} - d\frac{\partial\mathcal{L}}{\partial\, d\varphi^i}\right] + d\left(\sum_i \delta\varphi^i \frac{\partial\mathcal{L}}{\partial\, d\varphi^i}\right).$$

The Euler equations require that $[\ \ ] = 0$, and for $\delta\Phi$ the change $\Phi \to e^{-e\varepsilon\Lambda}\Phi$ yields $\delta\mathcal{L} = ed(\Lambda\varepsilon^j{}_k\varphi^k \wedge {}^*(D\varphi)^j)$ to first order in Λ. If we choose a global gauge transformation ($d\Lambda = 0$), the invariance of $\mathcal{L}$ guarantees $\delta\mathcal{L} = 0 = -d^*J$.

Next we shall extend these considerations to the non-Abelian gauge group $O(3)$. In this case Φ has three components, and one might think that in (10.2.2) one just has to take $\sum_i$ from one to three. However, now there are also three gauge potentials, and dA is not invariant under the gauge transformation $A \to L^{-1}AL + L^{-1}\,dL$ (see 10.1.12). This suggests replacing dA with $DA = \Omega$ (see 10.1.20; 1), which transforms as $(DA) \to L^{-1}AL$. Thus the bilinear expression $\operatorname{tr}\Omega \wedge {}^*\Omega = \Omega^i{}_k \wedge {}^*\Omega^k{}_i$ is gauge invariant and seems to be a good candidate for the Lagrangian of the gauge field. Such a Lagrangian was first proposed by Yang and Mills and later turned out to describe the strong and electroweak interactions if further fields are added.

(10.2.5) The Yang–Mills Lagrangian

$$\mathcal{L} = \tfrac{1}{2}\operatorname{tr} DA \wedge {}^*DA - \tfrac{1}{2}\langle D\varphi \wedge {}^*D\varphi\rangle - \tfrac{1}{2}m^2\langle\Phi \wedge {}^*\Phi\rangle.$$

(10.2.6) Remarks

1. We used the notation $\langle\Phi \wedge \Psi\rangle = \sum_i \varphi^i \wedge \psi^i$ for the scalar product in the fibers.
2. Although we were thinking of $O(3)$, the Lagrangian (10.2.5) has the same appearance for all $O(n)$.

The Euler equations derived from (10.2.5) are similar to (10.2.3) except that now the Maxwell field F is replaced by the curvature Ω. Here we just state the results, as we shall go through the details of the variational procedure in (10.2.15).

(10.2.7) The Yang–Mills Equations

$$d^*\Omega = -^*J := \frac{\partial\mathcal{L}}{\partial A}, \qquad D^*D\Phi = m^{2*}\Phi.$$

(10.2.8) **Remarks**

1. The derivation of (10.2.7) follows the lines of (8.1.7): one just notes that $^*\Omega = \partial\mathcal{L}/\partial\, dA$, since both factors in $\Omega \wedge {}^*\Omega = {}^*\Omega \wedge \Omega$ give the same contribution.
2. To get more explicit expressions one might use a basis b_i for the Lie-algebra = (3 × 3 antisymmetric matrices) which diagonalizes the trace. If $\operatorname{tr} b_\alpha b_\beta = -\delta_{\alpha\beta}$ and $A = \sum b_\alpha A^\alpha$, $A^\alpha \in E_1$, then the curvature $\Omega = \sum b_\alpha F^\alpha$ has the components
$$F^\alpha = dA^\alpha + \tfrac{1}{2} c_{\beta\gamma}{}^\alpha A^\beta \wedge A^\gamma$$
according to (10.1.20; 1). For $O(n)$ α goes from 1 to $n(n-1)/2$, whereas for φ^i i goes from 1 to n. The $c_{\beta\gamma}{}^\alpha$ are the structure constants of $O(3)$ with this basis: $b_\beta b_\gamma - b_\gamma b_\beta = c_{\beta\gamma}{}^\alpha b_\alpha$. With this decomposition the Lagrangian of the gauge field becomes $\mathcal{L}^{\mathrm{gf}} = -\frac{1}{2} F^\alpha \wedge {}^*F^\alpha$.
3. Since the F^α do not depend on dA only, but also on A, there is a contribution $-\partial\mathcal{L}^{\mathrm{gf}}/\partial A^\alpha = c_{\alpha\beta}{}^\gamma A^\beta \wedge {}^*F^\gamma$ to the current $^*J$. This does not appear in Maxwell's theory and means that the Yang–Mills equations are nonlinear even in the absence of the scalar field. On the other hand, the contribution of the scalar field to the current $(b^\alpha)_{kl}\varphi^k \wedge {}^*(D\Phi)^l$ has the same structure as previously.
4. The consistency of the Yang–Mills equations requires $d^*J = 0$. This follows indeed from the variational principle, as can be seen by considerations similar to the preceding ones. A gauge transformation $L = 1 + b_\alpha \Lambda^\alpha$ with $d\Lambda^\alpha = 0$ and $\Lambda^\alpha \to 0$ generates $\delta A^\alpha = c_{\beta\gamma}{}^\alpha A^\beta \Lambda^\gamma$, and since $\delta\mathcal{L} = 0$ the generalization of (10.2.4) tells us that
$$\Lambda^\alpha d(c_{\beta\alpha}{}^\gamma A^\beta \wedge {}^*F^\gamma + (b^\alpha)_{mn}\varphi^m \wedge {}^*(D\varphi)^n) = 0.$$
Thus the general conclusions which we drew from Maxwell's equations, such as the vanishing of the total charge in a closed universe or a plane wave still hold.
5. There is an essential difference from Maxwell's theory, where the current was gauge invariant. Here the current does not transform like Ω under a gauge transformation, $J \not\to L^{-1}JL$, but instead transforms inhomogeneously. Indeed, $J \to L^{-1}JL$ would be incompatible with (10.2.7). Although $^*\Omega$ transforms as $^*\Omega \to L^{-1}{}^*\Omega L$, $d^*\Omega$ does not. Therefore it has no gauge-invariant meaning to say that one has a certain amount of current at a point x in space—one can always find a gauge such that $J(x) = 0$. The total charges $Q_\alpha = \int_{N_3} {}^*J_\alpha = -\int_{\partial N_3} {}^*F_\alpha$ are more robust. Since they can be expressed as boundary integrals, they transform homogeneously under all gauge transformations which have $dL = 0$ on ∂N_3.

The culprit responsible for the inhomogeneous transformation law of J is the A contained in the contribution from the gauge field. The contribution $^*J(\Phi) = b\Phi \wedge {}^*(D\Phi)$ from the scalar field transforms homogeneously. Indeed, if we put the current from the gauge field on the other side of (10.2.7), we obtain the combination $d^*F^\alpha + c_{\beta\gamma}{}^\alpha A_\beta \wedge {}^*F^\gamma$ on the left side, since the structure constants are totally antisymmetric, $c_{\beta\gamma}{}^\alpha = c_{\alpha\beta}{}^\gamma$. The extension (10.1.6) of the covariant derivative to

$S_p(L(F))$ shows (compare Problem 7) that these are just the components of the covariant exterior derivative D which is constructed such that it transforms with $L^{-1}D^*\Omega L$. Thus the Yang–Mills equation can also be written $D^*\Omega = -{}^*J(\Phi)$. ${}^*J(\Phi)$ alone is not conserved, however: it turns out that D applied twice to ${}^*\Omega$ gives zero (Problem 7), and thus $d^*J \neq 0$ but $D^*J(\Phi) = 0$. Together with (10.1.26) we find a form for (10.2.7) which looks like Maxwell's equation except that d is replaced by D.

(10.2.9) The Covariant Form of the Yang–Mills Equations

$$D\Omega = 0, \qquad D^*\Omega = -{}^*J(\Phi) \quad \Rightarrow \quad D^*J(\Phi) = 0.$$

Since the gauge theories discussed so far seem to be the appropriate description for the electroweak and strong interactions, it is tempting to think that a theory of gravitation should follow the same pattern. According to the equivalence principle discussed in §7.3, gravity is related to the metric structure of space and time M. This suggests taking as fiber bundle $T(M)$, as fiber metric the pseudo-Riemannian metric, as gauge group the local Lorentz transformations, and as gauge potential the connection ω. Thus a possible Lagrangian would be $-\frac{1}{2}R^\alpha{}_\beta \wedge {}^*R^\beta{}_\alpha$ + the matter Lagrangian (9.1.20). However, this does not give a theory of gravitation, because the connection ω does not appear in (9.1.20)), everything being expressible in terms of * and d. Thus the current ${}^*J = \delta\mathcal{L}/\delta\omega$ does not get a contribution from the Maxwell or scalar fields. Only if we had a spinor field would its spin density [40] contribute to the current. Since we know that all matter acts as source of the gravitational potential, this theory does not seem to work. More specifically it is believed that all energy and momentum currents are sources of gravitation. Since they are the generators of translations one might think that it is more the translation part of the Poincaré group and not the rotations which are relevant for gravity. If this is so, we lose the strict analogy with the previous development where the gauge groups were orthogonal (or unitary) groups. Therefore the construction of the gravitational Lagrangian will involve some guess-work. In any case the energy and momentum currents of matter turn out to be $\partial\mathcal{L}^{\text{matter}}/\partial e^\alpha$. Here the tetrads $e^\alpha \in E_1$ are a basis and therefore it is reasonable to look at them as the gauge potentials replacing the A's. In this case the ω's play the rôle of a field strength and not a potential, as actually happens in the equations (1.1.6) for the motion of a particle in a gravitational field. If we use an orthogonal basis they contain all the information of the metric $g = \sum \eta_{\alpha\beta} e^\alpha \otimes e^\beta$ and its influence on matter comes about only through the *-map contained in $\mathcal{L}^{\text{matter}}$. We shall represent matter by a scalar and the electromagnetic field and take the Lagrangian from (9.1.20) (in units $\rho e^2/m = 1$):

(10.2.10) $$\mathcal{L}^{\text{matter}} = -\tfrac{1}{2}J \wedge {}^*J - \tfrac{1}{2}F \wedge {}^*F.$$

When choosing the gravitational Lagrangian $\mathcal{L}^{\text{gf}}$ some points should be borne in mind:

(10.2.11) **Remarks**

1. At the time of the birth of gravitational theory, the requirement of general covariance provided some relief from the labor pains, but later on it was more often a source of confusion. The concept of a manifold incorporates it automatically when the definition uses equivalence classes of atlases, and hence only chart-independent statements are regarded as meaningful. This program is by no means unique to gravitational theory—we have also followed it in classical mechanics and electrodynamics. The big difference is that now the metric g on M is not determined a priori.
2. There are some coordinate systems in which Einstein's equations simplify, just as Maxwell's equations are easier to work with in the Lorentz gauge $\delta A = 0$. For example, some formulas are shorter (cf. (10.2.20)) when written in the popular "harmonic coordinates," which satisfy $\delta\, dx = 0$ [45]. It is a matter of opinion, which we leave to the reader, whether this fact is of fundamental importance.
3. Even when the chart is fixed, there still remains the choice of a basis e^i. Orthogonal bases are special in that they standardize g_{ik} as η_{ik}. They still leave open the possibility of a Lorentz transformation $e^\alpha \to L^\alpha{}_\beta(x)e^\beta$, $L^t(x)\eta L(x) = \eta$ for all x. $\mathcal{L}^{\text{matter}}$ in (10.2.10) is invariant under this transformation, which means that matter does not define a "teleparallelism." Looking just at matter one cannot observe the orientation of the local Lorentz systems relative to each other. It is thus reasonable to postulate that gravity likewise has no preferred frame, and therefore $\mathcal{L}^{\text{gf}}$ should be invariant under this transformation too.

We now have to select $\mathcal{L}^{\text{gf}} \in E_4$ from the material gathered in §10.1 for $T(M)$ and pose the requirements that it:

(a) is invariant under a change of basis; and

(b) is quadratic in the derivative of the e^α.

This leads uniquely, up to a factor, to

(10.2.12)

$$\mathcal{L}^{\text{g}} \simeq {}^*R = R_{\alpha\beta} \wedge {}^*e^{\alpha\beta} = 2d(e^\alpha \wedge {}^*de_\alpha) - (de^\alpha \wedge e^\beta) \wedge {}^*(de_\beta \wedge e_\alpha) + \tfrac{1}{2}(de^\alpha \wedge e_\alpha) \wedge {}^*(de^\beta \wedge e_\beta)$$

(for the last formulation see Problem 8). Thus the total Lagrangian will have three contributions one of which arises from a scalar field $\mathcal{L}^{\text{s}} = -\frac{1}{2} J \wedge {}^*J$ representing matter. For simplicity we choose units where $\rho e^2/m = 1$. Secondly, the electromagnetic field adds $\mathcal{L}^{\text{e}} = -\frac{1}{2} F \wedge {}^*F$, and finally comes $\mathcal{L}^{\text{g}}$ equipped with a factor to be adjusted later.

(10.2.13) **The Lagrangian for the Total System**

$$\mathcal{L} = -\tfrac{1}{2} J \wedge {}^*J - \tfrac{1}{2} F \wedge {}^*F + \frac{1}{16\pi\kappa} R_{\alpha\beta} \wedge {}^*e^{\alpha\beta}, \quad F = dA, \quad J = dS + eA.$$

(10.2.14) Remarks

1. If we consider the e^α as the counterpart to A, then the simplest analogue to $dA \wedge {}^*dA$ which is invariant under global Lorentz transformations would be $de^\alpha \wedge {}^*de_\alpha$. This leads to a theory with teleparallelisms and contradicts our knowledge gained from experiments on the bending of light rays by the Sun. The somewhat different form of (10.2.13) ensures invariance under local Lorentz transformations. Because of this invariance we may use an orthonormal basis for the derivation of Euler's equation. The variation of the metric g results from our making the variation of e without imposing any orthogonality constraints. Since the metric η is constant we may freely pull indices up and down under the derivative and use the notation $\omega_{\alpha\beta} = \eta_{\alpha\gamma}\varepsilon^\gamma{}_\beta$, etc., which amounts only to a change of sign for each subscript zero.
2. Another quantity invariant under local gauge transformations would be simply ^{*1}. It contains no derivatives and by itself cannot lead to a differential equation. In an early stage of the theory Einstein added it in as a "cosmological term." Later it fell into disfavor, and there is no empirical evidence for it. Yet its absence is still mysterious since there are many possible sources which could contribute such a term. We shall occasionally resurrect it for the purpose of comparison.
3. In an orthogonal basis we find that

$$\begin{aligned} {}^*e^{\alpha\beta} \wedge d\omega_{\alpha\beta} - d({}^*e^{\alpha\beta} \wedge \omega_{\alpha\beta}) &= -(d{}^*e^{\alpha\beta}) \wedge \omega_{\alpha\beta} \\ &= -\omega^\alpha{}_\gamma \wedge {}^*e^{\gamma\beta} \wedge \omega_{\alpha\beta} + \omega^\beta{}_\gamma \wedge {}^*e^{\alpha\gamma} \wedge \omega_{\alpha\beta} \\ &= -2{}^*e^{\beta\gamma} \wedge \omega_\gamma{}^\alpha \wedge \omega_{\alpha\beta}, \end{aligned}$$

and so we would simply use

$$\mathcal{L}' = \frac{-1}{16\pi\kappa} {}^*e^{\beta\gamma} \wedge \omega_\gamma{}^\alpha \wedge \omega_{\alpha\beta}$$

for gravitation. This equivalent $\mathcal{L}'$, when taken by itself, is altered by a change of basis. It is useful for showing why we get an equation of second order: it contains only ω, i.e., derivatives of e, but not $d\omega$.

(10.2.15) Derivation of the Euler–Lagrange Equations

(a) Variation[3] of the Lagrangian of S: From $e^\alpha \wedge {}^*J = J \wedge {}^*e^\alpha$ (see (7.2.18(a))), we conclude that

$$e^\alpha \wedge \delta^* J = \delta J \wedge {}^*e^\alpha + J \wedge \delta^* e^\alpha - (\delta e^\alpha) \wedge {}^*J.$$

If we successively vary the e's occurring in ${}^*e^\alpha$, we discover that

$$\delta^* e^\alpha = \delta e^\beta \wedge (i_\beta{}^* e^\alpha), \qquad i_\beta := i_{e_\beta}.$$

[3]Here δ denotes the variation, not the codifferential.

Using this in the above formula, multiplying by J_α, and summing:

$$J \wedge \delta^* J = \delta J \wedge {}^*J - \delta e^\alpha \wedge [J \wedge i_\alpha {}^*J + (i_\alpha J) \cdot {}^*J] \quad \Rightarrow$$
$$\delta(-\tfrac{1}{2} J \wedge {}^*J) = -\delta J \wedge {}^*J + \tfrac{1}{2}\delta e^\alpha \wedge [J \wedge i_\alpha {}^*J + (i_\alpha J) \cdot {}^*J].$$

(b) Variation of the electromagnetic part of $\mathcal{L}$: As in part (a), we conclude from $e^{\alpha\beta} \wedge {}^*F = F \wedge {}^*e^{\alpha\beta}$ that

$$e^{\alpha\beta} \wedge \delta^* F = (\delta F) \wedge {}^*e^{\alpha\beta} + F \wedge \delta^* e^{\alpha\beta} - (\delta e^{\alpha\beta}) \wedge {}^*F.$$

As before, multiplication by $F_{\alpha\beta}$ yields

$$\delta(-\tfrac{1}{2} F \wedge {}^*F) = -\delta F \wedge {}^*F + \tfrac{1}{2}\delta e^\alpha \wedge [(i_\alpha F) \wedge {}^*F - F \wedge i_\alpha {}^*F].$$

(c) Variation of the gravitational part: First note that by (7.2.18(c)),

$$\delta^* e^{\alpha\beta} = \delta e^\gamma \wedge i_\gamma {}^*e^{\alpha\beta} = \delta e^\gamma \wedge {}^*e^{\alpha\beta}{}_\gamma, \qquad e^{\alpha\beta}{}_\gamma := \eta_{\gamma\sigma} e^{\alpha\beta\sigma}, \text{ etc.}$$

Although $R_{\alpha\beta}$ is itself constructed from the ω's and a variation of e induces a variation of ω, we need not calculate these variations, because

(10.2.16) $$^*e^{\alpha\beta} \wedge \delta R_{\alpha\beta} = d({}^*e^{\alpha\beta} \wedge \delta\omega_{\alpha\beta})$$

(Problem 1), and hence the variation of R does not affect the Euler–Lagrange equations.

The combination of these three results produces

$$\delta\mathcal{L} = -\delta J \wedge {}^*J - \delta F \wedge {}^*F + \delta e^\alpha$$
$$\wedge \left[{}^*t_\alpha + {}^*\mathcal{T}_\alpha + \frac{1}{16\pi\kappa} {}^*e_{\alpha\beta\gamma} \wedge R^{\beta\gamma}\right] + \frac{1}{16\pi\kappa} d({}^*e^{\alpha\beta} \wedge \delta\omega_{\alpha\beta})$$
$${}^*t_\alpha = \tfrac{1}{2}((i_\alpha J)^* J + J \wedge i_\alpha {}^*J),$$
$${}^*\mathcal{T}_\alpha = \tfrac{1}{2}((i_\alpha F) \wedge {}^*F - F \wedge i_\alpha {}^*F),$$

or, writing $J = dS + eA$ and $F = dA$,

(10.2.17) $$\delta\mathcal{L} = \delta S\, d^*J - \delta A \wedge (d^*F + e^*J) + \delta e^\alpha \wedge \left[{}^*t_\alpha + {}^*\mathcal{T}_\alpha + \frac{1}{16\pi\kappa} {}^*e_{\alpha\beta\gamma} \wedge R^{\beta\gamma}\right] - d\left(\delta S^* J + \delta A \wedge {}^*F - \frac{1}{16\pi\kappa} {}^*e^{\alpha\beta} \wedge \delta\omega_{\alpha\beta}\right).$$

The requirement that $\delta \int \mathcal{L} = 0$ therefore results in the

(10.2.18) **Field Equations of the Total System**

(a) $d^*J = 0$;

(b) $d^*F = -e^*J$; and

(c) $-\frac{1}{2}{}^*e_{\alpha\beta\gamma} \wedge R^{\beta\gamma} = 8\pi\kappa({}^*t_\alpha + {}^*\mathcal{T}_\alpha)$.

(10.2.19) Remarks

1. These calculations generalize those of §9.1, because the metric has also been varied. The effect is to produce Einstein's equations (c) for the gravitational potential, in addition to the equations we got before.
2. The right sides of (c) are the energy-momentum currents of a scalar field and an electromagnetic field. Note that the gauge-dependent contribution (8.1.8; 1) of the canonical energy-momentum tensor does not appear, and that the Maxwellian contribution is coupled with gravity. These currents have the structure of the Hamiltonian $p\dot{q} - L$ in mechanics as anticipated in (7.3.22). They can be written
$$^*t_\alpha = (i_\alpha{}^*J) \wedge dS - i_\alpha\mathcal{L}^s, \qquad {}^*\mathcal{T}_\alpha = i_\alpha{}^*F \wedge dA - i_\alpha\mathcal{L}^e.$$
Now $^*J = -\partial\mathcal{L}^s/\partial\, dS$ (resp. $^*F = -\partial\mathcal{L}^e/\partial\, dA$) correspond to p; dS (resp. dA) correspond to $\dot{q}$ and the inner product picks out the right component.
3. Although we have used an orthogonal basis to derive (10.2.17), (c) has the same form in any basis, because of the homogeneous transformation law for $R_{\alpha\beta}$ (10.1.20; 3). Equation (c) seems to have a different structure than (a) and (b), it does not say that the derivative of a field strength is a current. (c) is the analogue of the covariant form of the Yang–Mills equations (10.2.9). We shall shortly cast it into the form (10.2.7), which is like (a) and (b), but thereby the two sides of the equation lose the property of transforming homogeneously under gauge transformations.

(10.2.20) Different Versions of Einstein's Equations

(a) The classical version. The 3-forms of energy, momentum, and electric current all occur in (10.2.8). In §9.1 we wrote down the equations for the corresponding 1-forms, and we can similarly rewrite (c). By Rule (7.2.18(b)), for all $\omega \in E_2$,
$$-{}^*e_{\alpha\beta\gamma} \wedge \omega = ({}^*e_\alpha i_\beta i_\gamma + {}^*e_\beta i_\gamma i_\alpha + {}^*e_\gamma i_\alpha i_\beta)\omega.$$
If $\omega = R^{\beta\gamma}$ and we observe that $i_\alpha i_\beta R^\beta{}_\gamma = i_\alpha R_\gamma = i_\gamma R_\alpha$ and $^*e^\gamma i_\gamma R_\alpha = {}^*R_\alpha$ (cf. (10.1.24; 3)), then we see that
$$R_\alpha - \tfrac{1}{2}e_\alpha R = 8\pi\kappa(t_\alpha + \mathcal{T}_\alpha),$$
i.e.,
$$R_\alpha = 8\pi\kappa(t_\alpha + \mathcal{T}_\alpha - \tfrac{1}{2}e_\alpha(t + \mathcal{T})), \qquad t = i_\alpha t^\alpha, \text{ etc.}$$
For the components of the Riemann–Christoffel tensor (10.1.24; 2) and the energy-momentum tensor $T_{\alpha\beta}$ such that $t_\alpha + \mathcal{T}_\alpha = T_{\alpha\beta}e^\beta$, this means that
$$R^\gamma{}_{\alpha\gamma\beta} - \tfrac{1}{2}g_{\alpha\beta}R^\gamma{}_{\sigma\gamma}{}^\sigma = 8\pi\kappa T_{\alpha\beta}.$$

(b) As a Yang–Mills type equation, $\mathcal{L}'$ of (10.2.14; 3) can be written (see (10.2.13))

$$\mathcal{L}' = -\tfrac{1}{2}(de^\alpha \wedge e^\beta) \wedge {}^*(de_\beta \wedge e_\alpha) + \tfrac{1}{4}(de^\alpha \wedge e_\alpha) \wedge {}^*(de^\beta \wedge e_\beta)$$

if we momentarily put $8\pi\kappa = 1$. Performing the variational procedure (10.2.17) with the e^α we get another version of (10.2.18(c))

(10.2.21)

$$\begin{aligned} d^*F_\alpha &= -{}^*J_\alpha, \\ {}^*F_\alpha &= \frac{\partial \mathcal{L}}{\partial de^\alpha} = e^\beta \wedge {}^*(de_\beta \wedge e_\alpha) - \tfrac{1}{2} e_\alpha \wedge {}^*(de^\beta \wedge e_\beta), \\ {}^*J_\alpha &= \frac{\partial \mathcal{L}}{\partial e_\alpha} = i_\alpha{}^*J \wedge dS + i_\alpha{}^*F \wedge dA + i_\alpha{}^*F_\beta \wedge de^\beta - i_\alpha \mathcal{L}'. \end{aligned}$$

They are now in the form of the exterior derivative of a field strength yielding the dual of a current and have the following features in common with the Yang–Mills equations:

(i) There is a contribution ${}^*t_\alpha = i_\alpha{}^*F_\beta \wedge de^\beta - i_\alpha \mathcal{L}'$ from the e^α to the currents. It can be interpreted as the energy-momentum currents of the gravitational field. It has exactly the same structure $p\dot{q} - L$ as the contributions from S and A.

(ii) Under $e^\alpha \to L^\alpha{}_\beta(x)e^\beta$ both d^*F_α and ${}^*J_\alpha$ transform inhomogeneously. From the three contributions to ${}^*J_\alpha$ it is only t_α which does not transform homogeneously. For any given point x we can find a frame such that $t_\alpha(x) = 0$. Conversely, on a flat space with non-Cartesian coordinates $t^\alpha \neq 0$. As discussed in §7.3 these facts reflect how the balance of energy-momentum is affected in accelerated reference frames. The total currents are conserved, $d^*J_\alpha = 0$ and thus we draw again the conclusions that the total energy-momentum of a closed universe or a periodic field is zero. If space is asymptotically flat and we restrict ourselves to transformations which are asymptotically constant Lorentz transformations, then the total energy and momentum transform like a vector.

To see the relation to the classical version we have to re-express de in terms of the ω's. This is done most easily if one remembers that the exterior product is the dual of the inner product

(10.2.22)

$$\begin{aligned} F^\alpha &= i_\beta(de^\beta \wedge e^\alpha - \tfrac{1}{2}\eta^{\alpha\beta} de^\gamma \wedge e_\gamma) \\ &= -\, i_\beta(\omega^\beta{}_\gamma \wedge e^{\gamma\alpha} - \tfrac{1}{2}\eta^{\alpha\beta}\omega^\gamma{}_\sigma \wedge e^\sigma{}_\gamma) \\ &= -\, \langle \omega_{\beta\gamma} | e^\beta \rangle e^{\gamma\alpha} + \tfrac{1}{2}\langle \omega_{\gamma\sigma} | e^\alpha \rangle e^{\sigma\gamma} \\ &= -\, \tfrac{1}{2} i_{\omega_{\beta\gamma}} e^{\alpha\beta\gamma} = -\tfrac{1}{2}{}^*(\omega_{\beta\gamma} \wedge {}^*e^{\alpha\beta\gamma}). \end{aligned}$$

Thus $d^*F^\alpha = \frac{1}{2}d(\omega_{\beta\gamma} \wedge {}^*e^{\alpha\beta\gamma})$ which has some similarity to the left side of (10.2.18(c)), the difference being ${}^*\iota^\alpha$. To get the latter in terms of the ω's we use ${}^*e^{\alpha\beta\gamma} = \varepsilon^{\alpha\beta\gamma\delta}e_\delta$,

$$d(\omega_{\beta\gamma} \wedge e_\delta) = d\omega_{\beta\gamma} \wedge e_\delta - \omega_{\beta\gamma} \wedge \omega_{\sigma\delta} \wedge e^\sigma$$

and

$$R_{\beta\gamma} = d\omega_{\beta\gamma} - \omega_{\sigma\beta} \wedge \omega^\sigma{}_\gamma.$$

Comparing (10.2.21) with (10.2.18(c)) we see

$${}^*\iota^\alpha = -\frac{\varepsilon^{\alpha\beta\gamma\delta}}{16\pi\kappa}(\omega_{\sigma\beta} \wedge \omega^\sigma{}_\gamma \wedge e_\delta - \omega_{\beta\gamma} \wedge \omega_{\sigma\delta} \wedge e^\sigma).$$

This form of the currents has been put forward by Landau and Lifshitz so that we shall call (10.2.21) the Landau–Lifshitz form of Einstein's equation. The basis they use is not orthogonal, but it is a natural basis. In the natural basis $e^\alpha = dx^\alpha$, $\omega_{\alpha\beta} := \Gamma_{\alpha\beta\mu}\,dx^\mu$ with $\Gamma_{\sigma\beta\mu} = \Gamma_{\sigma\mu\beta}$. Then the energy-momentum tensor of gravity, that is the components of ι^α in this basis, are symmetric:

$$\begin{aligned} dx^\rho \wedge {}^*\iota^\alpha &= -\frac{\sqrt{|g|}}{16\pi\kappa}(\Gamma_{\sigma\beta\mu}\Gamma^\sigma{}_{\gamma\nu}\varepsilon^{\rho\mu\nu}{}_\delta + \Gamma_{\beta\gamma\mu}\Gamma_{\sigma\delta\nu}\varepsilon^{\rho\mu\nu\sigma}) \cdot \varepsilon^{\alpha\beta\gamma\delta} \cdot {}^*1 \\ &= dx^\alpha \wedge {}^*\iota^\rho. \end{aligned}$$

This symmetry ensures local angular momentum conservation: $d({}^*t^\beta + {}^*T^\beta + {}^*\iota^\beta) = 0$ implies $d(x^\alpha({}^*t^\beta + {}^*T^\beta + {}^*\iota^\beta) - x^\beta({}^*t^\alpha + {}^*T^\alpha + {}^*\iota^\alpha)) = 0$ only if it is true for ι as well as for t and T (cf. (10.1.11; 2)) that $dx^\alpha \wedge {}^*\iota^\beta - dx^\beta \wedge {}^*\iota^\alpha = 0$. Here x^α is a local coordinate, so at this stage the theorem of conservation of angular-momentum is formulated strictly on the domain of a single chart. The appropriate 3-form is defined globally only on special manifolds.

From (10.2.23) one deduces immediately (Problem 8) the so-called ADM-expression for the total energy in an asymptotically flat space. It was a major discovery [41] when it was shown that it is positive provided the external energies T^0 and t^0 have this property. Thus the negative energy of gravitation can never exceed the energy of the sources if the space is to stay flat in the large.

(10.2.23) **Remarks**

1. The versions (10.2.20) do not exhaust the possibilities of writing Einstein's equations in terms of the exterior differentials of 2-forms. Numerous other variants have been proposed [29], [48], [58]; we just wanted to exhibit the analogy to non-Abelian Yang–Mills theories. The field equations are nonlinear because the gravitational field carries energy and momentum. However these quantities evade localization since they transform inhomogeneously under local Lorentz transformations.
2. It is not the curvature forms $R_{\alpha\beta}$ but rather their contractions R_α that are locally determined by Einstein's equations. However, if the Weyl tensor is known,

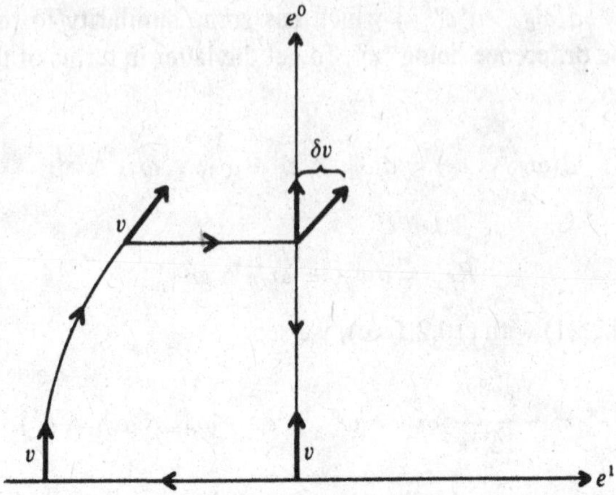

FIGURE 10.2. Parallel displacement of v along the streamlines of e^1 and e^0.

which happens, for instance, if the space is conformally flat ($C_{\alpha\beta} = 0$), then (10.2.11(c)) does determine $R_{\alpha\beta}$. In vacuo ($\mathcal{T}_\alpha = t_\alpha = 0$), $R_{\alpha\beta}$ and $C_{\alpha\beta}$ are the same and conformally flat solutions are flat. In two and three dimensions, where the Weyl-tensor vanishes, all solutions of Einstein's equations in vacuo are flat.

3. If one follows Cartan's suggestion and retains the torsion, generalizing the foregoing argument, then, like R_α, it is determined locally, by the spin density of the matter present [40]. In the absence of spin, the space becomes torsionless, and the theory reduces to the one presented above. Since we know of no objects with sufficiently high spin densities, this variant of the theory agrees with experiment as well as Einstein's.
4. The geometrical significance of the 1-forms R_α determined by (10.2.20(a)) is brought out through the following heuristic argument: Suppose that v points in the time direction e^0. Then if it is translated parallelly around the infinitesimal loop formed by $e^1 \wedge e^0$, it changes by $(\delta v)^1 = v^0 R^1{}_{010} \times$ the surface area, or, summing over the three spatial components,

$$\begin{aligned}(\delta v)^1 + (\delta v)^2 + (\delta v)^3 &= v^0 R^\alpha{}_{0\alpha 0} \times \text{surface area} \\ &= v^0 8\pi\kappa \int [T_{00} - \tfrac{1}{2} g_{00} T^\alpha{}_\alpha].\end{aligned}$$

If e^0 is a geodesic vector field, then v remains in the tangential direction of the geodesic vectors when translated parallelly along e^0. The positivity of the right side of the above equation indicates that the geodesic lines converge (Figure 10.2). This reflects the attractive character of gravity, and contains the seeds of the destruction of space and time that will be discussed in §10.6.

5. The e^i, and thus also the $\mathfrak{e}^\alpha$, cannot in general be defined globally, though they can if there exists a compact, spatially orientable submanifold N without a boundary. It is known that such a manifold is parallelizable, and hence e^1, e^2, and e^3 exist globally on N. If e^0 is taken as the timelike normal vector field,

then ι^α can be defined globally on N. Therefore the statement (7.3.35; 2) that an orientable, closed universe has zero total energy-momentum is always valid.

Now that the observables introduced in §7.3 have been identified, we are ready to derive the finite speed of propagation of the gravitational field, which, as anticipated in §7.1, equals the speed of light. Speaking mathematically, it is a matter of finding the characteristics of the equations (10.2.11(c)). In §7.3, the characteristics were defined as the possible

(10.2.24) **Surfaces of Discontinuity of the Solutions**

Let e^α have possibly discontinuous second derivatives with respect to a local coordinate u, and let t^α and T^α be continuous. Then either $R_{\alpha\beta}$ is continuous, or else $du = n_\alpha e^\alpha$ in an orthogonal basis e^α, where $n^2 := n_\alpha n_\beta \eta^{\alpha\beta} = \langle du|du\rangle = 0$.

(10.2.25) **Remarks**

1. If the e^α are allowed not to be C^∞, then it is of course possible to choose them with discontinuous second derivatives even in flat space, a fact which reflects only the choice of basis. Any genuine discontinuity would have to show up in the $R_{\alpha\beta}$, and (10.2.13) states that such discontinuities can only occur along surfaces with lightlike normals.
2. The equations $\delta\, dA = J$ also allow arbitrary discontinuities in A even if J is continuous; after all, they can be contained in a gauge potential $A = d\Lambda$. The analogous alternatives are that either $F = dA$ is continuous, or else $n^2 = 0$ (cf. (8.2.1)).

Proof of (10.2.24): The part of de^α that contains discontinuous first derivatives must be proportional to du:

$$de^\alpha = (A^\alpha{}_\beta + S^\alpha{}_\beta)\, du \wedge e^\beta,$$
$$A_{\alpha\beta} := \eta_{\alpha\gamma} A^\gamma{}_\beta = -A_{\beta\alpha}, \qquad S_{\alpha\beta} = S_{\beta\alpha}.$$

We have separated the coefficients into a symmetric and an antisymmetric part, because they act differently in $\omega_{\alpha\beta} = -\omega_{\beta\alpha}$. If we accept the following equations modulo continuous terms, then

$$\omega_{\alpha\beta} = -A_{\alpha\beta}\, du + S_\alpha n_\beta - S_\beta n_\alpha, \qquad S_\alpha = S_{\alpha\gamma} e^\gamma,$$

as can be verified by substitution into $de^\alpha = -\omega^\alpha{}_\beta \wedge e^\beta$. By assumption, any possible discontinuity in the curvature originates with $d\omega_{\alpha\beta}$, hence either with S'_α such that $dS_\alpha = du \wedge S'_\alpha$ or with $dA_{\alpha\beta} = A'_{\alpha\beta}\, du$. The latter possibility does not contribute to $d\omega_{\alpha\beta}$, so the discontinuous part of the curvature becomes

$$R_{\alpha\beta} = du \wedge (S'_\alpha n_\beta - S'_\beta n_\alpha),$$

and thus

$$R_\beta \wedge du = i_\alpha R^\alpha{}_\beta \wedge du = (n^\alpha S'_\alpha n_\beta - S'_\beta n^2) \wedge du.$$

But in that case,

$$(R_\beta n_\gamma - R_\gamma n_\beta) \wedge du = n^2(S'_\gamma n_\beta - S'_\beta n_\gamma) \wedge du = n^2 R_{\gamma\beta}.$$

Since R_β must be continuous according to Einstein's equations (10.2.11(c)), either $n^2 = 0$ or else $R_{\gamma\beta}$ stays continuous. □

(10.2.26) The Linear Approximation

The analogy we have just discovered between the characteristics of Einstein's and Maxwell's equations extends in the case of weak fields to a simple wave equation for the gravitational field. Even though we shall not show here when the contributions we drop are actually negligible, the approximation is of value as a first orientation to the problem, especially as the space around us is quite weakly curved.

Let

$$e^\alpha = dx^\alpha + \varphi^\alpha{}_\beta\, dx^\beta, \qquad \varphi_{\alpha\beta} := \eta_{\alpha\sigma}\varphi^\sigma{}_\beta = \varphi_{\beta\alpha},$$

be an orthogonal basis (the symmetry of φ will be justified later), and suppose $|\varphi^\alpha{}_\beta(x)| \ll 1$ for all x. Then from

$$de^\alpha = \varphi^\alpha{}_{\beta,\gamma}\, dx^\gamma \wedge dx^\beta$$

it follows to first order in φ that

$$\omega^\alpha{}_\beta = (\varphi^\alpha{}_{\gamma,\beta} - \varphi_{\beta\gamma,}{}^\alpha)\, dx^\gamma,$$

since (7.2.25) is satisfied to this order. In that case,

$$d\omega^\alpha{}_\beta = (\varphi^\alpha{}_{\gamma,\beta\rho} - \varphi_{\beta\gamma,\rho}{}^\alpha)\, dx^\rho \wedge dx^\gamma,$$
$$i_\alpha\, d\omega^\alpha{}_\beta = (\varphi^\alpha{}_{\gamma,\beta\alpha} - \varphi_{\beta\gamma,}{}^\alpha{}_\alpha - \varphi^\alpha{}_{\alpha,\beta\gamma} + \varphi_{\beta\alpha,}{}^\alpha{}_\gamma)\, dx^\gamma.$$

In harmonic coordinates, where $\delta\, dx^\alpha = 0$,

(10.2.27) $$\varphi_{\beta\alpha,}{}^\alpha = \tfrac{1}{2}\varphi^\alpha{}_{\alpha,\beta}$$

(Problem 2), and Einstein's equations (10.2.11(a)) become

(10.2.28) $$-\varphi_{\beta\gamma,}{}^\alpha{}_\alpha\, dx^\gamma = 8\pi\kappa(T_{\beta\gamma} - \tfrac{1}{2}\eta_{\beta\gamma}T^\alpha{}_\alpha)\, dx^\gamma$$

in this approximation. These can be solved with the Green function (8.2.3):

(10.2.29)

$$\varphi_{\alpha\beta}(\bar{x}) = \varphi^{\text{in}}_{\alpha\beta}(\bar{x}) + 8\pi\kappa \int d^4x\, D^{\text{ret}}(\bar{x} - x)(T_{\alpha\beta}(x) - \tfrac{1}{2}\eta_{\alpha\beta}T^\rho{}_\rho(x))$$
$$\varphi^{\text{in}}_{\alpha\beta,\rho}{}^\rho = 0, \quad \varphi^{\text{in}}_{\beta\alpha,}{}^\alpha = \tfrac{1}{2}\varphi^{\text{in}\alpha}_\alpha{}_{,\beta}.$$

(10.2.30) **Remarks**

1. The $T_{\alpha\beta}\, dx^\beta$ are the energy-momentum forms without the gravivational contribution. It is not inconsistent to neglect gravity in this approximation, because $\kappa T_{\beta\gamma.}{}^{\gamma}$ vanishes to zeroth order. This makes (10.2.29) and (10.2.27) consistent to first order.
2. In the static limit, $\int dt\, D^{\mathrm{ret}}(x-\bar{x}) = \dfrac{1}{4\pi|\mathbf{x}-\bar{\mathbf{x}}|}$, $T_{\alpha\beta} = M j_\alpha j_\beta$, where $j = (-1, \mathbf{v})\delta^3(\mathbf{x})$, $\varphi_{\alpha\beta}$ becomes
$$\varphi_{\alpha\beta}(\bar{x}) = \frac{2M\kappa}{r}(j_\alpha j_\beta + \tfrac{1}{2}\eta_{\alpha\beta});$$
i.e., the same result as stated in (5.6.2), because $g = (\eta_{\alpha\beta} + 2\varphi_{\alpha\beta})\, dx^\alpha \otimes dx^\beta$. This justifies the choice of the factor $8\pi\kappa > 0$. The sign is not dictated by the geometry, but is only found empirically.
3. The analogy with electrodynamics should not make us overlook that, as discussed in Chapter 6, the metric as measured is g and not $\eta_{\alpha\beta}\, dx^\alpha \otimes dx^\beta$, although the difference is not great if the fields are weak.
4. The symmetry of $T_{\alpha\beta}$ justifies our ansatz $\varphi_{\alpha\beta} = \varphi_{\beta\alpha}$ a posteriori.

We close this section by investigating whether the generalization of the calculations of §8.1 and §9.1 connected with conservation laws brings new insights to the case of a gravitational field. Since the metric is not fixed a priori, there are now more invariance properties, and one would expect to find additional conserved quantities.

Returning to equation (10.2.17) for the variation of $\mathcal{L}$, we start by looking at the new contribution from gravitation. If the variation comes from the Lie derivative L_X in the direction of the vector field X, then equation (10.2.17) implies that

(10.2.31)
$$\begin{aligned} L_X({}^*e^{\alpha\beta} \wedge R_{\alpha\beta}) - d({}^*e^{\alpha\beta} \wedge L_X\omega_{\alpha\beta}) &= (L_X e_\alpha) \wedge {}^*T^\alpha \\ &= (i_X\, de_\alpha + d i_X e_\alpha) \wedge {}^*T^\alpha \\ &= -(i_X\omega_{\alpha\sigma})e^\sigma \wedge {}^*T^\alpha + (i_X e^\sigma)\omega_{\alpha\sigma} \wedge {}^*T^\alpha - (i_X e_\alpha)\, d{}^*T^\alpha \\ &\quad + d((i_X e_\alpha){}^*T^\alpha), \end{aligned}$$

where ${}^*T^\alpha := {}^*e^{\alpha\beta\gamma} \wedge R_{\beta\gamma} \in E_3$ and e^α is an orthogonal basis.

If (10.2.31) is integrated over a four-dimensional manifold N without a boundary, and X has compact support in N, then from $\int L_X\omega = 0$ for all $\omega \in E_m$, the invariance of the integral under Lie differentiation (2.6.14), and from $(i_X\omega_{\alpha\sigma})e^\sigma \wedge {}^*T^\alpha = 0$ (because $e^\sigma \wedge {}^*T^\alpha = e^\alpha \wedge {}^*T^\sigma$) we infer that

$$\int_N x_\alpha(d{}^*T^\alpha + \omega^\alpha{}_\sigma \wedge {}^*T^\sigma) = 0, \qquad x_\alpha := i_X e_\alpha.$$

Since this must be true for all vector fields X of compact support in N—no invariance properties have been assumed of X—we obtain the

(10.2.32) **Contracted Bianchi Identity**

$$d^*T^\alpha = -\omega^\alpha{}_\sigma \wedge {}^*T^\sigma.$$

(10.2.33) **Remarks**

1. This fact follows from the more general equation (10.1.26) (Problem 3).
2. The invariance of the integral under general coordinate transformations is expressed by $\int L_X\omega = 0$. No new conservation theorems result from this general covariance, but only identities that hold independently of any field equation (cf. (8.1.8; 5) and (8.1.18; 3)).
3. Although (10.2.32) was derived for orthogonal bases, it has the same form in all bases, since the ${}^*T^\alpha$ transform as ${}^*e^\alpha$.
4. If Einstein's equations $T^\alpha = -16\pi\kappa(T^\alpha + t^\alpha)$ hold, then (10.2.32) implies equation (7.3.28):

$$d({}^*T^\alpha + {}^*t^\alpha) = -\omega^\alpha{}_\beta \wedge ({}^*T^\beta + {}^*t^\beta).$$

5. Because of the symmetry $i_\beta T_\alpha = i_\alpha T_\beta$, the 1-forms T_α have ten linearly independent components, among which (10.2.22) creates four differential identities. Thus only six of Einstein's equations are independent of one another. This is felt to be the correct number of equations for the ten components $g_{\alpha\beta} = g_{\beta\alpha}$ of $g = g_{\alpha\beta}\, dx^\alpha \otimes dx^\beta$: the equations ought not to fix the coordinate system, and so there must remain four arbitrary functions $\bar{x}^\alpha(x)$ to play with.
6. It is part of the relativity folklore that Einstein's theory differs from other field theories inasmuch as Einstein's equations also determine the equations of motion of matter. In particular (10.2.32) is supposed to imply that particles move on geodesics provided no other forces act on them. Whereas the proof of this claim for point particles encounters the difficulty that they generate singular gravitational fields we shall find a simple proof for continuous matter in (10.6.30; 4). There we will show that for an ideal fluid without pressure Bianchi's identity demands that the velocity field be geodesic. Similarly, if one has only one scalar field (10.2.32) implies the field equation. With two scalar fields the condition (10.2.32) for the sum of their energy-momentum tensors cannot imply the field equation because they may interact without changing the conservation of their total energy-momentum. Thus the situation is not too different from Maxwell's equations where the one real field describing charged matter $d^*J = 0$ is already the field equation $d^*(dS + eA) = 0$. For one complex field $\sqrt{\rho}e^{iS}$ the current is $\rho(dS + eA)$ and its conservation does not imply the field equation.

(10.2.34) **Problems**

1. Show that ${}^*e^{\rho\tau} \wedge \delta R_{\rho\tau} = d({}^*e^{\rho\tau} \wedge \delta\omega_{\rho\tau})$.
2. Show that $\delta\, dx^\alpha = 0$ implies that $\varphi_{\alpha\beta,}{}^\beta = \frac{1}{2}\varphi^\beta{}_{\beta,\alpha}$ for weak fields.
3. Derive (10.2.32) from (10.1.26).

4. What effect would a term $\Lambda {}^*\mathbf{1}$ in $\mathcal{L}$ have on Einstein's equations?

5. Calculate ${}^*t_0$ from (10.2.18(c)) on flat space in the orthogonal basis of polar coordinates.

6. Show that $R^\alpha{}_\beta \wedge R^\beta{}_\alpha$ is exact.

7. Use the natural extension of D for $V \in \mathrm{Sp}(L(F))$: $DV = dV + \omega \wedge V + (-1)^p V \wedge \omega$ to show $DD^*\Omega = 0$.

8. Show

$$\tfrac{1}{2} R_{\alpha\beta} \wedge {}^*e^{\alpha\beta} = d(e^\alpha \wedge {}^*de_\alpha) - \tfrac{1}{2}(de^\alpha \wedge e^\beta) \wedge {}^*(de_\beta \wedge e_\alpha)$$
$$+ \tfrac{1}{4}(de^\alpha \wedge e_\alpha) \wedge {}^*(de^\beta \wedge e_\beta).$$

Hint: Use (10.2.23) and (10.2.14; 3).

9. Show that the total energy ("ADM-energy")

$$\frac{-1}{8\pi\kappa}\int_N d^*F^0 = \frac{-1}{8\pi\kappa}\int_{\partial N} {}^*F^0,$$

where $\partial N = \{x \in \mathbb{R}^4 : |\vec{\mathbf{x}}| = R, x_0 = 0\}$ in an asymptotically flat space is given by

$$\lim_{R\to\infty} \frac{1}{16\pi\kappa}\int_{\partial N} {}^*e^{jo}(g_{jk,k} - g_{kk,j}) \qquad \text{with} \quad j, k = 1, \ldots, 3.$$

(Use a natural basis $e^k = dx^k$, (10.1.37) for ω and that $g_{ik} \to \eta_{ik}$ for $R \to \infty$.)

(10.2.35) Solutions

1.

$$d({}^*e^{\rho\tau} \wedge \delta\omega_{\rho\tau}) = -2\omega^\rho{}_\sigma \wedge {}^*e^{\sigma\tau} \wedge \delta\omega_{\rho\tau} + {}^*e^{\rho\tau} \wedge d\delta\omega_{\rho\tau}$$
$$= {}^*e^{\rho\tau} \wedge (d\delta\omega_{\rho\tau} + 2\omega_\rho{}^\sigma \wedge \delta\omega_{\sigma\tau}) = {}^*e^{\rho\tau} \wedge \delta R_{\rho\tau}.$$

2. To first order, $dx^\alpha = e^\alpha - \varphi^\alpha{}_\beta e^\beta$; so

$$0 = {}^*\delta\, dx^\alpha = d({}^*e^\alpha - \varphi^\alpha{}_\beta {}^*e^\beta) = -\omega^\alpha{}_\beta \wedge {}^*e^\beta - \varphi^\alpha{}_{\beta,\gamma} e^\gamma \wedge {}^*e^\beta$$
$$= (\varphi_{\beta\gamma,}{}^\alpha - \varphi^\alpha{}_{\gamma,\beta} - \varphi^\alpha{}_{\beta,\gamma}) e^\gamma \wedge {}^*e^\beta.$$

By (7.2.28), however, $e^\gamma \wedge {}^*e^\beta = \eta^{\gamma\beta}\,{}^*\mathbf{1}$.

3.

$$d^*T^i = d({}^*e^{imk} \wedge R_{mk}) = -\omega^i{}_j \wedge {}^*e^{jmk} \wedge R_{mk} - 2\omega^m{}_s \wedge {}^*e^{iskh} \wedge R_{mk}$$
$$+ 2{}^*e^{imk} \wedge \omega_m{}^s \wedge R_{sk} = -\omega^i{}_j \wedge {}^*T^j.$$

4. $\delta(\Lambda {}^*\mathbf{1}) = \Lambda \delta e^j \wedge i_j {}^*\mathbf{1} = \Lambda\, \delta e^j \wedge {}^*e_j$ (see (7.2.18(c)). Thus there is an additional term $\sim {}^*e_j$ in ${}^*T_j$.

5.

$$e^\alpha = (dt, dr, r\,d\vartheta, r\sin\vartheta\,d\varphi),\quad \omega_{12} = -d\vartheta,\quad \omega_{23} = -\cos\vartheta\,d\varphi,\quad \omega_{31} = \sin\vartheta\,d\varphi,$$
$$\tfrac{1}{2}\varepsilon_{0\beta\gamma\delta}d(\omega_{\beta\gamma}\wedge e_\delta) - d(-\cos\vartheta\,d\varphi\wedge dr + 2r\sin\vartheta\,d\varphi\wedge d\vartheta)$$
$$= -\sin\vartheta\,dr\wedge d\vartheta\wedge d\varphi.$$

6. $R^\alpha{}_\beta\wedge R^\beta{}_\alpha = d[\omega^\alpha{}_\gamma\wedge d\omega^\gamma{}_\beta + \tfrac{2}{3}\omega^\alpha{}_\beta\wedge\omega^\beta{}_\gamma\wedge\omega^\gamma{}_\alpha]$.

7.

$$\begin{aligned} D^*\Omega &= d^*\Omega + \omega\wedge{}^*\Omega - {}^*\Omega\wedge\omega,\\ DD^*\Omega &= d\omega\wedge{}^*\Omega - {}^*\Omega\wedge d\omega - \omega\wedge d^*\Omega - d^*\Omega\wedge\omega\\ &\quad + \omega\wedge(d^*\Omega + \omega\wedge{}^*\Omega - {}^*\Omega\wedge\omega) + (d^*\Omega + \omega\wedge{}^*\Omega - {}^*\Omega\wedge\omega)\wedge\omega\\ &= \Omega\wedge{}^*\Omega - {}^*\Omega\wedge\Omega = 0 \end{aligned}$$

according to (7.2.17).

8. According to (10.2.14; 3) ${}^*e^{\alpha\beta}\wedge R_{\alpha\beta} = d({}^*e^{\alpha\beta}\wedge\omega_{\alpha\beta}) + {}^*e^{\beta\gamma}\omega_\gamma{}^\alpha\wedge\omega_{\alpha\beta}$. Now always using (7.2.18),

$$e_\alpha\wedge{}^*de^\alpha = {}^*e^\beta\langle\omega^\alpha{}_\beta|e_\alpha\rangle$$

and

$$\begin{aligned} ({}^*e^{\alpha\beta})\wedge\omega_{\alpha\beta} &= {}^*i_{\omega_{\alpha\beta}}e^{\alpha\beta} = 2{}^*e^\beta\langle\omega^\alpha{}_\beta|e_\alpha\rangle \quad\Rightarrow\\ \tfrac{1}{2}{}^*e^{\alpha\beta}\wedge\omega_{\alpha\beta} &= e^\alpha\wedge{}^*de^\alpha. \end{aligned}$$

Next, with (10.2.23),

$$\begin{aligned} -de^\alpha\wedge{}^*F_\alpha &= \tfrac{1}{2}\omega_{\alpha\sigma}\wedge e^\sigma\wedge\omega_{\beta\gamma}\wedge{}^*e^{\alpha\beta\gamma}\\ &= \tfrac{1}{2}\omega_{\alpha\sigma}\wedge\omega_{\beta\gamma}[\eta^{\sigma\alpha}{}^*e^{\beta\gamma} - \eta^{\sigma\beta}{}^*e^{\alpha\gamma} + \eta^{\sigma\gamma}{}^*e^{\alpha\beta}]\\ &= {}^*e^{\gamma\alpha}\wedge\omega_\gamma{}^\beta\wedge\omega_{\alpha\beta}. \end{aligned}$$

Inserting the expression (10.2.22) for ${}^*F_\alpha$ completes the demonstration of the equality.

9. With (10.2.23) we have

$$-{}^*F^0 = -\tfrac{1}{2}(g_{ik,j} + g_{ij,k} - g_{jk,i})\tfrac{1}{2}e^j\wedge{}^*e^{ik0} = \tfrac{1}{2}(-g_{ii,k} + g_{ik,i}){}^*e^{0k},$$

since by symmetry the first term cancels out and the two others give the same contribution.

10.3 Maximally Symmetric Spaces

The spaces with the simplest structure, after flat spaces, are those of constant curvature. They are a generalization of the spherical surface and, though simple, have some physically interesting aspects.

Killing vector fields generate isometries (i.e., diffeomorphisms that leave g invariant) of the space and are bijectively related to the constants of motion and conserved currents. The fields need not be complete; it is possible for their flow to lead out of the manifold (see (2.3.7)). However, they imprint a local structure on the space even when they do not generate one-parameter groups of isometries. Generally there are none, but if there are enough of them around, then Einstein's equations become more tractable, and explicit calculations are possible.

The prototype of a Killing vector field is a rotation of $\mathbb{R}^m$; for some pair of indices (i, k), $v^i = x^k$ and $v^k = -x^i$, and the other components are zero. Note that:

(a) $v^i{}_{,k} + v^k{}_{,i} = 0$; and

(b) $v^l{}_{,jk} = 0$.

The generalizations of these facts to pseudo-Riemannian spaces are

(10.3.1) The Relationships Among the Covariant Derivatives of Killing Vector Fields v

We use the natural basis $e_\alpha = \partial_\alpha$ and the notation

$$\langle D_{e_\beta} v | e_\alpha \rangle =: v_{\alpha;\beta}, \quad \langle (D_{e_\gamma} D_{e_\beta} v - D_{D_{e_\gamma} e_\beta} | e_\alpha \rangle =: v_{\alpha;\beta;\gamma}.$$

Then, with the $R^\lambda{}_{\sigma\rho\mu}$ of (10.1.24; 2):

(a) $v_{\alpha;\beta} + v_{\beta;\alpha} = 0$; *and*

(b) $v_{\mu;\rho;\sigma} = R^\lambda{}_{\sigma\mu\rho} v_\lambda$.

Proof:

(a) Killing vector fields leave the scalar product invariant, $L_v\langle X|Y\rangle = \langle L_v X|Y\rangle + \langle X|L_v Y\rangle$. On the other hand, with Axioms (10.1.6(f)) and (g′),

$$\begin{aligned} L_v\langle X|Y\rangle = D_v\langle X|Y\rangle &= \langle D_v X|Y\rangle + \langle X|D_v Y\rangle \\ &= \langle D_X v|Y\rangle + \langle X|D_Y v\rangle + \langle L_v X|Y\rangle + \langle X|L_v Y\rangle. \end{aligned}$$

Combined, these make $\langle D_X v|Y\rangle + \langle X|D_Y v\rangle = 0$; i.e., (a), if X and Y are taken as the basis fields.

(b) Since a Killing vector field v preserves the metric structure and a torsion-free connection is uniquely determined by the metric, L_v commutes with the covariant exterior derivative D. To demonstrate this formally one has to consider the one-parameter (local) group Φ_t of isometries generated by v and observe that

$$\frac{\partial}{\partial t} D\Phi_t \bigg|_{t=0} = \frac{\partial}{\partial t} \Phi_t D \bigg|_{t=0}.$$

Thus we infer from (10.1.39) that $R(v, X)u = \langle D_X Dv|u\rangle$. Using the properties (10.1.44; 2) of R this becomes (b) when written in index notation.

(10.3.2) **Discussion**

Part (b) allows the second derivative of v to be written linearly in v. By carrying the procedure further, one can reduce all the higher derivatives to v and its first derivatives. If we assume analyticity, we can express v locally in terms of v and its first derivative at a single point, say 0:

$$v_\rho(x) = A_\rho{}^\lambda(x) v_\lambda(0) + B_\rho{}^{\lambda\sigma}(x) v_{\lambda;\sigma}(0).$$

This greatly restricts the number of possible Killing vector fields. In an m-dimensional space, $v_{\lambda;\sigma}(0) = -v_{\sigma;\lambda}(0)$ can assume $m(m-1)/2$ values, and $v_\lambda(0)$ can assume m values. Hence there are at most $m + m(m-1)/2 = m(m+1)/2$ independent Killing vector fields.

(10.3.3) **Remarks**

1. "Independent" means that they have no linear relationships with constant coefficients. They may satisfy equations with variable coefficients, just as rotations $x_j \partial_i - x_i \partial_j$ can be expressed in terms of the translations ∂_i. Since the Killing vector fields do not form a module (8.1.10; 1), this does not mean dependence.
2. On flat space the Euclidean group is the largest group of isometries (4.1.13; 4), and it has exactly $m(m+1)/2$ parameters. The statement that the group can be at most this large if space is curved is therefore quite plausible.

To classify the symmetric spaces, we begin with a

(10.3.4) **Definition**

(a) A space is **maximally symmetric** iff it possesses $m(m+1)/2$ independent Killing vector fields.

(b) A space is **isotropic about the point** x iff it has $m(m-1)/2$ Killing vector fields, for which x is a fixed point of the flow, and the $A^i{}_k$ of $(L_v e^i)(x) = A^i{}_k(x) e^k(x)$, where e^k are an orthogonal basis, generate the total Lorentz group of $(\mathbb{R}^m, \eta)$.

(c) A space is **isotropic** iff it is isotropic about all of its points.

(d) A space is **homogeneous** if it has a transitive[4] group of isometries.

(e) A space is **stationary** if it has a timelike Killing vector field. If the latter is orthogonal to a family of spacelike hypersurfaces the space is called **static**.

---

[4]Transitive means that any point can be reached from every point by a transformation from the group.

(10.3.5) Remarks

1. Definition (b) means that there are $m(m-1)/2$ Killing vector fields v^i such that $v^i{}_\alpha(x) = 0$, and the $v^i{}_{\alpha;\beta}(x)$ form a basis for the space of $m \times m$ antisymmetric matrices, where i runs from 1 to $m(m-1)/2$.
2. On a maximally symmetric space there are Killing vector fields as in (b) at every point. Therefore it is isotropic.
3. To appreciate the distinction made in (e) we remark that in general for a vector field X there is locally a family of transversal hypersurfaces N_α such that X and the vectors in N_α span the tangent space. However, this hypersurface will not be orthogonal to X in the sense of a given metric and there will be no orthogonal family of hypersurfaces. In contradistinction to the above, for a given spacelike hypersurface one can find, at least locally, an orthogonal geodesic vector field.

The curvature must be the same in every direction on an isotropic space. Because of this, there is an extremely simple

(10.3.6) Structure of the Curvature Forms of an Isotropic Space

On an isotropic space,

$$R^{ik} = K e^{ik}, \quad \textit{for some constant } K.$$

Proof: Let Φ be an isometry and Φ_* the translation it induces on tensors. We noted that Φ_* commutes with D and from $DD\Phi_* = \Phi_* DD$ we conclude $R\Phi_* = \Phi_* R$. Let Φ be an isometry with x as fixed point such that $(\Phi_* e^i)(x) = L^i{}_k e^k(x)$ with $L^t \eta L = \eta$. If R is decomposed in this basis, then its invariance under Φ_* means that the Riemann–Christoffel-tensor (10.1.42; 2) is invariant under $R_{ijkl} \to R_{mnop} L^m{}_i L^n{}_j L^o{}_k L^p{}_l$; it transforms as a matrix in the antisymmetric tensor product space. This representation is irreducible (except for $m = 4$) and the matrix group elements must be proportional to the unit matrix: $R_{ij}{}^{lm} = K(\delta_i{}^l \delta_j{}^m - \delta_i{}^m \delta_j{}^l)$ or, for the curvature forms, $R_{ij}(x) = K(x) e_{ij}(x)$. For $m = 4$ there would be the possibility $R_{ij} = K^* e_{ij}$, which is excluded by $R_{ij} \wedge e^j = 0$. In order to see why K has to be constant, consider the equation $dR_{ik} = dK \wedge e_{ik} + K\, de_{ik}$. By using Bianchi's identity (10.1.26), we discover that $dK = 0$, and hence K is independent of x. (For $m = 2$ another argument works.) □

(10.3.7) Remarks

1. If the curvature is independent of the direction, then it is also independent of the position. For that reason, one says simply that such spaces have constant curvature.
2. By (10.1.44; 3) and (10.3.6), isotropic spaces have vanishing Weyl forms, and hence they are conformally flat if $m > 3$.
3. It is even true that the existence of more than $m(m-1)/2 + 2$ Killing vector fields implies that $R_{ik} = K e_{ik}$ (cf. [23]).

(10.3.8) Construction of Isotropic Spaces with $m \geq 3$

Since such spaces are conformally flat, there is an orthogonal basis of the form $e^a = dx^a/\psi$, $\psi \in E_0$. Therefore $de^a = \psi_{,b} e^{ba}$, and thus $\omega_{ab} = \psi_{,a} e_b - \psi_{,b} e_a$. With the help of (10.1.45) again, this leads to the curvature forms

$$R_{ab} = \psi(\psi_{,ac} e^c{}_b - \psi_{,bc} e^c{}_a) - \psi_{,c} \psi^{,c} e_{ab}.$$

If this is to equal $K e_{ab}$, then $\psi_{,ab}$ must be zero for all $a \neq b$, and hence

$$\psi = \sum_{i=1}^{m} f^i(x^i).$$

If $f_a = \eta_{ai} f^i$, then (10.3.6) implies that

$$f_b'' + f_a'' = \psi^{-1}(K + f_c' f^{c\prime}).$$

Since the left side depends only on x^a and x^b, while the right side is the same for all a and b, both sides are in fact constant. This makes f a quadratic function, and therefore ψ can be put into the form

$$\psi = 1 + \frac{K}{4} x^a x^b \eta_{ab}.$$

Therefore, locally a space of constant curvature always has coordinates for which

$$g = \frac{dx^i\, dx^k\, \eta_{ik}}{(1 + Kx^2/4)^2}, \qquad e^i = \frac{dx^i}{1 + Kx^2/4},$$
(10.3.9)
$$\omega^{ik} = \frac{K}{2}(x^i e^k - x^k e^i)$$

(where $x^2 := x^a x^b \eta_{ab}$).

(10.3.10) The Killing Vector Fields of Isotropic Spaces

Because of the isotropy about the origin, the generators of rotations are Killing vector fields: Let $v = x_j \partial_k - x_k \partial_j$, where $x_j = \eta_{jk} x^k$ and ∂_k is the dual basis to $dx^k (i_{\partial_k} dx^i = \delta^i{}_k)$, and fix a pair of indices (j, k); then

$$\begin{aligned} L_v e^m &= d i_v e^m + i_v de^m \\ &= d\left(x_j \left(1 + \frac{Kx^2}{4}\right)^{-1} i_{\partial_k} dx^m\right) \\ &\quad - x_j \frac{K}{2}\left(1 + \frac{Kx^2}{4}\right)^{-2} x_l i_{\partial_k} dx^l \wedge dx^m - (j \leftrightarrow k) \\ &= d\left[x_j \delta^m{}_k \left(1 + \frac{Kx^2}{4}\right)^{-1}\right] \end{aligned}$$

$$+\frac{K}{2}x_j\left(1+\frac{Kx^2}{4}\right)^{-2}(\delta^m{}_k x_l\,dx^l - x_k\,dx^m) - (j \leftrightarrow k)$$

$$= dx_r\left(1+\frac{Kx^2}{4}\right)^{-1}(\delta^m{}_k\delta^r{}_j - \delta^m{}_j\delta^r{}_k).$$

Since $L_v e^m = A^{mr} e_r$ and $A^{mr} = \delta^m{}_k\delta^r{}_j - \delta^m{}_j\delta^r{}_k = -A^{rm}$, rotations about the origin are Killing vector fields. (See (8.1.10; 2).)

Moreover, there are generalizations of translations: If k is fixed, then $v = \partial^k(1 - x^2K/4) + (K/2)x^k x_j\partial^j$ is a Killing vector field. This can be verified as in (10.3.10), or, what is easier in this case, one can verify the equation of Problem 1 for the components v^i and g_{lm} of v and g in the natural basis,

$$0 \stackrel{?}{=} v^i g_{lm,i} + g_{im}v^i{}_{,l} + g_{il}v^i{}_{,m}$$

$$= \left(1+\frac{Kx^2}{4}\right)^{-3}\left\{-\eta_{lm}\left[Kx^k\left(1-\frac{x^2K}{4}\right)+\frac{x^kx^2K^2}{2}\right]\right.$$

$$+\left(1+\frac{Kx^2}{4}\right)\left[-\frac{K}{2}\delta^k{}_m x_l + \frac{K}{2}(x^k\eta_{ml} + x_m\delta^k{}_l) - \frac{K}{2}\delta^k{}_l x_m\right.$$

$$\left.\left. + \frac{K}{2}(x^k\eta_{ml} + x_l\delta^k{}_m)\right]\right\}$$

$$= 0.$$

These v's form a basis for small x, and so the group they generate in a neighborhood of the origin acts transitively: any point can be sent to any other. We can assemble our discoveries in a

(10.3.11) **Proposition**

For a pseudo-Riemannian manifold M, the following properties are equivalent:

(a) *M is maximally symmetric;*

(b) *M has constant curvature;*

(c) *M is isotropic; and*

(d) *M is homogeneous and isotropic about some point.*

(10.3.12) **Remark**

Our arguments have been strictly local, so nothing can be concluded about the global behavior—the Killing vector fields need not even be complete. Unions and pieces of spherical surfaces are also isotropic spaces. However, isotropic spaces with the same K are locally isometric.

Every m-dimensional manifold can be imbedded as a submanifold in $\mathbb{R}^{2m+1}$. Isotropic spaces can be imbedded in $\mathbb{R}^{m+1}$, whereby the metric is the restriction of a pseudo-Euclidean metric on $\mathbb{R}^{m+1}$. (Restricting the metric means that the scalar product determining it is restricted to the vectors in the tangent space of the submanifold.)

(10.3.13) **The Geometrical Imbedding of Isotropic Spaces**

Let $\eta_{\alpha\beta}$, $\alpha, \beta = 0, 1, \ldots, m-1$, fix the sign of the metric of an isotropic space, and choose the curvature $|K|$ as the unit of length, so that $K = \pm 1$. Let the η_{ik} on $\mathbb{R}^{m+1}$ equal $\eta_{\alpha\beta}$ for $i, k = 0, 1, \ldots, m-1$, and let $\eta_{mm} = K$. Then the isotropic space is locally isometric to the submanifold $H = \{\bar{x} \in \mathbb{R}^{m+1} : \bar{x}^i \bar{x}^k \eta_{ik} = K\}$, where g is the restriction of $d\bar{x}^i \otimes d\bar{x}^k \eta_{ik}$ to H.

Proof: The equation for H can be written as ($m = \dim H$)

$$K \bar{x}^m \bar{x}^m = K - \eta_{\alpha\beta} \bar{x}^\alpha \bar{x}^\beta \equiv K - \bar{r}^2.$$

Introduce the coordinates $x^\alpha \in \mathbb{R}^m$ on H by

$$\bar{x}^\alpha = \frac{x^\alpha}{1 + Kr^2/4}, \qquad \bar{x}^m = \frac{1 - Kr^2/4}{1 + Kr^2/4},$$

where $r^2 = x^\alpha x^\beta \eta_{\alpha\beta}$; then

$$d\bar{x}^\alpha = \frac{dx^\alpha (1 + Kr^2/4) - K x^\alpha r\, dr/2}{(1 + Kr^2/4)^2},$$

$$d\bar{x}^m = \frac{-Kr\, dr}{(1 + Kr^2/4)^2},$$

and the restriction of the metric

$$d\bar{x}^i\, d\bar{x}^k \eta_{ik} = d\bar{x}^\alpha\, d\bar{x}^\beta \eta_{\alpha\beta} + K\, d\bar{x}^m\, d\bar{x}^m = dx^\alpha\, dx^\beta\, \eta_{\alpha\beta} \left(1 + \frac{Kr^2}{4}\right)^{-2}$$

takes on the form (10.3.9). □

(10.3.14) **Remarks**

1. The Killing vector fields of (10.3.10) are simply the restrictions of the generators $\bar{x}^k \bar{\partial}^j - \bar{x}^j \bar{\partial}^k$ of the Lorentz group of $\mathbb{R}^{m+1}$ to H. Their flow leaves H and the metric $\eta_{ik}\, d\bar{x}^i\, d\bar{x}^k$ invariant, and hence so does their restriction to H. The group of isometries of spaces of maximal symmetry is therefore isomorphic to some Lorentz group $O(n, m+1-n)$.
2. The geodesics are the intersections of H with hyperplanes passing through the origin. They are the solutions of a variational problem $\delta \int ds\, \dot{\bar{x}}^i \dot{\bar{x}}^k \eta_{ik} = 0$ with the constraint $\bar{x}^i \bar{x}^k \eta_{ik} = K$. The introduction of a Lagrange multiplier λ leads to the equations $\ddot{\bar{x}}^i = \lambda \bar{x}^i$, making the

$$L^{ik} = \bar{x}^i \dot{\bar{x}}^k - \bar{x}^k \dot{\bar{x}}^i$$

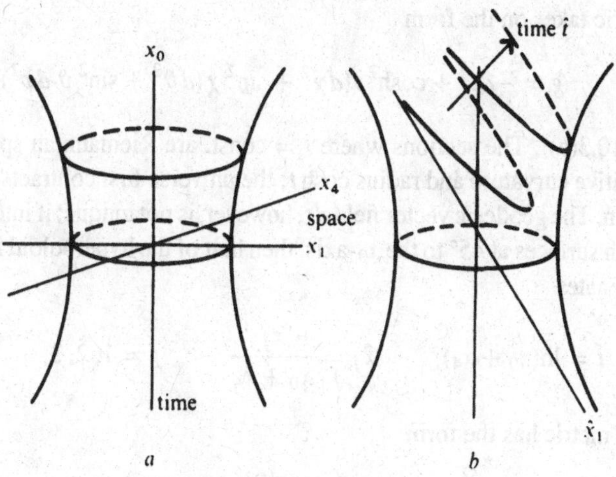

FIGURE 10.3. Various cross-sections of the de Sitter universe.

constant. Thus $\bar{x}$ lies on the plane of $\bar{x}(0)$, $\dot{\bar{x}}(0)$ since

$$0 = \varepsilon_{ike}\bar{x}^i L^{ke} = \varepsilon_{ihe}\bar{x}^i \bar{x}^k x^l.$$

(10.3.15) **The Physical Significance of Isotropic Spaces**

Spaces of a high degree of symmetry gratify the esthetic feelings of physicists, and are therefore popular as models of the world. Since the space around us is isotropic and homogeneous as far as we can see, its isotropy about every point is often elevated to a cosmological principle. Some theorists went beyond this and require a maximal symmetry for space and time, the "perfect cosmological principle." But aside from such cosmological speculation, maximally symmetric spaces are also important as solutions of Einstein's equations when the energy-momentum distribution is sufficiently symmetric. It is necessary to distinguish between the cases of positive and negative K ($K = 0$ is Minkowski space):

(10.3.16) **$K = 1$. The de Sitter Universe**

This can be represented as the hyperboloid

$$-x_0^2 + x_1^2 + x_2^2 + x_3^2 + x_4^2 = 1$$

in $\mathbb{R}^5$ with $g = -dx_0^2 + dx_1^2 + dx_2^2 + dx_3^2 + dx_4^2$. When reduced to $\mathbb{R}^3$, it looks as in Figure 10.3. The intersections with planes containing the x_0-axis are timelike geodesics. If these geodesics are introduced as coordinate lines with the proper time as a new coordinate t (these are known as **synchronous**, or **comoving**, coordinates; see Problem 5),

$$x_0 = \sinh t, \qquad x_1 = \cosh t \sin\chi \sin\vartheta \cos\varphi,$$
$$x_2 = \cosh t \sin\chi \sin\vartheta \sin\varphi, \qquad x_3 = \cosh t \sin\chi \cos\vartheta,$$
$$x_4 = \cosh t \cos\chi,$$

then the metric takes on the form

$$g = -dt^2 + \cosh^2 t\{d\chi^2 + \sin^2\chi(d\vartheta^2 + \sin^2\vartheta\, d\varphi^2)\} \tag{10.3.17}$$

(see Figure 10.3(a)). The sections where t = const. are Riemannian spaces with constant positive curvature and radius $\cosh t$; the universe first contracts, and then expands again. The geodesic vector field dt, however, is not unique; if intersections are taken with surfaces at 45° to the x_0-axis, then half of the hyperboloid is covered by the coordinates

$$\hat{t} = \ln(x_0 + x_4), \qquad \hat{x}_j = \frac{x_j}{x_0 + x_4}, \qquad j = 1, 2, 3,$$

in which the metric has the form

$$g = -d\hat{t}^2 + e^{2\hat{t}}(d\hat{x}_1^2 + d\hat{x}_2^2 + d\hat{x}_3^2) \tag{10.3.18}$$

(see Figure 10.3(b)). The intersections where $\hat{t}$ = const. are expanding Euclidean spaces, in the sense that the geodesics $\hat{x}_j$ = const., $j = 1, 2, 3$, grow steadily farther apart.[5] There are a great many facets of the de Sitter universe; we shall even be able to find coordinates in which g_{ik} does not depend on time at all. It will furthermore turn out in (10.4.42) to be a static space. In order to survey the causal relationships better, it is convenient to map the whole space into a compact set, in what is known as a **Penrose diagram.** To this end, we write the t of (10.3.17) as

$$t' = 2\arctan(\exp t) - \frac{\pi}{2},$$

which makes the metric

$$\begin{aligned} g &= \cosh^2(t)(-dt'^2 + d\chi^2 + \sin^2\chi\, d\Omega^2), \\ d\Omega^2 &= d\vartheta^2 + \sin^2\vartheta\, d\varphi^2, \quad 0 < \chi < \pi, \quad -\frac{\pi}{2} < t' < \frac{\pi}{2}. \end{aligned} \tag{10.3.19}$$

The conformal equivalence to Minkowski space is again evident if the coordinates

$$t + r = \tan\frac{t' + \chi}{2}, \qquad t - r = \tan\frac{t' - \chi}{2},$$
$$0 < \chi < \pi, \qquad -\pi + \chi < t' < \pi - \chi,$$

are used to turn the Minkowski metric into

$$g = \left[2\cos\frac{t' + \chi}{2}\cos\frac{t' - \chi}{2}\right]^{-2}(-dt'^2 + d\chi^2 + \sin^2\chi\, d\Omega^2). \tag{10.3.20}$$

---

[5]This is observable, since, for example, the proton and electron in a hydrogen atom do not move geodesically, but are electrically bound together, which keeps the Bohr radius from expanding with the geodesics.

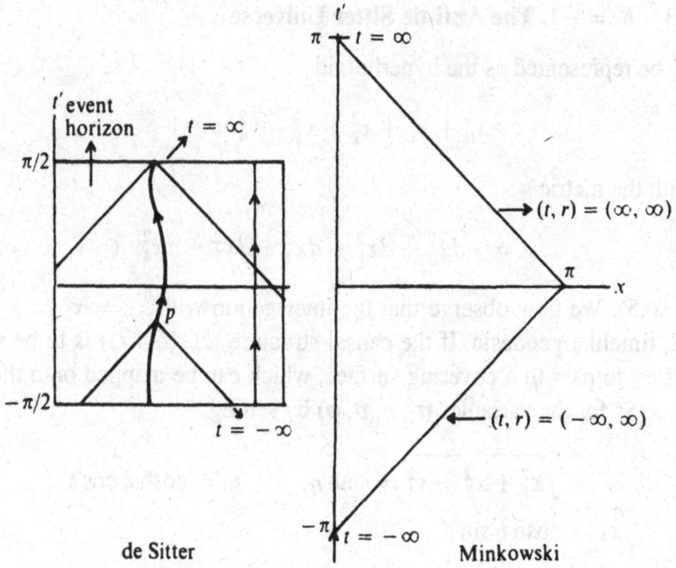

FIGURE 10.4. Penrose diagrams for flat and curved spaces.

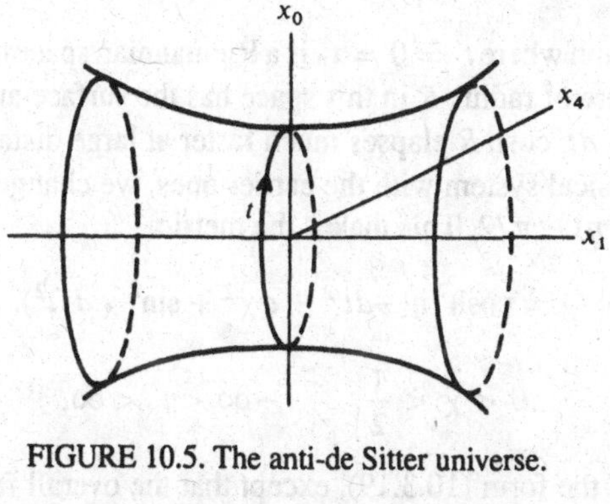

FIGURE 10.5. The anti-de Sitter universe.

Both the de Sitter universe and Minkowski space are mapped into a relatively compact part of $\mathbb{R} \times S^3$. The difference between the causal structures of the two spaces comes about because they cover different parts of the (t', χ)-plane (Figure 10.4). In de Sitter space, timelike geodesics begin on the lines $t' = -\pi/2$, $0 < \chi < \pi$, and end at $t' = \pi/2$, $0 < \chi < \pi$. There are some that do not intersect the past of a given point p, and so an observer at p would be unaware of them—they are "beyond the particle horizon." Conversely, the union of the past light-cones of a particle's trajectory does not fill the whole space; there are points that an observer on a geodesic would never see, "beyond the event horizon." This does not occur in Minkowski space, where a timelike geodesic begins at a point $(t', \chi) = (-\pi, 0)$ and ends at $(\pi, 0)$. Only an accelerated particle could emerge from within the line $(t, r) = (-\infty, \infty)$ of Figure 10.4, and in that case it is possible for two accelerated observers never to see each other (recall (6.4.10; 2)).

(10.3.21) $K = -1$. **The Anti-de Sitter Universe**

This can be represented as the hyperboloid

$$-x_0^2 + x_1^2 + x_2^2 + x_3^2 - x_4^2 = -1$$

in $\mathbb{R}^5$ with the metric

$$g = -dx_0^2 + dx_1^2 + dx_2^2 + dx_3^2 - dx_4^2$$

(Figure 10.5). We now observe that the intersection with $x_1 = x_2 = x_3 = 0$ is a closed, timelike geodesic. If the causal structure (cf. (6.4.7)) is to be saved, it is necessary to pass to a covering surface, which can be mapped onto the region $\mathbb{R} \times \mathbb{R}^+ \times S^2$ for the variables $(t', \eta, \vartheta, \varphi)$ by setting

$$r = \sqrt{x_1^2 + x_2^2 + x_3^2} = \sinh \eta, \qquad x_0 = \cosh \eta \cos t',$$
$$x_4 = \cosh \eta \sin t'.$$

The metric is turned into

(10.3.22) $$g = -\cosh^2 \eta \, dt'^2 + d\eta^2 + \sinh^2 \eta \, d\Omega^2.$$

Now the intersection where $t' = 0 = x_4$ is a Riemannian space of constant negative curvature. A sphere of radius R in this space has the surface area $4\pi \sinh^2 R$, and proper time $ds = dt' \cosh R$ elapses much faster at large distances. To be able to compare this physical system with the earlier ones, we change to radial variables $\chi = 2 \arctan(\exp \eta) - \pi/2$. This makes the metric

(10.3.23) $$g = \cosh^2 \eta(-dt'^2 + d\chi^2 + \sin^2 \chi \, d\Omega^2),$$
$$0 < \chi < \frac{\pi}{2}, \qquad -\infty < t' < \infty,$$

which is again of the form (10.3.19), except that the overall factor depends on χ rather than t'. The Penrose diagram is an infinite strip, which shows that in anti-de Sitter space there are no Cauchy surfaces at all. If we take the intersection with a spacelike surface at $\vartheta = 0$, then it must lie at an angle of less than 45 in the Penrose diagram. It is always possible to find a light ray along $\vartheta = 0$ which will never intersect this surface (Figure 10.6). Thus, in the covering surface, which has infinitely many sheets, time has the unusual property that for any infinite spacelike surface it is possible to find an event at a much later time having no causal connection with it.

(10.3.24) **Einstein's Equations for Isotropic Spaces**

If $R_{\alpha\beta} = K e_{\alpha\beta}$, then $R_\beta = i_\alpha R^\alpha{}_\beta = 3K e_\beta$, $R = i_\alpha R^\alpha = 12K$, and $R_\alpha - e_\alpha R/2 = -3K e_\alpha$, so the energy-momentum tensor is $T_{\alpha\beta} = -\eta_{\alpha\beta} 3K/8\pi\kappa$. In the phenomenological description (9.1.25; 3), this corresponds to a fluid at rest

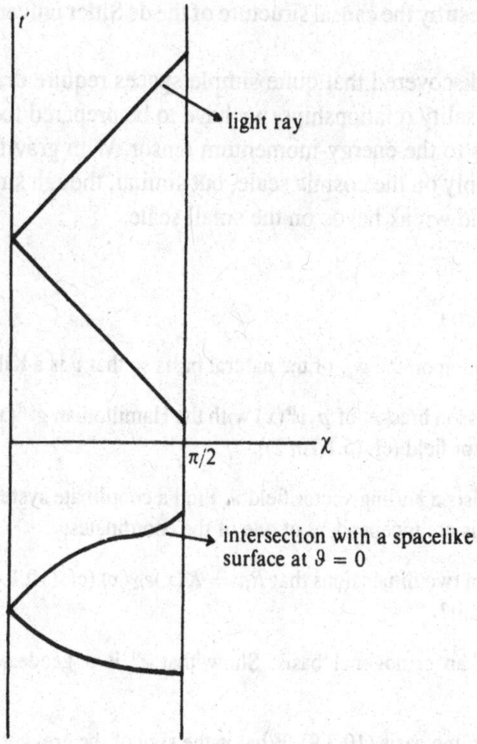

FIGURE 10.6. Penrose diagram for the covering surface of the anti-de Sitter universe.

with energy density = −pressure = $3K/8\pi\kappa$. A negative pressure is necessary to maintain the de Sitter universe ($K > 0$), and a negative energy is needed for the anti-de Sitter universe ($K < 0$); of course the distributions must also be homogeneous. Such unphysical $T_{\alpha\beta}$'s could be caused by:

(a) a contribution Λ^*1 in $\mathcal{L}$ (cf. (10.2.14; 2));

(b) the vacuum expectation value of the energy-momentum tensor of fields, which can be $\sim \simeq \eta_{\alpha\beta}$ for reasons of invariance; and

(c) extra terms in Einstein's equations [50].

There is no very persuasive reason to believe in any of these suggestions, so, despite their esthetic appeal, maximally symmetric spaces are not the front-running candidates for models of the universe; in fact, the empirical evidence goes somewhat against them.

(10.3.25) **Remarks**

1. Einstein's equations control only the local behavior of space, and not its global structure. They do not say whether it is necessary to enclose the anti-de Sitter universe in a covering surface to save the causal structure, or whether it is

permissible to destroy the causal structure of the de Sitter universe by identifying antipodal points.

2. Since we have discovered that quite simple spaces require drastic revisions of our familiar causality relationships, we have to be prepared for the worst when we couple fields to the energy-momentum tensor. With gravitation, these phenomena occur only on the cosmic scale, but similar, though strong, interactions for hadrons could wreak havoc on the small scale.

(10.3.26) Problems

1. What is the condition on the g_{lm} of the natural basis so that v is a Killing vector field?
2. Show that the Poisson bracket of $p_\alpha v^\alpha(x)$ with the Hamiltonian $g^{\alpha\beta}(x)p_\alpha p_\beta$ vanishes iff v is a Killing vector field (cf. (5.1.10; 2)).
3. Suppose there exists a Killing vector field v. Find a coordinate system in which the g_{lm} of the natural basis are independent of one of the coordinates.
4. It is always true in two dimensions that $R_{ik} = K(x)e_{ik}(x)$ (cf. (10.1.43)). Why is K not necessarily constant?
5. Let $e^0 = dt$ in an orthogonal basis. Show that e^0 is a geodesic vector field (see (10.1.41; 3)).
6. Calculate d^*F^α in the basis (10.3.9). What is the sign of the gravitational energy?
7. Calculate the integral $\int {}^*1$ over a Riemannian space of constant positive curvature ($K = 1$).

(10.3.27) Solutions

1. It is that $L_v\, dx^m = d(i_v\, dx^m) = dv^m = v^m{}_{,i}\, dx^i$ (cf. (2.5.12; 5)); thus

$$0 = L_v g_{lm}\, dx^l\, dx^m = v^i g_{lm,i}\, dx^l\, dx^m + 2g_{lm} v^l{}_{,i}\, dx^i\, dx^m$$
$$\Rightarrow v^i g_{lm,i} + g_{im} v^i{}_{,l} + g_{il} v^i{}_{,m} = 0.$$

2. $\{p_\alpha v^\alpha(x), g^{\beta\gamma}(x)p_\beta p_\gamma\} = -2p_\alpha v^\alpha{}_{,\gamma} g^{\gamma\beta} p_\beta + v^\alpha g^{\beta\gamma}{}_{,\alpha} p_\beta p_\gamma = 0$ for Killing vector fields: $g_{\alpha\beta} g^{\beta\gamma} = \delta_\alpha{}^\gamma \Rightarrow g^{\beta\gamma}{}_{,\alpha} = -g^{\beta\sigma} g_{\sigma\rho,\alpha} g^{\rho\gamma}$, which brings us to the condition of Problem 1.
3. By Theorem (2.3.12), it is always possible to find local coordinates for which the 1-component of v (in the natural basis) equals 1, and the others vanish. Then by Problem 1, $g_{lm,1} = 0$. If v is timelike, this coordinate can be treated as time, making the metric constant and $v = \partial_t$.
4. There are no 3-forms in two dimensions, and hence there is no Bianchi identity.
5. $0 = de^0 = \omega^0{}_k \wedge e^k$ implies that $\omega^0{}_k \sim e^k$, for $k = 1, 2, 3$. Hence (cf. (10.1.11) and (10.1.17)),

$$D_{e_0} e^0 = -e^k(\omega^0{}_k | e) = 0.$$

6. The basis is orthogonal, hence $\varepsilon_{\alpha\beta\gamma\delta} = -\varepsilon^{\alpha\beta\gamma\delta}$, and

$$-\tfrac{1}{2}\varepsilon^{\alpha\beta\gamma\delta}\omega_{\beta\gamma}\wedge e_\delta = -\frac{K}{2}\varepsilon^{\alpha\beta\gamma\delta}x_\beta e_{\gamma\delta};$$

$$-\tfrac{1}{2}\varepsilon^{\alpha\beta\gamma\delta}\,d(\omega_{\beta\gamma}\wedge e_\delta) = -\frac{K}{2}\varepsilon^{\alpha\beta\gamma\delta}\left[e_{\beta\gamma\delta}\left(1+\frac{Kr^2}{4}\right) - Kx_\beta x^\sigma e_{\sigma\gamma\delta}\right]$$

$$= -\frac{K}{2}\left[3!^*e^\alpha\left(1+\frac{Kr^2}{4}\right) - Kx_\beta x^\sigma 2! e_\sigma \wedge {}^*e^{\alpha\beta}\right]$$

$$= -3K^*e^\alpha + \frac{K^2r^2}{4}{}^*e^\alpha - K^2x^\alpha x_\beta{}^*e^\beta = 8\pi\kappa({}^*t^\alpha + {}^*T^\alpha).$$

The term linear in K is the same as the right side of Einstein's version, and the parts $\sim K^2$ are the gravitational contribution $\sim \omega\wedge\omega$, which is always negative for the energy density T_{00}: $-K^2((|\mathbf{x}|^2+3t^2)/4$. It is necessarily negative if $K>0$, since the integral of the energy density over a compact space must be zero (7.3.35; 2), and the part $\sim K$ is positive.

7. $\int {}^*\mathbf{1} = \int d^m x/(1+r^2/4)^m = S_m\int_0^\infty dr\, r^{m-1}/(1+r^2/4)^m$, where $S_m = 2\pi^{m/2}/\Gamma(m/2)$ is the surface area of the m-sphere. If $\beta = (1+r^2/4)^{-1}$, then, recalling that $\Gamma(m)/\Gamma(m/2) = (2^{m-1}/\sqrt{\pi})\Gamma(\frac{1}{2}(m+1))$,

$$\int {}^*\mathbf{1} = S_m 2^{m-1}\int_0^\infty d\beta\,\beta^{m-2}\left(\frac{1}{\beta}-1\right)^{(m-2)/2} = S_m 2^{m-1}\frac{\Gamma(m^2/2)}{\Gamma(m)}$$

$$= \frac{2\pi^{(m+1)/2}}{\Gamma\left(\frac{m+1}{2}\right)} = S_{m+1}.$$

Since the space is isometric to the $(m+1)$-sphere, the calculation of the volume is correct.

10.4 Spaces with Maximally Symmetric Submanifolds

The nonlinearity complicates Einstein's equations so much that the general solution lies beyond human capabilities. Explicit solutions can be written down only if the space is of sufficiently high symmetry.

If the symmetry of a maximally symmetric space is reduced, the variety of possible curvature forms becomes great enough to conceivably correspond to physically acceptable energy-momentum currents. It would, however, lead too far afield if we tried to classify all the possibilities exhaustively, so instead our plan will be to investigate the physically relevant metrics that come up when the symmetry is reduced in successive stages.

(10.4.1) Spaces with Six Killing Vector Fields

The interesting case is the **Friedmann universe**, with six spacelike Killing vector fields, generating a group isomorphic to $O(4)$. The trajectories of a point under the action of the group form a spacelike submanifold with six Killing vector fields, i.e., a three-dimensional Riemannian space of constant curvature. It is convenient to choose comoving coordinates, for which the geodesic vector field furnishes the coordinates lines $x =$ const. perpendicular to this space, and the proper time on these geodesic lines is the time-coordinate t (cf. (10.3.26; 5)). Writing $r^2 = |\mathbf{x}|^2$, the metric g is of the type of a

(10.4.2) **Robertson–Walker Metric**

$$g = -dt^2 + R(t)^2 \frac{|d\mathbf{x}|^2}{(1 + Kr^2/4)^2}.$$

(10.4.3) **Remarks**

1. $R(t)$ is an as yet unspecified function of time, something like $\cosh(t)$ in de Sitter space (see (10.3.17), and note that $|d\mathbf{x}|^2(1 + Kr^2/4)^{-1}$ is $d\chi^2 + \sin^2\chi\, d\Omega^2$ in the coordinates used there), or like $\exp(t)$ for $K = 0$ (see (10.3.18)). If $K > 0$, then the submanifold $t =$ const. has the finite volume $2\pi^2 R^3(t)/K^{3/2}$ (see (10.3.26; 7)).
2. If a new time-variable t' such that $dt'/dt = 1/R(t)$ is introduced as in (10.3.17), then g becomes conformally equivalent to that of de Sitter space, and consequently of Minkowski space,

$$g = R(t)^2 \left(-dt'^2 + \frac{|d\mathbf{x}|^2}{(1 + Kr^2/4)^2} \right)$$

(cf. (10.3.19)). It frequently happens that t' takes values only in a finite interval $t_0 < t < t_1$, as in de Sitter space. In that case the causality relationships turn out to be similar to those discussed in (10.3.16), and in particular there are particle and event horizons.

(10.4.4) **The Curvature Forms of the Friedmann Universe**

Let Greek indices run from 0 to 3 and Roman ones from 1 to 3. If we write the orthogonal basis as

$$e^\alpha = \left(dt, \frac{R(t)\, dx^a}{1 + Kr^2/4} \right),$$

then

$$de^\alpha = \left(0, \frac{\dot{R}}{R} e^{0a} - \frac{K}{2R} x_b e^{ba} \right),$$

and therefore

(10.4.5) $$\omega^a{}_0 = \omega^0{}_a = \frac{\dot{R}}{R} e^a, \qquad \omega_{ab} = \frac{K}{2R}(x_a e_b - x_b e_a).$$

Consequently,

$$d\omega^{0a} = \frac{\ddot{R}}{R} e^{0a} - \frac{K\dot{R}}{2R^2} x_b e^{ba},$$

$$d\omega^{ab} = \frac{K}{R^2} e^{ab}\left(1 + \frac{Kr^2}{4}\right) + \frac{K^2}{4R^2}(x^a x_c e^{bc} - x^b x_c e^{ac}),$$

and a similar calculation to that of (10.3.8) leads to

(10.4.6)
$$R^{0a} = \frac{\ddot{R}}{R} e^{0a}, \qquad R^{ab} = \frac{K + \dot{R}^2}{R^2} e^{ab}.$$

The contracted quantities become

(10.4.7)
$$R^0 = 3\frac{\ddot{R}}{R} e^0, \qquad R^a = \left(\frac{\ddot{R}}{R} + 2\frac{K + \dot{R}^2}{R^2}\right) e^a,$$

$$R = 6\left(\frac{K + \dot{R}^2(t)}{R^2(t)} + \frac{\ddot{R}(t)}{R(t)}\right).$$

(10.4.8) Remarks

1. If $R(t) = \text{const.}$, then the curvature is constant only in spatial directions,

$$R^{ab} = \frac{K}{R^2} e^{ab}, \qquad R^{0a} = 0.$$

The time-independence of R gives rise to an R^{0a} and contributes to R^{ab}.

2. A comparison of (10.4.7) and (10.1.44; 3) reveals that the Weyl forms are now zero, as required by the conformal equivalence to Minkowski space.

(10.4.9) Einstein's Equations in the Classical Form

In order to satisfy (10.2.20(a)), the energy-momentum forms of matter must be $e^\alpha \times$ (some function of t), because of (10.4.7). Therefore the energy-momentum tensor of matter is necessarily diagonal, and in the spirit of the phenomenological description (9.1.25; 3) we set $T_{00} = \rho =$ energy density and $T_{jj} = p =$ pressure. Einstein's equations then imply that

(10.4.10)
$$3\frac{\dot{R}^2 + K}{R^2} = 8\pi\kappa\rho,$$

$$-\frac{2\ddot{R}}{R} - \frac{\dot{R}^2 + K}{R^2} = 8\pi\kappa p.$$

(10.4.11) Remarks

1. The Bianchi identity (10.2.32) relates ρ and p to R, and of course the same relationship follows from (10.4.10). It implies that

$$d{}^*T_0 = d(\rho {}^*e^0) = d\rho \wedge {}^*e^0 + \rho\, d{}^*e^0$$
$$= -\omega^0{}_j \wedge {}^*T^j = -p\omega^0{}_j \wedge {}^*e^j,$$

and therefore

$$d\rho \wedge {}^*e^0 = (\rho - p)\omega^0{}_j \wedge {}^*e^j, \qquad \text{i.e.,} \quad \dot{\rho} = 3\frac{\dot{R}}{R}(\rho - p).$$

In the form

$$p = -\frac{\frac{d}{dt}(\rho R^3)}{\frac{d}{dt}R^3}$$

this has the interpretation that pressure = −(rate of change of energy)/(rate of change of volume). It is noteworthy that gravity does not appear in the total energy in comoving coordinates.

2. The static situation $\dot{R} = 0$ requires a negative pressure if $K > 0$, as in de Sitter space, and a negative energy if $K < 0$. This originally induced Einstein to include the cosmological term $\Lambda {}^*\mathbf{1}$ in his action principle. Friedmann later discovered the solution bearing his name, and the modern tendency is to accept the dynamical equations as is. In order to illustrate their significance better, let us examine

(10.4.12) Einstein's Equations in Landau and Lifshitz's Form

In (10.2.20(a)) and (10.2.22(b)) the energy-momentum forms for matter and gravitation were represented as exterior differentials of 2-forms. Since it is not yet apparent what the gravitational energy is, let us track it down. To this end, we calculate the restriction of the exterior differential of the 2-form

$$-\tfrac{1}{2}\varepsilon^{0bcd}\omega_{bc} \wedge e_d = -\frac{K}{2R}\varepsilon^{0bcd}x_b e_{cd},$$

$$-\tfrac{1}{2}\varepsilon^{0bcd}d(\omega_{bc} \wedge e_d)\Big|_{t=\text{const.}} = -\frac{K}{2R^2}\varepsilon^{0bcd}\left[e_{bcd}\left(1 + \frac{Kr^2}{4}\right) - Kx_b x_m e_{mcd}\right]\Big|_{t=\text{const.}}$$

$$= \frac{3K - K^2r^2/4}{R^2}{}^*e^0\Big|_{t=\text{const.}} = 8\pi\kappa({}^*T^0 + {}^*t^0) \qquad (10.4.13)$$

to $t = $ const.

(10.4.14) **Remarks**

1. Equation (10.4.13) states that

$$8\pi\kappa \cdot \text{the total energy density} = \frac{3K - K^2r^2/4}{R^2}$$

so $8\pi\kappa\cdot$ the gravitational energy density equals $8\pi\kappa\cdot$ (total energy density $-$ $\rho) = -(K^2r^2 + 12\dot{R}^2)/4R^2$. In the static situation ($\dot{R} = 0$) this is the energy density of a homogeneous mass distribution ρ according to Newton's theory, constructed just as in electrodynamics, but with the other sign and $4\pi\kappa \to e^2$. If we chose units where $R = 1$, then

$$\Delta V = e\rho = \frac{3K}{2e}, \qquad V = \frac{Kr^2}{4e},$$

$$\nabla V = \frac{\mathbf{x}K}{2e}, \qquad -\tfrac{1}{2}|\nabla V|^2 = -\frac{1}{2e^2}\frac{K^2r^2}{4},$$

and thus, as promised in §7.3, $-K^2r^2/4R^2$ corresponds exactly to $8\pi\kappa\cdot$ the Newtonian gravitational energy in the most nearly Cartesian coordinates. In the dynamical case, there is also a contribution $\dot{R}^2$.

2. The gravitational energy is exactly large enough so that if $K > 0$, the integral of the total energy over the whole space is zero:

$$3K\int_0^\infty \frac{r^2\,dr}{(1+Kr^2/4)^3} = \frac{K^2}{4}\int_0^\infty \frac{r^4\,dr}{(1+Kr^2/4)^3}.$$

3. If we write the first of equations (10.4.10) as

$$\frac{\dot{R}^2}{2} - 4\pi\kappa\rho\frac{R^2}{3} = -\frac{K}{2} = \text{const.},$$

then it has the form of the conservation equation for the energy of a (nonrelativistic) particle with coordinate R and speed $\dot{R}$. The kinetic energy plus the potential energy is constant, where the latter is taken as the potential energy on the surface of a ball of radius R and homogeneous density ρ:

$$-V(r) = \frac{4\pi R^3\rho}{3r}\Theta(r-R) + \frac{4\pi R^2}{2}\left(-1 + \frac{r^2}{3R^2}\right)\Theta(R-r).$$

Note that $V(0) - V(R)$ is smaller than $V(R) - V(\infty)$ by a factor of 2.

In the following section, we shall return to (10.4.10) when we study the collapse of stars, and we shall solve them for selected pressure-density ratios. A detailed discussion of their importance for cosmology may be found in [27], [28], and [29].

(10.4.15) **Spaces with Five Killing Vector Fields**

In §5.8 we learned that there are in general five constants of the motion in the field of a gravitational wave, which are linear in the momentum components. This motion accordingly allows five Killing vector fields, and the invariance group they generate is isomorphic to the invariance group of a gravitational wave, and consequently (5.8.4; 1) to that of an electromagnetic wave (5.5.3). The metric given in (5.8.1) is a special case of what we examined in (10.2.24), where it depended on only one coordinate u. The interesting situation is where discontinuities are allowed, in which $du = n_\alpha e^\alpha$, $0 = n_\alpha n^\alpha =: n^2$. If the coordinates are chosen so that

$$n = (-1, 1, 0, 0), \qquad u = x - t, \tag{10.4.16}$$

then the metric (5.8.1), with the required invariance structure, is of the form

$$g = -dt^2 + dx^2 + p(u)^2\, dy^2 + q(u)^2\, dz^2. \tag{10.4.17}$$

(10.4.18) **Remarks**

1. The form of g is that of a metric in a comoving coordinate system (cf. (10.3.26; 5)). Therefore the coordinate lines $\mathbf{x} =$ const. are possible particle trajectories.
2. The gravitational field described by g is a kind of transverse wave, which alters the distance perpendicular to the direction of propagation between particle trajectories. If $p - 1$ and $q - 1$ have compact support, then the pulse looks schematically as shown in Figure 10.7. For example, in the solution given in (5.8.7) $p \le 1$ and $q \le 1$, so the trajectories draw closer together in the y-direction and spread apart in the z-direction, as with a quadrupole field. This effect would not be measured by measuring rods, as they would be stretched in the same way. However, the deformation would be observable by measuring the time-delay of a reflected light signal.

(10.4.19) **The Curvature Forms**

If the orthogonal basis

$$e^\alpha = (dt, dx, p\, dy, q\, dz),$$
$$de^\alpha = (0, 0, p'\, du \wedge dy, q'\, du \wedge dz)$$

is used, then the affine connections become

$$\omega_{\alpha\beta} = S_\alpha n_\beta - S_\beta n_\alpha, \qquad S_\alpha = (0, 0, p'\, dy, q'\, dz),$$

as in (10.2.13). Since $n_\alpha S^\alpha = 0 = n_\alpha n^\alpha$, we find that $\omega_{\alpha\beta} \wedge \omega^\beta{}_\gamma = 0$ for all α and β, and the curvature forms are

$$\begin{aligned} R_{\alpha\beta} &= d\omega_{\alpha\beta} = du \wedge (S'_\alpha n_\beta - S'_\beta n_\alpha), \\ S'_\alpha &= (0, 0, p''\, dy, q''\, dz), \end{aligned} \tag{10.4.20}$$

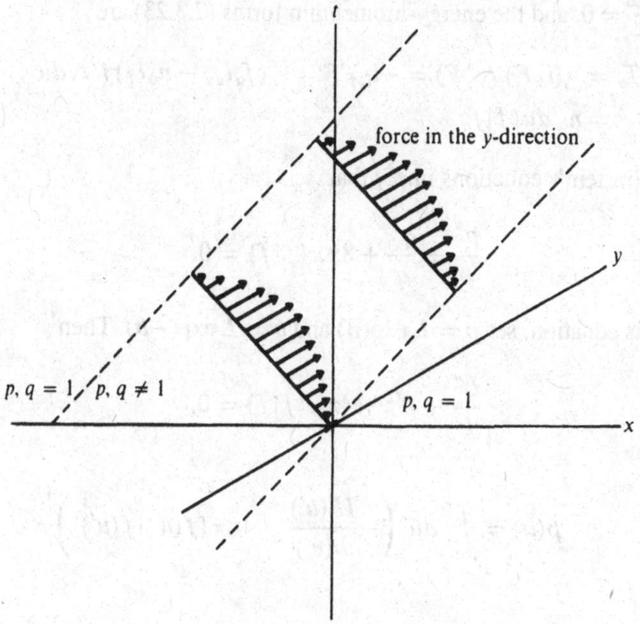

FIGURE 10.7. Schematic drawing of a gravitational pulse.

with the contractions

$$R_\beta = i_\alpha R^\alpha{}_\beta = -n_\beta\, du\, i_\alpha S'^\alpha = -n_\beta\, du \left(\frac{p''}{p} + \frac{q''}{q}\right), \tag{10.4.21}$$

$$R = 0.$$

(10.4.22) **Remarks**

1. In two dimensions, $g = -dt^2 + p^2(t)\,dy$, the curvature is $p''\,dt \wedge dy$, and thus $R_{\alpha\beta}$ are precisely the curvatures of the corresponding two-dimensional surfaces.
2. The Weyl forms are not of necessity zero, and the space need not be conformally flat. However, if $p'' = q'' = 0$, then it is always flat (cf. Remark (5.8.8; 2)).

(10.4.23) **Einstein's Equations in the Classical Form**

By equations (10.4.21),

$$R_\alpha - \tfrac{1}{2} e_\alpha R = -n_\alpha\, du \left(\frac{p''}{p} + \frac{q''}{q}\right),$$

and thus an energy current $\sim du$ could well be a source of the gravitational pulse, as long as it is accompanied by an equally large current of the 1-component of the momentum while the other components vanish. Such an energy-momentum current could be produced by an electromagnetic wave. If

$$F = f(u) \wedge du \qquad \text{with} \qquad \langle f|du\rangle = \langle du|du\rangle = 0,$$

then $F \wedge {}^*F = 0$, and the energy-momentum forms (7.3.23) are

$$\mathcal{T}_\alpha = {}^*((i_\alpha F) \wedge {}^*F) = -i_{i_\alpha F} F = -(f_\alpha i_{du} - n_\alpha i_f) f \wedge du$$
$$= n_\alpha \, du \langle f | f \rangle.$$

Therefore Einstein's equations imply that

$$\frac{p''}{p} + \frac{q''}{q} + 8\pi\kappa\langle f|f\rangle = 0.$$

To solve this equation, set $p = L\exp(\beta)$ and $q = L\exp(-\beta)$. Then

(10.4.24)
$$\frac{L''}{L} + \beta'^2 + 4\pi\kappa\langle f|f\rangle = 0,$$

(10.4.25)
$$\beta(u) = \int_0^u du' \left(-\frac{L''(u')}{L(u')} - 4\pi\kappa\langle f(u')|f(u')\rangle \right)^{1/2}.$$

(10.4.26) **Remarks**

1. In the approximation linear in κ,

$$L(u) = 1 - 4\pi\kappa \int_0^u du' \int_0^{u'} du'' \langle f(u'')|f(u'')\rangle,$$

while β remains arbitrary to first order. If the equations are homogeneous ($f = 0$), this provides us with a solution that could be used for φ^{in} in (10.2.18) because to first order, $\beta = \varphi_{22} = -\varphi_{33}$, and the other φ's are zero, so that $0 = \varphi_\alpha{}^\alpha = \varphi_{\beta\alpha,}{}^\alpha = \varphi_{\alpha\alpha,\rho}{}^\rho$.
2. If $L > 0$, then it is a concave function, because $\langle f|f\rangle > 0$, and f and β' contribute in similar ways to the curvature L'' of the function $L(u)$. In this situation, the trajectories of particles are focused in the (y–z)-plane. This is an effect of the gravitational field produced by the electromagnetic or gravitational wave.
3. If L' is ever negative, then L must sooner or later have a zero. This singularity in the metric might not be a genuine one, but may only indicate that the gravitational wave has disrupted the coordinate system. The space might appear as Minkowski space in some other chart (cf. (5.8.8; 2)), as soon as the wave has passed.

(10.4.27) **Einstein's Equations in Landau and Lifshitz's Form**

It remains to find out how well the interpretation (10.4.26; 2) of β'^2 as an energy density accords with the formulations (10.2.22(b)). In the latter formulation gravity contributes

$$t_\alpha = \frac{-1}{8\pi\kappa} \tfrac{1}{2} \omega_{\beta\gamma} \wedge \omega_{\alpha\rho} \wedge {}^*e^{\beta\gamma\rho}$$

to the energy, and the other summand is zero. Substitution from (10.4.19) yields

(10.4.28)
$$\begin{aligned} 8\pi\kappa t_\alpha &= {}^*(S_\beta n_\gamma \wedge S_\rho n_\alpha \wedge {}^*e^{\rho\beta\gamma}) \\ &= -n_\alpha i_{S_\beta} i_{S_\rho}(e^{\rho\beta} \wedge du) = -n_\alpha\, du\, 2\frac{p'q'}{pq} \\ &= 2n_\alpha\, du\left(\beta'^2 - \left(\frac{L'}{L}\right)^2\right), \end{aligned}$$

so that an additional negative term $-(L'/L)^2$ occurs along with β'^2.

(10.4.29) **Remarks**

1. In the linear approximation (10.4.26; 1), $L' = 0$, $2\dot\beta = \dot g_{22} = -\dot g_{33}$, and the result is that
$$\text{energy density} = \text{momentum density} = \frac{(\dot g_{22} - \dot g_{33})^2}{64\pi\kappa}.$$
2. Provided that L has a slowly varying amplitude in comparison with β, (10.2.22(b)) states that $\dot\beta^2$ creates gravity like any other kind of energy, and that the energy of a gravitational wave is positive.
3. Fictitious energies associated with fictitious forces also appear in (10.4.27); their origin is that the t_α do not vanish even in flat space ($p'' = q'' = 0$).
4. The speeding-up of a double star with a short period seems to be consistent with the energy loss due to gravitational radiation as calculated with this formula.

(10.4.30) **Spaces with Four Killing Vector Fields**

We shall consider the spaces that are the counterpart to the problem of a central force in mechanics. The energy and angular momentum will correspond to the operations of time-displacement and rotations that leave g invariant. In the polar coordinates for $M = \mathbb{R} \times \mathbb{R}^+ \times S^2$, the $g_{\alpha\beta}$ depend only on $r = |\mathbf{x}|$, and the metric can be written as
$$g = -dt^2 \exp(2a(r)) + dr^2 \exp(2b(r)) + r^2(d\vartheta^2 + \sin^2\vartheta\, d\varphi^2).$$

(10.4.31) **Remarks**

1. If Einstein's equations in vacuo hold, then it can be shown that the existence of the timelike Killing vector field follows simply from the spherical symmetry (Birkhoff's theorem, Problem 5).
2. In comoving coordinates, the $g_{\alpha\beta}$ are in general time-dependent.

(10.4.32) **The Curvature Forms**

In the orthogonal basis
$$\begin{aligned} e^\alpha &= (e^a\, dt, e^b\, dr, r\, d\vartheta, r \sin\vartheta\, d\varphi), \\ de^\alpha &= (a'e^a\, dr \wedge dt, 0, dr \wedge d\vartheta, \sin\vartheta\, dr \wedge d\varphi + r\cos\vartheta\, d\vartheta \wedge d\varphi), \end{aligned}$$

the affine connections turn out to be

$$\text{(10.4.33)} \qquad \omega^\alpha{}_\beta = \begin{array}{|c|c|c|}\hline e^{a-b}a'\,dt & 0 & 0 \\ \hline & -e^{-b}\,d\vartheta & -e^{-b}\sin\vartheta\,d\varphi \\ \hline & & -\cos\vartheta\,d\varphi \\ \hline \end{array}$$

(since $\omega_{\alpha\beta} = -\omega_{\beta\alpha}$, we write them only for $\alpha < \beta$). They make

$$d\omega^\alpha{}_\beta = \begin{array}{|c|c|c|}\hline e^{a-b}(a'' + a'(a' - b'))dr \wedge dt & 0 & 0 \\ \hline & e^{-b}b'\,dr \wedge d\vartheta & e^{-b}(-\cos\vartheta\,d\vartheta + \sin\vartheta b'\,dr) \wedge d\varphi \\ \hline & & \sin\vartheta\,d\vartheta \wedge d\varphi \\ \hline \end{array}$$

$$\omega^\alpha{}_\gamma \wedge \omega^\gamma{}_\beta = \begin{array}{|c|c|c|}\hline 0 & -e^{a-2b}a'\,dt \wedge d\vartheta & -e^{a-2b}a'\sin\vartheta\,dt \wedge d\vartheta \\ \hline & 0 & e^{-b}\cos\vartheta\,d\vartheta \wedge d\varphi \\ \hline & & e^{-2b}\sin\vartheta\,d\vartheta \wedge d\varphi \\ \hline \end{array}.$$

The term $\sim d\vartheta \wedge d\varphi$ cancels out of $d\omega^1{}_3$, and the $R^\alpha{}_\beta$ become proportional to $e^\alpha{}_\beta$:

$$\text{(10.4.34)} \qquad R^\alpha{}_\beta = \begin{array}{|c|c|c|}\hline (a'b' - a'' - a'^2)e^{-2b}e^0{}_1 & -\dfrac{a'e^{-2b}}{r}e^0{}_2 & -\dfrac{a'e^{-2b}}{r}e^0{}_3 \\ \hline & \dfrac{b'e^{-2b}}{r}e^1{}_2 & \dfrac{b'e^{-2b}}{r}e^1{}_3 \\ \hline & & \dfrac{(1 - e^{-2b})}{r^2}e^2{}_3 \\ \hline \end{array}$$

(10.4.35) **Einstein's Equations**

Since the remaining symmetry still suffices to make $R^\alpha{}_\beta$ of the form $R^\alpha{}_\beta = K^\alpha{}_\beta e^\alpha{}_\beta$ (no sum), the energy-momentum forms $\mathcal{T}^\alpha$ (letting this embrace everything coupled to gravity) must likewise be $\sim e^\alpha$. The coefficients are $-\sum_{\substack{\beta<\gamma \\ \beta\neq\alpha,\gamma\neq\alpha}} K_{\beta\gamma}$, and depend only on r:

$$\left(-\frac{1-e^{-2b}}{r^2} - 2b'\frac{e^{-2b}}{r}\right)e^0 = 8\pi\kappa\mathcal{T}^0, \tag{10.4.36}$$
$$\left(-\frac{1-e^{-2b}}{r^2} + 2a'\frac{e^{-2b}}{r}\right)e^1 = 8\pi\kappa\mathcal{T}^1,$$
$$\left(a'' + a'^2 - a'b' + \frac{a'-b'}{r}\right)e^{-2b}e^2 = 8\pi\kappa\mathcal{T}^2,$$
$$\left(a'' + a'^2 - a'b' + \frac{a'-b'}{r}\right)e^{-2b}e^3 = 8\pi\kappa\mathcal{T}^3.$$

(10.4.37) **Remarks**

1. Because of the spherical symmetry, there is an invariance under 2 ↔ 3, and, more specifically, $\mathcal{T}^2$ and $\mathcal{T}^3$ have the same factor in front of e^2 and, respectively, e^3.
2. $\mathcal{T}^\alpha$ is of the form $c^\alpha e^\alpha$ (no sum), and the contracted Bianchi identity (10.2.32) subjects the coefficients c^α to the equation

$$dc^\alpha \wedge {}^*e^\alpha = \sum_\beta \omega^\alpha{}_\beta \wedge {}^*e^\beta (c^\alpha - c^\beta).$$

3. The $\mathcal{T}^\alpha$ are written with the basis of (10.4.32), and thus the $\mathcal{T}^j$, $j = 1, 2, 3$, are obtained from Cartesian energy-momentum forms by local rotations (cf. (7.3.26)). If, for instance, $\mathcal{T}^j = pe^j$ in the Cartesian basis, then it is also true in this basis.

(10.4.38) **Special Cases**

1. If we make the phenomenological assumption that

$$\mathcal{T}_\alpha = (\rho e^0, pe^j)$$

(cf. (9.1.25; 3)), where ρ and p are not too singular and decrease sufficiently fast as $r \to \infty$, then the first of equations (10.4.37) is solved by

$$e^{-2b} = 1 - \frac{8\pi\kappa}{r}\int_0^r dr'\, r'^2 \rho(r') =: 1 - \frac{2\kappa M(r)}{r}, \tag{10.4.39}$$

and the second one determines a once b is known:

$$a = -b - 4\pi\kappa \int_r^\infty dr'\, r'\, e^{2b(r')}(\rho(r') + p(r')). \tag{10.4.40}$$

The last two identical equations relate ρ to p, and are equivalent to requiring the contracted Bianchi identity. This subject will be pursued in the following section; for the $\mathcal{T}^\alpha$ treated here it is satisfied.

If it happens that $\rho(r) = p(r) = 0$ for all $r > r_1$, then the metric in the region where $r > r_1$ is the

(10.4.41) **Schwarzschild Metric**

$$g = -\left(1 - \frac{r_0}{r}\right) dt^2 + \frac{dr^2}{1 - \frac{r_0}{r}} + r^2\, d\Omega^2, \qquad r_0 = 8\pi\kappa \int_0^\infty r^2\, dr\, \rho(r).$$

On the other hand, if $\rho = \text{const.} = -p$, then we return to the situation of (10.3.23) and obtain (cf. Problem 4) a

(10.4.42) **Static Form of the de Sitter Metric**

$$g = -(1 - Kr^2)\,dt^2 + \frac{dr^2}{1 - Kr^2} + r^2\, d\Omega^2, \qquad K = \frac{8\pi\kappa}{3}\rho.$$

2. Equations (10.4.37) allow the pressure in the radial direction to differ from the pressure in the ϑ and φ directions. This could happen for the Coulomb field of a point particle, for which the energy-momentum forms can be calculated as

(10.4.43) $$T^\alpha = \frac{e^2}{2r^4}[e^0, -e^1, e^2, e^3].$$

If we set $\exp(2a) = \exp(-2b) = \psi(r)/r$ in (10.4.37), then Einstein's equations read:

$$\left(\frac{1 - \psi'}{r^2} e^0, \frac{-1 + \psi'}{r^2} e^1, \frac{\psi''}{2r} e^2, \frac{\psi''}{2r} e^3\right) = 8\pi\kappa \mathcal{T}_\alpha.$$

If $\psi = r - r_0 + 4\pi\kappa e^2/r$, then this simply reproduces (10.4.43), and the resulting metric is called the

(10.4.44) **Reissner–Nordstrøm Metric**

$$g = -\left(1 - \frac{r_0}{r} + \frac{4\pi\kappa e^2}{r^2}\right) dt^2 + \frac{dr^2}{1 - \frac{r_0}{r} + \frac{4\pi\kappa e^2}{r^2}} + r^2\, d\Omega^2.$$

(10.4.45) **Remarks**

1. In the linear approximation, and with $|p| \ll \rho$, equation (10.4.40) becomes

$$a = -\frac{g_{00} - 1}{2} = -4\pi\kappa \left[\frac{1}{r}\int_0^r dr'\, \rho(r') r'^2 + \int_r^\infty dr'\, \rho(r') r'\right],$$

which is the Newtonian potential of the spherically symmetric energy density ρ, as it must be on account of (10.2.30; 2).

2. The pressure contributes to (10.4.40) as the density contributes to g_{00}. Hence the negative pressure $p = -\rho$ of the de Sitter universe in fact makes $a = -$the Newtonian potential, because for constant densities,

$$\frac{1}{r}\int_0^r r'^2 dr'\,\rho = -\frac{1}{2}\int_r^R r'\,dr'\,\rho + \text{const.}$$

3. If ρ is more singular than r^{-3} at $r = 0$, then one can write $\int_r^\infty$ instead of $-\int_0^r$ in (10.4.39). This is the case with the Reissner solution (10.4.44), so the positive field energy contributes with a reversed sign to the gravitational potential. The interpretation is that $M = M(\infty)$ represents the total energy, and the potential

$$-\frac{M\kappa}{r} + \frac{4\pi e^2\kappa}{2r^2}$$

shows that as one approaches the origin, part of the energy density is left behind, and the potential is effectively decreased compared with its asymptotic value, $-M\kappa/r$. If $M < \infty$, then it follows that the "naked mass" at the origin must be $-\infty$, since the electromagnetic mass

$$\frac{4\pi}{2}\int_0^\infty \frac{dr}{r^4}$$

is divergent. This has the paradoxical consequence that gravity is repulsive at short distances. Once again, the infinite electrostatic self-energy of a point charge is causing trouble.

4. The basis of (10.4.32) is less suitable for a discussion of gravitational energy using the version (10.2.20(b)) of Einstein's equations, since it is possible to simulate a gravitational t_α even in flat space, with polar coordinates. In Problem 6 the gravitational energy is discussed in the maximally Cartesian coordinates (see (5.7.17; 4)). It turns out that as long as there is asymptotically a Schwarzschild metric, the total energy including the gravitational energy is M. Note that

$$\int_{t=0} {}^*T^0 = 4\pi\int_0^\infty dr\,r^2 e^{b(r)}\rho(r) \neq M = 4\pi\int_0^\infty dr\,r^2\rho(r).$$

5. In Problem 7 the energy density of gravitation is calculated as $-\kappa M^2/8\pi r^4$ in these coordinates. This is equal to the negative of the energy density $(e/4\pi r^2)^2/2$ of a Coulomb field, where e^2 is replaced with $4\pi\kappa M^2$, analogously to Remark (10.4.14; 1). Since M is the integral of the total energy density, there can be an everywhere regular solution with $M > 0$ only if $\rho > 0$ counterbalances the negative gravitational energy. The increase of the Schwarzchild potential compared with the Newtonian potential at small r can be interpreted as the field produced by the negative gravitational energy.

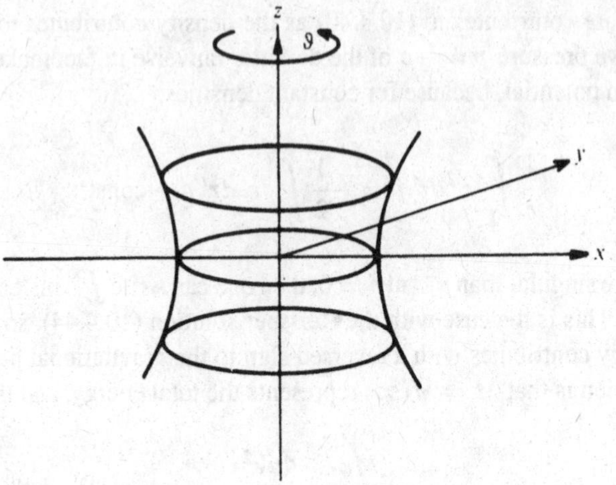

FIGURE 10.8. A choice of the metric on the surface t = const., φ = const.

(10.4.46) **Properties of Spherically Symmetric Fields**

1. The Geometric Interpretation of the Spatial Metric. The restriction of the metric to a plane passing through the origin,

$$g\Big|_{\substack{t=\text{const.}\\ \varphi=\text{const.}}} = e^{2b}\,dr^2 + r^2\,d\vartheta^2,$$

is the metric on a surface of rotation in $\mathbb{R}^3$. If it is written in cylindrical coordinates as $z(r)$, then

$$dz^2 + dr^2 + r^2\,d\vartheta^2 = dr^2(1+z'^2) + r^2\,d\vartheta^2 = e^{2b}\,dr^2 + r^2\,d\vartheta^2,$$

or, using (10.4.39),

$$z' = \sqrt{1-e^{2b}} = \sqrt{\frac{2\kappa M(r)}{r-2\kappa M(r)}},$$

$$z(r) = \int_0^r dr' \sqrt{\frac{2\kappa M(r')}{r'-2\kappa M(r')}}.$$

As a consequence, the Schwarzschild metric $M(r) =$ const. gives the intersection $\varphi =$ const. the geometry of a paraboloid of revolution,

$$z(r) = \sqrt{4r_0}\sqrt{r-r_0}$$

(Figure 10.8). The metric is singular at $r = 2\kappa M(r)$. The paraboloid of the Schwarzchild metric can be extended beyond that point, but, if so, r is no longer a monotonic function of z.

2. The Causal Structure. The Schwarzschild metric was extended beyond $r = r_0$ in (5.7.2; 5), with the aid of the coordinates

(10.4.47)
$$u = \sqrt{\frac{r}{r_0} - 1}\, \exp\left(\frac{r}{2r_0}\right) \cosh\left(\frac{t}{2r_0}\right),$$
$$v = \sqrt{\frac{r}{r_0} - 1}\, \exp\left(\frac{r}{2r_0}\right) \sinh\left(\frac{t}{2r_0}\right), \qquad u^2 - v^2 > -1,$$

in which

(10.4.48)
$$g = 4r_0^3 \exp\left(-\frac{r}{r_0}\right) \frac{du^2 - dv^2}{r} + r^2\, d\Omega^2.$$

There remains a singularity at $r = 0$, which is now a spacelike hypersurface $u^2 - v^2 = -1$. The quickest way to understand the causal relationships is to draw the Penrose diagram that results from using the coordinates

(10.4.49)
$$v + u = \tan\left(\frac{\psi + \xi}{2}\right), \qquad v - u = \tan\left(\frac{\psi - \xi}{2}\right),$$
$$-\pi < \psi \pm \xi < \pi, \qquad -\frac{\pi}{2} < \psi < \frac{\pi}{2}.$$

The metric

$$g = A^2(-d\psi^2 + d\xi^2 + R^2\, d\Omega^2),$$
$$A = \frac{r_0}{\sqrt{r}} e^{-r/2r_0} \cos^{-1} \tfrac{1}{2}(\psi - \xi) \cos^{-1} \tfrac{1}{2}(\psi + \xi),$$
$$R = \frac{r}{A},$$

reveals that timelike lines run in the (ψ–ξ)-plane at angles of at least 45°, and radial light rays run at 45°. The region covered by the new time and radial coordinates looks as shown below. Since the boundary contains the spacelike piece where $r = 0$, there is a horizon. Although $r = r_0$ is not a singularity, it is the event horizon for all trajectories that remain in Region I, where $r > r_0$. Regions II and III are invisible from Region I, which is itself invisible from Regions III and IV. Although nothing exceptional happens locally at $r = r_0$, the surface $r = r_0$ has a global significance.

The Reissner metric (10.4.44) becomes singular for some $r \in (0, \infty)$ if $4\pi\kappa e^2 < M^2\kappa^2$. For hadrons this inequality is far from being satisfied, since in natural units $e^2 = \frac{1}{137} \geq (\kappa/4\pi)M^2 \equiv$ (Planck length/Compton wave length)^2 = $(10^{-33}$ cm$/10^{-14}$ cm$)^2$. Thus no horizon prevents people from starting at the "naked singularity" at $r = 0$. The "cosmic censorship hypothesis" conjectures that naked singularities do not develop in reality. Indeed $4\pi e^2 \geq M^2\kappa$ means that the Coulomb repulsion of charged matter

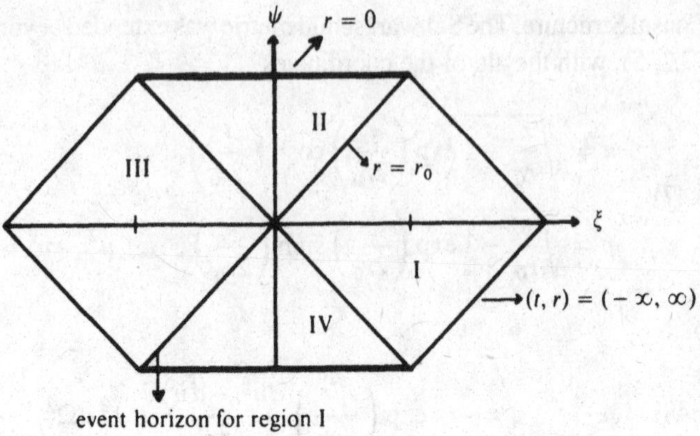

FIGURE 10.9. Penrose diagram for the Schwarzschild solution.

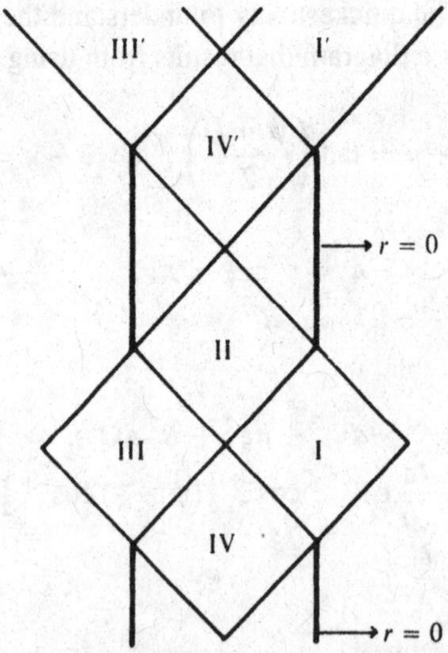

FIGURE 10.10. Penrose diagram for the Reissner solution.

would be stronger than the gravitational attraction, thus preventing the collapse to a singularity. If $M^2\kappa^2 > 4\pi\kappa e^2$, then the singularity of (10.4.44) at small r lies only in the choice of coordinates; with other coordinates it would be possible to continue to $r = 0$. In that case, the repulsive nature of gravity makes $r = 0$ a timelike line. The appropriate Penrose diagram, Figure 10.10, thus extends in the timelike direction to infinity, as with the anti-de Sitter universe ([28, p. 921]). Hence there are again no global Cauchy surfaces, but instead there is a bizarre possibility that one might crawl through the wormhole bounded by $r = 0$ into another universe just like ours (I′ in Figure 10.10).

3. Singularities. Now that the singularity at $r = r_0$ has successfully been removed from the Schwarzschild metric, the question arises of whether the

singularity at $r = 0$ is genuine. It is not necessarily significant that the g_{ik} are infinite there, since these quantities depend on the coordinates. However, the invariant $^*(R_{\alpha\beta} \wedge {}^*R^{\alpha\beta})$ goes as r^{-6}, and grows without bound as $r \to 0$ (Problem 8). This would not be possible at a regular singular point, so we conclude that the Schwarzschild metric cannot be extended regularly across the region $u^2 - v^2 > -1$.

(10.4.50) Problems

1. Construct the five Killing vector fields of the metric (10.4.17) (cf. (5.8.3)).

2. In the linear approximation, the metric at large distances

$$g_{\alpha\beta} = \eta_{\alpha\beta} + \frac{4\kappa}{r} \int d^3x'(T_{\alpha\beta} - \tfrac{1}{2}T\eta_{\alpha\beta})_{t-|x-x'|}$$

looks like the field of a plane wave. Using (10.4.29; 1), calculate the energy radiated in the 1-direction, and express it in terms of the quadrupole tensor

$$D_{ab} = \int d^3x\, T_{00}(3x_a x_b - \delta_{ab}|\mathbf{x}|^2).$$

3. Calculate the $\mathcal{T}_\alpha$ of (10.4.43) for $F = (e/r^2)e^{01}$. Verify that $i_{e_\alpha}\mathcal{T}_\alpha = 0$.

4. Reexpress (10.4.42) in the form (10.3.9).

5. Prove Birkhoff's theorem: If the a and b of (10.4.30) also depend on time, and Einstein's equations hold with $\mathcal{T}_\alpha = 0$, then there exists a time-coordinate t' such that g is of the form (10.4.30) with a and b independent of time. The metric g must then be of the form (10.4.41).

6. Use the Schwarzschild metric in the form

$$g = -h^2(r)\,dt^2 + f^2(r)|d\mathbf{x}|^2, \qquad f^2 = \left(1 + \frac{\kappa M}{2r}\right)^2,$$
$$h^2 = \left(1 - \frac{\kappa M}{2r}\right)\left(1 + \frac{\kappa M}{2r}\right)^{-1},$$

(cf. (5.7.17; 4)) and the version (10.2.20(b)) to identify the total energy

$$\int_{N_3} {}^*(T^0 + t^0)$$

with M. Use a ball of radius $R \to \infty$ for N_3.

7. Calculate the density of the total energy as in (10.4.45; 4) with the ω's of Problem 6.

8. Calculate $^*(R_{\alpha\beta} \wedge {}^*R^{\alpha\beta})$ for the Schwarzschild metric, and check that it is unbounded at $r = 0$.

(10.4.51) **Solutions**

1. The fields with components

$$v^i = (0,0,1,0), (0,0,0,1), (1,1,0,0), \left(z, z, 0, \int_0^{t-x} \frac{du}{q^2(u)}\right), \left(y, y, \int_0^{t-x} \frac{du}{p^2(u)}, 0\right)$$

satisfy

$$v^i g_{lm,i} + g_{im} v^i{}_{,l} + g_{il} v^i{}_{,m} = 0,$$

and are therefore Killing vector fields according to (10.3.27; 1).

2. From the continuity equation of the linear approximation, $T^{\alpha\beta}{}_{,\beta} = 0$, it follows that

$$\frac{\partial^2}{\partial t^2} T_{00} = \nabla_a \nabla_b T_{ab}, \qquad \text{i.e.,} \qquad \int d^3x\, T_{ab} = \frac{1}{2}\frac{\partial^2}{\partial t^2} \int d^3x\, x_a x_b T_{00}.$$

Therefore

$$\dot{g}_{ab} = \frac{2\kappa}{r}\frac{\partial^3}{\partial t^3} \int d^3x\, x_a x_b T_{00} + \eta_{ab} c.$$

The contribution $\sim \eta_{ab}$ is irrelevant, because we require only the square of the difference of the eigenvalues of the (2–3)-subspace $(\dot{g}_{33} - \dot{g}_{22})^2 + 4\dot{g}_{32}^2$. Substitution of D_{ab} yields

$$T_{10} = \left(\frac{\kappa}{144\pi r^2}\right)[(\dddot{D}_{33} - \dddot{D}_{22})^2 + 4\dddot{D}_{23}^2]$$

(cf. (7.1.13) and [51, §104]).

3.

$$^*F = -\frac{e}{r^2} e^{23}, \qquad T_\alpha = \tfrac{1}{2}[-i_{i_\alpha {}^*F} {}^*F - i_{i_\alpha F} F],$$

$$T_0 = -\frac{1}{2}\frac{e}{r^2} i_{e_1} \frac{e}{r^2} e^{01} = \frac{e^2}{r^4} e^0,$$

$$T_1 = \frac{1}{2}\frac{e}{r^2} i_{e_0} \frac{e}{r^2} e^{01} = -\frac{e^2}{r^4} e^1,$$

$$T_2 = -\frac{1}{2}\frac{e}{r^2} i_{e_3} \frac{e}{r^2} e^{23} = \frac{e^2}{r^4} e^2,$$

$$T_3 = \frac{1}{2}\frac{e}{r^3} i_{e_2} \frac{e}{r^2} e^{23} = \frac{e^2}{r^4} e^3.$$

4. Suppose $K = 1$, and introduce the coordinates

$$x_4 = (1 - r^2)^{1/2} \cosh t \qquad \text{and} \qquad x_5 = (1 - r^2)^{1/2} \sinh t$$

on the surface where $x_1^2 + x_2^2 + x_3^2 + x_4^2 - x_5^2 = 1$. Then

$$g = |d\mathbf{x}|^2 + dx_4^2 - dx_5^2 = |d\mathbf{x}|^2 - (1 - r^2)\,dt^2 + \frac{r^2\,dr^2}{1 - r^2}.$$

If $K = -1$, then take the coordinates

$$x_4 = (1 + r^2)\cos t \qquad \text{and} \qquad x_5 = (1 + r^2)\sin t$$

on the surface where $-|\mathbf{x}|^2 + x_4^2 + x_5^2 = 1$.

5. If a and b depend on time, then the only immediate change is $\omega^0{}_1$, by $\exp(b-a)\dot{b}\,dr$. This produces the following extra terms in the $8\pi\kappa T^\alpha$ of (10.4.35):

$$\alpha = 0\colon 2\dot{b}\frac{e^{-a-b}}{r}e^1,$$

$$\alpha = 1\colon -2\dot{b}\frac{e^{-a-b}}{r}e^0,$$

$$\alpha = 2, 3\colon -e^{-2a}(\ddot{b} + \dot{b}^2 - \dot{a}\dot{b})e^\alpha.$$

If $\alpha = 0$, then we also find that $\dot{b} = 0$ and hence $\exp(-2b) = 1 - r_0/r$. If, however, $\alpha = 1$, then $a' = -b'$, and thus $\exp(2a) = (1 - r_0/r)f^2(t)$. With the variables $dt' = f(t)\,dt$, we have the Schwarzschild metric. No new coordinates result if $\alpha = 2$ or 3.

6. $$e^\alpha = (h\,dt, f\,d\mathbf{x}), \qquad \omega^{0j} = \frac{h'}{f}\frac{x^j}{r}dt, \qquad \omega^{jk} = \frac{f'}{fr}(x^k\,dx^j - x^j\,dx^k),$$

$$\begin{aligned} 8\pi\kappa\int_{N_3}({}^*T^0 + {}^*t^0) &= -\tfrac{1}{2}\varepsilon^{0bcd}\int_{\partial N_3}\omega_{bc}\wedge e_d \\ &= -\varepsilon^{0bcd}\int_{\partial N_3}\frac{f'}{r}x^c\,dx^b\wedge dx^d \\ &= -\lim_{R\to\infty}\int_{r=R} d\Omega\, 2f'r^2 = 8\pi\kappa M. \end{aligned}$$

In the above equations, $d\Omega$ is the element of solid angle, and we have recalled that $-\varepsilon^{0123} = \varepsilon_{0123} = 1$.

7.

$$8\pi\kappa({}^*T^0 + {}^*t^0) = -\frac{1}{2}\varepsilon^{0bcd}\,d(\omega_{bc}\wedge e_d) = -\varepsilon^{0bcd},$$

$$\begin{aligned} d\left(\frac{f'}{r}x^c\right)dx^b\wedge dx^d &= -dx^1\wedge dx^2\wedge dx^3\left[6\frac{f'}{r} + 2r\left(\frac{f'}{r}\right)'\right] \\ &= -\frac{(\kappa M)^2}{r^4}dx^1\wedge dx^2\wedge dx^3. \end{aligned}$$

8. $R^{\alpha\beta} = c^{\alpha\beta}e^{\alpha\beta}$ (no sum);

$$c^{\alpha\beta} = \frac{\kappa M}{r^3}\begin{array}{|c|c|c|} \hline 2 & -1 & -1 \\ \hline & -1 & -1 \\ \cline{2-3} & & 2 \\ \cline{3-3} \end{array},$$

$${}^*(R_{\alpha\beta}\wedge{}^*R^{\alpha\beta}) = \sum_{\alpha,\beta}c_{\alpha\beta}{}^2 = \frac{24(\kappa M)^2}{r^6}.$$

10.5 The Life and Death of Stars

Gravity differs from other interactions by having a very small coupling constant, and by being universal. For cosmic bodies, the latter property

makes the action of gravity sum constructively to such an extent that it dominates all other interactions.

(10.5.1) The Orders of Magnitude

The gravitational energy of N protons (mass m) in a volume V is on the order of

$$E_G \sim -\frac{\kappa(Nm)^2}{V^{1/3}} = -\kappa m^2 N^{2/3} N \rho^{1/3}, \qquad \rho = \frac{N}{V}.$$

Although the Coulomb interaction is unimaginably stronger than this, $e^2 \sim 10^{36}$ κm^2, it is neutralized in normal matter, so that the electrical energy per particle is $\sim -e^2$/(the distance between nearest neighbors). This distance is $\sim \rho^{-1/3}$, so the total electrostatic energy is

$$E_e \sim -e^2 N \rho^{1/3} = \frac{e^2}{\kappa m^2 N^{2/3}} E_G. \tag{10.5.2}$$

We see that if $N \sim (e^2/\kappa m^2)^{3/2} \sim 10^{54}$, then gravity starts to dominate the electrical forces. The mass of Jupiter is about that of 10^{54} protons, which is the point at which the Newtonian potential supplants the Coulomb potential as the determiner of the structure. In a larger body, gravity crushes the atoms together, and the matter turns into a highly compressed plasma.

The Fermi energy, which is the origin of the solidity of matter, is (the number of electrons) × (the nearest-neighbor distance)$^{-2}$ × (the electron mass)$^{-1}$; in natural units ($\hbar = c = 1$):

$$E_F \sim \frac{N\rho^{2/3}}{m_e}. \tag{10.5.3}$$

The density ρ of an object adjusts so as to minimize the total energy. Whereas for the Coulombic energy (10.5.2) this makes the density independent of N,

$$\rho \sim (e^2 m_e)^3 = (\text{Bohr radius})^{-3}, \tag{10.5.4}$$

in the case of gravitation objects containing more particles are smaller:

$$\rho^{1/3} \sim m^2 \kappa N^{2/3} m_e, \qquad V \sim N^{-1} m_e^{-3} (\kappa m^2)^{-3}. \tag{10.5.5}$$

However, as soon as the separation between nearest neighbors is on the order of magnitude of the Compton wavelength $\rho^{1/3} \sim m_e$, the relativistic energy $|p|$ is to be used in calculating E_F instead of $|p|^2/2m_e$, and (10.5.3) is replaced by

$$E_F \sim N\rho^{1/3}. \tag{10.5.6}$$

The gravitational energy consequently dominates the Fermi energy when $\kappa m^2 N^{2/3} > 1 \Rightarrow N > (\kappa m^2)^{-3/2} \sim 10^{57}$, i.e., when the mass is somewhat greater than the mass of the Sun; and the minimum of the total energy is attained when $\rho = \infty$ and

$v = 0$. After that point, there is a process in nature that dramatically controls what happens. The rate of energy loss from stars is normally rather slow—one photon takes several million years to escape from the interior of the Sun—but sufficiently energetic electrons can create neutrinos by inverse beta decay $e^- + p \rightarrow \nu + n$, which, as they feel no strong interaction, leave the star immediately. This makes the transition to states of lower energy proceed at a much higher rate, and in a matter of seconds the star collapses to a neutron star, of nuclear density. Hence the energy released is on the order of the kinetic energy of neutrons at this density, about 10 MeV per particle, and thus as much energy is emitted as in the normal thermonuclear reactions, but much more rapidly. That is why it is assumed that the catastrophe just described is what takes place in a supernova, for which a single star may radiate with the brilliance of a whole galaxy for a week. The energy released would be the same, because a galaxy has typically 10^{10} stars, and normally a star takes 10^9 years $\sim 10^{10}$ weeks to burn up all its nuclear fuel.

This line of reasoning makes use of a naive, Newtonian picture of gravity, and it is interesting to see how it changes in Einstein's theory, with the help of the material developed in the preceding section. It might be hoped that a sufficiently great pressure could counteract the gravitational attraction and render the stars stable. This is not necessarily the case, however, because in the relativistic theory pressure can also produce gravity, which can aggravate the situation.

Recall, in the spirit of the phenomenological description of the energy and momentum of matter,

$$\text{(10.5.7)} \qquad \mathcal{T}_0 = -\rho e_0, \qquad \mathcal{T}_j = p e_j, \qquad j = 1, 2, 3,$$

that the energy density ρ and pressure p cannot be chosen completely arbitrarily, due to the contracted Bianchi identity (10.2.32) connecting them. For the special form (10.5.7) it requires that

$$\text{(10.5.8)} \qquad dp \wedge {}^*e^1 = \omega^1{}_0 \wedge {}^*e^0 (p + \rho)$$

(see (10.4.37; 2) with $\alpha = 1$ and $\omega^1{}_\beta$ only nonzero if $\beta = 0$).

(10.5.9) **Remarks**

1. We shall later be primarily interested in the static, spherically symmetric, case. Then both sides of the equation in (10.4.37; 2) vanish for $\alpha \neq 1$, with the e's and ω's of (10.4.30); hence (10.5.8) contains all the information of (10.2.32).
2. Since we have earlier expressed the metric in terms of ρ and p, equation (10.5.8) creates a relationship between ρ and p, which must be satisfied in order to have static equilibrium. If an equation of state is known for ρ and p, then there can be a static state only at the density distribution for which (10.5.8) agrees with the equation of state.

Taking the ω of (10.4.33),

$$\text{(10.5.10)} \qquad dp \wedge {}^*e^1 = \exp(-b)\left(\frac{\partial p}{\partial r}\right){}^*\mathbf{1} = -\exp(-b)a'(\rho + p){}^*\mathbf{1},$$

and according to (10.4.39) and (10.4.40) (with $\kappa M(r) \to M(r)$),

$$\begin{aligned} a' &= -b' + 4\pi\kappa r(\rho + p)e^{2b} \\ &= \left(1 - \frac{2M(r)}{r}\right)^{-1} \cdot \left[-4\pi\kappa r\rho + \frac{M(r)}{r^2} + 4\pi\kappa r(\rho + p)\right]. \end{aligned}$$

When this is substituted into (10.5.10), there results the

(10.5.11) **Tolman–Oppenheimer–Volkoff Equation**

$$-\frac{dp}{dr} = \frac{(\rho + p)[M(r) + 4\pi\kappa p r^3]}{r(r - 2M(r))}.$$

(10.5.12) **Remarks**

1. Of course, this also follows from (10.4.37), but the Bianchi identity does the trick without the extraneous information of (10.4.37).
2. Equation (10.5.11) generalizes the nonrelativistic fact that

$$-\frac{\partial p}{\partial r} = \frac{\rho M(r)}{r^2}.$$

The increase of the pressure for decreasing r is intensified by the following relativistic effects:

(a) There is an additional term $\sim p$ in $M(r)$, since pressure also produces gravity;

(b) it is necessary to add p to ρ, since the gravitational force also acts on p; and

(c) gravity increases faster than $\sim 1/r^2$ as $r \to 0$.

We saw at the outset that large, gravitating masses lose their stability in the special theory of relativity, because a relativistic electron gas is not as stiff as a nonrelativistic one, and does not stand firm against gravity. The general relativistic situation is even more precarious, because the solidity of matter also fails to help. In order to see this, we integrate (10.5.11) for the most extreme equation of state, viz., that of incompressible matter, which cannot be squashed to arbitrarily high density. If $\rho = \text{const.}$, and we require the boundary condition $p(R) = 0$, where R is the radius of the star, then in dimensionless variables we find

(10.5.13)

$$x = r\sqrt{\frac{8\pi\kappa\rho}{3}}, \quad x_0 = R\sqrt{\frac{8\pi\kappa\rho}{3}} = \sqrt{\frac{r_0}{R}},$$

$$p(x) = \rho\frac{\sqrt{1 - x^2} - \sqrt{1 - x_0^2}}{3\sqrt{1 - x_0^2} - \sqrt{1 - x^2}}$$

(Problem 1). As a consequence, we can read off the

(10.5.14) Maximal Pressure in Homogeneous Stars

$$p(0) = \rho \frac{1 - \sqrt{1 - r_0/R}}{3\sqrt{1 - r_0/R} - 1}.$$

(10.5.15) Consequences

1. Whereas $p(0)$ goes as $\rho r_0/4R$ for stars of homogeneous densities whose radii are much larger than the Schwarzschild radius, and thus $p(0)$ is normally much less than ρ, if $R \to r_0$ it increases rapidly and becomes infinite at $R = 9r_0/8$.

2. The pressure in matter comes from the electrons, while the protons give rise to the energy density. The relative orders of magnitude are that $p/\rho \sim$ (electron speed $v)^2 \times$ (electron mass)/(proton mass) $\sim v^2 \cdot 10^{-3}$, so that the pressure in the center of a star like the Sun, with $R \sim 10^5 r_0$, requires electrons to be moving at $\sim \frac{1}{10}$ the speed of light. The electrons must be relativistic in stars of the same mass but hundreds of times smaller (white and black dwarfs), and the situation becomes critical.

The next question to answer is how Einstein's theory affects the naive expression (10.5.1) for the gravitational energy. In §10.4 we saw that

$$\frac{r_0}{2\kappa} = 4\pi \int_0^\infty r^2\, dr\, \rho(r) =: M$$

is the total energy of the system, while

(10.5.16)
$$\int \rho^* e^0 = 4\pi \int_0^\infty r^2\, dr\, \rho(r) \left(1 - \frac{2M(r)}{r}\right)^{-1/2}$$

equals the total energy of the matter alone. If $\rho = \text{const.}$, then equation (10.5.16) can be evaluated easily, and there results the

(10.5.17) Gravitational Energy of a Homogeneous Star

$$E_G = M - \int \rho^* e^0 = M\left(1 - \frac{3}{2}\left[\arcsin\sqrt{\frac{r_0}{R}} - \sqrt{\frac{r_0}{R}}\sqrt{1 - \frac{r_0}{R}}\right]\left(\frac{R}{r_0}\right)^{3/2}\right);$$

if

$$R \gg r_0\colon E_G = -\frac{3}{5}\frac{\kappa M^2}{R},$$

$$R = r_0\colon E_G = -M\left(\frac{3\pi}{4} - 1\right).$$

When the density is small, this reduces to the Newtonian self-energy of a ball of uniform density, and as $R \to r_0$ this formula is of the same order of magnitude, but its numerical factor is somewhat different. At the limit of stability $R = 9r_0/8$, it reads $E_G = -0.37M$.

If the pressure called for in (10.5.11) cannot be provided, then it is not possible to have static equilibrium, and the star collapses. In order to pursue this drama analytically, let us consider only stars of uniform pressure and density distributions. The Friedmann solution (10.4.2) applies in the interior, while in the exterior the free Einstein equations hold, for which, according to Birkhoff's theorem (10.4.50; 5) the only available solution with spherical symmetry is the Schwarzschild solution. The problem of matching the solutions will be discussed later, after we study the dynamics in the interior.

The point of departure is equations (10.4.10), which will be used in the form

$$p = -\frac{\frac{d}{dt}(\rho R^3)}{\frac{d}{dt}R^3}, \tag{10.5.18}$$

$$\frac{\dot{R}^2}{2} - \kappa\frac{4\pi}{3}\frac{\rho R^3}{R} = -\frac{K}{2}. \tag{10.5.19}$$

We solve them first for the extremal equation of state, $p = 0$. In normal matter, p is always much less than ρ; the greatest pressure is that of massless particles, $p = \rho/3$. When either $p = \rho/3$ or $p = \rho$, analytic solutions can be written down (Problem 3), and we shall later figure out the qualitative behavior for all $p > 0$, which is generally similar.

(10.5.20) Solutions with $p = 0$

Equation (10.5.18) implies that $M = 4\pi\rho R^3/3 = \text{const.}$, making (10.5.19) of the form of the energy of the radial Kepler motion, with no angular momentum. This equation was integrated in §4.2, and the solution is most conveniently written in the form of Kepler's equation (4.2.24; 7). We can identify the variables used there as

$$m = 1, \qquad E = -\frac{K}{2}, \qquad \alpha = -\kappa M \quad \Rightarrow \quad a = \frac{\kappa M}{K}, \qquad \varepsilon = 1,$$

and must distinguish three cases:

(a) $K > 0$

$$R = \frac{\kappa M}{K}(1 - \cos u), \qquad t - t_0 = \frac{\kappa M}{K^{3/2}}(u - \sin u). \tag{10.5.21}$$

(b) $K = 0$

$$R = (t - t_0)^{2/3}\left(\frac{9\kappa M}{2}\right)^{1/3}. \tag{10.5.22}$$

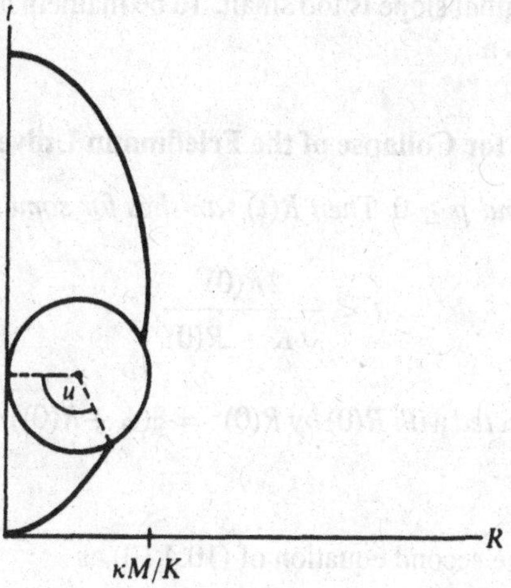

FIGURE 10.11. Cycloid for $R(t)$.

(c) $K < 0$

(10.5.23) $$R = \frac{\kappa M}{|K|}(\cosh u - 1), \qquad t - t_0 = \frac{\kappa M}{|K|^{3/2}}(\sinh u - u).$$

(10.5.24) **Remarks**

1. The case $K > 0$ in (10.5.19) corresponds to a negative energy in the Kepler problem. Accordingly, R equals zero when $u = 0$ ($t = t_0$) and when $u = 2\pi$ ($t = t_0 + 2\pi\kappa M/K^{3/2}$). In this case Kepler's equations (10.5.21) are the parametric representation of a cycloid, specifying how time elapses during a free fall into the center.
2. Case (a) describes a big bang at $t = t_0$, which was so weak that the particles flying along the geodesics $x =$ const. are eventually stopped by gravity, turn around, and all will eventually crash together again. In case (b), on the other hand, the initial velocity is high enough to send the particles to infinity. The space $t =$ const. is not compact, but in fact simply $\mathbb{R}^3$.
3. In case (c) the particles retain some kinetic energy when they reach infinity, and the space $t =$ const. has negative curvature.
4. If $t = t_0$, then $R = 0$, and we learn from (10.4.7) that the metric has a genuine singularity at that point, because the curvature scalar approaches infinity.

These results are probably not too surprising, since matter without pressure or angular momentum would be expected to fall into the center unless it has a large enough initial outward radial velocity. A positive pressure changes nothing, because the extra gravity it produces actually favors the collapse. The reason can be seen formally in (10.4.10), by which a positive p contributes negatively to $\ddot{R}$, thus increasing the concavity of the function $t \to R(t)$. In that case, $R(t)$ must approach zero if the initial slope is too small. To be mathematically precise, let us state these thoughts as a

(10.5.25) Criterion for Collapse of the Friedmann Universe

Let $\dot{R}(0)^2 < K > 0$ and $p \geq 0$. Then $R(t)$ vanishes for some

$$t < \frac{2R(0)}{\sqrt{K} - \dot{R}(0)},$$

in which $R(0)$ is connected with $\dot{R}(0)$ by $R(0)^2 = 3(K + \dot{R}(0))/8\pi\kappa\rho(0)$, according to (10.4.10).

Proof: Let us write the second equation of (10.4.10) as

$$\ddot{R}(t) = -\frac{\dot{R}(t)^2}{2R(t)} - \frac{K}{2R(t)} - \frac{R(t)8\pi\kappa p(t)}{2} \leq -\frac{K}{2R(t)}.$$

Then

$$R(t) = R(0) + t\dot{R}(0) + \int_0^t dt' \int_0^{t'} dt'' \ddot{R}(t'')$$
$$\leq R(0) + t\dot{R}(0) - \frac{Kt^2}{4a} = R(0) + \frac{\dot{R}(0)^2 a}{K} - \frac{K}{4a}\left(t - \frac{2\dot{R}(0)a}{K}\right)^2,$$

where $a := \sup_t R(t)$. If $\dot{R}(0)^2 < K$, then $a \leq R(0) + a\dot{R}(0)^2/K$ implies

$$a \leq R(0)\left(1 - \frac{\dot{R}(0)^2}{K}\right)^{-1},$$

and the zero for $R(t)$ happens before

$$t_0 = \frac{2\dot{R}(0)a}{K} + \sqrt{\frac{4aR(0)}{K} + \frac{4a^2\dot{R}(0)^2}{K^2}}, \quad \text{as} \quad R(0) + t_0\dot{R}(0) = \frac{Kt_0^2}{4a}.$$

The bound on a implies (10.5.25). □

(10.5.26) Remarks

1. If $p = 0$, then the condition $\dot{R}(0)^2 < K$ corresponds to the statement for the equivalent Kepler problem that the kinetic energy is less than minus the total energy. This obvious criterion preventing escape is valid for all $p \geq 0$.
2. The time t is that of a comoving coordinate system, and thus the space collapses to a point within a finite proper time for freely falling observers.

Finally, we construct a solution of Einstein's equations, which describes gravitational collapse. The physical picture of what takes place is as follows: If, after having exhausted its nuclear fuel, a star has shrunk down so far that the Fermi energy of the electrons has risen above the threshold for inverse beta decay $e^- + p \to \nu + n$, then the greater part of the matter is turned into neutrons. Since the star is supported against collapse mainly by the Fermi pressure of the electrons, it suddenly gives way. Thus the model would be that of a star in static equilibrium, whose pressure at some time is suddenly reduced to zero. The solution of Einstein's equations before that time is as in (10.4.30). Afterward, the solution in the interior is (10.4.2), and in the exterior it is the Schwarzschild metric. We now need to show that the solutions can be joined smoothly at the surface to satisfy Einstein's equations with $\rho =$ const. inside and 0 outside, and $p = 0$. Since the surface of the star falls freely, its radius in the co-falling coordinates (10.4.2) is $r = a =$ const. For simplicity we use units in which $r_0 = 8\pi\kappa a^3 \rho/3 = 1$ and consider the case $K = 0$. This makes the motion parabolic, with the surface of the star infinitely large at the beginning. Similarly, the solution with $K > 0$ is a Friedmann space in the interior, matched to a Schwarzschild metric. If $p > 0$, the calculation becomes much more complicated, because it cannot be constant inside the star, as otherwise there would be an infinite pressure gradient at the surface. However, the essential features are not greatly altered if $p > 0$ [55].

In order to join (10.4.30) to (10.4.2), we have to express the two metrics in the same coordinates. For this reason, we write the

(10.5.27) **Schwarzschild Metric in Co-Falling Coordinates**

It is convenient to introduce the coordinates $(\tau, \tilde{r})$ in place of (t, r), where τ is the proper time for radial parabolic motion, and $\tilde{r}$ is r at the time $t = 0$. Since the speed approaches zero asymptotically,

$$p^0 = \left(\frac{dt}{d\tau}\right)\frac{r-1}{r} = 1,$$

and thus

$$-1 = \frac{r}{r-1}\left(-1 + \left(\frac{dr}{d\tau}\right)^2\right) \quad \Rightarrow \quad \frac{dr}{d\tau} = -\frac{1}{\sqrt{r}} \quad \Rightarrow \quad \tau = \tfrac{2}{3}(\tilde{r}^{3/2} - r^{3/2}).$$

Consequently,

$$\frac{dt}{dr} = \frac{d\tau}{dr}\left(1 + \frac{1}{r-1}\right) \quad \Rightarrow \quad t = \tau - 2\sqrt{r} + \ln\frac{\sqrt{r}+1}{\sqrt{r}-1}.$$

This puts the metric in the normal form $g = -d\tau^2 + g_{ij}\, dx^i\, dx^j$, because

$$d\tau = \sqrt{\tilde{r}}\, d\tilde{r} - \sqrt{r}\, dr,$$

$$dt = d\tau - dr\frac{\sqrt{r}}{r-1} = d\tau\frac{r}{r-1} - d\tilde{r}\frac{\sqrt{\tilde{r}}}{r-1}$$

leads to

(10.5.28)

$$g = -dt^2\frac{r-1}{r} + dr^2\frac{r}{r-1} + r^2\,d\Omega^2$$
$$= -d\tau^2 + \left(1-\tfrac{3}{2}\tau\tilde{r}^{-3/2}\right)^{-2/3} d\tilde{r}^2 + \left(1-\tfrac{3}{2}\tau\tilde{r}^{-3/2}\right)^{4/3}\tilde{r}^2\,d\Omega^2.$$

(10.5.29) **Remarks**

1. This chart can be used for $3\tau/2 < \tilde{r}^{3/2}$ and becomes singular at $3\tau/2 = \tilde{r}^{3/2}$, which corresponds to $r = 0$ though. Equation (10.4.29) therefore extends the Schwarzschild metric beyond $r = r_0 = 1$, but it is not the maximal extension (10.4.48).
2. A particle falling freely from infinity travels from $r = \tilde{r}$ to $r = 0$ in proper time $2\tilde{r}^{3/2}/3$, in units where $r_0 = 1$.

To discover the proper Friedmann solution, note that equation (10.5.19), with $a = R(0)$, $r_0 = 8\pi\kappa\rho(0)a^3/3 = 1$, and $K = 0$, implies that

$$\dot{R} = -\frac{1}{\sqrt{R}},$$

and hence that

$$R(t) = a\left(1-\tfrac{3}{2}ta^{-3/2}\right)^{2/3}.$$

By redefining the coordinates $t \to \tau$, $ar \to \tilde{r}$, we come up with the

(10.5.30) **Oppenheimer–Snyder Solution**

The metric

$$g = -d\tau^2 + \begin{cases} (1-\frac{3}{2}\tau a^{-3/2})^{4/3}(d\tilde{r}^2 + \tilde{r}^2\,d\Omega^2) & \text{if } \tilde{r} \le a, \\ (1-\frac{3}{2}\tau\tilde{r}^{-3/2})^{-2/3}\,d\tilde{r}^2 + (1-\frac{3}{2}\tau\tilde{r}^{-3/2})^{4/3}\tilde{r}^2\,d\Omega^2 & \text{if } \tilde{r} \ge a, \end{cases}$$

satisfies Einstein's equations with $p = 0$,

$$\rho = \begin{cases} 3(a^{3/2} - 3\tau/2)^{-2}/8\pi\kappa & \text{if } \tilde{r} \le a, \\ 0 & \text{if } \tilde{r} > a. \end{cases}$$

Proof: Einstein's equations are satisfied for $\tilde{r} > a$ and $\tilde{r} < a$ by construction, and, as the curvature ought to be discontinuous at $\tilde{r} = a$ (cf. (10.2.24)), it is only necessary to check that it has no delta-function singularity there. Let us write the orthogonal basis for g,

$$e^\alpha = (d\tau, e^\nu \nu'\,d\tilde{r}, e^\nu\,d\vartheta, e^\nu \sin\vartheta\,d\varphi),$$

(10.5.31)
$$e^{\nu(\tau,\tilde{r})} := \begin{cases} \tilde{r}(1-\frac{3}{2}\tau a^{-3/2})^{2/3} & \text{if } \tilde{r} \le a, \\ \tilde{r}(1-\frac{3}{2}\tau\tilde{r}^{-3/2})^{2/3} & \text{if } \tilde{r} \ge a, \end{cases}$$

where $\nu' := \partial\nu/\partial\tilde{r}$ and $\dot{\nu} := \partial\nu/\partial\tau$; ν' is discontinuous and $\dot{\nu}$ is continuous with a discontinuous first derivative. When restricted to $r = a$ ($d\tilde{r}_{|\tilde{r}=a} = 0$), only the continuous parts of e^α remain, and even the de^α are continuous at $\tilde{r} = a$:

$$de^\alpha = (0, e^\nu(\dot{\nu}' + \dot{\nu}\nu')\, d\tau \wedge d\tilde{r}, e^\nu(\dot{\nu}\, d\tau + \nu'\, d\tilde{r}) \wedge d\vartheta,$$
$$e^\nu((\dot{\nu}\, d\tau + \nu'\, d\tilde{r}) \sin\vartheta + \cos\vartheta\, d\vartheta) \wedge d\varphi).$$

From this formula we get the affine connections,

(10.5.32) $\omega^\alpha{}_\beta =$

τ	$\tilde{r}$	ϑ	φ
0	$e^\nu(\dot{\nu}' + \dot{\nu}\nu')d\tilde{r}$	$e^\nu\dot{\nu}\, d\vartheta$	$e^\nu\dot{\nu} \sin\vartheta\, d\varphi$
	0	$-d\vartheta$	$-\sin\vartheta\, d\varphi$
		0	$-\cos\vartheta\, d\varphi$
			0

Observe that the discontinuous functions ν' and $\dot{\nu}'$ are multiplied by $d\tilde{r}$, and so no ν'' shows up in $d\omega^\alpha{}_\beta$ (see Problem 4). Hence, while $R^\alpha{}_\beta$ is discontinuous, it does not contain a delta function. □

(10.5.33) **Remarks**

1. As Einstein's equations do not allow delta-function singularities in the contractions R_α, the question arises of whether they can occur in $R^\alpha{}_\beta$. The answer is that they cannot, because the surface of discontinuity has a spacelike normal $d\tilde{r}$, and by (10.2.24), the regularity of R_α implies that of $R^\alpha{}_\beta$ in this case.
2. If the basis given in the proof is supplied with a more general function $\nu(\tau, r)$, then it is easy to find a solution of Einstein's equations that describes the gravitational collapse of a C^∞ density distribution. The discontinuous solution (10.5.30) can be considered as the limiting case of a C^∞ solution (Problem 4).
3. In order that the solutions join seamlessly, the Schwarzschild radius of the outer solution must be
$$8\pi\kappa \int_0^a d\tilde{r}\, \tilde{r}^2 \rho(\tilde{r}, \tau)_{|\tau=0}.$$
This is to be expected, because when $\tau = 0$, $dr = d\tilde{r}$. In the static coordinates (t, r), the dramatic action in the interior is not detectable from outside. The Schwarzschild solution is as silent as the grave about what it hides within.

(10.5.34) **The Geometric Significance of the Oppenheimer–Snyder Solution**

The chart of (10.5.30) is workable if

$$\tau < \tfrac{2}{3} \min[a^{3/2}, \tilde{r}^{3/2}],$$

though the metric becomes singular at the boundary. The lines $\tilde{r} =$ const. are the world-lines of freely falling observers, and τ measures their proper time. In this chart the light-cones inside the star flatten out as they approach the singularity,

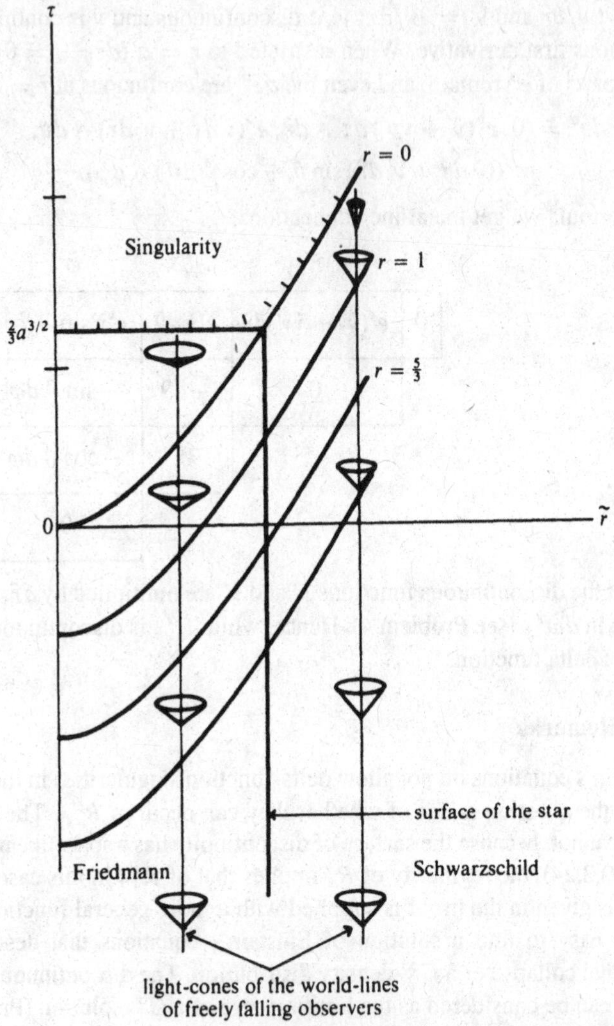

FIGURE 10.12.

while on the outside they narrow down (see Figure 10.12). Note that the significance of $r = 1$ as a horizon for $\tilde{r} > a$ can thereby be expressed as the fact that from this point on, the light-cones remain completely on one side of the curve $r = \text{const}$.

(10.5.35) Problems

1. Integrate (10.5.11) for $\rho = \text{const}$. (cf. (10.4.12)).

2. Show that $\rho = 3/56\pi\kappa r^2$ is a solution of (10.5.11) for the equation of state of radiation, $p = \rho/3$.

3. Solve (10.4.10) for $K > 0$, if $p = \rho/3$ and if $p = \rho$.

4. Use the basis (10.5.31)

$$e^\alpha = (d\tau, e^\nu \nu' \, d\tilde{r}, e^\nu \, d\vartheta, e^\nu \sin\vartheta \, d\varphi), \qquad \text{with} \quad \nu(\tau, \tilde{r}),$$

to find a solution of Einstein's equations for $\mathcal{T}_0 = \rho(\tilde{r}) e^0$, $\mathcal{T}_j = 0$.

(10.5.36) Solutions

1. Let x be the dimensionless variable given in (10.5.13), and $y = p/\rho$; then this equation becomes

$$-\frac{dy}{dx} = (1+y)(1+3y)\frac{x/2}{1-x^2}, \quad \text{i.e.,} \quad dy\left(\frac{-3}{1+3y} + \frac{1}{1+y}\right) = \frac{dx\,x}{1-x^2}$$

$$\Rightarrow \quad \ln\frac{1+3y}{1+y} = \frac{1}{2}\ln\frac{1-x^2}{1-x_0^2} \quad \Rightarrow \quad \frac{1+3y}{1+y} = \sqrt{\frac{1-x^2}{1-x_0^2}}$$

$$\Rightarrow \quad \frac{\sqrt{1-x^2} - \sqrt{1-x_0^2}}{3\sqrt{1-x_0^2} - \sqrt{1-x^2}} = y.$$

2.

$$p = 1/56\pi\kappa r^2, \qquad M(r) = 4\pi\kappa \int_0^r dr' \, r'^2 \rho(r') = 3r/14,$$

$$-\frac{\partial p}{\partial r} = \frac{1}{28\pi\kappa r^3} = \frac{4}{56\pi\kappa r^3}\frac{\frac{3}{14}r + r/14}{r(1-\frac{3}{7})} = \frac{\rho + p}{r}\frac{M(r) + 4\pi\kappa p r^3}{r - 2M(r)}.$$

3.

$$p = \frac{\rho}{3}: \; \frac{K + R^2}{R^2} = -\frac{2R\ddot{R} + \dot{R}^2 + K}{R^2} \quad \Rightarrow \quad \dot{R}^2 + R\ddot{R} + K = 0 \quad \Rightarrow \quad \frac{d^2}{dt^2}R^2 = -2K$$

$$\Rightarrow \quad R = (ct - Kt^2)^{1/2}.$$

$p = \rho$: Let $d\tau/dt = 1/R$. Then $4K + 4\dot{R}^2 + 2R\ddot{R} = 0$ becomes the oscillator equation,

$$\frac{d^2R^2}{d\tau^2} = -4KR^2 \quad \Rightarrow \quad R = R_{\max}[\sin 2\sqrt{K}\tau]^{1/2};$$

consequently t is given by

$$R_{\max}\int_0^t d\tau' [\sin 2\sqrt{K}\tau']^{1/2}.$$

In both cases we have chosen $R(0) = 0$, and we observe that R decreases again to zero after a finite time.

4. By (10.1.28) the affine connections (10.5.32) lead to the curvature forms

$$\begin{aligned}
R^0{}_{\tilde{r}} &= e^\nu(2\dot{\nu}\dot{\nu}' + \dot{\nu}^2\nu' + \ddot{\nu}' + \ddot{\nu}\nu')\,d\tau \wedge d\tilde{r}, \\
R^0{}_\vartheta &= e^\nu(\dot{\nu}^2 + \ddot{\nu})\,d\tau \wedge d\vartheta, \\
R^0{}_\varphi &= e^\nu(\dot{\nu}^2 + \ddot{\nu})\sin\vartheta\,d\tau \wedge d\varphi, \\
R^{\tilde{r}}{}_\vartheta &= e^{2\nu}(\dot{\nu}' + \dot{\nu}\nu')\dot{\nu}\,d\tilde{r} \wedge d\vartheta, \\
R^{\tilde{r}}{}_\varphi &= e^{2\nu}(\dot{\nu}' + \dot{\nu}\nu')\dot{\nu}\sin\vartheta\,d\tilde{r} \wedge d\varphi, \\
R^\vartheta{}_\varphi &= e^{2\nu}\dot{\nu}^2\sin\vartheta\,d\vartheta \wedge d\varphi.
\end{aligned}$$

Once again, $R_{\alpha\beta} \sim e_{\alpha\beta}$, so $T_{\alpha\beta}$ is diagonal. Einstein's equations require that

$$8\pi\kappa\rho = 3\dot{\nu}^2 + 2\frac{\dot{\nu}'\dot{\nu}}{\nu'} = 8\pi\kappa T_{00},$$
$$0 = e^{\nu}(3\dot{\nu}^2 + 2\ddot{\nu}) = T_{\tilde{r}\tilde{r}},$$
$$0 = \left(1 + \frac{1}{2\nu'}\frac{d}{d\tilde{r}}\right)(3\dot{\nu}^2 + 2\ddot{\nu}) = T_{\vartheta\vartheta} = T_{\varphi\varphi}.$$

The last two of these equations are solved by $e^{\nu} = (F(\tilde{r})\tau + G(\tilde{r}))^{2/3}$. Since the basis is invariant under a change of charts $\tilde{r} \to \tilde{r}(\tilde{r})$, we may set $G = \tilde{r}^{3/2}$, and are left with only one function, $F(\tilde{r})$. If $\tau = 0$, then the first of Einstein's equations becomes

$$FF' = 9\pi\kappa\tilde{r}^2\rho(\tilde{r}, 0) \quad \Rightarrow \quad F(\tilde{r}) = -\left[18\pi\kappa\int_0^{\tilde{r}} dr\, r^2\rho(r, 0)\right]^{1/2}.$$

In the case $\rho(r, 0) = 3/8\pi\kappa a^3$, we revert to (10.5.30).

10.6 The Existence of Singularities

The solutions of nonlinear differential equations have a tendency to develop singularities; in particular this is true of Einstein's equations, where the attractive nature of gravity reveals its physical origins.

The models we have considered of gravitational collapse, in which a singularity develops, were all radially symmetric. It is not at all surprising that a fall directed right at the center will end in a catastrophe. The one new feature of Einstein's theory is that the catastrophe cannot be prevented by any pressure, no matter how strong, because the pressure itself produces more gravity. There is a question, however, whether the situation is qualitatively changed by a perturbation of the radial symmetry, just as the angular momentum in the Kepler problem prevents the plunge into the center. In the relativistic Kepler problem, the effective gravitational potential goes as $-1/r^3$, which is stronger than the centrifugal potential (cf. (§5.7)), but it is conceivable that other mechanisms might impede the growth of a singularity. It is often claimed ([51, §1.14]), on account of this observation, that normally the solutions are free of singularities, which are pathologies afflicting spaces of high symmetry. It was the accomplishment of R. Penrose and others of the school of D. Sciama to disprove this claim: as long as energy and pressure are positive in some reasonable sense, and at some instant there exists the kind of geometry set up by a large mass, then the formation of a singularity is unavoidable, regardless of any symmetry.

Let us agree at this stage what we mean by a singular space. Regularity is incorporated in the concept of a manifold, and any singular points are removed. It might be suggested that unbounded growth of $R^{\alpha}{}_{\beta}$ could be taken as a sign of a singularity in the vicinity. We shall see shortly that the $R^{\alpha}{}_{\beta}$ describe the tidal force, and consequently this conjectured indicium has a direct physical significance, as it can be observed as bodily discomfort. Unfortunately, it is difficult to express this mathematically, since the components of $R^\alpha{}_\beta$ depend on the basis, and could also become infinite in the absence of a genuine singularity. Conversely, it is possible for all 14 of the invariants that can be constructed from $R^\alpha{}_\beta$ to vanish without $R^\alpha{}_\beta$ itself vanishing. For example, this happens for plane gravitational waves, and is analogous to a nonzero vector in Minkowski space having zero length.

Hence we resort to a different feature of the solutions we have discussed as the criterion for a singularity, viz., that an observer falls into the singularity in a finite proper time, thus leaving the manifold. There is, of course, the trivial possibility that the manifold has simply been chosen too small—if the manifold were only a piece of Minkowski space, then one could leave it in a finite time, although there is not necessarily any singularity outside the piece. In order to exclude such cases, we make a

(10.6.1) **Definition**

A pseudo-Riemannian manifold M is **extensible** iff it is a proper subset of a larger manifold M', i.e., its metric is the restriction to M of the metric on M'.

(10.6.2) **Remarks**

1. M' is not, of course, uniquely determined by M, so our criterion cannot involve examining an extensible manifold to see where there are singularities. For instance, the Schwarzschild metric for $r > 5r_0$ can be extended either to the regular solution with a continuous mass distribution for $r < 5r_0$ or to the singular solution.
2. When confronted with an extensible manifold, one gets the feeling that something has been intentionally left out. Therefore we postulate that the physical space–time continuum is nonextensible.
3. There are examples ([52, p. 58]) of nonextensible manifolds that can be escaped from, so it is actually necessary to postulate a more refined property, local nonextensibility. However, the examples seem rather artificial, so we shall content ourselves with the primitive definition.

The next step is to decide what observers we will grant an unlimited stay in the manifold.

(10.6.3) **Definition**

A pseudo-Riemannian manifold is said to be **geodesically complete** in timelike directions iff every timelike geodesic can be extended to an arbitrarily long proper-time parameter.

(10.6.4) **Remarks**

1. A positive metric g defines a metric for the topology of a Riemannian space M, and then geodesic completeness means the same thing as completeness in the sense of a metric topological space.

2. An affine parameter could also be defined on lightlike geodesic lines, and one can speak of lightlike and spacelike geodesic completeness. These conditions are not equivalent (Problem 1); but at any rate (10.6.3) must be required on physical grounds.
3. Geodesic incompleteness puts an observer who can stay in the manifold for only a finite time into a predicament, but is not necessarily evidence of any kind of infinity. This is shown by the example

$$g = -dt^2\left(1-\frac{h}{2}\right)+dx^2\left(1+\frac{h}{2}\right)+dx\,dt\,h,$$

$$h = \frac{\lambda}{2}(\cos^4(t-x)-1), \qquad \lambda \in (0,2),$$

on $\mathbb{R}^2$ (if desired, dy^2+dz^2 can be added in). If λ is small, this is only a weak gravitational wave that spreads throughout the flat space, but nonetheless the space fails to be geodesically complete in timelike directions, even for arbitrarily small λ (Problem 2). The reason is that a particle of the right initial velocity rides the crests of the waves, as in a linear accelerator, and reaches nearly the speed of light. Its proper time runs ever more slowly and never exceeds some finite value. There is no singularity, and the only $R^i{}_j$ that does not vanish is $R^0{}_1 = e^0{}_1 h''/2$. Since h is periodic in $u = t-x$, g can be used as a pseudometric on T^2, in which case even this compact set is geodesically incomplete, although it is certainly not a piece of a larger connected manifold. Despite that, we follow common usage and refer to the space as singular.
4. Even in Minkowski space it is possible to reach the end of the manifold after a finite proper time on certain timelike lines. If, for instance, $x = t + 1/t^2$ for $t > 1$, then

$$\int ds = \int \frac{ds}{dt}dt = \int_1^\infty dt\sqrt{1-\left(\frac{dx}{dt}\right)^2} < 2\int_1^\infty dt\, t^{-3/2} = 4.$$

It is only the choice of coordinates that makes the end at $x = \infty$, and it can be transformed to any finite point, just as the end lay at $-\infty$ in the Schwarzschild metric with the variable $\ln r$.
5. One might require that timelike lines with bounded acceleration $\ddot{z}^\alpha \ddot{z}^\beta g_{\alpha\beta}$ can be continued to arbitrarily long proper times. If this were not so, then the crew of a rocket with a finite supply of fuel could conceivably find themselves at the edge of the Universe, and would not know what to do. Yet geodesic completeness leaves this possibility open [53].

Geodesic lines are the world-lines of freely falling observers (cf. Problem 4). The nonrelativistic analogue of a geodesic vector field is the velocity field v_i of an ideal fluid with no pressure in a gravitational potential Φ. For stationary fluid flow, the equations of hydrodynamics require that $v_k v_{i,k} = -\Phi_{,i}$. Let n be the distance-vector field between nearby fluid particles, which is carried along with the stream. Its Lie derivative with respect to v vanishes, so

(10.6.5) $$v_k n_{i,k} = n_k v_{i,k}$$

(recall (2.5.12; 5)), which makes the second derivative along the streamlines

(10.6.6) $$v_k \frac{\partial}{\partial x^k}\left(v_j \frac{\partial}{\partial x^j} n_i\right) = -n_k \Phi_{,ik}.$$

Thus the gradient of the field $\Phi_{,i}$ affects the distance between two particles, and in fact the effect of the second derivative of Φ is to focus them together: Since Φ satisfies the equation

(10.6.7) $$\Phi_{,jj} = \rho \geq 0,$$

the net effect of the gravitational field, when averaged over all directions, is to focus particles. For irrotational fluid flow, $v_{i,k} \equiv v_{k,i}$, this can be expressed as an increase in the rate of convergence $c = -v_{i,i}$ of the flow along the streamlines:

(10.6.8) $$v_i \frac{\partial}{\partial x_i} c = -v_i v_{k,ik} = v_{k,i} v_{i,k} + \Phi_{,kk} \geq \frac{c^2}{3}.$$

This equation used (10.6.7) and irrotationality, which entered through the trace inequality for symmetric $(n \times n)$-matrices

(10.6.9) $$(\mathrm{Tr}\, M)^2 \leq n\, \mathrm{Tr}(M^2)$$

(Problem 3). If c is positive at some point, then it increases so rapidly by (10.6.8) that it soon reaches infinity, and the streamlines meet. If s is the parameter on a streamline, given as $x(s)$, $v_i(x) = dx_i/ds$ (cf. (10.1.42; 1)), then (10.6.8) implies that

(10.6.10) $$\frac{dc}{ds} \geq \frac{c^2}{3} \quad \Rightarrow \quad c(s) \geq \frac{c(0)}{1 - sc(0)/3},$$

and thus c gets arbitrarily large before $s = 3/c(0)$. This elementary property of gravity contains the essential features of the relativistic theory discussed below.

The relativistic generalizations of (10.6.6) are

(10.6.11) The Equations of Geodesic Deviation

Let $v = v^\alpha e_\alpha$ be a geodesic vector field and n a vector field such that $L_v n = 0$. Then

$$D_v D_v n = -e_\alpha (R^\alpha{}_\beta | n \otimes v) v^\beta.$$

Proof: By (10.1.7(g′)), $D_v n = D_n v$, and because of (10.1.19), (10.1.33; 2), and the equation $D_v v = 0$,

$$\begin{aligned} 0 = D_n D_v v &= D_v D_v n + (D_n D_v - D_v D_n) v \\ &= D_v D_v n + e_\alpha (R^\alpha{}_\beta | n \otimes v) v^\beta. \end{aligned}$$ □

(10.6.12) Example

Consider the Friedmann universe (10.4.2). Let the fields v and n be the natural contravariant basis elements ∂_t and ∂_x. Their Lie brackets with each other vanish, and ∂_t is geodesic, since we are using co-falling coordinates. The contravariant components of the metric (10.4.2) are

$$g^{00} = -1, \qquad g^{jj} = \left(\frac{1 + Kr^2/4}{R}\right)^2,$$

so v and n can be written in the orthogonal basis as

$$v = \partial_0 = e_0, \qquad n = \partial_1 = \frac{e_1}{\sqrt{g^{11}}} = e_1 \frac{R}{1 + Kr^2/4}.$$

With the affine connections of (10.4.5), we see that

$$D_v n = \frac{\dot{R}}{1 + Kr^2/4} e_1 + \frac{R}{1 + Kr^2/4} D_v e_1 = \frac{\dot{R}}{1 + Kr^2/4} e_1,$$

$$D_v D_v n = \left(D_v \frac{\dot{R}}{1 + Kr^2/4}\right) e_1 = \frac{\ddot{R}}{1 + Kr^2/4} e_1 = \frac{\ddot{R}}{R} n,$$

which, because of (10.4.6), is precisely

$$-e_1(R^1{}_0 | n \otimes v).$$

(10.6.13) Remarks

1. Since $R(t)$ describes how the distance between neighboring world-lines $x =$ const. varies, we perceive that $D_v D_v n$ has the significance of a relative acceleration.
2. Proposition (10.6.11) shows that from the physical point of view it is $R^\alpha{}_\beta$ rather than $\omega^\alpha{}_\beta$ that takes over the role of the electric field strength. Because of the principle of equivalence (5.6.11), there is no trace of the ω's; freely falling observers can only notice the gradient of the field R, specifying the corrections to the principle of equivalence, which holds only in the infinitesimal limit.

The curvature forms may have either sign, either focusing or defocusing. The contractions R_α are immediately determined by the energy and momentum, from which they inherit the positivity (8.1.13). As with (10.6.8), this leads to an

(10.6.14) Increase in the Rate of Convergence of Geodesic Vector Fields

Let v be a timelike geodesic vector field perpendicular to a hyperplane $t = 0$, and assume $i_v R_0 \geq 0$. Then the rate of convergence $c = -\sum_\alpha \langle e^\alpha | D_{e_\alpha} v\rangle$ (cf. Problem 4) satisfies the differential inequality

$$D_v c \geq \frac{c^2}{3}.$$

Proof: We work in the natural basis of a co-falling coordinate system, so that $v = \partial_t$ and $g = -dt^2 + g_{ab}\, dx^a \otimes dx^b$. As in (10.6.11), $D_v v = 0$, and $D_v \partial_a = D_{\partial_a} v$. By also recalling that

$$0 = D_v \delta^\alpha{}_\beta = \langle D_v\, dx^\alpha | \partial_\beta \rangle + \langle dx^\alpha | D_v \partial_\beta \rangle,$$

we find

$$\begin{aligned} D_v c &= -D_v \langle dx^\alpha | D_{\partial_\alpha} v \rangle = -\langle D_v\, dx^\alpha | D_v \partial_\alpha \rangle - \langle dx^\alpha | D_v D_{\partial_\alpha} v \rangle \\ &= -\langle D_v\, dx^\alpha | \partial_\beta \rangle \langle dx^\beta | D_v \partial_a \rangle + \langle R^\alpha{}_\beta | \partial_\alpha \otimes v \rangle v^\beta \\ &= \langle dx^\alpha | D_v \partial_\beta \rangle \langle dx^\beta | D_v \partial_\alpha \rangle + i_v R_0 \geq \frac{c^2}{3}, \end{aligned}$$

since the trace inequality (10.6.9) is again applicable:

$$M_{\alpha\beta} := \langle \partial_\alpha | D_v \partial_\beta \rangle = \langle \partial_\alpha | D_{\partial_\beta} v \rangle \equiv \Gamma^0_{\alpha\beta}$$

is symmetric in α and β and vanishes when α or β is zero, because of $D_v v = 0$. In the space orthogonal to v, g is positive and $c = \mathrm{Tr}(Mg) = \mathrm{Tr}(\sqrt{g} M \sqrt{g})$, while $\mathrm{Tr}(\sqrt{g} M \sqrt{g} \sqrt{g} M \sqrt{g})$ occurs in the above equation. □

(10.6.15) **Example**

Let us take another look at the Friedman universe, for which

$$D_{\partial_a} v = D_v \partial_a = \frac{\dot{R}}{R} \partial_a,$$

and thus $c = -3\dot{R}/R$. Proposition (10.6.14):

$$D_v c = -\frac{\partial}{\partial t} \frac{3\dot{R}}{R} = \frac{3\dot{R}^2}{R^2} - \frac{3\ddot{R}}{R} \geq \frac{c^2}{3} = \frac{3\dot{R}^2}{R^2}$$

holds if $\ddot{R} \leq 0$, which amounts to the condition that $i_v R_0 \geq 0$. From equations (10.4.10), this condition is met if $\rho + 3p \geq 0$.

(10.6.16) **Remarks**

1. According to Einstein's equations (10.2.20(a)), the condition $i_v R_0 \geq 0\ \forall v$: $\langle v | v \rangle < 0$ implies for the total energy-momentum tensor that $T_{00} + T_{11} + T_{22} + T_{33} \geq 0$ ("positivity of the energy"). This is true for all sensible models of matter. The reason for the positivity condition is that negative energy produces a repulsive gravitational force, which could prevent the convergence of the geodesics.
2. Generalizing the nonrelativistic result (10.6.10), we see that if c is ever positive, then it must become infinite after a finite time under the circumstances of (10.6.14). Let s be the proper-time parameter (cf. (10.1.42; 1)) on the geodesics of v, $v^i(x(s)) := \dot{x}^i(s)$, and $c(s) := c(x(s))$. If $c(0) > 0$, then c becomes infinite for some s such that $0 \leq s \leq 3/c(0)$.

3. If $N_{ab} := \langle e_a | e_b \rangle$, then $(\det N)^{1/2}$ is the volume spanned by the spacelike basis vectors. Since

$$c = -\langle e^a | D_v e_a \rangle = -\tfrac{1}{2} D_v \ln(\det N),$$

$c \to \infty$ would mean that the volume approaches zero, and neighboring geodesic lines meet. Hence there exists a basis field n, $L_v n = 0$, which becomes zero.

It is not necessarily a sign of a disaster if c does not remain bounded; (10.6.14) made no use of the strict positivity of $i_v R_0$, which means that the same conclusions could be reached in flat space, even though the metric is everywhere regular.

(10.6.17) Example

Let us introduce the coordinates $\tau = -\sqrt{t^2 - x^2}$ and $u = x/t$ on $M = \{(t, x) \in \mathbb{R}^2 : t < 0, x < 0, t^2 > x^2; g = -dt^2 + dx^2\}$, and let the vector fields

$$v = \partial_\tau = -\frac{t}{\sqrt{t^2 - x^2}} \partial_t - \frac{x}{\sqrt{t^2 - x^2}} \partial_x,$$

$$n = \partial_u = \frac{x t^2}{t^2 - x^2} \partial_t + \frac{t^3}{t^2 - x^2} \partial x,$$

serve as the basis; it is easy to verify that $D_v v = 0$, $D_v \tau = 1$, and $D_v n = D_n v = -n/\tau$. The streamlines of v are geodesic since they are straight lines through the origin, and the streamlines of n are normal to them in the sense of the metric (Figure 10.13).

$$c = -\langle v | D_v v \rangle - \frac{\langle n | D_v n \rangle}{\langle n | n \rangle} = \frac{1}{\tau},$$

and in fact

$$D_v c = -\frac{\partial}{\partial \tau} \frac{1}{\tau} = \frac{1}{\tau^2} = c^2,$$

and in one space dimension (10.6.14) can be strengthened to $D_v c \geq c^2$. At the origin c becomes infinite, although the space is not singular; it is only that the chart (τ, u) is unsuitable there.

When geodesics cross, they lose the property of being extremal, which can bring about some contradictions. This will lead to the conclusion that one cannot make the assumption that geodesics are extensible past the points where they cross. Let us distinguish the different possible states of affairs by making

(10.6.18) Definition

(a) The **future** $J^+(x)$ (resp. **past** $J^-(x)$) of $x \in M$ consists of the points of M that can be connected to x by causal curves (see (8.1.15)) directed toward the past (resp. future).

(b) Let S be a spacelike hypersurface and $x \in D^+(S)$. Then the set of causal (resp., differentiable causal) curves from x to S is denoted by $C(x, S)$ (resp. $C^1(x, S)$). The set $C(x, S)$ is topologized as follows: A basis of neighborhoods of the curve λ consists of all curves that stay in a neighborhood of

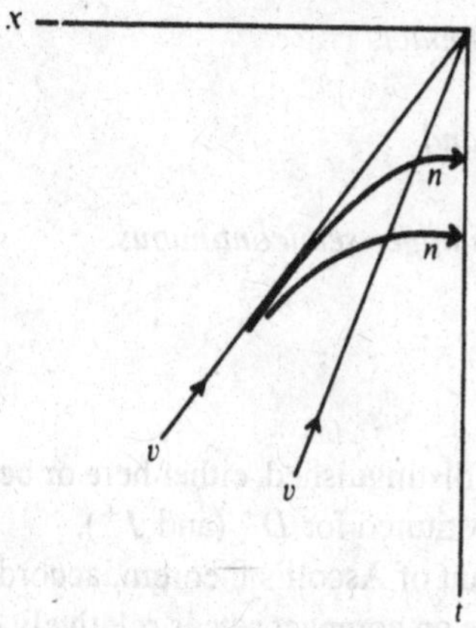

FIGURE 10.13. The convergence of a geodesic vector field on flat space.

λ, in the sense of the topology of M. We let C^1 have the topology induced by C.

(c) The **length**, or **proper time**, of $\lambda \in C^1(x, S)$ is defined by

$$d(\lambda) = \int_{s_0}^{s_1} ds \sqrt{-g_{\alpha\beta}\dot{\lambda}^\alpha \dot{\lambda}^\beta}, \qquad \lambda(s_0) \in S, \qquad \lambda(s_1) = x.$$

(10.6.19) **Remarks**

1. The topology on $C(x, S)$ is that of uniform convergence. It is metrizable, since the topology of M is metrizable: the Hausdorff distance function between two subsets of M is defined in terms of the topology of M, and the distance function of two causal curves produces the metric on $C(x, S)$. Consequently, compactness becomes synonymous with sequential compactness.
2. C^1 is dense in C, and so $d(\lambda)$ can be extended to C (Problem 5).

To simplify the next chain of reasoning, let us eliminate pathologies at the outset by making for the duration the

(10.6.20) **Assumptions**

(a) that M is orientable with respect to time (8.1.15); and

(b) that if x and y vary over M, then the interiors of $J^-(x) \cap J^+(x)$ form a basis for the topology of M (cf. (6.4.10; 3)).

The above assumptions imply the

(10.6.21) **Propositions**

For all x in the interior of $D^+(S)$:

(a) *$J^-(x) \cap D^+(s)$ is compact;*

(b) *$C(x, S)$ is compact; and*

(c) *$d\colon C^1(x, S) \to \mathbb{R}^+$ is upper semicontinuous.*

(10.6.22) **Remarks**

1. Neither time direction is distinguished, either here or below. Where appropriate, D^- (and J^-) can be substituted for D^+ (and J^+).
2. Proposition (b) is a variant of Ascoli's theorem, according to which any family of equicontinuous curves on compact sets is relatively compact in the topology of uniform convergence. However, a set of curves of arbitrary gradient is not compact; for example, $x = \sin nt$, $n = 1, 2, \ldots$ is not uniformly convergent to anything. The requirement that a curve never gets out of a light-cone prevents this from happening in (10.6.21).
3. Proposition (a) is a necessary condition for (b); for example, $x = \sin t/n$, $n = 1, 2, \ldots$, does not converge uniformly to zero on $-\infty < t \leq 0$.
4. The function d is not continuous, because in any neighborhood of $\lambda \in C(x, S)$ it is possible to reflect lightlike curves back and forth to make d vanish.
5. The extension of d to $C(x, S)$ is likewise upper semicontinuous (Problem 5). We let it define the proper time on a nondifferentiable causal curve.

The proofs of these propositions are rather technical, and are left for Problem 6. An important consequence of them is

(10.6.23) **Theorem**

Let S be a spacelike hypersurface and p be in the interior of $D^+(S)$. Then there is a curve of greatest proper time from p to S, and it is the geodesic through p that is orthogonal to S in the sense of the pseudometric g.

Proof: It follows from the compactness of C and the upper semicontinuity of d that d achieves its supremum ([41, 12.7.9]). The maximal curve must be geodesic, because otherwise one could find a nearby curve of greater proper time to the same point of intersection with S. The orthogonality follows from the requirement that geodesics to other nearby points of S take less time: according to (3.2.18; 6), the change in the time taken on a geodesic line having endpoint x is

$$\delta x^\alpha \frac{\partial}{\partial \dot{x}^\alpha} \sqrt{-\dot{x}^\beta \dot{x}^\gamma g_{\beta\gamma}} = \frac{\delta x^\alpha \dot{x}^\beta g_{\alpha\beta}}{\sqrt{-\dot{x}^\beta \dot{x}^\gamma g_{\beta\gamma}}}.$$

If $\dot{x}$ were not perpendicular to all the tangent vectors of S, then one could find a way to increase the time. □

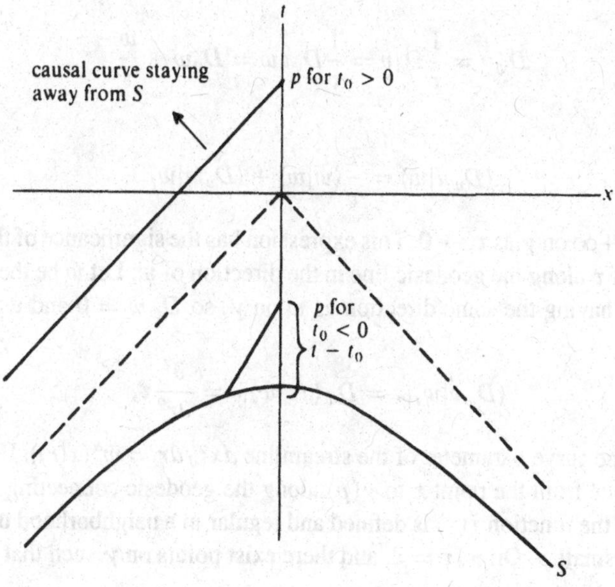

FIGURE 10.14. Geodesic lines perpendicular to S in (10.6.24).

(10.6.24) **Example**

As in Example (10.6.17), let $g = -dt^2 + dx^2$, $p = (t_0, 0)$, and $S = \{(t, x) \in \mathbb{R}^2 : t^2 - x^2 = 1, t \leq -1\}$. The straight lines through the origin are the geodesics perpendicular to S. The straight lines through p, $x = v(t - t_0)$, intersect S where $x^2 = v^2(t - t_0)^2 = t^2 - 1$, so the distance along them to S is $(t_0 - t)\sqrt{1 - v^2} = (1 + t_0^2 - 2t_0 t)^{1/2}$ (see Figure 10.14). If $t_0 \leq 0$, the maximum is achieved at $t = -1$, since $t \leq -1$, and the geodesic orthogonal to S is then the longest line. If $t_0 > 0$, then the distance grows without bound as $t \to -\infty$; there exists no maximum, and the line through $(-1, 0)$ is the shortest. This does not contradict (10.6.23), since in this case $p \notin D^+(S)$.

It is intuitively reasonable that one could obtain a more nearly extremal curve from two crossing geodesics, by rounding them off near the intersection. This expectation is confirmed by the

(10.6.25) **Theorem**

Let v: $D_v v = 0$ and $\langle v|v \rangle = -1$ be the geodesic vector field perpendicular to a hypersurface S, let $\gamma(\tau)$ be a streamline of v and suppose n with the properties $L_v n = 0$ and $\langle n|v \rangle = 0$ vanishes at $\gamma(0)$ but not on all of γ. Then for all $p > 0$, γ is not the curve of greatest proper time from $\gamma(p)$ to S.

Proof: We choose comoving coordinates $g = -d\tau^2 + g_{ik}\, dx^i\, dx^k$, and let τ be the proper time along $-v = \partial_\tau$. Since n satisfies (10.6.11) and vanishes at $\tau = 0$, it is impossible for $D_v n$ to vanish at that point, as it would otherwise be zero on γ. Therefore, letting $w := n/\tau$, $\lim_{\tau \to 0} \langle w|w \rangle$ is positive (i.e., n is spacelike).

Because

$$D_w v = \frac{1}{\tau} D_n v = \frac{1}{\tau} D_v \tau w = D_v w + \frac{w}{\tau}$$

the quantity

$$\langle D_w u | w \rangle = \frac{1}{\tau} \langle w | w \rangle + \langle D_v w | w \rangle$$

approaches $+\infty$ on γ as $\tau \to 0$. This expression has the significance of the second derivative of τ along the geodesic line in the direction of w: Let $\bar{w}$ be the geodesic vector field having the same direction as w on γ, so $D_{\bar{w}} \bar{w} = 0$ and $w_{|\gamma} = \bar{w}_{|\gamma}$; then

$$\langle D_w v | w \rangle_{|\gamma} = D_{\bar{w}} \langle d\tau | \bar{w} \rangle_{|\gamma} = \frac{\partial^2}{\partial r^2} \tau,$$

where r is the curve parameter of the streamline $dx^\alpha / dr = \bar{w}^\alpha(x(r))$. If $p - \bar{\tau}(x)$ is the distance from the point x to $\gamma(p)$ along the geodesic connecting these two points, then the function $\bar{\tau}(x)$ is defined and regular in a neighborhood of $\gamma(0)$ for sufficiently small p. On γ, $\tau = \bar{\tau}$, and there exist points on γ such that

$$\frac{\partial^2}{\partial r^2} \tau > \frac{\partial^2}{\partial r^2} \bar{\tau};$$

therefore there exists a point q in the neighborhood of $\gamma(0)$, for which $\tau_1 > \bar{\tau}_1$. If γ intersects S at $\tau = \tau_0 < 0$, then the distance from $\gamma(p)$ to S along γ equals $-\tau_0 + p$. The distance along the geodesics from p to q is $p - \bar{\tau}_1$, and the distance from there to S is $\tau_1 - \tau_0$, so that the total proper time along this path γ' is $p - \bar{\tau}_1 + \tau_1 - \tau_0 > p - \tau_0$ (Figure 10.15). □

(10.6.26) Example

In the case of straight lines (10.6.17) and (10.6.24) with $S = \{(t, x) \in \mathbb{R}^2 : t^2 - x^2 = 1, t \le -1\}$, $\gamma =$ the t-axis, $\tau = \sqrt{t^2 - x^2}$, and $p = (t_0, 0)$ for $t_0 > 0$, we have $\bar{w} = \partial_x$, and thus $r = x$ and

$$\frac{\partial^2}{\partial r^2} \tau_{|\gamma} = \frac{1}{|t|} > \frac{\partial^2}{\partial r^2} \bar{\tau} = \frac{\partial^2}{\partial x^2} \sqrt{(t - t_0)^2 - x^2}_{|x=0} = \frac{1}{|t - t_0|}$$

for all $t < 0$. The explicit calculation of (10.6.24) confirms the conclusions reached earlier.

Finally, let us collect our results in a

(10.6.27) Theorem

Let (M, g) be orientable with respect to time, $i_v R_0 \ge 0$ for timelike vectors v, and let $S \subset M$ be a spacelike hypersurface on which the rate of convergence of the orthogonal geodesic vector field v is always $\ge c_0 > 0$. Then there cannot exist a point p in $D(S)$ at a distance greater than $3/c_0$ from S.

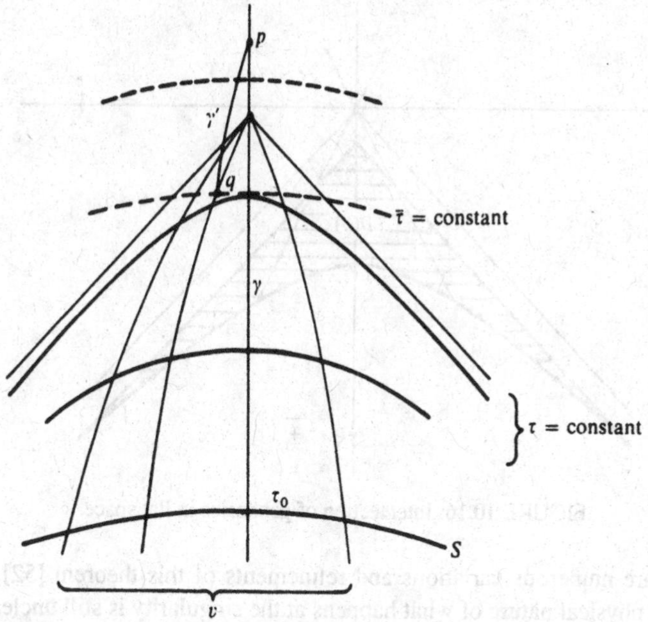

FIGURE 10.15. A curve γ' having longer proper time than the geodesic γ.

Proof: Proposition (10.6.14) states that the rate of convergence of v becomes infinite within a distance $\leq 3/c_0$ of S. Hence, given any geodesic through p perpendicular to S, there exists a field n that vanishes before p (cf. (10.6.16; 3)), and thus no geodesic through p is the curve of greatest proper time from S. This contradicts (10.6.23), and the only remaining logical possibility is that the geodesics cannot be extended to p. □

(10.6.28) **Remarks**

1. If $D(S)$ is sufficiently large, for instance all of M if S is Cauchy, and c is positively bounded on S, then M cannot be geodesically complete; there must be a singularity somewhere. The shortcoming of this statement is that we do not learn what happens physically at the singularity—whether there are infinite tidal forces, or a "quasiregular singularity" for which everything remains finite, as in (10.6.4; 3).
2. The convergence of v leads one to suppose that the rate of convergence of the streamlines of matter also becomes infinite, and that the energy density is divergent at some point. The difficulty in proving this is that there might exist an earlier, quasiregular singularity, and the time-evolution might stop before reaching an infinite density.
3. In order to draw conclusions from (10.6.23) about the existence of singularities, it is necessary to know something about the size of $D(S)$. If there were a Cauchy surface S with $c_0 > 0$, then singularities would be unavoidable.

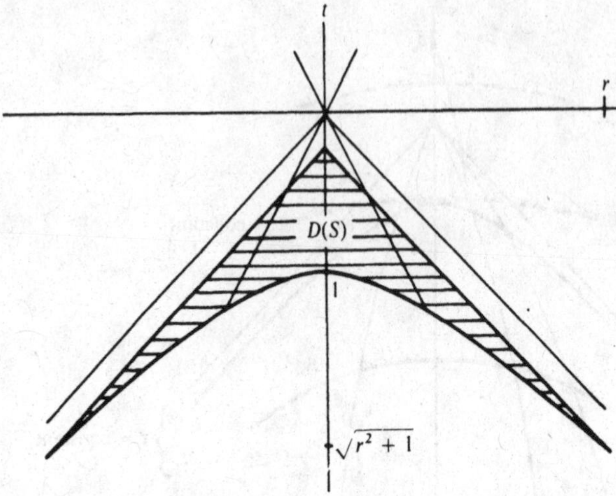

FIGURE 10.16. Intersection of geodesics in flat space.

4. There are numerous variations and refinements of this theorem [52], yet the precise physical nature of what happens at the singularity is still unclear.

(10.6.29) **Examples**

1. The Friedman universe (10.6.15) with S equal to the hypersurface at $t =$ const. This is a Cauchy surface, so every later point lies in $D(S)$. Hence, if $c = -3\ddot{R}/R > 0$, then no pressure, however great, can prevent the formation of a singularity.
2. Let (M, g) be $(\mathbb{R}^2, -dt^2 + dx^2)$ and $S = \{(t, x): t^2 - x^2 = 1, t \leq -1, |x| \leq r\}$. Then it can be calculated that

$$D(S) = \{(t, x): t^2 - x^2 > 1, t < x + r - \sqrt{r^2+1}, t < -x + r - \sqrt{r^2+1}\}$$

(see Figure 10.16). In this case, $c_0 = 1$ (cf. (10.6.17)), and the geodesics perpendicular to S leave $D(S)$ at the latest at $r + 1 - \sqrt{1+r^2} < 1 < 3/c_0$, so (10.6.27) predicts no singularity.
3. The Oppenheimer–Snyder Solution (10.5.30). With S such that $\tau = 0$ and $v = \partial_0$, we calculate that

$$c = {}^*di_v{}^*\mathbf{1} = -\left(3\dot{\nu} + \frac{\dot{\nu}'}{\nu'}\right) = \begin{cases} \frac{3}{a^{3/2} - 3\tau/2} & \text{if } \tilde{r} < a, \\ \frac{3/2}{\tilde{r}^{3/2} - 3\tau/2} & \text{if } \tilde{r} > a. \end{cases}$$

Therefore c is again infinite at the singularity at $r^{3/2} = \tilde{r}^{3/2} - 3\tau/2 = 0$. In the static coordinates, $c \sim r^{-3/2}$, because the radial speed

$$v_r \sim |\text{potential energy}|^{1/2} \sim r^{-1/2}$$

means that $v_{r,r} \sim r^{-3/2}$.

The great interest in singularities is clearly because what is at stake is whether most stars end up as black holes or black holes exist only in peculiar circumstances, and whether the Universe originated in a point and will some day return to a point. The question of singularities reveals how incomplete our understanding of natural laws is—are they ever superseded by something beyond human understanding?

(10.6.30) Problems

1. Construct an example of a pseudo-Riemannian manifold that is geodesically complete in spacelike and lightlike directions, but not in timelike directions. Do this by choosing $g = \Omega(t, x)(dx^2 - dt^2)$ on $\mathbb{R}^2$, with a suitable function Ω.

2. Find timelike geodesics for the metric of (10.6.4; 3), on which $x + t$ becomes infinite after a finite proper time.

3. (a) Prove (10.6.9). (b) Find an example of a nonsymmetric, real (2×2)-matrix that violates (10.6.9).

4. In the notation of (10.3.1), for the energy-momentum tensor (9.1.25; 3) with $p = 0$, $T^\alpha = \rho v v^\alpha$; the contracted Bianchi identity implies that $(\rho v^\beta v^\alpha)_{;\beta} = 0$. Show that this makes v geodesic. Identify the rate of convergence of an arbitrary vector field as $c = -v^\alpha{}_{;\alpha} = -\delta v = {}^*(L_v{}^*1) = {}^*(d^*v)$.

5. Extend the definition of d from the dense set $C^1(p, S)$ to $C(p, S)$ upper semicontinuously. Give an example of a densely defined continuous function which cannot be extended upper semicontinuously, and several examples of upper semicontinuous extensions of continuous functions.

6. Prove (10.6.21).

(10.6.31) Solutions

1. Let $\Omega = 1$ for $|x| \geq 1$, $\Omega_{,x}(t, 0) = 0$, and $\lim_{t\to\infty} |t|^{2+\varepsilon}\Omega(t, 0) = 0$ for some $\varepsilon > 0$. Then the time axis is geodesic, and the proper time on it is

$$\int_{-\infty}^{\infty} ds = \int_{-\infty}^{\infty} dt\sqrt{\Omega(t, 0)} < \infty.$$

However, light rays and spacelike lines leave the strip $|x| < 1$ and continue on as in Minkowski space.

2. Let $u = t - x$ and $v = t + x$; then the Lagrangian for the motion becomes

$$\mathcal{L} := \frac{h}{2}\dot{v}^2 - \dot{u}\dot{v}.$$

Consequently

$$\frac{h}{2}\dot{v}^2 - \dot{u}\dot{v} = -1 \quad \text{and} \quad P := \frac{\partial \mathcal{L}}{\partial \dot{v}} = h\dot{v} - \dot{u} = \text{const.}$$

Therefore we must integrate

$$\dot{u} = \sqrt{P^2 + 2h}, \qquad \dot{v} = \frac{1}{h}(P + \dot{u}) = \frac{2}{\dot{u} - P}.$$

If $P^2 := \lambda$, then

$$\dot{u} = P\cos^2 u \quad \Rightarrow \quad Ps = \tan u \quad \Rightarrow \quad \dot{u} = \frac{P}{P^2 s^2 + 1}$$
$$\Rightarrow \quad \dot{v} = -2\frac{1 + 1/P^2 s^2}{P} \quad \Rightarrow \quad v = -\frac{2s}{P} + \frac{2}{sP^3} + \text{const.}$$

3.

(a) If $M = TmT^{-1}$, where m is diagonal, with eigenvalues m_i, then (10.6.9) is the Cauchy–Schwarz inequality

$$\left(\sum_i m_i\right)^2 \leq \sum_i 1 \cdot \sum_i m_i^2.$$

(b) If $M = \begin{pmatrix} 1 & 1 \\ -1 & 0 \end{pmatrix}$, then $(\mathrm{Tr}\, M)^2 = 1$, but $M^2 = \begin{pmatrix} 0 & 1 \\ -1 & -1 \end{pmatrix}$, and $\mathrm{Tr}(M^2) = -1$.

4. Multiply $0 = (\rho v^\beta)_{;\beta} v^\alpha + \rho v^\beta (v^\alpha{}_{;\beta})$ by v_α. From $\langle v|v\rangle = -1$ it follows that $v_\alpha (v^\alpha{}_{;\beta}) = 0$, and we conclude that $0 = (\rho v^\beta)_{;\beta}$. In that case, $v^\beta (v^\alpha{}_{;\beta}) = \langle e^\alpha | D_v v\rangle = 0$. The equivalence of the expressions for c follows from

$$L_v{}^*\mathbf{1} = di_v{}^*\mathbf{1} = d^*v,$$
$$d({}^*e^\alpha v_\alpha) = -v_\alpha \omega^\alpha{}_\beta \wedge {}^*e^\beta + dv_\beta \wedge {}^*e^\beta = {}^*\langle e^\beta | dv_\beta - v_\alpha \omega^\alpha{}_\beta\rangle$$
$$= {}^*\langle e^\beta | D_{e_\beta} v\rangle = {}^*(v^\beta{}_{;\beta}).$$

5. Let $d(\lambda) = \inf_{C^! \supset U \ni \lambda} \sup_{\bar{\lambda} \in U} d(\bar{\lambda})$. This is upper semicontinuous and workable as long as the supremum is finite for sufficiently small U, which is the case as a corollary of the proof of (10.6.21) (see Problem 6(c)). If $f: \mathbb{R}\setminus\{0\} \to \mathbb{R}$ sends $x \to |x|^{-1}$, then this is not the case at $\{0\}$, and this function cannot be extended upper semicontinuously to a function $f: \mathbb{R} \to \mathbb{R}$. Incidentally, the above extension is maximally continuous; for example, to $f: \mathbb{R}\setminus\{0\} \to \mathbb{R}$, $x \to |x|$ it ascribes the value $f(0) = 0$, whereas $f(0) = a > 0$ would make the extension only upper semicontinuous.

6.

(a) If $J^-(x) \cap D^+(S)$ were not compact, then there would exist an infinite, locally finite covering with relatively compact neighborhoods U_i with $a_i \in U_i$ for $\{a_i\}$ without a point of accumulation. Let $x \in U_1$ and γ_i be a family of causal curves from x to a_i. Then $\gamma_i \cap \partial U_1$ has a point of accumulation h_1. If c_1 is a causal curve from x to h_1, then c_1 contains a point x_1 that lies not only in U_1 but also in another set U_2. Since $J^-(x_1)$ contains a neighborhood of h_1 by Assumption (10.6.20(b)), it also contains an infinite subfamily $\{\gamma_{1_i}\}$ of the $\{\gamma_i\}$, and consequently infinitely many a_{1_i}. There is a point of accumulation h_2 for $\{\gamma_{1_i}\} \cap \partial U_2 \cap J^-(x_1)$, and there exists a causal curve c_2 from x_1 to h_2, and so on (Figure 10.17). This procedure yields a causal line connecting $x, x_1, x_2, \ldots$, which cannot be extended father downward, since the a_i have no point of accumulation. However, for the same reason, it cannot intersect S, as otherwise one of the relatively compact U_i would contain infinitely many a_j. The existence of a nonextensible causal curve not meeting S contradicts the definition of $D^+(S)$, and therefore $J^-(x) \cap D^+(S)$ must be compact.

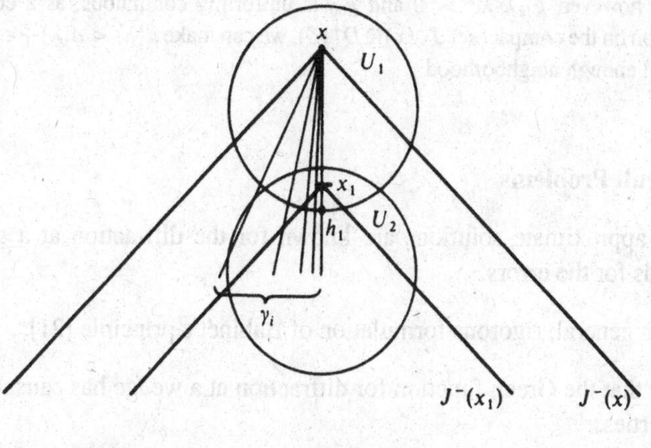

FIGURE 10.17. Construction of a nonextensible, causal curve that does not intersect S.

(b) $C(x, S)$ is compact as a metric space if it is complete and precompact. A uniform limit of causal curves is continuous ([41, 7.2.1]), and, because of (10.6.20(b)), causal. Therefore completeness follows immediately. Precompact means that for all ε there exists a finite covering of $C(x, S)$ with neighborhoods of diameter ε. Such a neighborhood of a curve γ is the set of curves

$$\left\{\gamma' : \sup_{y\in\gamma'} \inf_{x\in\gamma} \rho(x, y) < \varepsilon\right\},$$

where ρ is a distance function for the metric on M. Since $J^-(x)\cap D^+(S)$ is compact, we can cover it with finitely many A_i = the interior of $J^-(a_i)\cap J^+(a_i')$, $i = 1, \ldots, n$, with diameter $< \varepsilon$. Let x be in A_1, and form all unions

$$B_j = \bigcup_{k=1}^{n_j} A_{jk},$$

with

$$A_{j_1} = A_1, \qquad A_{j_{n_j}} \cap S \neq \emptyset,$$
$$A_{jk} \cap A_{jk+1} \neq \emptyset, \qquad J^-(A_{jk}) \cap A_{jk+2} \neq \emptyset.$$

The $B_j' = \{\gamma \in C(x, S) : \gamma \subset B_j\}$ are a covering for $C(x, S)$, since every causal curve of x must be in some such union. The number of the B_j is finite, and their diameter $< \varepsilon$.

(c) We need to show that for all ε there exists a neighborhood U of λ such that $d(\bar{\lambda}) < d(\lambda) + \varepsilon$ for all $\bar{\lambda} \in U$. For this purpose, we use comoving coordinates moving with λ, so that λ is one of the time axes x^j = const., $j = 1, 2, 3$, and the g_{0j} vanish. Since d does not depend on the choice of the curve parameter, we can take this parameter as x^0; we then have to compare

$$d(\lambda) = \int_\lambda dx^0 \sqrt{-g_{00}}$$

with

$$d(\bar{\lambda}) = \int_{\bar{\lambda}} dx^0 \sqrt{-g_{00} - g_{jk}\dot{\bar{\lambda}}^j \dot{\bar{\lambda}}^k}.$$

Since, however, $g_{jk}\dot{\lambda}^j\dot{\lambda}^k > 0$ and g_{00} is uniformly continuous, as a continuous function on the compact set $J^-(x)\cap D^+(S)$, we can make $d(\bar{\lambda}) < d(\lambda)+\varepsilon$ by taking a small enough neighborhood.

Some Difficult Problems

1. Only approximate solutions are known for the diffraction at a slit. Find bounds for the errors.

2. Give a general, rigorous formulation of Babinet's principle [21].

3. Show that the Green function for diffraction at a wedge has causal support properties.

4. Find a solution of Einstein's equations that describes the emission of gravitational waves.

5. Harmonic coordinates are used to prove that Einstein's equations are hyperbolic. Give a purely geometric formulation of this state of affairs, without reference to particular coordinates.

6. Discover singularity theorems that show that M is not only geodesically incomplete, but that the curvature invariants are in fact unbounded, in the right circumstances.

7. Solve the general relativistic two-body problem.

Bibliography

Works Cited in the Text

[1] J. Dieudonné. Foundations of Modern Analysis (four volumes). New York: Academic Press, 1969–1974.

[2] E. Hlawka. Differential Manifolds. In: *Acta Phys. Austriaca Suppl.*, vol. 7, pp. 265–307. Vienna: Springer-Verlag, 1970.

[3] R. Abraham. Foundations of Mechanics. New York: Benjamin, 1967.

[4] H. Flanders. Differential Forms. New York: Academic Press, 1963.

[5] S. Sternberg, Lectures on Differential Geometry. Englewood Cliffs, NJ: Prentice-Hall, 1964.

[6] C.L. Siegel and J. Moser. Lectures on Celestial Mechanics. Berlin: Springer-Verlag, 1971.

[7] V. Szebehely. Theory of Orbits, the Restricted Problem of Three Bodies. New York: Academic Press, 1967.

[8] V. Szebehely. Families of Isoenergetic Escapes and Ejections in the Problem of Three Bodies. *Astronom. and Astrophys.* **22**, 171–177, 1973.

[9] V.I. Arnold and A. Avez, Ergodic Problems of Classical Mechanics. New York: Benjamin, 1968.

[10] N. Kerst and R. Serber. Electronic Orbits in the Induction Accelerator. *Phys. Rev.* **60**, 53–58, 1941.

[11] V. Arnold. O Povedeniye Adiabaticheskogo Invarianta pri Medlennom Periodicheskom Izmeneniye Funktsiye Gamil'tona. *Dokl. Akad. Nauk.* **142**, 758–761, 1962.

[12] A. Schild. Electromagnetic Two-Body Problem. *Phys. Rev.* **131**, 2762–2766, 1963.

[13] Y. Sinai, *Acta Phys. Austriaca Suppl.*, vol. 10. Vienna: Springer-Verlag, 1973.

[14] J. Moser. Stable and Random Motions in Dynamical Systems. Princeton, NJ: Princeton University Press, 1973.

[15] R. McGehee and J.N. Mather. Solutions of the Collinear Four Body Problem which Become Unbounded in Finite Time. In: Lecture Notes in Physics, vol. 38, J. Moser, ed. New York: Springer-Verlag, 1975. (Entitled: Battelle Rencontres, Seattle 1974. Dynamical Systems: Theory and Applications.)

[16] G. Contopoulos. The "Third" Integral in the Restricted Three-Body Problem. *Astrophys. J.* **142**, 802–804, 1965.
G. Bozis. On the Existence of a New Integral in the Restricted Three-Body Problem: *Astronom. J.* **71**, 404–414, 1966.

[17] V. Arnold. Small Denominators and Problems of Stability of Motion in Classical and Celestial Mechanics. *Russian Math. Surveys* **18**, 85–191, 1963.

[18] R.C. Robinson. Generic Properties of Conservative Systems. *Amer. J. Math.* **92**, 562–603 and 897–906, 1970.

[19] M. Breitenecker and W. Thirring. *Suppl. Nuovo Cimento*, 1978.

[20] J.D. Jackson. Classical Electrodynamics. New York: Wiley, 1975.

[21] W. Franz. Theorie der Beugung elektromagnetischer Wellen. Ergebnisse der angew. Math., vol. 4. Berlin: Springer-Verlag, 1957.

[22] T. Fulton and F. Rohrlich. Classical Radiation from a Uniformly Accelerated Charge. *Ann. Phys.* **9**, 499–517, 1960.

[23] K. Yano. The Theory of Lie Derivatives and Its Applications. Amsterdam: North-Holland, 1955.

[24] H.M. Nussenzveig. High Frequency Scattering by an Impenetrable Sphere. *Ann. Phys.* **34**, 23–95, 1965.

[25] F. London. Superfluids, vol. I: Macroscopic Theory of Superconductivity. New York: Wiley, 1950.

[26] J.L. Anderson. Principles of Relativity Physics. New York: Academic Press, 1967.

[27] R.U. Sexl and H.K. Urbantke. Gravitation und Kosmologie, BI-Hochschultaschenbuch. Mannheim: BI-Wissenschaftsverlag, 1974.

[28] C.W. Misner, K.S. Thorne, and J.A. Wheeler. Gravitation. San Francisco: Freeman, 1973.

[29] S. Weinberg. Gravitation and Cosmology. New York: Wiley, 1972.

[30] J.M. Souriau. Géometrie et Rélativité. Paris: Hermann, 1964.

[31] M.D. Kruskal. Maximal Extension of Schwarzschild Metric. *Phys. Rev.* **119**, 1743–1745, 1960.

[32] J.C. Graves and D.R. Brill. Oscillatory Character of Reissner–Nordström Metric for an Ideal Charged Wormhole. *Phys. Rev.* **120**, 1507–1513, 1960.

[33] B. Carter. The Complete Analytic Extension of the Reissner–Nordström Metric in the Special Case $e^2 = m^2$. *Phys. Lett.* **21**, 423–424, 1966.

[34] A. Trautman. Theory of Gravitation. In: The Physicist's Conception of Nature, J. Mehra, ed. Boston, MA: Reidel, 1973.

[35] C.N. Yang. Integral Formalism for Gauge Fields. *Phys. Rev. Lett.* **33**, 445–447, 1974.

[36] W. Rühl. Finite Conformal Transformations in Local Quantum Field Theory. In: Electromagnetic Interactions and Field Theory. *Acta Phys. Austriaca Suppl.* vol. 14, pp. 643–646. Vienna: Springer-Verlag, 1975.

[37] M. Schönberg. *Rev. Brasileira de Fisica* **1**, 91, 1971.

[38] A. Uhlmann. *Wiss. Z. Friedrich-Schiller-Univ.* **8**, 31, 1958.

[39] M. Abramowitz and I.E. Stegun, eds. Handbook of Mathematical Functions. Applied Mathematics Series, vol. 55, Washington, DC: National Bureau of Standards, 1964.

[40] F. Hehl, P. von der Heyde, and G.D. Kerlick. General Relativity with Spin and Torsion: Foundations and Prospects. *Rev. Mod. Phys.* **48**, 393–416, 1976.

[41] J. Dieudonné. Foundations of Modern Analysis, vols. III and IV. New York: Academic Press, 1972 and 1974.

[42] A. Schild. Electromagnetic Two-Body Problem. *Phys. Rev.* **131**, 2762–2766, 1962.

[43] F. Rohrlich. Classical Charged Particles: Foundations of Their Theory. Reading, MA: Addison-Wesley, 1965.

[44] L.P. Eisenhart. Riemannian Geometry. Princeton, NJ: Princeton University Press, 1949.

[45] V. Fock. The Theory of Space, Time and Gravitation. New York: Macmillan, 1964.

[46] A. Lichnerowicz. Théories Rélativistes de la Gravitation et de l'Electromagnetisme Rélativité Générale et Théories Unitaires. Paris: Masson, 1955.

[47] J. Wess and B. Zumino. Superspace Formulation of Supergravity. *Phys. Lett.* **66B**, 361–364, 1977.

[48] A. Trautman. Conservation Laws in General Relativity. In: Gravitation, an Introduction to Current Research, L. Witten, ed. New York: Wiley, 1962.

[49] E. Pechlaner and R. Sexl. On Quadratic Lagrangians in General Relativity. *Comm. Math. Phys.* **2**, 165–175, 1966.

[50] F. Hoyle and J.V. Narlikar. Cosmological Models in Conformally Invariant Gravitational Theory–II. A New Model. *Mon. Nat. Roy. Astr. Soc.* **155**, 323–335, 1972.

[51] L. Landau and E. Lifshitz. The Classical Theory of Fields. Reading, MA: Addison-Wesley, 1977.

[52] G.F.R. Ellis and S.W. Hawking. The Large Scale Structure of Space–Time. Cambridge: Cambridge University Press, 1973.

[53] R.P. Geroch. What Is a Singularity in General Relativity? *Ann. Phys.* **48**, 526–540, 1968.

[54] M. Fierz and R. Jost. Affine Vollständigkeit und kompakte Lorentz'sche Mannigfaltigkeiten. *Helv. Phys. Acta* **38**, 137–141, 1965.

[55] F. Hoyle et al. In: Quasi-Stellar Sources and Gravitational Collapse, I. Robinson, ed. Chicago: University of Chicago Press, 1965.

[56] M. Reed and B. Simon, Methods of Modern Mathematical Physics (four volumes). New York: Academic Press, 1974–1979.

[57] H. Bondi and T. Gold. The Field of a Uniformly Accelerated Charge, with Special Reference to the Problem of Gravitational Acceleration. *Proc. Roy. Soc. London* **A229**, 416–424, 1955.

[58] W. Thirring and R. Wallner. The Use of Exterior Forms in Einstein's Gravitation Theory. *Rev. Brasileira de Fisica*.

Further Reading

Chapter 2

W.M. Boothby. An Introduction to Differentiable Manifolds and Riemannian Geometry. New York: Academic Press, 1975.

Th. Bröcker and K. Jänich. Einführung in die Differentialtopologie. Heidelberger Taschenbüchner, vol. 143. Heidelberg: Springer-Verlag, 1973.

Y. Choquet-Bruhat, C. DeWitt-Morette, and M. Dillard-Bleick. Analysis, Manifolds, and Physics. Amsterdam: North-Holland, 1977.

V. Guillemin and A. Pollack. Differential Topology. Englewood Cliffs, NJ: Prentice-Hall, 1974.

R. Hermann. Vector Bundles in Mathematical Physics, vol. 1. New York: Benjamin, 1970.

H. Holman and H. Rummler. Alternierende Differentialformen. Bibliographisches Institut, 1972.

S. Kobayashi and K. Nomizu. Foundations of Differential Geometry, vol. 1. Interscience Tracts in Pure and Applied Mathematics, No. 15, vol. 1. New York: Interscience, 1963.

L.H. Loomis and S. Sternberg. Advanced Calculus. Reading, MA: Addison-Wesley, 1968.

E. Nelson. Tensor Analysis. Princeton, NJ: Princeton University Press, 1967.

M. Spivak. Calculus on Manifolds: A Modern Approach to Classical Theorems of Advanced Calculus. New York: Benjamin, 1965.

Chapter 3

R. Barrar. Convergence of the von Zeipel Procedure. *Celestial Mech.* **2**, 494–504, 1970.

N. Bogoliubov and N. Krylov. Introduction to Non-linear Mechanics. Princeton, NJ: Princeton University Press, 1959.

J. Ford. The Statistical Mechanics of Classical Analytic Dynamics. In: Fundamental Problems in Statistical Mechanics, vol. III, E. Cohen, ed. Amsterdam: North-Holland, 1975.

G. Giacaglia. Perturbation Methods in Non-linear Systems. New York: Springer-Verlag, 1972.

M. Golubitsky and V. Guillemin. Stable Mappings and Their Singularities. New York: Springer-Verlag, 1973.

V. Guillemin and S. Sternberg. Geometric Asymptotics. Providence, RI: American Mathematical Society, 1977.

M. Hirsch and S. Smale. Differential Equations, Dynamical Systems, and Linear Algebra. New York: Academic Press, 1974.

W. Hunziker. Scattering in Classical Mechanics. In: Scattering Theory in Mathematical Physics, J.A. Lavita and J. Marchand, eds. Boston, MA: Reidel, 1974.

R. Jost. Poisson Brackets (An Unpedagogical Lecture). *Rev. Mod. Phys.* **36**, 572–579, 1964.

G. Mackey. The Mathematical Foundations of Quantum Mechanics. New York: Benjamin, 1963.

J. Moser, ed. Dynamical Systems: Theory and Applications. New York: Springer-Verlag, 1975.

J.-M. Souriau. Structure des Systèmes Dynamiques: Maîtrises de Mathématiques. Paris: Dunod, 1970.

Chapters 4 and 5

A. Hayli, ed. Dynamics of Stellar Systems. Boston, MA: Reidel, 1975.

L. Landau and E. Lifschitz. The Classical Theory of Fields. London: Pergamon Press, 1975.

H. Pollard. Mathematical Introduction to Celestial Mechanics. Englewood Cliffs, NJ: Prentice-Hall, 1966.

S. Sternberg. Celestial Mechanics. New York: Benjamin, 1969.

K. Stumpff. Himmelsmechanik. Berlin: Deutscher Verlag der Wissenschaften, 1959.

Chapter 6

J. Ehlers. The Nature and Structure of Spacetime. In: The Physicist's Conception of Nature, J. Mehra, ed. Boston, MA: Reidel, 1973.

E.H. Kronheimer and R. Penrose. On the Structure of Causal Spaces. *Math. Proc. Cambridge Philos. Soc.* **63**, 481–501, 1967.

C. Misner, K. Thorne, and J. Wheeler. Gravitation. San Francisco: W.H. Freeman, 1973.

S. Nanda. A Geometrical Proof that Causality Implies the Lorentz Group. *Math. Proc. Cambridge Philos. Soc.* **79**, 533–536, 1976.

R. Sexl and H. Urbantke. Relativität, Gruppen, Teilchen. Vienna: Springer-Verlag, 1976.

A. Trautman. Theory of Gravitation. In: The Physicist's Conception of Nature, J. Mehra, ed. Boston, MA: Reidel, 1973.

S. Weinberg. Gravitation and Cosmology. New York: Wiley, 1972.

E.C. Zeeman. Causality Implies the Lorentz Group. *J. Math. Phys.* **5**, 490–493, 1964.

Chapter 10

Section 10.1

R.L. Bishop and S.I. Goldberg. Tensor Analysis on Manifolds. New York: Macmillan, 1968.

N.J. Hicks. Notes on Differential Geometry. New York: Van Nostrand-Reinhold, 1971.

S. Kobayashi and K. Nomizu. Foundations of Differential Geometry, vols. I and II. New York: Interscience, 1963 and 1969.

B. Schmidt. Differential Geometry from a Modern Standpoint. In: Relativity, Astrophysics, and Cosmology, W. Israel, ed. Boston, MA: Reidel, 1973.

Section 10.2

I.M. Gel'fand and S.V. Fomin. Calculus of Variations. Englewood Cliffs, NJ; Prentice-Hall, 1963.

P. Havas. On Theories of Gravitation with Higher Order Field Equations. *Gen. Relativity Gravitation* **8**, 631, 1977.

D. Lovelock and H. Rund. Variational Principles in the General Theory of Relativity. *Jahresber. Deutsch. Math.-Verein.* **74**, No. 1/2, 1972.

A. Trautman. Conservation Laws in General Relativity. In: Gravitation, an Introduction to Current Research, L. Witten, ed. New York: Wiley, 1962.

Sections 10.3 and 10.4

S. Helgason. Lie Groups and Symmetric Spaces. In: Batelle Rencontres: 1967 Lectures in Mathematics and Physics, C. DeWitt and J.A. Wheeler, eds. New York: Benjamin, 1968.

Section 10.5

B.K. Harrison, K.S. Thorne, M. Wakano, and J.A. Wheeler. Gravitational Theory and Gravitational Collapse. Chicago: University of Chicago Press, 1965.

H. Scheffler and H. Elsässer. Physik der Sterne und der Sonne. Mannheim: BI-Wissenschaftsverlag, 1974.

Ya.B. Zel'dovich and I.D. Novikov. Relativistic Astrophysics, vols. I and II. Chicago: University of Chicago Press, 1971.

Section 10.6

C.J.S. Clarke. The Classification of Singularities. *Gen. Relativity Gravitation* **6**, 35–40, 1975.

C.J.S. Clarke. Space–Time Singularities. *Comm. Math. Phys.* **49**, 17–23, 1976.

G.F.R. Ellis and B. Schmidt. Singular Space–Times. *Gen. Rel. Grav.* **8**, 915–953, 1977.

R.P. Geroch. Singularities in the Spacetime of General Relativity, Their Definition, Existence, and Local Characterization. Dissertation, Princeton University, 1967.

R.P. Geroch. What Is a Singularity in General Relativity? *Ann. Phys.* **48**, 526–540, 1968.

S.W. Hawking. Singularities and the Geometry of Spacetime. Essay submitted for the Adams Prize, Cambridge University, 1966.

S.W. Hawking. The Occurrence of Singularities in Cosmology I, II, III. *Proc. Roy. Soc. London* **294A**, 511–521, 1966; *Ibid.* **295A**, 490–493, 1966; *Ibid.* **300A**, 187–201, 1967.

W. Kundt. Recent Progress in Cosmology, Springer Tracts in Modern Physics, vol. 47. New York: Springer-Verlag, 1968.

R. Penrose. Gravitational Collapse and Space–Time Singularities. *Phys. Rev. Lett.* **14**, 57–59, 1965.

Alternating Differential Forms

H. Cartan. Differential Calculus. Paris: Hermann, 1971.

H. Cartan. Differential Forms. Paris: Hermann, 1970.

G.A. Deschamps. Exterior Differential Forms. In: Mathematics Applied to Physics, E. Roubine, ed. New York: Springer-Verlag, 1970.

H. Flanders. Differential Forms with Applications to the Physical Sciences. New York: Academic Press, 1963.

S.J. Goldberg. Curvature and Homology. New York: Academic Press, 1962.

W. Greub, S. Halperin, and R. Vanstone. Connections, Curvature, and Cohomology. New York: Academic Press, 1972.

H. Holmann and H. Rummler. Alternierende Differentialformen. Mannheim: BI-Wissenschaftsverlag, 1972.

Tensor Analysis and Geometry of Manifolds

L. Auslander and R.E. MacKenzie. Introduction to Differential Manifolds. New York: McGraw-Hill, 1963.

R.L. Bishop and R.J. Crittenden. Geometry on Manifolds. New York: Academic Press, 1964.

R.L. Bishop and S.I. Goldberg. Tensor Analysis on Manifolds. New York: Macmillan, 1968.

F. Brickell and R.S. Clark. Differential Manifolds. New York: Van Nostrand-Reinhold, 1970.

T. Bröcker and K. Jänich. Einführung in die Differentialtopologie, Heidelberger Taschenbuch, vol. 143. Heidelberg: Springer-Verlag, 1968.

Y. Choquet-Bruhat, C. DeWitt-Morette, and M. Dillard-Bleick. Analysis, Manifolds, and Physics. Amsterdam: North-Holland, 1978.

D. Gromoll, W. Klingenberg, and W. Meyer. Riemannsche Geometrie im Grossen. Lecture Notes in Mathematics, vol. 55. New York: Springer-Verlag, 1968.

N.J. Hicks. Notes on Differential Geometry. New York: Van Nostrand-Reinhold, 1971.

S. Kobayashi and K. Nomizu. Foundations of Differential Geometry, vols. I and II. New York: Interscience, 1963 and 1969.

A. Lichnerowicz. Elements of Tensor Analysis. New York: Wiley, 1962.

C.W. Misner. Differential Geometry. In: Relativity, Groups, and Topology, C. DeWitt and B.S. DeWitt, eds. New York: Gordon and Breach, 1964.

E. Nelson. Tensor Analysis. Princeton, NJ: Princeton University Press, 1967.

S. Sternberg. Lectures on Differential Geometry. Englewood Cliffs, NJ: Prentice-Hall, 1964.

T.J. Willmore. An Introduction to Differential Geometry. Oxford: Oxford University Press, 1959.

J.A. Wolf. Spaces of Constant Curvature. New York: McGraw-Hill, 1967.

General Relativity

R. Adler, M. Bazin, and M. Schiffer. Introduction to General Relativity. New York: McGraw-Hill, 1965.

A. Einstein. The Meaning of Relativity. Princeton, NJ: Princeton University Press, 1955.

W. Pauli. Theory of Relativity. New York: Pergamon, 1958.

W. Rindler. Essential Relativity. New York: Springer-Verlag, 1977.

R.U. Sexl and H.K. Urbantke. Gravitation und Kosmologie, BI-Hochschultachsenbuch. Mannheim: BI-Wissenschaftsverlag, 1975.

J.L. Synge. Relativity, the General Theory. Amsterdam: North-Holland, 1965.

A. Trautman, F. Pirani, and H. Bondi. Lectures on General Relativity. Englewood Cliffs, NJ: Prentice-Hall, 1972.

Global Analysis

Y. Choquet-Bruhat and R. Geroch. Global Aspects of the Cauchy Problem in General Relativity. *Comm. Math. Phys.* **14**, 329–335, 1969.

G.F.R. Ellis and D.W. Sciama. Global and Nonglobal Problems in Cosmology. In: General Relativity, Papers in Honor of J.L. Synge, L. O'Raifeartaigh, ed. Oxford: Clarendon Press, 1972.

D. Farnsworth, J. Fink, J. Porter, and A. Thomson, eds. Methods of Local and Global Differential Geometry in General Relativity. Lectures Notes in Physics, vol. 14. New York: Springer-Verlag, 1972.

R.P. Geroch. Topology in General Relativity. *J. Math. Phys.* **8**, 782–786, 1967.

R.P. Geroch. Domain of Dependence. *J. Math. Phys.* **11**, 437–449, 1970.

R.P. Geroch. Space–Time Structure from a Global Point of View. In: General Relativity and Cosmology, R.K. Sachs, ed. New York: Academic Press, 1971.

ICTP, Global Analysis and its Applications, vols. I, II, and III. Lectures Presented at an International Seminar Course at Trieste from 4 July to 25 August, 1972. New York: Unipub, 1975.

W. Kundt. Global Theory of Spacetime. In: Differential Topology, Differential Geometry and Applications, J.R. Vanstone, ed. Montreal: Canadian Mathematical Congress, 1972.

A. Lichernowicz. Topics on Space–Time. In: Batelle Rencontres: 1967 Lectures in Mathematics and Physics, C. DeWitt and J.A. Wheeler, eds. New York: Benjamin, 1968.

R. Penrose. Structure of Space-Time. *Ibid.*

Proceedings, Summer Schools, and Collected Papers

P.G. Bergmann, E.J. Fenyves, and L. Motz, eds. Seventh Texas Symposium on Relativistic Astrophysics. *Ann. New York Acad. Sci.* **262**, 1975.

M. Carmeli, S. Fickler, and L. Witten, eds. Relativity. New York: Plenum, 1970.

H.-Y. Chiu and W.F. Hoffman, eds. Gravitation and Relativity. New York: Benjamin, 1964.

C. DeWitt and J.A. Wheeler, eds. Batelle Rencontres: 1967 Lectures in Mathematics and Physics. New York: Gordon and Breach, 1964.

C. DeWitt and J.A. Wheeler, eds. Batelle Rencontres: 1967 Lectures in Mathematics and Physics. New York: Benjamin, 1968.

Editorial Committee. Recent Developments in General Relativity. New York: Macmillan, 1962.

J. Ehlers, ed. Relativity Theory and Astrophysics. Providence, RI: American Mathematical Society, 1967.

W. Israel, ed. Relativity, Astrophysics, and Cosmology. Boston, MA: Reidel, 1973.

C.W. Kilmister, ed. General Theory of Relativity, Selected Readings in Physics. New York: Pergamon, 1973.

C.G. Kuper and A. Peres, eds. Relativity and Gravitation. New York: Gordon and Breach, 1971.

L. O'Raifeartaigh, ed. General Relativity, Papers in Honor of J.L. Synge. Oxford: Clarendon Press, 1972.

R.K. Sachs, ed. General Relativity and Cosmology. Proceedings of Course 47 of the International School of Physics "Enrico Fermi." New York: Academic Press, 1971.

G. Shaviv and J. Rosen, eds. General Relativity and Gravitation. New York: Wiley, 1975.

P. Suppes, ed. Space, Time, and Geometry. Boston, MA: Reidel, 1973.

J.R. Vanstone, ed. Differential Topology, Differential Geometry and Applications. Proceedings of the Thirteenth Biennial Seminar of the Canadian Mathematical Congress. Montreal: Canadian Mathematical Congress, 1972.

L. Witten, eds. Gravitation, an Introduction to Current Research. New York: Wiley, 1962.

Index

C^r vector field, 138
N-body problem, 204
p-forms, 56

action, 100, 329
action-angle variables, 116
adiabatic theorem, 235
adjoint, 46
ADM energy, 459
affine connections, 300
almost-periodic, 8
asteroids, 192, 197
asymptotic completeness, 128
asymptotic fields, 347
atlas, 12
automorphism, 7

basis, 294
Betatron, 232
Bianchi's identity, 442
Birkhoff's theorem, 489
black hole, 252, 292
boundary, 20
Bruns's theorem, 205
bundle atlases, 30

canonical flow, 97
canonical forms, 89
canonical transformation, 61, 89, 90
Cartan's structure equation, 440
Cauchy surface, 303
causal curve, 336
causal space, 276
causal structure, 276
center-of-mass, 173
Cesàro average, 102
characteristics, 301, 340, 463
chart, 12
Christoffel symbols, 446
circular polarization, 240
closed p-form, 67
closed form, 296
codifferential, 299
collinear equilibrium, 192
comparison diffeomorphism, 41
compatible, 12
complete, 39
components, 48
configuration space, 43
conformal transformations, 335
conservation of angular momentum, 98
conservation of momentum, 98
constant acceleration, 102
constant of motion, 42, 122

Euler–Lagrange, 43
Lagrangian, 42
constraints, 6
contracted Bianchi identity, 466
contraction, 273
contravariant, 48
convergence of flow, 515
convergence of perturbation theory, 157
Cornu's spiral, 414
cosmic censorship, 495
cosmological principle, 475
cosmological term, 457, 480
cotangent bundle, 55
cotangent space, 46
Coulomb field, 226
Coulomb potential, 348
covariant, 48
covariant differentiation, 433
covariant Lie derivative, 437
cross-angle, 137
current density, 317
curvature form, 440
cyclotron frequency, 222

de Sitter metric, 482
deflection angle, 122
deflection of light by the Sun, 254
delay time, 131
delta function, 305
derivation, 33
derivative, 12, 25, 30
diffeomorphism, 19
differentiable, 11, 19
differential, 65
differentiation, 11, 65
diffraction, 405
dilatation, 275
dilatations, 172
dimension, 12
dipole radiation, 289
discontinuity surface, 303, 463
distribution, 305
domain, 12
domain of influence, 336
double pendulum, 165
double stars, 292
driven oscillator, 154
dual basis, 46
dual space, 46
duality map, 298

effective potentials, 184
Einstein's equations, 459
Einstein's synchronization, 271
elapsed time, 176
electric field, 214
electrodynamic equations of motion, 214
electromagnetic radiation, 285, 287
electromagnetic waves, 286
elliptic fixed point, 144
energy momentum loss by radiation, 350
energy shell, 106
energy-momentum currents, 320
energy-momentum forms, 361, 389
energy-momentum loss by radiation, 375
energy-momentum tensor, 331, 389
equilateral equilibrium, 192
equilibrium positions, 106
equivalence principle, 323
ergodic theory, 83
escape criterion, 200, 209
escape trajectory, 82, 230
Møller transformations, 188
Euclidean group, 169
Euler–Lagrange equations, 43
exact p-form, 67
exact form, 296
extended configuration space, 101
extended phase space, 101
exterior differential, 295
exterior product, 49, 295

fiber metric, 434
fibers, 29
fictitious forces, 322
field strength form, 312
fine structure, 229
fixed point, 140
flat, 441
flow, 36, 39
form, 295
form-valued sections, 435
frame, 295
Frauenhofer region, 429
free fall, 102

free particles, 169, 217
Fresnel's integral, 413, 428
Friedmann Universe, 482
future, 518

Galilean group, 171
gauge group, 435
gauge theory, 452
gauge transformation, 216, 452
Gauss's theorem, 297
general covariance, 16
generator of a canonical transformation, 91
geodesic deviation, 515
geodesic vector field, 447
geodesically complete, 513
geodesics, 246
geodetic form of the equations of motion, 245
geometric optics, 408
gravitational radiation, 290
gravitational red-shift, 279
gravitational wave, 257, 485
Green function, 305
Green's formula, 306
group velocity, 399

half-space, 19
Hamilton's equations, 96
Hamilton–Jacobi equation, 103
Hamiltonian, 43
    vector field, 61, 93
harmonic coordinates, 456, 464
Heaviside step function, 303
hedgehog, 85
Helmholtz's circulation theorem, 384
Hertz dipole, 365
homogeneous, 470
homogeneous stars, 503
horizon
    event, 477
    particle, 477
hyperbolic fixed point, 144
hyperbolic motion, 357
hyperbolic trajectory, 223
hypersurface, 303

imbedding theorem, 15
impact parameter, 138
incompressible, 82
infinitesimal variations, 6
integrable system, 113
integral, 77, 78, 296
    curve, 36
    invariant, 92
    of motion, 105
interior is a manifold, 20
interior product, 50, 52, 297
isometries, 61
isotropic, 470

Jacobi's constant, 191
Jacobi's identity, 74

K–A–M theorem, 163
Kepler problem, 173
Kepler's equation, 180
Kepler's Third Law, 177
Killing vector field, 61, 333, 447
Kirchhoff's theory of diffraction, 415

Lagrangian, 330
Landau–Lifshitz form, 461
Laplace–Beltrami operator, 299
Larmor orbits, 221
Larmor's formula, 288, 372
Legendre transformation, 43
length of a curve, 519
Lenz vector, 174
Levinson's theorem, 137
Liénard–Wiechert potentials, 348
Lie bracket, 72
Lie derivative, 33, 69, 70, 299
lightlike, 238
lightlike coordinates, 306
linear approximation, 464
linear motion, 38
Liouville measure, 81
Liouville operator, 33
Lissajou figure, 8, 110
local canonical transformation, 90
local coordinate, 15
local flow, 39
local Lorentz transformation, 456
locally Hamiltonian vector field, 93
London's equations, 384
Lorentz force, 214, 320
Lorentz gauge, 344

Lorentz transformations, 217, 455

Møller transformations, 189
Möbius strip, 29
Møller-transformations, 125
magnetic charges, 314
magnetic field, 214
manifold, 6, 11, 12
  with a boundary, 19
mass-renormalization, 378
maximally symmetric spaces, 468
Maxwell's equations, 215, 313
metallic boundary conditions, 390
minimal frequency, 402
Minkowski space, 271
mixed fixed point, 144

naked singularity, 495
natural basis, 34, 47, 296
neutron star, 292
Noether's theorem, 331
nondegenerate, 57
normal modes, 397

observables, 7
Oppenheimer–Snyder solution, 508
orientable, 56, 78, 298
orthogonal basis, 51, 300
oscillator, 98, 118
oscillator with a changing frequency, 155

parallel at a distance, 28
parallel transport, 435, 439
parallelizable, 29
partial differential equations, 300
past, 518
Peano curves, 110
pendulum, 118
Penrose diagram, 476
perfect cosmological principle, 475
perturbation series, 147
perturbation theory, 145
phase space, 43
plane wave, 257
plasma frequency, 386
Poincaré group, 217
Poincaré transformations, 317
Poincaré's lemma, 67
Poincaré's recurrence theorem, 82
point-particle, 321
Poisson bracket, 94
Poynting's vector, 335
precession, 219, 252
principle of equivalence, 323
product manifold, 15
projection to a basis, 29
proper time, 519
pseudo-Riemannian, 57
pull-back, 62

quadrupole oscillations, 261
quasi-periodic orbits, 110
quasiregular singularity, 523

radiation field, 374
red-shift, 279
reduced mass, 174
regularizations, 6
Reissner–Nordstrøm metric, 492
Reissner–Nordstrom metric, 482
renormalized equation of motion, 378
resonant cavity, 401
restricted three-body problem, 190
restriction, 296
retarded Green function, 342
reversal of the motion, 125
Riemann normal coordinates, 247
Riemann–Christoffel tensor, 449
Riemannian space, 57
Riemannian structure, 297
rotating basis, 322
rotating charges, 361
rotating coordinates, 103
run-away solution, 379

saddle-point method, 410
scattering angle, 429
scattering cross-section, 137, 288, 430, 431
scattering transformation, 129
Schwarzchild's capture theorem, 83
Schwarzschild metric, 492
Schwarzschild radius, 248
section, 433
secular terms, 153
shadow, 414, 421, 427
signal velocity, 399

small denominators, 149
small oscillations, 118
soldering form, 444
sphere, 14
stable, 140
star (*) mapping, 53
starlike, 67
static, 470
stationary, 470
steepest descent, 410
step function, 303
stereographic projection, 28
Stokes's theorem, 79, 297
submanifold, 16
superconductor, 383
supernova, 501
surface area of the m-sphere, 481
surface tensor, 56
symplectic matrix, 93
symplectic space, 57
synchrotron radiation, 373

tachyons, 401
tangent
    bundle, 28
    space, 23, 24
tangential, 23
TE solutions, 403
tensor, 48
tensor algebra, 48
tensor fields, 294
tensor product, 48, 295
tensors, 45
tetrad, 294
Thompson's theorem, 85
tidal force, 247, 523
TM solutions, 403
Toda molecule, 119
Tolmann–Oppenheimer–Volkoff
    equation, 502
torsion, 444
torus, 14
total charge, 318
trajectory, 17, 36
traveling plane disturbance, 237
trivial, 30
trivializable, 30
Trojans, 192
two centers of force, 182

unbounded trajectories, 187
uniform acceleration, 357
uniform motion, 355
universality of gravitation, 245
unstable, 140

vector bundle, 29
vector fields, 31, 294
vector potential, 314
virial theorem, 206
virtual displacements, 6

wave-front, 401
wave-guide, 396
wedge, 49
wedge product, 295
Weyl forms, 449

Yang–Mills theory, 455
Yukawa potential, 386

Zeeman's theorem, 277